1 MONTH OF
FREE
READING

at
www.ForgottenBooks.com

By purchasing this book you are eligible for one month membership to ForgottenBooks.com, giving you unlimited access to our entire collection of over 1,000,000 titles via our web site and mobile apps.

To claim your free month visit:
www.forgottenbooks.com/free911546

ISBN 978-0-265-93070-0
PIBN 10911546

東 京 帝 國 大 學

理 科 大 學 紀 要

第 貳 拾 九 冊

THE

JOURNAL

OF THE

COLLEGE OF SCIENCE,

IMPERIAL UNIVERSITY OF TOKYO.

VOL. XXIX.

東 京 帝 國 大 學 印 行

PUBLISHED BY THE UNIVERSITY.

TOKYO, JAPAN.

1912—1914.

MEIJI XLV—TAISHO III

Publishing Committee.

Prof. **J. Sakurai,** *LL. D., Rigakuhakushi,* Director of the College, (*ex officio*)

Prof. **I. Ijima,** *Ph. D. Rigakuhakushi.*

Prof. **F. Ōmori,** *Rigakuhakushi.*

Prof. **S. Watasé,** *Ph. D. Rigakuhakushi.*

—•➤◆⧏•—

9279

CONTENTS.

Art. 1.—A Descriptive Monograph of Japanese Asteroidea. I. Archasteridæ, Benthopectinidæ, Porcellanasteridæ, Astropectinidæ, Luidiidæ, Pentagonasteridæ, Oreasteridæ, Gymnasteriidæ, Asterinidæ. By SEITARO GOTO. With 19 Plates.—Publ. Dec. 17th, 1914.

Art. 2.—Studies on Actinopodous Holothurioidea. With 8 Plates. —Publ. July 10th, 1912.

—————>+<+———

PRINTED BY THE SANSHUSHA, TOKYO.

東 京 帝 國 大 學
理 科 大 學 紀 要
第貳拾九冊 第壹編

THE

JOURNAL

OF THE

COLLEGE OF SCIENCE,

IMPERIAL UNIVERSITY OF TOKYO.

VOL. XXIX., ART. 1.

東 京 帝 國 大 學 印 行

PUBLISHED BY THE UNIVERSITY.

TOKYO, JAPAN.

1914.

TAISHO III.

Publishing Committee.

Prof. **J. Sakurai,** *LL. D., Rigakuhakushi,* Director of the College, (*ex officio*)

Prof. **I. Ijima,** *Ph. D. Rigakuhakushi.*

Prof. **F. Ōmori,** *Rigakuhakushi.*

Prof. **S. Watasé,** *Ph. D. Rigakuhakushi.*

Vol. XXIX., Art. 1.

A Descriptive Monograph of Japanese Asteroidea.

I. Archasteridæ, Benthopectinidæ, Porcellanasteridæ, Astropectinidæ, Luidiidæ, Pentagonasteridæ, Oreasteridæ, Gymnasteriidæ, Asterinidæ.

With 19 Plates.

By

SEITARO GOTO.

Published

DEC. 17TH, 1914.

————•>+<•————

PRINTED BY THE SANSHUSHA TOKYO.

JOURNAL OF THE COLLEGE OF SCIENCE, TOKYO IMPERIAL UNIVERSITY.

VOL. XXIX., ARTICLE 1.

A Descriptive Monograph of Japanese Asteroidea.

I. Archasteridæ, Benthopectinidæ, Porcellanasteridæ, Astropectinidæ, Luidiidæ, Pentagonasteridæ, Oreasteridæ, Gymnasteriidæ, Asterinidæ.

By

Seitaro Goto.

Professor of Zoology in the Science College, Imperial University of Tokyo.

With 19 Plates.

This monograph is based on a study of the following collections of Asteroidea :

1). Collection of the Science College, Imperial University of Tokyo. (S. C.)

2). Collection of the Imperial Museum at Ueno, Tokyo. (I. M.)

3). Collection of the Fisheries Bureau, Department of Agriculture and Commerce, Tokyo. (F. B.)

4). Collection of the Higher Normal School of Tokyo. (H. N. S.)

5). Collection of the Higher Normal School of Hiroshima. (H. N. S. H.)

6). Collection of the First High School, Tokyo. (I. H. S.)

7). Collection of the Higher Normal School for Women in Tokyo. (H. N. S. W.)

8). Collections under the care of Prof. Nozawa and Prof. Hatta of the Agricultural College of the Tôhoku Imperial University, Sapporo, Hokkaido. (S.)

9). Collection of Mr. Alan Owston, Yokohama, donated to the Science College.

These collections will be referred to in the following pages by the abbreviations placed in parentheses, as given above.

At the outset it may be advisable to say a few words of explanation as to the plan on which this work is written. My primary object is to give succinct descriptions of all the species represented in the collections just enumerated. I have however endeavoured under each species to bring together previous descriptions and references, so far as they are important for specific determination—with translations in the case of languages less in vogue in scientific circles—, so as to give what may be called the taxonomic history of the species under consideration; and since it is believed to be nearly exhaustive it will be of some help to those future students who may not enjoy the good fortune of having an adequate library within easy reach. The tracing of previous descriptions sometimes consumes much time and may be disproportionately expensive, and any contribution that will lighten the work should therefore be welcome, since it will enable zoologists to turn their attention to questions which are of more fundamental importance. This plan also serves incidentally as a means of doing justice to the first describers of species without a too narrow adherence to the rule of priority. The generic and specific names I have adopted in the following pages may not be satisfactory from the standpoint of a purist, but I have endeavoured to make clear which is the oldest name according to the Linnæan criterion. I have not undertaken any revision of the genera in the present contribution, since such a work can not in my opinion be done with profit without an examination of a larger number of species than have come into my hands, and above all without a study of the type specimens. I have also included in this monograph historical reviews of those species which have been reported from Japanese waters but which are not represented in any of the collections enumerated at the outset.

The abbreviations used in the text are I believe self-explanatory, except perhaps " MS ", which denotes the number of superomarginal plates.

It may be mentioned in passing that the Asteroidea of the " Albatross " Expedition in the north-western Pacific in 1906 are being worked out by the writer, and that several species new to science or to the seas covered by the Expedition will be added.

Finally, it is a pleasure to me to acknowledge my indebtedness to the late Prof. H. LUDWIG of Bonn, Prof. E. PERRIER of Paris, Dr. F. A. BATHER of the Natural History Department of the British Museum, Dr. W. K. FISHER of Stanford University, Dr. J. A. GRIEG of Bergen, Dr. T. MORTENSEN of Copenhagen, Prof. G. H. PARKER of Harvard University, and Prof. C. HARTLAUB of Helgoland.

ARCHASTERIDÆ.

Archaster typicus MÜLLER & TROSCHEL.

This species is not represented in any of the collections examined by me, and its occurrence in Japanese waters is not very certain, being founded on a specimen (or specimens) in the Berlin Museum referred to by VON MARTENS ['65, p. 353]. Its first description by MÜLLER and TROSCHEL ['40a, p. 323] is not accessible to me. GRAY mentions it in the same year under the name of *Astropecten stellaris* ['40, p. 181]. It is the third of his species of *Astropecten*. Its locality is unknown but its characteristics may be gathered from the following two extracts ['40, p. 181]: "Body 5-rayed, arms depressed; the upper series of marginal tubercles broad, rounded or shelving towards the edge." "The dorsal tubercles between the angles of the arms on the centre of the back and on the lines down the centre of the arms the largest."

According to SLUITER ['95, p. 52] *Archaster typicus* and *Archaster angulatus* are to be merged into one species, but as I am not able to examine specimens of either species I will confine myself to a review of the literature on *A. typicus* alone.

In their " System " MÜLLER and TROSCHEL give the following description of this species ['42, p. 65], referring to GRAY's *Astropecten stellaris* as synonym :

"Fünf Arme, seltener 4 oder 6. Verhältniss des kleinen Radius zum grossen wie 1 : 5. Arme viermal so lang wie breit. Die Winkel zwischen den Armen scharf. Auf der innern Reihe der Furchenpapillen bilden immer 3 auf einer Platte einen Keil; in der zweiten Reihe stehen zwei grössere platte neben einander, oder eine solche und eine Pedicellarie. Die ventralen Randplatten mit kleinen Schuppen besetzt, am äussern Rande mit einer Reihe platter Stacheln, deren je einer auf einer Platte steht. Die dorsalen Randplatten, 36 an jedem Arme, einfach graunlirt, sind viel höher als breit. Die Paxillen des Rückens sind so geordnet, dass sich auf der Mitte jedes Armes, vom After bis zur Spitze, eine Längsreihe stärkerer plattenartiger Fortsätze mit Borsten bildet, von denen aus jederseits schräge Reihen kleinerer Paxillen zum Rande verlaufen. Die Paxillen auf dem Rücken der Scheibe sind ähnlich denen der Mittelreihe. Vier bis fünf von diesen grösseren umgeben den zwischen ihnen versteckten After. Zwischen den Furchenpapillen stehen lange zangenartige Pedicellarien. Die Madreporenplatte liegt etwa in der Mitte zwischen dem Centrum und dem Rande.

"Farbe : an trockenen und Weingeistexemplaren dunkelbraun.

"Grösse : $4\frac{1}{2}$ Zoll.

"Fundort : Indischer Ocean. Im Museum zu Berlin aus der SCHOENLEINschen Sammlung ; im Museum zu Leyden durch MACKLOT ; in Paris durch PERON und LESUEUR und QUOY und GAIMARD."

The species is illustrated with two figures, Taf. V, Fig. 2, *a* and *b*.

Under the name of *Archaster* M. T. *nicobaricus* n. sp. MÖBIUS gives the following description of this species ['59, p. 13]:

"Er hat fünf Arme. Das Verhältniss des kleinen Radius zum grossen beträgt 1 : 6. Die Höhe der dorsalen Randplatten ist beinahe ⅓ des kleinen Radius. Durchschnittlich sind diese Platten doppelt so hoch wie breit.

"Zwei Reihen Saumstacheln. In der innern Reihe bilden je 3 einen Keil, indem die mittlere gegen die Furche vorspringt; in der äussern stehen auch 3 auf jeder Platte in schiefer Richtung gegen die Furche, so dass der aborale Stachel einer Gruppe den adoralen der folgenden theilweis deckt. Sowohl in der äussern, wie in der innern Reihe ist der Mittelstachel grösser als die beiden seitlichen.

"Die ventralen Randplatten sind dicht mit Schuppen besetzt; an ihrer äussern Fläche ist ein flacher Stachel, dessen Basis beinahe die ganze Breite der Platte einnimmt; hier und da sind jedoch statt eines zwei kleinere; im mittleren Theile des Armes misst das frei hervortretende Stück desselben ungefähr ¾ so viel wie die Breite seiner Randplatte.

"Die dorsalen Randplatten tragen dichte, schiefe Reihen kleiner Stacheln, die schräg nach oben aufgewachsen sind und eine eilanzettförmige Endfläche haben.

"Die Mittellinie des Armrückens wird von Paxillen eingenommen, die alle breiter als lang sind und um so deutlicher sechseckig erscheinen, je näher sie der Scheibe stehen. An jeder Seite der Mittelreihe ist eine Reihe Paxillen mit weniger Borsten; die übrigen, noch borstenärmern (aber untereinander gleichen) laufen in schiefen (der Mittellinie des Nachbararmes parallelen) Reihe nach dem Rande. In der Mitte der Scheibe sind unregelmässig abgerundete Paxillen von dem Durchmesser der grössten sechseckigen.

"Die Madreporenplatte ist dem Centrum etwas näher als dem Armwinkel. Der grosse Radius misst 81 mm.

"Bezeichnung: 'Kieler Museum. Durch Prof. BEHN von den Nikobarischen Inseln. 1846.'"

In a foot-note MÖBIUS states that the specific name is due to Professor BEHN, who left the work of describing to him.

DUJARDIN and HUPÉ describe this species as follows, with *Astropecten stellaris* GRAY as synonym ['62, p. 411] :

"Espèce ordinairement à cinq bras, rarement quatre ou six, environ quatre fois aussi longs que larges, et dont la longueur, à partir du centre, est quintuple du plus petit rayon du disque. Les angles rentrants inter-brachiaux ont leurs côtés rectilignes.

"Le sillon ambulacraire est pourvue de deux rangées de piquants ; la rangée interne offre, sur chaque plaque, trois piquants qui sont réunis en manière de coin ; la rangée externe n'a que deux piquants aplatis, plus grands ; quelquefois il n'y en a qu'un seul avec une pédicellaire. Les plaques marginales ventrales portent des petites écailles et, sur leur bord externe, une rangée de piquants aplatis, un pour chaque plaque. Les plaques marginales dorsales, au nombre de trente-six sur chaque bras, sont beaucoup plus hautes que larges et n'ont qu'nn seul rang de granules. Les papilles des dos sont disposées de telle sorte qu'elles forment, sur le milieu de chaque bras, une rangée longitudinale de plaques plus grandes et garnies de soies, d'où partent, de chaque côté, des rangées de papilles plus petites, dirigées vers le bord. Les papilles du disque, sur la face dorsale, sont analogues à celles du milieu des bras, et quatre ou cinq, plus grandes que les autres, entourent l'anus, qui forme une saillie au milieu d'elles. La plaque mad-réporique est située à peu près au milieu de la distance du centre au bord. Des pédicellaires allongées, en pince, se trouvent entre les piquants du sillon ambulacraire.

"Coloration d'un brun foncé. Dimension : largeur totale 120 mm.

"Habite la mer des Indes (Mus. Paris)."

LÜTKEN establishes the identity of *Archaster nicobaricus* and *Archaster typicus* ['64, p. 135] :

"*Archaster nicobaricus* BEHN, beskrevet af MÖBIUS efter Exemplarer hjembragte af BEHN fra Nikobar-Öerne, hvorfra ogsaa Prof. REIN-HARDT har hjembragt en större Række, er ikke i sin Forekomst indskræn-

Archaster nicobaricus BEHN, described by MÖBIUS after examples brought home by BEHN from the Nicobar Islands, whence also Pof. REINHARDT has brought it home in large series, is not confined to this

ket til denne Ögruppe; ogsaa fra Billiton har Hr. v. HEDEMANN hjemsendt den i betydelig Antal. Den er imidlertid ikke forskjellig fra *A. typicus* M. T. De Forskjellig-heder, som man kan faae frem ved at sammenholde MÖBIUS's Beskrivelse med den i 'System der Asteriden', turde indskrænke sig til följende to Punkter: 1) at BEHNS Exemplar er forholdsvis mere langarmet, en naturlig Fölge af at det er större (ældere) (R=81 mm). 2) At det har 3 ydre Fodpapiller istedenfor 2, af hvilke den ene hos *A. typicus* endda ofte skal erstattes[1] af en Pedicellarie; det er nu ganske rigtigt, at det normale Tal er 3, af hvilke den ene dog, især hos större Exemplarer, ofte erstattes af en tvegrenet Tang, hvis Forekomst ikke omtales hos MÖBIUS; men paa den anden Side finder man virkelig hist og her kun 2, saa at MÜLLERS og TROSCHELS Beskrivelse i dette Punkt vel er ufuldstændig, men ikke aldeles urigtig.—At Madrepor-pladen i det ene Værk siges at ligge 'noget nærmere ved Centrum end ved Armvinklen,' i det anden 'om-trent midt imellum Centrum og Randen,' har maaskee ikke engang til Hensigt at antyde en faktisk Forskjellighed.

"Da Arten uden Vanskelighed vil erkjendes af begge de citerede Skrifter, vil det ikke være nödventigt her at levere en ny Beskrivelse af

group of islands in its occurrence; Hr. v. HEDEMANN has also brought it back in considerable numbers from Billiton. It is not however different from *A. typicus* M. T. The dif-ferences, which one can obtain by comparing MÖBIUS' description with that in the 'System der Asteriden,' may be reduced to the following two points: 1) that BEHN's example is relatively longer armed, a natural result of its larger size (older) (R= 81 mm). 2) That it has 3 outer foot-papillæ instead of two, of which however one is said to be often replaced by a pedicellaria in *A. typicus;* now it is quite correct, that the normal number is 3, of which one is, especially in larger examples, replaced by a bifurcated for-cep, whose occurrence is not mention-ed by MÖBIUS; but on the other hand, one finds here and there only 2, so that MÜLLER and TROSCHEL'S description is indeed incomplete on this point, but not entirely wrong.— That the madreporic plate is stated in the one work to lie 'somewhat nearer to the centre than to the arm angle,' in the other 'nearly midway between the centre and the margin,' is not perhaps to be taken as indicat-ing an actual difference.

"Since the species will be re-cognised without difficulty from both of the writings cited, it will not be necessary to furnish here a new

---

1) There is a misprint here in the original.

den. Jeg kan indskrænke mig til to Bemærkninger : Den ene er, at allerede hos Exemplarer af c. 4″ i Tvermaal söges Pedicellarierne forgjæves, hvoraf man maa slutte, at disse Redskaber her, som hos adskillige andre Söstjerner, först træde op hos aldeles udvoxne Exemplarer; den anden, at der gives en ret hyppig Form af denne Art, som har hist og her en temmelig stor lodret Pig paa de övre Randplader; af 33 Exemplarer finder jeg den hos 10, som oftest dog kun een paa en enkelt Arm, kun hos eet et större antal, fra 4‑13, paa hver Arm. Man kunde let forledes til i et saadant Exemplar at formode en egen Art, og det er derfor ikke overflödigt at bemærke, at det kun er en individuel Varietet af den typisk piglöse Art."

description of it. I can confine myself to two · remarks : One is that already in examples of ca. 4″ in diameter one looks for the pedicellariæ in vain, hence one must conclude that these organs first appear in this case, as in several other starfishes, in fully grown up examples; the other, that there is a quite common form of this species, which has here and there a tolerably large vertical spine on the superior marginals; of 33 examples I find it in 10, most frequently, however, only one on a single arm, and only in one example in a greater number, from 4–13, on each arm. One might be led to suppose such an example to be a distinct species, and it is therefore not superfluous to remark that it is merely an individual variation of a typically spineless species.

VON MARTENS refers to this species as follows ['65, p. 353] : "*Archaster typicus* M. TR. findet sich mit der Angabe : Japan, vom Leidner Museum erhalten, in der Berliner Sammlung. Mir ist er in Japan nicht vorgekommen." Again in the continuation of his paper published in 1866, VON MARTENS mentions this species from the Indian Archipelago ['66, p. 83] :

"*Archaster typicus* MÜLL. TROSCH. l. c. S. 66. Taf. 5. Fig. 2.

"Unterscheidet sich von *Astropecten*, dem er im Habitus ähnelt, neben den wichtigeren Unterschieden des Afters und der Saugscheibe an den Ambulakren auch dadurch, dass die oberen Randplatten wesentlich an der Seite des Armes liegen und nur mit ihrem obersten Theil den Rücken erreichen, so dass die Kante, welche die Rücken- und Seitenfläche bildet, in den oberen Rand der Randplatte fällt und nicht in deren Mitte. Diese oberen Randplatten sind mit stumpfen keulenförmigen Fortsätzen dicht besetzt, ähnlich denen der Rückenfläche, aber gleichmässig dicht gedrängt und im Allgemeinen feiner ; wo sie abgefallen sind, erscheint die Platte mit feinen in etwas wellige Radialreihen gestellten Narben bedeckt. Die unteren Flatten sind mit kurzen schuppenförmigen Stacheln bekleidet, welche (bei trockenen Exemplaren) alle nach oben sich anlegen ; nach dem Abfallen derselben bleibt eine schuppig-höckerige weit gröbere Sculptur. Die Stelle des Afters im Centrum der Scheibe ist öfters deutlich durch längere und etwas zusammenneigende Fortsätze bezeichnet. Der Rücken ist mit Reihen von Höckern besetzt, deren jeder eine Anzahl kurzer keulenförmiger Fortsätze trägt ; die Höcker der mittleren Reihe jedes Armes sind breiter als die übrigen. Furchenpapillen der inneren Reihe 3 auf jeder Platte, die mittleren etwas grösser. Zahl der (unteren) Randplatten an einer Armseite 40–45, bei *A. Mauritianus* GRAY sp. (*angulatus* Müll. Tr.) 70.

"Farbe während des Lebens oben aschgrau, mit mehr oder weniger bestimmten schwärzlichen Querbändern auf den Armen ; unten blassgelb.

"Armradius bis 69 Mill. Scheibenradius zum Armradius wie 1 : 4—4½. Armbreite zum Armradius wie 1 : 3—4. Armhöhe zur Armbreite wie 6 : 11.

"Insel Batjan (Molukken) häufig an sandigen Stellen, 2–3 Fuss unter Wasser bei Ebbezeit, Insel Timor und Adenare neben Flores. Amboina nicht häufig.—Java, FRANK im Berliner Museum. Von denselben Orten, aber anderen Findern auch im Leidner Museum."

In his "Synopsis" ['66, p. 3] also GRAY mentions *Astropecten stellaris* without adding anything new to its description, but he cites "*Archaster typicus*, MÜLL and TROSCH. Monatsb. Berl. 1840" as its synonym. He also cites "Ast. 65, t. 5. f. 2 (B. M.)" as referring to the same species.

PERRIER describes the pedicellaria as follows ['69, p. 95]:

"*Archaster typicus*, M. et T.—Chez l'*Archaster typicus*, nous n'avons trouvé de Pédicellaires que dans les angles des bras et dans le sillon ambulacraire.

"Les branches de ces Pédicellaires[1] ne sont pas toujours parfaitement égales entre elles, elles ne sont pas non plus parfaitement régulières. La longueur

---

1) "Pl. 2, fig. 13."

de chacune d'elles est un peu plus de quatre fois sa plus grande largeur. Cette plus grande largeur est atteinte à peu près au sommet de l'échancrure musculaire interne. A partir de ce point le bord interne de chaque branche devient légèrement oblique de manière à rétrécir graduellement la pince qui se termine par un sommet arrondi. Le bord interne de chaque mâchoire est très-finement denté sur tout son pourtour à partir du sommet de l'échancrure."

PERRIER mentions the specimens examined by him in his "Revision" ['76, p. 265] :

"Quatre individus desséchés, dont un à six bras, de Tongatabou (MM. QUOY et GAIMARD) 1829. Trois individus desséchés de l'expédition de 1842 de DUMONT D'URVILLE, dont le naturaliste était M. LEGUILLOU. Un individu desséché sans indication de provenance. Trois individus de la Nouvelle-Calédonie, donnés en 1875, par M. GERMAIN. Dans l'alcool, un individu receuilli en 1829, par MM. QUOY et GAIMARD (probablement à Tongatabou). Un individu receuilli en 1803, par PÉRON et LESUEUR, dans le voyage du capitaine BAUDIN. Deux receuillis en 1842, par HOMBRON et JACQUINOT, dans la baie de Raffles (expédition DUMONT D'URVILLE). Un individu des 'Indes orientales' donné en 1870 par M. le professeur PAUL GERVAIS ; un individu receuilli en 1872, à la Nouvelle-Calédonie, par M. BALANSA.

"*Observation.*—Deux individus desséchés, désignés dans la collection sous le nom d'*Archaster typicus*, n'appartiennent certainement pas à cette espèce. Nous les rapportons à l'espèce suivante."[1]

VIGUIER gives a detailed description of the skeleton of this species and illustrates it with six figures ['78, p. 237] :

"Les figures 1 et 2 (pl. XVI) représentent les deux faces de l'*Archaster typicus*.

"Sur le bord des bras se trouve une double rangée de plaques marginales, qui donnent à cet animal une certaine ressemblance extérieure avec l'*Astropecten*. On peut remarquer de suite cependant que, si les marginales inférieures sont disposées comme chez ce dernier, les marginales sont en contact, par toute l'étendue de leurs faces correspondantes, avec leurs voisines

---

1) *Archaster angulatus.*

de série, et ce n'est qu'à la face ventrale que l'on trouve les sillons séparant les surfaces de deux plaques voisines.

"Les marginales supérieures atteignent leurs plus grandes dimensions précisément dans l'angle interbrachial, où les deux du sommet sont taillées en coin ; elles dècroissent ensuite régulièrement jusqu' à la plaque ocellaire, qui est bien loin de présenter ici le volume que nous lui avons vu chez l'*Astropecten*. Elle est au contraire très-petite, légèrement elargie à son extéremité libre,. et taillée en coin par l'autre bout. A chacun des côtés de ce coin correspondent plusieurs plaques marginales très-petites, on en compte jusqu' à quatre chez l'*Archaster angulatus* et non plus une seule, comme chez l'*Astropecten*. Le sommet tronqué du coin est en rapport avec la dernière plaque de la rangée médiane, qui arrive seule jusqu' à ce niveau avec les deux premières latérales.

"Cette rangée médiane, qui se distingue très-nettement tout le long du bras, et dont les ossicules vont en grandissant de l'extremité du bras vers la base, se compose de plaques hexagonales dont deux des côtés sont transversaux à la direction du bras. De chaque côté de cette série on en distingue une autre de plaques plus petites, régulièrement emboitées avec les premières, mais dont la forme commence déja à se dégrader un peu. Enfin, de chacune de ces plaques partent des lignes d'autres plaques plus petites et de forme irrégulièrement losangique qui se portent, parallèlement les unes aux autres, vers les plaques marginales, en divergeant de la ligne médiane comme les barbes d'une plume. Au sommet des angles interbrachiaux ces séries deviennent rayonnantes pour passer régulièrement d'un bras à l'autre. En outre, dans ces séries transversales, les petites plaques sont disposées de façon à former des lignes longitudinales parallèles à la médiane, et qui se raccordent par des arcs de cercle d'un bras à l'autre.

"Toutes ces petites plaques sont imbriquées par de petits prolongements de leur base, chacune d'elles recouvrant celle qui lui est immédiatement externe, jusqu' à la série médiane qui recouvre les deux latérales.

"Sur le disque les pièces deviennent beaucoup plus larges, et leurs contours s'arrondissent. On distingue très-bien au centre une grande plaque, à gauche de laquelle se trouve l'anus, en supposant la plaque madréporique en arrière. Celle-ci repose sur les ossicules dorsaux qui sont échancrés au-dessous d'elle pour le canal hydrophore, absolument comme nous l'avons vu chez les *Goniasteridæ*. Il n'y a rien là qui rappelle l'*Astropecten*. Cette plaque est épaisse, convexe, arrondie, et marquée de plis assez droits qui convergent vers le centre.

"La face ventrale n'est constituée, comme chez l'*Astropecten* et la *Luidia*, que par les plaques marginales inférieures et les séries adambulacrair-es. Comme chez ces deux genres on trouve dans les coins buccaux quelques pièces complémentaires, qui sont ici plus grosses, et ordinairement au nombre de deux paires seulement, dont la plus interne est la plus volumineuse. Il n'y a que les deux plaques marginales du sommet de l'angle qui soient raccourcies.

"Les plaques adambulacraires ont une forme assez singulière ; leur surface, qui est à peu près plane, au lieu d'être arrondie, comme chez l'*Astropecten*, a la forme d'un pentagone irrégulier, à côtés échancrés, et dont le sommet arrondi fait saillie dans le sillon.[1]"

"Les pièces ambulacraires ont une forme différente de celle des *Astro-pecten* et qui se rapproche de celle des *Goniasteridæ*. Il n'existe pas de soutiens ambulacraires. Enfin, nous trouvons des systèmes interbrachiaux très-puissants,[2] composés de deux grosses pièces placées verticalement, et reliées à l'odontophore.

"Les gros ossicules du dos sont creusés en fossette pour recevoir la pièce supérieure, et forment une sorte de bourrelet tout autour de son insertion.

"La bouche diffère absolument de ce qu'elle est chez l'*Astropecten*. Les dents, larges, triangulaires, contiguës sur la ligne interbrachiale, ne font pas saillie à la face ventrale, et s'avancent jusqu'au milieu de la bouche, qu'elles ferment presque complètement. La première pièce ambulacraire a son apo-physe en aile très-développée.[3]

"L'odontophore présente une forme très particulière : il est grossi cinq fois sur la figure 4, et l'on voit qu'il présente deux facettes articulaires bien détachées en apophyses, et que sa face ventrale n'est pas rétrécie en tri-

---

1) "Fig. 2, 3 et 7." 2) "Fig. 6, *i.*" 3) "Fig. 3."

angle, comme dans l'*Astropecten*. Cette face ventrale pent être considérée la base d' une pyramide quadrangulaire, dont deux des arêtes sont courbes. L'épaisseur de la pièce est considerable."

STUDER mentions it in the collection of the "Gazelle" ['84, p. 48]. "Diese weit verbreitete Art lebt in mässiger Tiefe zwischen Corallen. Sie fand sich in Neu-Guinea, MacCluergolf, Mataku, Fidji, Neu-Irland und Neu-Britannien."

BELL mentions "three dried specimens from Port Denison, 4 fms." as well as from the Seychelles ['84a, p. 133, 510]. He also mentions it from the Andaman Islands ['87, p. 140] and the sea of Bengal ['88, p. 388].

SLUITER mentions it from Java and Mauritius ['89, p. 309]:

"*A. typicus* (M. und TR.). E. PERRIER, Revision des Stellérides," pag. 265. Mehrere Exemplare (No. 240, 254, 575) aus der Bai von Batavia, von den Tausend Inseln im Java-Meer und von Mauritius. Die Tiere sind auf dem sandigen Strande der Korallen Inseln des Java-Meeres sehr gemein. Bei tiefer Ebbe fallen sie sogar trocken, halten sich sonst 1—3 Fuss unter Wasser. Immer sind sie mit einer dünnen Schicht Sand bedeckt, so dass die Stelle, wo sie liegen, nur durch den fünf strahligen Sandhaufen leicht zu entdecken ist. Sie leben gewöhnlich in grösseren Gesellschaften beisammen. Die Farbe ist ein ziemlich helles Graugelb, die dunkleren graubraunen Flecken vereinigen sich zuweilen zu unregelmässigen Querbändern, bleiben aber auch öfters gesondert. Im Aquarium konnte ich die Tiere nur während einiger Monate am Leben halten. Sie bedecken sich sogleich mit Sand und kriechen, gewöhnlich nur des Nachts, langsam über dem Sande umher."

This species is also mentioned in the Challenger Report, where some important critical remarks are found [SLADEN, '89, p. 123]:

"*Localities.*—Samboangan, Philippine Islands. Depth 10 fathoms.

"On the reefs at Zebu, Philippine Islands.

"Station 200. East of Samboangan, Philippine Islands. October 23, 1874. Lat. 6° 47′ 0″ N., long. 122° 28′ 0″ E. Depth 250 fathoms. Green mud. Surface temperature 85°.5 Fahr.

"On the reefs at Kandavu, Fiji Islands. August 1874.

" *Remarks.*—Attention may be drawn to the occurrence of this species at 250 fathoms (Station 200), for it is, so far as I am aware, the greatest depth at which *Archaster typicus* has been found. I can detect no differences worthy of remark between these examples and specimens from shallow water.

"It is interesting to note that amongst this series from station 200 there are two examples which are provided here and there with short, conical, robust, stunted spinelets, standing upright on the upper margin of the supero-marginal plates. These spinelets are quite irregular in their occurrence. In one of the examples less than a dozen are present on the whole starfish, but in the other case they are much more numerous. In this example it is also to be remarked that the lateral walls of the rays are much more vertical than in the other specimens from this locality, the supero-marginal plates being less bevelled or arched towards the abactinal surface, with which the lateral walls consequently forms a more angular junction, resembling in this respect the character of *Archaster angulatus*. In all other respects this interesting specimen is an extremely well-marked example of *Archaster typicus*.

" LÜTKEN [1] has placed on record the presence of occasional spinelets on the supero-marginal plates of this species, and it appeared to be of frequent occurrence in the large series of examples from the Nicobar Islands studied by him. Through his kindness I had the opportunity of examining a remarkably fine example from Billiton in the Natural History Museum of Copenhagen, in which from four to six spinelets were present on each side of a ray. In the Museum at Leyden are examples from Java and the Togean Islands (N. E. of Celebes) also similarly characterised, and this form has been named in manuscript by Professor C. K. HOFFMANN *Archaster typicus*, var. *multispina*. The presence of these spinelets on the supero-marginal plates is so very irregular and sporadic, and seems to me to be unaccompanied by any other character of sufficient importance, that I fail to appreciate the necessity of ranking the examples in question as a named variety.

"The normal composition of the adambulacral armature in this species is :—(1.) A furrow series of three spines, the middle one much in advance

1) LÜTKEN '64, p. 135.

regelmässig entwickelt zwischen den beiden Reihen Furchenpapillen, wie
das typisch für den *A. angulatus* von MÜLLER und TROSCHEL ist. Wie aber
DE LORIOL[1] schon hervorhebt sind die Pedicellarien sehr ungleich zahlreich entwickelt. Andererseits besitze ich Exemplare von *A. typicus* von den Molukken, welche das Verhältniss der Arme als $1 : 6\frac{1}{2}$ besitzen, mit 55 dorsalen
Randplatten versehen sind, und die Anordnung der Pedicellarien genau so
aufweisen wie die Tiere von Mauritius. Alle mögliche Uebergänge zwischen
den beiden Formen sind zu finden, und ich glaube nicht, dass beide als
gesonderte Arten beibehalten werden können. Auch SLADEN[2] erwähnt schon
einige Uebergänge zwischen beiden Formen."

KŒHLER mentions it from the Islands of the Sunda ['95, p. 386]:

"Un seul échantillon, ayant 18 centimètres de distance entre les extrémités des deux bras opposés (R = 12). Un certain nombre de plaques
marginales dorsales portent, surtout vers le milieu des bras, chacune un
piquant conique de hauteur variable. Cette disposition a été observée à
différents reprises, et certains auteurs ont fait une variété (*A. typicus* var.
multispina) de ces échantillons à plaques marginales armées.

"P. SLADEN et DE LORIOL, qui ont en entre les mains de pareils échantillons, estiment que ce caractère est trop irrégulier pour autoriser la création
d'une variété.

" Je n'observe pas sur cet échantillon de pédicellaires parmi les piquants
du sillon ambulacraire.

" Le nombre des plaques marginales est de 50."

DÖDERLEIN mentions this species from SEMON's collection ['96, p. 305],
and remarks, " Mehrere Exemplare von Amboina, mit und olme Stacheln
auf den Supramarginalplatten."

BEDFORD mentions this species from Singapore. After giving references
to MÜLLER and TROSCHEL ['40, and '42], v. MARTENS ['66], SLUITER ['89],
DE LORIOL ['93], CUÉNOT [Arch Biol. '91], and LUDWIG [BRONN], he makes
the following notes [: 00, p. 289]:

" R = 5.3—7.0 × r.

---

1) DE LORIOL '85, p. 79.
2) Challenger Report.

"*Locality and Habitat.* This species was very abundant on a sand-flat just exposed at low tide on Po Senang, Singapore ; I did not meet with any examples elsewhere ; its habitat appears to be identical on the islets of the Java Sea (*cf.* SLUITER).

"*Distribution.* Extends from the Nicobars and Andamans through the Mergui Archipelago, Malay Archipelago, and N. Australia, as far as the Fiji and Tonga Is. It seems doubtful whether it occurs in Mauritius.

"Out of five specimens brought back the superomarginals varied in number from 45 to 50 on each side of interbrachial arch (v. MARTENS gives 40–50, while MÜLLER and TROSCHEL give 36). There were no traces of superomarginal spines.

"In one specimen at the base of one of the arms there is a constriction somewhat similar to that which occurs in *Linckia* before schizogony of the arm takes place ; in the present case the superomaginals meet across the abactinal surface, but the actinal plates are unaffected.

"The anus is central, without any tendency towards CUÉNOT'S interradius BC in any of the five specimens brought back : Prof. LUDWIG, in his definition of the genus *Archaster*, says, 'After central' (*loc. cit.* p. 667) ; whereas in the body of the work (p. 587) he corroborates CUÉNOT by stating that it lies 'stets mehr oder weniger excentrisch in der Richtung einer interradialen Hauptebene,' and he then continues to describe the particular interradius in agreement with CUÉNOT'S notation BC, although he adopts a somewhat different mode of orientation.

"Measurements of two extreme individuals :—

No. of marginals.	R.	r.	Arm-breadth.
46	64	12	13
46	71	10	13."

It may be mentioned in passing that the apparent contradiction in BRONN'S 'Klassen u. Ordn.' referred to by BEDFORD in the above extract is fully explained by LUDWIG (*vide infra*).

PFEFFER simply mentions it from Ternate [: 00, p. 83].

It is also described by LUDWIG from the Albatross collection of 1899–1900 from the South Pacific [: 05, p. 53 and Taf. XXI, fig. 116] :

"Von dieser durch das indopacifische Gebiet weit verbreiteten und insbesondere von den Fidschi-Inseln längst bekannten Art hat die Albatross-Expedition des Jahres 1899 von dem Hafenort Suva (Viti-Levu, Fidschi-Inseln) neun jugendliche Exemplare mitgebracht, die dort am Strande gesammelt wurden.

"Die Masse der neun Examplare sind die folgenden:

Nr.	R in mm	r in mm	r : R	ZoR[1]
1	34	9	1 : 3.78	29
2	24	6	1 : 4.00	26
3	23	6	1 : 3.83	25
4	22	5.5	1 : 4.00	24
5	20	5.5	1 : 3.63	23
6	16	4.5	1 : 3.55	21
7	16	4.5	1 : 3.55	21
8	15	4.5	1 : 3.33	21
9	15	4.5	1 : 3.33	10

"Das aus sogenannten Pseudopaxillen gebildete Rückenskelett ist bei diesen jugendlichen Tieren (ich untersuchte namentlich das kleinste Exemplar Nr. 9) von grosser Regelmässigkeit (Taf. XXI, Fig. 116) und lässt sowohl die primären Interradialplatten als auch die primären Radialia und die Centralplatte deutlich erkennen. Das Armrückenskelett beginnt in der Scheibe des Exemplares Nr. 9 mit neun Längsreihen von Flatten, nämlich einer radialen und jederseits einer adradialen und drei dorsolateralen. Die ersten Adradialia je zweier benachbarten Arme stossen nach aussen von der betreffenden primären Interradialplatte zusammen und es folgt dann auf sie in interradialer Richtung eine zweite Interradialplatte, an welche sich innen das verkalkte interradiale Septum ansetzt. Jede primäre Interradialplatte schliesst mit den an sie anstossenden beiden ersten Adradialia und diese wieder mit der zweiten Interradialplatte dicht zusammen, während zwischen allen anderen Platten mehr oder weniger grosse, schmale Lücken für den Durchtritt der Papulä zu bemerken sind.

1) Number of superomarginals.

" Die Papulä selbst sind einfach schlauchförmig und finden sich in den
Skelettlücken des Scheitels wie der übrigen Rückenseite, fehlen also nur in
den erwähnten, aus je vier Platten gebildeten, interradialen Plattengruppen ;
gewöhnlich zählt man in Umkreis einer Platte sechs (selten sieben oder
acht) Papulä.

" Die Madreporenplatte ist ein selbständiges Skelettstück, das sich
zwischen eine primäre Interradialplatte und die beiden angrenzenden ersten
Adradialplatten einschiebt ; die Art gehört demnach sicher zu den Euplacoten.

" Die Regelmässigkeit des ganzen Dorsalskelettes wird nur an wenigen,
in der Figur (Taf. XXI, Fig. 116) mit * bezeichneten Stellen durch secundä-
ren Einschub von Platten gestört.   Die Adradialplatten und Dorsolateralplatten bilden ausser Längsreihen zugleich schiefe Querreihen.   Untersucht
man die Verbindungsweise der Dorsalplatten miteinander, so ergibt sich,
dass keine Connectivplättchen vorhanden sind, sondern die Platten mit
kurzen, lappenförmigen Fortsätzen ihrer Basis sich untergreifen und dadurch
in festeren Zusammenhang kommen, wie das übrigens schon durch VIGUIER[1]
festgestellt worden ist.   Ich möchte aber die Bemerkung hinzufügen, dass
die zu einer Querreihe gehörigen Dorsolateralplatten sich nur in der Richtung
der Querreihe miteinander durch einen an ihrem oberen Rande befindlichen
Fortsatz verbinden, während sie mit den Dorsolateralplatten der vorhergehenden
und der folgenden Querreihe eine derartige Verbindung nicht eingehen.

" BEDFORD[2] behauptet, dass der After genau central liege, ohne irgend
eine Verschiebung nach dem vorderen Interradius zu zeigen.   Dabei sucht er
mich als mir selbst widersprechend hinzustellen, weil er mir eine Angabe
auf p. 667 der Echinodermenbearbeitung in BRONN's Klassen und Ordnungen
des Tierreiches zuschiebt, die allerdings in Widerspruch mit meiner Angabe
auf p. 587 desselben Werkes steht.   Hätte er aber die Notiz auf seite V
desselben Buches gelesen, sowie meine Namensunterzeichnung auf. p. 623
beachtet, so könnte er wissen, dass ich von p. 623 an keinerlei Verantwor-
tung für den Text der von da ab von HAMANN fortgeführten Bearbeitung
habe.   Was die Sache selbst angeht, so zeigen die neun mir hier vorliegen-

1) '78' p. 238.
2) : 00, p. 290.

den jugendlichen Exemplare sowie auch mehrere erwachsene Tiere der hiesigen Sammlung, dass der After, trotz der BEDFORD'schen Behauptung, keineswegs genau central liegt. Seine Excentrizität ist freilich nur eine sehr geringe. Stets liegt der After umgeben von in der Regel vier, seltener fünf Platten so, dass er der vorderen primären Interradialplatte ein wenig genähert ist. Von den ihm umgebenden Flatten ist bei den jugendlichen Tieren immer eine und zwar stets dieselbe etwas grösser als die anderen. Diese grössere Platte, die ich für die Centralplatte halte, nimmt das eigentliche morphologische Centrum der Scheibe ein und der After liegt neben derselben in der Richtung nach vorn. Die Stachelchen, welche den an das Afterfeld angrenzenden Rändern der vier oder fünf Platten aufsitzen, sind etwas länger und dicker als die übrigen Stachelchen der Rückenplatten und legen sich über dem After, denselben beschützend, zusammen.

"An der Ventralseite der jungen Tiere ist in den Armwinkeln nur ein Paar von kleinen Ventrolateralplatten vorhanden, an dessen adoraler Seite die nach innen von den Mundeckstücken gelegene unpaare Interoralplatte (sog. Odontophor) zum Teile sichtbar wird. Die Mundeckstücke der kleinsten Exemplare besitzen am ambulacralen Rande erst fünf in einer Reihe stehende kleine Stacheln, ferner dem suturalen Rande entlang vier und am distalen Rande zwei noch kleinere Stacheln. Bei dem Exemplar Nr. 2 hat sich die Zahl der Stacheln am ambulacralen Rande der Mundeckstücke schon auf sechs oder sieben vermehrt. Die Adambulacralplatten, deren man in der proximalen Armhälfte elf auf die Länge von acht unteren Randplatten zählt, tragen auf dem in die Armfurche vortretenden Vorsprung bereits dieselben drei Stacheln wie die alten Tiere.

"Zweiteilige zangenförmige Pedicellarien, wie sie PERRIER[1] von dieser Art beschrieben hat, finden sich bei allen diesen neun jungen Exemplaren und sind so verteilt, dass erstens jede Ventrolateralplatte eine (bei dem Exemplar Nr. 1 eine oder zwei) und zweitens die meisten Adambulacralplatten ebenfalls je eine auf ihrer ventralen Oberfläche tragen. Bei dem Exemplar Nr. 2 sind überdies an einer der zehn Mundeckplatten zwei der aufeinanderfolgenden ambulacralen Stacheln zu einer Pedicellarie umgewan-

---

1) '69, p. 95.

delt. Bei dem Exemplar Nr. 1 sind ferner auch eine Anzahl der Rückenplatten, insbesondere auf dem proximalen Armabschnitt und dort wieder
namentlich den oberen Randplatten entlang, mit einer Pedicellarie ausgerüstet, die durch Umbildung je zweier Stachelchen der Platte entstanden zu
sein scheint.

"Von demselben Fundorte (Suva) liegen auch noch zwei trockene Exemplare vor, von den bei dem einen R=33 und bei dem anderen R=22 mm.
misst. Beide sind auf den Adambulacral- und Ventrolateralplatten mit
Pedicellarien ausgestattet. Dagegen kann ich zwei grösseren Tieren von R=39
und R=55 mm. der Bonner Sammlung, die gleichfalls von den Fidschi-Inseln
stammen, keine einzige Pedicellarie finden. Auf dieses bemerkenswerte
individuelle Schwanken im Besitze oder Fehlen der Pedicellarien bei *Archaster*
typicus und dem nahe verwandten, vielleicht damit identischen *Archaster*
angulatus haben schon LORIOL[1] und SLADEN[2] aufmerksam gemacht.

FISHER refers to this species in his paper on the Hawaiian starfishes
[: 06, p. 1045] :

"This well known and widely distributed species was not taken by the
naturalists of the *Albatross*, although considerable work was done on reefs and
dredging was carried on in shallow water. A specimen collected by W. H.
PEASE is recorded by J. E. IVES (op. cit., p. 175) from the 'Sandwich Islands'.
He says : 'The specimen from the Sandwich Islands differs from the others
by its narrower arms and greater number of arm plates, having about 40
to each side of an arm, whereas others have only about 35. It differs also
in colour, being of a very light cream colour instead of light or dark umber.
This however may be due to fading of the original tint.'[3]

"An examination of several Samoan specimens reveals the fact that

---

1) '85, p. 79 (*Archaster angulatus*) ; '93, p. 378—379 (*Archaster typicus*).

2) '89, p. 125 (*Archaster typicus*).

3) "After this paper was ready for the press I received from Dr. H. A. PILSBRY, of the
Philadelphia Academy of Sciences, the above specimen. I have nothing to add to Mr. IVES's
observations, except that very few of the superomarginals bear a small upright tubercular spinelet. In one of the Samoan specimens these are conspicuous. The number of superomarginals
is not great for the size (R=60 mm., r=10.5 mm.), since a specimen from Samoa, with R=65
mm., has 48 superomarginals. The armature of the adambulacral plates is the same in Hawaiian and Samoan examples."

there is some variation in the length and breadth of the rays, while the number of superomarginal plates may be as great as 52.

"This species has a certain resemblance to some forms of *Astropecten*, but may be readily distinguished from any member of that group by the absence of superambulacral plates, and by the presence of a very well defined median radial series of paxillæ."

CLARK has the following notes on this species [: 08, p. 280] :

"60 specimens, 60–125 mm. in diameter. Saonek, Waigiou Island, New Guinea.—45 specimens, 50–120 mm. in diameter. Amboina. BARBOUR collection.

"According to Mr. BARBOUR's notes, these specimens were taken in very shallow water on a bottom of white sand. The colour in life was orange-red, but in drying the specimens nearly all trace of this colour was lost, and they became pale yellowish, with only here and there patches of orange-red. One of the specimens from Amboina has 6 rays, while two of those from Saonek have only 4 rays each."

BROWN reports this starfish from the Mergui Archipelago [: 10, p. 28] :

"Locality.—XVI., Alligator Rock, 8 to 18 fathoms, rock and sand.

"Two specimens in which R = 44 and 46 and r = 10 and 9 [mm.] respectively. In the smaller specimen a single spine appears on one superomarginal; otherwise the supero-marginals have no trace of spines. A similar occurrence in this species is noted by LÜTKEN (Vidensk. Medd. (1864), p. 136), and by SLADEN ('Challenger' Reports, XXX., p. 124).

"This species is also recorded from the Mergui Archipelago by Dr. ANDERSON in 1882. Widely distributed in Eastern Indian Ocean and Western Pacific."

KŒHLER· mentions this species from the collection of the Indian Museum, with two figures [: 10, p. 9] :

"Iles Andaman. Un échantillon.

"No. 2231. Profondeur 26 brasses et demie. Quelques échantillons.

"Dans l'individu des îles Andaman, qui est très bien conservè, R = 36 mm. Les autres sont dans un état de conservation plus ou moins satisfaisant ; R = 30 mm. dans le plus grand et 19 mm. dans le plus petit. J'ai représenté ce dernier exemplaire Pl. I, fig. 1 et 2, pour servir de point de comparaison avec une petite asterie provenant des îles Andaman et qui appartient certainement au genre *Archaster* (Pl. I, fig. 3 et 4). Cet individu, tres jeune, a un diamètre maximum de 17.5 mm. : R=9 mm. et r=3 mm. On peut s'assurer, en le comparant au petit *Archaster typicus* des fig. 1 et 2, qu'il n'appartient pas à la même espèce. Je ne pense pas non plus que ce soit un *A. tenuis* BELL, d'autant plus qu'il n'est pas bien certain que ce dernier appartienne réellement au genre *Archaster*. Je me contente de représenter ici cet échantillon trop jeune pour être étudié en detail."

KŒHLER again mentions it from the south-eastern Moluccas, as follows [: 11, p. 266] :

"20 Avril 1908. Dobo. Profondeur 4 m. Quatre échantillons.

"15 Mai 1908. Dobo. Profondeur 4 m. Quatre échantillons.

"La valeur de R varie entre 50 et 75 mm. L'un des exemplaires a quatre bras et un autre en a sept, tous égaux. Dans quelques individus, les plaques marginales dorsales portent un nombre variable de tubercules coniques, disposition qui a été souvent signalée chez cette espèce."

BENTHOPECTINID.E.

Benthopecten spinosus VERRILL.

This starfish is not represented in any of the collections mentioned at the outset. It is perhaps better known by the name used by SLADEN in the Challenger Report, *Pararchaster semisquamatus*. VERRILL's original description runs as follows ['84, p. 218] : "*Benthopecten spinosus* V., sp. nov. Rays five, long, rather slender, except at base flat, and gradually tapered to long narrow tips. Disk not very large and like the arms covered with a smooth skin and covered plates, each with a slender central spine ; these become larger toward the centre, where there is a group of about 20 long, erect, tapering, sharp spines. A circle of 4 to 6 papillæ, placed singly, surrounds each plate. There are neither paxillæ nor granules. Upper marginal plates rather small, elevated in the middle, more than 40 on each side ; each bears several minute spines and a single central, large and long, tapered, sharp spine, the basal ones and those bordering the disk becoming larger and exceeding those of the centre of the disk. Lower marginal plates bear each a vertical row of about 3 spines, the upper one much the longest, and along the middle of the arm as long as those of the upper plates, but becoming much smaller at the base of the arm and along the disk, where the latter increase in size. The ventral plates form small triangular areas; they are rounded, raised in the middle, and each bears one or two small acute spines, but no granules. The adambulacral plates project strongly inward, and each bears an inner convex row of 4 or 5 small, slender spines and a transverse outer row of much larger ones, of which one or two inner are much the largest. Suckers large, tapered, pinched up at tip. Larger radius, 150 mm; smaller, 22 mm. Still larger examples were taken. Station 2035, in 1362 fathoms, etc.

"*Benthopecten*, gen. nov., resembles *Archaster*, but differs in having no paxillæ; the dermal plates usually bearing a single spine, sometimes two or three."

In his paper of 1885 VERRILL refers to this species as follows ['85, p. 519]:

"A very interesting new form, taken in many localities, is related to *Archaster* and *Astropecten*, closely resembling some of the spinose species of these genera in general appearance. It represented a new genus (*Benthopecten spinosus* V.). The flat dorsal surface is closely covered with tesselated, angular plates, having single, definite, small pores for solitary branchial papulæ between them, while there are no true paxillæ, the small spinules arising singly, or two or three together, directly from the plates. The marginal plates, above and below, bear single large, sharp spines, the five largest ones occupying the central interbrachial plates, on the upper margin. The disk is of moderate or rather small size, but the arms are long and tapered. It occurred in 855 to 1,917 fathoms, in 1883, but is most abundant in 1,200 to 1,500 fathoms."

Further on in the same paper VERRILL remarks as follows ['85, p. 543]: "B[athymetrical] range, 855 to 1,917 fathoms, 1883. Common; locally abundant."

SLADEN's description is as follows ['89, p. 7] :

" *Pararchaster semisquamatus*, n. sp. (Pl. II. figs. 1 and 2 ; Pl. IV. figs. 7 and 8).

" Rays five. R=166 mm.; r=15 mm. R=11 r. Breadth of a ray near the base, 14.5 mm.

" Rays very elongate, comparatively narrow and flat, tapering gradually and slowly from the base to the extremity, the outer part being very attenuate. Disk very small. Abactinal surface plane, feebly convex or subcarinate along the median line of the rays. Lateral walls of the ray low and vertical. Actinal surface of the disk prominent at the mouth-angles, and sloping thence to the margin and very slightly along the rays. Interbrachial arcs widely rounded.

" The abactinal surface of the disk and rays is covered with small, uniform, subcircular scale-like plates, which are overlaid with a delicate membranous tissue. The plates bear on their centre a single minute subconical or cylindrical spinelet; along the rays these are quite microscopic thornlets, but upon the disk and at the base of the rays there are a number of much larger spinelets ; the largest are elongate, about 7 to 8 mm. in length, robust, tapering, and sharply pointed, and their position probably marks the primary apical plates ; the primary radials and basals being especially distinguishable, and perhaps also the dorso-central and the underbasals ; other spinelets rather smaller are present in the vicinity of these, but they rapidly decrease in size as they recede from the central area ; and really definite spinelets do not extend further along the base of the ray than the third or fourth supero-marginal plate.

" The supero-marginal plates, fifty-five in number from the median interradial line to the extremity, are elongate and suboval in form, and are confined entirely to the lateral wall of the ray ; their posture appears slightly oblique when viewed from the side, the aboral end of one plate standing over the adoral end of the next outward; their height is less than half their length ; and the upper margin of the plate forms the boundary of the abactinal surface of the ray. On the centre of each plate is a large well-defined tubercle, on which is articulated a robust, cylindrical, tapering spine, the ninth from the interradial line being about 8 mm. in length. The surface of the plate is covered with membrane and bears no spines, except occasionally a minute thornlet, irregularly placed near the base of the large spine.

" In the median interradial line there is one large odd supero-marginal plate, developed abactinally into a prominent, truncate, conical tubercle, upon which is borne a powerful robust spine about 8 mm. in length, directed vertically, and thicker than any of the other spines on this species.

" The infero-marginal plates are similar in form and character to the superior series, each being nearly exactly beneath its corresponding upper companion. Like them, each has a prominent tubercular eminence, upon which is borne a straight, robust, tapering, and sharply pointed spine, the sixth or seventh from the interradial line measuring about 10 mm. in length; the length decreasing slightly as they proceed along the ray. A second similar but rather smaller spine, articulated on a tubercle, stands close to the lateral spine on the inner side. The surface of the plate is covered with membrane; and two or three irregularly placed microscopic thornlets may be present. On the two or three innermost plates on each side of the median interradial line, the lateral spines and their companions are greatly reduced in size, being little more than mere miliary spinelets.

" In the interbrachial arcs, the marginal plates have the appearance of being brought over upon the disk; and their surface forms a peculiar bevelled area sloping outwards and downwards, the lower margin only of the infero-marginal plates falling in the outline of the interbrachial arc.

" The adambulacral plates are large and massive, of great breadth, and rather broader than long. They are comparatively widely separate, and the interspaces are filled up with ligament; the furrow margin is slightly convex. Their armature consists of :—(1.) a furrow series of five very small, short, equal, cylindrical, obtusely pointed spinelets, which radiate apart and form a small isolated semicircular comb directed over the furrow, the successive combs being well spaced apart. (2.) On the actinal surface of the plate are two large, robust, tapering, but obtusely pointed spinelets, placed one behind the other, which are subequal in length, or occasionally the outer-

most is the longest; this measures 4 mm. on the sixth or eighth adambulacral plate, the longest spinelets of the furrow series exceeding 1 mm. One
minute thornlet usually stands on the adoral side of the outer of the two
superficial spines; no other spinelets are present, and the surface of the
plate is covered with a membranous tissue. The ambulacral furrows are
comparatively widely open. The tube-feet are large, and have a rounded,
well-developed, knob-like termination.

"The mouth-plates are large, prominent and convex actinally, presenting a broad and rounded, but centrally rather flattened, margin towards the
actinostome; and the median suture of the pair is imperfectly closed. Their
armature consists of a marginal series of five mouth-spines on each plate,
the innermost one being twice or three times as long and robust as the
others, which are subequal and rather widely spaced. On the actinal surface
of each plate are three (or sometimes only two) robust tapering spinelets,
standing wide apart and forming a line parallel to the median suture; these
do not appear to attain the size of the superficial actinal spinelets on the
adambulacral plates. The surface of the plates is covered with membrane,
and no other spinelets are present.

"The actinal interradial areas are of very small extent, and do not
reach beyond the fifth adambulacral plate. The size and number of the
intermediate or ventral plates is undeterminable in spirit specimens on account of the membrane with which they are overlaid. So far as can be
judged, they appear to be comparatively large, and are certainly few in
number. A few of these bear one, or occasionally two, small spinelets 1.5–
2 mm. in length.

"The anal aperture is subcentral but indistinct, surrounded by large
spinelets.

"The madreporiform body, which is large and oval, is situated midway
between the centre of the disk and the odd inter-radial supero-marginal
plate. Its surface is slightly convex, finely grooved with numerous highly
convoluted striation furrows, and has on that account a remarkable resemblance to the 'brain' coral. Several large spinelets surround its margin, the
largest being at its adcentral end.

" The papulæ are very small and numerous, but confined to an area at the base of the rays, which does not extend beyond the fourth supero-marginal plate. The area occupies nearly the whole breadth between the supero-marginal plates, and extends well upon the disk, but no papulæ are present along a broad band which traverses the median interradial line.

" No pedicellariæ of any kind are found upon this species.

" Colour in alcohol, a dirty bleached ashy grey.

" *Individual variation.*—There is a fragmentary specimen from Station 235, in a very bad condition, of a much smaller example than the type above described, in which the spines of the actinal surface generally are proportionately much more delicate. From their length and character, I am disposed to consider this a feature of individual or even locational variation, rather than a phase of growth; a remark, however, which is merely conjectural, for its validity could only be proved by the examination of a large series of specimens.

" *Localities.*—Station 237. Off the coast of Japan, south of Kawatsu. June 17, 1875. Lat. 34° 37' 0" N., long. 140° 32' 0" E. Depth 1875 fathoms. Blue mud. Bottom temperature 35°.3 Fahr.; surface temperature 73°.0 Fahr.

" Station 235. Off Japan, south of Omae Saki. June 4, 1875. Lat. 34° 7' 0" N., long. 138° 0' 0" E. Depth 565 fathoms. Green mud. Bottom temperature 38°.1 Fahr.; surface temperature 73°.0 Fahr.

" *Remarks.—Pararchaster semisquamatus* and *Pararchaster antarcticus* are readily distinguished from the other species of the genus by the absence of pedicellariæ and the simplicity of the spinulation of the abactinal plates. The differences between the two forms are discussed in detail ·in the description of *Pararchaster antarcticus.*

" If these two species and the variety are viewed together as an independent or specially characterised type of the genus, their distribution is very remarkable and instructive, *Pararchaster semisquamatus* of the North Pacific being represented by a variety in the North Atlantic, whilst the closely-allied species *Pararchaster antarcticus*, from the Southern Ocean, presents some of the characters of the typical or Pacific form of *Pararchaster semisquamatus*, as well as some of its Atlantic variety *occidentalis.*

"*Pararchaster semisquamatus* var. *occidentalis*, nov.

"There is a single specimen from the western side of the North Atlantic, off the east coast of North America, which, although agreeing in a remarkable way in all essential points with the type just described, presents a number of variations which render it worthy in my opinion of nominal recognition,—in fact, it may ultimately prove to be a distinct species. At present, however, I hesitate from according it that rank on the slender evidence of a solitary and imperfect specimen, although the widely separated geographical position of the two dredging stations would certainly favour the adoption of such a course.

"The two forms are almost exactly of the same size. In the Atlantic example—the variety under notice—the spines on the supero-marginal plates are distinctly thicker and more robust at the base, while those on the infero-marginal plates are relatively smaller than in the Pacific form (the type). The two large spines on the actinal surface of the adambulacral plates are also smaller and shorter. On the abactinal surface the single minute thornlet which springs from the centre of the abactinal plates is distinctly shorter and thicker—a circumstance which gives at first sight a finely tuberculate character to the abactinal area when viewed from above. The large spines in the central region of the disk are smaller and much less numerous than in the Japanese form, and do not extend to the base of the rays. The lateral wall at the summit of the interbrachial arc is much less bevelled towards the abactinal surface of the disk, and the marginal plates do not bend over so conspicuously as in the type figured. The infero-marginal plates appear proportionately smaller in their transverse dimensions; and the marginal and furrow series of spines on the adambulacral plates are slightly more delicate and elongate. The roughening of the surface of the large spines on the marginal plates and elsewhere is more conspicuous and decided in the variety. The madreporiform body has not the slightly convex character noticed in the type, and less convolution is present in the striations, which have a more or less regular appearance of centrifugal radiation. The tube-feet have a smaller terminal knob.

"*Locality.*—Station 44 or 45. Off the coast of North America, east of Delaware and Maryland.

"Station 44. May 2, 1873. Lat. 37° 25' 0" N., long. 71° 40' 0" W. Depth 1700 fathoms. Blue mud. Bottom temperature 37°.2 Fahr.; surface temperature 49°.5 Fahr."

Pararchaster armatus, which is looked upon by VERRILL as another synonym of the present species, is described by SLADEN as follows ['89, p. 19]:

"*Pararchaster armatus*, n. sp. (Pl. I, figs. 5 and 6; Pl. IV, figs. 5 and 6).

"Rays five. R = 37 mm.; r = 6 mm. R > 6 r. Breadth of a ray between the second and third supero-marginal plates, 4 mm.

"Rays elongate, narrow and attenuate outwardly, tapering from the base to the extremity; subdepressed, with low vertical walls, and consequently nearly rectangular in section. Disk small. Abactinal surface flat and level. Actinal surface very slightly prominent at the mouth-angles. Interbrachial arcs wide and well-rounded.

"The abactinal surface of the disk and rays is covered with a plating of small, thin, subcircular spicules, overlaid with a delicate membranous tissue. The spicules bear centrally one, or sometimes two, and rarely three, very small papilliform thornlets, of uniform character throughout, excepting a few· on the disk which are slightly larger, but still quite inconspicuous and unnoticeable without close examination. Occasionally in some examples two or three pedicellarian apparatus may be present,—these are of the double comb or spiracle-like form; their structure is very simple, and the spinelets composing them are comparatively large, in fact, slightly larger than the small thornlets usually borne on the spicules.

"The supero-marginal plates, twenty-three to twenty-five in number from the median interradial line to the extremity, are elongate and low. The length is about three times greater than the height, and the upper margin of the plate, though actually confined to the lateral wall of the ray, is very slightly bevelled upon the abactinal surface. Each plate bears centrally on this margin a rather prominent tubercle, upon which is articulated an elongate, tapering, sharply· pointed spine, the third or fourth from the interradial line measuring about 4.5-5 mm. in length. There are

generally one or two small thornlets near the base of the spine, otherwise
the surface of the plate is simply covered with a very thin membranous
tissue. In the median interradial line is a high, odd, supero-marginal plate,
thick and tubercular abactinally, which bears an elongate, cylindrical, tapering spine, about 8 mm. in length, directed vertically. On the vertical wall
of the plate which stands in the interbrachial arc are a number of minute,
conical, sharply-pointed granules, usually along the median area ; and the
one or two neighbouring plates on each side are likewise similarly, but
irregularly and more sparsely, granulated.

"The infero-marginal plates are elongate and low like the supero-
marginal series, to which they correspond. Each plate bears a straight,
tapering, lateral spine articulated on a tubercle, and usually more delicate
and often shorter than the spine on the accompanying supero-marginal
plate. A second but much smaller spine is sometimes present on the inner
side, usually in the large examples, but it is often wanting altogether, and
seldom appears on the inner part of the ray and towards the extremity.
Occasionally one or two microscopic miliary thornlets are present on the
plate, and the surface of the plate is covered with a delicate membranous
tissue.

"In the median interradial line there is a large, broad, odd infero-
marginal plate, larger and broader than any of the others ; it is placed
immediately beneath the odd supero-marginal plate. It bears a small
representative of the lateral spine, and in addition several delicate miliary
spinelets and thornlets. The one or two next plates on each side are
likewise frequently more spinulate than any of the others.

"The adambulacral plates are large in proportion to the size of the
starfish, their length slightly exceeds their breadth, and they present a
prominent and rather abrupt convexity into the furrow. Their armature
consists of :—(1.) a furrow series of five to seven small, short, cylindrical,
obtusely tipped spinelets, subequal excepting the extremities of the series,
which are rather shorter ; all are closely placed and form compact isolated
little combs. (2.) On the actinal surface of the plate is one comparatively
large spinelet, robust at the base, and tapering to a pointed extremity ; on

the outer side of this a second, but smaller and more delicate, spinelet is present, and frequently also a small miliary spinelet on the adoral side of the large spinelet. The presence, however, of the second spine and of the miliary is by no means constant. They are often (perhaps usually) absent in small examples and on the inner and outer parts of the ray even of large specimens. The larger actinal spine measures about 1.7 mm. in length on the fifth or sixth plate from the mouth-angle, the marginal or furrow series being normally less than half this length.

"The mouth-plates are large, and convex actinally, and the united pair form a broad projection towards the actinostome, the free margin being more than a semicircle, and bulging laterally. The armature consists of a marginal series of five, or sometimes six, short, slightly tapering mouth-spines on each plate, the innermost being twice as large as the others. On the actinal surface of the plate are three subequal tapering spinelets, forming a line parallel to the imperfectly closed median suture; sometimes a fourth is present on the outer side of the series, and sometimes its place is occupied by a miliary thornlet only; but often it is wanting altogether. No other spines are present, and the surface of the plates is covered with thin membrane.

"The actinal interradial areas are very small indeed, not more than three or four intermediate plates being present in young examples, and in larger ones not more than eight to ten. These form only a single series between the marginal plates and the adambulacral and mouth-plates; their surface is covered with membrane and the larger plates bear two or three small thornlets. There is usually one small pedicellarian apparatus in each area, which does not stand in the median interradial line, but is usually separated therefrom by one plate. It is of the 'spiracle'-like or double-comb form, previously described. Rarely two are present. There are no pedicellarian apparatus between the infero-marginal plates.

"The anal aperture is subcentral and distinct, and usually there are two or three spinelets in its neighbourhood larger than those borne on the spicules generally.

"The madreporiform body is comparatively large and oval, placed about midway between the centre of the disk and the margin, and its surface is conspicuously convex and is grooved with deeply cut and highly convoluted striation furrows.

"The papulæ are tolerably numerous, but are large and distinct and limited to a small area at the base of the ray, which does not extend outwardly beyond the first supero-marginal after the odd interradial plate, and its inward extent on the disk would be bounded by a circle drawn upon the disk, with its margin touching the inner edge of the madreporiform body.

"Colour in alcohol, a bleached ashy white.

"*Individual variation.*—In some examples a second infero-marginal spine, standing below the true lateral spine, is much more strongly developed than in others, and this in specimens of the same size and from the same locality. In others again it is represented only by a small miliary spinelet, or may be entirely absent altogether. I have only found them well-developed in one case, and that not the largest specimen in the series. In one example I notice a tendency towards diminution in the number of spinelets in the furrow series on the adambulacral plates, and this is shown in the abortion or total absence of the outer spinelets at either extremity of the series; as a result of this reduction there may be only three, four, or five spinelets in place of six, which appears to be the normal number, and the central spinelets of the series appears comparatively long. The occasional presence of an additional spine on the actinal surface of the adambulacral plates has already been remarked upon.

"*Young phase.*—The smallest example in the collection (from Station 46) has a minor radial measurement of 3.5 mm., and the rays appear to be comparatively robust. This specimen presents in a most unequivocal manner all the characters of the type, and even though so young there need not be the slightest hesitation about referring the form to this species. So small indeed are the differences between the juvenile and the adult stages, that the mature form in this species may well be said to exhibit on a large scale all the features of the embryonic phase.

"In this juvenile example there are no actinal intermediate (ventral) plates. The disk and the base of the rays have a somewhat villous or subpapillose appearance, the spinelets on the abactinal plates being decidedly robust for the size of the animal, and rather thickly covered with membrane. The papulæ are well-developed and distinct; and there are three large spiracle-formed pedicellarian apparatus on the disk. The odd interradial plates and spines are very large, the latter being about 5 mm. long; and are strongly denticulate along the shaft, suggesting to a certain degree the miniature of a *Cidaris*-spine. The knob-like terminations of the tube-feet are large and button-shaped. The genital foramina are discernible on each side of the odd interradial plate, opposite the first supero-marginal plates and near their inner edge.

"*Localities.*—Station 50. South of Halifax, Nova Scotia. May 21, 1873. Lat. 42° 8' 0" N., long 63° 39' 0" W. Depth 1250 fathoms. Blue mud. Bottom temperature 38°.0 Fahr.; surface temperature 45.°0 Fahr.

"Station 46. Off the coast of North America, east of New Jersey. May 6, 1873. Lat. 40° 17' 0" N., long. 66° 48' 0" W. Depth 1350 fathoms. Blue mud. Bottom temperature 37.°2 Fahr.; surface temperature 40.°0 Fahr.

"Station off the coast of Portugal. January 1873. (Exact date and station not recorded.)

"*Remarks.*—*Pararchaster armatus* is characterised by the presence of comb-formed pedicellariæ on the abactinal area and in the actinal interradial areas, and by their absence between the infero-marginal plates. The simplicity of the armature of the infero-marginal plates (one lateral spine and sometimes a small companion), and the comparatively large number of six or seven spines in the furrow series on the adambulacral plates, also serve to readily distinguish this interesting form."

With regard to the possible identity of PERRIER's *Archaster simplex* with *P. armatus* SLADEN makes the following remarks ['89, p. 6]: " The asterid obtained during the ' Blake' dredgings and described by PERRIER ['81, p. 28] under the name of *Archaster simplex*, is without doubt a *Pararchaster*. It is, however, an immature form, and I am unable to say from the description given whether it belongs to an independent species or is the young of one of those herein described. The figure given ['84, p. 264] is altogether unlike the smallest example of *Parachaster armatus* in the character of the armature both of the adambulacral and infero-marginal plates, and this appears to be the only form with which a direct comparison can be instituted. The description is too short to render any assistance in this case."

In his paper of 1894 VERRILL makes some important critical remarks on this species, putting down *Pararchaster semisquamatus* var. *occidentalis* SLADEN and *Pararch. armatus* SLADEN as synonyms of his *Benthopecten spinosus*. He says ['94, p. 245] :

" A comparison of a large series of this species, of various sizes from those that are 15 mm. up to large ones 260 mm. in diameter, shows that the two forms described by SLADEN from off the Atlantic coast are probably both identical with that described by me.

" This species varies considerably in several details of its structure, according to its age. None of SLADEN's specimens were full grown (largest size given is 74 mm. in diameter). Moreover there is often considerable variation in specimens of the same size and from the same locality, in the size of the disk, number, size, and arrangement of the spines on the marginal plates, etc. Some few examples have the disk at least one-third broader than others having the same length of rays, and such specimens naturally have large inferior interradial areas, with the plates more numerous than usual, as many as twenty to twenty-five being present in some cases. The papulæ often extend out on the rays, in large examples, as far as the fifth pair of marginal plates ; they cease sooner in the median line than to either side of it. They are often present on the central area of the disk, among the large primary spines. The actinal and adambulacral spines on the largest specimens are more numerous and longer than SLADEN's descriptions indicate, but the half-grown specimens agree well with his examples, in most respects.

" The pectinate pedicellariæ described by SLADEN as characteristic of *P. armatus* are commonly lacking entirely on our specimens, or exist only in very small numbers. The dorsal plates of the rays are rounded and ovate, unequal, and most commonly isolated in the integument. They usually bear only a single, small, slender, acute spine, rarely two. The large disk-spines are variable in number and length, but they are always restricted to the central area of the disk, and the largest are borne on the primary plates.

The large single spines on the odd interradial marginal plates are usually long, tapered, acute, and distinctly larger and longer than those on the disk. The lower marginal plates generally bear, in large specimens, one large, primary, acute spine, and one or two, rarely three, secondary ones below it, besides several small, slender, divergent, rough spinelets scattered around their bases. The adambulacral plates, in such specimens, generally have two or three long, slender, rough spines on the actinal side, besides several small, slender, spinelets, on the actinal margin ; the angular and salient inner margin usually bears about seven slender spines in a V-shaped group.

"I have seen a few regular four-rayed specimens, and also one peculiar monstrosity, in which a small supplementary ray buds out from the side of the regular ray, near the base. This species occurred at many stations in 721 to 2021 fathoms."

Wood-Mason and Alcock mention this species under Sladen's name from the dredgings of the "Investigator," and adds ['91, p. 428], "One specimen from station 111, 1664 fathoms. Colour in fresh state uniform salmon-red." Alcock ['93, p. 75] again refers to it as follows : "Bay of Bengal, on a bottom of *Globigerina*-ooze, in 1664 fathoms. Colour uniform salmon-red."

In his paper of 1895 Verrill makes some further remarks on this species as follows ['95, p. 129] :

"Bathymetrical range, 721–2021 fathoms. Most common in 1200 to 1600 fathoms. It was taken at 60 stations, between N. lat. 42° 47' and 35° 10', by the U. S. Fish Commission. Off the coast of Portugal (t. Sladen).

"The genus *Pararchaster* Sladen (1889) is synonymous with *Bentho-pecten* (1884). This is a strictly deap-sea genus, none of the species occurring in less than 400 fathoms. It is found in all the oceans. The following additional species were described by Mr. Sladen :—

B. spinosissimus (Sl.) V., Atlantic................ 425 fath.
B. simplex (Perrier) V., Caribbean .... 1323 „
B. antarcticus (Sl.) V., Southern Ocean .... 1675 „
B. pedicifer (Sl.) V.,   „   „   ...... 1600–1900 „
B. semisquamatus (Sl.) V., Pacific .... 565–1875 „

S. GOTO :

" All the above species are closely related. Mr. SLADEN also described a single young specimen, taken off Delaware Bay by the Challenger, as a variety (*occidentalis*) of *B. semisquamatus*. The type of the latter was from off Japan.

" This supposed variety appears to me to agree in all respects with many young specimens of our *B. spinosus*, judging from Mr. SLADEN's detailed description. The two species are evidently very closely allied."

Some further remarks on this species are made by VERRILL as follows ['99, p. 217] :

" This species was taken north of Cape Hatteras, at 62 stations, between N. lat. 42° 47′ and 35° 10′, in 721 to 2021 fathoms, by the U. S. Fish Commission. Most common in 1200 to 1600 fathoms.

" It was also taken in the Gulf of Mexico, station 2380 and station 2381, in 1430 and 1330 fathoms, and off Jamaica, station 2127, 1639 fathoms."

LUDWIG in his recent revision of the *Notomyota* [:10, p. 464] looks upon *B. spinosus*, *B. armatus* and *B. semisquamatus* as distinct species. Further, he regards *B. armatus* as identical with *Archaster simplex* PERRIER (= *B. simplex*). According to him *B. spinosus* and *B. simplex* occur in the Atlantic, and *B. semisquamatus* in the Indian Ocean and the West Pacific. He says [:10, p. 464], " *Pararchaster armatus* SLADEN 1889 fehlt in dieser Liste und in der folgenden Bestimmungstabelle, weil ich den früher aufgestellten *Archaster simplex* PERRIER 1881 für die Jugendform der von SLADEN als *Pararchaster armatus* bezeichneten Art halte." Again a little further on [p. 466] he says, " Mit VERRILL stimme ich darin überein, dass SLADENs *semisquamatus* var. *occidentalis* identisch ist mit VERRILLS *spinosus*. Dagegen scheint mir *armatus* SLADEN in Gegensatz zu VERRILLS Ansicht nicht mit *spinosus* vereinigt werden zu können ; *armatus* muss aber seinen Namen zugunsten von PERRIERS *simplex*[1] einbüssen, da *simplex* der ältere Name für

---

[1] PERRIER's full description of *A. simplex* runs as follows ['84, p, 264] :

" No. 40.—Profondeur, 1,323 brasses.—Lat. N., 23° 26′.—Long. O., 84° 02′ (1 exemplaire).

" Point de plaques ventrales entre les plaques marginales ventrales et les pièces dentaires.— Une plaque marginale impaire dans l'angle des bras supportant les pièces dentaires qui sont arrondies au sommet.—Quatorze plaques marginales plus longues que larges, légèrement convexes sur le bord externe, de manière à faire paraître le bord des bras un peu festonné. Ces plaques supportent directement les plaques ambulacraires qui sont plus petites, polygonales, saillantes

eine Form ist, die ich glaube mit Sicherheit als eine Jugendform von *armatus* ansprechen zu können. *B. antarcticus* ist offenbar dem *semisquamatus* nahe verwandt. Auch *acanthonotus* nähert sich dem *semisquamatus* und wird vielleicht einmal damit vereinigt werden müssen, was aber bei dem derzeitgen Stande unserer Kenntnisse eine offene Frage bleibt."

FARRAN reports *B. armatus* (SLADEN) from the west coast of Ireland and makes the following remarks [:13, p. 2] :

" Helga.

" S. R. 944—17 v '10· 51° 22′ N., 12° 41′ W., soundings 982 fms., ooze. Trawl.—Thirty.

" This species is now added to the British and Irish deep-water fauna. It was originally described by SLADEN in 1889, under the name *Pararchaster armatus*, from *Challenger* specimens taken off Nova Scotia and New Jersey and off the coast of Portugal. In the Irish specimens there are usually 2–3 subequal spinules on each plate in the papular area at the base of the arms, but towards the extremity of the arms there is rarely more than one. SLADEN describes and figures these spinules as single or rarely in groups of two or three, but an examination of the *Challenger* specimens showed that the arrangement in groups of 2–4 was frequent. In the largest Irish specimen R=ca. 60 mm. (tips of arms broken), and in the smallest R=11 mm., the average value of R being about 35 mm. Authorities differ as to the synonymy of this species. VERRILL (1895) regards it as a synonym of *Benthopecten spinosus* described by him in 1884. LUDWIG, in a recent revision of the species and genera belonging to the order Notomyota (1910), holds

dans le sillon ambulacraire. Quelques petites épines et une beaucoup plus grande, fine et pointue, situées tout à fait sur le bord de chacune des plaques marginales ventrales.—Plaques adambulacraires portant sur leur bord 4 ou 5 piquants divergents et un autre très grand à leur centre. Tubes ambulacraires terminés par une ventouse bien distincte.

" Quatorze plaques marginales dorsales, plus longues que larges, de forme presque ovale, plus une impaire triangulaire, portant chacune un long piquant pointu au sommet de l'angle interbrachial.

" Piquant de la plaque impaire ayant une longueur au moins double de celles des autres Une autre épine près de l'anus.—Plaque du disque petites, très peu apparentes.—Dos présentant de nombreuses petites épines, espacés, sans plaques calcaires apparentes. Plaque madréporique petite, arrondie, à surface irrégulièrement vermiculée, presque au contact de la plaque impaire.

" E=35 mm ; R=18 mm ; r=3 mm ; R=3 r ; d=4 mm."

that *Benthopecten spinosus* is distinct, but that SLADEN's *Pararchaster armatus* is the same as *Archaster simplex* described by PERRIER in 1881, and should consequently be known as *Benthopecten simplex*. According to SLADEN, however, PERRIER's figure 'is altogether unlike the smallest example of *P. armatus* in character of armature both of adambulacral and infero-marginal plates.' Under these circumstances it seems best, provisionally, to retain SLADEN's name.

"It may be noted, however, that the elongated disc spines which are so noticeable in *B. spinosus* are not to be found in any of the Irish specimens of *B. armatus*. In the largest Irish specimen (R=ca. 60 mm.) the largest disc spine measures only 1.5 mm., while in a specimen of *B. spinosus* in the British Museum (R=95 mm.) the disc spines are as much as 7 mm. long, and in a smaller specimen (R=50 mm.) they measure 2.5 mm. The large interbrachial spines in the latter specimen measure only 6 mm., whereas in Irish specimens of R=40 mm. they reach a length of 8 mm. · It was only in the largest Irish specimen (R=ca. 60 mm.) that the comb-shaped pedicellariæ could be found."

Cheiraster oxyacanthus (SLADEN).

This species is not represented in any of the collections studied by me. The following is the original description of SLADEN ['89. p. 38]:

"*Pontaster oxyacanthus*, n. sp. (Pl. IX. figs. 1 and 2; Pl. XII. figs. 7 and 8).

"Rays five. R=73 mm.; r=11 mm. R>6.5 r. Breadth of a ray near the base, 11 mm.

"Rays elongate, tapering continuously from the base to the extremity, the outer part narrow and attenuate, and of great flexibility, the outer part in the specimen under notice curled round with an abactinal recurvature. Interbrachial arcs well-rounded. Abactinal surface plane. Actinal surface subplane. Lateral walls of the rays comparatively high and vertical.

"The general paxillæ of the abactinal surface are comparatively small and simple, closely placed on the disk and inner part of the rays, but diminish greatly in size as they proceed outward, and are more widely spaced on the outer half of the ray. The crown consists of four to seven short, delicate, tapering spinelets, which radiate outward nearly horizontally, and appear to proceed almost from the centre of the tabulum. A few have a small central spinelet, more elongate and robust than the surrounding series. Upon the disk and at the extreme base of the rays a number of larger and specially-armed paxillæ are distributed amongst the general paxillæ above described. Each of these bears an elongate conical, tapering, robust, vertical spinelet, surrounded at the base by a ruff-like collarette of twenty or more minute ciliary thornlets. The central spine is powerful, and may measure 2 to 3 mm., but the length decreases as the paxillæ recede from the centre of the disk. The disposition of these armed paxillæ is somewhat irregular; they are, however, confined to a median radial area, and two or perhaps three irregularly defined longitudinal lines may be distinguished; their presence amongst the small and comparatively inconspicuous general paxillæ of the abactinal surface forms a striking feature in the species.

"The supero-marginal plates, thirty-eight in number from the inter-radial line to the extremity, form a well-defined though rather narrow border to the rays. The breadth, height, and length are about subequal, the last dimension being slightly in excess, and the plates are slightly convex and subtubercular abactinally. Each plate bears a robust conical, tapering spine, about 4 mm. in length midway along the ray, directed perpendicularly to the abactinal plane of the ray. These spines diminish a trifle in length as they approach the base of the ray, but increase in robustness; whilst they decrease both in length and robustness on the outer part of the ray. On the outer side of this spine is usually a smaller and more delicate spinelet directed outward at an angle of about 45°, and occasionally two may be present, and sometimes they are little more than elongate miliary spinelets. The rest of the plate is covered with numerous very short, delicate, pointed miliary thornlets.

"The infero-marginal plates alternate with the superior series, and their dimensions are subequal. Each plate bears three robust, conical tapering, pointed spines, similar to those on the supero-marginal plates. These spines form a series along the median transverse line of the plate; the outermost or lateral spine is the longest, and slightly greater than the supero-marginal spine ; the innermost spine of the three is the smallest, and is about half the length of the lateral one. The longest lateral spine is the third or fourth from the interradial line, and measures about 5 mm., and the succeeding ones decrease slightly in length as they proceed outward. The rest of the plate bears a few widely spaced, small, miliary thornlets, some of which at the base of the large spines are more elongate than the others. There is a naked suture-line between each plate.

"The adambulacral plates are slightly longer than broad, and with only a slightly convex margin towards the furrow. Their armature consists of :—
(1.) A furrow series of six short, rather thick, cylindrical, obtusely pointed spinelets, the inner pair slightly longer, and the outer one at each extremity very much smaller, than the rest ; their posture resembles that of the fingers of a hand held slightly concave. (2.) On the actinal surface of the plate are two robust, slightly tapering but obtusely pointed spinelets, one behind the other in the transverse median line. The innermost stands close behind the marginal series, the outermost is slightly larger and is subequal in size to the innermost of the three spines on the infero-marginal plate. Two or three minute miliary spinelets may be present on the outer margin of the plate, but the rest of the plate has a naked appearance, and is only covered with membrane.

"The mouth-plates are powerful and slightly convex ; each plate of the united pair bears on its free margin six mouth-spines, the innermost being the longest, most robust, and slightly compressed ; the others decrease as they recede from the mouth and become very short. On the actinal surface of each plate are three robust, tapering, secondary mouth-spines, two placed so that a line joining them would run parallel to the median suture, and this line is continued on the outer part of the plate by one or two smaller spinelets. The third large spinelet is placed opposite the interspace between the two large spinelets above mentioned, midway between them and the outermost of the marginal mouth-spines.

"The actinal interradial areas are very small, not more than eight to ten intermediate plates being present in each. The two innermost may bear a small central conical spinelet surrounded by a few minute miliary thornlets only. There are three complex pedicellarian apparatus in each area, situated in the lateral sutures which separate the two innermost intermediate or ventral plates; these organs consist of an oval cavity equally scooped out of the margins of the two adjacent plates, each margin beset with about five short, compressed, pointed 'dog-tooth' shaped spinelets, directed over the cavity, and frequently turned upwards into the same. The major axis of the cavity measures about 1 mm. There are also structures which I take to be very minute pedicellariæ present on a number of the adambulacral plates, appearing to protrude through the membrane, usually on the outer part of the adoral margin.

" The anal aperture is subcentral and distinct, its margin being surrounded by a close circlet of small spinelets longer than the small spinulation of the paxillæ. At a little distance from the aperture is a circlet of the large armed paxillæ, standing more or less regularly in the radial and interradial lines.

" The papulæ, though confined to the base of each ray, occupy a much greater area than in the other members of the genus, and are probably not comprised in a specially constituted papularium. They are small and widely spaced, more than fifty may be counted in each area, and isolated ones extend as far as the fourth marginal plate.

" The madreporiform body, which is small, circular, and convex, is situated close to the marginal plates, and its surface is striated with rather fine convoluted furrows. One of the large powerfully spined paxillæ stands on its adcentral side.

" Colour in alcohol, a bleached ashy white.

" *Locality.*—Station 232. South of Yeddo (Japan). May 12, 1875. Lat. 35° 11' 0" N., long. 139° 28' 0" E. Depth 345 fathoms. Green mud. Bottom temperature 41°.1 Fahr.; surface temperature 64°.2 Fahr.

" *Remarks.*—This is, perhaps, the handsomest species in the genus, at any rate the most striking, and is at the same time remarkably well characterised. Without referring to minor points of difference, it will suffice to say that the form is at once distinguished from all others by the group

of large conical spines on the abactinal area of the disk, and by the presence
of more than one large spine arranged in transverse series on the infero-
marginal plates. Even without these striking features, *Pontaster oxyacanthus*
would be well marked."

LUDWIG [: 03, p. 18] gives a synoptical key of the *Cheiraster* species
known down to the date of his writing, in which this species is also
included.

LUDWIG [:10] has lately referred this species to the genus *Luidiaster* as
defined by him, and he further thinks that *Archaster dawsoni* VERRILL '80
(= *Acantharchaster dawsoni* VER.) and *Cheiraster horridus* FISHER :06 are
different names for the same species, that therefore, according to the law of
priority, *L. dawsoni* should be the correct name. FISHER [:11, p. 127]
however, is of the opinion that both *horridus* and *oxyacanthus* are distinct
from *dawsoni*.

The genus *Luidiaster* is characterised as follows [LUDWIG :10, p. 451] :

"Diagnose: *Cheirasteridæ* mit flachen, in ihrem distalen Teile zwei-
lappig umgrenzten Papularien ; die Pedicellarien sind kammförmig und stehen,
falls sie nicht ganz fehlen, auf den Ventrolateralplatten oder auch auf
anderen Platten, stets über je zwei Platten ; die Paxillen tragen eine grössere
Zahl von Stachelchen, die ein häufig verlängertes Zentralstachelchen umstellen;
die Adambulacralplatten besitzen zwei oder mehr subambulacrale Stacheln.

"Die Paxillen haben bald einen ganz niedrigen, bald einen erhöhten
Schaft, der sich aber dann von der Basis des Paxillus nicht scharf absetzt.
Das Zentralstachelchen der Paxillenkrone ist besonders lang auf den grösseren
Paxillen des *L. dawsoni*. Die unteren Randplatten besitzen bei *L. hirsutus*,
gerlachei und *vincenti* nur einen, bei *teres* zwei, bei *dawsoni* mehrere Stacheln.
Die Ventrolateralstachelchen sind bald von gleicher, bald von ungleicher
Grösse. Die Pedicellarien können bei jungen Tieren noch ganz fehlen, z.
B. bei *L. hirsutus* und *dawsoni*, oder sind bei den jungen Tieren weniger
zahlreich, z. B. bei *L. teres*, und sind auch bei erwachsenen oft sehr
unbeständig, z. B. bei *L. hirsutus*."

Cheiraster yodomiensis, n. sp.

(Pl. I, figs. 1—7.)

I have only one dried specimen of this new species, which however is in excellent condition. The measurements are as follows :

r	R	R : r	MS
23 mm.	170 mm.	7.4	43

The disk is comparatively large, considering the slenderness of the arms; the interbrachial angle is very nearly a right angle. The actinal side is plane, and the long spines of the adambulacral and inferomarginal plates are very conspicuous. The abactinal side of the arms is for the most part perfectly plane, but that of the disk and the basal parts of the arms is slightly concave in the dried specimen, and must have been somewhat inflated in life. There is also, in my specimen, a circular depressed area of an orange red colour in the centre of the disk, in which the anus is situated. The paxillæ with large central spines extend into the arms about as far as the fifth or sixth superomarginals, beyond which the spines are considerably smaller and form small groups of two or three arranged in a row on either side of the arm. The spines near the outskirts of the spinous area are also smaller than those nearer the centre. The spines of the marginals are very long and conspicuous.

Superomarginals.—The first superomarginal is nearly twice as high as broad, and is more prominent than the others, on the abactinal side. It carries a single long, slender, conical, pointed spine which may be as long as 14 mm., and one or two very much smaller spines near the base of the former, usually on the

outer side. In one interradius, one of the first superomarginals
carries two spines of nearly equal size and much smaller than
the principal spine of its fellow. The first superomarginal presents
a wart-like form on the abactinal side, and a bluntly keeled ridge
on the outside which is covered with small rough conical granulations. The remaining superomarginals each carry a large
conical spine, exactly similar to that of the first superomarginal
and hardly smaller as far as about the middle of the arms, but
gradually decreasing in size from thence towards the tip. Some
of the plates may carry in addition one or two, or exceptionally
three, smaller spines. There is besides a more or less complete
circle of very small spines around the base of the principal
spine, to which the accessory spines usually belong. Each
superomarginal presents, like the first plate, a wart-like form
on the abactinal side, and is separated from the next one
by a more or less soft tissue, when seen from the outside.
The number of superomarginals is 43 for one of the complete
arms. The superomarginals as a series are more elevated than
the paxillar area and form a conspicuous rim on the abactinal
side.

Inferomarginals—Only the first inferomarginals are coincident
with the superomarginals, all the rest are alternate, and from the
second or third plate on they are directly contiguous to the
adambulacral plates to which they are closely applied. Each plate
carries a transverse series of three or four large spines, of which
the one nearest the abactinal side is largest, being only slightly
smaller than the principal spines of the superomarginals, and the
one at the opposite end is the smallest. The modal number of
spines for the inferomarginals is three. There is a more or less

complete enclosure of very small spines around the base of the
principal spines, which is more distinct on the arms but much
less so or atrophied in the interbrachial angles of the disk.
There are besides some very small spines either forming a series
near the margin of the plate or more irregularly scattered on its
surface. Once in a while, some of the spines near the margin
are better developed than usual and together with the similarly
developed spines of the next plate form what have been called
" pectinate pedicellariæ," which are characteristic of this genus, and
appear to have been described so far only from the ventrolateral
plates. These structures will be mentioned again later. The
lateral margin of the arms is formed more by the inferomarginals
than by the superior series.

Adambulacrals.—The adambulacral plates (Pl. I, fig. 4) are
nearly twice as broad as long and present each a wedge-shaped
surface towards the furrow. At the base of the arms there are
two adambulacral plates to each inferomarginal, but further out
the adambulacrals become relatively more numerous, although the
ratio of the two may not become as great as three to one. Each
adambulacral plate bears at the furrow margin a series of two to
three, very lightly curved spines, which may be as long as
3.5 mm. at the central end of the furrow. The modal number of
these furrow spines is two, but three is not very rare. On the
actinal surface of the plate there are 2–4 large spines, forming a
single row at right angles to the furrow and making a single
transverse series with the inferomarginal spines, where the two
sorts of plates happen to coincide. These spines may be as long
as 7 or 8 mm., and unlike those of the marginals are always
rounded or truncated at the end. These actinal spines of the

adambulacral plates are mostly two in number, occasionally three
and very rarely four. In the last case the spines are of subequal
size and much smaller than when there are less. There are in
addition some exceedingly small spines on the actinal surface of
the adambulacral plates.

Mouth-plates.—The mouth-plates (Pl. I, fig. 3) are comparatively large and are similar in general form to the adambulacral
plates, the two plates of a pair together forming a trapezoid with
the base turned towards the mouth and separated from each
other by a space covered over with a membrane. Each plate has
the form of a nearly right-angled triangle with the right angle
apposed to that of its fellow plate. On the side turned towards
the month there is a seires of 4–6 spines, gradually decreasing in
size towards the first adambulacral plate ; on the actinal surface
of the plate there is a more or less regular series of some five or
six spines similar to the large spines of the adambulacral plates
and gradually becoming smaller away from the mouth.

Ventrolaterals.—The ventrolateral plates (Pl, I, fig, 5, 6) are
confined to the disk, and there are only about a dozen plates in
each interradius, forming two arc-shaped series between the inferomarginals and the adambulacral plates. The inner series i.e.
the one nearer the centre of the disk, consists of some nine plates,
of which the two on either side of the interradial line are larger
than the others and bear each one or two long spines similar to
those of the inferomarginals but smaller. The next pair of plates
are also tolerably large and bear each a similar spine ; the
remaining plates of the inner series are small and bear a number
of exceedingly small spines without any regular arrangement ;
these small spines being also found sometimes on the larger

plates. The outer series of ventrolaterals consists of two or three very small plates confined to the inner side of the first infero-marginals; they may bear a few exceedingly small spines. The pectinate pedicellariæ which are characteristic of this genus are found between the first and second plates of the inner series of ventrolaterals, either one on either side or only one in an inter-radius. Each pedicellaria consists of 4–6, somewhat flattened, slightly curved spines, about a millimeter in length, situated on the margin of the first ventrolateral turned towards the second ventro-lateral plate, and about as many spines on the latter projecting against the spines of the first plate, the spines of the two plates alternating with one another. Without seeing a live specimen it is hard to say how these spines act during life, but the propriety of applying the name of pedicellaria to them appears to me open to doubt. In all the Phanerozonia I have examined, the pedicel-laria is accompanied by a small hole in the plate, which gives exit to the muscle fibres from the subjacent tissue to the base of the pedicellaria-valves; so that even after the pedicellariæ have been abraded, their presence can be detected by means of the small holes left in their places. Moreover the valves of a pedicellaria always belong in other forms to the same plate. Now in the case of the pectinate pedicellariæ of *Cheiraster* these two characteristics do not apply, the constituent spines belong to different plates and they do not leave any hole after them, showing that the muscle fibres, which are no doubt present around the bases of the spines, must be merely superficial and do not pass out from the deeper layer, as in the typical pedicellariæ. As mentioned above these pectinate pedicellariæ are also present sometimes on the infero-marginals, and in such a case the plates may be so far apart from each other that the comparatively short spines of the two adjacent plates do not come together at all. The exact function of these pectinate pedicellariæ has yet to be determined by observations on living specimens.

Paxillæ.—The paxillæ are comparatively small and those with a large central spine are confined to the disk and the basal parts of the arms. There are also more or less irregular series of paxillæ bearing small spines on either side of the arms at a short distance from the superomarginals, each member of the series being separated from the next by four or five paxillæ without the large spine. These latter (Pl. I, fig. 7) are in general more or less roundish in shape, except where the papular pores are comparatively numerous, when the plates tend to assume more or less a stellate form. The tabulum is very low, with a simply rounded summit covered with some six to ten exceedingly small granuliform spines without any distinction into centrals and peripherals. In the paxillæ with a central large spine the top of the tabulum is more or less elevated and nearly hemispherical; the central spine is sharply pointed and in the-disk may be as long as 7 mm., but only 2 mm. or so at the outskirts of the spinous area. The central spine is surrounded by a circlet of very small spines, placed at a short distance from its base. The papular pores are absent from the immediate vicinity of the anus, but are quite numerous elsewhere in the spinous area. An irregular series of these pores also extends along either side of the arms close to the superomarginals, about three-fourths the length of the entire arm.

Structures similar to the pectinate pedicellariæ of the ventro-lateral and inferomarginal plates are also found on the abactinal surface. They are smaller in size, and are usually formed by spines belonging to three paxillæ, there being three or four spines to each paxilla. They may also be formed by spines belonging to only two paxillæ.

Madreporite.—The madreporite is fairly large, rounded polygonal in outline, elevated above the general surface of the abactinal side, close to the margin of the disk; the grooves are mostly radiating.

Locality.—My specimen is from Yodomi, off Misaki, from a depth of 160 m. The colour of the dried specimen is brownish yellow all over, except the central perianal area which is orange red.

A single dried specimen in S.C.

This species is closely related to *Ch. oxyacanthus* (SLADEN), but is distinguished from it by several characters.

Dr. W. K. FISHER points out to me that this species perhaps belongs to *Luidiaster*, in which special dorsal muscles are attached to either one or two of the proximal ambulacral plates by a distinct tendon, while in *Cheiraster* they are not. This and the allied genera have recently been subjected to revision by LUDWIG [: 10], and according to his synoptical key the present species appears to fall in well with the genus *Luidiaster* as defined by him. Although owing to the dried condition of my only specimen I am not able to settle the anatomical characteristic mentioned by FISHER, there is no doubt that the present species stands close to *Ch. oxyaçanthus;* and as the latter is referred to *Luidiaster* by LUDWIG *yodomiensis* may perhaps also be included in that genus.

PORCELLANASTERIDÆ.

Ctenodicus crispatus (RETZIUS).

(Pl. VII, figs. 113-119; Pl. VIII, figs. 120-126).

This species appears to have been first figured by LINCK in 1733 under the name of *Astropecten corniculatus*, and was then described by RETZIUS in 1805 as *Asterias crispata*. PERRIER also gives references to SABINE ['24] and DEWURST, the former figuring it under the name of *Asterias polaris*, the latter referring to it as *Asterias aurantiaca*. GRAY in his paper of 1840 [p. 180] includes two species in the first section of his genus *Astropecten*, viz. *Astropecten corniculatus* and *Astr. polaris*, both now considered as synonyms of the present species. The passages relating to the species under review are as follows :

" 1. Body pentagonal; rays short.

" 1. *Astropecten corniculatus*, LINCK. t. 27. & t. 36. f. 63.

" Inhab.— Perhaps a variety of the next.

" 2. *Astropecten polaris* = *Asterias polaris*. SABINE, PARRY'S Voy. 223. t. 1. f. 2, 3.

" Inhab. North Sea."

The genus *Ctenodiscus* is due to MÜLLER and TROSCHEL, who describe two species in their " System " as follows [p. 76] :

" Species 1. *Ctenodiscus polaris* Nob.

(References to LINCK with? SABINE and GRAY ['40]).

" Der kleine Halbmesser verhält sich zum grossen wie 1 : 2. Körper pentagonal, die Arme dreieckig, weniger breit am Grunde als lang. Furchenpapillen 3-4 nebeneinander auf jeder Platte, fast gleich, conisch, spitz; nach aussen davon trägt die Platte meist noch einen Stachel. Die Bauchschienen bestehen aus zwei Reihen glatter Schuppen, an den längsten Reihen sind 7 Schuppen vorhanden. Randplatten 12 an jedem Arme. Die unpaarige Platte an der Spitze ist fast doppelt so lang wie breit, trägt 2 Stacheln am Ende und dahinter noch eine kleinere Erhöhung. Sie ist auch am

Rande gewimpert. Die Madreporenplatte liegt in der Nähe des Randes und hat parallele Furchen.

"Grösse : 1½ Zoll.

"Fundort: Groenland, mitgetheilt durch Prof. ESCHRICHT. Im Museum zu Stockholm.

"Species 2. *Ctenodiscus pygmaeus* Nob.

"In allen Punkten dem vorigen gleich; er unterscheidet sich nur durch die Zahl der Randplatten, 6 an jedem Arme, deren unpaarige letzte verhältnissmässig weniger lang und grösser ist. Dann zeichnet sich diese Art durch eine auf der Mitte des Rückens befindliche beutelförmige Erhebung der Rückenhaut aus, die ebenfalls mit Paxillen besetzt ist. Dieser Vorsprung bildet eine warzenartige conische Erhebung, die noch einmal so lang wie breit ist. Geschuppte Schienen der Bauchseite befinden sich bloss in den Winkeln zwischen den Armen, und enthalten nur zwei Schuppen in einer Reihe.

"Grösse : 5 Linien.

"Fundort: Groenland, mitgetheilt druch Prof. ESCHRICHT."

In an appendix to the same work (p. 129), we find the following addenda :

"*Ctenodiscus polaris* p. 76.

"Hierher als synonym *Asterias crispata* RETZ. Diss. p. 17. Das Originalexemplar von RETZIUS im Museum zu Lund hat nur vier Arme, die sehr spitz sind, und soll aus dem Indischen Ocean sein, was wahrscheinlich ein Irrthum ist.

"*Ctenodiscus pygmaeus* p. 76.

"Es ist wahrscheinlich nur das Junge des vorhergehenden."

This conclusion of MÜLLER and TROSCHEL has been endorsed by subsequent writers.

The species is mentioned by DÜBEN and KOREN ['46[1]], FORBES ['52[1]], STIMPSON ['53[1]], LÜTKEN ['57[1]], and SARS ['50,[1] '61[1]].

DUJARDIN and HUPÉ describe it as follows ['62, p. 431] :

"Ctenodisque crépu. *Ctenodiscus crispatus.*—LÜTKEN.

---

1) *Fide* LUDWIG.

(References omitted.)

" Espèce à cinq bras, dont la longueur est double du plus petit rayon du disque : ces bras sont triangulaires, un peu élargis à leur base. Les piquants du sillon ambulacraire, au nombre de trois ou quatre sur chaque plaque, sont coniques, pointus, presque égaux et réunis ; un autre piquant existe ordinairement, plus en dehors, sur la même plaque. Les bandes ventrales consistent en deux rangées d'écailles lisses, lesquelles sont au nombre de 7 dans les plus longues rangées. Les plaques marginales sont au nombre de 12 à chaque bras, et la plaque terminale du sommet est presque deux fois aussi longue que large ; elle porte deux piquants à l'extrémité et une petite saillie en arrière ; enfin, elle est ciliée sur son bord. La plaque madréporique est située sur le disque, près du bord ; elle est couverte de sillons parallèles.

" Coloration jaunâtre. Dimension : largeur totale 40 mm.

" Il serait possible que cette espèce fut la même que celle qui est figuré dans LINCK, sous le nom d'*Astropecten corniculatus*, mais comme il reste encore quelques doutes à cet égard, il vaut mieux, à l'exemple de M. LÜTKEN, l'assimiler à l'espèce de RETZIUS (*A. crispata*). MM. MÜLLER et TROSCHEL avaient d'abord décrit comme une deuxième espèce, sous le nom de *Ctenodiscus pygmœus*, un petit exemplaire large de 11 mm., et n'ayant que six plaques marginales à chaque bras ; mais plus tard, dans l'Appendice de leur ouvrage (Syst. der Aster., p. 129), ces auteurs regardent comme très-vraisemblable, que c'est le jeune âge de la première espèce (*Ct. polaris = Ctenod. crispatus*), LÜTKEN."

According to LUDWIG [:00] this species is again mentioned by SARS ['65, p. 56].

In his paper of 1866, GRAY adds some remarks to his brief descriptions of 1840 ['66, p. 3] :

" 1. Body pentagonal ; rays short. GRAY, l.c. *Ctenodiscus*, MÜLL. and TROSCH. Ast. 76.

" 1. *Astropecten polaris*, GRAY, Ann. N. H. 1840, p. 180. *Asterias polaris*, SABINE, Append. PARRY'S Voy. 233, t. 1. f. 2, 3. *Ctenodiscus polaris*, MÜLL. and TROSCH. Ast. 76, t. 5. f. 5, 129. *Ctenodiscus crispatus*,

LÜTKEN (B. M.). *Asterias crispata* RETZ. Diss. 17. Inhab. North Sea. "*Astropecten corniculatus*, LINCK, t. 27 and t. 36. f. 63 ; perhaps a variety of the former ; and *Ctenodiscus pygmœus*, MÜLL. and TROSCH. Ast. 76, p. 129, is the young."

VERRILL mentions it from New England ['66, p. 345, 355].

PERRIER has a short remark on *Ctenodiscus* in his work on pedicellariæ ['69, p. 106]. He says, "On ne connaît encore q'une seule espèce de *Ctenodiscus*, le *Ctenodiscus* ou *Anodiscus crispatus*. La collection du Muséum en possède de fort beaux échantillons ; nous avons constaté qu'ils sont, comme les *Astropecten*, dépourvus de Pédicellaires."

According to LUDWIG this species is mentioned by SARS ['69]. WHITEAVES mentions it from the Gulf of St. Lawrence and says ['72 p. 346], "*Ctenodiscus crispatus*, DÜBEN and KOREN, was very abundant in every haul at depths greater than 100 fathoms." According to LUDWIG it is also mentioned by VERRILL ['73] and WHITEAVES ['74].

In his "Révision" PERRIER makes some critical remarks on the synonymy of the species as ·follows ['76, p. 301] :

"*Ctenodiscus corniculatus*.

(References omitted.)

"J'avoue ne pas pouvoir partager les doutes emis par MÜLLER et TROSCHEL, puis par DUJARDIN et HUPÉ relativement à l'identité entre l'*Astropecten corniculatus* de LINCK et l'espèce dont il s'agit ici. Les differences qu'on pourrait signaler entre la figure de LINCK et l'*Asterias crispata* de RETZIUS sont tout à fait de l'ordre de celles que comporte la représentation d'un animal incomplétement etudié et dessiné après dessication. Cependant la forme générale de l'animal, l'ornamentation de son disque, ses grandes plaques marginales, ses pièces maxillaires sont représentées d'une manière suffisante pour lever tous les doutes et ne peuvent se rapporter, parmi les Astéries connues, qu' à celles-ci, et il est tout á fait improbable que l'Astérie figurée par LINCK, si elle est différente, n'ait pas été retrouvée depuis lui. Je crois donc que les scrupules qu'on a pu avoir à cet égard sont tout à fait exagérés ; on devrait en avoir de tout aussi grands en ce qui concerne la plupart des autres espèces figurées par LINCK. L'*Asterias crispata* de RETZIUS doit, en conséquence, reprendre le nom que LINCK lui avait imposé et ce sera pour nous le *Ctenodiscus corniculatus.*

"Un échantillon du Groënland, donné en 1870 par M. le professeur GERVAIS, et quatre, donnés en 1845 par M. le professeur LOVÉN. Ces derniers portent Copenhague comme indication de provenance ; mais cela se renouvelle pour toutes les espèces données par M. LOVÉN, et l'indication plus vague : mer du Nord, serait sans doute ici plus convenable.

"Tous ces individus sont en très-bon état et conservés dans l'alcool.

"Une deuxième espèce de ce genre, appartenant aux régions australes, a été décrite en 1871 par le docteur LÜTKEN sous le nom de *Ctenodiscus australis* (Vidensk. Meddel., p. 258)."

MARENZELLER mentions this species as follows ['78, p. 385] : "Gefunden am 3 April ; geogr. Breite 79° 4′ 9, geogr. Länge 66° 42′ 3 und am 26. Juni 1873 ; geogr. Breite 79° 13′ 3, geogr. Länge 59° 55′ 3.   Tiefe 220 Meter. Meeresgrund Schlamm.

"Verbreitung : Norwegen (bis 300 Faden) (M. SARS), Spitzbergen, Grönland, England (Porcupine-Expedition), Wellington-Canal, St. Lorenzbucht, Neu-England."

STUXBERG also reports it as follows ['78, p. 30] :

"Murmanska hafvet : lat. 71° 39′ × long. 48° 12′, 80 famnar, sand och lerbotten ;

Matotschkin scharr : östra mynnigen, 90 famnar ; mellan Gubin och Beluscha vikarne, 60–70 famnar, lerbotten ;

Kariska hafvet : lat. 70° 30′ × long. 62°, 60 famnar, lerbotten ;

lat. 71° × long. 65° 30′, 12 famnar, sandbotten ;

lat. 71° 5′ × long. 63° 20′, 90 famnar, lerbotten ;

lat. 72° 5′ × long. 67° 30′, 36 famnar, lerbotten ;

"The Murman Sea : lat. 71° 39′ × long. 48° 12′, 80 fathoms, sand and clay bottom ;

Matochkin Strait : the eastern entrance, 90 fathoms ; between the Gubin and Beluscha coves, 60–70 fathoms, clay bottom ;

The Kara Sea : lat. 70° 30′ × long. 62°, 60 fathoms, clay bottom ;

lat. 71° × long. 65° 30′, 12 fathoms, sand bottom ;

lat. 71° 5′ × long. 63° 20′, 90 fathoms, clay bottom ;

lat. 72° 5′ × long. 67° 30′, 36 fathoms, clay bottom ;

lat. 75°× long. 75° 20′, 22 famnar, lerbl. sandb ;	lat. 75°× long. 75° 20′, 22 fathoms, bottom sand mingled with clay ;
lat. 75° 40′× long. 78° 40′, 26 famnar, lerbl. sandb.	lat. 75° 40′× long. 78° 40′, 26 fathoms, bottom sand mingled with clay.

It is also mentioned by Storm ['78, *fide* DANIELSSEN and KOREN].

VIGUIER gives a detailed description of the skeletal system of this species ['78, p. 226] :

"Il n' existe, de ce genre, que deux espèces : le *Ctenodiscus australis* décrit en 1871 par M. LÜTKEN, et qui ne figure pas au Muséum, et le *Ctenodiscus corniculatus*, que LINCK avait déjà figuré, dans la planche XXXVI de son ouvrage, sous le nom d'*Astropecten corniculatus*.

"On comprend très-bien que LINCK ait pris le *Ctenodiscus* pour un *Astropecten*. Le dos est constitué, en effet, dans ces deux types, d'une façon tellement semblable, que tout ce que je dirai sur la forme et l'arrangement des paxilles chez l'*Astropecten*, s'applique au *Ctenodiscus*, et je crois inutile d'en parler maintenant ; mais nous trouvons ailleurs d'importantes différences.

"Les plaques marginales sont réduites ici à de simples lames disposées verticalement, et dont le milieu porte une crête saillante. Ceci peut être considéré comme une exagération de ce qui voit chez les *Astropecten* et les *Luidia*, comme on peut s'en rendre compte en comparant entre elles les coupes des ces trois genres, qui se trouvent sur la planche XV. La face ventrale présente une différence bien plus grande. Ici, en effet, à cause de la forme du corps, les plaques marginales ne sont en rapport avec les séries adambulacraires que tout à fait à l'extrémité des bras, et tout l'espace triangulaire laissé entre la série marginale inférieure et les deux séries adambulacraires est occupé par des lames très-minces, imbriquées de la periphérie au centre. Ces lames forment des séries rayonnantes du sommet de l'angle interbrachial, et sont sans rapport régulier · de nombre avec les plaques marginales ou les pièces adambulacraires. Ces pièces adambulacraires sont singulièrement aplaties.

"Quant aux pièces ambulacraires, elles ressemblent à ce que nous verrons dans la *Luidia ;* mais elles sont relativement beaucoup plus fortes, et leur direction est presque horizontale.

"Les soutiens ambulacraires $s^{1)}$, beaucoup plus petits que dans les types suivants, sont cependant parfaitement distincts.

"La bouche du *Ctenodiscus* est largement ouverte, et limitée par des dents grosses, saillantes, un peu écartées dans la même paire ; et se rapproche par tous ces caractères de ce que nous verrons plus loin ; mais on remarque bientôt de grandes différences. Les dents ne sont pas amincies comme dans l'*Astropecten* ou la *Luidia*, mais larges et epaisses ; elles se projettent aussi beaucoup plus loin vers le centre de la bouche. Ceci est dû à une disposition particulière, déjà citée plus haut, et dont je ne connais que ce seul exemple. Il est facile de constater, sur la figure 19, que la première pièce ambulacraire est constituée par la coalescence de trois pièces, tandis que la dent ne répond qu'à deux pièces adambulacraires. Le pore ambulacraire très-réduit, que l'on voit tout à fait en avant, ne peut laisser de doutes à cet égard. L'apophyse, en aile, est extrêmement réduite.

"On doit s'attendre, avec une modification aussi grande dans la bouche, à trouver un odontophore singulier ; en effet, il ne rappelle aucun des types que nous avons vu jusqu'ici, comme on peut s'en assurer en regardant la figure 18. Les apophyses articulaires sont excessivement petites, pour répondre aux plus petits trous ambulacraires, et la pièce elle-même est large, pour maintenir en situation les dents d'une même paire.

"Cette singulière modification de la bouche m'a paru constante chez le *Ctenodiscus* ; mais je n'ai pu m'en assurer que sur mon échantillon, et peut-être ne faut-il pas généraliser ce fait, qui sera toujours, du moins, fort intéressant au point de vue de la constitution de la bouche.

"La plaque madréporique, située près du bord du disque, comme dans les deux genres[2)] suivants, est volumineuse, ovale, convexe et marquée de sillons longitudinaux et sinueux.

"La plaque ocellaire, très-grosse également, présente une forme assez singulière. Elle est marquée, près de son bord libre, de deux petites fossettes séparées par une crête médiane. En arrière de ces fossettes, la pièce est fortement convexe, presque globuleuse.

"Il n'existe pas de système interbrachial, non plus que dans les genres[2)]

---

1) Superambulacral plates.   2) *Luidia* and *Astropecten*.

suivants. On n'a pas signalé jusqu'ici de pedicellaires chez le *Ctenodiscus*, qui en paraît complétement dépourvu, comme l'*Astropecten*.

"Le *Ctenodiscus corniculatus* est un animal des mers du Nord."

D'URBAN mentions this species from Barents Sea and says ['80, p. 256], " *Ctenodiscus crispatus* was also very generally distributed at all depths from 62 to 210 fathoms." According to LUDWIG it is also mentioned by VERRILL ['80] and STUXBERG ['80].

DUNCAN and SLADEN give a very good description of this species together with a comprehensive list of previous references ['81, p. 49] :

" *Ctenodiscus corniculatus* (LINCK), PERRIER. Plate III, Figs. 17–20.

(References omitted.)

" Body depressed and goniodiscoid in outline ; radii five in number, with the arm-angles well rounded ; proportion of greater to lesser diameter 2 : 1. The calcareous elements of the abactinal surface form a compact network, similar to *Astropecten*, in which the interspaces are but very small. A great number of small closely-placed paxillæ are borne upon this frame-work, each carrying 5–10 round, blunt spinelets, the whole so densely crowded together that the spinelets are normally directed upward from the pedicle. The abactinal surface is frequently puffed up and more or less convex in profile (owing probably to the quantity of sand or clay with which this Starfish fills its stomach) ; whilst a small peak-like protuberance rises from the centre, around which the paxillæ rapidly diminish in size. The sides of the disk are perpendicular, and formed of two series of marginal plates—one ventral, the other dorsal. Each dorsal plate is ankylosed to a corresponding ventral plate—the pair thus formed being separated from the neighbouring pair on either hand by a deep furrow, which follows the lateral suture of the plates, the margins being fringed with a series of fine, compress-ed, cilia-like spinelets, which arch over the furrow. Each of the dorso-lateral plates bears a small compressed, but pointed, spinelet, which stands erect on its upper margin ; and the lower or ventro-lateral series likewise carry a similar spinelet, which is placed near the junction of the ventral and dorsal plates, and projects at right angles to the side walls of the test. The last or terminal dorso-lateral plates of each side of a ray are ankylosed together, and form a large arched or tubercular plate, indented on its outer margin, and bearing three more or less prominent tubercles—the rudiments of dorso-marginal spines. The furrows between the marginal plates are continued onto the actinal surface of the animal and extend to the ambulacral furrow, cutting up the ventral interradial areas into band-like spaces, each of which is tessellated with irregular, subquadrate, scale-like plates that imbricate upon one another, and form normally, in large adult examples, a double alternating series behind each adambulacral plate. The innermost band, however, of each area comprises two adambulacral plates; and the trapezoid tessellating scales, which here always form a regular double alternating series, are, in consequence, twice as large in the neighbourhood of the furrow as the scales in the other bands. All these plates bear on the margin that opens on the sutural furrow a series of papillæ that form a continuation with the papillæ above-noted on the sides of the lateral plates, from which they differ only in being not flattened, nor are they at the same time so regular and closely placed. The adambulacral plate represents a wedge-shaped projection into the furrow, and carries five or six papillæ, three only of which usually stand on the margin of the ambulacral furrow, the remaining two or three (which are generally much smaller) being situated on the aboral margin opening on the sutural furrow of the interbrachial area; not unfrequently, however, one of them is as large as the ambulacral spinelets, and is placed somewhat inward upon the plate, away from the sutural fringe and behind the am-bulacral series. Towards the extremity of the ray the adambulacral plates stand next to the ventro-marginal plates, and are not separated from them by the trapezoid imbricating scales above described.

"Each pair of mouth-plates forms an ovoid mass, the inner or apposed margins of the plates being elevated into a prominent keel. The inner-most pair of mouth-papillæ are very large and thick, and taper to a point—the remainder, from 7 to 9 in number, being considerably smaller and arranged round the free margin of the plate. Along, or near to, the median keel of the mouth-plate are 3–5 coarse spinelets, the innermost being large and thick, and are much less pointed than the marginal series. The mad-reporiform body is frequently not more than its own diameter distant from the margin, and is generally oval in outline and covered with elongate striæ running in the direction of the major diameter.

" The entire body and all its appendages are covered with an investing leathery skin.

" *Size.*—Ordinary specimens are about 30–40 millims. in diameter, the largest recorded by SARS, from Tromsö, being 65 millims.

" *Colour.*—The colour is recorded as brown-red; specimens preserved in spirit are either black, greenish, or various shades of drab.

" *Habitat.*—*Ctenodiscus corniculatus* is found in mud or soft clay bottoms at very various depths, being dredged by SARS, at Finmark, in 40–200 fathoms depth; and further south, at Christiansund, in 40–80 fms., by Insp. MÖLLER (LÜTKEN).

" *Premature form.*—The young form of this species was described by MÜLLER and TROSCHEL under the name of *Ct. pygmæus.* Small individuals, of about half an inch in diameter, are characterized by the flatter test, the comparatively greater prominence of the latero-dorsal spinelets, and the three large, conspicuously-developed spinelets which are present on the terminal plate of ray. The upper margin of this plate, which lies towards the centre of the disk, is fringed with a series of papillæ similar to those on the sides of the lateral plates; and these papillæ, as well as the spinelets, appear to be subject to a greater or lesser degree of obliteration (or resorption) during the progress of the growth of the Starfish; in fact in old specimens the spinelets become reduced to mere tubercles.

" According to LÜTKEN, the apical prominence in the centre of the disk is more prominent and characteristically developed in young forms; but in the specimens which we have examined (from Novaya Zemlya) it would appear to be quite the reverse, for we have been unable to detect any difference, except a proportional diminution in size, from the condition presented by the mature animal. The adambulacral plates in these specimens bear their papillæ or ' ambulacral spines' on the furrow-margin, with one large one placed thumb-like behind them on the aboral side.

" At the extremity of the ray there are only two of the ambulacral spinelets; and the thumb-like spinelet is larger than either of them, and is persistently present on every plate. The mouth-plates are small and simple, having only three or four mouth-papillæ on the margin of each plate, and two only on the median ridge (or at most three), the innermost of these being very large and prominent, and standing perpendicular to the plane of the plate, midway between the extremities. When the young Starfish is examined under the microscope from above, it will be seen that in the lateral sutural furrows, which open on the dorsal surface, there are, in addition to the marginal fringe of compressed spinelets, an inner series of fine, pointed, cilia-like spinelets, at least at the upper portion.

" In the early stages of this species, the lower or ventral series of lateral plates lies much more upon the actinal or ventral surface of the Starfish than it generally does in the fully grown form. In the examples above mentioned the row of single or true lateral spinelets, borne by the ventro-lateral plates, stands almost at the angle formed by the vertical side of the test and the actinal surface, the plates themselves arching sharply under into the actinal surface. The spinelets also seem to occupy a position relatively nearer to the centre of the plate than at a later stage. In large specimens no portion of the lateral plates curve onto the ventral area, and the spinelets are situated much nearer to the upper extremity of their respective plates.

" *Variations.*—Dr. Lütken records examples from a station off Norway, which probably lies near the southern limit of the area of distribution, that differ strikingly from the normal short-armed form by their longer and more pointed rays, whereby the contour approaches that of *Archaster Parelii ;* but in other respects no differences were noticed. We have observed a similar variation in the radial proportions amongst a series of specimens from Barents Sea, but not developed to such a marked degree as in those examined by our learned comtemporary. The relative proportions of the greater to the lesser radius in two specimens are respectively 17 millims. to 9.75 millims. in the one, and 17 millims. to 7 millims. in the other ; two smaller examples measured similarly 10 millims. to 6 millims., and 10 millims. to 5 millims. —the difference in character presented by the wide and gentle curve of the arm-angle of the one in comparison to the more acute and angular outline of the other being much more striking to the eye that the figures which indicate the actual proportions would seem to imply. From the occurrence of both these forms together, as well as the identity of their general structure, it is, perhaps, not improbable that we have here nothing more than a sexual character.

" Upon the whole, this species would seem to be remarkably constant. After a careful study, however, of specimens from Greenland, Novaya Zemlya, and North America, we are inclined to believe that a certain amount of variation does occur (probably of locational permanence) in the features of the ventro-lateral plates and of the ambulacral spines, after the manner indicated whilst treating of the phases of growth; and although this would seem rather like a confusion of the stages characteristic of growth with the features presented by circumstantial variation, the evidence has been such as to lead to the inference that certain characters of early growth-phases are, in some localities, retained until a much later period of growth,—perhaps even becoming a permanency through life—a state of things which is perfectly explicable on the not improbable assumption that the exigencies of arctic existence have acted in retarding the progress of growth-characters and in the maintenance of the youthful or more simple form. The spinulation of the paxillæ is similarly subject to variation.

" Distribution.

" a. *Greenland* : Hare Island, Waigat Strait, lat. 70° 30′ N., 175 fms. ('*Valorous*' *Exped.*).

" b. *North of American Continent* : Melville Island, about lat. 74° 47′ N., long. 110° 48′ W. (*PARRY's Exped.*) ; Assistance Bay, 7–15 fms. (*PENNY's Exped.*) ; Newfoundland (*SARS*) ; Bay of Fundy, 50–60 fms. (*STIMPSON*) ; Maine.

"'c. *North of European Continent* : Spitzbergen (*LÜTKEN*) ; Barents Sea, lat. 76° 58 N., long. 45° 40 E., 110 fms. ('*Willem Barents*' *Exped,*), the most northern locality on record ; Finmark ; Scandinavian coast."

According to LUDWIG this species is mentioned by HOFFMANN ['82]. VERRILL mentions it from the New England coast from depths of 182–310 fathoms ['82, p. 218].

DANIELSSEN and KOREN mention this species under the name of *Cteno-*

discus corniculatus, (LINCK) PERRIER, with a very full list of previous references ['84, p. 83] :

" Wherever this starfish was found, it was generally very abundant. It was collected at the following stations :

" Salten-fjord, Nordland, in great abundance.

" Station No. 257. Many specimens, but small ones.

—— „ 260. Extremely abundant.

—— „ 261. Extremely abundant.

—— „ 262. A few specimens.

—— „ 273. A few specimens.

—— „ 275. A few specimens.

—— „ 323. Frequent.

—— „ 326. Immense quantities.

—— „ 336. A few specimens.

—— „ 338. Several specimens.

—— „ 357. Immense quantities.

—— „ 363. A few specimens.

" Distribution.

" Along the North coast of Norway. The Murman coast. The North American coast. Greenland. Spitzbergen and Nova Zembla."

VERRILL mentions this species from the dredgings of the ' Fish Hawk,' 1880–1882. He says ['85, p. 551], " B. [athymetrical] range, 182 to 321 fathoms, 1880, 1881, 1882. Local; abundant north of Cape Cod." Accordinging to LUDWIG it is also mentioned by JARZYNSKY ['85].

In his paper on the Echinoderms of the Bering Sea, LUDWIG describes a new species, *Ct. krausei*, which he, however, subsequently concedes to be identical with *Ct. corniculatus*. *Ct. krausei* n. sp. is described as follows ['86, p. 290] :

" Von der Gattung *Ctenodiscus* M. u. TR. sind bis jetzt nur zwei Species bekannt : *Ct. corniculatus* (LINCK) PERRIER und *Ct. australis* LÜTKEN. Der erstere ist eine der charakteristischsten Formen des nordatlantischen Meeres ; sie erstreckt ihr Wohngebiet von der Ostküste Nordamerikas über Grönland, Spitzbergen, Novaja Semlja östlich bis ins Karische Meer und ist zuletzt ausführlich von DUNCAN u. SLADEN (Echinod. Arctic Sea, 1881; p. 49, tab. III, fig. 17–20) beschrieben worden (genauere Angaben über die Verbreitung finden sich auch bei DANIELSSEN u. KOREN, Norske Nordhavs-Expedition, Asteroidea, 1884; p. 83). *Ct. australis* hingegen gehört dem antarctischen Meere an; (vergl. LÜTKEN, Vidensk. Meddelels. Naturh. Forening, 1871; p. 238; ferner STUDER, Verzeichn. d. von der 'Gazelle' gesamm. Asteriden u. Euryaliden, 1884; p. 42). Von beiden unterscheidet sich die neue in zwei Exemplaren vorliegende Art aus dem Beringsmeere, welche ich zu Ehren der beiden Forschungsreisenden, der Herren Dr. ARTHUR und AUREL KRAUSE benenne. Beide Exemplare wurden südöstlich von St. George in einer Tiefe von 30 m erbeutet.

"Da *Ct. australis* sich nur durch die gröberen, weniger zahlreichen und mit mehr (15–20) Stachelchen besetzten Paxillen von *Ct. corniculatus* unterscheidet, so beschränke ich mich im Folgenden auf eine Vergleichung meiner neuen Art mit der zuletzt genannten. Dabei muss ich die Bemerkung voranschicken, dass mir keine Exemplare von *Ct. corniculatus* zur Verfügung stehen und ich mich deshalb nur auf die in der Litteratur vorhandenen Beschreibungen und Abbildungen beziehen kann.

"In der Grösse stimmen beide Exemplare von *Ct. krausei* miteinander überein, sie haben eine grösste Länge von 34 mm; der grosse Radius misst 18 mm, der kleine 9.5 mm. Sie schliessen sich also in dieser Hinsicht an die von *Ct. corniculatus* bekannten Verhältnisse an.

"Die Paxillen des Rückens tragen gewöhnlich 6–8 kurze Stachelchen, von denen eines oben auf der Mitte der Paxille, die übrigen rings um jenes mittlere angeordnet sind. Bei *Ct. corniculatus* geben DUNCAN u. SLADEN die Zahl der Stachelchen auf den Paxillen auf 5–10 an. Ganz ähnlich wie bei *Ct. corniculatus* sind die Paxillen auf der Rückenmitte kleiner und dichter zusammengedrängt, und es erhebt sich an dieser Stelle die Rückenhaut zu einem kurzen, kegelförmigen Vorsprung. Auch darin stimmt der Rücken von *Ct. krausei* mit *Ct. corniculatus* überein, dass er im Ganzen etwas gewölbt ist. Ein bemerkenswerter Unterschied bezüglich der Rückenhaut beider Arten ergibt sich erst dann, wenn man die Insertion der Paxillen untersucht. Bei *Ct. corniculatus* geben nämlich DUNCAN u. SLADEN an, dass sich in der Rückenwand ein compactes Netzwerk kalkiger Skeletteile befinde, welches die Paxillen trägt. Diese Angabe steht ·nicht ganz im Einklang mit dem Bau der Rückenhaut, wie man denselben bei anderen Astropectiniden zu finden gewohnt ist (vergl. VIGUIER, Squelette des Stellérides, Arch. de zool. expér. VII, 1878 ; p. 225–234) ; denn dort sind es die Paxillen selbst, welche durch Verbreiterung und Aneinanderlagerung ihrer basalen Ende das Netzwerk in der Rückenhaut zu Stande bringen. Indessen wenn man auch annimmt, dass DUNCAN u. SLADEN nicht behaupten wollen, es sei jenes Netzwerk bei *Ct. corniculatus* aus besonderen Skeletstücken aufgebaut, so geht doch aus ihrer Beschreibung hervor, dass ein Netzwerk in der Rückenhaut von *Ct. corniculatus* überhaupt vorhanden ist. Bei *Ct. krausei* aber fehlt ein solches vollständig ; die Paxillen verbreitern sich zwar an ihrem basalen Ende zu einer Fussscheibe, welche etwa doppelt so breit wie die Paxille selbst ist ; aber diese Fussscheiben berühren sich nicht, sondern bleiben stets durch skeletfreie Zwischenräume von einander getrennt.

" Die Zahl der oberen (und unteren) Randplatten beträgt an jedem Arme der beiden Individuen jederseits 10. Bei *Ct. corniculatus* zeichnen DUNCAN u. SLADEN 11–13 ; MÜLLER u. TROSCHEL (System der Asteriden 1842 ; p. 76 ; *Ct. polaris = corniculatus*) und LÜTKEN (l.c.) geben 12 an ; M. SABS dagegen fand dass bei besonders grossen Exemplaren die Zahl der Randplatten auf 16–18 steigt (Overs. af Norges Echinod., 1861 ; p. 26 ; *Ct. crispatus = corniculatus*). Unter diesen Umständen ist es leicht möglich, dass auch bei *Ct. krausei* mehr als 10 Randplatten auftreten können und sonach die Zahl der Randplatten kein sicheres Merkmal zur Unterscheidung beider Arten abgiebt. Dies gilt um so mehr, wenn man die Beobachtungen v. MARENZELLER's über die mit dem Wachstum Hand in Hand gehende Vermehrung der oberen Randplatten bei der Gattung *Astropecten* mit in Betracht zieht (vergl. v. MARENZELLER, Revision adriat. Seesterne in : Verhdl. zool. bot. Gesellschaft Wien, 1875 ; p. 364). Anders liegt die Sache, wenn man die Bestachelung der Randplatten ins Auge fasst. Der grössere Stachel, welcher auf dem oberen Rand der oberen Randplatten steht, sowie der ähnliche Stachel, welcher auf der untern Randplatte dort eingelenkt ist, wo ihre

Aussenfläche sich ventralwärts zu wenden beginnt, verhalten sich zwar wie bei *Ct. corniculatus.* Jedoch die kleineren Stachelchen, welche rechts und links von der mitteleren Längsleiste der Randplatten in einer Reihe übereinander stehen, sind viel kleiner und zahlreicher als bei *Ct. corniculatus ;* in den Armwinkeln zählt man zwischen je einem oberen und unteren Stachel in jeder Reihe etwa 20 kleinere Stachelchen, also etwa doppelt so viel als bei *Ct. corniculatus ;* von diesen 20 Stachelchen gehören etwa 15 der oberen Randplatte, 5 der unteren Randplatte an (vergl. Fig. 14 und die citirten Abbildungen von DUNCAN u. SLADEN). Hinter der soeben besprochenen, von aussen mit blossem Auge sichtbaren Stachelchenreihe besitzen die Randplatten eine abgeschrägte Fläche, welche von einer weichen Haut bekleidet ist, die sich in 3–4 parallelen Längsfalten legt. Jede derartige Längsfalte umschliesst nun in ihrem Inneren eine Längsreihe sehr feiner, flach zusammengedrückter, kleinster Stachelchen, deren Form und Grösse aus Figur 15 und 16 erhellt. Ob diese Längsfalten mit ihren kleinsten Stachelchen auch bei *Ct. corniculatus* vorkommen, bedarf einer weiteren Untersuchung. Eine solche würde auch zu zeigen haben, ob der schon von AGASSIZ versuchte Vergleich der zwischen den Randplatten der Astropectiniden befindlichen Strassen mit den Saumlinien (Semiten) der Spatangiden sich durchführen lässt (vergl. A. AGASSIZ, North American Starfishes, 1877, p. 119.)

"An der Spitze der Arme schiebt sich zwischen die letzten Randplatten eine grosse Terminalplatte ein, welche in ihrer Gestalt ganz mit derjenigen von *Ct. corniculatus* übereinstimmt. Ich verstehe aber nicht, weshalb DUNCAN u. SLADEN die Terminalplatte von *Ct. corniculatus* aus einer Verschmelzung oberer Randplatten entstehen lassen. Nach Allem, was wir über die Entstehung der Skeletteile der Seesterne bis jetzt wissen, scheint mir gar kein Grund zu einer solchen Annahme vorhanden zu sein; vielmehr muss man, solange nicht bestimmte Beobachtungen uns eines anderen belehren, annehmen, dass auch bei *Ctenodiscus*, wie bei anderen Seesternen, die Terminalplatte von Anfang an als ein unpaares Gebilde auftritt. Vielleicht ist die betreffende Angabe von DUNCAN u. SLADEN nur eine unbeabsichtigte Reminiscenz an eine Stelle bei LÜTKEN, wo derselbe gleichfalls die Terminalplatte für eine Vereinigung von zwei oberen Randplatten erklärt (vergl. LÜTKEN, Overs. over Grönlands Echinoderm., 1857 ; p. 46).

"Bei *Ct. corniculatus* geben die verschiedenen Autoren nirgends etwas über die Lage der Geschlechtsöffnung an. Ich möchte aber vermuthen, dass dieselben (an Spiritusexemplaren) ebensoleicht· zu sehen sind, wie das bei *Ct. krausei* der Fall ist. Hier liegen sie in der Rückenhaut der Scheibe, dicht über den oberen Randplatten, und zwar befinden sich deren in jedem Interradius zwei. Eine jede ist genau über dem Zwischenraum zwischen der ersten und zweiten Randplatte angebracht, also von ihrem Partner durch die Breite zweier oberen Randplatten getrennt.

"An der Unterseite (vergl. Fig. 13) unterscheidet sich *Ct. krausei* fast nur darin von *Ct. corniculatus*, dass die Schuppenreihe, welche die inter-brachialen Felder zwischen den unteren Randplatten und den Adambulacral-platten einnehmen, schon an der vierten unteren Randplatte aufhören, während sie bei *Ct. corniculatus* sich viel weiter nach der Armspitze hin erstrecken (vergl. DUNCAN u. SLADEN, l.c., fig. 18, 19 und VIGUIER, l.c., tab. XV, fig. 15).

"Nach VIGUIER (l.c., p. 79, p. 227) soll bei *Ct. corniculatus* der erste Wirbel nicht aus der Verwachsung der zwei, sondern der drei ersten Ambulacralstücke gebildet sein. Da ich schon früher in einem anderen Zusammenhang (Z. f. wiss zool. XXXII, p. 678) die Ansicht vertreten habe, dass diesem Befunde nicht entfernt diejenige Bedeutung beizumessen ist, welche der genannte Forscher ihm beilegt, so unterliess ich es nicht, an einem der beiden Exemplare von *Ct. krausei* die Zusammensetzung des sog. ersten Wirbels zu untersuchen. Ich konnte mit Leichtigkeit feststellen, dass bei *Ct. krausei* das dritte Ambulacralstück ebensowenig wie bei anderen Asterien in die Bildung des sog. ersten Wirbels eintritt. Die von VIGUIER für *Ct. corniculatus* angegebene Verwachsung zwischen dem ersten (aus den beiden ersten Ambulacralstücken gebildeten Wirbel) und dem dritten Ambulacralstück ist bei *Ct. krausei* nicht vorhanden."

According to LUDWIG this species is mentioned by AURIVILLIUS ['86], LEVINSEN ['86], STUXBERG ['86] and RUIJS ['87.]

In the Challenger Report *Ctenodiscus corniculatus* (LINCK) PERRIER is mentioned and the localties are given as follows [SLADEN, '89, p. 171] :

" *Localities.*—' Porcupine ' Expedition :

" Station 82. In the Faeröe Channel. Lat. 60° 0′ 0″ N., long. 5° 13′ 0″ W. Depth 312 fathoms. Bottom temperature 5°.2 C.; surface temperature 11°.2 C.

" Station 57.[1] In the Faeröe Channel. Lat. 60° 14′ 0″ N., long. 6° 17′ 0″ W. Depth 632 fathoms. Bottom temperature −0°.8 C. ; surface temperature 11°.1 C.

" Station 58.[1] In the Faeröe Channel. Lat. 60° 21′ 0″ N., long. 6° 51′ 0″ W. Depth 540 fathoms. Bottom temperature −0°.6 C. ; surface temperature 10°.6 C.

" *Other Localities.*—This species also occurs off the coast of Greenland, off the eastern coast of North America, off the Scandinavian coast, off Spitzbergen and Nova Zembla, and in the Barents Sea.

" *Ctenodiscus corniculatus* has been found in the fossil state by the late M. SARS ['61, p. 144] in the older beds of the Postpliocene or Glacial formation of Norway, near Christiania."

In the same report is described and figured *Ctenodiscus procurator* as a new species, but it appears to be specifically identical with *Ct. crispatus* ['89, p. 173]

" *Ctenodiscus procurator*, n. sp. (Pl. XXX. figs. 7–12).

" This form has so many points of close resemblance to the North-Atlantic *Ctenodiscus corniculatus* that examples might be selected which at first sight would easily be mistaken for that species. A number of small differences, however, present themselves when a large series is examined, which appear sufficiently constant to warrant the recognition of this form as a distinct species. Under these circumstances the description of *Ctenodiscus procurator* will probably be most intelligible if it takes the form of a comparative review of the characters of this species in relation to those of the two previously known species of *Ctenodiscus*, viz., *Ctenodiscus corniculatus* of the North Atlantic, and *Ctenodiscus australis*, LÜTKEN, from the East of Patagonia.

---

1) " This occurrence is recorded in Sir WYVILLE THOMSON's Depths of the Sea, but I have not seen a specimen."

" When these three species are compared *inter se* it is evident that in many respects *Ctenodiscus corniculatus*, though so widely separated geographically, appears to occupy an intermediate classificatory position between *Ctenodiscus australis* and *Ctenodiscus procurator*, which inhabit the eastern and western sides respectively of South America. In *Ctenodiscus procurator* the rays are generally a trifle longer, and, even when not actually so, have at least that appearance in consequence of being slightly narrower at the base and more attenuate and pointed outwardly. The abactinal area is plane, its union with the lateral wall, especially in the region of the disk and the base of the rays, forming a sharp angle in consequence of the rapid adoral slope of the whole lateral wall; the supero-marginal plates being also affected in the majority of cases. This feature at once strikes the eye in comparison with the usually vertical and actinally well-rounded margin of *Ctenodiscus corniculatus* and the thick and tumid one of *Ctenodiscus australis*.

" The paxillæ of the abactinal area are small and crowded, similar to those in *Ctenodiscus corniculatus*. The madreporiform body is distinct and not hidden by paxillæ as in *Ctenodiscus australis*. The marginal plates appear to be invariably rather more numerous than in *Ctenodiscus corniculatus*, and consequently still more so than in *Ctenodiscus australis*;—for example, in a specimen of *Ctenodiscus procurator*, measuring $R = 28.5$ mm., there are eighteen supero-marginal plates counting from the median interradial line to the extremity; whereas in *Ctenodiscus corniculatus* of exactly the same radial dimensions ($R = 28.5$ mm.) there are only fifteen. *Ctenodiscus corniculatus*, with $R = 27$ mm., has fourteen supero-marginal plates; *Ctenodiscus procurator*, with $R = 27$ mm., has seventeen. *Ctenodiscus procurator* appears to have generally one or more spines less on the adambulacral plates than in *Ctenodiscus corniculatus*, three only being actually marginal or furrow spines, and a fourth standing backward and on the actinal surface of the plate at the aboral end. Very rarely indeed are four furrow spines present; whereas four and five are general in *Ctenodiscus corniculatus*.

" From the foregoing remarks it will be seen that *Ctenodiscus procurator* is much more closely allied to the North-Atlantic *Ctenodiscus corniculatus* than to the comparatively neighbouring form *Ctenodiscus australis*, from which it is readily distinguished. On the other hand the individual points of difference between the Chilian and the Northern forms are small and trifling, but when taken as a whole may be regarded as sufficient to differentiate them specifically, especially when the constancy of the characters in question and the widely separated geographical position of the two forms are taken into account.

" Colour in alcohol, a bleached yellowish white.

" *Localities.*—Station 303. Off the western coast of South America, off the Chonos Archipelago. December 30, 1875. Lat. 45° 31′ 0″ S., long. 78° 9′ 0″ W. Depth 1325 fathoms. Blue mud. Bottom temperature 36°.0 Fahr. ; surface temperature 54°.8 Fahr.

" Station 306. In the Messier Channel, between Wellington Island and the west coast of Chili, off Port Grappler. January 4, 1876. Lat. 49° 24′ 30″ S., long. 74° 23′ 30″ W. Depth 140 fathoms. Blue mud. Surface temperature 53°.0 Fahr.

" Station 309. Off Puerto Bueno. January 8, 1876. Lat. 50° 56′ 0″ S., long. 74° 15′ 0″ W. Depth 40 fathoms. Blue mud. Bottom temperature 47°.0 Fahr. ; surface temperature 50°.5 Fahr.

" Station 311. Off the entrance to Smyth Channel. January 11, 1876. Lat. 52° 45′ 30″ S., long. 73° 46′ 0″ W. Depth 245 fathoms. Blue mud. Bottom temperature 46°.0 Fahr. ; surface temperature 50°.0 Fahr."

FEWKES ['91, p. 64] gives a brief diagnosis of *Ctenodiscus* and mentions this species. BELL describes it as follows, with 3 figures ['92, p. 64] :

" R = 2 r.

" General form stellate, with rather deeply incurved sides, flattened ; a large terminal plate. Ambulacra wide, with large suckers, and bordered by a row of spines, three to each plate, and externally to them one or two others, which are generally rather smaller. The ventral surface spineless. The marginals with spines ; the superomarginals are elongated from above downwards and form a vertical wall on every side of the disk ; neither on them nor on the inferomarginals is a spine always developed, but it is

generally; it is never large, and there is never more than one. There
are about fifteen marginals on either side of each arm. The paxilliform
plates of the upper surface are very delicately stellate and closely packed.
The madreporite, which is distinct and rather deeply grooved, is not far
from the margin of the disk.

"Colour in spirit creamy yellow.

"R=28.5   26   24.

"r= 14   14   11.

"*Distribution.* Both sides of the northern part of the Atlantic, and
Arctic Ocean, 7–632 fms.

"A.   Faeroe Channel, 60° 0' 0" N., 5° 13' 0" W., 312 fms. 'Porcupine'
Exp."

It is mentioned by NORDGAARD ['93, p. 10] as being very common in
Beitstad Fiord. PFEFFER mentions it from Spitzbergen. In the first of his
papers referred to here ['94, p. 98] he mentions "6 erwachsene und 1
junges Stück," and in the second ['94a, p. 102] he says, "Auf Steingrund;
auch Mudd bez. Lehm mit Steinen. 15–140 Faden," besides giving the
stations where the specimens were obtained.

VERRILL refers to this species as follows ['95, p. 132]; "B [athymetri-
cal] range, 5 to 632 fath. Most abundant from 50 to 150 fath. Taken at
numerous stations in Massachusetts Bay, Gulf of Maine, Bay of Fundy, etc.
It extends to Greenland, Spitzbergen, and Northern Europe. Circumpolar.
Allied species occur in the South Atlantic and South Pacific." SLUITER
mentions it in the collection of the Amsterdam Museum as follows ['95, p.
52]: "*Ctenodiscus corniculatus* (LINCK) PERRIER. Mehrere Exemplare aus der
Barents See, 70° 17' NB., 46° 31' OL. Gr. und zahlreiche Exemplare aus
der Kara See, 71° NB., 63' OL. aus einer Tiefe von 45–57 Faden. Alle in
Alcohol, und drei von Bodö, Norwegen (WEBER)." It is also mentioned by
VANHÖFFEN ['97, *fide* LUDWIG]. BIDENKAP mentions it as occurring in
Lyngenfjord 'not seldom, but never in large numbers, in 30–50 fathoms'
['99, p. 100], as well as at 'Hukøbotn in 20–25 fath. and Ramfjord in 80
fath. on soft muddy bottom; being common along Lofoten and Finmarken'
['99a, p. 107].

DÖDERLEIN in his short communication on Arctic starfishes, considers *Ct. krausei* as a synonym of *Ct. corniculatus*. He says ['99, p. 337], "Bei *Ctenodiscus corniculatus*, einem der häufigsten arktischen Seesterne, liess sich nachweisen, dass *Ctenodiscus krausei* LUDWIG (Zool. Jahrb. Syst. Bd. 1. p. 290) aus dem Behringsmeer damit identisch sein dürfte, da die für *Ct. krausei* als specifisch angegebenen Merkmale auch bei *Ct. corniculatus* nachgewiesen werden können."

LUDWIG in his work on Arctic starfishes uses the name *Ct. crispatus* and after giving an exhaustive list of references makes the following remarks [: 00, p. 451]:

"DÖDERLEIN hat unlängst (1890) die von mir (1886) aus dem Beringsmeere aufgestellte Art *Ctenodiscus krausei* ganz mit Recht für identisch mit *Ct. crispatus* erklärt; wie ich mich durch eine vergleichende Untersuchung nunmehr selbst überzeugt habe, war meine frühere Aufstellung des *Ct. krausei* als besonderer Art nur möglich, weil es mir damals an Vergleichsmaterial fehlte und die älteren Beschreibungen des *Ct. crispatus* in manchen Punkten unzulänglich waren. Dies vorangeschickt, ergiebt sich für *Ct. crispatus* heim heutigen Stande unserer Kenntnisse ein Verbreitungsgebiet, das in der Richtung von West nach Ost von 170° w. L. bis ca. 79° ö. L., also durch 249 Längengrade reicht. Aus dem Beringsmeere (LUDWIG 1886) geht die Art der Nordküste Amerika's entlang an der Melville-Insel (SABINE 1824) vorbei durch die Barrow-Strasse (FORBES 1851) und von da an Grönland (MÜLLER und TROSCHEL 1842; LÜTKEN 1857; PERRIER 1875; NORMAN 1877; DUNCAN und SLADEN 1881; VANHÖFFEN 1897) hinunter zur Ostküste Amerikas, wo sie von Neufundland (M. SARS 1861) und dem St. Lorenz-Golf (WHITEAVES 1872, 1874) bis Cap Cod vorkommt (STIMPSON 1853; VERRILL 1866, 1873, 1880, 1882, 1885, 1895; FEWKES 1891). Nordatlantisch kommt sie weiterhin an Spitzbergen (LÜTKEN 1857; PFEFFER 1894; DÖDERLEIN 1899) vor und erreicht westlich von Spitzbergen unter 80° 3' n. Br. ihren nördlischsten Fundort (DANIELSSEN und KÖREN 1884). Weiter südlich und östlich kennt man sie aus dem Färöer-Kanal (SLADEN 1889; BELL 1892) und dann an der norwegischen Küste von Christiansund bis Finmarken (RETZIUS 1805; DÜBEN und KOREN 1846; M. SARS 1850,

1861, 1865, 1869 ; Storm 1878 ; Danielssen und Koren 1884 ; Aurivillius .1886 ; Nordgaard 1893 ; Sluiter 1895 ; Bidenkap 1899), ferner aus der Bärents-See (Stuxberg 1878, 1886; D'Urban 1880; Hoffmann 1882; Da-nielssen und Koren 1884; Sluiter 1895), von der Murmanschen Küste .(Jarzynsky 1885), aus der Matotschkin-Strasse (Stuxberg 1878, 1886) und nördlich von Nowaja Semlja (v. Marenzeller 1877), sowie aus dem Ka-rischen Meere, bis zum 79° ö. L. (Stuxberg 1878, 1880, 1886; Levinsen 1886; Ruijs 1887; Sluiter 1895). Dagegen ist sie noch weiter östlich bis zum Ostcap noch nicht angetroffen worden, so dass man sie trotz ihrer weiten Verbreitung dennoch nicht als völlig circumpolar bezeichnen kann. Von Süd nach Nord reicht ihr Gebiet westatlantisch von 42–75° n. Br., ostatlantisch von 60–80° n. Br.

"Sie findet sich vorzugsweise auf lehmigem, schlickigem und schlamm-igem Boden, seltener auf harter, steiniger Unterlage und bewohnt Tiefen von 9–1156 (meistens 30–400) m.

"Die Römer-Schaudinn'sche Sammlung enthält 95 Exemplare von Spitzbergen und von der Murmanschen Küste von den Stationen 6, 11, 17, 18, 19, 21, 26, 35 und 57. Station 6 (78° 15′ n. Br., 105–110 m, Lehm mit enizelnen kleinen Steinen) liegt im Storfjord, Station 11 (79° 2′ n. Br., 250–395 m, feiner Schlick mit Steinen) und 21 (78° 12′ n. Br., 210–240 m, Mud, wenig kleine Steine) an der Westseite von Westspitzbergen, die Stationen 17 (79° 44′ n. Br., 430–450 m, feiner Mud, wenig kleine Steine, viele Wurmröhren), 18 (80° 8′ n. Br., 480 m, feiner Mud, wenig kleine Steine) und 19 (79° 34′ n. Br., 112 m, Mud mit Steinen) an der Nordseite von Westspitzbergen, Station 26 (78° 5′ n. Br., 290 m, Schlick, wenig kleine Steine) in der Olga-Strasse und Station 35 (79° n. Br., 195 m, Lehm, wenig kleine Steine) zwischen König-Karls-Land und Nordostland. Demnach ist die Art rings um Westspitzbergen verbreitet und geht hier nördlich bis 80° 8′ n. Br. Ausserdem wurde sie von Römer und Schaudinn an der Mur-manschen Küste (Station 57 : 69° 36′ n. Br., 128 m, wenig Steine, viele Algen und Laminarien) erbeutet. Die Tiefen jener spitzbergischen Fundorte betragen 105–480 m."

Döderlein describes some specimens from the " Olga " Expedition and discusses the identity of. *Ct. krausei* LUDWIG with the present species [: 00, p. 222]:

'	"Exemplare dieser Art fanden sich auf

"Station 16 : 75° 40' N, 17° 30' O, 179 m Tiefe, blauer Schlick mit Muscheln.

"Station 26 : 78° 5' N, 14° 13' O, 145 bis 180 m Tiefe, schlickig.

"Station 33 : 78° 23' N, 16° 20' O, 190 m Tiefe, zäher Schlick; zahlreich.

"Station 55 : 75° 40' N, 17° 1' O, 190 bis 200 m Tiefe, grüner Schlick.

"Station 58 : 76° 27' N, 21° 24' O, 160 m Tiefe, grüner Schlick.

	a.	b.	c.	d.	e.	f.	Drontheim. g.	Drontheim. h.	i.
"Scheibenradius	10 mm.	11.5	13	13	16	16	14	14.5	12
"Armradius	18 ,,	21	26	27	29	31	30	37	30
"Zahl der Dorsomarginal-platten	11 ,,	11	12	14	14	15	16	17	14

"Die von der Olga-Expedition gesammelten Exemplare gehören sämtlich zu der normalen kurzarmigen Form, bei der das Verhältnis von Scheiben-radius zum Armradius nahezu 1 : 2 ist. Wie schon LÜTKEN erwähnt, finden sich an den Küsten von Norwegen Formen, die sich auszeichnen durch ihre längeren und schmäleren Arme, so dass r : R nahezu $1 : 2\frac{1}{2}$ ist, wie das die zum Vergleiche herangezogene Exemplare h. und i. von Drontheim zeigen. Die Zahl der oberen Randplatten nimmt mit dem Grössenwachstum des Seesternes ziemlich gleichmässig zu, sodass die kleinsten der gesammel-ten Exemplare von 18 mm Armradius deren 11 jederseits besitzen, die grössten von 31 mm deren 15 ; das Drontheimer langarmige Exemplar von 37 mm Armradius besitzt 17 obere Randplatten.

"Unter dem Namen *Ctenodiscus krausei* hat LUDWIG (Zool. Jahrbücher, System., Bd. I, pag. 290, Taf. 6 Fig. 13–16) zwei Exemplare aus dem Beringsmeer (SO. von St. George in 30 m Tiefe) beschrieben, die ich nach genauer Vergleichung dieser Beschreibung mit zahlreichen mir zugänglichen *Ctenodiscus crispatus* als unzweifelhaft dieser Art angehörig bezeichnen muss. Der Autor von *Ct. krausei*, dem Exemplare von *Ct. crispatus* nicht zugänglich waren, sah sich durch die sehr eingehende Beschreibung und Abbildungen dieser Art bei DUNCAN und SLADEN (Echinoderma of the Arctic Sea) veranlasst, Unterschiede zwischen der ihm vorliegenden Form und *Ct. crispatus* anzunehmen, die tatsächlich nicht vorhanden sind oder höchstens in so unbedeutendem Masse bestehen, dass sie zur Aufstellung einer besonderen Art nicht zu verwenden sind. Der wichtigste Unterschied zwischen beiden Arten soll in Bau des Rückenskeletts liegen, das nach DUNCAN und SLADEN bei *Ct. crispatus* ein kompaktes Netzwerk kalkiger Teile bilden soll (jedoch ausdrücklich ' ähnlich dem von *Astropecten* ' !), ' welches die Paxillen trägt,' während LUDWIG bei *Ct. krausei* fand, dass ' ein Netzwerk gänzlich fehlt, dass die Paxillen sich zwar an ihrem basalen Ende zu einer Fussscheibe verbreitern, welche doppelt so breit ist, wie die Paxille selbst ; dass aber diese Fussscheiben sich nicht berühren, sondern stets durch skelettfrei Zwischenräume von einander getrennt bleiben.' Tatsächlich zeigt *Ct. crispatus* genau das von LUDWIG bei der Form aus dem Beringsmeer geschilderte Verhalten.

"Den zweiten wichtigeren Unterschied findet LUDWIG in der Zahl der kleineren, wimperartigen Stachelchen an den Seiten der Randplatten, welche die cribriformen Organe bedecken ; ' in den Armwinkeln zählte er zwischen je einem oberen und einem grossen Randstachel in jeder Reihe etwa 20 kleinere Stachelchen bei *Ct. krausei,* also etwa doppelt so viel als bei *Ct. crispatus,* von diesen 20 Stachelchen gehören etwa 15 der oberen und 5 der unteren Randplatte an.' An der bezeichneten Stelle zähle ich nun bei drei trockenen Exemplaren von *Ct. cripatus* (an Spiritusexemplaren sind diese Stachelchen kaum sicher zu zählen) 16, 17 und 18 Stachelchen und zwar 12+4, bezw. 12+5 und 13+5 nach ihrer Zugehörigkeit zur oberen und unteren Randplatte.

"Nachdem sich diese beiden für wichtg gehaltenen Unterschiede zwischen der nordatlantischen und nordpacifischen *Ctenodiscus*form als nicht stichhaltig erwiesen haben, muss *Ct. krausei* als Synonym von *Ct. crispatus* angesehen werden.

"Die 10 oberen Randplatten, welche an der pacifischen Form bei einem Scheibenradius von 9.5 mm und einem Armradius von 18 mm beobachtet wurden, entsprechen durchaus der Zahl, die bei gleich grossen atlantischen Form zu erwarten ist. Die cribriformen Organe haben bei dieser den von LUDWIG an der pacifischen Form geschilderten Bau. Die Lage der Geschlechtsöffnungen fand ich bei norwegischen Exemplaren an der von LUDWIG bei seiner Form beobachteten Stelle, dicht über den Zwischenraum zwischen der 1. und 2. oberen Randplatte. Die Ventrolateralplatten trennen bei meinen kleineren Exemplaren 5 der unteren Randplatten von den Adambulacralplatten, bei den grösseren wächst diese Zahl; bei der Form aus dem Beringsmeer fand LUDWIG an seinen noch kleineren Exemplaren nur 4 Randplatten von den Adambulacralplatten getrennt.

"Aus dem Obigen geht die interessante Tatsache hervor, dass *Ctenodiscus crispatus* eine circumpolare Art ist, die sich aus dem nördlichen Eismeer nach dem Nord-Atlantic und dem Nord-Pacific verbreitet."

According to the Zoological Record this species is mentioned by BRØGGER [: 00, p. 157], KNIPOVITSCH (in Russian) and WHITEAVES [: 01, p. 48]. GRIEG, in his paper on the echinoderms of northern Norway, makes the following remarks on this starfish [: 02, p. 19]:

"Denne art forekommer meget talrig paa evjebund i fjordbasinerne mellem Skjaerstadfjord og Porsangerfjord, 30–530 m. Eksemplarerne, der havde en skiveradius af indtil 17 mm. og armradius af 40 mm, tilhørte dels den normale kortarmede, dels den langarmede form.

"Ved den norske kyst er den sydlig udbredt til Christiansund."

This species occurs in large numbers on clayey bottom in the fjord-basins between Skjaerstad Fjord and Porsangerfjord, 30–530 m. The examples, which had the disc radius of up to 17 mm. and arm radius of 40 mm., belonged partly to the normal form with short arms, and partly to that with long arms.

On the Norwegian coast it is distributed southwards to Christiansund.

Under the name of *Ct. corniculatus* (LINCK) MICHAILOVSKIJ has some remarks on this species [: 02, p. 487]:

"Bei der überwältigenden Mehrheit der zu meiner Verfügung gewesenen spitzbergenschen Exemplare dieser Art ist das Verhältnis $r : R = 1 : 2$, oder doch nahe daran. Nur bei einigen ging es bis $1 : 1.8$ einerseits und $1 : 2.3$ ja auch $1 : 2.4$ andrerseits. Letztere Zahlen nähern sich sehr der von LÜTKEN angegebenen Proportion für die langarmige Varietät. In enger Beziehung zur relativen Armlänge steht auch die Zahl der dorso-marginalen Plättchen, was man leicht aus folgender Zusammenstellung ersehen kann :

r =	11.5 mm.	15 mm.	12 mm.	12 mm.	11 mm.
R =	21 mm.	29 mm.	26 mm.	28 mm.	26 mm.
r : R =	1 : 1.8	1 : 1.9	1 : 2.1	1 : 2.3	1 : 2.4

Zahl der dorso-mar-
ginalen Plättchen. 12   12   13   16   17

"Weniger genau, aber noch offenkundiger tritt die Abhängigkeit in der Zahl der Plättchen von Alter und Grösse des Tieres aus folgender Tabelle hervor :

$$r = \begin{cases} 3 \text{ mm.} \quad 6 \text{ mm.} \quad 6 \text{ mm.} \quad 7 \text{ mm.} \quad 7 \text{ mm.} \quad 8 \text{ mm.} \quad 9 \text{ mm.} \\ 10 \text{ mm.} \quad 11 \text{ mm.} \quad 11 \text{ mm.} \quad 12 \text{ mm.} \quad 15 \text{ mm.} \quad 16 \text{ mm.} \end{cases}$$

$$R = \begin{cases} 6 \text{ mm.} \quad 12 \text{ mm.} \quad 12 \text{ mm.} \quad 14 \text{ mm.} \quad 15 \text{ mm.} \quad 16 \text{ mm.} \quad 19 \text{ mm.} \\ 21 \text{ mm.} \quad 22 \text{ mm.} \quad 22 \text{ mm.} \quad 26 \text{ mm.} \quad 29 \text{ mm.} \quad 33 \text{ mm.} \end{cases}$$

$$r : R = \begin{cases} 1 : 2 \quad 1 : 2 \quad 1 : 2 \quad 1 : 2 \quad 1 : 2.1 \quad 1 : 2 \quad 1 : 2.1 \\ 1 : 2.1 \quad 1 : 2 \quad 1 : 2 \quad 1 : 2.1 \quad 1 : 1.9 \quad 1 : 2 \end{cases}$$

Zahl der dorso-mar- (6   8   9   9   9   10   11
ginalen Plättchen. (13   11   12   13   12   14

"Die gleiche Anzahl von dorso-marginalen Plättchen bei Exemplaren, die verschiedene Dimensionen haben, oder eine verschieden grosse Zahl davon bei Individuen derselben Grösse, bedingen ihrerseits wieder die allgemeine Form der Arme, denn die enger und in grösserer Menge auf einer und derselben Fläche befindlichen Plättchen treten bedeutend stärker am Ende des Armes auf die obere Seite über, wobei sie die zwischen ihnen liegende freie Fläche verengern und verkleinern. Eines der aus Nord-Spitzbergen stammenden Exemplare, zeigt die bemerkenswerte Eigentümlichkeit, dass hinter der gewöhnlichen ventro-marginalen Stachelreihe bei ihm noch eine zweite, nicht weniger deutlich ausgesprochene Reihe ebensolcher Stacheln sich vorfindet, welche unterhalb der ersten, in ungefähr derselben Entfernung von ihr, wie sie von dem oberen Rande der entsprechenden Plättchen gelegen ist. In allen übrigen ist dieses Stück ganz normal.

"Die russischerseits constatierten Fundorte dieser Art unter 80° 57' und

81° 1′ n. Br. sind die allernördlichsten aller bis dato von Spitzbergen bekannten.

"A. BIBULA. 1899. St. 23=23[1] (13)= [Storfjord, lat. 76° 42 N., long. 17° 28′ E., 139–131.5 m., coarse stone fragments, bottom temperature 0.7° C.].

"Dr. TSCHERNYSCHEV. 1899. St. 27=50[1](1)=[Lat. 80° 57′ N., long. 20° 51′ E., 190 m., mud, bottom temperature 0.7° C.] u. 28=56[1](1)=[Lat. 81° 1′ N., long. 18° 28′ E., 180 m., mud, bottom temp. 0.3° C.].

"Dr. A. WALKOWITSCH. 1900. St. 1=61[1](2)=[Storfjord, near Whaleshead, lat. 77° 27′ N., long. 18° 45′ E., 120.5 m., mud, bottom temp. –1.9°– –2° C.], 2=62[1] (sehr zahlreich)[=Storfjord, near Whaleshead, lat. 77° 28′ N., long. 18° 40′ E., 108–117 m., mud, bottom temp. –2° C.], 5=65[1] (10)[= Icefjord, opposite Advent Bay, lat. 78° 22′ N., long. 15° 25′ E., 243 m., mud, gravel and 'small algae' (?), bottom temp. 0.8° C.] u. 9=69[1](8)[= Icefjord; Klassbilin Bay, 5 nautical miles from Nordenskjöld Glacier, lat. 78° 37′ N., long. 16° 40′ E., 142–133 m., thin mud with numerous stones, bottom temp. –1.9° C.].·1901. St. 6=76[1](16)[=Storfjord, between Whaleshead and Cape Agardh, lat. 77° 47′ N., long. 19° 07′ E., 101 m., soft mud with fragments of annelid tubes, bottom temp. –1.8° C.] u. 14=84[1](2)[= Storfjord, lat. 78° 3′ N., long. 20° 5′ E., 75.5 m., thin sandy mud with gravel, bottom temp. 2.5 C.]."

NORMAN refers to this species as follows [: 03, p. 407] : "*Ctenodiscus crispatus*, RETZIUS. Inner reach of Lang Fiord, in 5–30 fathoms; Bög Fiord, in 100–120 fathoms; and Varanger Fiord, 100–125 fathoms."

MICHAILOVSKIJ mentions *Ct. corniculatus* from the vicinity of Nova Zembla as follows [: 04, p. 171] : "Das Verhältnis r : R schwankt bei den gemeinsamen Exemplaren von 1 : 1.8 bis 1 : 2.2. Von einer Station liegen zwei unregelmässig gebildete Formen vor, bei welchen je zwei Arme einander beträchtlich genähert und auf eine gewisse Strecke hin gleichsam mit einander verschmolzen sind." The specimens were obtained at the following stations: St. 41=lat. 72° 51 N., long. 37° 52′ E., 240 m., mud with stones, bottom

---

1) These numerals show the numbers of the stations in the whole series; those in parentheses the number of specimens taken.

temperature 0.9° C., 11 specimens; St. 48=lat. 74° 31′ N., long. 53° 20′ E.,
211 m., sand, bottom temp. 1.9° C., 1 spec.; St. 48b=lat. 74° 28′ N., long. 54°
18′ E., 160 m., mud, 3 young spec.; St. 48c=lat. 74° 32′ N., long. 54° 20′ E.,
150 m., gravel, 1 young spec.; St. 52b=lat. 74° 30′ N., long. 54° 10′ E.,
175 m., mud, 13 spec.; St. 58=lat. 75° 13 N., long. 53° 23′ E., 179 m.,
mud, bottom temp. 1.9° C., 1 spec.; St. 60=lat. 77° 11′ N., long. 53° 27′
E., 312 m., mud, bottom temp. 1.6° C., 1 spec.; St. 61=lat. 78° N.,
long. 52° 57′ E., 308 m., mud, bottom temp. 0.6° C., 7 spec.; St. 72=lat. 76°
35′ N., long. 61° 11′ E., 111 m., mud, bottom temp. 1.7° C., 15 spec.; St. 76
=lat. 77° 53′ N., long. 61° 29′ E., 356 m., mud, bottom temp. 1.2° C., 1
spec.; St. 79=lat. 79° 30′ N., long. 60° 50′ 30″ E., 225 m., sand, bottom
temp. 0.6° C., 3 spec.; St. 80=lat. 79° 51′ 18″ N., long. 60° 44′ E., 323
m., mud, bottom temp. 1.1° C., 6 spec.; St. 82=lat. 80° 26′ N., long. 64°
14′ E., 204 m., mud, 1 spec.; St. 94=lat. 73° 53′ N., long. 52° 55′ E.,
162 m., mud, bottom temp. 1.3° C., 10 spec.; St. 95=lat. 73° 30 N., long.
50° 12′ N., 275 m., mud, bottom temp. 1.3° C., 4 spec.; St. 98=lat. 71° 38′
N., long. 40° 30′ E., 320 m., mud, bottom temp. 1.4° C., 3 spec." MEISSNER
mentions *Ct. procurator* from the collection of the Hamburg Magellan Expedition [: 04, p. 18, *fide* Zool. Record].

In his work on the "Albatross" Asteroidea LUDWIG, using the name
Ct. crispatus (RETZIUS), describes specimens from the tropical Pacific and
brings out some new points in its synonymy [: 05, p. 104]:

"SLADEN's *Ctenodiscus procurator* von der Westseite Südamerika's (zwisch-
en 45° und 53° s. Br., in 75–2423 m. Tiefe) unterscheidet sich von der lang-
armigen Form des nordischen *crispatus* nur durch eine geringe Differenz in der
Zahl der Randplatten und etwas schwächere Bewaffnung der Adambulacralplatten. Die hier vorliegenden Exemplare gleichen aber diese Unterschiede[1]
völlig aus und sind überdies durch ihre Grösse bemerkenswert, in welcher
sie die grössten bis jetzt bekannt gewordenen nordischen Exemplare (R= 40
mm.) zum Teil erheblich übertreffen. Es liegen vier grosse Exemplare mit
folgenden Maassen vor:

---

1) "Für die Identität des *Ct. procurator* mit *Ct. crispatus* ist auch ein Vergleich der beiden
Abbildungen: SLADEN l.c. Taf. XXX, Fig. 7 und DÖDERLEIN, l.c. Taf. IX, Fig. 2 sehr lehrreich."

Nr.	R in mm.	r in mm.	r : R	ZoR	ZuR
1	56	19	1 : 2.95	20	21
2	49	18	1 : 2.72	18	18
3	45	17	1 : 2.65	16	17
4	38	15	1 : 2.53	15	15

"Diese vier grossen Exemplare stammen aus dem Golf von Panama, Station 3393; 10. März 1891; 7° 15′ n. Br.; 79° 36′ w. L.; Tiefe 1865 m.; Bodentemperatur+2.67° C, oberflächliche Temperatur+23.33° C; Bodenbeschaffenheit: grüner Schlamm.

"Ausserdem liegt ein junges Exemplar aus dem Eingang des Golfes von Californien vor, Station 3430; 19. April 1891; 23° 16′ n. Br.; 107° 31′ w. L.; Tiefe 1558 m.; Bodentemperatur +3.28° C, oberflächliche Temperatur +22.78° C; Bodenbeschaffenheit: schwarzer Sand; mit den Maasen: R=11 mm., r=5 mm., r : R=1 : 2.2, ZoR=8, ZuR=9.

"Da ich durch die vorliegenden Stücke zu der Auffassung komme, dass SLADEN'S *procurator* mit der nordischen Art zusammengehört,[1] so ergiebt sich aus den Fundstellen des Challenger und des Albatross für die geographische Verbreitung die sehr bemerkenswerte Tatsache, dass an der Ostseite des pacifischen Meeres eine nordische, annähernd circumpolare und bereits in Norden von geringer Tiefe (9 m.) bis in Tiefen von 1156 m. herabsteigende Art, die man im pacifischen Gebiete bis jetzt nur aus dem Beringsmeere kannte, ihr Verbreitungsgebiet südwärts bis zum 53° s. Br. (Eingang des Smyth Channel) fortsetzt und dabei überall da, wo das Oberflächenwasser eine höhere Temperatur besitzt in grössere kältere Tiefen hinabgeht.

"Eines der vorliegenden grösseren Exemplare (Nr. 2 der Tabelle) liess bei der anatomischen Untersuchung das Folgende feststellen. Die Füss-

---

1) "Dagegen kann ich mich der PERRIER'schen Meinung (Echinodermes de la Mission scientifique du Cap Horn, T. VI, Paris 1891, p. K 142–144), dass *Ctenodiscus australis* LÜTKEN, dessen Verbreitungsgebiet dem des *Ct. procurator* benachbart ist, als Varietät zu *Ct. crispatus* zu ziehen sei, nicht ohne Weiteres anschliessen. Denn, abgesehen davon, dass mir keine Vergleichsobjekte vorliegen, hat SLADEN (l.c. 1889, p. 172, Taf. XXX, Fig. 1-6) für *Ct. australis* einige unterscheidende Merkmale betont, die PERRIER nicht berücksichtigt zu haben scheint."

chenampullen sind einfach. Die apicale Erhebung ist gut ausgebildet und endigt blindgeschlossen; ihr der Leibeshöhle angehöriger Innenraum wird ebenso wie bei *Porcellanaster waltharii* durch ein senkrechtes Längsseptum in zwei Halbkanäle zerteilt. Superambulacralstücke sind nicht vorhanden. Die interradialen Septen sind dünnwandig und unverkalkt. In jedem Interradius fand sich eine, im Interradius des Steinkanales aber zwei (jederseits vom Steinkanal eine), grosse Poli'sche Blasen. Die Tiedemann'schen Körperchen sind sehr gut ausgebildet. Interradiale Blinddärme sind nicht vorhanden; dagegen gut ausgebildete, gesondert aus dem Magen entspringende radiale Blinddärme, die bis zur siebten oder achten Randplatte in die Arme hinein-reichen. Die verästelten, dicken, in ein Bündel zusammengedrängten Genital-schläuche reichen nicht bis in die Arme und sind ausschliesslich mit ver-schiedenen Grössenstadien kleiner Eizellen erfüllt.

"An dem kleinen Exemplar von R=11 mm. hat der Apicalfortsatz eine Länge von 4 mm. Während bei den grossen Tieren auch im Armrücken Papulä bis fast zur Armspitze vorkommen, sind sie bei dem kleinen Exemplar noch auf die Rückenhaut der Scheibe beschränkt."

Clark [: 05, p. 1], Nordgaard [: 05, p. 160, 235, 241] and Schmidt [: 05, p. 18, 19] mention this species (*fide* Zool. Rec.).

Under the name of *Ctenodiscus crispatus* Retzius, Grieg mentions this species as follows [: 06, p. 9]:

"Fundstätte: 1900 Stat. 10 (1); 15 (zahlreich); 52 (sehr zahlreich); 53 (do.).

"1901 Stat. 50 (sehr zahlreich); 56 (einige Exemplare); 83 (1); 87(3); 96 (zahlreich).

"'Heimdal' 1900 Stat. 14 (zahlreich).

"Diese Art ist sowohl in der kalten wie warmen Area verbreitet. Dem Material im Museum zu Bergen nach zu urteilen ist die Kaltwasserform kurzarmiger als die Warmwasserform. Die Individuen aus der kalten Area hatten einen Armradius, der höchstens doppelt so gross war als der Schei-benradius. So war bei den Exemplaren vom 'Heimdal' Stat. 14 r. 3-11 mm., R. 4.5-22 mm., r. : R.=1 : 1.5-2, die Dorsomarginalplatten 4-13, und bei den Exemplaren vom 'Michael Sars' 1900 Stat. 52 r. 4.5-14 mm., R.

8–23 mm., r.: R.=1 : 1.7–2, die Dorsomarginalplatten 6–11. Bei den Individuen aus der warmen Area dagegen fand ich häufig einen Armradius, der über doppelt so gross war als der Scheibenradius. So war bei den Exemplaren vom 'Michaël Sars' 1900 Stat. 15 r. 5–16 mm., R. 9–35 mm., r.: R.=1 : 1.7–2.3, die Dorsomarginalplatten 7–16. An der norwegischen Küste findet man nicht selten das Verhältins r.: R.=1 : 2.5, ja M. Sars erwähnt eines Exemplares von Reine, Lofoten, wo es 1 : 2.7 war. Es sei doch bemerkt, dass andere Forscher Exemplare dieser Art aus der kalten Area erwähnen, die ebenso langarmig sein dürften wie die Warmwasserform. Bei dem von Michailovskij in 'Zool. Ergebnisse der russischen Expeditionen nach Spitzbergen' beschriebenen Materiale variierte das Verhältnis zwischen Scheiben- und Armradius zwischen 1 : 1.8 und 1 : 2.4, am häufigsten war es : 1 : 2. Die Bodentemperatur der Oertlichkeiten, wo das sehr reiche Material gesammelt worden war, war + 2.5 ÷ 2°, nur vier Exemplare aus der warmen Area, weshalb man sagen darf, dass das Material der kalten Area angehört. Es scheint jedoch, dass der einzige Unterschied zwischen Warm- und Kaltwasserform dieser Art eben der ist, dass selbst die langarmigsten Individuen der kalten Area nie oder höchst selten die relative Armlänge derer aus der warmen erreichen können."

This species is mentioned by Derjugin [: 06, p. 149] from the Murman Sea and by Grieg [: 07, p. 134] from the North Sea. The latter had ' two long-armed examples (r : R=1 : 2 and 1 : 2.27). In the larger example the disk radius was 15 mm., arm radius 34 mm., 15 dorsomarginal plates.' The station was 75° 58′ 5″ N., 14° 8′ W., depth 300 m., brown and grey clay, bottom temperature 0.40° C.

Kalischewskij makes some critical remarks on this species and proposes to distinguish two varieties [: 07, p. 25, 9 figs.] :

"Die Vergleichung der Exemplare *Ctenodiscus crispatus* (Retzius), welche von der Russischen Polar-Expedition in dem Karischen und Nordsibirischen Meere erbeutet worden sind, mit den Exemplaren derselben Art, die von der Murman-Küste herstammen, lässt die Vermutung aufkommen, dass in den arktischen Meeren zwei Varietäten dieses Sternes existieren ; die sibirische Kaltwasser- und die europäische Warmwasser-Varietät.

" Dank der Liebenswürdigkeit der Herrn A. S. SKORIKOW und K. M.
DERJUGIN, hatte ich 45 Exemplare *Ctenodiscus crispatus* (RETZIUS) von der
Murmanküste, aus der Umgebung der murmanschen zoologischen Station'
zu meiner Verfügung, wofür ich ihnen hier meinen tiefgefühlten Dank
ausspreche. Ausserdem erwies sich es, dass im zoologischen Museum der
Universität zu Odessa noch 6 Exemplare dieses Sternes vorhanden sind,
welche mit der Etiquette ' Eismeer, A GRAFTIO ' versehen sind und augenscheinlich auch aus dem Murmanmeere stammen.

   " Obgleich die Vergleichung des erwähnten Materials mit dem der
Sammlung der Expedition keine Möglichkeit gab irgend welche feste Unter-
scheidungsmerkmale wahrzunehmen, so geben doch die Durchschnittsziffern
ein sehr gutes Resultat.

   " Das Verhältnis R : r bei den karischen und nordsibirischen Exemplaren
des *Ctenodiscus crispatus* (RETZIUS) ist durchschnittlich (von 14 Beobachtungen) gleich 1.96 bei einer Amplitüde von 2.1-1.9=0.2, während bei den
murmanschen (durchschnittlich von 45 Messungen) es nur 1.87 bei einer
Amplitüde von 2.2-1.6=0.6 ist. Bemerkenswerth ist auch die Verteilung
der Dornen auf der abaktinalen Seite der Arme und des Diskus bei den
karisch-sibirischen und murmanschen Exemplaren : während bei der ersteren in
jeder Gruppe gewöhnlich 8-9 und auch mehr Dornen vereinigt sind und
sogar manchmal die sehr solide Zahl 16 erreichen, fällt diese Zahl bei den
letzteren bis auf 5-6-7 Dornen ; nur ausnahmsweise erreicht sie 8-9 (s. Fig.
76 u. 8 b auf der Tabelle I).   Die Mundplättchen sind überhaupt bei beiden
Varietäten auch nicht gleich gestaltet.   Die Figuren 7 a und 8 a auf der
Tabelle I stellen die gewöhnliche Form dieser Plättchen der beiden oben
erwähnten *Ctenodiscus*-Varietäten dar; übrigens sind auch Uebergänge
möglich ; das massive langgestreckte Mundplättchen, welches gewöhnlich
der karischsibirischen Form *Ctenodiscus crispatus* (RETZIUS) eigen ist, kann
auch bei der murmanschen Form vorkommen, und das breite dünne
Plättchen wird, als Ausnahme, bei den karisch-sibirischen Exemplaren
wahrgenommen.

   " Auf Grund der oben angegebenen Merkmale halte ich es für nötig in
den Gewässern der eurasiatischen Arctis mindestens zwei Varietäten *Cteno-*

discus crispatus (RETZIUS), die verhältnismässig kurzarmige murmansche und die langarmige karisch-sibirische, zu unterscheiden.

"Die Sammlungen der Russischen Polar-Expedition rückt die östliche Verbreitungsgrenze des *Ctenodiscus crispatus* (RETZIUS) bis 138° 47' Ost vor, was diesen Stern als circumpolare Art anerkennen lässt.

"Fundorte: 1. 13 (26) viii 1900, Karisches Meer, am Westufer der West-Taimyr gegenüber dem Cap-Sterlegow, Lt. 75° 49' N., Lg. 89° 35' Ost.; Tiefe 38 Mtr.; Boden—Schlamm; zool. Trawl (St. 14c); 1 Exemplar.

"2. 21 viii (3 ix) 1901, Nordenskiöldmeer, östlich von der Ost-Taimyr-Halbinsel, Lt. 77° 1' N,, Lg. 114° 35' Ost.; Tiefe 60 Mtr.; Boden—Schlamm mit Steinen; zool. Trawl (St. 46); 5 + 1 juv. = 6 Exemplare.

"3. 24 viii (6 ix) 1901, ebendaselbst, offenes Meer, Lt. 75° 42' N., Lg. 124° 41' Ost.; Tiefe 51 Mtr., Boden—Schlamm; zool. Trawl (St. 49); 7 Exemplare.

"4. 28 viii (10 x) 1901, Eismeer nördlich von den Neusibirischen Inseln, Lt. 77° 20' 30" N., Lg. 138° 47' Ost.; Tiefe 38 Mtr.; Boden—Schlamm; zool. Trawl (St. 50); 1 Exemplar.

"Tabelle der Messungen des *Ctenodiscus crispatus* (RETZIUS) aus der Sammlung der Russischen Polar-Expedition:

St.	Exempl.	R in mm.	r in mm.	R : r	Anzahl d. Seitenplat. des Armes.	Anzahl der Dornen in den Dornenbündeln.									
14c	I	30	15	2	13	12	11	14	10	7	13	10	6	15	14
46	I	31	15.5	2	13	10	10	11	11	6	7	13	12	7	9
46	II	31	14	2.1	14	6	10	12	13	10	8	11	4	8	9
46	III	28	15	1.9	12	13	12	12	13	8	7	11	10	15	9
46	IV	28	15	1.9	12	6	4	11	11	11	10	8	11	11	10
46	V	29	15	1.9	12	9	11	10	11	9	7	10	10	11	14
49	I	30	15	2	15	14	14	10	9	9	10	9	7	5	12
49	II	24.5	12	2	11	13	8	15	8	11	6	5	9	9	13
49	III	21	10.5	2	11	7	9	8	9	8	6	5	9	9	8
49	IV	21	11	1.9	12	11	10	10	6	8	10	9	4	16	13

49	V	20	10	2	12	10	8	10	11	9	5	6	8	9	10
49	VI	19	10	1.9	11	7	8	9	6	4	9	7	9	9	6
49	VII	17	9	1.9	11	8	7	8	8	8	9	9	4	6	7
50	I	25	13	1.9	13	11	14	11	15	8	14	6	12	8	8

" Tabelle der Messungen des *Ctenodiscus crispatus* (Retzius) aus dem Murmanschen Meere.

Exempl.	R in mm.	r in mm.	R : r	Anzahl d. Seitenplatt. des Armes.	Anzahl der Dornen in den Dornenbündeln.									
I[1]	20	9.5	2.1	14	11	8	8	10	7	9	6	6	9	7
II	17	10	1.7	11	6	6	6	7	7	6	5	4	7	4
III	25	13	1.9	13	8	9	8	8	2	7	6	9	4	8
IV	28	14	2	13	10	9	8	7	18	7	9	7	7	6
V	20	10	2	11	9	7	7	7	7	8	7	6	6	10
VI[2]	19	9.5	.2	11	6	8	8	10	7	6	9	5	8	8
VII	19	9.5	2	11	6	9	6	9	6	7	6	9	9	7
VIII	18	9	2	12	6	6	7	9	6	6	7	7	6	7
IX	19.5	9.5	2	12	4	5	5	5	7	6	4	7	5	5
X	16.5	9.5	1.7	10	7	7	5	5	5	5	8	9	5	7
XI	18	10	1.8	11	6	6	5	7	6	6	6	5	6	4
XII	24	11.5	2	12	8	6	6	8	8	8	6	9	8	3
XIII	23.5	12	2	13	7	7	6	9	7	9	4	5	6	9
XIV	22	11	2	13	6	5	7	7	5	6	6	6	8	4
XV	24	15	1.6	10	6	6	8	9	7	4	10	4	6	8
XVI	23	12	1.9	12	8	8	7	8	5	9	7	7	8	9
XVII	24	12	2	12	6	6	8	6	6	5	8	5	5	8
XVIII	22	11	2	12	6	6	7	7	7	5	5	5	7	6
XIX[2]	22	13.5	1.7	11	7	7	7	8	7	7	5	9	6	7
XX	16	9	1.7	9	7	8	6	7	5	6	5	5	8	8
XXI[2]	19	11.5	1.7	10	6	6	5	5	5	3	6	5	6	6
XXII[2]	19.5	12	1.6	11	6	7	6	7	6	5	7	5	5	4
XXIII[2]	15	7	2.1	11	6	6	6	4	5	5	6	4	6	8

XXIV[2]	27	16	1.7	15	8	9	8	8	8	5	9	9	8	8
XXV	21	12	1.8	11	6	8	8	9	4	4	7	8	5	5
XXVI	27	13	2.1	15	7	6	8	6	7	7	6	4	7	6
XXVII	19	10	1.9	11	7	6	7	7	7	8	9	4	5	7
XXVIII	14	7	2	9	6	4	4	5	6	6	7	4	5	4
XXIX	21	11.5	1.8	11	6	8	7	7	6	8	6	9	9	8
XXX	25	13	1.9	13	8	6	4	7	7	6	7	10	7	8
XXXI	20.5	11	1.9	11	6	6	6	7	6	4	7	6	7	4
XXXII	23	12	1.9	13	9	8	6	9	8	8	7	5	7	8
XXXIII	19	10	1.9	11	6	7	6	6	5	7	5	8	7	11
XXXIV	18	9.5	1.9	11	6	5	5	6	4	8	5	4	6	5
XXXV	22	11	2	12	13	10	7	6	4	5	11	12	8	12
XXXVI	15	7.5	2	9	5	5	4	5	4	5	5	6	5	4
XXXVII	21	10.5	2	12	7	9	8	7	6	5	7	7	7	5
XXXVIII	20.5	9.5	2.2	12	7	6	7	6	6	7	8	8	4	7
XXXIX[2]	18	10	1.8	12	4	6	6	5	6	4	5	5	5	6
XL	18	11	1.6	11	8	5	9	7	6	6	8	5	8	7
XLI[2]	21	12.5	1.7	12	6	8	6	6	4	6	5	4	7	7
XLII[2]	22.5	14	1.6	11	10	11	8	8	8	8	9	9	6	5
XLIII[2]	18	10	1.8	12	6	6	7	4	7	5	5	4	5	4
XLIV	23	14.5	1.7	12	8	5	6	7	8	7	6	6	3	6

"Ausser diesen 44 Exemplaren, befindet sich in dem mir zur Verfügung stehenden Material aus dem Murmanschen Meere noch ein Exemplar des *Ctenodiscus crispatus* (RETZIUS), welches eine interessante Missgestalt, eine Asymmetrie darstellt. Das Centrum des Diskus ist bei ihm auf die Seite gerückt, infolgedessen ist in einem Falle R=24.5 mm. und im anderen= 20.5 mm.; r schwankt von 15.5 mm. bis 12 mm., R : r=1.7–1.6 Die Zahl der Seitenplättchen ist aber an allen Armen gleich, nämlich 12; die Zahl der Stacheln in einzelnen Gruppen 9, 7, 5, 6, 7, 5, 4, 8, 7, 6.

1) "Ein zusammengepresstes Exemplar."
2) "Ein aufgetriebenes Exemplar mit spärlich sitzenden Dornenbündeln."

"Die Messungstabelle der Exemplare *Ctenodiscus crispatus* (RETZIUS) des zoologischen Kabinets der Universität Odessa.

Exemp.	R in mm.	r in mm.	R : r	Zahl der Seitenplatt. des Armes.	Zahl der Dornen in Dornenbündel.								
I	20	11.5	1.8	12	6	4	6	5	10	11	7	6	5	6
II	22	13.5	1.6	11	6	8	3	6	6	6	9	7	4	8
III	17.5	10	1.75	11	5	6	6	4	5	5	6	7	5	5
IV	15	8	1.9	11	6	6	5	5	4	6	7	7	5	4
V	15	9	1.7	10	4	4	6	7	5	6	4	5	3	7
VI[1)	15	9	1.7	10	7	4	6	8	4	6	6	6	7	5

KŒHLER describes some specimens from the dredgings of the "Princesse Alice" [: 09, p. 28]:

"Campagne de 1898 : Stn. 939, profondeur 177 m. Nombreux échantillons.—Stn. 976, profondeur 186 m. Quelques échantillons.—Stn. 1012, profondeur 430 m. Plusieurs échantillons.

"Campagne de 1906 : Stn. 2495, profondeur 251 m. Un échantillon.

"Campagne de 1907 : Stn. 2619, profondeur 50–20 m. Un échantillon.

"Certains auteurs ont distingué une forme à bras courts et une forme à bras longs et KALISCHEVSKY a même cru reconnaître, en outre, que chaque de ces formes différait aussi par le nombre des granules des plaques : cet anteur ajoute cependant qu'il y a des passages entre les deux formes.

"Dans nos exemplaires recueillis par la Princesse-Alice, le rapport $\frac{R}{r}$ est égal à 2 ou il est très voisin de ce chiffre. Je trouve par exemple que dans le plus grand individu, R=38 et r=18 mm; dans d'autres, R=30 et r=14 mm.; R=22 et r=10 mm., etc.

"DÖDERLEIN notamment a publié [: 00, p. 221] une description détaillée accompagnée d'excellentes photographies du *Ct. corniculatus*.

"*Distribution géographique.*—Le *Ct. corniculatus* est une espèce presque complétement circompolaire, car sa forme pacifique, que LUDWIG avait d'abord séparée du *Ctenodiscus* atlantique, doit lui être réunie, de l'avis même de ce savant. Sur les côtes de l'Amerique du Nord, le *Ct. corniculatus* s'étend

1) "Ein aufgeblasenes Exemplar."

entre le 42° et le 75° Lat. N. On le connaît dans toutes les régions septen-trionales de l'Europe et de l'Asie, et ses limites d'extension vers l'Est ont été reculées par l'Expédition polaire russe jusqu'au 138° Long. E. Il s'étend donc du 170° Long. O. jusqu'au 138° E., soit sur un espace de plus de 300°. Il peut vivre dans les mers boréales, entre 9 et 1156 m de profondeur.

"On aurait pus croire que le *Ct. corniculatus* restait localisé dans les régions septentrionales de notre globe, et ce n'est pas sans quelque surprise qu'on le voit figurer parmi les espèces draguées par l'Albatross dans le golfe de Panama et à l'entrée du golfe de Californie, à des profondeurs de 1865 m et de 1558 m. LUDWIG, qui a étudié ces échantillons, les trouve absolument identiques à ceux des mers boréales. Quatre individus dragués dans le golfe de Panama sont particulièrement remarquables par leur taille et, dans le plus grand, R=56 et r=19 mm.

"Quelle est au juste la valeur des différences qui séparent les *Ct. australis* LÜTKEN et *procurator* SLADEN des mers australes, du *Ct. corniculatus* boréal? Il me paraît pour le moment, impossible de répondre à cette question. Il est incontestable que les trois espèces sont très voisines et que le *Ct. procurator* se rapproche beaucoup du *Ct. corniculatus*. Si leur identité venait à être démontrée, le *Ct. corniculatus* devrait être noté comme une des très rares formes bipolaires actuellement connues."

KŒHLER again mentions it from the collection of the French expedition to the Arctic [: 09b, p. 122]: "*Ctenodiscns corniculatus* (LINCK). Quelques échantillons de taille plutôt petite. Mer de Barentz." VERRILL mentions some four rayed specimens from New England and figures one beside a regular five rayed example [: 09, p. 548, 549].

MORTENSEN describes a specimen from north-east Greenland as follows [: 10, p. 256]:

"Stat. 98 (77° N. 17½° W.) 300 m. 1 specimen. R=39 mm., r=18 mm.

"It is noteworthy that the paxillæ in this species are connected at their bases by radiating muscles, very beautifully arranged (Pl. XVI. Fig. 5). The spinelets of the paxillæ are invested in a thick skin, and generally they are united in bundles of two or three, as is especially seen very beautifully on staining the skin. The wall of the stomach is full of small, irregular specules (Pl. XVI. Fig. 4).

"In one of the interradia I found on this specimen a cluster of eggs, attached to one of the genital openings. The eggs are rather large, 0.6–0.7 mm in diameter. The ovaries are seen to contain eggs in all stages of development, from quite young to such as are quite mature, ready to be laid. These facts make it very probable that this species has not pelagic larvæ ; evidently the eggs must be laid at different epochs, since they do not become ripe at the same time, as is generally the case in those forms, where the eggs develop into pelagic larvæ.

"The present specimen differs not inconsiderably from the typical form in the marginal plates being somewhat spinous and not so regular in shape (Pl. XIV, Fig. 4, to compare with Pl. XIV, Fig. 6, representing the corresponding plates from a specimen of the typical form). The limit between the upper and lower marginal plates is indistinct, on account of a thick covering skin. Whether this specimen represents a distinct variety or perhaps an individual abnormality can, of course, not be settled on the evidence of the present material. But I am inclined to regard it as an abnormal specimen. In a specimen from the Kara Sea I have found traces of a similar irregularity in the marginal plates, but much less developed than in the present specimen.— Otherwise it agrees with KALISCHEWSKIJ's cold water variety.

"Concerning the most interesting geographical distribution of this species, see below, p. 294." The wide distribution of this and several other species of starfishes is cited as an argument in favour of A. H. CLARK's "Polar-Pacific" region.

FISHER adds to the list of synonyms the *Ctenodiscus australis* of LOVÉN MS. LÜTKEN, a point on which I would not express my own opinion without examining the original specimens. He then gives a very exhaustive account of *C. crispatus* [FISHER, :11, p. 31]:

"*Diagnosis.*—Rays five, exceptionally four or six. $R = 1.66$ r to 3.16 r. General form stellato-pentagonal, to stellate, extremely variable. Abactinal surface more or less tumid and usually with an elevated cone in centre of disk; paxillæ variable, usually low, with few to many spinelets which are short, clavate and skin-covered. Marginal, actinal intermediate, and adambulacral plates obscured by a thin soft skin. Continuous narrow deep grooves extend between marginal plates, across intermediate area, and between consecutive adambulacral plates. These are overhung on either side by a fold of skin embedded in which are numerous flattened spinelets, as in the lamellæ of typical cribriform organs. Between special raised ridges of marginal plates these furrows are deeper and V-shaped with five to seven superimposed lamellæ on either side. Marginal plates eleven to twenty, in each series, from median interradial line to extremity of ray. A single short conical spine at upper end of each superomarginal, and another similar one on each inferomarginal on the actinolateral margin of ray. Adambulacral plates with an oblique series of three to five sharp short skin-covered spines, and on aboral outer corner a similar, usually shorter spine, covered with the general investment of actinal surface. Mouth plates prominent; along free margin, about six spines like those of adambulacral and at inner end of plate a single more prominent spine; two or three short conical tubercles usually stand in a series on either side of median suture, these sometimes as long as furrow spines. Superambulacral plates present.

"*Description.*—No adequate description of this species is readily available. With the ample material at my disposal it seems well to give a description and at the same time to point out some of the most prominent of the variations. Instead of placing this last, a separate subhead is given to each category of characters and the variations considered at once.

"This a remarkably variable species, especially in the length of the rays, and their width, and in the general facies of the animal. Practically all of the most diverse variations may occur in examples from a single station, and they are thus not due to locality. As slender and broad armed forms occur among the smaller specimens this difference is not due to age.

"*Proportions.*—A striking series of four nearly equal sized specimens from stations 4235 illustrates admirably the difference in form, measurements being given in the accompanying table.

"Measurements of *Ctenodiscus crispatus.*

Station.	R.	r.	R : r	Number of supero-marginal plates.[1]	Width of ray at base.[2]	Number of furrow spines.[3]	Width of madre-poric plate.	Remarks.
4235[1]	38	12.0	3.16 : 1	20	14	5	4.0	Actinal area very narrow; paxillæ medium sized.
4235[2]	39	16.0	2.45 : 1	18	20	3–5	3.5	Paxillæ medium sized.
4235[3]	36	14.0	2.57 : 1	16–18	18	3–4	4.0	Do.
4235[4]	40	17.0	2.35 : 1	17–18	21	4–5	4.0	Paxillæ smaller.
4223[1]	35	19.0	1.8 : 1	13	23	3–5	3.0	Paxillæ medium sized; average proportion.
4223[2]	33	18.5	1.84 : 1	11	22	4–5	2.5	Paxillæ larger (Pl. 4, fig. 5).
4286	29	17.5	1.66 : 1	12	20	3–4	2.75	Paxillæ small.

"Specimen 4235[1] has slender, narrow rays (from above resembling a *Psilaster* somewhat) and numerous marginal plates with fairly wide fasciolar furrows. The intergradation is perfect, through 4235[4] to 4286, a very short-rayed form almost arcuately pentagonal, with few superomarginal plates. If, in the above table, the width of ray at base is compared with R, and the proportion R : r taken into account, the great difference in proportion is at once evident.

"*Abactinal surface; paxillæ.*—In some specimens the epiproctal cone is *inverted*, in others very inconspicuous. In specimens with rays nearly the same length the width of the paxillar area varies considerably, especially at the end of ray, thereby giving some specimens a much more robust appearance. The fact that in some examples the abactinal wall is nearly plane, except for the epiproctal cone, and much inflated in others is due, of course, to the condition of the animal at the moment of death, but is important as magnifying one or two trivial characters, such as the angle of marginal

1) "From interradial line to tip of ray."

2) "Measured from one interradial line, across abactinal surface to adjacent interradial line."

3) "Spines on furrow edge of adambulacral plate; one or more spines on surface of plate not counted."

plates and compactness of paxillæ. Two extremes, both from station 4223, will serve to illustrate variability of paxillæ. Each specimen has R=34 mm.; in A, paxillæ in neighbourhood of madreporic body have seven to twelve spinelets, occasionally as few as four on very small ones; of these never more than one is situated in centre of tabulum and rather more than half the paxillæ have no central spinelet at all; in B the paxillæ ordinarily have from twelve to twenty-two spinelets, of which three to five occupy centre of tabulum, and all or very nearly all have central spinelets, those paxillæ on outer part of ray having so few as one central spinelet.[1] When these two specimens are placed side by side the difference is very striking. Many of the specimens, both small and large, lack. a central spinelet to paxillæ altogether, thus resembling exactly some Atlantic examples. There is as much if not more difference in the extremes of these specimens than is shown by SLADEN'S two figures illustrating paxillæ of *Ctenodiscus australis* and *C. procurator*,[2] while in the extremes of body-form the difference is greater than between figures 1 and 7, illustrating the same two species. I thought at first that the difference in size of madreporic body might furnish a character of some constancy, to separate *krausei* from *crispatus*, the latter having the larger body. This character also is very variable in Pacific specimens, some examples having fairly large, others small madreporic bodies. In four rather poorly preserved Atlantic specimens the madreporic body is more constant and is one and one-half times greater in diameter than in Pacific examples of the same size.

"The shaft of the paxilla varies in length. In specimens from very deep water it is longer than in shallow water specimens.

"The bases of paxillæ, or the abactinal 'plates,' are circular and rather closely placed, usually not quite touching. They are largest about one-half r from centre of disk, decreasing in size toward tip of ray and centre of disk. Along mid-radial line where there are no papulæ the plates are smaller and usually broadly elliptical. Papulæ are not regularly arranged, four to six usually occurring about a plate. They are single, and are lacking on a circular

1) "Atlantic specimens show the same range of variation."
2) "Challenger Asteroidea, pl. 30, figs. 4 and 9."

area in centre of disk (including the central cone) and on five narrow radial and five narrow interradial areas extending from the centre like the spokes of a wheel. On either side of the radial areas, papulæ extend to tip of ray.

" *Marginal plates.*—Superomarginal plates exactly opposite inferomarginals. The former are thin and confined to side wall of ray. The actinolateral border of ray with its series of spines is slightly nearer upper than lower border of inferomarginals. The part of inferomarginal below the actinolateral spine has a rather broad specialised ridge, which is broader than intervening fasciolar furrow. This furrow is roofed over by a single row of spinelets immersed in a continuous web, eight or nine of these spinelets occurring between lower edge of plate and actinolateral spine. When dried the surface of plate shows often numerous minute bosses. Directly above the actinolateral spine the specialised ridge narrows abruptly and joins without a conspicuous break, that of superomarginal. The fasciolar channel is several times broader than the specialised ridge and is a V-shaped trough, on whose sides are five to seven parallel superimposed but distinctly spaced series of delicate spinelets in a continuous web. Each web extends from the upper end of a superomarginal plate to the actinolateral spine. The whole apparatus of five to seven webs or lamellæ on either side of the V-shaped trough, extending over its cavity, forms a very delicate cribriform organ or filter. The series of spinelets roofing the furrow are much flattened and the tip ends in several minute points. Owing to the greater width of furrow, these spinelets are much longer than those of lower end of inferomarginals. The other spinelets are narrow and more delicate and decrease in size in each successive tier toward bottom of furrow. In alcoholic specimens the membrane investing the spinelets is usually so thick that they are only seen with difficulty. The spinelets act only as a support for the membrane and do not themselves function as strainers, as in *Astropecten*. The superomarginal spine which stands at the top of the plate is either terete, tapering, and pointed, or broadly lanceolate and acute like a spear tip. On outer part of ray there are sometimes two spines on a few plates.

" The variation in number of marginal plates has already been indicated in the table. The margin of ray and. disk is thicker in some examples than in others, the appearance being heightened by recurved rays, when the actinolateral angle is rounder and less abrupt. The exposed surface of the specialised ridge of superomarginal plates is thin or narrow, but varies more or less, being slightly broader in four Atlantic specimens; but in another Atlantic example from off Newfoundland they are as narrow as in Alaskan specimens. The height of superomarginals is variable, specimens from 1,033 fathoms having much lower ones. This is readily appreciated by noting the distance between the two series of marginal spines in the interbrachial arc.

" *Actinal surface.*—The adambulacral armature is essentially alike in both Atlantic and Pacific specimens, some examples showing a preponderance of three or four spines, others of four and five. Besides the spine (not of furrow series) which usually stands on the outer aboral corner, there is usually one to several very much smaller and more delicate spinelets along adoral and outer edge of plate. These can not be distinguished readily unless specimen is dry.

" Actinal interradial area, like marginal and adambulacral plates, are overlaid by membrane through which the plates are scarcely visible until dried. Plates are arranged in series running from marginals to adambulacrals. Deep channels also follow the same course. These are overhung by a series of spinelets embedded in membrane, being a continuation of the marginal fascioles. The interradial channel splits and runs on either side of mouth plates. The first on ray always runs between second and third adambulacral, the second between the third and fourth, and so on. The photographic figure will show the arrangement of plates.

" *Anatomical notes.*—No intestine, no intestinal cœcum, no anus. The conical eminence in centre of disk contains a prolongation of the cœlom and is divided by a vertical septum. Stomach large, single, firmly moored to abactinal wall. Hepatic cœca large, extending nearly to end of ray, and with spacious cavity. Gonads interradial, one on either side of the membranous interradial septum. Ampullæ single; tube feet large, conical, pointed; no deposits in walls; one Polian vesicle to each interradius.

" Superambulacral plates present; absent from first and sometimes second ambulacral ossicle and also from last six or eight at end of ray. These plates are not very conspicuous, being dorsoventrally flattened and overlaid by membrane. Although LUDWIG[1] states that they are absent from his specimens (Panama region), I think he must have overlooked them. They are very easily seen if a portion of the ray is treated with caustic potash solution. They are present in a specimen examined, from station 3307, from the great depth of 1,033 fathoms. I also dissected a specimen from station 2452, off Newfoundland, 89 fathoms, and the superambulacral ossicles are present.

" In the centre of the conical abactinal prolongation one can easily distinguish in many specimens a small ' pore ' evidently connecting with the body cavity. This is also present in many *Eremicaster tenebrarius*, and is what SLADEN took to be an anus in *Porcellanaster*. I think it must be an artificial opening caused by a stretching of the abactinal membrane at the summit of the cone, and possibly subsequent wearing, as the rudimentary paxillæ are usually more or less worn down here.

" The walls of the stomach contain numerous small straight or irregular rods and grains from 0.01 to 0.175 mm. in length. They are sometimes provided with irregularities on sides or are irregularly triradiate. They are found also in the walls of the hepatic cœca, but are not so numerous. On the lips of the peristome they are transformed into broader irregular flattened rods with a few perforations, but in the peristome itself are comparatively few scattered rods like those of stomach walls, and only near the lip they are perforated. In the walls of stomach near mouth the rods are usually simple and very regular, and tend to arrange themselves in close meridional series.

" *Japanese specimens.*—I have eight specimens from station 4818, Sea of Japan, 225 fathoms. One of the largest of these is figured. All have very small low paxillæ with comparatively few spinelets. The superomarginals are also slightly narrower than in shallow-water Alaskan examples, but about as in specimens from 1,033 fathoms, Bering Sea. (The latter have large paxillæ, with usually high pedicels, which lack entirely the central spinelets). The Japanese specimens have four, but occasionally also three and five furrow spines.

---

1) " Mem. Mus. Comp. Zool., vol. 32, 1905, p. 105."

"Another specimen from station 5039, south east of Hokushu, 326 fathoms, is very different, having narrower and longer rays (R=3 r), large paxillæ, with comparatively long pedicels, the summit of which lacks central spinelets, and small actinolateral spinules. There are four furrow spines, as in the Japan Sea examples. This is also figured.

"Superambulacral plates are present.

"*Distribution.*—Bering Sea, along the north coast of America to Melville Island; through Barrow Strait to Greenland; south along the east coast of North America. to Cape Cod; west. and north of Spitzbergen to latitude 80° 3' N.; south to Faroe Islands, and on the Norwegian coast from Kristiansund to Finmark; Spitzbergen, Barents Sea, Murman coast, Matochkin Strait, Nova Zembla, and northward; Kara Sea as far as longitude 79° E. From here to East Cape the species has not yet been recorded.[1] From Bering Sea the species ranges south into the Sea of Japan, and on the American side to California, and is recorded from the mouth of the Gulf of California (LUDWIG) and Gulf of Panama (LUDWIG), and under the name *procurator* from off the Chonos Archipelago, Chile, south to entrance of Smyth Channel (SLADEN); off the east coast of southern South America as *australis* (SLADEN).

"*Specimens examined.*—About eight hundred and eighty-three.

"Specimens of *Ctenodiscus crispatus* examined.

Station.	Locality.	Depth. Fathoms.	Nature of bottom.	Number.	Collection.
2848	Near Shumagin Islands, Alaska.	110	green mud	85	U. S. Nat. Mus.
2849	"	69	"	12	"
2852	"	58	black sand	45	"
2855	Off Sitkalidak Island, near Kadiak Island, Alaska.	69	green mud	14	"
2860	South end Queen Charlotte Islands.	876	green mud	5	"
3075	Off Sea Lion Rock, Washington.	859	"	3	"
3076	Off Washington.	176	"	3	"
3077	Near Prince of Wales Island, southeastern Alaska.	322	"	58	"

1) "The above is condensed from LUDWIG, Fauna arctica, vol. 1, p 451, where authorities are cited."

3128	Off Monterey Bay, California.	627	blue mud	1	U. S. Nat. Mus.
3216	South of Alaska Peninsula.	61	black sand, mud	66	,,
3217	,,	42	black gravel	1	,,
3307	Bering Sea	1,033	green ooze	33	,,
3494	,,	65	green mud, fine sand	41	,,
3530	,,	59	dark green mud	16	,,
3532	,,	77	,,	66	,,
3538	,,	59	green mud	4	,,
3550	,,	76	brown mud	104	,,
3551	,,	74	green mud	38	,,
3552	,,	54	black sand	2	,,
3607	Bering Sea (North of Unalaska)	987	green mud, black lava sand	156	,,
3610	Bering Sea	75	green mud	many; all spoiled	,,
	Alaska (exact locality unknown)	—	....	52	,,
4194	Gulf of Georgia, British Columbia	111–170	soft green mud	1	Albatross, 1903.
4197	,,	31–90	....	1	,,
4223	Boca de Quadra, southeastern Alaska	48–57	soft green mud	21	,,
4228	Near Naha Bay, Behm Canal, southeastern Alaska	41–134	gravel and sponges	2	,,
4229	,,	198–256	soft gray mud	1	,,
4231	,,	113–82	green mud, fragments of slate	2	,,
4235	Near Yes Bay, Behm Canal, Alaska	130–181	gray mud, black specks	8	,,
4246	Kasaan Bay, Prince of Wales Island, Southern Alaska	42–47	green mud	1	,,
4274	Alitak Bay, Kadiak Island	35–41	....	1	,,
4281	Chignik Bay	42–43	green mud	1	,,
4286	,,	57–63	green mud, rocks	1	,,
4287	Uyak Bay, Kadiak Island	66–67	gray mud	2	,,
4292	Shelikof Strait	102–94	blue mud	2	,,
4768	Bowers Bank, Bering Sea	764	greenish brown mud	23	Albatross, 1906.
4775	,,	584	green mud	11	,,

"*Remarks.*—Attention is again called to the marvellous variation exhibited by this species and to its very extended range. It is extremely doubtful if *Ctenodiscus australis* LÜTKEN, 1871, from off the east coast of southern South America is a distinct species. The differences noted by SLADEN are among the most variable characters. Compare the various figures of undoubted *crispatus* published herewith, and then the figures published by SLADEN. It would not be difficult to make at least two species in Bering Sea and a third in the Sea of Japan with greater differences than seem to exist between '*australis*' and '*procurator.*' PERRIER[1] has already expressed the same doubt as to the difference between *australis* and European *crispatus*.

"This starfish, judging by its wide distribution, seems well adapted to life on soft mud. The creatures are usually gorged with mud, from which they evidently derive their food materials."

CLARK mentions it from Lower California as follows [: 13, p. 188]:

"A single small specimen ($R=15$ mm.) is all the collection contains of this common and widespread species.

"Sation 5686. Off Ballenas Bay, west coast of Lower California, 930 fms. Bottom temp., 37.3°."

MORTENSEN reports this starfish from Greenland as follows [: 13, p. 330]:

"Forekomst (Occurrence). Vest-Grønland : Igaliko (MØLLER) Bredefjord, 10–ca. 500 m. ; Skovfjord, 65–ca. 400 m. ; Kvan-Fjord, 115–420 m. (STEPHENSEN, 1912) ; Arsuk (BARRETT) ; Ivigtut (MØLLER) ; Sukkertoppen (BISTRUP, 1894) ; Nordre-Strømfjord 50–380 m. (Dr. NORDMANN, 1911) ; Akudlek, Bredebugt, Jacobshavn (TRAUSTEDT, 1892) ; Hare-Ø ('Valorous' Exped.) ; Karajak-Fjord (VANHÖFFEN) ; 63° 56′ N. 53° 12′ V. ; 64° 53′ N. 53° 06′ V., 203 Fv. (WANDEL, 1889) ; 66° 45′ N. 56° 30′ V., ca. 200 Fv. ; 68° 20′ N. 54° 03′ V., 220–280 Fv. ; 69° 17′ N. 52° 50′ V., ca. 225 Fv. ; ('Tjalfe,' 1908–9).

"Øst-Grønland : 74° 24′ N. 19° 42′ V., 130 Fv. ; 72° 25′ N. 19° 33′ V., 140 Fv. (RYDERS Exp. 1891) ; 75° 58′ N. 14° 08′ V., 300 m. (DUC D'ORLÉANS, 1905) ; 77° N. 17½ V., 300 m. (Danmark-Exped. 1908).

"Dybde (Depth) : ca. 10–1940 m.

1) "Exp. sci. Cap Horn, 1891, pp. 143, 144."

"Udbredelse : Circumpolar. I Atlanterhavet gaar den mod Syd til Færø-Kanalen og til Kap Cod paa Nord-Amerikas Ostkyst; i Stillehavet til det Japanske-Hav og langs hele Amerikas Vestkyst til Magellanstræde (*Ctenodiscus procurator* SLADEN); ogsaa ved Syd-Amerikas Ostkyst (*Ctenodiscus australis* LÜTKEN).

"Distribution : Circumpolar. In the Atlantic Ocean it goes towards the south to the Færoë Canal and to Cape Cod on the east coast of North America; in the Pacific Ocean to the Japan Sea and along the whole west coast of America to the Strait of Magellan (*Ctenodiscus procurator* SLADEN); also on the east coast of South America (*Ctenodiscus australis* LÜTKEN).

That the *Ctenodiscus krausei* of LUDWIG is probably identical with *C. crispatus* has been suggested by DÖDERLEIN and admitted by LUDWIG himself. The principal points of difference which led the last mentioned authority to make a new species for his specimens from the Bering Sea was the absence of a skeletal reticulum in the dorsal wall of the disk, the greater number of spinelets on either side of the keel of the superomarginal plates, and the presence of skin folds containing small flattened calcareous rods behind the row of lateral spinelets just mentioned. A dorsal skeletal reticulum has been described as being present in *Ct. crispatus* by DUNCAN and SLADEN; but an examination of a couple of specimens of this species from Casco Bay, Maine, has shown beyond doubt that no such reticulum is present, the bases of the paxillæ expanding somewhat but remaining entirely separate from one another. As to the presence of longitudinal skin folds in the fasciolar grooves of the marginal plates, they are very well developed in all the specimens of *Ct. crispatus* examined by me, while the number of the lateral spinelets on the keel of the superomarginal plates is subject to much variation. It appears then that one is completely justified in regarding *Ct. krausei* as a synonym of *Ct. crispatus*.

This species is one of the commonest starfishes in the Japan Sea, and it was dredged up at certain stations by the bushelful during the Albatross expedition of 1906. Leaving a detailed report on these specimens to the future I may here give the results of the measurements of some specimens taken at random. It may be mentioned that a specimen of this species from the Japan Sea as well as some from the Pacific were lying in the collection of the Science College, for some time. I at first thought that the Pacific specimens presented some differences from those of the Japan Sea, but they are evidently of minor importance and do not justify one in making them the basis for a new species. The dimensions of the specimens are as follows :

Specim.	r mm.	R mm.	R : r	MS	Locality
1	8	17	2.1	9	Japan Sea
2	8.3	16	1.9	9	,,
3	8.5	18	2.1	8	,,
4	9	16	1.8	9	,,
5	9	20	2.2	—	,,
6	9	21	2.3	10	,,
7	10	19	1.9	9	,,
8	10	19	1.9	9	,,
9	10	20	2.0	10	,,
10	10	21	2.1	10	,,
11	11	23	2.1	11	,,
12	11	20	1.8	10	,,
13	11	23	2.1	13	Misaki
14	11	25	2.3	—	Japan Sea
15	11	26.5	2.4	—	,,
16	12	24	2.0	—	..

17	12	22	1.8	9	Japan Sea
18	12	23	1.9	11	„
19	12	23	1.8	—	..
20	12	24	2.0	11	„
21	12	25	2.1	—	„
22	12	26	2.2	—	..
23	13	24	1.8	11	„
24	13	24	1.8	10	„
25	14	25	1.8	11	„
26	14	27	1.9	14	Misaki
27	14	27	1.9	12	Japan Sea
28	15	28	1.9	12	„
29	15	32	2.1	15	,,
30	16	36	2.2	13	„
31	19	36	1.9	13	„

It thus appears that both the radial ratio and the number of marginal plates are subject to a good deal of variation, the latter evidently depending primarily on the size of the specimens. It may also be mentioned that many of the long armed specimens of the Albatross collection (1906) can not be distinguished specifically from the *Ct. procurator* of SLADEN, as figured in the Challenger Report.

The abactinal surface is either plane or convex, depending upon the amount of mud contained in the stomach when the specimen was put in preservative fluid, and the central conical protuberance is very conspicuous. In some specimens, especially of larger size, five radial lines can be seen extending from the central process to the apices of the arms. The actinal surface is nearly plane and covered over with a humid membrane, which masks the individual plates of this side, so that they can usually be distinguished only after treatment with caustic potash. The lateral margins are also covered with a similar membrane and the two series of marginal plates can be distinguished from each other only with difficulty. The individual plates of each series are distinctly separated from one another by deep grooves.

Superomarginals.—The superomarginal plates are covered with a tumid membrane continuous with that of the actinal side, and in surface view their boundaries towards the inferomarginal plates are either invisible or indistinct. The individual plates of the series are however very distinct and are separated from one another by fasciolar grooves, which are in reality very spacious but are largely covered over by the series of delicate spinelets standing out at right angles from the keel of the plate and connected together by a fold of the superficial membrane. The outer face of the superomarginals is perfectly vertical or very slightly inclined towards the actinal surface. In the interbrachial angles the superomarginals are nearly one and a half times as high as they are wide, but near the tip of the arms the width and height are about equal. The lateral spinelets on either side of the keel are 10–12 for each plate in the interradii (Pl. VII, fig. 115). At the abactinal end of each superomarginal plate there is a pointed, somewhat flattened, conical spine, which becomes smaller towards the tip of the arms and is covered over by the same membrane as the plate itself. On the lateral face of the keel, i.e. in the fasciolar grooves behind the row of delicate spine-lets above mentioned, there are three or four skin folds parallel to the outer face of the superomarginal plate and to the row of lateral spinelets, and they contain numerous flattened calcareous rods of microscopic size (Pl. VII, fig. 118; Pl. VIII, fig. 125). There appear to be some differences in the size and form of these calcareous rods between the specimens from the Japan Sea and those from the Pacific side, but they are of too trifling a character to be made a basis for specific distinction, and my Pacific specimens are too few. The superomarginal plates show very little on the abactinal side, especially in the interbrachial arcs, where only the spines at the aboral end are visible, but further out in the arms the plates show distinct faces on this side. The keel of the superomarginal plates become relatively broader towards the apex of the arms.

Inferomarginals.—The inferomarginal plates are strictly coincident with the superomarginals and are nearly triangular in surface view in the interbrachial arcs. Further out in the arms the plates are more or less rectangular; and when viewed from the interior of the body all the plates are either rectangular or nearly square. The portion visible from the surface is the keel, which is articulated with that of the corresponding superomarginal. The individual plates of the inferomarginal series become distinct from the superomarginal and the ventrolateral plates only after maceration or treatment with caustic potash, and it is then seen that the outer surface of each inferomarginal plate is curved and that the sharply pointed, somewhat flattened actinolateral spine is twice as far removed from the actinal end of the plate as from the abactinal. The surface of the plate on the actinal side of the spine slants obliquely towards the oral surface, while the part lying on the other side stands up vertically towards the superomarginal. The margins of the inferomarginals facing the fasciolar grooves are armed with a row of 9–11 subcapillary spinelets similar to those of the superomarginal series and connected together by a skin fold. 'Of these spinelets those that lie on the abactinal side of the external spine are similar in shape to those of the corresponding series of the superomarginals, while those that lie on the actinal side of the same spine are like those of the ventrolateral plates, being shorter, flattened and rounded at the top. On the plate lying next the interradial line there are in a specimen of R=20 mm. some fifteen of these spinelets on either side, of which four or five lie between the abactinal end of the plate and the actinolateral spine. The delicate skin folds that cover the lateral faces of the superomarginal plates are continued downwards on to the inferomarginals to the level of the actinolateral spine. The external surface of the inferomarginal plates is usually more or less finely granulated.

Adambulacrals.—Of these there are for each side of an arm one to three or four more than the inferomarginals. The first two plates appear always to correspond to the first marginal plate, being connected with the latter by the series of ventro-lateral plates lying between them. As to the remaining adambulacral plates each one may correspond to a marginal plate or some may be intercalary in position. The adambulacral plates stand out very distinctly on treatment with caustic potash and are then seen to present each a rounded border towards the ambulacral furrow. On this border there is a row of three to five, short, sharply pointed spines, of which the one at the adcentral end is usually the largest on the first three or four plates, while on the remaining plates these spines are more nearly equal and spread out in the form of a fan. There may be one or more additional small spines on the actinal surface of the plate, usually close to the adcentral or abcentral border (Pl. VII, fig. 115 ; Pl. VIII, fig, 122).

Mouth-plates.—The mouth-plates are relatively large and conspicuous and the pair taken together is almost regularly elliptical or rhomboidal in outline. On each plate there is at the mouth end one or two pointed spines notably larger than the rest, and 4–6 pointed spines of unequal sizes on the furrow border (Pl. VII, fig. 115 ; Pl. VIII, fig. 122). On the border of the plate facing the first adambulacral plate there is a row of four or five inconspicuous stunted spines nearly covered over by the humid membrane of the body surface, while on the actinal surface of the plate there may be 1–8 short spinelets of unequal sizes, of which the one next the large oral spine may be subequal to the latter. In some specimens the inner intermediate plate at the abcentral end of each pair of mouth-plates comes out distinctly to view and in such cases its abcentral border may be armed with six to eight indistinct spinelets mostly covered over by the superficial membrane.

Ventrolaterals.—The ventrolateral plates are very well developed and appear to be subject to a great deal of variation. They are arranged in obliquely transverse rows parallel to the interradial line and are divided into groups covered over by the same humid membrane as that of the marginals and the adambulacral plates ; each group is separated from the next by a deep furrow continuous with the fasciolar grooves of the marginals. In most cases there are two transverse rows of ventrolateral plates corresponding to each marginal, but occasionally one or more plates may be intercalated between the two rows of a single group (Pl. VIII, fig. 122). In larger specimens the ventrolateral plates extend outwards as far as the sixth marginal, but in smaller specimens they may terminate with the fourth. Again, the number of plates that form a single transverse row varies a good deal ; e.g. in a specimen with $R=27$ mm. there were eight plate in a single row next the interradial line, while in another with $R=23$ mm. there were only six. These plates are imbricated like the scales of a teleost and bear on their outer border, i. e. the border facing the transverse furrow dividing a group from the next, delicate flattened spines, two to five for each plate, which sometimes become apparent only after treatment with caustic potash.

Paxillæ.—The paxillæ may be well spaced or closely set according as the dorsal membrane of the animal is convex or flat, and this depends again upon the amount of mud contained in the stomach. In some specimens the paxillæ are so closely pressed against one another as to impart a comparatively smooth appearance to the abactinal surface. The paxillæ are all very small but especially so are those that are found on the central conical prominence. There is no distinction between the central and peripheral coronal spinelets except in their positions, and in larger paxillæ there may be as many as eight peripherals and one or two centrals, while in smaller ones the centrals are usually absent and the peripherals may be only three or four in number. The tabulum is comparatively short and has an expanded base of rounded outline when viewed from the interior of the body and remains entirely separate from the neighbouring ones (Pl. VII, fig. 116, 117 ; Pl. VIII, fig. 124). The papulæ are distributed singly between the paxillæ, mostly one between any three contiguous paxillæ. They are however absent in the narrow band-shaped areas along the crest of the arms and the interradial lines. The central conical prominence and its immediate vicinity are also destitute of papulæ.

Madreporite.—The madreporite shows on the surface, although the peripheral parts may be slightly covered over by the surrounding paxillæ, and may be from twice to four times as far removed from the centre of the disk as from the margin, this variation in position being largely due to the degree of inflation of the dorsal body wall. The madreporite is either circular or elliptical in outline and covered with subparallel grooves showing a tendency to converge towards the inner end of the plate (Pl. VII, fig. 119 ; Pl. VIII, fig. 126). In some specimens there were a few granules near the margin of the plate.

Terminal plate.—The terminal plate of the arms is comparative· ly large and conspicuous and nearly quadrangular in form. The surface may be finely granulated and there are a pair of short spines at the apex, which are apt to fall off.

Locality.—Japan Sea ; Misaki. It occurs on fine mud, with which its stomach is completely filled.

Specimens in S. C.

Porcellanaster tuberosus SLADEN.

This species is not represented in any of the collections on which this paper is based. The following is the first description of it [Sladen, '83, p. 223] :

" *Porcellanaster tuberosus*, n. sp.

" Rays five, interbrachial angles well rounded, the minor radial porportion being 32 per cent., R=18.5 millim., r=6 millim. The rays spring gradually from the angle and taper moderately towards the extremity, maintaining a robust character throughout. Disk not high, and very slightly inflated. Dorsal area covered with a rather fleshy integument beset with simple spinelets somewhat closely placed ; these are short, cylindrical, obtuse ; covered with membrane, and occupy the whole of the surface excepting only the extreme angle at the base of the ray. A well-developed epiproctal tubular prolongation rises from the centre of the dorsal area, and is nearly equal in length to the distance between the centre and the inner edge of the marginal plates in the arm-angle ; it tapers very slightly towards its extremity and is indurated with spicular spinelets like the rest of the dorsal membrane.

"The marginal plates form a deep margin and curve over roundly in the interbrachial angles, the inferior as well as the superior series being visible from above. Upon the rays the superior series arch well over and almost meet in the median dorsal line, giving to the ray a more or less subcarinate character. The supero-marginal plates are four in number from the median interbrachial line to the extremity, exclusive of the large terminal plate, and all are distinctly longer than high. The second and third supero-marginal plates from the interbrachial line bear short conical upright spine-lets; but all the rest are unarmed excepting the terminal plate, which carries three spines—one at the extremity in the median line of the ray, and one on either side at the anterior extremity of the inferior margin of the plate. The terminal plate is swollen and prominently tubercular dorsal-ly, and is excavated on its outer extremity for the passage of the terminal ambulacral tube. In one ray of the specimen under notice, the penultimate supero-marginal plates are also swollen and ankylosed, in such a manner as to resemble the terminal plate, and bear a single spinelet. The infero-marginal plates are five in number, and are much shallower than the superior series and also shorter. The two series consequently do not cor-respond, a result probably brought about by the extreme development of the terminal plate, which occupies the space both of superior and inferior plate. Cribriform organs one in each angle, rather broad and with a deep depres-sion down the median line; structure lamelliform.

"Ambulacral furrows wide and open, occupying nearly the whole of the actinal surface of the ray. Adambulacral plates small, and form regular triangular prominences which indent, as it were, the margins of the furrow. Ambulacral spines two on each plate, short, subconical, sharply pointed or thorn-like, placed side by side on the aboral side of the projecting angle; they are consequently directed aborally and at an angle towards the furrow, diverging also slightly from one another.

"Mouth-plates rather large, forming an acute angle adorally, with an elevated angular ridge along the line of suture, each plate being strongly bent downwards and having the upturned edges compressed together to form the keel. The aboral extremity is more elevated than any other part and presents a sharp angular peak, the mouth-plates sloping down therefrom with a graceful inward curve to the level of the interbrachial area. A single short conical mouth-spine is placed at the extremity of the adoral peak; and two others, about equal in size to the ambulacral spines, stand on the lateral margins of each plate, the most adoral of the two being situated nearly midway between the extremities of the margin.

"The actinal interbrachial areas are small and sagittiform in outline, and do not extend beyond the third adambulacral plate. The plates are small, subregular, transversely elongate on the outer portion of the area, and with a tendency to imbricate, this character, however, being so faintly presented that it is difficult to say whether imbrication really exists or not.

"Colour, in alcohol, greyish white, rather darker over the dorsal area of the disk.

"Station 237. Lat. 34° 37' N., long. 140° 32' E. Depth 1875 fms.; bottom temperature 1°. 0 C.; red clay."

Substantially the same description is reproduced in the full report of the "Challenger," with addition of the locality and some remarks [SLADEN, '89, p. 140] :

"*Porcellanaster tuberosus*, SLADEN (Pl. XXIII. figs. 1–4; Pl. XXVII. figs. 13–16).

[Reference to the foregoing description omitted.]

"Rays five. R=18.5 mm.; r=6 mm. R=3 r.

"The rays spring gradually from the angles of the disk and taper moderately towards the extremity, maintaining a robust character throughout; the minor radius is in the proportion of 32 per cent. The disk is not high, and very slightly inflated. The interbrachial arcs are well rounded.

"The abactinal area is covered with a rather fleshy integument beset with simple spinelets somewhat closely placed; these are short, cylindrical, obtuse, covered with membrane, and occupy the whole of the surface excepting only the extreme angle at the base of the ray. A well-developed epiproctal tubular prolongation rises from the centre of the abactinal area, and is nearly equal in length to the distance between the centre and the inner edge of the marginal plates in the interbrachial arc; it tapers very slightly towards its extremity, and is indurated with spicular spinelets like the rest of the abactinal membrane.

"The marginal plates form a deep margin and curve over roundly in the interbrachial arcs, the inferior as well as the superior series being visible from above. Upon the rays the superior series arch well over and almost meet in the median dorsal line, giving to the ray a more or less subcarinate character. The supero-marginal plates are four in number from the median interradial line to the extremity, exclusive of the large terminal plate, and all are distinctly longer than high. The second and third supero-marginal plates from the median interradial line bear short, conical, upright spinelets; but all the rest are unarmed excepting the terminal plate, which carries three spines—one at the extremity in the median line of the ray, and one on each side at the anterior extremity of the inferior margin of the plate. The terminal plate is swollen and prominently tubercular abactinally, and is excavated on its outer extremity for the passage of the terminal ambulacral tube. In one ray of the specimen under notice, the penultimate supero-marginal plates are also swollen and ankylosed in such a manner as to resemble the terminal plate, and bear a single spinelet.

"The infero-marginal plates, which are five in number, are much shallower than the superior series, and also shorter. The two series consequently do not correspond, a result probably brought about by the extreme development of the terminal plate, which occupies the space of both superior and inferior plate.

"One cribriform organ is present in each interbrachial arc; it is rather broad and has a deep depression down the median line. The structure is lamelliform. (See Pl. XXVII.)

"The ambulacral furrows are wide and open, occupying nearly the whole of the actinal surface of the ray. The adambulacral plates are small, and form regular triangular prominences, which indent, as it were, the margins of the furrow. Their armature consists of two short, subconical, sharply-pointed, or thorn-like spinelets, placed side by side on the aboral side of the projecting angle; they are consequently directed aborally and at an angle towards the furrow, diverging also slightly from one another.

"The mouth-plates are rather large, forming an acute angle adorally, with an elevated angular ridge along the line of the suture, each plate being strongly bent downwards, and having the upturned edges compressed together to form the keel. The aboral extremity is more elevated than any other part, and presents a sharp angular peak, the mouth-plates sloping down therefrom with a graceful inward curve to the level of the actinal interradial area. Their armature consists of a single short conical mouth-spine, placed at the extremity of the adoral peak; and two others, about equal in size to the spinelets of the adambulacral armature, stand on the lateral margins of each plate, the most adoral of the two being situated nearly midway between the extremities of the margin.

"The actinal interradial areas, which are small and sagittiform in outline, do not extend beyond the third adambulacral plate. The intermediate plates are small and subregular, transversely elongate on the outer part of the area, and with a tendency to imbricate; this character, however, being so faintly presented that it is difficult to say whether imbrication really exists or not.

"Colour in alcohol, greyish white generally, but darker over the abactinal area of the disk.

"*Locality.*—Station 237. Off the coast of Japan, south of Kawatsu. June 17, 1875. Lat. 34° 37′ 0″ N., long. 140° 32′ 0″ E. Depth 1875 fathoms. Blue mud. Bottom temperature 35.°3 Fahr.; surface temperature 73.°0 Fahr.

"*Remarks.*—*Porcellanaster tuberosus* is distinguished from the other species of *Porcellanaster* with only one cribriform organ in each interbrachial arc, by its broad and robust rays, with a large and tubercular terminal plate armed with three spines, and by having only two supero-marginal plates on each side of a ray armed with spines, which are stout. Other points of difference are noticed in the description."

Hyphalaster inermis SLADEN.

This species is not represented in any of the collections mentioned at the beginning of this paper. Its first description by SLADEN is as follows ['83, p. 239]:

" *Hyphalaster inermis,* n. sp.

" Marginal contour stellato-pentagonoid. Rays five, well developed, slender, round, and tapering but slightly. Interbrachial angles very wide and expansive, the curve slightly flattened in the immediate angle, thereby emphasizing the marked pentagonal contour of the body-disk. The lesser radius is in the proportion of 42.5 per cent.; $R=20$ millim., $r=8.5$ millim. Disk depressed, not inflated ; both dorsal and actinal surfaces stand on a level with the edges of the marginal plates.

" Dorsal area covered with closely crowded paxillæ, the whole disk as well as the base of the rays being uniformly packed. The paxillæ are very fine and small, and are made up of about 5 to 10 spinelets ; towards the margin of the disk they become smaller and also in the centre, where they are very compact, a slightly prominent peak being formed as in *Ctenodiscus.* A slight elevation of the surface is present in the median radial line, oppo- site the base of each ray, and at about one third of the distance from the margin to the centre.

" Marginal plates occupy the entire margin and represent the whole thickness of the animal, forming perpendicular walls regularly rounded above and below. Along the rays the supero-morginal plates meet in the median dorsal line and form a complete casing to the ray, which is well rounded, small, and tapers but slightly. The supero-marginal series are 8 in number (or, with a very small aborted one, 9), exclusive of the terminal. The plates which fall in the margin of the disk proper have the length about equal to their height, but in those along the ray the height is greater than the length. The infero-marginal plates correspond in number and in length with the superior series. In the arm-angle, along the disk proper, the height is about equal to the length and the plates are uniform in size with the superior series ; towards the extremity of the ray the height diminishes gradually and the length is greater than the height—a reversal of the relative proportions presented by the plates of the superior series. The marginal plates are smooth and bear no spines, but when examined microscopically have the appearance of being subgranular and built up of a rather open network. The plates of both series are convex outwardly or tumid in a very slight degree, by which means the sutural divisions of the segments are clearly marked out, and a somewhat annulated appearance is given to the ray. The terminal plate is large and conspicuous, appearing somewhat tubercular and directed slightly upwards when viewed in profile, and oval in contour when seen from above. This plate bears three short and rather robust spinelets— one at the terminal extremity of the plate, situated in the median dorsal line, pointing in the direction of the prolongation of the ray, and diverging but little from the horizontal. Below this spine, and at either side of it, at the angle formed by the ventral edge of the plate and the external extremity, is a somewhat smaller spinelet, pointing in the direction of the prolongation of the ventral margin of the plate. Cribiform organs 7 in number, narrow and well defined; structure papilliform.

"Ambulacral furrows narrow and straight, almost completely closed-in by the over-arching adambulacral plates and spines, the sucker-feet, which are arranged in simple pairs, being entirely concealed from view. The adambulacral plates are about half as broad as long, but diminish in size as they proceed outwards; and form along the ray triangular prominences projecting into the furrow. Each plate bears 3 to 4 spines, rather short, rapidly pointed, more or less compressed, invested with membrane, arranged in line along the furrow-margin of the plate and sometimes slightly oblique to the course of the furrow. The row of spinelets can be raised at a right angle to the surface of the plate, so as to allow the sucker-feet to be protruded. Traces of an aborted secondary or external spinelet, represented by a mere granule, may be detected at the adoral extremity of the adambulacral plate, away from the furrow-series.

"Mouth-plates moderately large, the inner margins which fall in the median suture being elevated so as to form a rounded elongate tubercular protuberance, the lateral margins being flattened out. Mouth-spines 7 or 8 on each side, similar to the ambulacral spines, excepting the innermost one, which is much larger and stouter. Two large spines are thus conspicuous at each mouth-angle and are directed towards the centre, the series entirely closing the peristome, which is remarkably small. The small mouth-spines upon the margin of the plate interlock with those of the adjacent mouth-angle, and form a continuous series with the ambulacral spines. The rudiments of a secondary mouth-spine, represented by a thorn-like granule, occur on each plate near the median suture and at the highest portion of the keel.

"The interbrachial areas are triangular in outline, flat, extensive, and covered with imbricating scales of more or less regularly symmetrical hexagonal form. These plates are broader than long, and arranged in regular series of single columns extending from the margin of the disk to the ambulacral furrow; their breadth diminishes somewhat as they approach the margin, and consequently that of the column also. The adambulacral plates join up to the infero-marginal plates along the whole length of the free portion of the ray, and there is consequently no extension of the interbrachial area along the ray. The imbricating plates bear a few widely-spaced miliary tubercles or large granules upon their surface, usually 4 or 5 to a plate, but upon which they have no definite arrangement.

"Colour, in alcohol, grey, the paxillar area being a much darker shade, which shows a strong contrast with the greyish white of the marginal plates.

"Station 237. Lat. 34° 37′ N., long. 140° 32′ E. Depth 1875 fms.; bottom temperature 1°.7 C.; mud."

Essentially the same description is reproduced in the full report of the "Challenger," with a few additional observations, as may be seen from the following [SLADEN, '89, p. 162] :

"*Hyphalaster inermis*, SLADEN (Pl. XXV. figs. 4–6 ; Pl. XXVIII. figs. 5–8). (Reference to the foregoing description.)

"Rays five. R=20 mm.; r=8.5 mm. R<2.5 r.

"Marginal contour stellato-pentagonoid. Rays well developed, slender, round, and tapering but slightly. The disk is depressed, not inflated, and both the abactinal and actinal surfaces stand on a level with the edges of the marginal plates. The minor radius is in the proportion of 42.5 per cent. The interbrachial arcs are very wide and expansive, the curve being slightly flattened at the summit of the arc emphasises the marked pentagonal contour of the body-disk.

"The abactinal area is covered with closely crowded paxillæ, the whole disk as well as the base of the rays being uniformly packed. The paxillæ are very fine and small, and are made up of about five to ten spinelets. Towards the margin they become smaller, and also in the centre, where they are very compact—a slightly prominent peak being formed as in *Ctenodiscus*. A slight elevation of the surface is present in the median radial line, opposite the base of each ray, and at about one-third of the distance from the margin to the centre.

"The marginal plates occupy the entire lateral region, and represent the whole thickness of the animal, forming perpendicular walls regularly rounded above and below. Along the rays, the supero-marginal plates meet in the median radial line, and form a complete casing to the ray, which is well rounded, small, and tapers but slightly. The supero-marginal plates are eight in number (or, counting a very small aborted one, nine), exclusive of the terminal. The plates which fall in the margin of the disk proper have the length about equal to their height, but in those along the ray the height is greater than the length.

"The infero-marginal plates correspond in number and in length to the superior series. In the interbrachial arc, along the disk proper, the height is about equal to the length, and the plates are uniform in size with the superior series. Towards the extremity of the ray the height diminishes gradually, and the length is greater than the height—a reversal of the relative proportions presented by the plates of the superior series. The marginal plates are smooth and bear no spines; but when examined microscopically have the appearance of being subgranular, and built up of a rather open network. The plates of both series are convex outwardly or tumid in a very slight degree, by which means the sutural divisions of the segments are clearly marked out, and a somewhat annulated appearance is given to the ray. The terminal plate is large and conspicuous, appearing somewhat tubercular when viewed in profile, and oval in contour when seen from above. This plate bears three short and rather robust spinelets—one at the terminal extremity of the plate, situated in the median radial line, pointing in the direction of the prolongation of the ray, and diverging but little from the horizontal. Below this spine, and at each side of it, on the angle formed by the actual edge of the plate and the terminal extremity, is a somewhat smaller spinelet, pointing in the direction of the prolongation of the actinal margin of the plate.

"Seven cribriform organs are present in each interbrachial arc. They are narrow and well defined, and their structure is papilliform. (See Pl. XXVIII. fig. 8.)

"The ambulacral furrows are narrow and straight, almost completely closed in by the overarching adambulacral plates and spines, the tube-feet, which are arranged in simple pairs, being entirely concealed from view. The adambulacral plates are about half as broad as long, but diminish in size as they proceed outwards, and form along the ray triangular prominences projecting into the furrow. Each plate bears three or four spines, rather short, rapidly pointed, more or less compressed, invested with membrane, arranged in line along the furrow margin of the plate, and sometimes oblique to the course of the furrow. The row of spinelets can be raised at a right angle to the surface of the plate, so as to allow the tube-feet to be protruded. Traces of an aborted secondary or external spinelet, represented by a mere granule, may be detected at the adoral extremity of the adambulacral plate, away from the furrow series.

"The mouth-plates are moderately large, the inner margins which fall in the median suture being elevated so as to form a rounded elongate tubercular protuberance, the lateral margins being flattened out. Their armature consists of seven or eight mouth-spines on each plate, similar to those constituting the armature of the adambulcral plates excepting the innermost one, which is much larger and stouter. Two large spines are thus conspicuous at each mouth-angle, and are directed towards the centre, the series entirely closing the actinostome, which is remarkably small. The small mouth-spines upon the margin of the plate interlock with those of the adjacent mouthangle, and form a continuous series with the armature of the adambulacral plates. The rudiments of a secondary mouth-spine, represented by a thornlike granule, occur on each plate, near the median suture, and at the highest portion of the keel.

"The actinal interradial areas are triangular in outline, flat, extensive, and covered with imbricating scales of more or less regularly symmetrical hexagonal form. These plates are broader than long, and arranged in regular series of single columns extending from the margin of the disk to the ambulacral furrow. Their breadth diminishes somewhat as they approach the margin, and consequently that of the columns also. The adambulacral plates join up to the infero-marginal plates along the whole length of the free portion of the ray, and there is consequently no extension of the interradial area along the ray. The imbricating plates bear a few widely spaced miliary tubercles or large granules upon their surface, usually four or five to the plate, upon which, however, they have no definite arrangement.

"Colour in alcohol, grey; the paxillar area being a much darker shade, which shows a strong contrast with the greyish white of the marginal plates.

"*Young phase.*—There is a small example of this species, which, though measuring only R=10 mm., r=5 mm., so closely resembles in all respects the characters of the adult, that there is not the slightest hesitation in determining it specifically. Beyond the fact that the rays are shorter, the terminal plate more tubercular and broader, and that a less number of supero-marginal plates on the two sides of a ray meet in the median radial line, I can scarcely detect a feature worthy of mention as differentiating the immature from the adult form; excepting the changes in size, proportion, and number which affect plates and appendages normally. There are five supero-marginal plates between the median interradial line and the terminal in the small specimen.

"*Locality.*—Station 237. Off Japan, south of Kawatsu. June 17, 1875. Lat. 34° 37' 0" N., long. 140° 32' 0" E. Depth 1875 fathoms. Blue mud. Bottom temperature 35.°3 Fabr.; surface temperature 73.°0 Fabr.

"*Remarks.*—*Hyphalaster inermis* is distinguished from the other species with seven cribriform organs by its robust and rigid body-frame, by the supero-marginal plates meeting in the median radial line, by these being devoid of spinelets, by the fully developed paxillæ, and by the narrow cribriform organ."

ASTROPECTINIDÆ.

Astropecten scoparius VALENCIENNES.

(Pl. III, figs. 34–41.)

I look upon *Astrop. hemprichii* M. u. TR. and *Astrop. japonicus* M. u. TR. as synonyms of the present species. According to BELL [: 04, p. 149] *Astrop. zebra* SLADEN should be united with *Astrop. hemprichii*, and in that case SLADEN's species would also come in the range of the following review. But as SLADEN's *Astrop. zebra*, so far as can be judged from the published descriptions and figures, appears to me to be a good species, I have confined myself in the following to a review of the names mentioned at the outset. *Astrop. articulatus* SAY and *Astrop. acanthifer* SLADEN are also very nearly related to the present species.

The name of this species is due to VALENCIENNES, but its first description was published by MÜLLER and TROSCHEL. These authors divide the species of *Astropecten* known to them into four groups, as follows: (1) species with two or more spines on the superomarginals, (2) species with one spine on the superomarginals, (3) species with small tubercles instead of spines on the superomarginals, (4) species with uniformly granular superomarginals without spines or tubercles. The present species is of course included in the second group and described as follows [MÜLLER u. TROSCHEL, '42, p. 71].:

"*Astropecten scoparius* VAL. nov. sp.

"Fünf Arme. Verhältniss der Radien wie 1 : 4. Arme breit, an der Spitze verschmälert. Furchenpapillen in einem keilförmigen Haufen. Aus der Beschuppung der Bauchplatten zeichnen sich viele Stachelchen unter der feinern Beschuppung aus, nach dem Rande zu häufiger. Der äusserste Stachel ist sehr lang, breit, etwas schwertförmig gebogen. Die dorsalen Randplatten, 25–30 an jedem Arme, sind gleichförmig gekörnt; vom Ende der Arme zieht sich eine Reihe conischer Stachelchen hin, einer auf jeder Platte am Aussenrande. Diese Stachelchen fehlen jedoch den Platten in den Winkeln zwischen den Armen. Der Raum in der Mitte der Arme

zwischen den dorsalen Randplatten ist 2- bis 3 mal so breit wie die Rand-
platten.ˑ Madreporenplatte in der Nähe des Randes.

"Farbe : gelblich.

"Grösse : 7 Zoll.

"Fundort : unbekannt.   Im Museum zu Paris."

Astropecten hemprichii, which I include in the list of synonyms of the
present species for reasons given below, is described in the "System" as
follows [MÜLLER u. TROSCHEL, '42, p. 71] :

"*Astropecten hemprichii* Nob. nov. sp.

"Fünf Arme.   Verhältniss des kleinen Radius zum grossen wie $1 : 4\frac{1}{2}$.
Die Furchenpapillen in mehreren Reihen, auf jeder Platte steht ein keil-
förmiger Haufe, dessen Spitze von einer Papille gebildet wird.   Die Stachelchen auf den ventralen Platten sind äusserst zart, fast haarförmig ; erst in
der Nähe der Randstacheln zeigen sich feine, welche nur halb so lang sind
wie die Randstacheln.   Die Randstacheln sind platt, und mit Ausnahme
derjenigen in den Winkeln zwischen den Armen spitz.   Die dorsalen Randplatten sind viel breiter als hoch, am Grunde der Arme und in den Winkeln zwischen ihnen ohne Stacheln.   Allmählig entwickelt sich eine Reihe
kleiner Stachelchen bis gegen das Ende der Arme.   An jedem Arme 33
dorsale Randplatten.

"Grösse : $4\frac{1}{4}$ Zoll.

"Fundort : Rothes Meer.   Im Museum zu Berlin durch HEMPRICH und
EHRENBERG."

Astropecten japonicus, which is referred to the present species by
DÖDERLEIN, a procedure in which I entirely concur, was placed by MÜLLER
and TROSCHEL in their third group, with small tubercles instead of spines on
the superomarginals, and described as follows ['42, p. 73] :

"*Astropecten japonicus* Nob. nov. sp.

"Verhältniss des kleinen Radius zum grossen wie $1 : 4$.   An jedem Arme
30 Randplatten.   Die Furchenpapillen, fünf auf jeder Platte, bilden keil-
förmige Haufen.   Die Beschuppung der Bauchplatten ist sehr zart ; am
Rande steht auf jeder Platte ein grösserer Randstachel, unter dem in einer
mit dem Rande parallelen Reihe drei Stachelchen stehen, welche etwa ein

Drittel der Länge des grösseren erreichen. Die dorsalen Randplatten sind so breit wie hoch, granulirt, und tragen nur selten am äusseren Rande einen kleinen beweglichen Tuberkel. Die dorsalen Randplatten sind fast so breit wie das Mittelfeld auf der Mitte der Arme. Die Madreporenplatte nahe dem Rande. Das Endglied an den Armen ist sehr gross und braun.

"Grösse : 2⅓ Zoll.

"Fundort : Japan. Im Museum zu Leyden durch v. SIEBOLD."

According to DE LORIOL ['85, p. 74] and SLADEN ['89a, p. 324] MICHE-LIN mentions this species (*Astr. hemprichii*) from Mauritius under the name of *Astr. articulatus* [MICHELIN, '45, p. 24]. According to the first named authority it is also mentioned from Mosambique by PETERS ['52, p. 178].

DUJARDIN and HUPÉ apparently follow MÜLLER and TROSCHEL and describe the same species under the three different names mentioned above ['62, p. 418, 419, 423] :

"Astropecten balai. *Astropecten scoparius.*—MÜLLER et TROSCHEL.

"—VALENCIENNES. Coll. du Muséum.—MÜLLER et TROSCHEL, Syst. der Aster., p. 71.

"Espèce à cinq bras, dont la longueur, à partir du centre, est quadruple du plus petit rayon du disque ; ces bras sont larges à leur base et rétrécis au sommet. Les piquants du sillon ambulacraire forment des groupes en forme de coin. Parmi les écailles, dont les plaques ventrales sont revêtues, on distingue beaucoup de petits piquants qui deviennent plus fréquents vers le bord ; le piquant le plus extérieur est très-long, large et un peu courbé en lame d'épée. Les plaques marginales dorsales, au nombre de 25 à 30, sont uniformément granuleuses. De l'extrémité de chaque bras, part une rangée de petits piquants coniques ; il y en a un sur chaque plaque, au bord extérieur ; les piquants manquent aux plaques qui occupent l'angle rentrant interbrachial. L'aire dorsale des bras, entre les plaques marginales, est deux à trois fois aussi large que ces plaques.

"La plaque madréporique est rapprochée du bord.

"Coloration jaunâtre. Dimension : largeur totale 190 mm."

"Astropecten de Hemprich. *Astropecten hemprichii.*—MÜLLER et TROSCHEL.

"—MÜLLER et TROSCHEL, Syst. der Aster., p. 71.

"Espèce à cinq bras, dont la longueur, à partir du centre, égale quatre fois et un tiers le plus petit rayon du disque. Les piquants du sillon ambulacraire, en plusieurs rangées, forment, sur chaque plaque, un groupe en coin, dont le sommet est formé par un seul d'entre eux. Les petits piquants des plaques ventrales sont extrêmements fins, presque capillaires. Les plus rapprochés des piquants marginaux ne sont que moitié aussi longs ; ces derniers sont plats et pointus, à l'exception de ceux qui se trouvent dans l'angle rentrant interbrachial. Les plaques dorsales, au nombre de 33 à chaque bras, sont beaucoup plus larges que hautes. Celles de la base des bras et de l'angle interbrachial sont sans piquants ; mais une rangée de petits piquants se dirige vers l'extrémité des bras.

"Coloration jaunâtre. Dimension : largeur totale 122 mm.

"Habite la mer Rouge (Mus. Berlin)."

"Astropecten japonais. *Astropecten japonicus.*—MÜLLER et TROSCHEL.

"—MÜLLER et TROSCHEL, Syst. der Aster., p. 73.

"Espèce dont la longueur des bras est quadruple de celle du plus petit rayon du disque : chacun d'eux porte trente plaques marginales. Les piquants du sillon ambulacraire, au nombre de cinq sur chaque plaque, sont groupés en une sorte de coin. Les plaques ventrales sont revêtues d'écailles très-délicates, et sur chaque plaque, au bord, se trouve un piquant marginale plus gros, au-dessous duquel sont trois petits piquants formant une rangée parallèle au bord, et ayant à peu près le tiers de la longueur du plus grand. Les plaques marginales dorsales sont aussi larges que hautes, granuleuses, et portent rarement, à leur bord externe, un petit tubercule mobile. Sur le milieu des bras, leur largeur est égale à celle de l'aire médiane. L'article terminal des bras est très-gros et coloré en brun. La plaque madréporique est située près du bord.

"Coloration jaunâtre foncé. Dimension : largeur totale 67 mm.

"Habite les côtes du Japon. (Mus. Leyde)."

LÜTKEN, in a section of his paper of 1864 headed, "Om Slægten *Astropecten* samt om forskjellige Arter af Kamstjernernes Gruppe (*Astropectinidæ*)," remarks that, through an examination of a large number of species (ca. 20), he has come to the conviction that the last two groups of MÜLLER and TROSCHEL may best be united into one; so that his divisions are : (1) with a single series of high and strong spines on the superomarginals, (2) with a double series of smaller, both lower and feebler, spines, one along the outer edge, and the other along the inner edge of the superomarginals, (3) superomarginals simply granular, but not seldom one or two very low spines amid this granulation; strong spines however always wanting in the angles between the arms. *Astropecten japonicus* is placed in the last group, and the following remarks is made [Lütken, '64, p. 125, 127] :

" Jeg har endnu ikke seet nogen *A. pentacanthus* eller *A. japonicus* med Smaapigge paa Randpladerne.. ."	"I have not yet seen any *A. pentacanthus* or *A. japonicus* with small spines on the marginal plates...."

VON MARTENS, in a section of his paper of 1865, headed, "Japanische Seesterne," has the following ['65, p. 352] :

"*Astropecten scoparius* VAL. MÜLL. TROSCH. l. c. S. 71.

"Yokohama, auf Schlammgrund, nicht häufig, die Stacheln auf den dorsalen Randplatten sind sehr unbeständig, sie fehlen im ersten Drittel der Arme fast immer, bis zur Mitte kommen immer noch einzelne Platten ohne Stacheln vor."

At page 354 of the same paper, there is a reference which possibly has to do with the original specimen or specimens of MÜLLER and TROSCHEL.

"*Astropecten japonicus* MÜLLER und TROSCHEL S. 43—?

"? Herklots tab, inedit. X. fig. 2.

"Japan, Leidner Museum."

Again, in his second paper he says ['66, p. 87], "*Astropecten scoparius* MÜLL. TROSCH. Indischer Ocean, SCHÖNLEIN im Berliner Museum. Ist sicher in Japan zu Hause."

According to DE LORIOL ['85, p. 74] and SLADEN ['89a, p. 324] *Astropecten hemprichii* is referred to by v. MARTENS ['69, p. 131]. It is also mentioned from the Red Sea and the Gulf of Suez by GRAY ['72, p. 119]. The latter's specimens belong to the British Museum.

PERRIER mentions *Astropecten scoparius* in his "Révision," and it appears from his references that VALENCIENNES' manuscript is in possession

of the Muséum d'historie naturelle of Paris. He then refers to the original
specimens as follows ['76, p. 279] :

"Deux échantillons desséchés et en bon état, saus date d'entrée et
sans provenance. Ce sont les types de MÜLLER et TROSCHEL."

SLADEN describes a small specimen of *Astropecten japonicus* ['79, p. 427] :

"*Astropecten japonicus*, MÜLLER and TROSCHEL. 1842. *Astropecten japonicus*, MÜLLER and TROSCHEL, System der Asteriden, p. 73.

"Coll. ST. JOHN : Korean Straits, 9 fathoms.

"The arms are moderately long and narrow; $R = 11.25$ millims., $r = 4$
millims. The foot-papillæ, arranged in wedge-shaped groups of five, are
long, fine, and cylindrical. The first spinelet, which forms the apex of the
wedge, stands by itself, projecting inward upon the furrow, is thicker than
the rest and arched upwards at its base; the others stand external to this,
two and two together, the outermost pair being rather longer than the inner
pair; whilst on the inmost portion of the furrow the outer series of papillæ
are augmented by one or two additional spinelets. The adambulacral plates
which bear the foot-papillæ appear very much depressed, in consequence of
the gibbous character of the ventro-marginal plates—a feature which is very
striking when compared, for instance, with specimens of *Astr. formosus*,
mihi, of nearly equal size.

"The upper marginal plates are broader than long, and covered closely
with short stout granulose spinules of clavate form, and on the outer half of
the arm carry on their outer margin a small conical spinelet. In the
specimen under notice the nine outer, out of thirteen marginal plates, are
thus armed.

"The ventro-marginal plates project more outwardly than the upper
marginal plates, and bear one large, compressed, lanceolate spine at the
margin, which is generally followed by two smaller spines placed side by
side, not half its length, and very much finer and more cylindrical. The
rest of the spinulation consists of small, short, isolated, cylindrical spinelets.
In the present example these have been very much abraded; and little
further detail can be made out.

"The dorsal area or paxillar field is, in the middle of the arm, very little, if any, broader than the marginal plate.. The paxillæ are large and closely crowded—so much so that the radii (of which there are 8-9 and very robust) of a paxilla are directed upward, instead of at right angles to their pedicle; and this gives to the paxillay area a granulate rather than a stellate appearance to the naked eye, and without any indication of regular arrangement.

"Dr. Lütken remarks on never having seen an *Astr. japonicus* with the spines upon the dorsal marginal plates. On the specimen under consideration these are so small that they might easily be passed over without notice,—whilst, further, it is a character of such usual variability that I am fully prepared to believe in the existence of examples in which they are wanting altogether, their rudimentary state on the present specimen quite leading to that idea. A seemingly parallel instance may be pointed to in the case of *Astr. euryacanthus*, Ltk., in the premature stages of which small spines are present on the outer margin of the dorsal marginal plates towards the ends of the arms, but no trace of them remains in the adult.

"Our knowledge of this species at present is very scanty; and it may not be beyond the range of probability that a more extensive series of specimens will require the modification of our current ideas of the form altogether, and possibly even its amalgamation with such a species as *A. scoparius*, when more is definitely known about the premature stages of these Astropectens."

According to DE LORIOL ['85, p. 74] the form described by Möbius ['80, p. 50] under the name of *Astropecten mauritianus* is identical with *Astrop. hemprichii*.

SLADEN mentions *Astrop. japonicus* in his preliminary paper on the Challenger Asteroidea ['83, p. 255]:

"*Astropecten japonicus*, MÜLLER and TROSCHEL.

"Station. Off Yokohama, Japan. Depth 8–14 fms.; 5–25 fms.

"Station 233 A. Kobi, Japan. Lat. 34° 35′ N., long. 135° 10′ E. Depth 8–50 fms.; mud, sand.

"Station 233 B. Lat. 34° 20′ N.; long. 133° 35′ E. Depth 15 fms.; mud."

BELL mentions *Astropecten hemprichii* (?) from Mosambique and the Seychelles ['84a, p. 510].

DE LORIOL describes *Astropecten hemprichii* from Mauritius as follows
['85, p. 74] :

" Les exemplaires de cette espèce sont nombreux. Très constant dans
leurs caractères, ils varient dans leur taille, r=13 mm. à 22 mm. R=65 mm.
à 111 mm.

Ils me paraissent correspondre exactement à la description donnée par
MÜLLER et TROSCHEL, seulement ces auteurs disent que les plaques ventrales
portent des petits piquants très délicats, presque comme des cheveux, et
que ce n'est que près des piquants marginaux qu'il s'en trouve quelque-uns
plus longs, de moitié plus courts que ces derniers. Or, dans les individus
de Maurice, les piquants des plaques ventrales 'sont très fins à la vérité,
mais on ne peut pas dire d'eux qu'il sont *capillaires*, sauf pourtant de ceux
qui se trouvent sur le bord des plaques, de plus des piquants plus gros se
montrent beaucoup plus près des plaques adambulacraires. Il ne me semble
pas que cette légère différence, qui peut du reste n'être qu'une affaire d'appréciation, soit de nature à motiver une séparation spécifique. MÜLLER et
TROSCHEL disent encore que les piquants marginaux sont aigus, sauf dans
les angles interbrachiaux, ce n'est que dans les grands exemplaires que les
piquants des angles interbrachiaux sont coupé carrément à l'extrémité, il
sont à peu près tous aigus dans les individus de la taille de ceux que j'ai
fait représenter.

" Le nombre des bras est de cinq, dans un individu l'un d'entre eux
se bifurque vers la moitié de sa longueur. Près du bord externe 'des plaques
marginales ventrales, qui dépassent beaucoup les dorsales, les piquants plus
longs se multiplient, en devenant toujours plus longs jusqu' au dernier
piquant marginal qui est large, aplati, pointu, et aussi long que la plaque,
ceux qui se trouvent dans l'angle interbrachial sont généralement tronqués
dans les grands exemplaires. Les plaques adambulacraires portent, dans le
sillon, un groupe de trois piquants dont l'impair, le plus interne, est re-
courbé et notablement plus long que les autres, puis vient un petit groupe
de deux long piquants égaux et plats, enfin en dehors, un groupe de six ou
sept piquants serrés, très fins, cylindriques et assez longs. Les plaques
marginales dorsales, dont le nombre est de trente-trois de chaque côté des

bras, sont couvert de petits granules plats et écartés, et le pourtour de la plaque est garni de petites soies très courtes, très serrées, et d'une finesse extrême, un petit piquant conique, court, pointu, très fugace se montre presque sur chacune des plaques marginales dorsales, sur leur bord tout à fait externe ; mais il n'y en a aucun, au fond de l'angle interbrachial, sur un nombre de plaques variant de quatre à dix, et cela sur un même individu. L'aire paxillaire est large, à la base des bras elle à une largeur égale à un peu plus du double de la largeur d'une plaque marginale dorsale ; les paxilles sont relativement grandes avec une houppe d'une trentaine de petits piquants obtus. La plaque madréporique, logée tout à fait près du bord, dans un angle interradial, est très cachée par les paxilles et à peine apparente ; lorsqu'on la met à nu elle paraît assez grande. Couleur jaune-paille.

L'*Astr. hemprichii* ne peut être confondu avec l'*Astr. mauritianus*, comme le pense M. MÖBIUS, car, ainsi que nous l'apprend M. PERRIER (Révision des Stellérides du Muséum de Paris, p. 359.), cette dernière espèce porte un forte piquant interne sur les deux plaques marginales internes de chaque angle interradial, et ils manquent toujours dans l'*Ast. hemprichii*."

In the same work DE LORIOL describes a specimen of *Astropecten*, without naming it, which differs from *Astr. hemprichii* in having spines on all the superomarginals. There is no doubt in my mind that it is merely a variational form of *Astr. scoparius*. The description of this specimen is as follows ['85, p. 77] :

" *Astropecten sp.*

"Dimensions. r=21 mm., R=108 mm. Diamètre des bras à la base 25 mm.

" J'ai sous les yeux un échantillon d'un *Astropecten* de grande taille, que m'a envoyé M. DE ROBILLARD, et qui ressemble extrêmement, par la presque totalité de ses caractères, à l'*Astr. hemprichii*, mais qui en diffère par la présence d'un piquant conique, droit, sur toutes les plaques marginales dorsales, dans les angles interbrachiaux, et, par conséquent, sur toutes les plaques marginales dorsales sans exception ; de plus une ou deux plaques au milieu de l'angle portent même un second piquant semblable vers leur bord

interne ; il faut ajouter que les plaques marginales ventrales débordent relativement davantage les plaques marginales dorsales. Dans l'*Astr. hemprichii*
sept ou huit plaques marginales dorsales au moins, la plupart du temps
bien davantage, dans l'angle interbrachial, sont entièrement dépourvus de
piquants.   Faut-il voir dans ces différences des caractères suffisants pour
envisager cet échantillon comme appartenant à une espèce particulière ?
Comme je ne connais q'un seul exemplaire à lui rapporter, il m'a paru
préférable de ne pas résoudre cette question pour le moment, et d'envisager
provisoirement cet individu comme une variété de l'*Astr. hemprichii*. J'ai été
tenté, au premier abord, de la rapporter à l'*Astr. mauritianus*, GRAY, qui,
suivant les indications données par M. PERRIER, d'après l'examen qu'il a fait
des exemplaires types au British Museum, est pourvu de piquants coniques
sur les plaques des angles interbrachiaux (Révision des Stellérides, p. 359) ;
mais GRAY dit positivement que son *Astr. mauritianus* appartient à un
groupe dans lequel les plaques marginales ventrales *ne débordent pas* les
plaques dorsales, or, dans l'exemplaire en question, elles débordent au
contraire d'une manière très prononcée.

"Comme il est possible que cet individu devienne le type d'une nouvelle espèce, si d'autres observateurs découvrent de nouveaux caractères
distinctifs que je n'ai pas su voir, et ˙si de nouvelles découvertes montrent
que ceux que j'ai indiqués sont constants, j'ai pensé qu'il ne serait pas inutile
de le faire figurer."

DÖDERLEIN, in his paper on Ceylon Echinoderms, states ˙that he has
not found any specimens from Japan that can be referred to *Astropecten
japonicus*, but that he has several of *Astr. scoparius* [DÖDERLEIN, '88, p.
830].   BELL ['88, p. 384, 388] reports *Astr. hemprichii* from the Sea of Bengal.

In the Challenger Report, *Astropecten scoparius* is not mentioned, but
Astr. japonicus, of which SLADEN says ['89, p. 205], "A large series of this
species was obtained.   I have compared them with the original type-specimens
of MÜLLER and TROSCHEL belonging to the Leyden Museum."   The specimens
were obtained off Kobé,. in mud, 8 fathoms deep, near Awaji Shima, in sand,
50 fathoms, in Bingo Nada, in blue mud, 15 fathoms, and off Yokohama,
5–25 fathoms.

Astropecten hemprichii is reported by SLADEN from the Mergui Archipelago. He thinks that the *Astropecten articulatus*(?) of MICHELIN ['45, p. 24] and *Astropecten mauritianus* of MÖBIUS ['80, p. 50] refer to the same species. The following notes are then added [SLADEN '89a p. 324]:

" *Locality.* Sir William James Island; 7th. Dec. 1881.

" *Remarks.* I feel little hesitation in referring a single example to this species, which has recently been carefully described and figured by DE LORIOL on the basis of material obtained from Mauritius. The type specimen, preserved in the Berlin Museum, was collected by HEMPRICH and EHRENBERG in the Red Sea. The form appears to be closely allied to *Astropecten scoparius.*"

IVES ['91, p. 211] simply mentions this species from Japan and gives some figures of it. MEISSNER mentions it from Yokohama ['92, p. 188]: " 10 Exempl.—Yokohama, 10. VII. 84. (No. 3336.)"

Astropecten hemprichii is mentioned by PFEFFER ['96, p. 47] who remarks on its colour (*fide* Zoological Record).

LUDWIG mentions it from Zanzibar ['99, p. 539]: " *Astropecten hemprichii* MÜLLER und TROSCHEL, von Sansibar durch PFEFFER (1896), aus der südlich (c. 24° s. Br.) von Mozambique gelegenen Bai von Inhambane durch PETERS (1852)." It was not represented in the collection examined by LUDWIG.

DÖDERLEIN gives a very good description of this species [: 02, p. 328]:

" *Astropecten scoparius* MÜLLER u. TROSCHEL.

" Syn. *Astropecten japonicus* (MÜLLER u. TROSCHEL), SLADEN.

" Arme durchschnittlich etwas breiter und kürzer als bei *Astropecten kagoshimensis.* R : r=3–4.7.

" Dorsomarginalplatten am Aussenrande mit je einem sehr kurzen, spitzen Stachelchen, das aber den 3(2)–10 ersten Randplatten fehlt.

" Ventromarginalplatten fein und kurz bestachelt; dazwischen eine Querreihe verlängerter Stacheln, dünn und spitz; am Rand ein langer Stachel, platt, spitz, öfter gebogen, darunter zwei halb so lang, in die die übrigen allmählich übergehen.

" Adambulacralplatten mit drei inneren schlanken Furchenstacheln, zwei mittleren, ebenso dicken, von denen der adorale meist der kürzere ist, und zwei oder mehr äusseren, etwas kürzeren Furchenstacheln.

"Jederseits etwa 3 kleine Ventrolateralplatten. Bei drei verschieden grossen Exemplaren beträgt R=20, 40 u. 91 mm., r=7, 12 u. 19.5 mm., die Armbreite=7, 13 u. 21.5 mm, die Zahl der Randplatten=18, 29 u. 51.

"Diese Art ist einfarbig, graugelblich bis bräunlich.

"Ich fand A. scoparius in geringer Tiefe (bis ca. 20 m) sehr zahlreich auf Sand und Schlammboden in der Tokiobai, Sagamibai, bei Tango, Tagawa (am inneren Meere) und Kagoshima. Es ist der häufigste Seestern an den japanischen Küsten."

Astropecten hemprichii is mentioned by BELL [: 03, p. 244] from the "Zanzibar Channel, 8 fath." It is also mentioned from the Gulf of Manaar by the HERDMANS [HERDMAN, HERDMAN and BELL, : 04, p. 143] as follows :

"*Astropecten hemprichi*, M. and T.

"Stations I., IV., X., XIV., XX., XXVIII., XXII, XXXIX., XLII., XLVII., XLVII. and LIV., practically all round the island, at depths of 4 to 40 fathoms. It is especially abundant on some parts of the pearl banks."

In the same paper [p. 149] BELL has the following remarks on this species :

"I beg once more to offer an example of the variability of Echinoderms, and to call attention to the mode of distribution of' the spicules on the superomarginal plates of *Astropecten hemprichi*; the three figures here shown are taken from the three specimens found in the bottle to which the late Mr. SLADEN affixed the name of *Astropecten zebra*. In the Challenger Report *A. zebra* occupies the following position in the author's key :—

(A) With small spinelet, on the first four or five plates.

    (a) With four or five spinelets. A well-developed series of pseudo-pedicellariæ .... *zebra*.

    (b) With one spine only on the first plate. No pedicellariæ.... .... *velitaris*.

"Inspection of my photographs will show how little constant is the number of spinelets, and the uselessness of the character as an aid to specific distinction. *A. zebra* should, I think, be united with *A. hemprichi*[1]. Perhaps, also, SLADEN's *A. notograptus* is another synonym[2].

HERDMAN [: 06, p. 447] mentions *Astrop. hemprichi* as an enemy of the pearl oyster, prowling over the beds.

Astropecten hemprichii is reported from Portuguese East Africa by SIMPSON and BROWN [: 10, p. 48]:

"Several specimens of various ages and slightly different in superficial appearance represent this species. They were obtained in three separate localities, viz., Tunghi Bay, Mtundo Bay, and Montepes Bay. They agree exactly with those of the same species in the British Museum, collected by H. M. S. 'Alert' on the Mozambique coast. One specimen has been regenerating three arms from the disc.

"Localities.—Station I., Tunghi Bay; Station III., Mtundo Bay (Wamizi Is. to Kifuki Is.); Station X., Montepes Bay.

"Previously recorded from—Mozambique Coast; Red Sea; Mauritius; Ceylon; Tuticorin; Mergui Archipelago."

KOEHLER mentions it from the south-eastern Moluccas [: 11, p. 266]:

"Dragage No. 1. 18 Février 1908. Ngaiguli. Profondeur 14 m. Cinq échantillons.

"La longueur de R varie de 40 mm. à 12 mm. Tous les exemplaires sont bien caractérisés et ils sont exactement conformes à la description de DÖDERLEIN (Zoologischer Anzeiger, Bd. XXVI, p. 326). Les piquants marginaux commencent à se montrer soit sur la quatrième, soit sur la cinquième ou sur la sixième plaque marginale dorsale suivant la taille des exemplaires."

As may be seen from the above review, there are some differences in the descriptions of different observers, which may be brought out more clearly in a tabular form, as follows:

---

1) "I have taken the opportunity of examining the types of *A. velitaris*; they are all immature, and are, perhaps, not members of the same species."

2) "Describers of young starfishes should have their memoirs placed in some scientific Index Expurgatorius; they take no trouble, and give much."

Comparison of different descriptions of *Astropecten scoparius*.
(M=Number of marginals.)

Authors.	R:r	r	R	M	Adamb. spines.	Superomarg.	Inferomarg.	Paxil. area. Madrep.
MÜLLER and TROSCHEL	4	7 Zoll		25-30	In wedge shaped groups.	From end of arms runs a series of conical spinelets, one on each plate on the outer margin, absent in interbrachial angles.	From the general scaly covering of the ventral plates many spinelets can be distinguished by their larger size, become more numerous towards margin.	Paxil. area in middle of arm 2-3 times marginal. Madrep. near margin.
MÜLLER and TROSCHEL (*hemprichii*)	4.3	4½ Zoll		33	In several rows, in wedge shaped groups whose apex is formed by a single papilla.	Each with a spine. Much broader than high. Spines absent at base of arms and in interbrachial angles.	Spinelets very delicate, almost capillary. Near the marginal spines are found fine spinelets, only half as long as the marginal spines. The latter flattened and, except those of interbrachial angles, pointed.	
MÜLLER and TROSCHEL (*japonicus*)	4	2½ Zoll		30	5 on each plate, in wedge shaped groups.	As wide as high, granulated, bears only seldom a movable tubercle on outer margin.	Scaly covering very delicate; on the margin there is a large marginal spine, beneath which 3 spinelets form a row parallel with the margin, the spinelets being 1/3 as long as the marginal spine.	Paxil. area in middle of arm of about the same breadth as marginal. Madrep. near margin.
VON MARTENS						The spines on the dorsal marginals are very inconstant, they are almost invariably absent in the first third; up to middle occur single plates without spines between those with them.		

LÜTKEN (*japonicus*)							
SLADEN (*japonicus*)	2.8	4 mm. 11.25 mm.	In wedge shaped group of 5, long, fine, cylindrical.	Broder than long, on outer half of arm, with a small conical spinelet.	Never saw spines on (superior) marginals.	Projecting more than superomarg., with one large, compressed, lanceolate spine at margin, followed by two smaller, much finer and more cylindrical ones.	Paxil. area very little if any broader than marginal.
DE LORIOL (*hemprichii*)	5-5+	13.65 mm. 22 111 mm. 33	A group of 3 spines, of which the innermost and one is recurved and notably longer than the others, then follows a group of two equal flat spines, and lastly, on the outer side, a group of six or seven spines, close set, very fine, cylindrical and tolerably long.	Covered with small granules separated from one another. A small, short, conical, pointed, deciduous spine on each plate close to the outer margin; this spine absent in the interbrachial angles on plates varying in number from 4 to 10, and this in one and the same example.		More projecting than superomarg., near outer border some spines, which become more numerous towards margin, becoming longer towards last one, which is large, pointed and flattened, as long as the plate; those in the interbrachial angles generally truncated in larger specimens.	Paxil. area nearly double the marginal at base of arm. Paxil. large, with some 30 large spinelets. Madrep. large, near border, concealed by paxillae.
DE LORIOL (*Astrop. sp.*)	5	21 mm. 108 mm.			Spines present on all the superomarginals, those of the interbrachial angles bearing even a second spine.		Breadth of arm at base 25 mm.
DÖDERLEIN	3-4.7	7 12 19.5 mm. 20 40 91 mm. 18 21 51	3 inner slender spines, 2 middle, equally thick spines, of which the adoral one is mostly shorter, and 2 or more outer, somewhat shorter spines.		With a short pointed spine on outer border, absent on 3(2)-10 first plates.	Covered with fine and short spines; a transverse row of elongated, thin pointed spines; on margin a long, flat, pointed, frequently curved spines, beneath which are two spines half as long.	Breadth of arm. 7 mm. 13 ,, 21.5 ,,

I shall make a few remarks on some of the points brought
out in the foregoing table, basing them on a study of Japanese
specimens, of which I have examined a tolerably large series from
various localities.

Radial ratio.—This varies in general according to two factors,
locality and size. Taking specimens from the same locality we
find that the ratio R : r is greater for larger specimens than for
smaller ones. This has been brought out clearly by LUDWIG
['97, p. 6] for *Astropecten aurantiacus*; and indeed it may be
regarded as a general rule for all species of *Astropecten*, if not for
the starfishes in general, inasmuch as a starfish which has just
completed its metamorphosis has a pentagonal outline. The
radial ratio varies also in different localities; thus, my specimens
from Tomo and Miyazu have in general a greater ratio than those
of Misaki and other localities. At the same time a certain latitude
must be allowed on these points, as may be seen from the adjoined
table, in which the specimens are arranged in the order of r.

Specim.	r mm.	R mm.	R : r	MS	Loc.
1	5	13.5	2.7	. . . .	Tsu (Gulf of Isé).
2	6	17.5	3	16	Uraga.
3	6	18	3	. . . .	Misaki.
4	7	21	3	. . . .	Kanazawa (Bay of Tokyo).
5	7.5	24.5	3	. . . .	Misaki.
6	7.5	29	4	. . . .	Tomo (Inland Sea).
7	7.5	30	4	28	,,
8	8	21	2.6	. . . .	Otaru (Hokkaido).
9	8	25	3.1	. . . .	Misaki.
10	8.5	30	3.5	. . . .	Honmoku (Yokohama).
11	8.5	34	4	. . . .	Azamushi (Bay of Aomori).

Specim.	r mm.	R mm.	R : r	MS	Loc.
12	9	26	3	....	Otaru (Hokkaido).
13	9	30	3.3	....	Kanazawa (Bay of Tokyo).
14	9	32.5	3.6	....	"
15	9	34.5	3.8	....	Azamushi (Bay of Aomori).
16	10	33.5	3.35	....	Kanazawa (Bay of Tokyo).
17	10	34	3.4	....	Tsurumi (Bay of Tokyo).
18	10	34	3.4	....	Misaki.
19	10	34	3.4	....	Kanazawa (Bay of Tokyo).
20	10	34.5	3.45	....	Honmoku (Yokohama).
21	10	35	3.5	....	Kanazawa (Bay of Tokyo).
22	10	35	3.5	....	"
23	10	35.5	3.55	....	Tsu (Gulf of Isé).
24	10	36	3.6	....	Uraga.
25	10	44	4.4	....	Tomo (Inland Sea).
26	10	45	4.5	....	
27	10	46.7	4.67	....	
28	10	48	4.8	....	
29	10	48	4.8	....	"
30	10.5	41	4	....	Kanazawa (Bay of Tokyo).
31	10.5	50	4.7	....	Tomo (Inland Sea).
32	11	40	3.6	....	Kanazawa (Bay of Tokyo).
33	11	43	4	....	··
34	11	44.5	4	....	"
35	11	52	4.7	36	Tomo (Inland Sea).
36	11.5	37.5	3.3	....	Honmoku (Yokohama).
37	11.5	40	3.5	....	Misaki.
38	11.5	44.5	3.9	....	Wakanoura (Strait of Kii).
39	11.5	47	4	....	Misaki.
40	11.5	51	4.4	....	Tomo (Inland Sea).
41	12	41	3.4	....	Misaki.

Specim.	r mm.	R mm.	R : r	MS	Loc.
42	12	42,44	3.66	....	Hojo (Tateyama Bay).
43	12	44	3.7	....	Tomo (Inland Sea).
44	12	44.5	3.7	....	Misaki.
45	12	45	3.7	....	Azamushi (Bay of Aomori).
46	12	45	3.7	....	Misaki.
47	12	45	3.7	....	Bonin Islands.
48	12	46	3.8	....	Misaki.
49	12	50	4	....	,,
50	12	52	4.3	....	Wakanoura (Strait of Kii).
51	12	54	4.5	....	Tomo (Inland Sea).
52	12.5	41.5	3.3	....	Misaki.
53	12.5	48	3.8	....	Tsu (Gulf of Isé).
54	12.5	49	3.9	33	Tomo (Inland Sea).
55	13	44	3.4	....	Misaki.
56	13	46.5	3.6	....	,,
57	13	57.5	4.4	....	Miyazu (Japan Sea).
58	13.5	45	3.3	....	Misaki.
59	13.5	55	4	33	Tomo (Inland Sea).
60	14	45	3.2	....	?
61	14	53.5	3.8	....	Misaki.
62	14	55	4	....	Miyazu (Japan Sea).
63	14	60	4.3	....	Tomo (Inland Sea).
64	14	63	4.5	....	Shikajima (off Fukuoka).
65	14.5	59	4.1	....	Miyazu (Japan Sea).
66	15	48	3.2	....	?
67	15	48.5	3.2	....	Misaki.
68	15	52	3.4	....	,,
69	15	54	3.6	....	?
70	15	55	3.7	....	?
71	15	65	4.3	....	Miyazu (Japan Sea).

Specim.	r mm.	R mm.	R : r	MS	Loc.
72	15.5	62	4	....	Namerikawa (Japan Sea).
73	16	58	3.6	....	?
74	16	65	4	32	?
75	16.5	54.5	3.3	....	?
76	16.5	57	3.5	....	Misaki.
77	16.5	58.5	3.5	....	,,
78	18	72	4	....	Namerikawa (Japan Sea).
79	18.5	85	4.6	....	,,
80	20	84	4.4	....	?

The number of marginal plates is certainly of some systematic importance, but in my opinion only in a subordinate degree.

The relative extent to which the superior and inferior marginals project laterally has been taken hold of by some observers for distinguishing species. But this also appears to me to be hardly of systematic importance. It is true that in certain species the inferomarginals are always more projecting than the superomarginals; but in the present species, while the inferomarginals stand prominently outward in some specimens, they are flush with the superomarginals in others.

Again, the relative breadth of the paxillar area and the superomarginals is, in my opinion, only of a very secondary importance for specific distinction; much depending in this case on the contraction of the different muscles when the specimen was killed.

Most characteristic of all are the armature of the superomarginals, the lateral spines of the inferomarginals, the adambulacral spines, and the armature and form of the mouth-plates. The paxillæ are also characteristic of species, and the position and form of the madreporite may also be taken into account.

Bearing all these points in mind, the following description may be given.

Radial ratio.—As measured by me on Japanese specimens from different localities it is 2.7—4.8.

Superomarginals.—The number is given in the table at p. 135. They are much broader than long. At the base of the arm they are at right angles to the margin but very gradually become inclined further out, so that in the distal half of the arm, they may form an angle of about 15° or more with the horizontal (Pl. III, fig. 34, 39). The edge is perfectly rounded at the inter-brachial angle, but is less so along the greater part of the arm, although it never becomes angular. The abactinal surface is rather coarsely granulated, while the outer surface is covered with more pointed, conical granulations. I find that the description of SLADEN ['79], that these granulose spinules are of clavate form, applies only to those near the margins of the abactinal surface. Very characteristic are the short conical, pointed spines which are present one on each superomarginal, along the edge, and are directed obliquely towards the tip of the arm. In most cases these are absent in the interbrachial arc, the number of unarmed plates varying, according to my own observations, from 3 to as many as 16 for one side of an arm; the spines being present in the latter case only in the distal half of the arm. The specimens from Tomo show the least developed superomarginal spines, both in number and size. Excepting a very small specimen of r=3.5 mm., I have yet to see a specimen destitute of these spines, like those referred to by LÜTKEN. On the other hand I have a specimen from the Bonin Islands in which these spines are continuous from arm to arm through the interbrachial angle; not only that, but one or two of the superomarginals at the angle bear two or even three spines, though of smaller size. Again I have a few specimens from Misaki, in which several superomarginals of the interbrachial arc bear from one to four small spines, although these are not always continuous from arm to arm. Thus, in an interradius of one of these specimens only the first pair of supero-marginals are destitute of spines, while in another interradius the superomarginals all bear very short, blunt elevations hardly appearing as spines. In another large specimen (R=58.4 mm.) the first two pairs of superomarginals of an interradius bear an irregular group of 2-4 spines. These specimens recall the one described by DE LORIOL as *Astropecten sp.* from Mauritius cited above. There is no doubt in my mind that they are all only variational forms of the present species. As to my specimen from the Bonin Islands, it may be remarked that I have only one, and am therefore unable to say whether or not this variational form is of common occurrence there. The transverse grooves between the superomarginals are covered with capillary spinelets.

Inferomarginals.—These are usually one less than the supero-marginals. The general actinal surface of the plates is armed with pointed spines, which are more or less flattened and of various sizes (Pl. III, fig. 35, 36, 38). The more prominent, longer spines form one or two irregular rows, and the 2-4 next the lateral spine are espe-cially longer, being sometimes more than $\frac{2}{3}$ as long as the latter. The single lateral spine is very prominent, flattened, sharply pointed, and usually more or less curved towards the tip of the arm. The proximal, inner, and distal margins of the inferomarginal plate are armed with a single row of capillary spinelets. The transverse grooves appear to be naked on the actinal side; but

on the external side, they are covered with capillary spinelets
similar to those of the superomarginals.

Adambulacral plates.—These are much broader than long for
the greater part of the arm, and very much inclined to the furrow
at the base of the arm (Pl. III, fig. 36, 38). About the middle
of the arm there are 5 of these plates to every 3 of the inferomarginals. The adambulacral spines on one of the proximal plates
are as follows : one spine longer than the others at the apex
of each plate, then one pair, then a second pair, then a third
pair or three in a row, and a few smaller spines. In a more
distal plate the first five (one apical and two pairs) are usually
present, but the rest are subject to variation. The relative size
of the first pair mentioned by DÖDERLEIN is frequently, but not
always, true ; the proximal and distal spines are often of the same
length, or the former may even be longer. These spines are more
or less flattened, and have rounded apices.

Mouth-plates.—These are very prominent in specimens from
which the actinal spines have been removed ; they are narrow,
slightly curved, and ankylosed with each other at the two ends,
leaving a small space between covered over with a membrane
(Pl. III, fig. 36). Each plate has a main crest and an accessory
ridge, the former extending through the whole length of the plate,
the latter about one-third as long, and situated on the furrow
side of the mouth end, lying in fact in a line with the furrow
margins of the adambulacral plates (Pl. III, fig. 36, 37). The
main crest bears about 10 short, robust, conical spines ; and the
accessory ridge bears 4–5 smaller spines, of which the most
proximal (next the mouth) is of nearly equal size with those of
the main crest. A few distal spines of the main crest are more

slender, and those next the mouth more robust, than the rest, although the more robust ones near the mouth are not so prominently different from the others as in some other species. In smaller specimens however, the few spines at the mouth end are much longer than the others and form an oral armature. In addition there may be two or three very small, almost capillary spines in the distal part of the main ridge, on the furrow side.

Ventrolaterals.—The actinal interambulacral area is very small, and the ventrolateral plates are also very small, there being only 2 or 3 one either side of the interradial line (Pl. III, fig. 36).

Paxillæ.—At the middle of the arm the paxillar area is mostly twice or more as wide as the superomarginals of one side; but it is subject to considerable variation, and in some examples it is only about as broad as the superomarginals. The paxillæ are of various sizes in different parts of the abactinal surface (Pl. III, 41). At and near the centre they are always smaller and form here a circular area of a different external appearance from the surrounding parts but passing on to them without any demarkation. This part is sometimes elevated into a conical prominence. The paxillæ are also small along the superomarginals and bear only a few coronal spinules, presenting a certain external resemblance to the pedicellariæ. The paxillæ are largest at about the middle of the base of the arm, and in these there are as many as 10 or even up to 15 central blunt spinules surrounded by some 15 peripherals which are smaller. In some of the smaller paxillæ from the centre of the disk, there may be only one central and about 8 or 10 peripherals. The pedicel of the paxillæ is flattened from side to side; the base and top are expanded, and the latter is circular or elliptical in an apical view.

Madreporite.—The madreporite sometimes show well, but is more often hidden from view by the crowns of the surrounding paxillæ. When freed and exposed it is tolerably large, with strongly lobed borders, and half as far removed from the margin of the disk as from the centre (Pl. III, fig. 40).

Terminal plate.—Bilobed.

Locality.—*Astropecten scoparius* is the commonest starfish in Japan. It is especially common on sandy or muddy flats which are exposed at low tide, but is also found at greater depths (20–30 fathoms). Its colour varies from light greyish blue to light brown. I have examined specimens from Otaru, Shiribeshi (Hokkaido), Aomori Bay, Misaki and vicinity, Tokyo Bay, Uraga Channel, Zushi, Tateyama Bay, Wakanoura (Kishyu), Tomo (Bingo, on the Inland Sea), Shikajima (off Fukuoka), Gulf of Isé, Miyazu (Tango), Namerikawa (Etchu), Enoura (Suruga Bay), Gulf of Kagoshima. Specimens in S.C., I.M., H.N.S., H.N.S.W., S., F.B., I.H.S., O.

Remarks.—It only remains for me to add a few remarks on the synonymy of this species. That *Astropecten japonicus* should perhaps be amalgamated with *Astr. scoparius* occurred to SLADEN, and DÖDERLEIN took this step later. The specimen described by MÜLLER and TROSCHEL as *Astr. japonicus* evidently had weakly developed superomarginal spines. In my opinion *Astr. hemprichii* should also be united with the present species. The differences that deserve consideration are the radial ratio and the projecting character of the inferomarginals. As mentioned above, the latter character varies to some extent, and taken alone can not, in my opinion, be regarded as of specific importance. As to the radial ratio, it varies in Japanese specimens from 2.7–4.8. Calculating from the dimensions given by DE LORIOL, it is very nearly 5 for the Mauritius specimens ; but MÜLLER and TROSCHEL give it as 4.3. It is also to be remarked that in Mauritius this starfish appears to grow to a larger size than in Japan ; for the specimen which is figured by DE LORIOL and referred to as of " petite taille," would rank among the larger ones of my specimens. With this more luxuriant growth the radial ratio would be increased and the spines also would be larger, especially the lateral ones of the inferomarginals ; and the apparent difference between the descriptions of MÜLLER and TROSCHEL on the one hand and DE LORIOL on the other as to the armature of the inferomarginals may also be explained. As to the single specimen referred to by DE LORIOL as *Atropecten sp.*, it need hardly be remarked that the only point of importance to be considered is the continuous character of the superomarginal spines, which has been shown to occur also among Japanese specimens of *Astropecten scoparius*. I think therefore that I am justified in concluding that *Astropecten hemprichii* and *Astropecten scoparius* are one and the same species.

Astropecten polyacanthus MÜLLER & TROSCHEL.

(Pl. III, figs. 42–51.)

Next to *Astropecten scoparius* this is probably the commonest species of the genus in Japan. It is placed by MÜLLER and TROSCHEL in their second group, in which also come the two synonyms mentioned below. According to DE LORIOL ['85, p. 76] this species was illustrated for the first time by SAVIGNÝ ['03, pl. iv, fig. 1], but its first description was published in the " System der Asteriden " as follows [MÜLLER u. TROSCHEL, '42, p. 69, 2 figs.] :

"*Astropecten polyacanthus* Nob. nov. sp.

"Fünf Arme. Verhältniss des Scheibenradius zum Armradius wie 1 : 6. Die inneren Furchenpapillen stehen je drei auf einer Platte in der Figur eines Winkels, so dass die grösste mittlere vortritt. Hinter ihnen steht ein Häufchen kleinerer Papillen. Die Stacheln in der Beschuppung der Bauchseite sind alle gross, 4–5 auf einer Platte und erreichen vor den Randstacheln schon die Grösse derselben. Die dorsalen Randplatten, 33 an jedem Arme, sind höher als breit. Die Stacheln auf ihnen sind lang, conisch und spitz, auch diejenigen in den Winkeln stehen senkrecht und bilden eine Reihe am Rande. Auf der zweiten, oft auch auf der dritten Platte vom Grunde der Arme aus fehlen immer die Stacheln. Selten zeigt sich nach aussen die Anlage einer zweiten Reihe. Die Granula auf der dorsalen Randplatten sind conisch. Die Rückenfelder zwischen den dorsalen Randplatten sind breiter als bei der vorhergehenden Art.[1]

"Grosse : bis 1 Fuss.

"Fundort : Rothes Meer. Im Museum zu Berlin durch HEMPRICH und EHRENBERG ; im Museum zu Paris durch BOTTA."

On the next page is the first description of *Astropecten hystrix* [Müller u. Troschel, '42, p. 70] :

"*Astropecten hystrix* VAL. nov. sp.

"Fünf Arme, mit 30 Randplatten. Die Furchenpapillen, 9–10 auf jeder Platte, stehen in 2 Reihen quer auf die Arme ; nach der Furche bildet die grösste die Spitze. Auf den zart beschuppten Bauchplatten zeichnen sich 4–5 wenig platte spitze Stacheln aus, die nach aussen an Grösse zunehmen. Die äussersten sind bis gegen 3 Linien lang und dabei verhältnissmässig dick. Die dorsalen Randplatten springen an den Armen, besonders in der Mitte, schräg vor und sind daher durch starke Vertiefungen von einander getrennt. Sie sind höher als breit und tragen auf ihrem Gipfel einen kegelförmigen Stachel. Die beiden Platten in jedem Winkel zwischen den Armen stehen höher als alle übrigen. Die Madreporenplatte ist um weniger als ihren Durchmesser von den Randplatten entfernt. Das

---

1) *Artropecten bispinosus.*

Feld des Rückens oder der Paxillen auf der Mitte der Arme ist mehr als 3–mal so breit wie die schmalen Randplatten.

"Grösse: 6 Zoll.

"Fundort: Ceylon. Im Museum zu Paris durch REYNAUD."

Immediately next follows the description of *Astropecten armatus*[1] [MÜLLER u. TROSCHEL, '42, p. 71]:

"*Astropecten armatus* Nob. nov. sp.

"Fünf Arme. Verhältniss des Scheibenradius zum Armradius wie 1 : 3¾. Viele lange Furchenpapillen, ohne sich in bestimmte Reihen zu ordnen. Aus der Beschuppung der Bauchplatten tritt in der Nähe der Furchen eine Reihe kleiner Stachelchen hervor, die zuweilen sich verdoppelt. Am Rande finden sich auf jeder Platte zwei grosse Stacheln, von denen der oberste der grösste ist. Die dorsalen Randplatten, 20–22 an jedem Arme, sind höher als breit, jede trägt einen runden, senkrechten, spitzen Stachel, kürzer als die Randstacheln der ventralen Platten. Die Stacheln nehmen vom Grunde der Arme aus an Grösse ab. Auf der zweiten, dritten und vierten Platte steht jedoch nie ein Stachel, so dass zwei Stacheln in jedem Scheibenwinkel durch einen Zwischenraum jederseits von den übrigen Stacheln getrennt sind.

"Grösse: 4 Zoll.

"Fundort: Japan. Im Museum zu Leyden durch v. SIEBOLD."

The first description of *Astropecten vappa* MÜLL. & TROSCHEL, which SLADEN regards as a synonym of the present species comes in here chronologically ['43]; but I have not been able to see the original paper.

DUJARDIN and HUPÉ describe this species as follows, closely following MÜLLER and TROSCHEL ['62, p. 417]:

"Astropecten polyacanthe. *Astropecten polyacanthus.*—MÜLLER et TROSCHEL.

"—MÜLLER et TROSCHEL, Syst. der Aster., p. 69, pl. v, f. 3.

"Espèce à cinq rayons ou bras, dont la longueur, à partir du centre, est sextuple du plus petit rayon du disque. La rangée interne du sillon

---

1) The *Astropecten armatus* of GRAY ['40, p. 181; '66, p. 3] is referred to *Astrop. erinaceus* GRAY by PERRIER ['76, p. 278] and SLADEN ['89, p. 735].

ambulacraire présente, sur chaque plaque, trois piquants formant un angle,
de telle sorte que celui du milieu est le plus grand et le plus saillant;
derrière eux se trouve un groupe de piquants plus petits. Les piquants de
la face ventrale, au nombre de quatre ou cinq sur chaque plaque, sont tous
gros et atteignent la longueur des piquants marginaux. Les plaques dorsales
marginales, au nombre de trente-trois sur chaque bras, sont plus hautes que
larges, et portent des piquants long et coniques qui, dans les angles interbrachiaux, ont une attitude perpendiculaire et forment une rangée sur le
bord. Les piquants manquent toujours sur la deuxième et souvent aussi sur
la troisième plaque, à partir de la base des bras. Rarement on voit en
dehors l'indication d'une deuxième rangée. Les granules sur les plaques
marginales dorsale sont coniques. Les aires comprises entre ces dernières
plaques sont plus larges que dans l'espèce précédente.[1]

"Coloration jaunâtre foncé. Dimension : largeur totale 200 à 325 mm.

"Habite la mer Rouge (Mus. Paris)."

The same authors describe *Astrop. hystrix* as follows ['62, p. 418]:

"Astropecten porc-épic. *Astropecten hystrix.*—MÜLLER et TROSCHEL.

"MÜLLER et TROSCHEL, Syst. der Aster., p. 71, no. 8.

"Espèce pourvue de trente plaques marginales à chaque bras. Les
piquants du sillon ambulacraire, au nombre de neuf ou dix sur chaque
plaque, sont disposés sur deux rangées, le plus grand d'entre eux forme une
pointe vers le sillon. Les plaques ventrales, finement écailleuses, présentent
quatre ou cinq piquants un peu plats et pointus, qui vont en augmentant de
longueur, en dehors; le plus externe ayant environ 7 millimètres de longueur,
est proportionellement épais. Les plaques marginales dorsales sont obliquement saillantes sur les bras, particulièrement au milieu, ce qui fait qu'elles
sont séparées par une forte dépression; elles sont plus hautes que larges et
portent sur leur sommet un piquant conique.

"Les deux plaques de l'angle rentrant interbrachial, sont plus hautes
que toutes les autres.

"La plaque madréporique est éloignée de la plaque marginale d'un peu
moins que son diamètre. Les aires dorsales des bras, entre les plaques

---

1) *Astropecten bispinosus*

marginales, sont un peu plus de trois fois |aussi larges que ces plaques.

" Coloration brunâtre. Dimension : largeur totale 160 mm.

" Habite les côtes de Ceylan (Mus. de Paris)."

Immediately following is the description of *Astrop. armatus* [DUJARDIN et HUPÉ, '62, p. 418] :

Astropecten armé. *Astropecten armatus.*—MÜLLER et TROSCHEL.

" —MÜLLER et TROSCHEL, Syst. der Aster., p. 71 (non GRAY).

" Bras dont la longueur, à partir du centre, égale trois fois et un quart le plus petit rayon du disque. Les piquants du sillon ambulacraire sont longs, nombreux, et ne forment point de rangées distinctes ; une rangée de petits piquants, quelquefois doubles, s'élève au-dessus des écailles qui garnissent les plaques ventrales ; près de sillon ambulacraire, et sur chaque plaque, se trouvent deux grands piquants, dont l'inférieur est un peu moindre. Les plaques marginales dorsales, au nombre de 20 à 22 sur chaque bras, sont plus hautes que larges, et chacune porte un piquant rond, perpendiculaire, plus court que les piquants du bord ventral ; les piquants diminuent de grosseur à partir de la base des bras. Sur les deuxième, troisième et quatrième plaque, il n'y a jamais de piquant, de telle sorte que deux piquants dans chaque angle rentrant interbrachial se trouve séparés de part et d'autre des autres piquants.

" Coloration jaunâtre. Dimension : largeur totale 108 mm.

" Habite les côtes du Japon.

" M. GRAY a également établi une espèce sous ce même nom, mais nous ne savons si elle est la même que celle-ci."

Astropecten vappa MÜLLER & TROSCHEL which is regarded by SLADEN as a synonym of the present species is thus described by DUJARDIN and HUPÉ ['62, p. 421] :

" Astropecten vappa. *Astropecten vappa.*—MÜLLER et TROSCHEL.

" —MÜLLER et TROSCHEL, WIEGM. Archiv, 1843, p. 119.

" Espèce ayant les bras dont la longueur, à partir du centre, égale trois fois et demie le plus petit rayon du disque. Les piquants du sillon ambulacraire forment deux rangées présentant chacune trois piquants de même forme sur chaque plaque.

"Les plaques marginales sont au nombre de 21 à chaque bras, les ventrales présentent de petites écailles isolées, saillantes, en forme de piquants, dont la plus externe est plus longue que le piquant marginal, lequel est d'ailleurs pointu et un peu aplati. Les plaques marginales dorsales, au milieu des bras, sont aussi larges que hautes ; elles portent, en dehors, un petit piquant conique. L'aire paxillaire des bras, entre les plaques marginales, est très-large, elle égale plus de trois fois la largeur de ces plaques, vers le milieu des bras. La plaque madréporique est très-près du bord.

"Coloration jaunâtre. Dimension : largeur totale 40 mm.

"Habite la côte sud-ouest de la Nouvelle-Hollande (Mus. de Berlin)."

LÜTKEN, in the section of his paper of 1864 treating of the genus *Astropecten*, makes the following remarks bearing on this species ['64, p. 132] :

"I Beskrivelsen af *Astropecten armatus* M. Tr. (fra Japan) hedder det, at den anden, tredie og fjerde övre Randplade fra Armvinklen altid er piglös. Paa et for mig liggende Exemplar slaaer dette kun til for de to Armvinklers Vedkommende, men ikke for de tre andres ; i den ene af disse mangle kun Nr. 2 og 3 paa den ene og Nr. 2 paa den anden Side Piggen ; i den anden ere vel alle tre Plader piglöse paa den ene Side, men kun Nr. 2 og 3 paa den anden ; og i den tredie Armvinkel faae de alle hver sin Pig, men denne er dog paa Nr. 2 paa den ene Side og paa Nr. 3 og 4 paa den anden ikke lidet mindre end de, mellem hvilke de sidde, ligesom ogsaa den inderste Plade i hver Række, som ellers pleier at have en meget stor Pig, her kun har en meget lille, der endda paa den ene Plade ikke sidder paa sit sædvanlige Sted. Hos

In the description of *Astropecten armatus* M. Tr. (from Japan) it is stated that, the second, third and fourth upper marginal plates from the arm angle are always without spines. In the example lying before me, this is true only in so far as it relates to two of the arm angles, but not for the other three ; in one of these only _ No. 2 and 3 are destitute of spines on the one side and No. 2 on the other ; in the second all the three are indeed unarmed on one side, but only No. 2 and 3 on the other ; and in the third arm angle they all bear spines, but these are, for No. 2 on one side and No. 3 and 4 on the other not a little smaller than those between which they are found ; moreover, the innermost plate of each series, which elsewhere is apt to bear a very large spine, has here only a very small one, which however is not found at its customary spot on one of the

tre Exemplarer fra Hongkong i China er gjennemgaaende kun 2-den og 3-die, et Par Steder kun anden og et enkelt Sted ikke engang denne piglös; men intetsteds udvides Piglösheden til den fjerde Randplade. Af to andre foreliggende Exemplarer holder det ene sig Regelen ganske efterrettelig, medens det andet kun viser den ubetydelige Afvigelse derfra, at det i den ene Armvinkel paa den ene Side har to piglöse Randplader istendefor 3.—Hvorvidt denne Art er vel adskilt fra *A. vappa* M. Tr. (Nyholland) og *A. polyacanthus* M. Tr. (Röde Hav), turde derfor endnu være, tvivlsomt; mellem de Exemplarer af alle tre Former, som have staaet til min Raadighed, har jeg ikke kunnet finde rigtig gode Artsmærker, men jeg indrømmer, at der udfordres större Suiter for at kunne udtale sig med Bestemthed mod deres Forskjellighed. Af den store geografiske Afstand vil der neppe kunne hentes nogen Understöttelse for denne sidste, da der er mange andre Arter, der optræde med ligesaa stor Udbredning."

plates. In three examples from Hongkong in China, only the second and third are destitute of spines throughout, and in a couple of places only the second and in a single place none was without a spine; but nowhere does the spineless condition extend to the fourth marginal plate. Of two other examples lying before me, one faithfully keeps to the rule, while the other presents only an inconsiderable deviation from it, inasmuch as there are on one side of one of the arm angles two spineless marginal plates instead of 3.—How far this species is well separate from *A. vappa* M. Tr. (New Holland) and *A. polyacanthus* (Red Sea), it may therefore still be doubtful; among the examples of all the three forms, which have stood at my disposal I have not been able to find really good specific characters, but I concede that larger series are necessary in order to express oneself definitely against their distinctness. From the great geographical separation one would hardly derive any support for this last (their specific distinctness), since there are many another species which have as wide a distribution.

VON MARTENS makes the following simple remarks ['65, p. 352]:

" *Astropecten armatus* MÜLL. et TROSCHEL l. c. S. 71., non GRAY. HERKLOTS fn. jap. tab. inedit. X. fig. 1.

" Ein starker aufrechtstehender Stachel auf jeder Randplatte."

It is a rather strange coincidence that both MÜLLER and TROSCHEL and VON MARTENS had specimens with spines on all the superomarginals, while the break in the series of superomarginal spines on either side of the interbrachial angle is a very characteristic feature of the species.

GRAY mentions this species from the Gulf of Suez; the specimens belong to the British Museum ['72, p. 119].

PERRIER, under the heading of *Astrop. polyacanthus*, after mentioning *Astr. hystrix.* VAL., *Astr. armatus* MÜLL. & TROSCH., *Astr. vappa* M. & T., and *Astr. aster* DE FILIPPI in the list of synonyms, gives the results of a critical examination of the specimens studied by him ['76, p. 275]:

"Le docteur LÜTKEN a exprimé, en 1864, l'opinion que l'*Astropecten armatus* du Japon et l'*Astropecten vappa* de la Nouvelle-Hollande pourraient bien être la même espèce que l'*Astropecten polyacanthus* de la mer Rouge.

"Il existe dans la collection du Muséum des échantillons portant ces noms et dont la provenance est bien celle des types de MÜLLER et TROSCHEL; mais aucun d'eux n'a été examiné par les auteurs du System der Asteriden; l'*Astropecten armatus*, de Hong-Kong, représenté par un seul échantillon, a été donné au Jardin des Plantes, en 1864, par le Musée de zoologie comparative de Cambridge (Massachusetts), et cinq *Astropecten* rapportés de la Nouvelle-Hollande, en 1844, par M. JULES VERRAUX, ont été déterminés comme étant des *Astropecten vappa*, sans doute par VALENCIENNES et très-probablement, comme le précédent, sans comparaison directe avec le type de MÜLLER et TROSCHEL, qui est au Musée de Berlin. On peut donc concevoir quelques doutes à l'égard de l'authenticité de ces déterminations, d'autant plus qu' entre ces dernières espèces, MÜLLER et TROSCHEL signalent dans les proportions, dans le nombre des plaques marginales, dans la disposition des piquants ambulacraires, des différences qui ne se retrouvent pas sur nos échantillons. Il est juste, cependant, de faire rémarquer à cet égard que ces différences sont de celles qui sont sujettes au plus grand nombre de variation chez les Astérides. Sous le bénéfice de ces réserves, il n'est pas douteux que les *Astropecten armatus* et *vappa* du Muséum ne doivent être considérés comme de véritables *Astropecten polyacanthus*, et l'on doit encore réunir à cette espèce tous les individus de la collection du Muséum qui portent le nom *Astropecten hystrix* et dont plusieurs ont été sans doute déterminés par TROSCHEL. Si la détermination du musée de Cambridge et du Muséum sont exactes, il a donc là quatre espèces à rèunir en une seule. Il est certain, en tous cas d'après l'examen des échantillons du Muséum de Paris, que l'*Astropecten polyacanthus*, de la mer Rouge, se trouve encore à Zanzibar, Mascate, Ceylan, Hong-Kong, en Chine, Port-Jackson et divers autres points du littoral de la Nouvelle-Hollande ; enfin, M. Filhol vient encore d'en rapporter un exemplaire des îles Fidji. C'est là une aire de répartition très-vaste, mais dont plusieurs autres espèces nous ont déjà offert des exemples remarquables, et qui concorde, du reste, avec la grande uniformite de la faune des régions chaudes du Pacifique.

"En ce qui concerne les *Astropecten polyacanthus* et *hystrix*, au sujet desquels nous avons des éléments authentiques d'appréciation, nous retrouvons les mêmes dispositions des piquants marginaux et ventraux, les mêmes limites des variations dans le nombre des plaques marginales, dont la deuxième et quelquefois la troisième dorsale manquent ordinairement de piquant. Müller et Troschel attribuent aux piquants de la gouttière ambulacraire des dispositions assez différentes dans ces deux espèces. Dans l'*Astropecten polyacanthus*, 'les piquants ambulacraires, disent-ils, sont au nombre de trois sur chaque plaque et sont disposés de manière à figurer un angle et de manière que le plus grand des trois soit placé entre les deux autres. En arrière se trouve un petit groupe de piquants plus petits.' Dans l'*Astropecten hystrix*, 'les piquants ambulacraires sont au nombre de neuf ou dix sur chaque plaque et disposés en rangées obliques par rapport aux bras, le plus gros piquant se trouve le plus près du sillon.' En réalité dans les deux cas, les piquants ambulacraires sont disposés sur le bord des plaques interambulacraires et il en existe un en général plus grand que les autres, sur le milieu de la face de la plaque qui regarde la gouttière ambulacraire, lorsque ce piquant et le premier de chaque côté des deux séries marginales sont seuls très-développés, on a les trois piquants figurant un angle de l'*Astropecten polyacanthus ;* en arrière d'eux, les piquants moins développés forment le petit groupe dont parlent Müller et Troschel ; c'est toujours le cas dans les petits échantillons. Lorsque les piquants bordant les plaques interambulacraires se développent davantage et presque également, ils forment alors les deux séries obliques de l'*Astropecten hystrix* et le piquant médian isolé constitue le grand piquant le plus rapproché de la gouttière. Cette disposition se trouve plus particulièrement sur les grands individus, dont les plaques interambulacraires sont aussi plus allongées transversalement que chez les petits.

"La disposition fondamentale des piquants ambulacraires comme tous les autres caractères, est donc la même chez l'*Astropecten polyacanthus* et l'*Astropecten hystrix* de MÜLLER et TROSCHEL qui constituent une seule et même espèce. On a donné le premier nom aux individus de la mer Rouge et le second à ceux de toutes les autres provenances et elles sont, comme nous l'avons vu, assez nombreuses, puisque de cette dernière espèce le Muséum possède trois individus desséchés et trois dans l'alcool de Ceylan (M. REYNAUD, expédition de *la Chevrette*, 1829) ; un individu de Mascate et un autre de Zanzibar, rapportés en 1841, par M. LOUIS ROUSSEAU ; un individu de Port-Jackson (dans l'alcool), rapporté en 1829, par QUOY et GAIMARD ; enfin cinq individus desséchés, rapportés en 1841, de la Nouvelle-Hollande, par M. JULES VERREAUX, constituent les *Asterias vappa*, du Muséum. Il faut ajouter à cette liste un individu, dans l'alcool, rapporté cette année même (1876) par M. FILHOL des îles Fidji, et l'*Astropecten armatus*, de Hong-Kong, représenté par un individu desséché, du musée de Cambridge (Massachusetts). Quant aux individus portant le nom d'*Astropecten polyacanthus* et qui proviennent de la mer Rouge, le Muséum en a reçu, en 1833, de M. BORÉ, sept conservés dans l'alcool ; en 1834, de M. BOTTA, quatre desséchés ; deux individus de grand taille, également desséchés, proviennent de la collection du prince BONAPARTE. Enfin, trois individus dont l'un a ses deuxième et troisième plaques marginales armées, comme les autres d'un piquant, ne portent pas d'indication de provenance, mais se rattaches encore à ce type.

"Conformément à la règle que nous nous sommes imposée dans cette révision, nous avons conservé sur les étiquettes du Muséum, avec le nom que nous adoptons, toutes les dénominations reçues par les divers échantillons dont nous venons de parler, afin de laisser entre les mains de chacun tous les éléments possibles d'appréciation."

As to the *Astropecten aster* mentioned by PERRIER in his list of synonyms, it may be remarked that it has been referred lately by LUDWIG ['97, p. 50] to *Astropecten johnstoni*.

SLADEN, in his paper of 1879, makes some critical remarks of importance on this species. In the following I omit the list of synonyms given [SLADEN, '79, p. 429].

"*Astropecten polyacanthus* (of *armatus*-type) M. and T.

"Coll· St. John : Yedo Bay.

"So far back as 1864, Dr. LÜTKEN raised the question as to the validity of the separation of *A. armatus*, M. and T., from Japan, and *A. vappa*, M. and T., from Australia, as species distinct from the typical form of *A. polyacanthus* from the Red Sea, asserting his inability to detect in the material he had examined any characters of specific value to warrant such a division. M. PERRIER, after studying the large collections in Paris, concurs in these views, and maintains the consolidation of the above-mentioned forms,[1] including also *A. hystrix* (Val.), M. and T.

"M. PERRIER further expresses his opinion that the differences upon which the separation has stood are nothing more than conditions of age and locality—the series of specimens which the French savant has had the opportunity of examining being procured from stations as widely distant as Zanzibar, Muscat, Ceylon, Hong-Kong, Fiji Islands, Port Jackson and several other localities in Australia, thus indicating a very extensive distribution of the *A. polyacanthus* type.

"Although the present specimen is in a somewhat weathered condition, it can unmistakably be assigned to the varietal group formerly described under the name of *A. armatus*, M. and T. In each ray the three marginal plates which succeed to the innermost in the arm-angle are destitute of tubercles and dorsal marginal spines. This character is regular, and accords with the typical description given in the System der Asteriden. LÜTKEN (Vidensk. Medd., 1864, p. 132) chronicles the occurrence of considerable irregularity and variation in the number of these spineless plates in different rays of the same individual, and cites examples from Hong-Kong having only one, or two, or even none of the undeveloped spineless plates on different rays of the same specimen. This starfish measures R=35 millims., r=9.6 millims.

---

1) *Astrop. polyacanthus, hystrix, armatus, vappa.*

"Without calling in question the accuracy of M. PERRIER's determination, the occurrence of such instances as this of a form presenting strongly marked variations at different stations within the area of its distribution, urges upon naturalists the necessity of exercising extreme caution against being led away by a tendency to group too comprehensively the forms which may be included within a large and widely distributed genus ; for however seriously the multiplication of frivolous 'species' may embarrass a classification, the wholesale grouping, or, in other words, the unbounded extension of the limits of specific character, is productive of much more injurious results, in that it curtails the precision of definition, and, whilst ignoring environment as a factor, divests nomenclature of one of its highest and most important qualities.

" From the fact that forms are separated by much smaller and less striking differences in an extensive genus than in one of more limited scope, . 'species' in the larger group have often not such clearly marked or conspicuous characters as those which are presented by 'varieties' in a less comprehensive genus. It follows that the judgment should be very cautiously excercised when tempted to embrace within a single species all the strongly marked distributional extremes of any widely-spread type, however closely their connexion may seem to be preserved through intermediate forms ; for in many cases these gradations are nothing more or less than the links which indicate to us the development of 'species,' and are, in short, the stages with which generally we are unacquainted, owing either to the imperfection of knowledge, or more frequently by reason of their destruction through the hostility of unfavourable conditions.

" Taking into consideration the advance which knowledge is continually making by means of the addition of new material from hitherto unexplored fields, the process of too comprehensive grouping would ultimately result in the fomation of series which, from their unwieldiness, would require arbitrary division for the mere purposes of classification and comprehension, if the ordinary natural distinctions be ignored. Of course it will be acknowledged that 'species' are but arbitrary divisions after all, and that a *nomen triviale* serves but to register the state of information and our opinions upon certain

forms of life; but since under such an aspect the organisms themselvs stand
as the outcome of adaptation and the conditions of existence, the latter
factor being thus synonymous with *habitat* or geographical position, taken
in its widest sense, it would evidently be a disadvantage to science to lose
the record of the influence which has been exerted, and to sacrifice so simple
an indication of the relative position of a modified type within the area of
its general occurrence."

This species is mentioned by name by SLADEN ['83, p. 251] in his
preliminary paper on the Challenger Asteroidea as follows:

"*Astropecten polyacanthus*, MÜLLER and TROSCHEL.

"Station. Port Jackson, Australia. Depth 2–11 fms.

"Station. Admiralty Island. Depth 16–25 fms.

"Station. Yokohama, Japan. Depth 5–25 fms.

"Station. Kobi, Japan. Depth 8–50 fms."

STUDER mentions this species from Timor (Pariti) ['84, p. 43]. BELL
mentions it from the collections of the 'Alert' ['84a, p. 133]:

"A fine series of . ten specimens were taken in 0–5 fms., at Port
Jackson. In the case of two examples there are three spineless ossicles
succeeding the plate at the angle of the arm; in all the other cases the
more ordinary condition of two such ossicles only is found to ·obtain. The
smallest specimen has R equal to 15 millim., and the largest R equal to
36 millim. Some variability is to be noted in the tenuity of the arms."

The same author also mentions it from the Seychelles and Darros Island
[BELL, '84a, p. 510].

WALTER describes a specimen from Ceylon as follows ['84, p. 368]:

"*Astropecten armatus*. M. et TR. (=*hystrix* und *polyacanthus* M. et TR.)

"Ein getrocknetes Exemplar aus Ceylon ergiebt als Grössenverhältnisse:
Armlänge 89 mm., Scheibendurchmesser ohne die Randplatten 27, mit denselben 31 mm. Die Breite der Arme an der Basis: oben 18, unten 21 mm.
Vollkommen stimmt dies Exemplar zu M. et TROSCHEL's typischem *Astropect.
armatus* von Java, nicht zur var. *hystrix*, die schon im System der Asteriden
von Ceylon angeführt wird. Die einzige Abweichung von jener Beschreibung des *Astrop. armatus* besteht bei unserm Exemplar, neben grösserer

… die 2 mittleren … Stacheln … schliessenden, mit … Seitenplatten … nach M. et … alle ähnliche … schon an einer … sich finden. … Merkmal, die 2., und 4. … an einem … liegt an 2 Armen … Leinen nur noch an … stehen, gegen **den ceyloner und** dem von M. et … Exemplar. Im letzten rmwinkel endlich stehen 5 solcher, indem … Platte sich … die vorletzte **Armplatte, die vor dem** … Winkelbiegung aber trägt. Im Gegensatz zur … Zahl der Winkelstacheln scheint … dagegen ihre Stellungs … und ihr Verhältnis zu denen der eigentlichen Armplatte … und … stantes zu sein. Die Stacheln der im rmwinkel … nämlich auf dem oberen Ende dieser … der Oberfläche des Scheibenrandes … lich an … angebracht sind …

s autres, et
rouvent deux
a q'une, mais
de l'autre. Avec
omparé avec soin
int de Hong-Kong,
quels, à la vérité, R
que M. PERRIER, je ne
ce spécifique appréciable,
M. SLADEN, mais, pour
tes, il faudrait trouver des
voir que des variations dans

u ad adds ['87a, p. 648], "I
specie name of *armatus* in pre-
DERLIN also mentions it from the
388] again mentions it from the

va specimen ['89, p. 310]:
RRIER 'Revision des Stell.', pag.
i von Batavia. Das einzige Ex-
da de Armradius nur 15 m. M.
Beschreibung, welche MÜLLER und
weiter mit *A. hystrix* (M. und
n, idetisch seien. Ich bemerke
HEL besonders erwähnte Eigentüm-
der zweiten, oft auch auf der dritten
ie Stacheln auf den dorsalen Rand-
M men unes Exemplares vorkommt,
wie bei M und TR. angegeben wird."
includes in the list of synonyms, besides
iters, *Astropecten chinensis* GRUBE and

Zahl der Arm-Randplatten, hier 50 jederseits, darin, dass die 2 mittleren
Glieder der Armwinkel, deren jedes ein kleines kegelförmiges Stachelchen
trägt, von den weiteren, den Arm in seiner ganzen Länge einfassenden, mit
gleichgestalteten, nur etwas kleineren Stachelchen besetzten Seitenplatten
durch nur 2 stachellose Platten getrennt werden, deren es nach M. et
TR. 3–4 geben sollte. Indess scheint dies Verhalten wie alle ähnliche
Merkmale, ungemein wechselnd. So finde ich eine Brücke schon an einer
Seite eines Armes des gleichen Thieres, wo 3 stachellose Felder sich finden.
Noch deutlicher ergiebt sich jenes von M. et TR. betonte Merkmal, 'die 2.,
3. und 4. Armplatte trägt nie einen Stachel', als absolut unsicher an einem
chinesischen Exemplar der hiesigen Sammlung. Dieses zeigt an 2 Armen
jederseits nur je eine stachellose Platte; an 2 anderen Armen nur noch an
einer Seite eine solche, so dass in den Armwinkeln 4 Stacheln stehen, gegem
2 des ceyloner und dem von M. et TR. beschriebenen Exemplar. Im letzten
Armwinkel endlich stehen 5 solcher, indem dort keine stachellose Platte sich
findet, die vorletzte Armplatte, die vor dem Beginn der Winkelbiegung, aber
2 trägt. Im Gegensatz zur wechselnden Zahl der Winkelstacheln scheint
mir dagegen ihre Stellungsweise und ihr Verhältniss zu denen der eigentlichen Armplatten ein durchaus constantes zu sein. Die Stacheln der im
Armwinkel stehenden Platten sind nämlich auf dem oberen Ende dieser
angebracht. Ihre Basis also ruht auf der Oberfläche des Scheibenrandes,
während die Stacheln der Armplatten seitlich an diesen angebracht sind.
Dazu sind die Winkelstacheln stärker als die der Arme. Auf der Scheibe
nehmen die am Rande ziemlich gleichen Paxillen zum Centrum hin rasch
an Grösse ab, bis sie, im Centrum punktförmig geworden, nur noch wenige
Granula tragen. Die Granula der oberen Randplatten sind sowohl an den
Armen, wie der Scheibe grösser, als die der Rückenpaxillen."

Specimens from Mauritius are described by DE LORIOL ['85, p. 76]:

"De nombreux individus appartenant à l'*Astrop. polyacanthus* typique,
exactement indentiques à la figure citée de SAVIGNY, ont été envoyés de
Maurice. Dans le plus petit r[1]=50 mm., dans le plus grand R=100 mm.
Dans tous ces individus les deux plaques au sommet de chacun des angles

1) This must be a misprint of R.

interbrachiaux portent chacune un piquant droit plus long que les autres, et pans presque tous, de chaque côté de ces deux plaques, se¦ trouvent deux plaques entièrement dépourvues de piquants, parfois il n'y en a q'une, mais c'est rare, et parfois aussi il y en a une d'un côté et deux de l'autre. Avec ces exemplaires très typiques de l'*A. polyacanthus*, j'ai comparé avec soin des exemplaires de l'*A. armatus* de ma collection provenant de Hong-Kong, et des individus de l'*A. vappa* de l'Australie, dans lesquels, à la vérité, R ne dépasse pas 60 mm. Je dois dire que, de même que M. PERRIER, je ne sais trouver, entre tous ces individus aucune différence specifique appréciable. J'ai lu avec soin les observations très judiceuses de M. SLADEN, mais, pour pouvoir maintenir ces trois espèces comme distinctes, il faudrait trouver des caractères ayant quelque constance, et je ne sais voir que des variations dans l'aspect général des individus."

BELL mentions this species from Ceylon and adds ['87a, p. 648], "I cannot imagine why Dr. WALTER uses the specific name of *armatus* in preference to the familiar *polyacanthus*." DÖDERLEIN also mentions it from the same locality ['88, p. 830]. BELL ['88, p. 388] again mentions it from the Sea of Bengal.

SLUITER makes some remarks on a Java specimen ['89, p. 310]:

"*A. polyacanthus* (M. und TR.). PERRIER, 'Revision des Stell.', pag. 275. Ein Exemplar (No. 257) aus der Bai von Batavia. Das einzige Exemplar, das ich besitze, ist nur klein, da der Armradius nur 15 m. M. misst. Das Tier stimmt genau mit der Beschreibung, welche MÜLLER und TROSCHEL für ihren *A. polyacanthus* geben, weniger mit *A. hystrix* (M. und TR.), welche beide Arten, nach PERRIER, identisch seien. Ich bemerke noch, dass die von MÜLLER und TROSCHEL besonders erwähnte Eigentümlichkeit für *A. polyacanthus*, dass auf der zweiten, oft auch auf der dritten Platte, vom Grunde der Arme aus die Stacheln auf den dorsalen Randplatten fehlen, konstant bei den fünf Armen meines Exemplares vorkommt. Auch die Furchenpapillen sind genau, wie bei M. und TR. angegeben wird."

In the Challenger Report SLADEN includes in the list of synonyms, besides the names mentioned by previous writers, *Astropecten chinensis* GRUBE and *Astr. ensifer* GRUBE ['89, p. 201]:

"*Localities.*—Port Jackson, Australia. Depth 2 to 11 fathoms, 6 to 15 fathoms.

"Admiralty Islands. Depth 16 to 25 fathoms.

"Yokohama, Japan. Depth 6 to 25 fathoms.

"Kobé, Japan. Depth 8 to 50 fathoms.

"Hong-Kong. Beach.

"*Remarks.*—The examples from the Admiralty Islands are remarkable for the robustness and size of the large spines on both the supero-marginal and infero-marginal plates. This is especially conspicuous in one example in which the lateral spines are unusually broad and flattened, the breadth being maintained until near the tip, when it abruptly forms a broad lancet-like point. The same example is also noteworthy from the fact that each supero-marginal plate is armed with a spine; whereas another dredged at the same time follows the usual rule in this species in having the second, or second and third, supero-marginal plates on each side of the median inter-radial line unarmed. These points are very interesting, but I scarcely feel justified in giving a name to the variety on the strength of the material at command. This course, however, may perhaps be found desirable in the future, when more examples from this locality are available for study.

"There is a small *Astropecten* also from the Admiralty Islands, which, from having been dredged in association with the above-mentioned speci-mens, might not unreasonably be looked upon as the young phase of this species. I consider, however, that it is distinct, and should be referred to *Astropecten velitaris*, VON MARTENS. The example in question, which has a major radial measurement of 11 mm., and sixteen supero-marginal plates, has only the two innermost plates in each interbrachial arc armed with spines, no trace of their presence being found on any of the others. This view is strengthened by the fact that in young examples of *Astropecten polyacanthus* from Port Jackson of even smaller size than this, the supero-marginal plates are characteristically and powerfully armed; and the characters of the paxillæ and of the spinulation of the infero-marginal plates are also different.

"The type-specimen of MÜLLER and TROSCHEL's *Astropecten vappa* in the Berlin Museum is quite a young form and in bad condition. After the study of the large series of specimens from Australia, I have no hesitation whatever in regarding the name as a synonym of the present species, although the type-specimen could independently be scarcely held as available for specific recognition. I have also examined the types of GRUBE's *Astropecten chinensis* and *Astropecten ensifer* in Breslau, and I consider that both of them are with little doubt young stages of *Astropecten polyacanthus.*'

IVES simply mentions the name of *Astropecten armatus* without figure or any definite locality ['91, p. 211].

DE LORIOL ['93, p. 379] mentions "un exemplaire très typique dans lequel $R = 57$ mm." from the Bay of Amboina. It is mentioned by BELL from North-west Australia (32–34 fms.), Arafura and Banda Seas (12–15 fms.) and Macclesfield Bank (30–41 fms.) ['94, p. 394, 395, 396].

SLUITER mentions this species from the Red Sea and the Moluccas ['95, p. 53]: "*Astropecten polyacanthus* M. und TR. Ein Exemplar von dem Roten Meer und zwei von den Molukken (v. D. HUCHT), alle getrocknet."

LEIPOLDT refers to this species as follows ['95, p. 645]: "Ein Exemplar von Beilul (nordwestlich Assab) aus einer Tiefe von 5 m auf Algengrund; ein Exemplar von Assab.

"Das erstere Exemplar zeichnet sich dadurch aus, dass sämmtliche obere Randplatten einen grösseren Stachel tragen, während bei dem anderen, kleineren Exemplare die zweite obere Randplatte ihn nicht besitzt."

KŒHLER mentions this species from the Sunda Isles ['95, p. 387]: "Plusieurs échantillons de petite taille, mais très typiques." PFEFFER ['96, p. 47, *fide* Zool. Rec.] remarks on its colour. FARQUHAR ['98, p. 310] gives its distribution as follows: New Zealand, Australia, Admiralty and Fiji Islands, Japan, China, Ceylon, Andaman Islands, Bay of Bengal, Mauritius, Red Sea.

It is mentioned by LUDWIG from Zanzibar, although not represented in the collection reported on ['99, p. 538]:

"*Astropecten polyacanthus* MÜLLER und TROSCHEL, von Sansibar durch PERRIER (1875), von PFEFFER (1896), von den Amiranten und Seychellen durch BELL (1884)."

DÖDERLEIN mentions only *Astropecten armatus* MÜLLER & TROSCHEL as the synonym of this species, and makes the following remarks [: 02, p. 329]:

"Diese wohlbekannte, im wärmeren Indo-Pacific überall vorkommende Art fand ich nicht selten in der Tokio-Bai, Sagamibai, bei Tagawa und bei Kagoshima in geringerer Tiefe (ca. 20 m). Die vorliegenden Exemplare zeigen einen grossen Radius von 30–36 mm."

This species is mentioned by BELL [:03, p. 244] from "Zanzibar, 3–5 fath.," and by HUTTON [:04, p. 291] from New Zealand.

The HERDMANS mention it from the Gulf of Manaar [HERDMAN, HERD-MAN and BELL, :04, p. 143]:

"Astropecten polyacanthus, M. and T.

"Station XXVIII., Trincomalee, 7 to 14 fathoms; Station XXXVIII., off Galle, 9 to 22 fathoms; Station LX., outside Muttuvaratu Paar, 20 to 30 fathoms; Station XLIII., west of Kaltura, 22 fathoms; and Station LXIII., west of Periya Paar, 36 fathoms."

LUDWIG refers to this species as follows [:05, p. 76] :

"Ein Exemplar (R=49 mm.) von Suwa (Viti Levu, Fidschi-Inseln), am 13. December 1899 an der Küste aufgelesen.

"Die weit verbreitete und variabele Art is von den Fidschi-Inseln bereits bekannt."

FISHER gives a detailed description of this species [:06a, p. 1004]:

"Rays 5. R=47 mm.; r=10.5 mm. R=4.3 r. Breadth of ray at base, 12 mm.

"Arms rigid, very gently tapering to a bluntly pointed extremity. Sides of arms rather high, perpendicular. Disk of medium size. Paxillar surface somewhat inflated. Interbrachial arcs acute but rounded.

"Abactinal paxillar area is rather compact, the paxillæ being large and arranged in definite transverse rows, there being about 3 series to each superomarginal plate. Paxillæ are largest in the interradial areas, midway between centre and margin of disk, and also along median line of ray. Each paxilla consists of one or two central papilliform granules, surrounded by a radiating series of from 5 to 8 slightly longer ones, the whole crowning a rather long pedicel. On disk the largest paxillæ have upward to 5 central granules of unequal size, surrounded by 10–12 longer marginal ones, and occasionally one of the central granules is enlarged into a cylindrical pointed spinule. Less often one of the marginal spinelets is similarly enlarged (pl. II, fig. 1·a), but the two never occur on a single paxilla.

"Superomarginal plates, 22 in number from interradial line to extremity of ray, are much higher than broad and do not encroach conspicuously upon paxillar area. They form a rectangular edge to the abactinal surface, and the first plate is raised slightly above the level of the others. Except the second plate, and on one ray the third, each bears a perpendicular, stout, pointed, conical spine situated on the abactinal face, slightly nearer aboral than adoral margin. The spine on first plate is longest and stoutest (as long as the longest spine of any inferomarginal, 4.5 mm.), the series decreasing in length toward tip of ray. The second superomarginal plate, which does not bear a spine, is smaller than either the first or third and is crowded by them. Plates are covered with small cylindrical papilliform spinelets, which become stouter and squamiform toward base of spines. Wide fasciolar grooves between the plates (i.e., between the specialized, elevated exposed surfaces).

"The inferomarginals, which are broader than high, correspond to superomarginals in number and do not extend beyond them laterally. Each plate bears a transverse series of 3 stout and relatively long, tapering, slightly flattened, sharp-pointed spines, the upper longest; to which is added a fourth spinule at inner end of series on third to seventh plates. First 2 plates usually have only 2 shorter, widely spaced spines. Plates are covered with slender papilliform spinelets in the fasciolar grooves and at upper end, these becoming longer, strongly flattened, and bluntly rounded or chisel-shaped at tip, in the vicinity of spines, and on actinal surface generally.

"Adambulacral armature is in 3 series. The furrow series consists of 3 long, stout spinelets, the median longest, blunt, somewhat triangular in cross section at its base, the lateral spinelets flattened and truncate. Second series consists of 2 stouter, much flattened, truncate spinelets, the one nearest aboral margin being the larger. Third series consists of 3 blunt, flattened spinelets somewhat smaller than the furrow series, the median being usually slightly the longest of the 3 and most flattened. There is an odd spine, smaller and pointed, situated behind the third series. In all there are usually 9 spines on each adambulacral plate.

"Actinal interradial areas are so much reduced and are paved each with 4 small roundish plates, which bear spinelets very similar to those covering adjacent inferomarginals.

"Mouth plates are prominent, the armature, unfortunately, having been largely destroyed. The marginal spinelets are rather slender, slightly flattened, the innermost 2 or 3 of each plate forming at each mouth angle a horizontal fan of 4 to 6 teeth, of which the median are longest.

"Tube feet large, with an incipient conical sucker at end, easily distinguishable from the rest of the foot.

"Madreporic body is not visible superficially; hidden by the paxillæ.

"Colour in life: Paxillar area of distal half of arms vinaceous cinnamon; remainder of arms, and disk, fawn colour. The dorsal integument, largely hidden by the regular and ornate paxillæ, is bright vermilion, the colour being visible between the spinelets of the paxillæ. Spines of superomarginal plates, orange buff. Marginal plates, inferomarginal spines, and entire actinal surface, light buff pink. Colour in alcohol bleached yellowish.

"Locality: Station 4168, vicinity of Bird Island, 20 fathoms, coral, shells, and foraminifera. Bottom rough.

"Only a single specimen of this handsome species was secured, and that unfortunately, is not perfect. I have felt some misgivings in referring it to polyacanthus, having been obliged to depend wholly on the original description. In proportions the specimen agrees most nearly with MÜLLER and TROSCHEL's description of armatus (System der Asteriden, p. 71), from Japan, which SLADEN and others consider the same as polyacanthus, the type of which came from the Red Sea. The descriptions of these two species certainly differ in many points, and presumably the types do also, but in view of the opinion of SLADEN and PERRIER I have accepted the present name. I have given a full description, with figures, that there may be no mistaking the particular form referred to, whether the name be correct or not.

"This species may be readily distinguished from others of the genus inhabiting Hawaiian waters by the row of erect superomarginal spines, the second and sometimes the third superomarginal lacking the spine; and by the stout spines of the inferomarginals, arranged on each plate in a series of three or four.

"*Astropecten polyacanthus* has a wide distribution, extending from the Red Sea to Zanzibar, Ceylon, Hongkong, the coasts of China and Japan (Kobe, Yokohama), New Holland, Admiralty Islands, Fiji Islands, and Port Jackson, Australia. It is a shallow water species exclusively, ranging from 2 to 50 fathoms, the usual depths at which it is found being 25 fathoms and under. The station at Bird Island is the most eastern record for the Pacific, very materially extending the known range of the species."

BROWN reports this starfish from the Mergui Archipelago [:10, p. 30] :

" Locality.—I., Tavoy Island, 8 fathoms, shelly sand and mud.

" R=15.5   r=4 [mm.].   Breadth of arm at base=4.5.

" A single specimen apparently a young form, in which the spines of the ventral surface are scarcely developed. One supero-marginal on either side of the median interradial line bears a distinct tooth-like spine inclined slightly inwards.

" The specimen seems to be referable to this species.

" Recorded previously from the Red Sea, Zanzibar, and Mozambique, the Seychelles and Ceylon to Hong-Kong, the Fiji Islands, and Port Jackson."

SIMPSON and BROWN reports it from Portuguese East Africa [:10, p. 48] :

" In the specimens, which we have identified with this species, the first supero-marginal plate is strongly armed with a vertical spine. The second supero-marginal on each side of the median interradial line is devoid of a spine, while the remainder have the same structure as the first.

" SLADEN draws attention to a specimen in which this characteristic absence is not pronounced.

" This species is fairly abundant on the coast, and its distribution is interesting. It is a typically shallow-water species, as the following bathymetrical notes will show :—

" China, .Beach.

Japan, ...5 to 25 fathoms.

Admiralty Islands, ...16 to 25 fathoms.

Port Jackson, ...2 to 11 fathoms, and 6 to 15 fathoms.

Mergui Archipelago, ...8 fathoms.

East Africa, ...6 to 12 fathoms.

"Locality.—Station VI., Kero-Nyuni Bay, near Ras Pekawi.

"Previously recorded from—Japan ; China ; Fiji Is.; Admiralty Is.; Port Jackson ; Banda Sea ; Arafura Sea ; Macclesfield Bank ; N.W. Australia ; New Zealand ; Andaman Is.; Mergui Archipelago ; Ceylon ; Seychelles ; Amirante Is.; Mauritius ; Zanzibar ; and the Red Sea."

Kœhler mentions it from the collection of the Indian Museum as follows [:10, p. 41] :

"Iles Andaman. Profondeur 15–35 brasses. Quelques petits échantillons.

"Station 59. Côte S. de Ceylan, au large de Great-Basses. Profondeur 32 brasses. Quelques petits échantillons.

"Station 175.   8° 51′ 30″ Lat. N. 81° 11′ 52″ Long. E. Profondeur 28 brasses. Deux échantillons de taille moyenne.

"6° 01′ Lat. N. 81° 16′ Long. E. Profondeur 34 brasses. Un petit échantillon.

"No. 2234. Profondeur 26 basses et demi. Deux échantillons très incomplets.

"Dans certain individus, R ne dépasse pas 7 à 8 mm. mais les caractères de l'*A. polyacanthus* sont déjà bien indiqués, contrairement à ce que l'on observe d'habitude dans le genre *Astropecten* où les jeunes sont en général très mal caractérisés et fort difficiles à déterminer."

Kœhler also mentions it from the south-eastern Moluccas [:11, p. 266] :

"Dragage No. 4. 20 Mars 1908. Wammer. Profondeur 40 m. Un très petit échantillon.

"Dragage No. 6. 28 Mars 1908. Sungi Manumbai. Profondeur 23 m. Un échantillon de petite taille."

From the above review it may be seen that the following names are generally looked upon as synonyms of the present species, viz., *Astrop. armatus*, *Astrop. hystrix* and *Astrop. vappa ;* and to these we must add according to Sladen the *Astrop. chinensis* and *Astrop. ensifer* of Grube. I regret to say that I have not been able to see the original description of *Astrop. vappa*, but it may be presumed from what we know about the descriptions of some other species by DUJARDIN and HUPÉ that the description of *Astrop. vappa* by these authors is a close reproduction of the original of MÜLLER and TROSCHEL. The principal differences in the descriptions of these different forms may be brought out more clearly in a tabular form (p. 167).

It will be seen from an inspection of this table that the radial ratios as given by MÜLLER and TROSCHEL are very different for *Astrop. polyacanthus* and *armatus*; but as will be shown later on, this is subject to a good deal of variation in different specimens. The type specimen of *polyacanthus* appears to have been very large; and although I have not seen one of equal size, it appears from the table of radial ratios given below, that these are, in general, larger for large individuals than for small ones, so that the figures given by MÜLLER and TROSCHEL can not be looked upon as militating against the specific identity of the two species. Again, the descriptions of the adambulacral armature is very different for the two species; but it is perfectly possible that the specimen of *Astrop. armatus* examined by MULLER and TROSCHEL was not in a very good condition of preservation, so that the original arrangement of the furrow spines had become obscured. As to the descriptions of the armature of the inferomarginals for the two species in question, the differences appear to me too trifling to be considered as of specific value. Moreover, as *Astrop. polyacanthus* is one of the commonest species of the genus in Japan and there are no others that may possibly be confused with it, I have no hesitation in concluding that MULLER and TROSCHEL had this species before them in describing their *Astrop. armatus*,

Forms	R:r	r	R	M	Adambul. armature.	Superomarg.	Inferomarg.	Paxill. area, Madrep.
polyacanthus (Müller and Troschel)	6	9.6 mm. | 35 mm. (Sladen) 1 Fuss (London) 50-100 30-36 (Döderlein).		33	Inner spines forming a group of 3 in the form of an angle, the ... one largest and most prominent. Behind ... a group of smaller spines.	Spines long, conical, those at inter-brachial angle standing upright and forming a marginal series. The spines always absent on 2, often also on 3. ... rudiments of a second series are present on ... of the preceding. ... conical.	General armature scaly, spines all ..., 4-5 on ach plate, sub-terminal ones as large as the terminal.	Paxillar area much broader than supero-marginals.
hydria (Müller and Troschel)		6 Zoll		30	9-10 on each plate, in two rows transverse to length of arm; the largest spine forming the apex towards furrow.	Obliquely projecting, especially in middle of arm, separated from each other by deep furrows; higher than broad, with a conical spine at top. The two plates at arm angle higher than the others.	General armature scaly, with 4-5, not particularly flattened, pointed spines, which are larger towards margin. The outermost up to three lines long, and also relatively thick.	Paxillar area more than three times as broad as supero-marginals.
armatus (Müller and Troschel)	3 1/7	4 Zoll		20-22	Many long spines without any regular arrangement.	Higher than broad, with a round, upright, pointed spine, shorter than marginal spine of inferomarginal. The superomarginal spines decrease in length from base of arm outwards; always absent on 2, 3, and 4. plate.	General armature scaly, near ambulacral furrow a row of prominent spines which sometimes becomes double. On margin there are two large spines, of which the upper is the largest.	
trappa (after Dujardin and Hupé)	3 1/2	40 mm.		21	Two rows of spines, each consisting of 3.	As broad as high at middle of arm, with one small conical spine.	General armature consisting of small, isolated, projecting scales of spinous form, of which the outermost is longer than the marginal spine which is pointed and slightly flattened.	Paxillar area of arm very wide, more than 3 times as broad as the marginals in middle of arm. Madreporite very near the margin.

and as it accords very well with the description of *Astrop.* *polyacanthus* we may follow previous authors in uniting the two species. As for *Astrop. hystrix,* the question turns on the presence or absence of unarmed superomarginals next to the prominent interbrachial pair; but this has already been shown by previous observers to be subject to variation, and we shall see more of it later.

Basing my observations on a goodly number of specimens from different localities, I may give the following description of this species.

Radial ratio.—This is best seen from the following table.

Specim.	r mm.	R mm.	R : r	MS	Locality
1	3	6	2	. . . .	Enoura (Suruga).
2	3.5	8.3	2.4	. . . .	,,
3	3.5	9	2.6	10	
4	4	9	2.3	. . . .	
5	4	10	2.5	. . . .	
6	4.2	11	2.6	. . . .	
7	4.6	11	2.4	. . . .	
8	5	13.5	2.7	. . . .	
9	5	13.5	2.7	. . . .	,,
10	5.5	16	3	. . . .	Misaki.
11	6	13.5	2.2	. . . .	Enoura.
12	6	16	2.7	13	,,
13	6.5	14	2.1	. . . .	,,
14	7	23.5	3.4	. . . .	Misaki.
15	8	23	2.9	16	Enoura.
16	8	31	3.9	. . . .	?
17	10	37	3.7	. . . .	Wakanoura.
18	10	43	4.3	. . . .	Tosa.

Specim.	r mm.	R mm.	R : r	MS	Locality
19	10.5	44.5	4.2	....	Bay of Tokyo ?
20	11	37	3.4	20	Misaki.
21	11	41	3.7	....	,,
22	11	46.5	4.2	....	,,
23	11	47	4.3	....	Satsuma.
24	11.5	46	4	....	Tosa
25	11.5	55	4.8	27	,,
26	12	45	3.8	....	Misaki.
27	12	50	4.1	23	Wakanoura.
28	12	52	4.3	....	Iyo.
29	12.5	52	4.1	....	Hôjô.
30	12.5	53	3.8	....	Misaki.
31	13	44.5	3.3	22	,,
32	14	53	3.8	....	,,
33	14	57	4	25	,,
34	14	59	4.2	....	,,
35	14	59	4.2	....	Ohama (Wakasa).
36	14	68	4.9	....	Iyo.
37	14	71.5	5.1	23	Tosa.
38	15	56.5	3.7	....	Satsuma.
39	15.5	71	4.6	....	Misaki.
40	17	76	4.4	....	Tosa.
41	17	82	5	....	Misaki.
42	18	76	4.2	....	Hyuga.
43	18.5	92	5	28	Nagasaki.
44	19	72	3.8	....	Kagoshima Bay.
45	19	97	5.1	29	Nagasaki.

As in *Astrop. scoparius* the radial ratio is, in general, greater
for large specimens than for small ones, and although in

certain cases a tendency to hold by the same ratio may be detected in specimens from the same locality, it appears that this is more a matter of individual variation than a geographical habitus.

Superomarginals.—As may be seen from the foregoing table, the number of superomarginals is not necessarily proportional to the radial ratio, i.e., individuals with relatively large R do not always have more superomarginals. The most prominent feature of this species is the presence of a large, prominent, conical spine on the abactinal surface of the superomarginals (Pl. III, fig. 42, 47, 48). The series begins with the first superomarginal, which is usually more prominent than the others and bears an especially large spine. The next two or three are usually without spines, but it appears from the statements of previous observers that these also may be armed with spines, although I have not seen a specimen, in which the series of the superomarginal spines is continuous throughout. The following variations have, however, been observed. These formulæ are written so that the figures show the ordinal numbers of the unarmed superomarginals, the first set representing the madreporic interradius, and the following the other interradii in counter-clockwise order, the dash the interradial line.

Specim. A   3. 2.—2. 3., 3. 2.—2. 3., 4. 3. 2.—2. 3., 3. 2.,—2. 3.,
                 3. 2.—2. 3.

Specim. B   4. 3. 2.—2. 3. 4., 4. 3. 2.—2. 3. 4., 4. 3. 2.—2. 3. 4.,
                 4. 3. 2.—2. 3., 4. 3. 2.—2. 3. 4.

Specim. C   4. 3. 2.—2. 3., 3. 2.—2. 3. 4., 4. 3. 2.—2. 3.,
                 3. 2.—2. 3. 4, 4. 3. 2.—2. 3. 4.

Specim. D   3. 2.—2. 3. 4., 4. 3. 2.—2. 3. 4., 3. 2.—2. 3. 4.,
                 4. 3. 2.—2. 3. 4., 4. 3. 2.—2. 3.

Specim. E   3. 2.—2. 3., 3. 2.—2. 3., 3. 2.—2. 3., 3. 2.—2. 3.,
                 2.—2. 3.

Specim. F 4. 3. 2.—2. 3. 4., 3. 2.—2. 3. 4., 4. 3. 2.—2. 3. 4.,.
3. 2.—2. 3., 3. 2.—2. 3. 4. 6.

These formulæ represent fairly well the modal condition of the superomarginal spines of this species; but in a few extreme cases the number of unarmed spines next the first superomarginal may amount to as many as five or six. Again in specim. F of the examples above cited the fourth superomarginal on the last side of the last interradius had an exceedingly small spine barely distinguishable from the neighbouring granules.

In the basal portion of the arms, there is generally a small naked zone on the inner side of the abactinal surface of the superomarginals (Pl. III, fig. 47). The proximal and distal faces of the fasciolar grooves are covered with capillary spinelets, and the remaining portions are covered with granules, which are large and conical immediately on the outer side of the large spine and gradually become slender and cylindrical towards the margins and finally pass over into the capillary spinelets of the fasciolar grooves.

The superomarginals are higher than broad, and stand up erect, i.e. their external surfaces are nearly perpendicular. The abactinal surface is at right angles to the margin at the base of the arms, but becomes inclined more distally, the inclination sometimes amounting to about $30°$ at the middle of the arm.

Inferomarginals.—The inferomarginals are very much broader than long, and make an angle of about $30°$ with the ambulacral furrow (Pl. III, fig. 44, 46). The general armature may be described as coarse and spiny. The spinelets along the margins are subcapillary, but they form a less regular row than in *Astrop. scoparius* and are less fine. An irregular row of 4-6 large,

flattened spines is always present on each inferomarginal, the
spines becoming longer toward the margin, the last, usually the
largest, being found very close to the abactinal margin of the
plate. There may sometimes be detected a second row
of smaller spines, but more commonly these are not developed
well enough to show distinctly. The outer half of the fasciolar
grooves between the inferomarginals is covered with capillary
spinelets similar to those of the superomarginals, and continuous
with them.

Adambulacrals.—The adambulacrals are about one-third as
broad as the inferomarginals, and there are usually two of them
to each inferomarginal, but in a few places in each arm there
may be five of them to two inferomarginals. The spines may be
represented by the following formulæ.

Specim. 1 1 : 2 : 2, which means that there is an apical spine,
followed by a pair, and then a second pair. This
was found in some small specimens with r=5–8 mm.,
and represents probably the simplest case of adambulacral armature in this species. The following
variations have been observed.

Specim. 2 1 : 2 : 2 : 3 or 1 : 2 : 2 : 2 : 1
,, 3 1 : 2 : 2 : 4
,, 4 1 : 2 : 2 : 5
,, 5 1 : 2 : 2 : 2 : 3
,, 6 1 : 2 : 3 : 2 : 3
,, 7 1 : 2 : 2 : 2 : 2

the total number of spines amounting up to ten or eleven (Pl. III,
fig. 46). The first or apical spine is always much longer than
the rest; of the first pair the two may or may not be of the

same length; of the second pair the abcentral spine is almost invariably much the larger of the two, and in a few cases there were three in a row instead of two, when the middle one was the largest; there may be a third pair and an odd spine or other combinations which are shown by the preceding formulæ.

Mouth-plates.—When denuded of the spines, the mouth-plates are very prominent, and are more pointed at the distal end (Pl. III, fig. 44, 45). There is a principal crest running through the entire length, and an accessory ridge running up nearly half as far from the proximal end. There are about eight or nine spines on the principal crest, increasing in length towards the mouth, and about six on the accessory ridge, which are also longer towards the mouth.

Ventrolaterals.—These are very small, and there are 1–2 on either side of the interradial line (Pl. III, fig. 44). They are covered with spines similar to the smaller spines of the inferomarginals.

Paxillæ.—The paxillar area at the middle of the arms is more than twice as broad as the superomarginals of one side. There is an area of small paxillæ at the centre of the disk, which, however, is not so very prominent or so large as in *Astrop. scoparius* (Pl. III, fig. 42). The coronal spinelets are less numerous and more divergent than in *Astrop. scoparius*, there being some 27 or so in one of the largest paxillæ, which are usually found at about the middle of the base of the arm (Pl. III, fig. 49). The most striking difference in the paxillæ of this species, as compared with *Astrop. scoparius*, is, however, the fact that there is not such a well marked difference between the centrals and peripherals. All the spinelets look very similar to one another, the more central ones are only shorter and blunter than the more peripheral ones. In the smaller paxillæ found along the inner margin of the superomarginals there are no centrals and the coronal spinelets, of which there may be five or more, simply diverge from the apex of the pedicel.

Madreporite.—When exposed this is tolerably large. It is, however, more frequently hidden from view for the greater part by the surrounding paxillæ; but even when it is projecting, it is not so conspicuous as in *Astrop. scoparius*, since the exposed surface is covered with granules of irregular form (Pl. III, fig. 50). The margin is crenate, and the surface covered with irregularly radiating furrows. It is nearer the margin than the centre of the disk.

Terminal plate.—This is quite conspicuous and bilobed at the apex.

Locality.—My specimens are from the following localities: Akune (Satsuma), Kagoshima Bay, Hososhima Bay (Hyuga), Nagasaki, Kashiwa Island (off the coast of Tosa), Gogo Island (off the coast of Iyo), Ohama (Wakasa, Japan Sea), Wakanoura (Kishu), Tsu (Gulf of Isé), Enoura (Suruga), Misaki, Awa. According to a note in my possession the ambulacral feet of this species are red. The whole abactinal side of the body is of a murky brownish colour.

This species appears to live on sandy or muddy bottoms and is found at greater depths than *Astrop. scoparius*. I have however no definite data to fix its bathymetrical range. The records of the Challenger are perhaps the fullest on this point.

Specimens in S.C., H.N.S,, H.N.S.W., H.N.S.H., F.B., I.H.S., I. M., O.

Astropecten ludwigi DE LORIOL.

(Pl. IV, figs. 70–79.)

This species appears to have been first described from Japan by IVES, who mentions it under the name of *Astrop. japonicus* MÜLLER & TROSCHEL and gives drawings of various parts [IVES, '91, p. 211]. According to DE LORIOL it is again mentioned by the same writer under the same name in the following year ['92, p. 2]. Its first description as a distinct species is however due to DE LORIOL ['99, 21]:

" Diamètre total .... 117 mm.

" Diamètre des bras à la base .... 19 mm.

" R = 60 mm.   r = 17 mm.   R = 3½ r.

" Disque assez grand, très déprimé sur sa face dorsale qui est uniformément couverte de paxilles étoilées, relativement petites, couronnées d'une houppe de granules cylindriques, allongés, au nombre de 8 à 12 au pourtour, avec deux ou trois plus courts au centre.

" Cinq bras larges à la base et rapidement effilés. Plaques marginales dorsales au nombre de 35 de chaque côté avec une interradiale cunéiforme impaire, notablement plus larges que hautes, convexes, et rapidement arquées pour former le bord ; elles sont couvertes de granules spiniformes fins et serrés, et très finement ciliées sur leur pourtour de manière à couvrir les larges intervalles qui les séparent dans le squelette ; vers le milieu de la plaque se trouve une série transverse de granules plus gros que les autres qui, sur le côté des bras, deviennent de petits piquants très courts, mais bien distincts, au nombres de 2 à 4, formant parfois une série irrégulière sur le bord distal. Il n'y a aucun piquant dorsal proprement dit. L'aire paxillaire est large, elle atteint, à la base du bras, presque quatre fois la largeur de l'une des plaques marginales. Les paxilles ne sont pas très serrées, elles forment des séries transverses assez régulières, on en compte une quinzaine par série à la base des bras ; elles sont semblables à celles du disque avec un nombre de spinules un peu plus faible.

" Plaques marginales ventrales beaucoup plus larges que hautes ; elles ne débordent pas les marginales dorsales ; l'impaire, au fond de l'angle interra-

dial, est fortement cunéiforme, de même que ses deux voisines. Leur revêtement se compose de petites écailles peu serrées, redressées, un peu spatuli-
formes, arrondies au sommet ; sur leur bord distal se trouve une série de
petits piquants aigus, dont les 3 ou 4 premiers sont fort petits, les suivants
s'allongent rapidement et la frange marginale se compose, pour chaque
plaque, de quatre à cinq piquants relativement très courts, quoique bien plus
longs que les premiers, superposés, aciculés, à peu près egaux entre eux.
Dans l'exemplaire décrit, ces piquants marginaux sont presque tous appliqués
contre le bord du bras, ce qui, ajouté à leur brièveté relative, fait que la
frange marginale est à peine apparente ; dans le fond des arcs interbrachiaux
les piquants marginaux sont encore un peu plus courts, aplatis, élargis à
leur base et rapidement acuminés au sommet. La plaque terminale ou
ocellaire est peu développée et largement sillonée.

" Les plaques adambulacraires sont allongées, quadrangulaires, elles portent, dans le sillon, une série interne de trois à quatre piquants assez longs, fins,
aplatis, tronqués à l'extrémité, et sensiblement egaux entre eux, en dehors une
seconde série de trois piquants divergeants, un peu plus courts, et, enfin, une
double série de trois ou quatre piquants bien plus petits, cylindriques, très
serrés, qui tendent à se confondre avec le revêtement des plaques marginales,
tout en restant cependant bien distincts, en étant séparés par un léger sillon.

" Plaques buccales relativement courtes, entièrement couvertes de petits
piquants aplatis et entourées d'une frange de piquants qui, d'abord très
petit, s'allongent graduellement en s'aplatissant et forment, dans le péristome,
un éventail terminal de cinq ou six piquants allongés, aplatis et tronqués.
Celle des plaques adambulacraires qui touche la plaque buccale, de chaque
côté, a l'apparence d'une crête étroite surmontée d'une double série très
regulière de piquants aplatis, tronqués, tous égaux entre eux, chaque série en
a 15 ou 16. Le revêtement de la plaque adambulacraire voisine n'est qu'en
partie semblable, dans le sillon les piquants reprennent leurs caractères
normaux. M. SLADEN (Challenger Asteroidea, p. 210, pl. XXXV, fig. 1–2,
pl. XXXVIII, fig. 10–12) signale une disposition tout à fait semblable
dans l'*Astropecten pontoporæus*. En arrière des plaques buccales se trouvent
quatorze petites plaques ventrales disposées en chevron de 7 de chaque côté ;

les plus rapprochées de l'angle sont étroites et allongées, les postérieures sont presques arrondies et diminuent graduellement de grosseur ; elles sont couvertes de petits piquants égaux, courts et serrés.

" Plaque madréporique très petite, très finement sillonnée, située tout près du bord du disque et presque cachée par les paxilles.

" *Rapports et différences.* L'exemplaire décrit correspond parfaitement à l'espèce qui a èté figuré par M. Ives (*loc. cit.*) sous le nom d'*Astropecten japonicus*, sans y ajouter une description ; il en presente fort exactement tous les caractères, seulement les deux petites séries de piquants externes, sur les plaques adambulacraires, ne sont pas indiquées sur la figurè, ce qui peut provenir d'une imperfection de dessin, car elles tendent à se confondre avec les piquants des plaques marginales, par contre on discerne, bien que grossièrement indiquées, les doubles séries des piquants, si particulières, que portent les plaques adambulacraires adjacentes aux plaques buccales, de même que la sériè transverse de gros granules allongés en forme de piquants que portent les plaques marginales dorsales. D'un autre côté cette figure donnée par M. Ives ne me semble pas correspondre à la diagnose de l'*Astr. japonicus*, donnée par MÜLLER et TROSCHEL. En effet, d'après cette description du type, il n'aurait que 30 plaques marginales dorsales, et cinq piquants seulement pour chaque plaque adambulacraire ; sur chaque plaque marginale ventrale un grand piquant précédé, du côté ventral, par trois autres qui n'ont que le tiers de sa longueur, ses plaques marginales dorsales sont aussi larges que hautes, et ne ' portent que rarement un petit tubercule mobile sur le bord externe,' enfin, au milieu du bras, l'aire paxillaire n'a que la largeur d'une plaque marginale, le diamètrè total serait de 65 à 70 mm. M. SLADEN (The Asteroidea and Echinoidea of the Korean Seas, Journal of the Linnæan Society, Zoology, vol. XIV, p. 427) a décrit avec plus de détail un petit exemplaire dans lequel R=11.25 mm., et r=4 mm., les caractères qu'il énumère sont conformes à la courte description de MÜLLER et TROSCHEL, mais pas à la figure donnée par M. Ives, il mentionne aussi l'étroitesse de l'aire paxillaire des bras (*very little*, dit-il) ainsi que la grandeur des paxilles, si serrés qu'elles n'ont plus l'apparence étoilée. Ni MÜLLER et TROSCHEL, ni M. SLADEN, n'ont mentionné le revêtement particulièr des plaques adambulacraires adjacentes aux plaques buccales. Si cette particularité s'était montrée sur leurs types de l'*Astr. japonicus,* elle n'aurait pas manqué d'attirer leur attention ; ils ne parlent pas non plus de la série particulièrement nombreuse des petites plaques ventrales. Malgré mes recherches je n'ai trouvé ancune espèce décrite dont celle-ci pourrait être utilement rapprochée, elle ne saurait être confondue avec *l'Astr. scoparius* VAL. commun dans les mers du Japon. L'*Astropecten pontoporœus,* SLADEN, dont il a été question plus haut, se distingue sans peine par l'armature de ses plaques adambulacraires, en particulier, et par le revêtement de ses plaques marginales ventrales qui sont tout différents.

"Localité. Tago (Japon). Ma collection."

De LORIOL's description. is accompanied by one figure of the aboral aspect of the animal. From the statement of DÖDERLEIN quoted below, it is very probable that the name of the locality mentioned above is a misprint of "Tango," although there is a place named "Tago-no-ura." The former is on the Japan Sea, the latter in the Bay of Suruga.

DÖDERLEIN gives a succinct description of this species [: 02, p. 328] :

"*Astropecten Ludwigi* DE LORIOL 1899.

"Syn. *Astropecten japonicus* IVES.

"Die Arme sind gleichmässig verjüngt bis zur Spitze. R : r = 3.3–4.

"Dorsomarginalplatten, mindestens von der 6. Platte an, mit je einer Querreihe von 3–8 kurzen, mitunter etwas verkümmerten Stachelchen, von denen meist eines auf dem Innenrande der Platte steht.

"Ventromarginalplatten mit schuppenförmigen Granulä, einer Querreihe von 3–6 kürzeren und am Rande mit 3–5 längeren spitzen Stachelchen versehen.

"Adambulacralplatten mit einer inneren Längsreihe von 3–5 gleich oder ungleich langen, schlanken Furchenstacheln ; nach aussen davon ein dichter Büschel kürzerer Stacheln, von dem sich mitunter eine mittlere Längsreihe von Furchenstacheln deutlich ablöst.

"Jederseits sind 5–9 Ventrolateralplatten vorhanden.

"Beim kleinsten und grössten der mir vorliegenden Exemplare beträgt R = 20 u. 90 mm, r = 6 u. 23 mm, die Zahl der Marginalplatten 19 u. 43.

" Die Farbe ist rothbraun.

" Ich erhielt die Art zahlreich aus der Tokio- und Sagamibai und fischte sie bei der Provinz Tango in einer Tiefe von 10 m auf schlammigem Boden, zugleich mit *Temnopleurus toreumaticus ;* kleinere Exemplare erhielt ich in der Sagamibai aus ca. 100 m Tiefe.

" Die Art ist von DE LORIOL nach einem, wahrscheinlich von mir selbst gesammelten, nicht sonderlich gut conservierten Exemplare beschrieben worden; unpaare Marginalplatten konnte ich bei keinem meiner zahlreichen Exemplare beobachten."

FISHER has recently formed a new genus, which he calls *Ctenopleura*, to receive the present species and *Ct. astropectinides*, a new species from the Philippines. The new genus is characterised as follows [FISHER, : 13, p. 608]:

" Allied to *Astropecten*, but differing in having the gonads in a crowded series parallel to the marginal plates, and extending about a third the length of ray ; inferomarginal plates with a lateral, oblique, compact comb of 3 to 5, usually 4, slender appressed spines, closely resembling the lateral comb of *Persephonaster*, and in addition 1 to 5 flattened appressed spines on the actinal surface ; adambulacral plates with usually 4 or 5 furrow spines (or, on the second and third plates, sometimes 6 or 7) instead of 3, the usual number in *Astropecten* ; subambulacral spines small, none .enlarged ; often a fasciculate subambulacral pedicellaria is present ; Polian vesicles 5. Other characters as in *Astropecten*.

" Type of the genus.—*Ctenopleura astropectinides*, new species.

" This genus includes also *Astropecten ludwigi* DE LORIOL, of Japan, in which the gonads are arranged in series extending about a third the length of the ray. In *Astropecten* the gonads form a single tuft on either side of the interbrachial septum. The arrangement of the gonads in series will separate *Ctenopleura* from *Leptychaster, Bathybiaster, Psilaster, Blakiaster, Astromesites, Ctenophoraster, Persephonaster, Tritonaster,* and *Patagiaster*, while the very restricted development of the actinal intermediate plates will distinguish it from other Astropectinidæ having serially arranged gonads, such as *Anthosticte, Tethyaster, Thrissacanthias, Dipsacaster,* and *Plutonaster*. The armature of the marginals will separate *Ctenopleura* from *Lonchotaster* and

Ripaster in which the gonads are not described."

This is one of the best characterised species of the genus on our coasts and is not uncommon in certain localities at certain depths, its most characteristic feature being the presence of a series of 3–9 or 10 flattened spines on the superomarginals. As to the unpaired interradial marginals mentioned by DE LORIOL, my own observations agree with those of DÖDERLEIN, and in none of the specimens that have passed through my hands were they present. A careful examination of the figure given by DE *Loriol* also fails to bring them to light. It is true that in one or two of the interradii, a · wedge shaped superomarginal is represented in the figure, but its position is easily seen not to be strictly interradial. Again with regard to the armature of the first adambulacrals, which is emphasized by DE *Loriol* as a very characteristic feature of the species, it must be remarked that a similar comb-like arrangement of the spines can be observed in several other species of the genus. It may, however, be admitted that it is somewhat more conspicuous in the present species, owing to the relative thickness of the spines concerned, and the strict regularity of their arrangement.

Radial ratio.—The radial ratio of this species appears to be less subject to variation than in those described above, although it must be borne in mind that the range of the variation may be found to be wider with the study of more numerous specimens. The following table gives a clearer idea of the radial ratio than any general descriptions can.

Specim.	r mm.	R mm.	R : r	MS	Locality.
1	6.5	20.5	3.2	20	Misaki.
2	11	35	3.2	22	,,
3	14	48	3.4	25	,,
4	14	48	3.4	28	,,
5	15	58	3.9	28	,,
6	16	63	3.9	. . . .	,,
7	16.5	52	3.2	27	,,
8	16.5	57	3.5	30	,,
9	17	60	3.5	29	,,
10	18	57	3.2	. . . .	,,
11	18	69	3.9	. . . .	,,
12	19	68	3.6	30	,,
13	19.5	66	3.3	29	,,
14	20	62	3.1	28	,,
15	20	62.5	3.1	30	,,
16	21	82	3.9	38	,,
17	22	76	3.5	35	Namerikawa (Etchu).
18	22	76	3.5	33	,,
19	22	81.5	3.7	39	,,
20	22	88	4	38	Ôita (Bungo).
21	22.5	102	4.5	41	Misaki.
22	23	87	3.8	35	,,
23	24	100	4	39	Ôita (Bungo).
24	25	101	4	40	Hôjô (Awa).
25	26.5	91	3.4	39	Namerikawa (Etchu).
26	27	86	3.2	34	Misaki.

Superomarginals.—As above remarked the armature of the superomarginals is perhaps the most striking feature of the species. There is on each superomarginal a series of flattened conspicuous spines, the number of which may vary from as few as three or four up to as many as nine or ten. When there are only three or four, it frequently happens that two or three of these spines stand close together on the outer part of the plate, and one somewhat apart on the inner part. When they are more numerous, however, they usually form a continuous series, leaving only the outermost part of the plate free, as represented in fig. 76. It may also be added that the individual spines are larger when there are fewer than when they are more numerous.

The series of conspicuous spines above described divides, in a general way, the surface of a superomarginal into two portions which show a difference in their armature. In the abcentral part of each superomarginal the armature consists of very small, thickly set, cylindrical spinelets, while in the adcentral portion, which is the larger one, the armature immediately contiguous to the series of conspicuous spines may best be described as scaly, and the scaly form gradually passes on to a cylindrical towards the margin. On the outer face of the plate the conico-cylindrical spinelets are somewhat better spaced. The fasciolar grooves are thickly covered with capillary spinelets. It must finally be remarked that in the interbrachial angles the conspicuous spines are less numerous.

At the interbrachial angles the superomarginals are higher than broad, but nearly half as high as broad at the middle of the arms, where they are also half as long as broad. On the outer side the superomarginal plates are perfectly rounded, so that we can not well speak of an external edge. The number of superomarginals is given in the foregoing table.

Inferomarginals.—The inferomarginals are much broader than long, and bear each a transverse series of spines, which are very much flattened and almost spatulate in the inner, but much longer and slender in the outer part (Pl. IV, fig. 75). The whole series may contain from 5 to 10 spines in all, of which the outermost 3–5 are long and slender. These spines, but especially the marginal longer ones, are attached close to the distal border of the inferomarginals, and are generally closely appressed to the next distal plate, so that the spiny character of the inferomarginals is not so apparent as in some other species, when the animal is viewed as a whole. This series of spines is more distinct than in *Astrop. scoparius* and *Astrop. polyacanthus*, owing to the fact that they are more regularly arranged and better differentiated from the general armature of the inferomarginals. The general surface of the latter is covered with comparatvely large and flat scales. There is a series of small cylindrical spinelets along the margins of each plate.

Adambulacrals.—In the greater part of the arms there are three adambulacrals to every two inferomarginals. The first ad-ambulacrals are very narrow and the surface of each plate is for the greater part divided into two ridges separated from each other by a narrow longitudinal groove (Pl. IV, fig. 74). The spines, which are of equal length and short, are arranged regular-ly in two rows corresponding to the two ridges, and only in the outermost portion, where the crest is single, is there a slight tendency to an irregular arrangement. There are from a dozen to about twenty-five spines in each series, according to the size of the specimen (Pl. IV, fig. 73).

From about the fourth adambulacrals on there are, on each plate, three thick, somewhat flattened, prismatic truncated spines of nearly equal length on the inner border, then follow, separated by a rather conspicuous space, two or three spines arranged almost in a line parallel to the furrow, also flattened and more squarely truncated at the top, slightly shorter than the first three. Then follows a third irregular group of half a dozen to about fifteen spines of different sizes and shapes, the inner ones being like the foregoing spines, while the outer ones are more like the inner marginal spinelets of the inferomarginals. When there are only two spines in the second group, they form a horse-shoe shape with the spines of the first group, and enclose a central space, on the outside of which the remaining spines form a closely set group.

Mouth-plates.—The mouth-plates, though not particularly larger than in *Astrop. scoparius* and *Astrop. polyacanthus* in denuded specimens, bear a larger number of spines, and are consequently more conspicuous. As usual, there is a main ridge running through the whole length of the plate, and a secondary ridge which is nearly half as long. The secondary ridge runs along the ambulacral furrow and bears about five flattened spines, of which the innermost is conspicuously larger (Pl. IV, fig. 73). On the main ridge, there are at least two rows of spines, in the inner of which the spines are stout and more or less flattened, and in larger specimens tend to crowd out one another, in consequence of which they assume a more or less irregular arrangement. The spines at the central end of this row is very thick and longer than the others, and together with the innermost spine of the secondary ridge, constitute the oral armature. There may be as many as fifteen or sixteen spines in this row, of which a few at the peripheral end are conspicuously larger in smaller specimens. The spines of the outer row are smaller than those of the inner and are 8–15 in number, according to the size of the specimen. The third row, when present, consists only of about a dozen very

small spines lying between the outer of the two above mentioned
rows on the one hand and the secondary ridge and the first
adambulacral plate on the other.

Ventrolaterals.—These are very small, as in *Astrop. scoparius*
and *Astrop. polyacanthus*, but they are more numerous, there being
from twelve to fifteen of them in one interradius (Pl. IV, fig. 72).
They are not strictly paired on either side of the interradial line,
but more or less alternate, so that there is usually one wedged in
at the distal end of the mouth-plate.

Paxillæ.—The paxillar area at about the middle of the arms
is about two and a half times as wide as the superomarginals of
one side. The paxillæ are tolerably large, and in exceptionally
large ones there may be as many as seventeen or eighteen peripheral coronal spinelets and about fifteen centrals. The latter are
similar in shape to the peripherals, but shorter (Pl. IV, fig. 77).
The area of small paxillæ at the centre of the disk is very small
and not very conspicuous.

Madreporite.—The madreporite is comparatively small, and
nearly hemispherial in shape (Pl. IV, fig. 79). It is half as far
removed from the inner margin of the superomarginals as from
the centre of the disk. It is usually more or less exposed, and
the exposed surface is covered with granules, which are generally
minute in smaller specimens and are therefore not apparent, but
which may grow tolerably large in some large specimens, and
show a more or less definitely radial arrangement (Pl. IV, fig.
78). In a few specimens observed by me the madreporite was
nearly midway between the margin and the centre of the disk.

Terminal plate.—The terminal plates are large and distinctly
bilobed.

Locality.—As already stated this species is found at greater depth than *Astrop. scoparius* or *Astrop. polyacanthus*, the records in my possession giving the bathymetrical range as 30–480 metres. DÖDERLEIN however obtained specimens from a depth of 10 metres. It appears to be less common than the species described in the preceding pages, but is not at all uncommon. I have specimens from the following localities : Misaki, Hôjô (Awa), Ôita (Kyūshu), Namerikawa (Japan Sea), Iyo (Inland Sea).

Specimens in S. C., H. N. S., H. N. S. H., F. B., O.

Astropecten kagoshimensis DE LORIOL.

(Pl. IV, Figs. 62–69.)

This species was first described by DE LORIOL as follows ['99, p. 23] :

" Diamètre total .... 80 mm. à 85 mm.

" Largeur des bras à leur base .... 10 mm.

" R = 43 mm. (maximum) r = 10 mm. R = 4.3 r.

" Disque étroit, très aplati sur la face dorsale, entièrement couvert de paxilles très fines et très serrées, particulièrement au centre, qui masquent entièrement la plaque madréporique ; elles sont couronnées, au pourtour, par une houppe de 8 à 11 granules allongés, avec un à cinq granules centraux.

" Cinq bras très plats, assez larges à la base, et graduellement effilés, avec des angles interbrachiaux aigus. Sur la face dorsale ils sont bordés, de chaque côté, par une série de plaques marginales, au nombre de 28 à 29, avec une impaire au fond de l'angle interradial ; elles présentent peu de surface sur la face dorsale, mais s'arquent promptement pour former le bord latéral ; elles sont largement séparées dans le squelette, notablement plus larges que hautes, et, relativement, de faibles dimensions. Les granules qui les recouvrent sur leur face dorsale ont la forme de petits cylindres courts et arrondis au sommet ; sur leur face latérale ils sont plus aigus et plus écartés. Ces plaques portent, sur le bord externe du bras, et sur leur bord distal, un petit piquant très fin, aigu, dont la longueur, de 1 à 2 mm. environ sur les

premières, à la base du bras, diminue graduellement ; les quatre à six dernières plaques, vers l'extrémité des bras, en sont tout à fait dépourvues,
mais ils ne manquent jamais sur les plaques du fond de l'arc interbrachial. La
plaque terminale ocellaire est grande, renflée et divisée par un sillon. L'aire
paxillaire est large, trois ou quatre fois autant que l'une des séries de plaques
marginales et presque de niveau avec elles. Les paxilles forment des séries
transverses presque rectilignes, vers la base on en compte 12 par série dans
le petit exemplaire ; elles sont semblables à celles du disque, mais, ordinairement, avec un peu moins de granules dans la houppe. En général la surface
paxillaire paraît comme uniformément granuleuse, et, sur le disque, en
particulière, il est difficile d'observer la disposition étoilée des granules
paxillaires.

"Les plaques marginales ventrales ne débordent pas les dorsales ; elles
sont larges, bien plus larges que hautes, et également au nombre de 29 avec
une impaire dans l'angle interradial ; elles sont comme ciliées sur leur
pourtour par des spinules très fines, d'autres spinules, un peu plus robuste,
mais très écartées, couvrent la surface ; une série de cinq à six piquants très
fins, aigus, augmentant graduellement de longueur, occupe leur bord distal,
le dernier, le piquant marginal, au moins deux fois aussi long que le pénul-
tième est un peu plus robuste, tout en demeurant très fin, arqué et aigu ;
sa longueur est de 3 mm. Les plaques adambulacraires portent 6 piquants,
très fins, assez longs, cylindriques, obtus, à peu près égaux entre eux, une
série de trois, dont le médian est un peu plus long, se trouve dans le sillon,
et les trois autre forment une série externe contiguë aux plaques marginales.

"Plaques buccales étroites et allongées ; elles portent 5 ou 6 piquants
relativement longs, robustes, et obtus, avec une série de petits piquants très
courts au pourtour, un éventail terminal de 5 ou 6 longs piquants s'avance
dans le péristome. En arrière des plaques buccales se trouve quatre petites
plaques arrondies assez renflées, disposées en arc très peu cintré, et couvertes
d'un faisceau de très petits piquants.

"*Rapports et différences.* Je ne connais pas d'espèces avec lesquelles
celle-ci pourrait être confondue, elle a un aspect assez particulier dû à la
finesse de ses piquants, en général ; l'absence constante de piquants sur les

plaques marginales dorsales, vers l'extrémité des bras, est un caractère qui ne manque pas de valeur. Dans l'*Astr. javanicus*, LÜTKEN, le piquant des plaques marginales dorsales est fixé sur leur bord interne et non sur leur bord externe, les piquants marginaux sont larges, plats et tronqués, le revêtement des plaques marginales ventrales est différent; il en est de même pour l'*Astr. Orsinii*, LEIPOLDT, dans lequel le piquant marginal est très large, le piquant de ses plaques marginales dorsales se trouve sur leur milieu, de plus sa forme et ses proportions sont différentes, et il y a trois séries de piquants adambulacraires. L'*Astr. tamilicus*, DÖDERLEIN, a un piquant *tronqué* sur ses plaques marginales dorsales, l'aire paxillaire de ses bras est plus étroite, le revêtement des plaques marginales ventrales est différent, de même que le piquant marginal, enfin la série externe des piquants adambulacraires n'en a que deux dont l'un est plus épais que l'autre. Dans l'*Astr. scoparius*, il n'y a pas de piquants sur les cinq premières plaques marginales dorsales à partir de l'angle interradial, mais il s'en trouve jusqu'à l'extrémité des bras, les plaques marginales dorsales sont plus larges, le revêtement des plaques marginales ventrales est differents, les piquants marginaux sont plus robustes et il y a trois séries de piquants sur les plaques adambulacraires.

"Localité. Kagoshima (Japon). Deux exemplaires. Ma collection."

A succinct description of this species is given by DÖDERLEIN, who adds a new variety [: 02, p. 329]:

"*Astropecten kagoshimensis* DE LORIOL 1899.

"Arme lang und schmal; R:r=4.3–4. Mitte des Scheibenrandes [-rückens?] meist papillenartig erhöht. Dorsomarginalplatten vom Armwinkel ab mit je einem sehr deutlichen kleinen Stachel (selten auf der 1. Randplatte fehlend).

"Ventromarginalplatten fein bestachelt, mit einer Querreihe von nach aussen immer mehr an Länge zunehmenden Stacheln, deren äusserster als langer, schlanker, spitzer und etwas gebogener Randstachel auftritt.

"Adambulacralplatten mit 3 inneren dünnen, etwa gleich langen Furchenstacheln, und 2–3 mittleren, etwas kürzeren, zu denen aussen noch einige weitere treten können.

"Jederseits 2 kleine Ventrolateralplatten. Beim kleinsten und grössten

der mir vorliegenden Exemplare beträgt R = 36 u. 48 mm, r = 8.4 u. 12 mm,
die Armbreite 9 u. 13 mm, die Zahl der Marginalplatten 25 u. 29.

" Die Farbe der lebenden Exemplare ist dunkelrothbraun mit etwa 3
dunklen Binden über die Arme.

" Ich fischte diese Art in geringer Anzahl in der Sagamibai, bei Tango
und bei Kagoshima, durchgehends aber in grösserer Tiefe als *Astropecten
scoparius*, nämlich in etwa 40 bis 100 m.

" Die der Beschreibung DE LORIOL'S zu Grunde, liegenden Exemplare
gehören wahrscheinlich zu denen, die ich selbst bei Kagoshima gesammelt
hatte ; auch bei dieser Art kann ich unpaare Marginalplatten nicht finden.

" *Astropecten kagoshimensis* var *kochiana* nov. var.

" Vor Kochi (Shikoku) fieng ich in einer Tiefe von etwa 100 m ein
grosses Exemplar, das in den mittleren Furchenstacheln von der typischen
Art abweicht ; von solchen finden sich hier zwei, wie bei manchen anderen
Exemplaren, aber das aborale ist doppelt so gross und viel breiter als das
adorale, auch viel grösser und breiter als die inneren Furchenstacheln ; die
nach aussen davon stehenden Stachelchen zeigen gern ein etwas verbreitertes
Ende.   Die oberen Randplatten zeigen im Armwinkel beträchtlich grössere
Stacheln als auf dem übrigen Theile des Armes.   Im Uebrigen entspricht das
Exemplar ganz der typischen Form.

" R = 87 mm, r = 16 mm, Armbreite = 16 mm, Zahl der Marginalplatten = 44."

This species is tolerably common in certain parts of our
coast, and is a well defined species somewhat resembling the
Astr. bispinosus of the Mediterranean and the *Astr. pleiacanthus*
BEDFORD from Singapore. The colour of the abactinal side is
a murky reddish brown with a few irregular cross patches of a
darker line on each arm. In many specimens there is a mesial
dark coloured band on the dorsal surface of each arm radiating
from the centre of the disk, which is of the same dark colour.
The specimens I have examined so far all belong to DÖDERLEIN's
typical form. The most prominent characteristic of this species

is the presence of a small spine on the superomarginals ; but, as will be shown later on, this is subject to considerable variation, and these spines may even be entirely absent in some specimens.

Radial ratio.—The radial ratio has a larger range of variation than might be inferred from the previous descriptions. We again see that, as a general rule, it is less for specimens with smaller radii. The measurements are tabulated below.

Specim.	r mm.	R mm.	R : r	MS	Locality
1	7.5	23	3	....	Enoura.
2	7.5	30.5	4	25	Misaki.
3	8.5	38.5	4.5	27	,,
4	8.5	40	4.7	27	,,
5	9	39	4.3	....	,,
6	9.5	38.5	4	26	,,
7	9.5	40	4.2	29	,,
8	9.5	41	4.3	....	,,
9	9.5	41.5	4.4	....	,,
10	9.5	41.5	4.3	....	,,
11	9.5	44	4.6	31	,,
12	10	39	3.9	....	,,
13	10	43	4.3	....	,,
14	10	51	5	....	,,
15	10.5	41	3.9	...	,,
16	10.5	42	4	....	,,
17	10.5	44	4.2	....	,,
18	10.5	44.5	4.2	....	,,
19	10.5	45.5	4.2	....	,,
20	11	44	4	....	,,
21	11	44	4	28	,,
22	11	47.5	4.3	30	,,

Specim.	r mm.	R mm.	R : r	MS	Locality
23	11	49.5	4.5	32	Misaki.
24	11	51	4.6	32	,,
25	11	51	4.6	34	,,
26	12.5	50	4	30	,,

Superomarginals. — The number of superomarginals where counted is shown in the foregoing table.

Each superomarginal is armed with a small conical spine situated on its outer border, and directed obliquely outwards and distally (Pl. IV, fig. 62, 67). The spines on the first two plates at the interbrachial angle are usually situated a little more internally than the rest, and are sometimes found even close to the inner border; they may also be perceptibly larger than the others. The spines are usually absent from a few plates at the extreme end of the arms, and in one or two specimens that have come under my observation the superomarginals were entirely unarmed. In such cases we have to depend entirely on the combination of other characters for identification. An interradial plate may or may not be present. The first superomarginals, as also the second and third, may also be destitute of spines in some specimens, showing the unsafeness of depending for determination on a single character, however striking it may usually be. In a small specimen from Enoura, No. 1 of the foregoing table, the unarmed condition went so far as to extend to the first 3–11 superomarginals on either side of the interradial line.

The general covering of the superomarginals is granular (Pl. IV, fig. 67). The granules are short cylindrical and rather well-spaced on the abactinal surface, but on the outer side they are more conical and more separated from one another. The surfaces of the fasciolar grooves are covered with capillary spinelets.

The superomarginals themselves have a more decidedly angular external edge than in *Astr. scoparius* and many other species, and the spines are arranged along this edge, forming a conspicuous row. At the interbrachial angles the superomarginals are twice as high as broad, but at the middle of the arms the height and breadth are very nearly equal.

Inferomarginals.—The general armature of the inferomarginals is fine and consists of short, flattened, conical spinelets, from which a series of 4–6 large spines and one or two others stand out conspicuous (Pl. IV, fig. 66). The most external, or marginal spine is conspicuously larger than the others of the series, and is followed by one spine or by two forming a pair. The series of conspicuous spines lies close to the distal edge of the infero-marginals.

Adambulacrals.—There are about six adambulacrals to four of the inferomarginals in the greater part of the arms. There are three spines on the furrow border of each plate, then two and finally one usually. This may be shown by the formula 3 : 2 : 1, and may be regarded as the typical case; but the following variations occur very frequently, viz., 3 : 2 : 2, 3 : 2 : 3. Of the first three the middle one is the longest and the adcentral one is mostly slightly longer than the distal; of the next two, which are slightly flattened and of nearly uniform thickness along their length, the distal one is thicker than the other and frequently also longer; the remaining spines are irregular both in size and form (Pl. IV, Fig. 66).

Mouth-plates.—The mouth-plates are rather long and narrow,

and there is a principal and a secondary crest on each, the latter
being nearly half as long as the former (Pl. IV, fig. 64). On the
principal crest there is an irregular row of 12–15 spines, of which
the one at the mouth end is conspicuously larger than the others,
and the distal ones are mostly larger than the middle ones. There
are moreover some half a dozen spines of moderate size on the
adcentral part of each plate between the principal and the secondary crest, and an irregular row of very small spines on the outer
surface of the distal half of the plate. The secondary crest bears
some six or seven spines, the most proximal of which is larger
than the others and lies close on the outside of the large oral
spine of the principal crest (Pl. IV, fig. 65).

Ventrolaterals.—The ventrolaterals appear to be constant in
number in this species, there being two on either side of the interradial line. They are small, and bear each some ten or a dozen
spines. When denuded of the spines they usually show a hollow
portion at the centre (Pl. IV, fig. 64).

Paxillæ.—The paxillæ are fine, and there is in the disk a
central area of very small and thickly set paxillæ, which is very
conspicuous in some specimens (Pl. IV, fig. 62). Of the coronal
spinelets the centrals and peripherals are distinct, the former being
thicker and shorter than the latter (Pl. IV, fig. 68). In one of
the largest paxillæ, which are found about the middle of the base
of the arms, there may be some 15 or 16 peripherals and some
8–10 centrals. The paxillar area at the middle of the arm is
nearly two and a half times as broad as the superomarginals of
one side.

Madreporite.—The madreporite is very small and is almost
entirely hidden by the crowns of the surrounding paxillæ. It lies

close to the margin of the disk and is almost hemispherical in shape. The grooves are not very complicated, and in some parts there may be seen a tendency to assume a honey-combed structure (Pl. IV, fig. 69).

Terminal plate.—The terminal plate is comparatively large and deeply bilobed on the dorsal surface.

Locality.—This species is not rare in depths of 39–80 m. and DÖDERLEIN obtained his new variety from a depth of 100 m. My specimens are from Misaki, Enoura (Suruga) and Shimabara (Hizen). Specimens in S.C., I.M.

Astropecten latespinosus MEISSNER.

(Pl. III, fig. 52–61.)

The original description of MEISSNER is as follows ['92, p. 188]:

" *Astropecten latespinosus* n. sp.

" Exempl.—Mitsuga Hama (Japan, Inland Sea), 18, VI. 84.

" [Taf. XII, Fig. A—C.]

" Es liegen 3 Exemplare vor, die Maasse derselben sind :

a) R=35 ; r=13 ; Breite d. Armes a. d. Basis=16 ; längst. Stachel= 3.5 mm.,

b) R=36 ; r=13.5 ; Breite d. Armes a. d. Basis=17 ; längst. Stachel=3.8 mm.,

c) R=21.5 ; r=9 ; Breite d. Armes a. d. Basis=11 ; längst. Stachel =3 mm.

" Bei den beiden grösseren Sternen ist also R=2. 7 r, bei dem kleineren dagegen R=2.4 r.—Die Art gehört zu denjenigen *Astropecten*, deren obere Marginalplatten granulirt sind und keinerlei Armatur tragen (Fig. A). Die unteren Randplatten sind mit feinen, fast härchenartigen Stachelchen bekleidet, aus denen je 2 oder 3 grössere hervorragen, am Rande tragen sie je einen ca. 1 mm. *breiten, flachen, am Rande meist gezähnten Stachel* (Fig. A), nur die Eckplatten der Arme haben einen kürzeren zugespitzten Stachel (Fig. B). Der grosse, zusammengedrückte Stachel ist an seiner Basis von einem Kranze kleiner Stachelchen umgeben, aus dem an der einen Seite des grossen Stachels ein etwas grösserer auffällt. Die Armatur der Adambulacralplatten (Fig. C) besteht aus folgenden Theilen : Den Füsschen liegen drei dünne, lange, handförmig verbundene Stachelchen an, zu welchen noch ein vierter gleichartiger gehört, der jedoch meist seitwärts von jenen isolirt auf der Platte steht. In der Mitte der Adambulacralplatte erhebt sich ein kurzer, aber breiter, oben dolchartig zugespitzter Stachel und hinter diesem stehen noch auf der den Bauch-Randplatten zugewandten Seite 2 Paare dünnerer, keulenförmiger Stachelchen.—Die Zahl der Marginalplatten des Rückens ist 24 (bei dem kleinsten) bis 30 (bei dem grössten). Diese Species ist nahe verwandt mit *A. regalis* GRAY, doch trennt sie davon nicht nur die Form der Rand-Stacheln, sondern auch die Vertheilung dieser. Bei *regalis* stehen 2 Stacheln auf jeder unteren Marginalplatte, bei *latespinosus* nur einer.— Die Farbe der Exemplare ist hellgelb.—(No. 3337.)"

DÖDERLEIN redescribes it as follows [:02, p. 330] :

" *Astropecten latespinosus* MEISSNER 1892.

"Die Arme sind sehr kurz, spitz und breit, gleichmässig verjüngt. R : r = 2.5 – 3.

"Dorsomarginalplatten nicht besonders breit, ohne Stacheln. Ventromarginalplatten viel breiter, ragen weit über die Dorsomarginalplatten vor und bilden den Rand ; zwischen den kleinen Schüppchen einige verlängerte, spitze Stachelchen ; am Rande ein sehr breiter, kurzer, ziemlich abgestutzter Stachel.

"Adambulacralplatten mit drei inneren Furchenstacheln, 2 mittleren, von denen der adorale sehr klein, der aborale sehr gross ist, und 3–4 äusseren kleinen Stachelchen.

"Jederseits etwa 2 kleine Ventrolateralplatten.

"Bei dem kleinsten und grössten der vorliegenden Exemplare beträgt R = 27 u. 58 mm, r = 10.5 u. 19 mm, die Armbreite = 13 u. 30 mm, die Zahl der Randplatten = 29 u. 42.

"Die trockenen Exemplare sind einfarbig, ledergelb. Eine kleine Anzahl von Exemplaren erhielt ich in der Tokiobai und der Sagamibai."

This is a very well defined species with prominent characteristics, and in certain localities nearly as common as *Astrop. scoparius*. Some of the characters are fairly variable, and this fact should be borne in mind in the determination of the species. The body is very flat, and is of a greyish brown colour on the upper side.

Radial ratio.—So far as I have observed, the radial ratio does not appear to be so variable as in some other species, as may be seen from the following table. It may also be remarked that this ratio is more constant in relation to the size of the specimens.

Specim.	r mm.	R mm.	R : r	MS	Locality
1	3	6.5	2.2	....	Fushiki (Japan Sea).
2	3.4	6.5	1.9	....	,,
3	3.5	7	2	....	,,
4	3.5	8.5	2.4	....	
5	4	9	2.3	....	
6	4	9.5	2.4	15	
7	4	9.5	2.4	....	
8	4	10	2.5	....	
9	4	10	2.5	....	
10	4.5	9.7	2.2	....	,,
11	4.5	11	2.4	....	
12	4.5	11.5	2.6	....	
13	4.8	11	2.5	....	
14	5	11	2.5	....	
15	5	11	2.2	....	
16	5	11.4	2.3	....	
17	5	11.5	2.3	....	
18	5	11.6	2.3	17	
19	5	12	2.4	....	,,

Specim.	r mm.	R mm.	R : r	MS	Locality
20	5	13	2.6	....	Fushiki (Japan Sea).
21	5.5	12	2.2	....	,,
22	5.5	14	2.5	18	
23	6	13	2.1	....	,,
24	11	28	2.5	....	Kachiyama (Tokyo Bay).
25	11	29	2.6	30	Nabuto (Awa).
26	12	36	3	34	,,
27	12	30	2.5	....	Hôjô.
28	12	30	2.5	....	,,
29	12.5	30	2.4	....	Kachiyama.
30	12.5	31.5	2.4	....	Misaki.
31	13	30	2.3	....	Hôjô.
32	13	32	2.5	....	,,
33	13.5	36	2.7	31	Nabuto.
34	14	35	2.5	....	Kachiyama.
35	14	35.5	2.5	....	,,
36	14.5	34.5	2.4	....	Misaki.
37	15	37.5	2.5	....	Kachiyama.
38	15	45	3	....	Misaki.
39	15.5	38	2.5	31	Nabuto.
40	16	40	2.5	....	Kachiyama.

Superomarginals.—The number of superomarginals is shown
in the foregoing table. They are very flat and well rounded, so that
the external edge is wholly inconspicuous, except in those specimens in which there is a series of short rounded spines along it.
In by far the larger number of the specimens which have come
under my observation, the superomarginals are covered with
rounded granules, which are larger near the inner edge of the
plate and gradually diminish in size outwards (Pl. III, fig. 59).

Around the edges of the plate there are one or two rows of short cylindrical granules of very small size. In a few specimens, one of the granules of the central area near the external edge of the plate is larger and very prominent, so as to bring about a row of small rounded spines along the external edge of the superomarginals. This series may or may not be continuous through the interbrachial arcs. At the middle of the arms the superomarginals are nearly as high as long, and twice as wide as long. The fasciolar grooves are covered with capillary spinelets.

Inferomarginals.—The inferomarginals are very wide and the most prominent feature is that each plate bears at its external edge a single, stout, flattened spines, nearly as wide as the plate is long, and with a constricted base and flattened serrated edge (Pl. III, fig. 52, 53, 58, 59). The interbrachial pair of inferomarginals bear very small flattened spines with pointed edges, and the next pair usually bear spines intermediate in form between those of the first pair and the remaining plates. The inferomarginals project laterally much more than the superomarginals, and the margins of the body are entirely formed by them. The abactinal surface of the inferomarginals is closely covered with very fine, spiniform granulations, and there is a row of more conspicuous spines next the lateral spine. On the actinal side, the inferomarginals are covered with small flattened, erect scales, which stand well apart from one another (Pl. III, fig. 58). There is also an irregular transverse series of 2–4 slender spines, besides a closely set group of larger and more conspicuous spines at the base of the lateral spine, of which the two or three immediately next the latter are conspicuously larger. The transverse grooves between the inferomarginals are naked.

Adambulacrals.—There are about 5 adambulacrals to every 4 of the inferomarginals in the greater part of the arms. The ambulacral groove is rather shallow and well open. Of the ad-ambulacral spines, three stand at the furrow edge, the middle one being the longest and all three fairly long and slender (Pl. III, fig. 56). Next come a pair, of which the abcentral one is very much larger than the other, and more or less inflated at the base, so that the general form is somewhat triangular. Then follow a variable number (3–7) of small but rather thick spines, of which one may be larger than the others.

Mouth-plates.—In denuded specimens the mouth-plates are very prominent, and the secondary ridge is barely noticeable. On the principal ridge at least two more or less irregular rows of tubercles are present (Pl. III, fig. 56). In very small specimens of 12 mm. R the principal ridge bears two rows of spines, six or seven large ones on the crest, and five or six on the outer side ; the one at the mouth end being conspicuously larger than the others. The secondary ridge bears at this stage one, two or three spines. In larger examples the general arrangement of the spines remains the same, but the two rows on the main ridge become more or less duplicated and irregular, and the spines are more numerous, there being some twenty larger spines and some sixteen smaller ones. On the secondary ridge there are now 6–8 spines, but they always lie in a single row. The spines of the secondary ridge and the terminal ones of the main crest form a row along the ambulacral furrow.

Ventrolaterals.—The ventrolaterals are small, and there are usually 2 or 3 on either side of the interradial line (Pl. III, fig. 56).

Paxillæ.—The paxillæ are small and closely set, and the coronal spinelets are also finer than in most other species of the genus here described, so that the aboral surface looks more smooth. The largest paxillæ are, as usual, found at about the middle of the base of the arms, and there is usually a conspicuously large one on the inner side of the madreporite. The distinction between the peripherals and the centrals of the coronal spinelets is less sharp than in most other species, the centrals being only shorter and more rounded (Pl. III, fig. 60). It also happens very commonly that one or more of the centrals are conspicuously larger than others and almost spherical, and in such cases it looks as if there were one or more large centrals and numerous peripherals; but such a description would be totally incorrect. It also happens in some specimens that the single granule near or at the centre of the crown is not only conspicuously larger, but is produced into a spine. This I have found to be the case near the apex of the arms in a few of my larger examples; but the same condition is especially striking in several of the smaller examples from the Japan Sea, in which there is a row of such spines along the mesial line of the arms through nearly their whole length (Pl. III, fig. 54). This feature was, at first glance, so conspicuous that I provisionally made a new species; but further studies have left no doubt that it is either a local or individual variation. In one of the larger paxillæ there are some thirty coronal spinelets. The area of small paxillæ at the centre of the disk is large and conspicuous.

Madreporite.—The madreporite is large and well exposed, and is close to the margin. It is usually of a rounded kidney shape, the indentation being turned towards the centre of the disk.

Terminal plate.—The terminal plate is tolerably large and bilobed.

Locality.—This species occurs on the same sand flats with *Astrop. scoparius* and is in some localities nearly as common as the latter. I have specimens from the following localities : Hôjô (Awa), Zushi (Sagami), Nabuto (Awa), Kachiyama (Awa), Misaki, Fushiki (Japan Sea).

Specimens in S.C., H.N.S., I.M., I.H.S.

Astropecten formosus SLADEN.

This species is not represented in any of the collections mentioned at the beginning of this paper. The following is its original description by SLADEN ['79, p. 424] :

"*Astropecten formosus*, sp. nov. Pl. VIII. figs. 1, 2, 3, 4.

"Coll. ST. JOHN : Korea, 36 and 54 fathoms (young) ; W. Coast of Nippon, 60 fathoms.

"Disk large, rays short, arm-angles widely rounded ; the greater and lesser radii of the largest specimen measure respectively 14.5 millims. and 5.2 millims., or in the proportion of $2\frac{3}{4} : 1$ approximately. The foot-papillæ, which are all cylindrical and taper towards the tip, form two series : the inner one, which spreads out into a comb overhanging the ambulacral furrow, is composed of three papillæ, the middle one being longer than the others ; the outer series, which radiates towards the ventral plates, consists in the middle of the arm of three papillæ, whilst along the inner fourth of the furrow there are four or five, these being arranged two and two, or two and three together, one pair opposed to the inner series, the others placed more external and nearer together. On the innermost plates of the ray this external series of foot-papillæ is further augmented by two or three additional spinelets, and which form an almost imperceptible transition into the scuticles of the ventral plates. The ventral marginal plates bear three spines—the uppermost, or that nearest the margin, being the smallest; the second is large, compressed and acuminate, twice the size of the marginal spine, and is succeeded by another almost as large. The spines are arranged obliquely across the plate, except in the arm-angle, where they form a straight series along with two or three additional spinelets which lie between them and the furrow. The main spines of these inmost plates of the arm-angle are also somewhat smaller than their successors. The rest of the ventral plate is covered with numerous small compressed and finely acuminate scuticles, standing erect and fairly well spaced, which present quite a different appearance to the flat, closely-packed, spatulate scales which so frequently clothe the under surface of *Astropecten*. The furrows between the plates are wide, having the margins set with fine setæform spinules, very different from the armature of the plate just described. The upper marginal plates, which are broader than long, number about sixteen on each side, exclusive of the tip. They are closely papillate; and the spine-like papillæ are cylindrical, with radiate tips more or less expanded and quite clavate. There are about five rows of these spinelets upon a plate, exclusive of the marginal setæform series, the middle ranges being larger than the rest; whilst the setæform spinelets which fringe the furrows are much longer and more delicate; and present in a more marked degree the clavate character of the tip.

"The paxillary area is, at the middle of the ray, a little broader than the marginal plate (though not twice as broad); and the paxillæ are large and very distinctly stellate, 5–6-radial, with a ray springing from the centre as well, though sometimes this is wanting. The madreporiform body is situated close to the marginal plates.

"Two very young *Astropectens*, measuring respectively 10.5 millims. and 8 millims. in their greatest diameters, seem to belong to this species. The relative characters of the disk and rays, the arrangement of the foot-papillæ, the armature of the ventro-marginal plates, and the paxillæ of the dorsal surface present only such differences as might be expected in the premature conditions of the *Astropecten* above described. The inner row of foot-papillæ consists of three spinules as in the adult form; but in the outer series there are only two on the outer portion of the furrow, and three on the inner; their arrangement, however, being such as to leave little doubt, when comparison is made with the different portions of the furrow in the largest specimen, that they belong to one and the same species. On the marginal plates of the smaller specimen there is only one spine; but on the larger there seems indications here and there of the future development, out of the plate armature, of the larger companion spines. The paxillæ are large and much simpler than in the adult, having fewer radii.

"Although these juveniles were dredged on different occasions—one being taken off the Korea at the depth of 54 fathoms, and the other off W. coast of Niphon, 60 fathoms—they both agree in the singular circumstance of having gorged a small bivalve! and in each case apparently of the same species. In the larger of the two young starfish the distention of the test and the position of the shell lead to the supposition that the diminutive *gourmand* had fallen a martyr to the indulgence of its appetite!

"This *Astropecten* bears some resemblance to certain examples of the northern form known as *Astr. Mülleri*, M. & T.; regarding that, however, as an extreme variation of *A. irregularis*, the differences presented by the Asteroids at present under consideration are such as to justify the opinion that they should be classed (provisionally at any rate) as distinct from that species. It would not, however, be surprising to find, from the examination of a larger supply of material from this and other localities than is at present available, that the above specific determination would require to be included within the *extended* diagnosis of the type of *A. irregularis*, although the distribution as at present known of the varietal forms of that species (e.g. *A. Mülleri*, *A. echinulatus*, etc.) would hardly lead to such a supposition."

Persephonaster asper, n. sp.

(Pl. I, figs. 8–13.)

This is a large species presenting characters intermediate between *P. misakiensis* and *P. triacanthus*. Viewed from the actinal side it looks much like *P. misakiensis*, especially in the armature of the inferior marginals, but on the abactinal side it is more like *P. triacanthus* in the spiny character of the supero-marginal armature, although the paxillæ are more like those of *P. misakiensis*. The large spines are, however, much stouter than those of the latter species, and they are also not or hardly flattened. I have only dried specimens, and judging from their appearances, there must have been a great deal of slimy matter on the surface of the body when fresh. The smaller spines, but especially those of the adambulacral plates, are agglutinated together by a gelatinous substance, which forms in the dried specimens a distinct web-like membrane between the spines of the adjacent adambulacrals. The specimens measured as follows :

Specim.	r mm.	R mm	R : r	MS
1	34	154	4.5	36
2	33	155	4.7	36
3	43	205	4.8	38

Superomarginals.—The superomarginals bear a transverse series of three or four, or rarely five large spines (Pl. I, fig 9). When there are only three, either the outermost two are con-spicuously large, or the outermost one only is so and the other two much smaller. When there are four or five, the outermost two are conspicuously large and the rest much smaller. These spines are stout, pointed and conical, and the largest one may be as long as 9 mm. The two superomarginals on either side of the interradial line usually carry only one spine near the abactinal end forming the same series with the innermost spine of the

other superomarginals. The entire external surface of the supero-
marginal plates is thickly covered with very slender, almost subcapillary, pointed, conical spinelets, which are coarser around the
base of the conspicuous spines. The transverse grooves between
the successive superomarginals are absent. The plates in the
interbrachial angles are very high, facing towards the outside and
presenting a small surface on the abactinal side; but as they
proceed further into the arms, the plates become more and more
horizontal and nearly the whole of their surface is turned
towards the abactinal side. At about the middle of the arms
the plate is nearly one-half wider than long. Except in the
interbrachial angles where the superior and inferior marginals
stand flush, the lateral margin of the arms is formed by the
inferior series.

Inferomarginals.—The most striking feature of the inferomarginals is an obliquely transverse series of large, very slightly
flattened spines, mostly four or five but occasionally six in
number. Aside from occasional irregularities, the innermost three
spines usually form a strictly transverse series near the abcentral
margin of the plate, and the rest an oblique series running towards
the outside and centre. The two portions sometimes make an
angle, but usually they are continuous and pass gradually the one
into the other. The spines generally increase in length outwards,
but the most external one may be shorter than the next. The
longest spine may measure as much as 9 mm. or a little more.
The surface of each inferomarginal plate may be distinguished
into two sections which are covered with different spines. The
abcentral two-thirds or a little more of the surface, to which the
series of conspicuous spines belongs, is covered with well-spaced,

somewhat flattened, straight or slightly curved, conical spines. The adcentral one-third or a little less is covered with subcapillary spinelets, which are more closely set towards the centre. It must, however, be remarked that the transition between the two sections is gradual, and on the two plates on either side of the interradial line, the subcapillary spinelets are found near both the proximal and the distal margin. In all my specimens the conspicuously large spines are all directed obliquely outwards and towards the tip of the arms. There are no transverse grooves between the plates. The inferomarginal plates are strictly contingent with the superomarginals throughout the length of the arms.

Adambulacrals.—Except a few plates next the month, each adambulacral plate presents a sharply pointed wedge-shaped edge towards the furrow, and along this edge is a series of six or seven subequal spines, one of which invariably occupies the apex of the wedge and is lightly curved and flattened (Pl. II, fig. 11). On the actinal surface of the plate are 6–11 spines, a few of which are conspicuously larger than those of the furrow series. Of the spines of the actinal surface three or four always lie in the angle formed by the furrow series, and are, in the dried specimens, agglutinated together and with the furrow spines by a gelatinous substance. This circumstance makes the detection of the furrow series from the actinal side somewhat difficult, but a view from the furrow at once brings them out in relief. The first adambulacral plate is short and wide, and carries two somewhat irregular, parallel series of spines which become smaller towards the furrow; the second is intermediate in character between the first and the third which usually presents the peculiarities described above.

Mouth-plates.—The mouth-plates are comparatively long and narrow, being about 13 mm. in a specimen with R=205 mm., and enclose between them a tolerably large space coverved over with a membrane (Pl. I, fig. 10). Each plate bears some 32–40, irregularly arranged spines of various sizes. The exposed surface of each plate is a scalene triangle, of which the longest side faces the fellow plate and the shortest side the mouth. One can make out a series on the longest side of the triangle, con-sisting of some dozen, somewhat large spines, of which the one at the month end is usually larger than the others ; on the shortest side there are usually four or five spines, of which the innermost one is common to this and the first series ; the remaining side is subdivided into two portions by a short projection at a point corresponding to the furrow end of the first adambulacral plate, and on this side there are some twenty spines, most of which are small, and of which the one at the month end is common to this and the preceding series. The remaining spines, which are of various sizes, are mostly borne on the adoral half of the plate in the area enclosed by the three series mentioned above.

Ventrolaterals.—The ventrolateral area is very distinct, and there is a series of plates extending about two-thirds or more into the arms. In the area limited by the second inferomarginals on either side, there are some twenty-five plates,. which are mostly small and leave soft membranous areas between. The plates pre-sent convex surfaces covered with pointed conical spines, one or two of which in each plate are usually larger than the others. These larger spines occasionally tend to form a series with the large inferomarginal spines, especially at the base of the arms ; but they are very much smaller.

Paxillæ.—In the largest specimen the paxillæ are well spaced, but in the other two specimens, which are smaller, they are rather closely set. In all cases, however, there is a circular central area in the disk, where the paxillæ are very dense and just a trifle smaller than those of the immediately surrounding parts (Pl. I, fig. 9, 12). The paxillæ also gradually decrease in size towards the tip of the arms. The tabulum is rather long and cylindrical and is provided with an expanded base inbedded in the skin. The coronal spinelets are rather long, slender and pointed, the longest ones being slightly longer than the tabulum; their number varies according to the size of the paxillæ. In one of the largest I have counted as many as fifteen, but in a smaller one near the tip of the arms there are only some ten or less. There is no distinct difference in shape between the peripheral and central spinelets, but·a few around the centre are usually larger than the others. Moreover, the central part of the summit of the tabulum is more elevated than the peripheral part, so that the coronal spinelets of a paxilla usually assume the form of the painter's brush. In the largest of my specimens a good deal of the naked skin is visible between the paxillæ. None of my specimens is in a condition to show the papular pores.

Madreporite.—The madreporite is tolerably large, low, nearly circular, and covered with radiating grooves; the peripheral part is usually more or less overgrown by the surrounding paxillæ, but the madreporite itself is entirely naked; its inner margin is nearly twice as far removed from the centre of the disk as·from the superomarginals (Pl. I, fig. 13).

Terminal plate.—The terminal plate is comparatively large and elliptical, the major axis being at right angles to the length of the arm. The plates are much abraded in all my specimens, but they appear to be covered with subcapillary spinelets similar to those of the superomarginals.

Locality.—My specimens are from the vicinity of Misaki, two from Okinosé from depths of 480 m. and 1120 m., the other from Yodomi from a depth of 160 m.

Three dried specimens in S. C.

Dr. W. K. Fisher thinks that this species possibly belongs to his genus *Thrissacanthias*, in which the gonads extend some way into the rays, while in *Persephonaster* they are confined to the interradii. The dried condition of my specimens does not enable me to decide this point.

Persephonaster misakiensis, n. sp.

(Pl. II, figs. 23-33.)

I have two specimens of this species in alcohol. Both are of good size and look externally very much like *Astropecten ;* but the presence of the anus as well as certain features in the armature of the body make it necessary to refer them · to *Persephonaster*,[1] which is distinguished from *Astropecten* chiefly by the presence of

---

1) I may here express my opinion parenthetically that so far as can be judged from the published descriptions, the distinction between *Plutonaster* and *Persephonaster* appears to be vague and uncertain. The former genus was set up by SLADEN ['84, p. 81], to whose work one must refer back for the original diagnosis and the descriptions of typical species. The genus *Persephonaster* was created by WOOD-MASON and ALCOCK ['91, p. 430] and more precisely defined by ALCOCK ['93, p. 81]. As remarked by LUDWIG ['97, p. 117] in connection with his description of *Plutonaster subinermis* of the Mediterranean, the presence of the superambulacral plates is a character which does not agree with SLADEN's definition of the genus; but that aside from this and a certain peculiarity in connection with the genital organs, the Mediterranean species can most naturally be brought in that genus. In fact *Plutonaster subinermis* and *Plut. abyssicola* LUDWIG [:05, p. 30] appear to me to be intermediate forms between the typical species of the two genera in question. *Ctenophoraster hawaiiensis* FISHER [:06, p. 1015] seems to be nearly allied to *Persephonaster misakiensis* which is intermediate in character between the former and the typical *Persephonaster* species.

the anus and certain differences in the arrangement of the adambulacral spines, though closely similar to that genus in other respects. The superambulacral plates are very well developed, and the whole external characters appear to me amply sufficient for placing this species in the genus *Persephonaster* as defined by Alcock.

The colour of the alcoholic specimens is light straw yellow; the measurements are as follows:

Specim.	r mm.	R mm.	R : r	MS
1	26	91	3.5	35
2	26	44	3.6	33

The characteristic points in the general features of this species are the generally fragile character of the spines, the comparatively ample spaces left between the paxillæ, even in the subcentral area of small paxillæ around the anus, the comparative largeness of the disk, and the presence of a very slender terminal portion in each arm, where the superomarginals of the two sides are separated from each other only by about two rows of paxillæ, the arms undergoing at this point a comparatively sudden constriction from side to side (Pl. II, fig. 23).

Superomarginals.—The number of superomarginals is given in the foregoing table. At the middle of the arm they are nearly twice as broad as high, but at the interradial angles they are twice as high as broad, and their breadth gradually increases from the interradial angle to about the middle of the arm and then gradually decreases again towards the distal portion of the arm.

On each superomarginal plate (Pl. II, fig. 23, 28) can be distinguised a central transverse area of coarser armature. This is most conspicuous in the middle portion of the arm, but is less so on the first few plates, as well as on the plates belonging to the distal slender portion. Moreover, in the greater part of the arm there are in this area 3–5 conspicuous, pointed, somewhat flattened, subconical spines, forming a transverse series. These conspicuous spines are of unequal sizes even on the same plate, and are mostly present on the outer rather than on the inner portion. The remaining surface of the superomarginal is uniformly covered with very fine silky subcapillary spinelets, which gradually pass into the truly capillary spinelets clothing the surfaces of the transverse grooves, which are less deep and more open than in *Astropecten*. In the interradial angle the superomarginals stand out very slightly more than the inferomarginals, but in the larger part of the arms the latter are more projecting, so that the lateral margin is formed entirely by them.

Inferomarginals.—The inferomarginals (Pl. II, fig. 24, 26) are very broad in the greater part of the arm, but less so in the interradial angle and the distal slender portion. The first inferomarginal is very slightly higher than broad, the second and third are a little broader, while the next following seven or eight are of nearly equal breadth, the remaining plates gradually decreasing in breadth towards the slender portion of the arm, where they undergo again a sudden decrease in breadth. The general armature of the actinal surface of the inferomarginals may be described as scaly, the scales being of unequal lengths, directed obliquely outwards and distally, and overlapping one another. Towards the inner margin of the plates the scales are more or less slender and long. From the general armature of the infero-marginals a somewhat irregular series of long conspicuous spines distinguish themselves; it consists of 5–9 spines, runs usually obliquely from the outer proximal corner of the plate towards the diagonally opposite corner, the spines being particularly crowded along the margin of the arm, where they form a conspicuous feature when viewed from the side. They are not, however, so conspicuous in an oral or aboral view of the animal, since the spines near the margin are usually adpressed to the plates. The outermost of these conspicuous spines is very close to the aboral margin of the plate, and in a lateral view of the arm the conspicuous spines of the two marginals are seen to form a continuous series. Besides the series of conspicuous spines above mentioned, a few additional spines of conspicuous length may be present on each inferomarginal. The trans-verse grooves between the inferomarginals are covered with capillary spinelets.

Adambulacral plates.—In the greater part of the arms, there are 6 adambulacrals to 5 of the inferomarginals. In denuded specimens, the first 4–5 plates are rounded towards the ambulacral furrow, but the rest present a rather sharp wedge-shaped edge on this side, the edge becoming in general more acutely angled towards the tip of the arms. Along this edge is a regularly curved row of 5–7 tubercles supporting the ambulacral spines, and on the remaining actinal surface of the plate there are some ten or twelve tubercles, each supporting a spine (Pl. II, fig. 27). The furrow spines are very long, and usually flattened, with the sharp edge turned towards the furrow (Pl. II, fig. 26). The spines on the actinal surface of the plate are frequently furrowed longitudinally, and a row of them can sometimes be distinguished next to the ambulacral series. The total number of the spines on the actinal surface of each plate is some ten or twelve for the proximal two-thirds of the arms.

Mouth-plates.—In denuded specimens the mouth-plates project very prominently on the ventral surface. Each pair encloses a a capacious space between covered with a membrane. Each plate is long and bears on the actinal surface a conspicuous principal ridge extending along the whole length of the plate and two short accessory ridges near the mouth end, of which one is nearly parallel to the main ridge while the other makes an angle of nearly 60° with it. In both of my specimens the mouth-plates are much abraded in the distal part; but the arrangement of the spines on the main crest appears to be as follows (Pl. II, fig. 29). There are three irregular rows, of which two extend along the whole length of the plate, while the third extends only through the distal half of the plate and occupies the outermost margin of the crest. The first mentioned two rows are terminated at the mouth end by a common long, robust spine, which with its fellow forms a conspicuous pair in each interradius. The total number of spines on the main crest appears to be some 40–45. Of the two accessory crests, the one next the main crest extends through a little over one-third of the length of the mouth-plate and bears some 5 or 6 spines. The third crest which extends nearly vertically towards the mouth is nearly as long as the other secondary crest and bears about eight spines arranged in two rows. The three crests meet at the base of the conspicuous mouth spine, and the last mentioned crest can be seen clearly only in a lateral view of the mouth-plate (Pl. II, fig. 30).

Ventrolaterals.—The ventrolaterals are rather numerous, forming about 6 arc-shaped rows parallel with the inferomarginals. The outermost row extends for a little over two-thirds of the length of the arm, and in the interbrachial angle there are two of these plates corresponding to one inferomarginal (Pl. II, fig. 25). The second row may extend up to half the length of the arm, and the third row 2 or 3 plates into the arm, while the rest are wholly confined to the disk. It must at the same time be remarked that these rows are not perfectly regular. Regarding for convenience the second inferomarginals as the limit of the interradial area of the disk, there are some fifty ventrolateral plates in the enclosed area for each interradius. The ventrolaterals are covered with slender fragile spines, among which some ten or fifteen may be conspicuous by their size in a single interradial area of the disk. They form in general a transitional form between the outer spines of the adambulacral plates and those of the inferomarginals, those near the former being more like them, and those near the latter being more or less like the inner spines of the inferomarginals. In counting the spines of the actinal surface of the adambulacrals, some care is necessary not to include the spines on the ventrolaterals.

Paxillæ.—The paxillæ are well spaced, and in both of my specimens the coronal spinelets stand erect and parallel or even convergent, somewhat like the hairs of a brush, and they are comparatively long, being only slightly shorter than the height of the pedicel ; so that, when a single paxilla is detached and viewed from the side, it looks somewhat like a sea-anemone with its tentacles all extended vertically upwards (Pl. II, fig. 31). The paxillæ vary considerably in size in different parts of the body, being very small along the margin and in the distal part of the arms. There is in the disk a central area of smaller paxillæ around the anus, but it is not so well marked as in *Persephonaster triacanthus* or many species. of *Astropecten*; in fact the perianal paxillæ are not very numerous and pass on so gradually to those of ordinary size, that the area in question is not conspicuous (Pl. II, fig. 23). The tabula are more or less cylindrical, and not flattened as in many species of *Astropecten*. On one of the largest paxillæ I counted as many as some thirty coronal spinelets, while a small paxilla near the tip of the arms carries only five or six of them. The bases of the tabula of adjacent paxillæ are joined by slender, delicate, radiating calcareous pieces; and in the meshes thus formed are found the papulæ, one in each, of a conical form (Pl. II, fig. 32). In each arm there is a median longitudinal band where the papulæ are totally absent, as in *Astropecten*, but it is very narrow and inconspicuous.

Madreporite.—The madreporite (Pl. II, fig. 33) is rather conspicuous and is situated about midway between the centre of the disk and the margin. It is nearly round and in my specimens is 2.5 and 3 mm. in diameter respectively; its surface is covered with granules. The furrows are mostly radiating.

Terminal plate.—The terminal plates are small and about twice as long as broad.

Locality.—Both of my specimens are from the neighbourhood of Misaki. The only record I have about the habitat of this species is that one of the specimens was fished from a depth of about 550 m.

Two alcoholic specimens in S. C.

Persephonaster triacanthus, n. sp.

(Pl. II, figs. 14–22).

This species presents a very strong external resemblance to *Astropecten*, to which genus it will be referred by any casual observer; but the presence of the anus, together with a few other points of less importance, exclude it from that genus. The anus is very difficult to recognize, but dissection established its presence beyond doubt, and in certain favourable cases, when the speci-mens are dried, the anus can be recognized externally as a very small opening in the centre of a conspicuous conical prominence, where the paxillæ are considerably smaller than in the surrounding parts. I have four specimens of this species, all from the same part of the sea, and from nearly the same depth. This species is so closely related to the starfish described by SLADEN from the same region under the name of *Astropecten brevispinus*, that I should not wonder if a study of a large number of specimens should lead to an amalgamation of the two species. In any case I have no doubt that *Astropecten brevispinus* SLADEN should be referred to the genus *Persephonaster*. I shall come back to this question later.

The measurements of the four specimens are as follows :

Specim.	r mm.	R mm.	R : r	MS	Locality
1	18	74	4.1	35	Misaki.
2	22	92	4.2	....	,,
3	25	94	3.8	....	,,
4	15	67	4.5	31	Bay of Suruga.

The disk is tolerably large, and the arms gradually taper to a point from the base, the terminal part being especially slender, but not so conspicuous or long as in *Persephonaster misakiensis* (Pl. II, fig. 14, 15).

Superomarginals.—The most conspicuous feature of this species is the presence of three robust spines on the superomarginals, forming a transverse series (Pl. II, fig. 20). On a few plates in the interbrachial arc there are usually only two of these spines close to the inner edge of the plate, but there may also be four or even five, in which case one of them is usually much smaller than the others. In the terminal part of the arms there is usually only one spine on each plate. Again the relative size of the three spines is various, and any one of them may be the largest, although this appears to be most frequently the outermost one. The central part of each superomarginal is covered with short conico-squamiform spinelets, and these become subcapillary towards the transverse grooves, which are well developed and the faces of which are covered with capillary spinelets. The superomarginals are nearly as high as broad in the greater part of the arms, and their external face is perfectly rounded, so that the abactinal and lateral faces can not be distinguished sharply. The two marginals keep the same front on the outer side.

Inferomarginals.—The inferomarginals are about as broad as the superomarginals, and each plate carries three or sometimes four robust spines; the largest one close to the abactional end, and the other two (or three) in a series with it on the actino-distal face (Pl. II, fig. 19). When viewed from the side, the spines of the two marginals make a single series, but those of the infero-marginals are very apt to form a somewhat oblique row. Near the inner margin the inferomarginals are covered with conico-squamiform spinelets similar to those of the superomarginals. They become more slender and pointed towards the outer part, and towards the proximal and distal edges of the plate they gradually become subcapillary, and finally truly capillary on the faces of the transverse grooves. Some care is necessary to distinguish the armature of the inferomarginals from that of the ventrolaterals, which is intermediate in character between it and the outer spines of the adambulacral plates.

Adambulacral plates.—Of these there are, in the greater part of the arms, seven to every six of the inferomarginals ; and except the two or three at the mouth end, each plate presents a wedge-shaped margin towards the ambulacral furrow (Pl. II, fig. 16). The first plate is broad and short, and bears two parallel series of rather slender spines, each series consisting of some ten spines ; and when these stand up all erect, they present a comb-like appearance, which, however is not so conspicuous as in *Astropecten ludwigi*. From the third plate on there is a curved furrow series of 5–7 spines which are tolerably long, the longest ones being very nearly 3 mm. in a specimen whose r is 18 mm. (Pl. II, fig. 19). On the actinal face of the adambulacral plate there can usually be distinguished two somewhat irregular series of spines parallel to the furrow series, each of 3–5 spines, somewhat shorter and more pointed than those of the former. The spines of the furrow series are subequal in length.

Mouth-plates.—These are less prominent than in *P. misakiensis*, and are inclined at a rather sharp angle towards the mouth, the highest part being near the abcentral end of the plates, where they are firmly ankylosed. These points are clearly visible in denuded specimens, in which the two plates can also be seen to be separated at the mouth end and to leave a rather spacious cavity between, which is covered over with a membrane in alcoholic specimens. Each mouth-plate bears only one crest which runs along its whole length and is double, there being a shallow but distinct furrow on it (Pl. II, fig. 17). Corresponding to this double structure of the crest, there are two parallel series of spines on each plate. The spines of the inner series, i.e. the one on the side of the mouth-plate turned towards its fellow, are rather short but robust and rounded at the end ; the one at the mouth end is larger than the others but only slightly longer ; there are in all some ten or eleven spines in this series. The spines of the outer series are considerably more slender than those of the inner and are usually closely appressed to the latter ; the spine at the mouth end is, however, nearly as large as that of the inner series, so that there are in each interradius four spines of large size at the mouth end (Pl. II, fig. 18). The outer series may contain from eight or nine to some twelve spines, and may extend the whole length of the mouth-plate or may terminate at about two-thirds of its length.

Ventrolaterals.—The ventrolaterals are less numerous than in *P. misakiensis*, but extend outwards over two-thirds of the length of the arms (Pl. II, fig. 16). From the character and arrangement of the ventrolaterals of the arms we may say that the innermost series extends farthest into the arms, and that the outermost series terminates at the third or fourth inferomarginal. Again, if we adopt for the sake of comparison, the same limit to the interradial' disk area as in *P. misakiensis*, there are in the present species some 30–32 plates in each interradius. The ventrolaterals are covered with short, blunt, rough spines, which impart a coarse appearance to the area; they are intermediate in form between the outer spines of the adambulacrals and the coarser armature of the inferomarginals. In well preserved alcoholic specimens the spines of the ventrolaterals appear very distinct from those of the inferomarginals by their deeper color, as well as from the adambulacral spines which are almost colorless; moreover the ventrolateral spines present perfectly rounded tops.

Paxillæ.—In a well preserved alcoholic specimen (Pl. II, fig. 14), in which the aboral side is much inflated, the paxillæ are well spaced and clearly show the papulæ lying between them, but in specimens whose aboral side is perfectly level or slightly sunken they are less well spaced. There is a very prominent, central elevated area of small paxillæ, in which the anus opens in a subcentral position. The paxillæ are again very small in the terminal slender parts of the arms. They are also more like those of *Astropecten* than of *P. misakiensis*. The pedicel is flattened from side to side, and the coronal spinelets are hardly divergent (Pl. II, fig. 21). In the larger paxillæ there may be as many as four centrals and twelve peripherals, the former being slightly larger than the latter but otherwise similar. In the smaller paxillæ there may be one central or none at all. The coronal spinelets have truncated tops and are about half as long as the pedicel. In some of the well spaced portions there are found radiating calcareous pieces connecting the bases of the pedicels similar to those of *P. misakiensis;* but they do not appear to be present everywhere.

Madreporite.—The madreporite is tolerably conspicuous, nearly elliptical in outline, and covered with granules. It is situated slightly more outwards than midway between the margin and the centre of the disk. The grooves are irregular (Pl. II, fig. 22).

Terminal plate. — The terminal plate is rather large and elongated, and bears three spines at the end.

In a well preserved alcoholic specimen the colour is brownish grey for the marginals, the same color of a darker shade for the ventrolateral area, and the same colour of a lighter shade for the aboral surface. The terminal parts of the arms and the adam-bulacral portions are almost colourless. The papulæ are short conical processes of a deep brown colour, and are very numerous. They are, however, absent from the central part of the disk and a rather wide zone along the middle of the arms, exactly as in *Astropecten*.

Locality.—One of my specimens is from the Bay of Suruga, at the entrance of Enoura Cove, depth 560 m.; the others are from the Sagami Sea, depth 480 m.

Four alcoholic specimens in S. C.

In this species also the superambulacral plates are well developed.

Remarks.—As stated above this species is very closely related to *Persephonaster brevispinus* (SLADEN) mentioned below, with which further study may possibly prove it to be identical. So far as SLADEN's description goes, there are, however, well marked though not very great differences between the two species, which may be enumerated as follows :

Radial ratio 3.2 for *P. brevispinus* (R=32 mm., r=10 mm.) as against 3.8–4.5 for *P. triacanthus.*

The terminal slender portion of the arms present in *P. triacanthus* is not found in *P. brevispinus.*

Superomarginal spines three in *P. triacanthus*, and two in *P. brevispinus*. Only in the interbrachial arcs and the terminal portions of the arms are these spines less in *P. triacanthus*.

The furrow series of the adambulacral spines are four or five for *P. brevispinus*, and 5–7 for *P. triacanthus*. They are long and slender in the latter species but short, robust, and subpapilliform in the former. The spines on the actinal plates of *P. triacanthus* are also tolerably long and slender but small and stumpy in *P. brevispinus*.

The armature of the mouth-plates are similar in both; but the spines appears to be somewhat less numerous in *P. brevispinus*, there being about 16 (six pairs + 4) or 18 on each plate as against 18–23 for *P. triacanthus*.

The terminal plate bears three spines in *P. triacanthus*, but apparently none in *P. brevispinus*. This may however be due to abrasion.

Persephonaster brevispinus (SLADEN).

This is the *Astropecten brevispinus* of SLADEN. It is not represented in any of the collections studied by me, but there is no doubt in my mind that it must be referred to *Persephonaster*, because the character of its adambulacral armature shows it to be closely related to *triacanthus*. The only single character of decisive value that distinguishes this genus from *Astropecten* is the presence of the anus. SLADEN was not able to make out the anus, but this is very small in some species and is very apt to be overlooked, unless special pains be taken to prove its presence; and in dealing with a large mass of material such as SLADEN had before him one is but too liable to rely more on external characters that are comparatively easy of observation. It did not however escape the experienced eyes of SLADEN that the present species presents some characters that are not usually found in *Astropecten*, as is shown in the "remarks" at the end of the full description reproduced futher on.

The first description of this species, which is a preliminary one, is as follows [SLADEN, '83, p. 249] :

"*Astropecten brevispinus*, n. sp.

"Rays five. R=3.2 r; R=32 millim., r=.10 millim. Rays tapering regularly from the base to the tip and terminating in a point. Breadth of a ray at the base about 11 millim. Interbrachial angles slightly rounded.

"Supero-marginal plates 22 in number from the interbrachial line to the tip, higher than broad along the inner half of the ray, but broader than high on the outer portion. Each plate (excepting two or three in the arm-angle and a few at the extremity) bears two small, conical, sharply pointed spines. The inner series are placed close to the inner edge of the plate, and are continuous from the arm-angle until near the tip, decreasing in size as they proceed outward, until they disappear altogether. The outer series are slightly larger, and are placed at the extreme edge of the plate on the rounding where the dorsal and lateral superficies converge; they are continuous throughout the ray, excepting the innermost plate in the arm-angle.

"Infero-marginal plates higher than broad, and flush with the superior series. Each plate bears a single marginal spine, short, tapering continuously from base to tip, sharply pointed and slightly compressed. On the inner half of the ray, two similar and slightly smaller spines are situated on the median line of the plate—one, which is the smallest, not far from the inner edge of the plate adjoining the adambulacral plates, and the other about midway between this spine and the marginal spine, the three forming a lineal series transverse in relation to the direction of the ray. On the outer portion of the ray the inner spine is aborted or indistinguishable from the squamules of the plate. When the side or lateral wall of the ray is placed in direct view, the above-mentioned spines of the inferior plates are all visible, and they, together with the spinelets of the superior plates, appear to form a continuous vertical series. The marginal spine is very little, if at all, longer than the outer spine on the supero-marginal plate, and all these spines stand at an angle to the superficies of the plate and are directed upward and outward. Very short, widely spaced, papilliform squamules are distributed over the whole of the infero-marginal plates, and the granulation of the supero-marginal series partakes of the same character and is indistinguishable at the junction of the plates.

" The ambulacral spines are short and robust, subpapilliform, and do not taper, and they stand more or less perpendicular to the surface of the plate. The inner series are 4 or 5 in number, and their base-line forms a slight angle projecting into the furrow; the middle spinelets are a shade larger and more robust than the others. The ambulacral spinelets that occupy the rest of the plate behind the furrow-series are little more than elongate papillæ; they are small, stumpy, covered with membrane, and are rather widely spaced, no definite arrangement being discernible, although about two irregular rows may be traced in some instances. The spinelets on the ventral plates are similar in character and disposition to the foregoing, and they merge imperceptibly into the squamules of the infero-marginal plates. This uniformity in the dermal appendages imparts a characteristic appearance to the ventral aspect of the starfish.

" Mouth-plates elongate, each with two short, flattened, truncate spinelets at the inner extremity, then about six pairs of spinelets, short and robust, standing perpendicular on the surface of the plate and forming two series apposed to one another; and then about four rather broader, shorter, and more robust spinelets, forming a single series in continuation as it were of the two apposed series, at the outer extremity of the plate, towards which the spinelets decrease as they proceed outwards. Consequent on this method of arrangement there is a marked division of the mouth-plate armature into two narrow series along the median line of each mouth-angle.

" The paxillæ of the dorsal area are small and compact, and composed of six to nine spinelets, of which one is central. The spinelets are short and robust, and are directed upward, their radiation apart being very slight. No definite order is maintained in the arrangement of the paxillæ. The papulæ are small and dark brown, or almost black in colour, and a broad space occurs along the median line of the ray in which none are present.

In the centre of the disk there is a large, conspicuous, and well-developed conical prominence, upon and in the neighbourhood of which the paxillæ are greatly reduced in size. No anal puncture is traceable.

"The madreporiform body is small and situated at about one third of the distance from the margin to the centre of the disk. The terminal (ocular) plate, though small, is conspicuous and elongately oblong.

"Colour in alcohol, umber-brown, becoming lighter in shade towards the extremities of the rays. The spinelets white. Small specimens yellowish white.

"Station 232. Lat. 35° 11′ N., long. 139° 28′ E. Depth 345 fms.; bottom temperature 5°.0 C.; sandy mud."

SLADEN's final description of this species, which closely follows the preliminary, was published in the Challenger Report, as follows ['89, p. 198] :

"Astropecten brevispinus, SLADEN (Pl. XXXIII. figs. 1 and 2; Pl. XXXVII. figs. 1–3).

"Astropecten brevispinus, SLADEN, 1883, Journ. Linn. Soc. Lond. (Zool.), Vol. XVII, p. 249.

"Rays five. R = 32 mm.; r = 10 mm. R > 3 r. Breadth of a ray at the base, about 11 mm.

"Rays tapering regularly from the base to the tip, and terminating in a point. Interbrachial arcs slightly rounded.

"The paxillæ of the abactinal area are small and compact, and composed of six to nine spinelets, of which one is central. The spinelets, which are short and robust, are directed upward, and their radiation apart is very slight. No definite order is maintained in the arrangement of the paxillæ. The papulæ are small, and dark brown or almost black in colour; and a broad space occurs along the median line of the ray in which none are present. In the centre of the disk there is a large and conspicuous conical prominence, upon and in the neighbourhood of which the paxillæ are greatly reduced in size.

"The supero-marginal plates, which are twenty-two in number from the median interradial line to the extremity, are higher than broad along the inner half of the ray, but broader than high on the outer portion. Each plate, excepting two or three in the interbrachial arc and a few at the extremity, bears two small, conical, sharply pointed spines. The inner series are placed close to the inner edge of the plates, and are continuous from the arm-angle until near the tip, decreasing in size as they proceed outward, until they disappear altogether. The outer series are slightly larger, and are placed at the extreme edge of the plates on the curvature where the abactinal and lateral superficies unite : they are continuous throughout the ray, excepting the innermost plate on each side the median interradial line.

" The infero-marginal plates are higher than broad, and flush with the superior series. Each plate bears a single lateral spine, which is short, tapering continuously from base to tip, sharply pointed and slightly compressed. On the inner half of the ray, two similar and slightly smaller spines are situated on the median line of the plate—one, which is the smallest, not far from the inner edge of the plate adjoining the adambulacral plates, and the other about midway between this spine and the lateral spine, the three forming a lineal series transverse in relation to the direction of the ray. On the outer portion of the ray the inner spine is aborted or indistinguishable from the squamules of the plate. When the side or lateral wall of the ray is placed in direct view, the above-mentioned spines of the infero-marginal plates are all visible, and they, together with the spinelets of the supero-marginal plates, appear to form a continuous vertical series. The lateral spine is very little, if at all, longer than the outer spine on the supero-marginal plate, and all these spines stand at an angle to the superficies of the plate, and are directed upward and outward. Very short, widely spaced, papilliform squamules are distributed over the surface of the infero-marginal plates, and the granulation of the supero-marginal series partakes of the same character, and is indistinguishable at the junction of the plates.

" The armature of the adambulacral plates consists of short, robust, subpapilliform spinelets, which do not taper, and stand more or less perpendicular to the surface of the plate. The furrow series consists of four or five spines, and their base line forms a slight angle projecting into the furrow ; the middle spinelets are a shade larger and more robust than the others. The actinal surface of the plate behind the furrow series is occupied by spinelets which are little more than elongate papillæ ; they are small, stumpy, covered with membrane, rather widely spaced, and usually no definite order of arrangement is discernible, although about two irregular rows may be traced in some instances.

" The spinelets on the actinal intermediate plates are similar in character and disposition to the foregoing, and they merge imperceptibly into the squamules of the infero-marginal plates. This uniformity in the dermal appendages imparts a characteristic appearance to the actinal aspect of the starfish.

" The mouth-plates are elongate, each with two short, flattened, truncate spinelets at the inner extremity, followed by about six pairs of short robust spinelets, which stand perpendicular on the surface of the plate, and form two series apposed to one another ; these are succeeded by about four rather broader, shorter, and more robust spinelets, forming a single series in continuation, as it were, of the two apposed series, on the outer extremity of the plate, towards which the spinelets decrease as they proceed outward. Consequent on this method of arrangement there is a marked division of the mouth-plate armature into two narrow series separated by the median suture line of each mouth-angle.

" The madreporiform body is small, and situated at about one-third of the distance from the margin to the centre of the disk.

" The terminal (ocular) plate, though small, is conspicuous and elongately oblong.

" Colour in alcohol, umber-brown, becoming lighter in shade towards the extremities of the rays. The spinelets are white. Small specimens are yellowish white.

" *Young phase.*—A small example, which has a major radial measurement of 16.5 mm., may readily be distinguished as belonging to the species. It is to be noted, however, that the paxillæ of the abactinal area have quite a different character, the spinelets of the crown being long, and radiating nearly horizontally. The armature of the adambulacral plates and mouth-plates is comparatively longer, especially on the latter ; and on the actinal surface of the adambulacral plates behind the furrow series there are usually one or two larger spinelets, thickly invested with membrane, especially noticeable on the inner part of the ray, but of which no trace remains in the adult—that is to say, their prominence and juvenile robustness are altogether lost.

"*Locality.*—Station 232. Off the coast of Japan, south of Yeddo. May 12, 1875. Lat. 35° 11′ 0″ N., long. 139° 28′ 0″ E. Depth 345 fathoms. Green mud. Bottom temperature 41°.1 Fahr.; surface temperature 64°.2 Fahr.

"*Remarks.* — *Astropecten brevispinus* is distinguished from the other species in this section of the genus *Astropecten* by the single lateral spine, by the character of the adambulacral armature, and by the character of the paxillæ. It differs from *Astropecten antillensis* and *Astropecten brasiliensis* in each of these particulars. It is at once distinguished from *Astropecten aurantiacus*, which has a single lateral spine, by the general facies, and by the character of the adambulacral armature ; and from *Astropecten erinaceus* and *Astropecten duplicatus* by the single lateral spine and the different character of the spinulation of the infero-marginal plates."

As mentioned under *P. triacanthus* this species may prove to be identical with it.

· *Leptychaster arcticus* (Sars). ·

This species was first described by M. Sars under the name of *Astropecten arcticus* ['50, p. 161] and later figured by him ['56, p. 61, pl. IX, fig. 16–18]. It is also mentioned by M'Andrew and Barrett ['57, p. 45] and again by the same authors ['57, p. 45] as well as by Barrett ['57, p. 46, pl. IV, fig. 3] under the name of *Astropecten lütkeni*. Sars mentions it again under the original name ['61, p. 32]. All these papers are not accessible to me, and the above is given on the authority of Ludwig [: 00, p. 452].

Dujardin and Hupé describe this species under two names, *Astropecten arcticus* and *Astropecten lütkeni* ['62, p. 428] :

"Astropecten arctique. *Astropecten arcticus.*—Sars [References to Sars and Barrett].

"Espèce pentagone à cinq bras, dont la longueur est environ une fois et demie celle du plus petit rayon du disque. Plaques marginales au nombre de 25 à chaque bras. Face dorsale couverte partout de tubercules coniques, très-courts. Aire paxillaire quatre ou cinq fois aussi large que les plaques à la base des bras. Espaces interbrachiaux ou angles rentrants des bras, arrondis.

"Coloration jaune orangé, rougeâtre pâle. Dimension : 40 mm.

"Habite les mers du Nord."

"Astropecten de Lütken. _Astropecten lütkeni._—BARRETT.

"Disque pentagonal, portant cinq bras médiocrement allongés, élargis à leur base, acuminés vers leur pointe, pourvus de chaque côté d'environ vingt plaques couvertes d'épines et de tubercules sur les parties latérales. Les sillons ambulacraires sont bordés de deux rangées de petites épines, puis, en dehors, de cinq ou six autres rangées d'épines plus petites formant des sortes de touffes, et enfin, le reste de la face ventrale des bras porte des séries transversales de très-petites épines soyeuses, dirigées dans le sens des bras. La face dorsal est couverts de paxilles très-ténues, formant des tubercules, parmi lesquels on distingue un grand nombre d'épines semblables à celles des plaques marginales.

"Coloration jaunâtre. Dimension : largeur totale 50 à 60 mm.

"Habite les mers du Nord (Finmark)."

This species is mentioned under the name of _Archaster arcticus_ by SARS[1] ['69, p. 251] and VERRILL[1] ['73, p. 5, 14, 100]. Again VERRILL simply mentions it by name (_Archaster arcticus_) ['78, p. 214, 373]. It is also mentioned by PERRIER[2] ['78, p. 32, 88] and STORM[1] ['78, p. 252 ; ' 80, p. 119]. VERRILL again mentions it as occurring at depths of 183–310 fathoms ['82, p. 138, 218].

DANIELSSEN and KOBEN, after giving a list of previous literature, refers to this species under the name of _Astropecten arcticus_ as follows ['84, p. 82] :

"This starfish is also a genuine _Astropecten_ and no _Archaster_." Then

1) These citations are made on the authority of LUDWIG.
2) Cited after SLADEN and LUDWIG.

the stations where specimens were obtained are mentioned, and the distribution is given as follows: " The Drontheim fjord. The coast of Finmark. The Murman coast. The North American coast."

JARZYNSKY mentions this species under the name of *Astropecten arcticum* ['85, p. 170, *fide* LUDWIG] ; while VERRILL refers to it under the name of *Archaster arcticus* and adds ['85, p. 542] : " B [athymetrical] range, 113 to 547 fathoms, 1883. Frequent, but only in small numbers."

SLADEN mentions this species under the name of *Leptoptychaster arcticus* and sets up also a new variety as follows ['89, p. 189] :

" *Localities.*—' Porcupine ' Expedition :

" Station 65 (1869). Between the Shetland and the Faeröe Islands. Lat. 61° 10′ N., long. 2° 21′ W. Depth 345 fathoms. Bottom temperature —1°.1 C. ; surface temperature, 11°.1 C.

" Station 82 (1869). In the Faeröe Channel. Lat. 60° 0′ N., long. 5° 13′ W. Depth 312 fathoms. Bottom temperature 5°.2 C. ; surface temperature 11°.2 C.

" Station 3 (1870)." West of Ushant. Lat. 48° 31′ N., long. 10° 3′ W. Depth 690 fathoms.

" *Remarks.*—This form was originally described as an *Astropecten*. Some subsequent writers have retained it in that genus, whilst others have regarded it as an *Archaster*. I am unable to agree with either of these determinations. After careful study I consider that its structure accords in all points of generic import with the genus *Leptoptychaster* established by Mr. SMITH for the foregoing form[2] of the Southern Ocean, and I have therefore referred the species under notice to that genus. It is readily characterised by the aborted supero-marginal plates, the short but broad infero-marginals (both alike unarmed), the well-developed actinal interradial areas, the character of the paxillæ, and likewise that of the armature of the adambulacral plates. Finally, the species has more or less well-developed superambulacral plates ; the whole forming a combination of characters which accords neither with *Archaster* nor *Astropecten*.

---

1) " Recorded by Sir WYVILLE THOMSON in the Depths of the Sea ; but I have not seen a specimen from this locality."

2) *Leptychaster kerguelensis.*

"*Leptoptychaster arcticus,* var. *elongata* nov.

"The American examples of this species, of which a large series was taken at station 49, are all distinctly longer in the ray, and have the paxillæ of the abactinal area somewhat more delicate and less compact in character than in the European forms, although even in these some variation occurs. It would, however, be an easy matter to say which were the American examples out of a large number of mixed specimens, and on these grounds I consider that we are dealing with a well-marked variety.

"*Localities.*—Station 46. Off the coast of North America, east of New Jersey and Long Island. May 6, 1873. Lat. 40° 17' 0" N., long. 66° 48' 0" W. Depth 1350 fathoms. Blue mud. Bottom temperature 37°.2 Fahr.; surface temperature 40°.0 Fahr.

"Station 49. South of Halifax, Nova Scotia. May 20, 1873. Lat 43° 3' 0" N., long. 63° 39' 0" W. Depth 85 fathoms. Gravel, stones. Bottom temperature 35°.0 Fahr.; surface temperature 40°.5 Fahr."

BELL describes this starfish as follows ['92, p. 65]:

"R = 2.25 r.

"A small species, with a large disk and rapidly tapering arms, the angles between which are distinctly rounded; the rather narrow ambulacra are bounded by a row of long spines arranged by threes on each adambulacral, outside of which are other rows so closely packed as to be almost disorderly in their disposition; the intermediate plates which fill up the rather large interbrachial area, but do not extend far into the rays, are closely covered with sharp spiniform granules. The marginals are short but quite distinct, and the superomarginals, though much less wide than the inferomarginals, remain distinct to the top of the arm. They are both covered by large granular plates, which sometimes take on the appearance of spines; there are from 20 to 40 marginals on either side of each arm. The dorsal surface formed of the ends of fine paxillæ, the form of which is obscured by the integument which invests them.

"Colour: dry or in spirit, white; alive, cinnabar to orange.

"R =   18   15
"r =   8   7

"Breadth of arm at base   7   ,5.5

"*Distribution.* East and western sides of northern part of Atlantic to Arctic Ocean, 20–690 fms.

"a, b. Faeroe Channel, 1312 fms. 'Porcupine' Exp. (St. 82).

"c, d. Between Faeroe and Shetland, 640 fms. 'Porcupine' Exp. (St. 65)."

HERDMAN ['92, p. 89] probably refers to this species under the name of *Astropecten arcticus,* and NORMAN mentions it from the Trondhjem Fiord, Norway, under the name of *Leptophycaster arcticus* ['93, p. 346]. He adds, "This species has not, I think, been found to the south of the Trondhjem Fiord."

VERRILL refers to this species under the name of *Leptoptychaster arcticus* as follows ['94, p. 255] : " Our series of specimens show various gradations in the relative length of the rays, some of them agreeing in this and other respects with the form described as a variety by SLADEN. His variety was taken off New Jersey, in 1,350 fathoms. I am unable to make out any definite diagnostic characters for this form.

"This species has been taken at many stations off our coast, in 50 to 547 fathoms, but always in small numbers."

VERRILL again makes the following remarks ['95, p. 133] : " B [athy-metrical] range, 50 to 965 fath. (1350 fath ; SLADEN). Most common in 85 to 200 fathoms ; rare at greater depths. Taken at 23 stations, from N. lat. 45° 14' to 38° 29'. It always occurred in small numbers. It is also found off the northern coasts of Europe. Two closely allied species are found in the Antarctic Ocean."

SLUITER ['95, p. 53] mentions it in the collection of the Museum of Amsterdam : " *Leptoptychaster arcticus* (SARS) SLADEN. Sehr zahlreiche Exemplare in Alkohol von der Barents See, 72° 14'—72° 36 NB. 22° 30'—25° 38' OL. Gr. aus einer Tiefe von 140—155 Faden."

LUDWIG [: 00, p. 452] gives a very full list of previous literature and a good account of the distribution and habitat of this starfish, as follows :

"Westatlantisch findet sich diese subarktische Art an der Ostküste von Nordamerika vom 38–45° n. Br. (VERRILL 1873, 1878, 1882, 1885, 1894,

1895; SLADEN 1889). Ostatlantisch liegt ihr südlichster Fundort südlich von Irland unter 48° 31' n. Br., 10° w.L. (SLADEN 1889). Zwischen dieser vereinzelten Fundstelle und dem Färöer-Kanal ist sie noch nicht angetroffen worden. Im Färöer-Kanal aber wurde sie auf der Fahrt der 'Porcupine' erbeutet (SLADEN 1889, BELL 1892). Am zahlreichsten sind ihre Fundorte an der skandinavischen Küste von Throndhjem (STORM 1878, 1880; NORMAN 1893) bis nach Finmarken (M. SARS 1850, 1856, 1861, 1869; M'ANDREW und BARRETT 1857; BARRETT 1857; DANIELSSEN und KOREN 1884; HERDMAN 1892). Ferner kennt man sie westlich und nördlich von Norwegen (DANIELSSEN und KOREN 1884) bis etwa zum 73° n. Br. Von hier geht sie in östlicher Richtung bis in den westlichen Teil der Barents-See (SLUITER 1895) und bis zur Murmanschen Küste (JARZYNSKY 1885). Von West nach Ost reicht demnach ihr Verbreitungsgebiet von ca. 76° w.L. bis ca. 42° ö.L.=122 Längengrade, von Süd nach Nord westatlantisch von 38-45°, ostatlantisch 48-73° n. Br.

"Die Tiefen der Fundstellen betragen zwischen 37 und 1261 m, in einem Falle 2469 m. Sie lebt meistens auf reinem Lehmboden, kommt aber auch auf sandigem und steinigem Lehm, auf Schlamm und auf Kies und Steinen vor."

DÖDERLEIN mentions a small specimen of this species from the collection of the "Olga" [: 00, p. 224]. It is also mentioned by MICHAILOVSKIJ as follows [: 02, p. 486]:

"Zwei junge Exemplare, welche 1899 auf dem Wege nach Spitzbergen gefunden wurden, unterscheiden sich von den grösseren Stücken durch verhältnismässig weniger lange Arme (r : R=1 : 2, anstatt 1 : 2.14 bis 2.25 nach BELL und 1 : 2.2 bis 1 : 2.25 nach M. SARS), ferner durch die bedeutend kleinere Anzahl von Marginalplättchen. Ihre Anzahl ist nach BELL 20-40 für jede Seite des Armes, nach M. SARS aber 24; unsere Stücke haben nur 18 und 20 für den ganzen Armwinkel. Im Zusammenhang mit dieser und der anderen Zahl von Marginalplättchen ist r=3 und 3.5 mm., R dagegen 6 und 7 mm.

"Dr. A. TSCHERNYSCHEW. 1899. [St. 3=44[1)]" =Lat. 71° 21 N., long. 17°

---

1) The first numeral shows the station in a single voyage, the second that in the whole series.

32′ E., 278 m., stone fragments, bottom temperature 5° C., 2 young specimens.

GRIEG refers to it as follows [: 02, p. 19]:

"Ved siden af foregaaende art[1] er *leptoptychaster arcticus* en af de mest udbredte asterider i det nordlige Norge. NORDGAARD har den tildels i talrige eksemplarer fra ikke mindre end 14 stationer, 30–400 m.

"Sydgrænsen for dens udbredelse ved vor kyst er Trondhjemsfjorden."

Beside the foregoing species *leptoptychaster arcticus* is one of the most widely distributed asterids in northern Norway. NORDGAARD has it partly in numerous examples from no less than 14 stations, 30–400 m.

The southern limit of its distribution on our coast is Trondhjem Fiord.

NORMAN [: 03, p. 408] mentions it under the name of *Leptophychaster arcticus*, M. SARS from "Tana Fiord and off Vardö, in 127–148 fathoms (' Vöringen ')."

MICHAILOVSKIJ describes it from the north coast of Norway as follows [: 04, p. 170]:

"Die sechs, verschiedenen Altersstadien angehörenden Exemphare, weisen ein Verhältnis von r : R auf, welches zwischen 1 : 2.2 (bei den kleinsten) und 1 : 2.7 (ben grössten) schwankt. Die Zahl der Marginalplatten einer jeden Armseite beträgt bei den grössten Exemplaren 24, fällt rasch mit der Verminderung der allgemeinen Dimensionen des Tieres und beträgt bei Exemplaren mit r = 4.5 mm. und R = 11 mm. bereits nur etwa 16. Bei den allergrössten Exemplaren beträgt r = 9 mm. und R = 24 mm.

"St. 37 (6).[2] "

NORDGAARD mentions this species from the Norwegian Fiords [: 05, p. 160, *fide* KŒHLER]. GRIEG describes it from the collection of the "Michael Sars " as follows [: 06, p. 10]:

"*Leptoptychaster arcticus*, M. SARS.

"Fundstätte : 1900 Stat. 54 (3)[2]; 57 (1).[2]

"1901 Stat. 56 (einige Exemplare); 80 (1).[2]

"1902 Stat. 63 (1)[2]; 64 (1)[2]; 86 (1).[2]

---

1) *Ctenodiscus crispatus.*

2) The numerals in parentheses indicate the number of specimens taken. St. 37 = Lat. 71° 20′ N., long. 27° 49′ E., 413 m., mud, bottom temperature 3.1° C.

"M. SARS gibt in seiner ersten Beschreibung dieser Art das Verhältnis zwischen Scheiben- und Armradius als 1 : 2.2 an. In der 'Oversigt af Norges Echinodermer' ist das Verhältnis als 1 : 2.3–2.5 angeführt. Bei zwei jungen Exemplaren, welche die russische Gradmessungsexpedition auf dem Wege nach Spitzbergen nahm, fand MICHAILOVSKIJ [: 02, p. 27] das Verhältnis 1 : 2, und bei einigen Exemplaren, die der Eisbrecher 'Jermak' im Sommer 1901 ausserhalb Nordkyn nahm fand er, dass es zwischen 1 : 2.2 bei dem kleinsten Exemplar und 1 : 2.7 bei dem grössten variierte [: 04, p. 170]. Bei dem Material vom 'Michael Sars' variierte das Verhältnis zwischen 1 : 2.4 und 1 : 2.9. Bei einem Teil Exemplaren, von O. NORGAARD im nördlichen Norwegen genommen, hatten die kleinsten 1 : 2, die grössten bis an 1 : 3.13. Man wird aus diesen Angaben ersehen, dass junge Individuen dieser Art verhältnismässig kürzere Arme haben als ältere, übrigens ist auch dies grossen individuellen Variationen unterworfen.

"Mehre Exemplare meines Materials, sowohl von der norwegischen Küste als auch aus dem Nordmeere, zeichneten sich durch ihre langen, schlanken Arme aus. So massen zwei Exemplare des 'Michael Sars' 1901 Stat. 56 : r. 12 mm., R. 33 mm., Armbreite 13 mm., Dorsomarginalplatten 27–28 ; r. : R.=1 : 2.8, A. : R=1 : 2.5. Zwei Exemplare von den Færöbänken massen : r. 7 mm., R. 20 mm., Dorsomarginalplatten 19–22, r. : R.=1 : 2.9. Wie die von SLADEN in 'Challenger' Asteroidea (p. 189) beschriebene Varietät *elongata* von der nordamerikanischen Küste zeichneten sich diese Færöexemplare ausser durch ihre langen Arme auch durch feinere, weniger kompakte Paxillen auf der Rückenseite der Scheibe aus.

"*Leptoptychaster arcticus* ist eine echte Warmwasserform, die nur ein Mal in der kalten Area des Færökanals ('Porcupine' 1869, Stat. 65, 631 m., –1.1°) und möglicherweise ein Mal in der kalten Area des Barentsmeeres ('Willem Barents' 1879 Stat. 9, 372 m.–0.6°) genommen worden ist."

KŒHLER describes some specimens from the collections of the "Princesse-Alice" as follows [: 09, p. 58, *Leptoptychaster arcticus*] :

"Campagne de 1898 : Stn. 960, profondeur 394 m. Une vingtaine d'échantillons.

"Campagne de 1899 : Stn. 1052, profondeur 440 m. Un échantillon.

" Plusieurs auteurs ont mentionné les variations que présente la longueur des bras : le rapport entre R et r varie de 1/2 à 1/3.13. Des variations correspondantes se montrent dans le nombre des plaques marginales. SLADEN ['89, p. 189] mentionne une var. *elongata* des côtes americaines, mais il est assez vraisemble que cette variété n'est pas mieux distincte dans les individus d'Amérique que dans ceux d'Europe : VERRILL ['94, p. 255] dit, en effet, qu'il a observé toutes les transitions entre les formes à bras courts et les formes à bras longs.

" D'une manière générale, les jeunes ont les bras relativement plus courts que les adultes.

" Voici, à titre de renseignements, les mesures prises sur quelques exemplaires de différentes tailles choisis dans la collection de la Princesse-Alice depuis les plus grands jusqu'aux plus petits :

R = 26   23   18   15   13 mm.
r = 9   8   7   7   6 „

" *Distribution géographique.*—Le *L. arcticus* est plutôt une forme subarctique qu'on observe, sur les côtes américaines, entre le 38° et le 45° Lat. N. En Europe, sa station la plus méridionale se trouve sur la côte sud de l'Irlande, par 48° 31 Lat. N. et 10° Long. W. Il se rencontre aux îles Färoër, au large des côtes de Norvége, dans les mers de Barents et de Murman et il s'étend jusqu'au 42° Long E. Au Nord, il ne paraît pas dépasser le 73 Lat."

FISHER gives a very full account of this species as follows [: 11, p. 43, *Leptychaster arcticus,* 2 figs.] :

" *Diagnosis.*[1]—Rays five.   R = 32 mm. ; r = 10 mm. ; R = 3.2 r.   Breadth of ray at base, 12 mm. (measured from interradial line).   Disk medium sized, rays tapering gradually to a blunt extremity ; interbrachial areas rounded but rather abrupt ; abactinal surface slightly inflated ; sides of ray rounded ; superomarginal plates small, resembling enlarged paxillæ, not markedly wider at base of ray than at middle or on outer third, forming a narrow margin to abactinal area ; inferomarginals short, but fairly wide, placed very obliquely and forming the rounded edge to ray, most of the series being visible from

---

1) " From specimen taken at station 4792, in the vicinity of the Commander Islands."

above ; no specialised spines on either series ; adambulacral plates with a
furrow series of three or four spinelets, and on actinal surface two or three
longitudinal series (often irregular) of three to five spinelets each, with
sometimes two or three small spinelets out of line ; abactinal paxillæ very
compactly placed, longest on interradial area of disk and on either side of
median radial area along ray ; actinal interradial areas fairly large, the
intermediate plates extending far along the ray.

"*Description.*—The paxillæ are closely placed and. the difference in size
between those of the midradial region of ray and centre of disk, and those
along side of paxillar area (particularly in the interradial region) is very
marked—much more so than in *Leptychaster pacificus.* A large paxilla of the
interradial region of disk presents a slightly convex crown of small terete,
compactly placed spinelets, varying in number according to the size of the
paxilla, a fairly large one having twenty to twenty-five peripheral and about
the same number of central spinelets, a trifle shorter than column of paxilla.
Papulæ absent from centre of disk and along median radial area ; five or six
(sometimes four) about each plate irregularly, on interradial regions and
along border of paxillar area of ray. Abactinal plates slightly and irregularly lobed on papular areas, much less obviously so than in *pacificus.*

"Superomarginal plates, forty-one in number from interradial line to
extremity of ray, are much smaller than in either of the two other species[1]
described below, and have the form of large paxillæ, being not conspicuously
wider in interbrachial angle than midway along ray. They are irregular or
subcircular in outline, rather straight sided adjacent to inferior series, have
the appearance of being obliquely oriented and do not correspond exactly to
inferomarginals. Opposite the first ten superomarginals are seventeen abactinal paxillæ (or irregular transverse series). The superomarginals are
confined to the abactinal surface and their spinelets are heavier and slightly
more numerous than on the other paxillæ. Terminal plate wider than long,
with a rounded end ; notched adjacent to paxillæ.

"Inferomarginals much shorter than wide, obliquely oriented (about 45°)
to transverse axis of ray, and occupying entire side wall of ray. Each plate

---

1) *Pacificus* and *anomalus.*

is therefore strongly arched, this arch giving the rounded margin. Each plate is covered with spinelets, stouter than those on superomarginals, somewhat squamiform on exposed surface, and very slender in the fasciolar grooves. These fasciolar grooves are deep, and about twice as wide *at base of ray* (taking transverse axis as length of plate) as the adjacent specialised ridge of the plate. The latter is a thin, almost laminar ridge, very much thinner than in *pacificus*, where the specialised ridge is as wide or wider than the grooves. Farther along ray in *arcticus* the ridges become thicker, nearly or quite as wide as grooves and the inferomarginals are more massive. A longitudinal shallow furrow separates the two series of plates.

" Adambulacral plates with a curved furrow series of three or four rather long, slender, blunt, terete spinelets. Lateralmost spinelets slightly the shortest. On actinal surface of plate the spinelets decrease gradually in size outward, there being two or three longitudinal series of three or four spinelets, occasionally more. Sometimes a very few odd spinelets stand out of line. The first plate is wider than the rest, with more numerous spinelets.

" Mouth plates narrow, the combined pair forming a salient angle into actinostome. Furrow margin long, with about ten spinules like those of adambulacral plates, decreasing in size outward, and continued along adambulacral margin in several smaller spinelets ; innermost two spinules forming ' teeth ' at angle. An irregular series of spinelets smaller than furrow spinules stand along edge of median suture ; and at outer end of plate are a few intermediate spinelets between the two series, which throughout the length of plate stand rather close together.

" Actinal interradial areas fairly large, the plates being arranged in series between inferomarginals and adambulacrals. A single series of plates extends about half length of ray (from interradial angle) ; a second series, one-fourth ; a third series, one-eighth or less. Between first superomarginal and first adambulacral are about six plates in an interradial direction.

" Madreporic body nearer margin than centre of disk ; surrounded by six or seven large paxillæ which overhang the edges ; ridges coarse, transverse rather than centrifugal.

" *Variations.*—There are three specimens from station 3602 which are difficult to classify satisfactorily. Two resemble *arcticus* but have a larger disk, shorter and broader rays, and larger superomarginals. The third much resembles *anomalus*, and in the sum of its characters stands about midway between the two specimens, above referred to, and *anomalus*, which was taken at the same station. The aberrant specimens have inferomarginals resembling those of *arcticus*. R=37 mm. (largest specimen); r=15 mm.; R=2.5 r. Breadth of ray at base, measured from interradial line, 18 mm. It is not improbable that this species hybridizes with *anomalus* whenever the ranges overlap, and that the very aberrant specimens may be explained by such a theory.

" *Type-locality.*—Öxfjord, Finmark, 100 to 150 fathoms.

" *Distribution.*--The distribution of this species is evidently circumpolar. In the Atlantic hemisphere it is found along the east coast of North America from lat. 38° to 45° N., and on the coast of Europe from south of Ireland, the Faroe Channel, off Norway from Trondhjem to Finmark, and eastward to Barents Sea and the Murman coast.[1] In the north Pacific region the species ranges over Bering Sea and south on the Asiatic side to Yezo, Japan.

" *Specimens examined.*—One typical from station 4792, vicinity of Commander Islands, 72 fathoms, pebbles; two aberrant forms, 3602, vicinity Pribilof Islands, Bering Sea, 81 fathoms, green mud, sand; one from 5047; off Hokushu, Japan, 107 fathoms, dark gray sand, broken shells, pebbles.

" *Remarks.*—The specimen from near the Commander Islands agrees in most particulars with an example from the coast of Maine (station 21, Cashes Ledge). The Atlantic specimen has the raised ridges of inferomarginals at base of ray, slightly wider.

MORTENSEN reports this starfish from Greenland as follows [: 13, p. 328] :

" Forekomst (Occurrence). Vest Grønland : 66° 49′ N. 56° 28′ V., 235 Fv. (WANDEL, 1889, 1891). 66° 21′ N. 56° 54′ V., 400–600 m ; 66° 42′ N. 56° 12′ V., 130 Fv. ; 66° 44′ N. 56° 08′ V. ; ca. 175 Fv. ; 66° 45′ N. 56° 30′ V., ca. 200 Fv. ('Tjalfe ' 1908-09). Bredefjord, 225–290 m. (STEPHENSEN, 1912).

---

1) " Condensed from LUDWIG " [: 00, p. 452].

"Ikke kendt fra Øst-Grønland " [Not known from East-Greenland].

"Dybde " [Depth] : Ca. 40-1250 m. (2469 m).

"Udbredelse " [Distribution] : Nordlige-Atlanterhav, ned til [down to] 38° N. paa den amerikanske Side [on the American side]; paa den europaiske Side forekommer den fra Irlands Sydkyst til Murman-Kysten [on the European side it occurs from the south coast of Ireland to the Murman coast]. Nordlige-Stillhav, fra [from] Berings-Strædet til Japan (Yezo)."

This species is not represented in any of the collections named at the outset.

Leptychaster anomalus FISHER.

The first description of this species is due to FISHER and is as follows [: 06a, p. 115] ;

"Rays 5. R=27 mm.; r=17 mm.; R=1.6 r. Breadth of ray at base, 19 mm.

"In general form and ornamentation greatly resembling *Parastropecten inermis* LUDWIG. Disk broad, rays short, broad and blunt; interbrachial arcs shallow and wide; abactinal surface subplane, capable of slight inflation; marginal plates conspicuous, devoid of enlarged spines or spinelets, but covered with granules and granuliform spinelets; actinal intermediate areas broad; adambulacral plates with 3 or 4 furrow spines; small superambulacral plates present; a very tiny anal pore present.

"Abactinal paxillar area compact; paxillæ arranged in not very regular oblique transverse rows at sides of ray; without order in median radial areas and centre of disk. Paxillæ largest at base of ray and in interradial areas decreasing conspicuously in size toward centre of disk and tip of ray; large at sides of paxillar area than in mid-radial region. Paxillæ with subcircular bases having 5 or 6 very short irregular lobes, by which neighbouring plates touch, or even imbricate in centre of disk and mid-radial area. Papulæ in 5's and 6's (except in centre of disk and along mid-radial lines where they are absent). Column of paxillæ about as high as breadth of base, flaring at summit, the largest crowned with a coördinate floriform group of about 40 or 45 short, terete, often clavate, round-tipped spinelets; of these about one-half form a peripheral series and are a trifle slenderer and longer. On the smaller paxillæ the spinelets decrease markedly in size, but only slightly in number.

"Superomarginal plates, 15 in number from median interradial line to extremity of ray form an arched bevel to border of abactinal surface; plates shorter than wide, but increase in length on outer half of ray. Plates of both series separated by transverse narrow deep fasciolar grooves and a narrow deep groove (not so deep as transverse grooves) separates superomarginal from inferomarginal series. Superomarginal plates covered with short, terete, blunt granuliform spinelets, similar to but larger than paxillar spinelets, becoming well-defined slender spinelets in fasciolar grooves. Superomarginal covering is to be considered as a spinelet rather than granules.

"Inferomarginal plates much wider than long, encroaching more onto actinal area than do superomaginals onto abactinal, and corresponding in position to superomarginals. Spinelets, densely covering surface of plates, larger than those of superomarginals, and increasing in size toward outer end of plate which projects slightly beyond adjacent end of superomarginal, thus defining the ambitus. Inferomarginal spinelets granuliform in middle of plate, often attaining a squamiform appearance at outer end; spinelets in fasciolar furrows, slender. No enlarged spines of any sort on either marginal series. Terminal plate small, granulose, deeply notched below.

"Actinal interradial areas large; intermediate plates low-paxilliform, arranged in chevrons, the series adjacent to adambulacrals extending about three-fourths length of ray or to eighth inferomarginal. Plates decrease in size toward margin, are strongly imbricated internally, and the paxillar crowns which are composed of about 25 to 30 clavate obtuse, not very crowded, spinelets (slender when dry) surmount a low convex elevation or tabulum.

"Adambulacral plates about as wide as long, with a slightly rounded, angular furrow margin, the angularity being more pronounced in vicinity of mouth plates. Armature consists of (1) a furrow series of 4 (sometimes 3) terete or slightly flattened bluntly pointed tapering spinules about as long as plate and graduated in length orad, the longest spine being on aboral end of plate ; or the spinules may be disposed like rays of fan and graduated in length toward either end of series. (2) On actinal surface are about 3 longitudinal series of smaller spinelets, decreasing in length toward outer edge of plate where the spinelets are like those of actinal intermediate plates. Four spinelets commonly occur in the inner actinal series and about 3–5 in each of the outer ; or the 2 latter series may be wanting, the spinelets, instead, forming an irregular group, especially on outer part of ray where there are frequently upwards to 16 or 20 actinal spinelets.

" Mouth plates narrow, rather prominent actinally, the free margins of the combined plates forming a salient angle into actinostome ; free margin of each plate slightly angular near inner end and longer than the margin adjacent to first adambulacral. Armature consists of a furrow series of about 6 or 7 tapering spinules decreasing in length from the inner enlarged tooth, outward, and thence continued along margin adjacent to first adambulacral in about 9 much smaller spinelets similar to those of actinal intermediate plates. A superficial series of similar spinelets follows margin of median suture, increasing in size toward inner angle of plate, and an incomplete more or less irregular series often, but not always, occurs between marginal and superficial series. There is more or less variation in the details of dental armature.

" Madreporic body rather large, about midway between centre and extreme edge of disk. Striations coarse, centrifugal, very irregular ; madreporic body sometimes nearly hidden by 5 or 6 large paxillæ.

" Type, No. 21926, U. S. Nat. Mus. Type locality, Albatross Station 3310, Bering Sea, in 58 fathoms, on dark sand and mud.

" Remarks.—This species bears a close resemblance to Parastropecten inermis LUDWIG[1], and is probably congeneric with that form, although anomalus has a minute anal pore. The presence of an anal pore is, I believe, a character of scarcely more than specific importance. For instance one species of Astropecten has been shown by VERRILL to possess a minute anus. Although I have not yet had an opportunity to make serial sections of the

1) " Mem. Mus. Comp. Zool., xxxii, July, 1905, 76, pl. iv, fig. 21, 22 ; pl. xxi, fig. 117 pl. xxii, fig. 126. (Gulf of Panama and Cocos Id., 1,271 and 1,408 meters.) "

anal region of *anomalus*, I have been able to make out a tiny pore in 2 specimens, and the intestine leading to the pore is well developed. It may perhaps seem heretical to classify the present species with *Leptychaster*, but *anomalus* differs chiefly from *L. pacificus* in having a larger disk, shorter rays, broader actinal interradial areas, and a slightly different ornamentation on paxillæ and marginal plates. The superomarginals are only a trifle, if any, larger in *anomalus* although the inferomarginals are a little longer and not quite so broad. The chief differences are therefore. in proportion. But *pacificus* is an undoubted *Leptychaster*, an evident offshoot of *articus*, of the circumpolar fauna. It therefore follows in due course that *anomalus* is a *Leptychaster*, although superficially different enough from *kerguelensis*, perhaps to warrant another generic designation if we did not have the intermediate steps.

"Without having examined specimens of *Parastropecten inermis* I hesitate to further question the validity of the genus, although frankly I find no generic characters other than the size of the superomarginals that can separate the form from *Leptychaster*. At any rate, *L. anomalus* differs from *P. inermis* in having fewer furrow spines, more paxillæ spinelets, 5 and 6 papulæ about the very short-lobed roundish plates (instead of 4), and finally in possessing a minute anal pore. The general facies of the 2 forms is strikingly alike."

VERRILL makes a new genus, *Glyphaster*, for the present species, but does not give any generic diagnosis [: 09a, p. 553].

FISHER gives a very detailed description of it as follows [: 11, p. 48,4 figs.] :

"*Diagnosis.*—Rays five. R = 27 mm. ; R = 1.6 r. Breadth of ray at base, 19 mm. Usual form stellato-pentagonal ; disk broad, rays short, broad, and blunt ; interbrachial arcs shallow and wide ; general form depressed ; abactinal surface subplane, capable of slight inflation ; actinal surface slightly arched due to rays bending upward ; marginal plates conspicuous, few in number, devoid of any enlarged spines or specialised armature, but covered with short spinelets ; actinal intermediate areas broad ; adambulacral plates with three or four furrow spinules, and on actinal surface with three longitudinal series of smaller spinelets, decreasing in length toward outer edge of plate ; small superambulacral plates present ; a very small anal pore present.

" *Description.*—Abactinal paxillar area compact; paxillæ arranged in not very regular oblique transverse rows at sides of ray; without order in median radial area and centre of disk. Paxillæ largest at base of ray and in interradial areas, decreasing conspicuously in size toward centre of disk and tip of ray; larger at sides of paxillar area than in midradial region. Column of paxilla about as high as breadth of base, flaring at summit, the largest crowned with a coordinate floriform group of about forty-five short, terete, often clavate, round-tipped spinelets; of these about one-half form a peripheral series and are a trifle slenderer and longer. On the smaller paxillæ the spinelets decrease markedly in size, but only slightly in number.

" Abactinal plates in a prepared specimen, from inner or cœlomic side. Plates small, closer together along midradial line and in centre of disk where there are no papulæ than at sides of area where papulæ are present. Plates of latter region are circular in general shape, but nearly always more or less irregular; with four, five, or six short, abrupt, lobes irregular in length, thickness and in distribution on the periphery of plate. These plates of papular area are arranged in perceptible, although irregular, oblique trans-verse rows (parallel with interradial line). Usually six papulæ occur around each plate (often five, and rarely four) emerging between the irregular lobes by which plates usually touch. Plates along median area of ray are even less regular than others, and frequently there are no lobes at all. They are slightly smaller, and papulæ are absent from a strip about four plates wide. Toward centre of disk, plates decrease rapidly in size, become more crowd-ed, and lobes if present are very inconspicuous and of irregular occurrence. The large primary interradial plate placed just adcentrally to the madreporic canal is convex internally (bears a large paxilla externally), and is slightly concave on the outer edge, next to madreporic canal; three or four other less regular and smaller plates complete the circle about the madreporic canal.

" Superomarginal plates, fifteen in number from median interradial line to extremity of ray form an arched bevel to border of abactinal surface; plates shorter than wide, but increase in length on outer half of ray. Plates of both series separated by transverse narrow deep fasciolar grooves and a narrow deep groove from inferomarginal series. Superomarginal plates

covered with short, terete, blunt, granuliform spinelets, similar to but larger
than paxillar spinelets, becoming well defined, slender spinelets in the fasciolar
grooves. The superomarginal spinelets are close-set and small, forming an
even nap on the exposed surface of plates.

"Inferomarginal plates much wider than long, encroaching more onto
actinal area than do superomarginals onto abactinal, and corresponding in
position to superomarginals. Spinelets, densely covering surface of plate,
larger than those of superomarginals, and increasing in size toward outer
end of plate which projects slightly beyond adjacent end of superomarginal,
thus defining the ambitus. Inferomarginal spinelets somewhat granuliform
in middle of plate, often attaining a squamiform appearance near outer end ;
spinelets in fasciolar furrows, slender. Spinelets on extreme upper or outer
edge of plate (above the squamiform spinelets) are similar to those of adjacent portion of superomarginals, the true ambitus or edge of ray being a
little below the margin of plate. It is on this rounded edge of ray that the
largest spinelets occur. No enlarged spines or tubercles on either marginal
series. Terminal plate small, granulose, deeply notched below.

"Actinal interradial area large ; intermediate plates low-paxilliform, arranged in chevrons, the series adjacent to adambulacrals extending about three-
fourths length of ray or to eighth inferomarginal. Plates decrease in size
toward margin, are strongly imbricated internally, and the paxillar crowns
which are composed of about twenty-five to thirty clavate obtuse, not very
crowded, spinelets (slender when dry) surmount a very low convex elevation
or tabulum. Well-defined fasciolar channels separate these tabula.

"Adambulacral plates about as wide as long, with a slightly rounded,
angular furrow margin, the angularity being more pronounced in vicinity of
mouth plate. Armature consists of (1) a furrow series of four (sometimes
three) terete or slightly flattened bluntly pointed tapering spinules about as
long as plate and graduated in length orad, the longest spine being on
aboral end of plate ; or the spinules may be disposed like rays of fan and
graduated in length toward either end of series. (2) On actinal surface are
about three longitudinal series of smaller spinelets, decreasing in length toward outer edge of plate where the spinelets are like those of actinal inter-

mediate plates. Four spinelets commonly occur in the inner actinal series and about three to five in each of the outer ; or the two latter series may be wanting, the spinelets, instead, forming an irregular group, especially on outer part of ray where there are frequently upward to sixteen or twenty actinal spinelets.

" Mouth plates narrow, rather prominent actinally, the free margins of the combined plates forming a salient angle into actinostome ; free margin of each plate slightly angular near inner end and longer than the margin adjacent to first adambulacral. Armature consists of a furrow series of about six or seven tapering spinules decreasing in length from the inner enlarged tooth, outward, and thence continued along margin adjacent to first adambulacral in about nine much smaller spinelets similar to those of actinal intermediate plates. A superficial series of similar spinelets follows margin of median suture, increasing in size toward inner angle of plate, and an incomplete more or less irregular series often, but not always, occurs between marginal and superficial series. There is more or less variation in the details of dental armature. The exposed, outer, slightly convex surface of the combined plates has the appearance of being covered with a bristling armature of short papilliform spinelets, very similar to those on adjacent actinal intermediate plates.

" Madreporic body rather large, about midway between centre and extreme edge of disk. Striations coarse, centrifugal, very irregular ; madreporic body sometimes nearly hidden by five or six large paxillæ.

" Superambulacral plates present, though small. Absent from the first ambulacral plates, and from the distal six or seven, which are much reduced and crowded against the adjacent inferomarginals. Gonads forming a tuft of tubules on either side of the interradial septum, five or six tubules (two or three times dichotomously divided) to each tuft. The gonads do not consist of a series of tufts extending along the rays as in *Dipsacaster*. A Polian vesicle in each interradius. Interradial septa uncalcified.

" *Variations.*—The specimens assembled under this form present a very considerable amount of variation, and when the extremes are placed side by side it is hard to believe that there are not two valid species. But there

is such a bewildering number of more or less perfect intermediate stages
that one is forced to range them all under one head.

" The most important variations occurring in this species are in respect to
dimensions and the size of the marginal plates. Some examples have a
more stellate form, while others verge onto an arcuate pentagonal shape.
Again, one extreme has fairly massive superomarginals, which are large up
to the tip of the very blunt ray, while in the other extreme, the upper
marginals are more numerous and relatively smaller, and the tip of the ray
is not so blunt. The latter form is not so numerous as the first, which is
typical. The effect of this difference of shape on the relative dimensions is
shown in the following table :

" Table showing effect of shape on relative dimensions in *Leptychaster*
anomalus.

Station	R.	r.	R : r	Breadth of ray at base.	Number of supero-marginals.
3310	27	17	1.61 : 1	19.0	15
4281	21	13	1.62 : 1	16.0	10
3334	25	15	1.7 : 1	16.5	13
4281	29	17	1.7 : 1	19.0	13
3486	28	14	2 : 1	16.0 —	18

" The number of spinelets on the outer half of the adambulácral plates
(i.e., on actinal face) varies slightly, and they are a trifle longer and
slenderer in some individuals than in others. The tip of the ray in this
species is rounded and blunt, while it is sharp in *pacificus*. This, with the
fewer and more transversely placed inferomarginal plates of *anomalus*, will
serve to distinguish the longer rayed individuals from *pacificus*. Variation
within narrow limits occurs in the size of the paxillæ, in general the examples with more massive marginals having the larger paxillæ. The greatest
difference in general facies is caused, however, by the variation in size of
marginals referred to above. A slight difference in width causes a varying
amount of encroachment upon the abactinal paxillar area. In general the

shorter armed individuals have wider and more massive marginal plates, but this is not invariably the case. The extreme variants in proportions are found at the same station.

"*Young.*—The smallest specimen (station 4538) measures R = 8 mm., r = 6 mm.; rays broad, stout, blunt, with seven rather massive superomarginals, which are larger relatively than in adults. Paxillæ with nine or ten peripheral and two to five central spinelets; adambulacral plates with four furrow spines. Terminal plate small. This species has much larger marginal plates than equal sized *Pseudarchaster pusillus*, and the latter species has very large terminal plates and spiny inferomarginals, so that there is no danger of confusing the two forms in a superficial examination.

"*Type.*—Cat. No. 21926, U.S.N.M.

"*Type-locality.*—Albatross station 3310, vicinity of Unalaska, Alaska, 58 fathoms, on fine dark sand and mud.

"*Distribution.*—Bering Sea (vicinity of Pribilof Islands and west of St. Paul) to southeast Alaska, and off Monterey Bay, California; on the Asiatic side to the Sea of Japan. Bathymetrical range, 32 to 688 fathoms in Bering Sea and northern part of range, to 871 fathoms off Monterey Bay. Found on fine gray or black sand, green mud, or on pebbles.

"*Specimens examined.*—The following is a complete list of localities from which one hundred and seventy-eight specimens have been examined:

"Specimens of *Leptychaster anomalus* examined.

Station.	Locality.	Depth.	Nature of bottom.	Number of Specimens.	Collection.
2847	Vicinity of Shumagin Islands, Alaska..........	Fathoms 48	fine gray sand..........	1	U.S. Nat. Mus.
2848	do..........	110	green mud..............	13	do.
2849	do..........	69	do.	3	do.
2852	do..........	58	black sand..............	2	do.
2854	Off Kadiak Island, Alaska	60	do.	1	do.
2855	Off Sitkalidak Island, vicinity Kadiak Island Alaska....................	69	green mud..............	10	do.
3223	Bering Sea, vicinity of Unalaska, Alaska..........	56	black pebbles..........	1	do.

Station.	Locality.	Depth.	Nature of bottom.	Number of Speci-mens.	Collection.
		Fathoms			
3224	North of Unimak Island, Alaska...............	} 121	black sand and gravel ..	1	U.S. Nat. Mus.
3257	do............	81	gray sand and gravel....	2	do.
3263	do	61	black mud	12	do.
3310	Vicinity of Unalaska, A-laska	} 58	fine dark sand and mud	15	do.
3311	do	85	green mud	28	do.
3313	do............	68	fine black sand	50	do.
3334	do............	56	mud and sand..........	1	do.
3486	West of St. Paul Island, Bering Sea	} 150	green mud, fine sand....	2	do.
3488	do	106	green mud, gray sand....	5	do.
3501	South of St. George Is-land, Bering Sea	} 688	green mud	1	do.
3548	Near Unimak, Aleutians.	91	black mud	1	do.
3602	Bering Sea, vicinity of Pribilof Island	} 81	green mud, sand........	1	do.
	Shumagin Islands, Alaska	58	....	1	do.
	Unalaska...............	80	sand...................	1	U.S. Nat. Mus.,W.H. Dall, No. 6073.
	No locality [1]	....	....	9	U.S. Nat. Mus.
4230	Vicinity of Naha Bay, Behm Canal, southeastern Alaska................	}240-108	rocky.................	3	Albatross, 1903.
4233	Vicinity of Yes Bay, Behm Canal	39-45	grey mud, rocky........	1	do.
4265	Off Sitka Sound, Alaska..	590[2]	green mud, rocky	1	do.
4280	Chignik Bay, Kadiak Is-land...................	} 32	green mud, black sand..	2	do.
4286	do............	57-63	green mud, rocks	2	do.
4538	Off Monterey Bay, Cali-fornia	}871-795	gray sand and rocks	1	Albatross, 1904.
4768	Bowers Bank, Bering Sea	764	greenish brown mud....	3	Albatross. 1906.
4775	do............	584	green mud	1	do.
4784	Vicinity of Attu Island, Aleutians	} 135	coarse pebbles..........	1	do.
4818	Sea of Japan (38°08'N.; 138°31'E.)	225	fine brown mud........	1	do.
4867	Sea of Japan............	150	green mud	1	do.

1) " Bottles broken by earthquake, 1906."
2) " Not typical."

" *Remarks.*—The reasons which may be advanced for placing this species under the genus *Leptychaster* are the following : *Leptychaster arcticus*, which ranges into Bering Sea, has in the same region a close relative, *pacificus*, which differs in possessing broader and less numerous marginal plates. The superomarginals of *arcticus* are throughout the ray scarcely larger than paxillæ, but are of conspicuous size in *pacificus*, and in the interbrachial region are considerably wider than midway along ray. After a close comparison of the two forms it has not been possible to separate them generically. The character of the marginals, abactinal paxillæ, adambulacral plates and armature, distribution of gonads, and distribution of papulæ are essentially the same in the two forms. The width of the superomarginals remains the most conspicuous difference. By the same reasoning it is impossible to separate generically *anomalus* from *pacificus*. In the former the width of the superomarginals has been found to vary considerably ; hence if this character is unstable for a species it obviously can not be used to diagnose a genus. The wholly unarmed marginals remain, however, very characteristic of the genus. Thus *L. arcticus* and *L. anomalus* represent two extremes, with *pacificus* in between. *L. propinquus*, described below, somewhat resembles *L. kerguelensis* when viewed from above, and still more *Mimaster cognatus*, although the interbrachial arcs are rounder and the inferomarginals wider in proportion to length. As noted under *Gephyreaster swifti*, in view of the range of variation within the genus *Leptychaster*, it is very doubtful if *Mimaster cognatus* is a *Mimaster*. It seems to be a *Leptychaster*.

" *Leptychaster anomalus* bears a striking resemblance to *Parastropecten inermis* LUDWIG, which appears to be based on young specimens, however. From this species *anomalus* differs in having three or four furrow spinelets instead of six or seven, in having five or six papulæ about each plate or paxilla instead of four, and in having a minute anal opening.

" The diagnosis given by LUDWIG for *Parastropecten* is as follows [: 05, p. 76] : ' Die neue Gattung unterscheidet sich von *Astropecten* durch den völligen Mangel von unteren und oberen Randstacheln, durch kurze Arme und durch verhältnismässig grosse ventrale Interradialfelder, deren

Platten (= Ventrolateralplatten) eine paxilläre Form haben.' In other words, the genus is erected on the strength of the unarmed marginals and large actinal areas—just the features in which *Leptychaster* differs from *Astropecten*, for the size of the superomarginals has been shown to be variable within a species, and not diagnostic of a genus.

"The impossibility of recognising a separate genus *Parastropecten* for *inermis* and *anomalus* is the fact that *Leptychaster propinquus* would have to be ranked under *Leptychaster* on account of small superomarginals, *L. anomalus*, of course, under *Parastropecten*, *L. pacificus* probably under *Parastropecten*, on account of the larger superomarginals, although it is a close relative of *L. arcticus*. To divide the genera on the presence or absence of an anus would lead to the ranking of *Parastropecten inermis* and *Leptychaster arcticus* under one genus and the other forms under another, and would necessitate ignoring the very characters upon which we base genera.

"Consequently, although *Parastropecten* seemed fully warranted when described, it seems best now to merge it with *Leptychaster*.

"The genus *Glyphaster* VERRILL based (without diagnosis) on *L. anomalus* seems to be synonymous with *Parastropecten*, which would therefore have precedence if *anomalus* constituted a separate genus. Since there are intergrading forms with *arcticus*, I have no hesitation in saying that neither *Glyphaster* nor *Parastropecten* can be so diagnosed as to include *Leptychaster propinquus* and exclude typical *Leptychaster*.

"I have examined two small specimens from the Sea of Japan, collected by the *Albatross* in 1906. If the tiny specimen from off Monterey is certainly *anomalus*, the species probably has a continuous distribution by way of Bering Sea. It is of interest to note that the Japanese examples are fairly typical. Both of them have epiproctal cones, and in one the tiny anal pore is visible (in sunlight) with a strong glass."

This species is not contained in any of the collections mentioned at the outset.

Dipsacaster.

The genus *Dipsacaster* was set up by ALCOCK ['93, p. 87] with the following diagnosis:

"Disk large, with flat rigid rays of moderate length. Abactinal surface of disk and rays with compact definitely arranged paxillæ and numerous intervening papulæ.

"Marginal plates with highly developed ridges and fasciolar channels; the inferomarginals with enlarged spines.

"Actinal interradial areas large, with numerous intermediate plates extending far along the rays, and bearing compact rosettes of spinelets.

"Adambulacral plates with a palmate or pectinate furrow-series of spinelets, and actinally with several series of spinelets arranged in rosette-like groups.

"Madreporiform plate large but concealed.

"No anus exists, and though a minute pore is present in the centre of the disk, it has no communication with the lumen of the intestine. No pedicellariæ."

There have been described seven species, *D. sladeni* ALCOCK ['92, p. 87], *D. pentagonalis* ALCOCK ['93a, p. 172], *D. eximius* FISHER [: 05, p. 296 ; : 11, p. 86], *D. nesiotes* FISHER [: 06, p. 1026], *D. borealis* FISHER [: 10, p. 546 ; : 11, p. 91], *D. laetmophilus* FISHER [: 10, p. 546 ; : 11, p. 95], and *D. anoplus* FISHER [: 10, p.547 ; : 11, p. 97], the first two from the Indian Ocean and the rest from the Pacific.

In view of the characters presented by the species described below, it is necessary to modify the original generic diagnosis slightly by striking out the phrase "the inferomarginals with enlarged spines" and inserting "or more or less soft" between "rigid" and "rays of moderate length." It appears to me not necessary to set up a new genus for the sake of these characters. In having the inferomarginal plates destitute of any specialised spines this species is like *D. anoplus*.

Dipsacaster grandissimus, n. sp.

(Pl. VIII, figs. 136–139; Pl. IX, figs. 140–141.)

This is a very characteristic species, different from any one of the genus yet described, the most striking feature being the softness of the body, especially on the abactinal side. Its great size is also remarkable. I have only one specimen, in which the marginals are more or less disjointed, owing partly to maceration, but more, as I take it, to the softness of the intervening tissue. The whole body is very flat, and a slightly elevated carinal ridge is seen to be present in each arm, all meeting at the centre of the disk, so that there is left a depressed area in each interradius, similarly as in some species of *Pentagonaster*. The disk is very large; the arms are very broad at the base and rounded at the end; so that the body is regularly and broadly 5-radiate.

The mouth is largely open in my specimen, and the ambulacral furrows are also very wide; but these conditions evidently depend much on the contraction of the muscles. The ambulacral feet are very large and appear to be provided with a small sucker. My specimen measured as follows :

r mm.	R mm	R : r	MS
45	115	2.5	30

Superomarginals.—These are very conspicuous, but do not form the lateral forder of the body, except near the tip of the arms, where they are either flush with the inferomarginals or slightly more projecting. At the interbrachial arc the superomarginals are three times as broad as long, but they become relatively much longer distally. At a short distance from the tip of the arms the superomarginals are mostly seen to decrease in breadth rather suddenly, so that the paxillar area is more or less broadened out (Pl. IX, fig. 140). From about the middle of the arms outwards the superomarginals are very nearly twice as broad as long. In my specimen almost all of the superomarginals are denuded of their armature, but from what little scraps left on them I infer that they were uniformly and thickly covered with subcapillary spinelets of a silky appearance. I imagine that there were no prominent spines, as the surface of the superomarginals presents a very finely and uniformly granulated appearance. The fasciolar grooves are rather shallow but well developed, and are covered with capillary spinelets.

Inferomarginals.—The inferomarginals are much more projecting than the superior series at the interbrachial arc and the greater part of the arms, and form the lateral margin of the body, except for a short stretch near the tip of the arms. Like the superomarginals they are about three times as broad as long at the interbrachial angle, but become relatively much longer towards the tip of the arms. The two marginals are coincident along nearly half the length of the arms, but in the more distal part they are alternate. The inferomarginals are uniformly covered with a thick coat of very fine, somewhat flattened spines of a silky appearance. The fasciolar grooves are covered with very short capillary spinelets.

Adambulacrals.—The adambulacrals are more numerous than the inferomarginals by only four or five for the entire arm. As

seen from the surface the adambulacral plates are nearly square
near the mouth, but they soon become longer than broad towards the middle of the arms, and again become as broad as
long toward the tip of the arms. At the furrow margin on
each plate there is a series of very conspicuous, long spines, 7–9
in number, with the central ones longer and those towards the
two ends shorter (Pl. VIII, fig. 138). These spines are very
stout, pointed at the end, angular and flattened transversely to
the furrow, so that a cross-section of any one of them would be
a regular rectangle. On the actinal surface of the plate there
are some very sharply pointed spines with enlarged bases apparently without any regular arrangement, some twenty in number.
These spines are much shorter and more slender than those of
the furrow series.

Mouth-plates.—These are very large, and when denuded of
their spines are seen to be lightly ankylosed together for a short
distance at the mouth end. Each plate is comparatively long
and narrow, and the outline of its external margin is somewhat
like that of a brace (Pl. VIII, fig. 137). Along the edge bordering on the ambulacral furrow there is a conspicuous series of
very stout, angular, flattened spines, mostly with truncated
ends, ten to a dozen in number (Pl. VIII, fig. 136). The spine
at the mouth end is stout, but it does not project prominently
towards the mouth ; the spines at the farther end of the series
is smaller and more pointed in form. The above mentioned
furrow series of spines encloses with its fellow an area on the
mouth-plates, in which there are usually from a dozen to fifteen
sharply pointed spines, considerably smaller than those of the
furrow series. They are usually more or less conical in form,

but they may also be more or less like the oral spines in shape. The area bearing these conical spines, is generally depressed, and a distinct furrow is present on each plate (Pl. VIII, fig. 136, 137). The abcentral part of the surface of each mouth-plate is thickly covered with numerous spines, not very long, irregular in shape, some pointed but most of them with rough ends, and striated lengthwise. The spines along the border facing the first adambulacral plate are pointed and conical. The spines covering the abcentral part of each mouth-plate are some fifty or sixty in number, making from one hundred to one hundred and twenty for each pair of mouth-plates.

Ventrolaterals.—The ventrolateral areas are very large, and the plates very numerous, there being about 185 in each interradius. They are regularly arranged in obliquely transverse rows corresponding to each adambulacral plate, and reach out to near the end of the arms. The rows corresponding to the first adambulacrals do not reach the marginals, and there are five or six plates for these two adambulacrals in each interradius. The plates next the adambulacrals are tolerably large, but those away from them gradually diminish in size outwards, so that the plates adjoining the marginals in the interbrachial arc are very small. The ventrolateral plates are uniformly and thickly covered with very small spines exactly like those of the inferomarginals (Pl. VIII, fig. 138). They appear to be always directed towards the lateral margin of the body. Between the plates is left enough soft tissue to allow them a good deal of movement.

Paxillæ.—The paxillæ are well spaced and a good deal of soft tissue is present between their bases. When the coronal spinelets are removed and the tabulum is taken out, it is seen to be somewhat hour-glass shaped, the top of the tabulum being especially well rounded (Pl. VIII, fig. 139). The paxillæ are largest on the madreporite, where they are mostly flattened from side to side, so that on a top view they appear more or less ellipsoidal ; they are not, however, so tall as in the other parts of the body. Outside of the madreporite the largest paxillæ are found in the more central part of the disk and along the carinal ridges of the arms. Near the end of the arms and the margin of the body they are smaller. The coronal spinelets are very fine, pointed and slender ; there is no distinction of centrals and peripherals, but those borne along the periphery are usually shorter. The spinelets are close together, so that the general shape of the corona is that of a rounded cone (Pl. VIII, fig. 139). On one of the larger paxillæ near the centre of the disk 45 coronal spinelets were count-ed. There is no area of small paxillæ at the centre of the disk. The papular pores are very numerous and comparatively large, there being sometimes as many as four or five in an area enclosed by any four paxillæ ; so that in some places the dorsal wall of the body presents a reticulated appearance. The papulæ are distributed almost uniformly on the whole abactinal surface of the body.

Madreporite.—The madreporite is very large, being bout 12 mm. in diameter in the specimen here described, depressed, circular, of very uneven surface, the larger elevations and depressions being arranged radially. The whole surface is traversed by very minute grooves. It is situated a little nearer the margin than the centre of the disk. It bears on its surface about twenty large paxillæ, and is not conspicuous in an entire view of the animal.

Terminal plate.—The terminal plate is relatively very small, and nearly round.

Locality.—Off Misaki, from a depth of 640 m.
Specimen in S. C.

Craspidaster hesperus (Müller & Troschel).

This species was first described under the name of *Archaster hesperus* by Müller and Troschel ['40, p. 323], and redescribed in the "System." The following is extracted from the latter ['42, p. 66]:

" *Archaster hesperus* Nob.

" *Archaster hesperus* Müll. Tr. Monatsbericht der Acad. der Wissensch. zu Berlin 1840, April: Wiegmann's Archiv, Jahrgang VI, Band II, p. 323.

"Fünf Arme. Der kleine Halbmesser verhält sich zum grossen wie 1 : 3¼. Die Winkel zwischen den Armen sind scharf, oder kaum ausgerundet. Die Furchenpapillen bekleiden nicht bloss den Rand der Furchen, wo ihrer fünf grössere auf einer Platte stehen, sondern begränzen auch kleiner die übrigen Ränder ihrer Platte. An jedem Arme 25–30 Randplatten. Die Bedeckung derselben wie bei der vorigen Art.[1] Die Papillen des Rückens stehen auf den Armen in queren Reihen, ohne eine mittlere zusammenhängende stärkere Reihe zu bilden. Auf dem mittlern Theil der Scheibe, in der Nähe des Afters, sind die Paxillen sehr klein und zahlreich. Die Madreporenplatte liegt näher dem Rande als der Mitte.

"Farbe: im trocknen Zustande weiss.

"Grösse: 3 Zoll.

"Fundort: Japan. Im Museum zu Berlin durch Capitain Wendt; im Museum zu Leyden durch v. Siebold; auch im Museum zu Paris."

It is described by Möbius under a different name and as a new species ['59, p. 11]:

" *Stellaster* Gray *sulcatus* n. sp.

"Taf. IV. Fig. 1 und 2.

"Pentagonal und flach, mit ausgerundeten wenig (gegen 100°) stumpfen Armwinkeln. Das Verhältniss des kleinen zum grossen Radius ist gleich

---

1) *Archaster typicus* M.T.

1 : 3. Die Breite der Randplatten ist grösser als ihre Länge. Die innere (der Scheibe zugewandte) Kante der im Armwinkel liegenden dorsalen Randplatten misst $\frac{1}{4}$ des kleinen Radius. Die Randplatten und die Platten des dreieckigen Interambulacralfeldes sind durch tiefe Furchen von einander getrennt.

"Am innern Rande der Saumplatten stehen je 6 runde, dünne Stacheln und meistens ebenso viel halb so grosse auf dem äussern Rande in gebogener Linie. Mit ähnlichen Stacheln ist der obere Umfang der Rückenplatten umkränzt; noch dünnere stehen in den Furchen zwischen den Randplatten und zwischen den Interambulacralplatten.

"Auf der Scheibe treten die Rückenplatten hoch rund-warzenförmig, auf den Armen weniger hoch und eckig hervor. Die Rückenplatten sind unregelmässig vertheilt; die Armplatten ordnen sich in schiefe Reihen. Die Randplatten, die Interambulacralplatten und die Endfläche der Rückenplatten sind mit kleinen runden Körnern bedeckt.

"Die beweglichen Stacheln an den ventralen Randplatten sind platt und oberhalb ihres kurzen Stieles parallelrandig. Die Poren liegen einzeln zwischen den Rückenplatten. Der After ist im Centrum, die Madreporenplatte mitten zwischen diesem und dem Rande.

"Das Hamburger Museum besitzt zwei getrocknete Exemplare, deren grosser Radius 35 und 36 mm beträgt. Ihre Fundort ist nicht angegeben."

According to LUDWIG this species was described in 1862 by GIEBEL ['62] as a new species under the name of *Astropecten gracilis* (*vide infra*).

DUJARDIN and HUPÉ's description is as follows ['62, p. 412] :

"*Archaster hesperus.*—MÜLLER et TROSCHEL.

[References to MÜLLER and TROSCHEL '40 and '42].

"Espèce à cinq bras, dont la longueur, à partir du centre, égale trois fois et demie le plus petit rayon du disque. Les angles rentrants interbrachiaux ont leurs côtés rectilignes ou légèrement arrondis. Les piquants du sillon ambulacraire sont inégaux : cinq plus grands occupent le bord interne de chaque plaque ; les autres, plus petits, sont insérés sur le reste de cette plaque. Chaque bras porte 25 à 30 plaques marginales revêtues d'écailles comme dans l'espèce précédente." Les papilles du dos forment

1) *Archaster typicus.*

seulement des rangées transverses sur chaque bras, et non une rangée médiane, plus forte sur le milieu du disque, autour de l'anus. Les papilles sont très-petites et peu nombreuses.

" La plaque madréporique est située plus près du bord que du centre.

" Coloration blanchâtre (après dessication). Dimension : largeur totale 81 mm.

" Habite les côtes du Japon (Mus. de Paris)."

The same species is described by Lütken who refers it to *Archaster* ['64, p. 136] :

" I det nysnævnte Værk beskriver og afbilder Möbius fremdeles en Söstjerne under Navnet *Stellaster sulcatus* (p. 2.,[1] t. IV f. 1–2) efter to törrede Exemplarer, uden Lokalitetsangivelse, i Hamborgs Museum ; jeg gjenkjender denne Art i et ligeledes törret Exemplar, som længe har henligget i det tidligere ·kongelige Museum, ligeledes unden Lokalitetsangivelse, men har derved tillige havt Leilighed til at overbevise mig om, at det ikke er en *Stellaster* men en *Archaster*.[2] Der er ingen Lighed mellem de smaae Paxiller, hvormed hele Rygfladen er beklædt, og de store kornede Rygtavler hos en *Stellaster*; dens hele Bygning er

In the last mentioned work Möbius also describes and figures a starfish under the name *Stellaster sulcatus* (p. 2.,[1] t. IV, f. 1–2) after two dried examples, without statement of locality, in the Hamburg Museum; I recognise this species in a likewise dry example, which has lain for a long time in the former Royal Museum, also without statement of locality, but has thus at the same time given occasion to convince me that it is not a *Stellaster* but an *Archaster*.[2] There is no similarity between the small paxillæ, with which the whole dorsal surface is clothed, and the large granulated dorsal plates of a Stellaster ; its

1) Evidently a misprint of 11.

2) " Paa det foreliggende Exemplar kan Tilstedeværelsen af Anus ikke erkjendes, da det har været gjennemstukket af en Naal lige i Centrum, hvor Anus jo ligger hos Archasterne, og hvor den jo ogsaa efter Möbius's Beskrivelse har ligget (hvorimod Stellasterne have en subcentral Anus). Med *Archaster hesperus* M. Tr. har jeg ikke havt Leilighed til at sammenligne den og kan derfor ikke udtale mig om dens Forskjellighed fra eller mulige Overeensstemmelse med denne Art."

2) In the example lying before me the presence of the anus can not be recognised, as it had been pierced by a needle exactly in the centre, where the anus certainly lies in the Archaster-species, and where it lies also according to Möbius' description (on the contrary the Stellaster-species have a subcentral anus). I have not had occasion to compare it with *Archaster hesperus* M. Tr. and therefore can not pronounce on its difference from or possible coincidence with this species.

overhovedet aldeles *Archaster-* eller *Astropecten-*agtig, som man vil see af fölgende Charakteristik.

"Armene ere middellange, temmelig smalle og spidse, Armvinklerne runde, den store Radius 36 mm, den lille 11 mm, Randpladernes Antal 24-25. Midt paa Armene er Paxilfeltets Brede lig med Randpladernes; det bestaaer der af meget smaae Paxiller, paa Skiven blive disse större, endog temmelig store, men atter smaae mod Midten af Skiven. Jo större de ere, desto tydeligere seer man, at de dannes af to Slags Papiller, deels korte, tætpakkede, lige afskaarne, der beklæde Paxillens Endeflade, deels længere, smækkere, vandrette og mere cylindriske, der i Form af en Krands indfatte dens Rand. Madreporpladen formoder jeg at være skjult tilstede i en ved sin Störrelse udmærket Paxil, som ligger omtrent midtveis mellem Skivens Midtpunkt og Randpladerne. Disse ere tæt kornede af flade Korn, som ganske ligne dem, der beklæde Paxillernes Topflade; langs med de Furer, som adskille Pladerne, erstattes de af længere, finere, vandrette, börsteagtige Papiller.[1] De övre Randplader savne ethvert Spor til Pigge; paa hver af de nedre findes derimod i Nærheden af dens övre Rand en lille Knude, paa hvilken en bred, flad og but, men temmelig

whole structure is as a whole entirely like that of *Archaster* or *Astropecten,* as one can see from the following characterisation.

The arms are moderately long, rather slender and pointed, arm angles round, the great. radius 36 mm, the small 11 mm, number of the marginal plates 24-25. In the middle of the arms the breadth of the paxillar area is equal to that of the marginal plates; the area consists there of many small paxillæ, on the disk these become larger, even moderately large, but again small towards the centre of the disk. The larger they are, the more distinctly one sees, that there are two kinds of papillæ, those which are short, closely packed, squarely truncated, which clothe ·the top face of the paxilla, and those that are longer, slenderer, horizontal and more cylindrical, which border its margin in the form of a crown. The madreporic plate I suppose to lie hidden by a paxilla distinguished by its large size, and situated about midway between the centre of the disk and the marginal plates. These are thickly covered with flat granules perfectly similar to those which clothe the top surface of the · paxillæ; along the grooves that separate the plates they are replaced by longer, finer, horizontal, bristle-like papillæ.[1]   The

---

1) "Analogien mellem Randpladerne og Paxillerne er saaledes meget let at opfatte hos denne Art."

1) "The analogy between the marginal plates and the paxillæ is thus so evident in this species as to be noted [by any one]."

kort, tiltrykt Randpig [1] (kun et enkelt Sted to) er indleddet. Bugsidens övrige Beklædning bestaaer af : 1) Adambulacralpladerne, hvis Antal er omtrent dobbelt saa stort som Randpladernes, og som hver ere udstyrede med en Krands af 12–15 Papiller, af hvilke de 4–5 længste vende ind mod Fodgangen og de korteste deels mod Munden, deels mod Armens Spids ; 2) af de Smaatavler, som danne det her forholdsvis vel udviklede Bugfelt,[2] der dog ikke strækker sig ud paa selve Armene, men kun til den 10de Adambulacralplade omtrent. Der er i hvert Bugfelt 5–6 större Smaatavler paa hver Side, dannende een med Adambulacralpladerne parallel Række, samt en lille Gruppe, sammensat af en halv Snees meget smaae Tavler, mellem hine og Armvinklernes Randplader.[3] Alle Bugfeltets Smaatavler ere kornede paa deres Overflade og fryndsede i Randen ligesom Randpladerne og Paxillerne."

1) "Det er maaskee denne Pigs Lighed med Stellasternes 'hængende Randpig', som har forledt M. til at gjøre denne Søstjerne til en *Stellaster*."

,2) "Som bekjendt er dette overmaade lidet, næsten forsvindende hos de fleste Kamstjerner. Meest udviklet er det—paa Grund af de afrundede Armvinkler—hos *Archaster Parelii* og *Astropecten Andromeda*."

3) "Denne Bugfeltets Sammensætning er ikke heldig gjengivet paa Afbildningen."

upper marginals are destitute of all traces of spines; on each of the lower there is however in the neighbourhood of its upper margin a small node, on which a broad, flat and blunt, but rather short, adpressed marginal spine [1] (only in a single place two) is placed. The remaining covering of the ventral side consists of : 1) the adambulacral plates, whose number is nearly twice that of the marginal plates, and which are each covered with a crown of 12–15 papillæ, of which the 4–5 longest are directed inwards towards the ambulacral furrow and the shortest partly towards the mouth, partly towards the apex of the arm ; 2) of the tablets which form the ventral area,[2] relatively well developed in this species, which however does not extend into the arms, but only to about the 10th adambulacral plate. In each ventral area there are 5–6 larger tablets on each side, forming a series parallel with the adambulacral plates, together with a small group, composed of some half score very small plates between the adambulacral plates and the marginal plates of the arm angle.[3] All the tablets of the ventral area are granulated on their upper surface and fringed at the margin like the marginal plates and the paxillæ.

1) It is perhaps the similarity of this spine with the 'hanging marginal spine' of the Stellaster-species, which has misled M. to refer this starfish to *Stellaster*.

2) As is known this is exceedingly small, well-nigh *nil* in most Astropectens. It is most developed—on account of the rounded arm angles—in *Archaster Parelii* and *Astropecten Andromeda*.

3) This composition of the ventral area is not reproduced to advantage in the figure.

VON MARTENS mentions it from Japan and adds some important remarks ['65, p. 353]:

" *Archaster hesperus* MÜLL. TROSCH. l.c. S. 66. *Stellaster sulcatus* MÖBIUS Abhandl. der naturf. Gesellschaft zu Hamburg IV. Taf. 4. Fig. 1. 2.

" Japan, Berliner und Leidner Museum.

" Die Abbildung und Beschreibung bei MÖBIUS stimmt vollkommen zu den drei Originalexemplaren von MÜLLER und TROSCHEL im Berliner Museum. Selbst die Veränderlichkeit in Zahl und Stellung der kleinen Bauchplatten im Armwinkel bei einem und demselben Exemplar findet sich an den unsrigen ebenso wie in der von MÖBIUS gegebenen Figur. Die Rückenbedeckung verweist diese Art zu Archaster, indem sie aus stark convexen, kleinen und runden Stücken besteht, welche mit Stacheln besetzt sind oder doch waren. Die oberen Randplatten sind auch so schmal wie bei ben ächten *Archaster*, der Stachel der untern dagegen gleicht ganz dem von *Stellaster*, geht aber sehr leicht verloren. Die Spaltung der innersten Bauchplatte erinnert an *Ogmaster* (vergl. unten), ist aber ebenso auch bei *Archaster typicus* vorhanden. Beide *Archaster*[1] unterscheiden sich auch dadurch von *Stellaster* und überhaupt allen zu *Goniaster* gehörigen Gruppen, dass die Bauchplatten des Interbrachialraums, mit Ausnahme der innersten, paarig angeordnet sind, so dass eine Linie vom Mund zum Armwinkel stets zwischen den Platten verläuft, während sie bei Goniaster abwechselnd zwischen zwei Platten hinein und in die Mitte einer Platte trifft."

GRAY simply refers to this species for comparison under *Astropecten stellaris* (= *Archaster typicus*) in his " Synopsis " ['66, p. 3].

VON MARTENS again mentions this species under the name of *Archaster hesperus* from the South China Sea ['67, p. 112]:

" Ein kleines Exemplar (Scheibenradius 7, Armradius 23 Mill.) in der südchinesischen See auf dem Wege von Singapore nach Siam 14. November 1861 mit dem Schleppnetz aufgefischt.

" Zu der MÜLLER'schen Beschreibung kann nach diesem Exemplar sowohl als dem Originalexemplar der Berliner Sammlung hizugesetzt werden, dass die ventralen Randplatten ausser dem einen grossen platten Stachel

---

1) *A. typicus* and *A. hesperus*.

keine kleineren, sondern vier Körnchen tragen, und dass alle Randplatten durch merkliche Zwischenfurchen von einander getrennt sind, im Gegensatz zu *A. typicus*."

PERRIER, after giving references to the descriptions above reproduced, adds the following remarks ['76, p. 267]:

"Deux individus desséchés, du voyage de PÉRON et LESUEUR, en 1803, sans autre indication géographique que celle-ci: Mers australes.

"Cette espèce est en réalité japonaise (capt. WENDT, musée de Berlin; SIEBOLD, musée de Leyde)."

Under the name of *Astropecten macer* it is described as follows by SLUITER ['89, p. 311]:

"*A. macer* (n. sp.). Der Körper mit fünf Armen, sehr stark abgeplattet. Die Arme, an der Basis 12 mm. breit, verjüngen sich allmählich nach der Spitze zu, ohne eine Spur einer Anschwellung etwas von der Basis entfernt, wie eine solche bei den von PERRIER beschriebenen abgeplatteten Formen vorkommt (*A. Richardi*, *A. alatus* und *A. spatuliger*). R. = 50 m.M., r. = 15 m.M., also R = 3⅓ r. Die Höhe beträgt nur 5 m.M. Die Arme sind schlank.

"Die Platten, welche die Ambulacralfurchen begrenzen, tragen jede 7 Papillen, von welchen 5 etwa gleich lang, die zwei äussersten aber kürzer sind. Neben diesen verläuft eine Reihe fast viereckiger Plättchen, welche der ganzen Ambulacralfurche entlang bis an die äusserste Spitze des Armes zu unterscheiden sind. Die ventralen Randplatten erreichen also die Ambulacralfurchen nicht. Die der Ambulacralfurchen entlang verlaufenden Bauchplättchen tragen an drei Seiten kleine Dörnchen, nur nicht an der Seite, wo sie an die Ambulacralplättchen grenzen. Ausserdem kommen auf der Scheibe an der Bauchfläche, in jedem Dreieck zwischen den oben erwähnten Plättchen, neben den Furchen und den ventralen Randplatten, noch 20 oder 21 ziemlich regelmässig angeordnete Bauchplatten vor. Diese Platten tragen am Rande zahlreiche kleine Dörnchen oder Borsten, wie die Paxillen des Rückens, nur ist die centrale Scheibe viel grösser, und sind sie als riesige Paxillen zu betrachten.

"Die 36 ventralen Randplatten tragen zur Bildung der Ränder der

Arme für etwa die Hälfte der Höhe bei. In den Winkeln der Arme sind sie länglich, auf der Hälfte der Arme fast viereckig. Sie sind nur mit sehr weit auseinander stehenden, sehr kleinen Wärzchen versehen. Nur an den Armrändern finden sich einige grössere Tuberkelchen, und in den Armwinkeln noch ein oder zwei spitze Stacheln. Alle diese Platten tragen am Rande einen grössern platten Stachel, welcher aber sehr leicht abfällt. Ausserdem ist jede Platte noch an ihren Rändern, nach Art der Paxillen, mit kleinen Borsten versehen.

" Die dorsalen Randplatten sind ebenso zahlreich wie die ventralen, ohne Stacheln, aber fein granuliert. Die Granula werden nach dem äussern Rande zu etwas gröber. Jede dieser Platten ist wieder, nach Art der Paxillen, am Rande von kleinen Dörnchen umgrenzt. Auf der Mitte der Arme ist das Paxillenfeld am Rücken nur wenig breiter als die Dorsalplatten. Die Paxillen haben eine mehr oder weniger ovale centrale Scheibe, welche etwa 12–14 Stachelchen trägt. Die, welche den Randplatten am nächsten liegen, bilden aber öfters Verbindungen mit einander, wodurch ein festeres Skelett am Rande entsteht. Die Madreporenplatte liegt dem Rande etwas näher als dem Centrum der Scheibe. Die Farbe ist weisslich, der Rücken etwas mehr gräulich, der Bauch und die Ränder fast ganz weiss.

" Sechs Exemplare aus einer Tiefe von 10 Faden aus der Bai von Batavia (No. 255), Schlammboden. Bei dem grössten Exemplar war $R = 50$ m.M., bei dem kleinsten $R = 10$ m.M."

SLADEN refers this species to a new genus [1] and gives a very full description of it in the Challenger report ['89 p. 177] :

---

1) The new genus is characterised as follows [SLADEN, '89, p. 175] :

" Subfamily *Astropectininæ*, SLADEN, 1887.

" Genus *Craspidaster*, n. gen.

" *Archaster* (pars), MÜLLER and TROSCHEL, Monatsber. d. k. Akad. d. Wiss. Berlin, 1840, p. 104 ; System der Asteriden, 1842, p. 65.

" Rays five, tapering, subrigid. General form subdepressed and flat.

" Supero-marginal and infero-marginal plates largely developed, remarkably thick and massive. Covered with hyaline, deciduous granules ; devoid of spines, excepting one adpressed, flattened, lateral spine on the infero-marginal plates. Deep, well-defined channels along the sutures between successive plates, the margins bordered with a webbed fringe formed of small spinelets enveloped in a continuous membranous investment ; the fringe continuous round the inner end of the supero-marginal plates.

"*Craspidaster hesperus*, MÜLLER and TROSCHEL, sp. (Pl. XVII. figs. 5–7; Pl. XVIII. figs. 1–4).

"*Archaster hesperus*, MÜLLER and TROSCHEL, 1840, Monatsber. d. k. Akad. d. Wiss. Berlin, p. 104; System der Asteriden, 1842, p. 65.

"*Stellaster sulcatus*, MÖBIUS, 1859, Neue Seesterne des Hamburger und Kieler Museums, p. 11, Taf. iv. figs. 1 and 2 (Abhandl. a. d. Gebiete Naturw. hrsg. v. d. naturwiss. Verein, Hamburg, Bd. iv. Abth. 2, 1860).

"Rays five. $R = 53 + mm$. (the terminal plate being broken off in all the rays of the largest specimen); $r = 15$ mm. Breadth of the ray across the second supero-marginal plates, 14.5 mm.

"General form depressed and rigid. Rays moderately long and flat, tapering from the base to the extremity, which is not attenuated or sharply

"Abactinal area with paxillæ. Paxillæ with very massive basement plates, suboval internally, pedicle columnar, crown with one or more central granules on the tabulum, surrounded by a marginal series of short spinelets, which radiate horizontally, and are united, at least in part, by a membranous web.

"Adambulacral plates superficially subquadrangular or rhomboid; the furrow margin with a series of short, subcylindrical spinelets, five or six in number, forming a small radiating comb; the other three margins bearing small, skin-covered, papilliform spinelets, directed over a channel which intervenes between adjacent adambulacral plates, and also between the adambulacral and the marginal plates. Actinal area of the adambulacral plates covered with skin and devoid of spines. Ambulacral furrows entirely closed by the adambulacral plates and their armature, when contracted.

"Actinal interradial area well developed, with a few large plates, regular and pavement-like in their disposition, covered with hyaline deciduous granules, each plate margined with a webbed fringe like that on the marginal plates; well-defined channels along the suture lines of the plates.

"Superambulacral plates present. Tube-feet conically pointed.

"No anus. No pedicellariæ.

"*Remarks.*—The type of this remarkable genus is the starfish to which MÜLLER and TROSCHEL gave the name of *Archaster hesperus*. Specimens, nearly all in a dry state, are to be found in the British Museum, as well as in several of the Continental museums, but the form has nevertheless been left in its anomalous position, although other observers have noted some of its remarkable characters. Under these circumstances I have given below an account in detail of its general structure. It will be seen to have nothing of generic import in common with the two other members of MULLER and TROSCHEL's genus *Archaster*, *Archaster typicus* and *Archaster angulatus*, or indeed with the other forms which have been hitherto ranked as *Archaster*. The presence of the superambulacral plates, the conical pointed tube-feet, the absence of an anus, and also the absence of pedicellariæ, would seem naturally to associate this form with the Astropectinidæ, whilst the massive granulose plates, devoid of all spines excepting the lateral, with their singular marginal fringe, the character of the adambulacral plates and their armature, and likewise that of the actinal intermediate plates, constitute a series of structures that isolates the form very distinctly from other genera at present known."

pointed. Interbrachial arcs wide and well rounded ; lateral walls highest in the interbrachial arc, decreasing gradually towards the end of the ray.

" The abactinal surface of the disk and rays is covered with paxillæ of a rather peculiar form, which I have not observed in any other species. In the immediate centre of the disk the paxillæ are small, crowded, and individually indistinguishable ; they also diminish greatly in size as they proceed along the ray, but remain perfectly distinct, and though closely placed together, in no way interfere with one another's form by crowding. The largest paxillæ occur midway between the centre of the disk and the margin, and at the base of rays. These consist of a comparatively large, convex tabulum, coverd with rather coarse hemispherical granules, having more or less of a mulberry form, with a fringe-like series at the margin of the tabulum of short, equal, skin-covered papilliform spinelets, all directed horizontally. On the large paxillæ there may be from ten to twenty granules on the central area of the tabulum, and from sixteen to twenty in the marginal series. In the medium-sized paxillæ, beyond the base of the ray and in the neighbourhood of the margins throughout, there are not more than three or four of the central granules, and eight to ten of the marginal series, whilst in the smaller paxillæ on the outer half of the ray there is seldom more than one central granules and seven or eight appear to be the normal number of marginal spinelets ; in these the membranous investment is even more apparent than on the larger paxillæ, and is continuous or united for a considerable distance between adjacent spinelets.´ Upon the rays the paxillæ are arranged in remarkably regular, straight, transverse series, which extend uninterruptedly from one series of marginal plates to that on the other side of the ray, traversing the whole paxillar area of the ray. Although there is no formation of a definite median line, the paxillæ in the middle of the area on the inner third of the ray are distinctly larger than the others in the same transverse series.

" The marginal plates are large and remarkably massive, forming a broad border to the abactinal and actinal areas and are well rounded in the lateral wall. The superomarginal plates, thirty or thirty-one in number from the median interradial line to the extremity, have the breadth equal to about twice the length, the proportion diminishing slightly towards the extremity. Their height at the summit of the interbrachial arc is about twice the length, about midway on the ray it is nearly one and a half, and at the extremity subequal. The width of the paxillar area is equal to that of the superomarginal plate at the fifteenth plate from the median inter-radial line ; midway on the ray it is rather greater. The plates are slight-ly, but rather flatly, convex along their median transverse line (i.e., breadth), and are separated by well-defined channels. Their surface is covered with rather large, uniform, tolerably well-spaced, hyaline, hemispherical granules, which are abnormally deciduous, and around the margin of the plate is a fringe of small, uniform, papilliform spinelets, invested with a con-tinuous web-like membrane, directed horizontally in relation to the vertical plane traversing the breadth of the plate ; the fringe thus covers over the furrow between the superomarginal plates, and is continuous round the end of the plate abutting on the paxillar area. There are no spines on the superomarginal plates.

" The infero-marginal plates correspond exactly to the superior series, and their length is the same ; their height in the lateral view is nearly twice their length at the summit of the interbrachial arc, but diminishes along the ray, being subequal or even slightly less when midway. Their breadth on the actinal surface is about twice the length midway along the ray, but is considerably greater in the interbrachial arc, where the border formed by the infero-marginal plates occupies very nearly half the space be-tween the mouth-angle and the margin ; on the outer part of the ray the proportion of beadth to length diminishes gradually, but the breadth remains preponderant throughout. The surface of the plates is faintly but flatly convex, emphasized by the rounded bevel at the margin of the well-defined transverse channel between each successive plates. The surface of the plates is covered with uniform, well-spaced, hyaline, deciduous, hemispherical granules, similar to, but perhaps slightly smaller than, those on the supero-marginal plates, and the margins are furnished with a similarly webbed fringe of small spinelets directed horizontally over the transverse channels between the plates, the fringe increasing a little in breadth as it approaches the edge of the ray. On the infero-marginal plates which abut against adambulacral plates, the fringe is not present on that edge, but the four innermost infero-marginal plates, that is to say, two on each side of the median interradial line, which abut against the actinal intermediate (ventral) plates, have the fringe continuous round the inner end of the plates, and on a few plates next succeeding a trace of the fringe is discernible, the abortion being effected gradually. Each infero-marginal plate bears a single, small, compressed, comparatively broad, flat, truncate spinelet, scarcely longer than the length of the plate; it is articulated but adpressed to the ray, directed towards the extremity at a slight angle upwards, and it is placed at the extreme margin of the actinal surface, consequently a little below the upper edge of the infero-marginal plate, and stands close to its aboral margin. Occasionally on one or two plates in the interbrachial arc a second smaller and very much narrower spinelet may be present at some distance from the marginal one, on the actinal surface, and likewise close to the aboral margin of the plates.

"The adambulacral plates are small, and, as seen with their armature when viewed from above, appear subquadrate or rhomboid in form. Their armature consists of a furrow series of five or six short, cylindrical, slightly tapering spinelets, the outer ones rather smaller than the others, and all radiating slightly apart at an angle over the furrow. The actinal surface of the plate is covered with membrane, and is devoid of spinelets, but bears round its margin, that is to say, on the three remaining sides, a series of small, uniform, skin-covered, papilliform, obtuse spinelets, very much shorter than the furrow series, and directed at an angle of about 45° to the plane of the plate, towards the adjacent plate, whether this be an adambulacral or marginal one. There is thus the appearance of a straight channel intervening between the series of adambulacral plates and the marginal plates over which the series of skin-covered spinelets is directed, and the adambulacral plates are themselves distinctly and clearly spaced. Near the middle of the aboral margin of each adambulacral plate is one comparatively very robust, short, stumpy, subconical spinelet, its posture suggesting resemblance to a thumb in relation to the furrow series of spinelets, if these were considered as the fingers of an outstretched hand. It is directed at a slight angle to the vertical outwards (sometimes inwards on the outer part of the ray), and towards the extremity.

"The mouth-plates are small and elongate, the outline of the united pair being fusiform, and their surface is not convex or prominent actinally. Their armature consists of a marginal series of five or six small spinelets, which extend from the inner extremity to the junction with the adambulacral plate. Close within this series is a second which quite masks them, consisting of a lineal series of eleven or twelve short, robust, subconical and pointed spinelets, which extend from one extremity of the plate to the other, decreasing slightly in length, but less in robustness, as they recede from the mouth.

"The actinal interradial areas, which are comparatively small and triangular, are occupied by a few very large, regular; intermediate plates, arranged in definite order, mosaic-like and not imbricating. The first or innermost plate on each side of the median interradial line is considerably larger than the mouth-plates, and the pair together have an hexagonal outline; they occupy fully two-thirds of the distance between the mouth-plates and the marginal plates, and are separated by a median suture corresponding to the median interradial line. The space intervening between this pair and the marginal plates is occupied by a single odd oblong plate standing in the median interradial line. The other plates extend the whole way from the adambulacral to the marginal plates, and vary in shape according to their position. The number also varies from three to five in each half area, and may vary even in two halves of one and the same area. The second plate counting from within is in all cases the largest plate in the area. The surface of these plates is covered with small, hemispherical, deciduous granules, similar to those on the marginal plates, and round the margin of each plate is a fringe of small spinelets united by a membranous web, similar to that described on the marginal plates, which is directed horizontally over a channel running between the plates.

"I have been unable to detect the slightest trace of an anal aperture; indeed from the small and compact character of the paxillæ in the centre of the dorsal area, it might be said, reasoning from the analogy of *Astropecten*, that no such aperture existed. The region is sometimes slightly protruded in a low cone, sometimes slightly introverted in the centre.

" The madreporiform body is moderately large, subcircular, and situated midway between the centre of the disk and the margin. The central area of the body is abruptly elevated and occupied by one of the mulberry-like masses of hemispherical granules similar to those on the tabulum of the paxillæ, and this again is surrounded by the marginal fringe of spinelets (in fact a central, but sessile, paxilla); beneath this the striation-furrows, which are fine, may be seen radiating to the periphery of the body.

" The ambulacral furrows are completely closed in by the adambulacral plates and armature when contracted. The tube-feet have conical tips when extended, but which appear slightly knob-like when contracted.

" No pedicellariæ of any kind are present.

" Colour in alcohol, a brownish grey over the paxillar area; marginal plates and actinal surface a bleached yellowish white.

" *Young phase.*—In a small specimen from Station 203, measuring R=22 mm., r=6.5 mm., the number of marginal plates is twenty-four, and in general character this example accords in all points with the adult form, the species being unmistakable. The remarkable webbed fringes are fully developed, and the furrows between the marginal plates are very wide. The number of spinelets in the paxillæ and in the armature of the adambulacral plates is rather less, especially on the attingent sides of the latter, as might naturally be expected. The spinelets in the furrow series range in number from four to six, according to position. On the larger paxillæ of the disk, seldom more than six to nine marginal spinelets are present, and not more than one or two central granules; along the ray five or six appears to be the normal number. The madreporiform body is relatively nearer the margin than in the adult form. The centre of the abactinal region is slightly introverted as in many Astropectinidæ. The development of the terminal (or 'ocular') plate is interesting. In the young stage referred to, the plate is divided along its entire median line, forming a more or less wide and gaping channel, with the edges rounded, along which the abactinal membrane passes and more or less aborted paxillar spinelets; beneath the membrane lies the terminal tentacle, there being no completion of the calcareous ring on its abactinal side. In the larger example from this locality, thin, narrow, calcareous connection is developed, on the floor of the furrow, but only at the distal extremity of the plate, forming there a delicate arch over the terminal tentacle. This division of the terminal plate is full of significance as regards the formation of this so-called single plate.

" *Variation.*—In the specimen from Station 203 there is a small amount of variation which is worthy of notice, although it might readily be passed over. This occurs in the armature of the adambulacral plates, in prominent robust thumb-like spinelet on the aboral margin of the plate being wanting (see Pl. XVIII. fig. 2). The largest example is smaller than that from Hong-Kong, hence it is possible that the 'thumb' may be developed only after full maturity is attained. On the other hand it may be said that as the example under notice measures R=37.5 mm., r=9.75 mm., and has exactly the same number of marginal plates (thirty-one, exclusive of the terminal) as the larger Hong-Kong specimen, its normal adult characters may be considered to be present.

"On comparing the two forms it may further be remarked that the furrow series of spines on the adambulacral plates are comparatively longer in the specimens from Station 203, and also that the breadth of the adambulacral plates in relation to their length is slightly greater. The marginal · spines are comparatively longer and narrower, and they are frequently pointed and channelled along their length, or gouge-shaped. The spinelets in the webbed fringe· or the three attingent sides of the adambulacral plates are fewer in number, as also are the spinelets on the paxillæ of the abactinal area (see Pl. XVIII. fig. 1). These, however, are characters which I regard as attributable to the smaller size.

" *Localities.*—Challenger Expedition :

"Hong-Kong, 10 fathoms.

"Station 203, East of Panay Island (Philippine Group). October 31, 1874. Lat. 11° 6′ 0″ N. ; long. 123° 9′ 0″ E. Depth 20 fathoms. Mud. Surface temperature 85°.0 Fahr.

" *Other localities :* Japan (MÜLLER and TROSCHEL) ; Banka Straits (Stockholm Museum) ; Singapore (VON MARTENS).

" *Remarks.*—MÜLLER and TROSCHEL's type-specimen in Berlin from Japan collected by Captain WENDT, which I have examined, conforms in all points —in so far as the dry specimen can be compared—with the form above described from Hong-Kong. It is, however, somewhat smaller, measuring R=41 mm., r=11.5 mm. The thumb-like spinelet is present in the adambulacral armature. The thumb is also present in a specimen from Banka Straits preserved in the Stockholm collection.

" I have also seen the type of MÖBIUS's *Stellaster sulcatus.* It is un-questionably the same species. The two specimens are rather small, R=35 and 36 mm., with r=12 mm., in the example measured by me. This observer has noted[1] the peculiar bordering of spinelets on the marginal and actinal intermediate plates, also the granulate covering of these plates and the general character of the paxillæ.

" An example preserved in Copenhagen, which I have also studied, has been carefully described by LÜTKEN,[1] who at the same time pointed out that *Stellaster sulcatus,* MÖBIUS, was synonymous with *Archaster hesperus.*

" In the specimen at Leyden, seen by MÜLLER and TROSCHEL, I find that the thumb-like spine on the adambulacral plates is not so largely developed or so prominent as it sometimes is, nevertheless it is present. The locality of this example is unknown ; it is simply stated in the System der Asteriden to have been collected by VON SIEBOLD."

SLUITER mentions this species in the collection of the Museum of Amsterdam ['95, p. 52] :

" *Craspidaster hesperus* (M. u. TR.) SLADEN. Zahlreiche Exemplare in Alcohol von der Bai von Batavia aus einer Tiefe von 5–14 Faden (SLUI-TER). Nach der jetzt vorliegenden ausführlichen Beschreibung und Abbil-dung dieser Art von SLADEN, bin ich zu der Ueberzeugung gekommen, dass die früher von mir als *Astropecten macer* beschriebene Form mit dem alten *Archaster hesperus* M. TR. identisch ist. Die Zahl und Form der inter-

---

1) References are made to the papers above cited.

radialen Bauchplatten kann aber mehr oder weniger verschieden sein, und bei grösseren Exemplaren, bis 4 mehr betragen als von SLADEN beschrieben und abgebildet wird."

LUDWIG has the following remarks on this species under the heading, "Anmerkung zur Gattung *Astropecten*" ['97, p. 60]:

"Im Jahre 1862 hat GIEBEL eine angeblich neue *Astropecten*-Art aus dem 'Mittelmeer' unter dem Namen *A. gracilis* beschrieben. Angenommen, sie wäre wirklich neu, so müsste sie umgetauft werden, denn schon 1840 (p. 282) hat GRAY einen *Astropecten gracilis* von allerdings unbekannten Fundorte aufgestellt. Da die GIEBEL'sche Beschreibung auf keinen mir aus dem Mittelmeer bekannten Seestern passt, so wandte ich an den Director der zoologischen Sammlung in Halle, Herrn Prof. GRENACHER, der mir mit freundlicher Bereitwilligkeit das GIEBEL'sche Originalexemplar zur Ansicht schickte. Durch genaue Untersuchung desselben konnte ich bald feststellen, dass das Exemplar zwar eine Astropectinide ist, aber nicht mehr in die Gattung *Astropecten* in ihrem heutigen Sinne gehört, sondern identisch ist mit dem von SLADEN (1889) ausführlich beschriebenen *Craspidaster hesperus* (M. TR.). Was den Fundort angeht, so muss die Angabe, dass das Exemplar der Halle'schen Sammlung aus dem Mittelmeer herrühre, auf einem Irrthum beruhen, denn der *Craspidaster hesperus* ist bis jetzt nur von China, Japan, den Philippinen, aus der Banka-Strasse und von Singapore bekannt."

This starfish is reported from the Mergui Archipelago by BROWN [: 10, p. 28]:

"Locality.—XXXII., off Bentinck Island, 29 fathoms, soft mud and sand.

"Two specimens agree with the young phase described by SLADEN. The adpressed spinelets on the infero-marginal plates tend to fall off very readily, but their presence is quite evident in both specimens.

"R=22   r=6.5.

"Apparently not previously recorded from the Indian Ocean. Known from Japan to Singapore."

It is also mentioned by KŒHLER from the collection of the Indian Museum [: 10, p. 9]:

"Station 88. Vizagapatam. 4 miles au S.E. de Maurawalipur. Profondeur 9–13 brasses. Un échantillon.

" R=16 mm."

KŒHLER also mentions it from the south-eastern Moluccas [: 11, p. 268] :

"Dragage No. 15. 15 Avril 1908. Sungi Barkai. Profondeur 7 à 8 m. Un échantillon. R=28 mm. ; r=9 mm."

Psilaster gracilis SLADEN. ·

This species is not contained in any of the collections studied by me. SLADEN's original description is reproduced below ['89, p. 230] :

" *Psilaster gracilis*, n. sp. (Pl. XLI. figs. 5 and 6 ; Pl. XLII. figs. 9–11).

" Rays five. R=65 mm. r=12 mm. R<5.5 r. Breadth of the ray at the third supero-marginal plate, 11.5 mm.

" Rays elongate, narrow and tapering throughout to a finely pointed extremity, having a subcylindrical facies, slightly compressed. Abactinal area slightly convex and capable of inflation. Actinal area subplane. Lateral walls comparatively high and vertical. Interbrachial arcs open and widely rounded.

" The abactinal paxillar area of the disk and rays is covered with comparatively large and closely placed paxillæ. These are composed of very short, stumpy, papilliform spinelets. Three to five are central, more robust than the rest, often almost granuliform, and about a dozen or more form the marginal series, all very short and radiating outward. In some of the paxillæ the central spinelets are posed in such a way as to form incipient pedicellariæ, in others three or four of the central spinelets are slightly longer, and are distinctly pedicellarian in function. At the margin of the area an arrangement of the paxillæ in transverse series may be observed, but is not very conspicuous at first sight, on account of the crowding of the paxillæ, and is only well seen near the base of the rays.

" The supero-marginal plates, thirty-six in number from the median interradial line to the extremity, stand vertically, and, being confined entirely to the lateral wall of the ray, can hardly be said to have a distinct breadth on the abactinal area, excepting on the outer part of the ray, as they merge so gradually into the rounding of the lateral wall. On the outer part, however, they are more distinctly curved over and flattened on the abactinal area. The surface of the plates in relation to one another forms a continuous plane. On the inner part of the ray the height is about one-third greater than the length, and further outward than midway along the ray the dimensions are nearly subequal. The surface of the plates is closely covered with short obtusely tipped papillæ, equal in length but slightly more robust along the median region of the plate ; and all so closely placed as to give the appearance of coarse velvet pile. The supero-marginal plates bear no spinelets.

"The infero-marginal plates correspond to the superior series, each plate being equal in length to its companion in the upper series. Their height, however, is slightly less, and they are well curved upon the actinal surface. Their surface is covered with short close-set papillæ similar to those on the supero-marginal plates, which become slightly larger and more spiniform at the end of the plate adjacent to the adambulacral plates. Each plate bears along the upper half of the aboral margin a series of four (normally, but sometimes three) small tapering spinelets. The uppermost spine is the smallest, the rest subequal or with either the lowest or the median spine slightly longest. They are closely appressed to the side of the ray, and are directed at a slight angle upward : the position of the series on the plate being also sometimes very slightly oblique.

"The adambulacral plates are elongate, large, with a faintly convex margin towards the furrow. Their armature consists of a furrow series of six or seven rather long, delicate, cylindrical spinelets, equal in length, covered with membrane, and standing parallel to one another. The series or 'combs' thus formed are distinctly spaced from the neighbouring series on adjacent plates. On the actinal surface of the plate and near to the marginal series is a longitudinal series of three or four spines, slightly shorter and more robust, widely and irregularly spaced ; and external to these is another longitudinal series of four or five similar spines, but even more irregular in disposition : indeed, so far is this carried in both cases that it is often impossible to distinguish any regular serial arrangement at

all. Sometimes a few additional spinelets may be present, external to those
above mentioned, and the grouped character becomes then more marked.
These spinelets on the actinal surface of the adambulacral plate are covered
with membrane like the marginal series, and they have generally a more
or less straggling and irregular appearance.

"The mouth-plates are elongate and narrow, and the united pair form
a convex keel actinally. Their armature consists of a marginal series of
small, rather robust, and abruptly tapering spinelets, shorter than the
marginal series on the adambulacral plates, about six or seven on the free
margin of the plate, although others appear to continue the series up to the
outer extremity of the plate. At the innermost point of each mouth-plate
is one elongate, flattened and round-tipped mouth-spine, greatly exceeding
any of the other spines in size, and the pair of spines at each mouth-angle
are parallel to one another and directed towards the centre of the actinostome. At first sight these enlarged spines might be considered as the
innermost spines of the marginal series above described, but I am doubtful
whether this is really the case, as there is the singular occurrence in this
form of a small group of short spinelets present on each plate at a still
higher level than the foremost mouth-spines above mentioned, and this
little insignificant group is further peculiar from the fact that it is not
directed towards the centre of . the actinostome, but in the direction of a
line crossing the ambulacral furrow. If this group of small and abnormally
placed spinelets really belongs to the true marginal series it is probable
that the prominent and enlarged mouth-spine should be reckoned as the
foremost of the actinal or superficial series. On the actinal surface of each
plate is a longitudinal series of eight to twelve rather robust and conically pointed spinelets, parallel to the median suture; and an intermediate series, fewer
in number and more widely spaced, between these and the marginal series.

"The actinal intermediate plates are confined to a very small area in
the interradial region, but I am unable to say from superficial observation
whether their arrangement presents any regularity or not. Each of the
plates bears two or three short, robust, conical tipped spinelets, which have
a more or less marked tendency to form a group.

" The madreporiform body is entirely obscured by paxillæ.

" Colour in alcohol, a dirty greyish white.

" *Locality.*—Station 238. Off the coast of Japan, south of Kawatsu. June 17, 1875. Lat. 34° 37′ 0″ N., long. 140° 32′ 0″ E. Depth 1875 fathoms. Blue mud. Bottom temperature 35°.3 Fahr.; surface temperature 73°.0 Fahr.

" *Remarks.*—*Psilaster gracilis* has a very different facies from any of the other species described, and it is not without hesitation that I have included it in the genus. It is at once distinguished by its narrow subcylindrical rays and widely rounded interbrachial arcs; by the papilliform covering of the marginal plates; by the large inner pair of mouthspines; and by the general character of the actinal spinulation."

Nauricia pulchella GRAY.

I can not form any definite opinion as to the systematic position of this starfish, except that from a statement of GRAY quoted below it appears to be nearly related to *Craspidaster hesperus*. The following scanty notices are the only ones that I have been able to find out from the literature accessible to me.

Nauricia GRAY is placed by its author in the *Astropectinidæ*, which is characterised thus [GRAY, '40, p. 180] : " Back flattish, netted with numerous tubercles, crowned with radiating spines at the tip, called Paxilli." The first subdivision of the *Astropectinidæ* contains forms which have " the margin of the rays ciliated with a series of simple elongated spines, the paxilli or crowned tubercles regularly radiating " ['40, p. 180]. The first group under this subdivision, in which *Nauricia* is included, has " the rays edged with a series of large regular tubercles, which increase in number as the animal grows " ['40, p. 180].[1] The genus and species are then characterised as follows ['40, p. 180] :

" *Nauricia* GRAY.

" The ambulacral spines broad and ciliated; 2 series of tesseræ between the angles of the arms and the mouth beneath. Asiatic.

---

1) It must be remarked that this group contains *Astropecten corniculatus* (= *Ctenodiscus crispatus*) and other heterogeneous forms.

"1. *Nauricia pulchella*, GRAY. SEBA, iii. t. 8. f. 7. a. b. not good. Rays 5, half as long as the width of the body, gradually tapering, lower series of marginal tubercles with a series of broad flat spines on the upper margin of each.

"Inhab. China? Japan?"

DUJARDIN and HUPÉ are of opinion that this starfish should be referred to *Astropecten* ['62, p. 429].

VON MARTENS briefly refers to it as follows ['65, p. 354]:

Nauricia pulchella GRAY. Ann. Mag. n. h. VI. p. 180. Japan, ist mir unbekannt."

In his "Synopsis" GRAY reproduces essentially the same passages published in his paper of '40, but adds at the end, "The *Stellaster sulcatus*, MÖBIUS, l.c. t. 4. f. 1.2, appears to be from a specimen which has lost its marginal spines" ['66, p. 3].

As seen under *Craspidaster hesperus* the *Stellaster sulcatus* of MÖBIUS has been identified by LÜTKEN ['64] and SLADEN ['89a] with that species.

LUIDIIDÆ.

Luidia maculata MÜLLER & TROSCHEL.

(Pl. V, figs. 80–88.)

This species is due to MÜLLER and TROSCHEL, whose original description is as follows ['42, p. 77]:

"*Luidia maculata* Nob. nov. sp.

"Sieben bis neun Arme. Körper platt. Rücken der Arme und der Scheibe wenig gewölbt. Verhältniss des Scheibenradius zum Armradius wie 1:7. Die Arme sind 8 mal so lang wie breit. Furchenpapillen auf jeder Platte eine schwertförmig gekrümmte innen, und nach einem kleinen Zwischenraum noch 3–4 ebenfalls etwas gekrümmte aber nicht so platte Stachelchen. Zwischen diesen finden sich häufig lange zangenartige Pedicellarien, die zum Theil (vielleicht alle) dreizackig sind. Die längsten sind halb so lang wie die Stacheln in ihrer Nähe, und 3–4 mal so lang wie breit. Nun folgen nach einer Art von Furche die ventralen Platten. Diese sind stachelartig beschuppt und aus dieser Beschuppung erheben sich 2–3 etwas platte lanzettförmige grössere Stacheln, welche 2–3 Längsreihen auf den Armen bilden; die innersten scheinen die längsten zu sein. Die Paxillen des Rückens haben einen durchaus viereckigen Gipfel, und bilden an den Seiten des Rückens vier regelmässige Längsreihen. In der Mitte des Rückens sind sie kleiner und stehen viel weniger regelmässig, so dass man kaum sechs Reihen unterscheiden kann. Die Borsten auf den Paxillen sind in der Mitte der Paxille dicker und mehr granulaartig, mit abgestutztem Ende; am Rande werden sie dünner.

"Farbe: Die ganze Oberfläche ist im trocknen Zustande mit grossen grauen Flecken belegt, mit hellern abwechselnd.

"Grösse: 15 Zoll.

"Fundort: Japan. Im Museum zu Leyden durch v. SIEBOLD.

"Anmerkung. Das Exemplar mit sieben Armen unterschied sich von dem mit neun Armen durch die kleinere Scheibe, welche bei derselben Grösse des Thieres nur 1 Zoll im Durchmesser hatte, während sie bei dem Exemplare mit 9 Armen 2 Zoll mass; auch waren die Borsten der Paxillen am Ende nicht abgeschnitten, sondern abgerundet."

According to PERRIER ['76, p. 258] this species is mentioned by PETERS ['52, p. 178], but DE LORIOL ['85, p. 72] is of the opinion that the *Luidia maculata* of PETERS is to be referred (?) to *L. savignyi*.

DUJARDIN and HUPÉ ['62, p. 433] refer to this species as follows:

"Luidie tachetée. *Luidia maculata*.—MÜLLER et TROSCHEL.

"——MÜLLER et TROSCHEL, Syst. der Aster., p. .., sp. 2.

"Habite les mers du Japon."

Gray ['66, p. 4] simply mentions it for comparison under *Luidia ciliaris*.

VON MARTENS mentions this species from Japan and adds a new variety *quinaria*, which is however a distinct species, as explained under *Luidia quinaria*, where the whole passage bearing on the two species is reproduced (*vide infra*, p. 293). The same writer again mentions this species in his second paper as follows ['66, p. 84]:

"*Luidia maculata* MÜLL. TROSCHEL. l.c. S. 77.

"In der Bai von Manila ein achtarmiges Exemplar gefunden. Dieser Fundort verknüpft einigermassen die zwei bis jetzt bekannten entlegenen : Japan und Mosambique (A.)."[1]

PERRIER in his work on the pedicellariæ refers to this species as follows ['69, p, 110] :

"Nous n'avons pas vu les Pédicellaires de la *Luidia maculata*, M. et T., ni ceux de la *Luidia senegalensis*, bien que la collection du Muséum possède ces deux espèces."

According to the same writer ['76, p. 258] this species is mentioned by VON MARTENS from east Africa ['69, p. 131].

PERRIER in his "Révision" makes some critical remarks on the species ['76, p. 258] :

"Je trouve dans la collection du Muséum deux échantillons desséchés désignés sous ce nom. L'un provient de Batavia (île de Java), où il a été recueilli par M. RAYNAUD en 1829; l'autre, de la côte de Coromandel, d'où il a été rapporté par M. DUSSUMIER en 1830. Ces échantillons ne paraissent pas avoir été vus par MÜLLER et TROSCHEL, qui donnent leur type comme originaire du Japon. Leur provenance n'a rien qui puisse faire mettre en doute la détermination du Muséum, car VON MARTENS, dans le mémoire que nous citons dans la synonymie de celle espèces, indique que la *Luidia maculata*, M. et T., a été trouvée non-seulement au Japon mais encore à Manille et dans le détroit de Mozambique, localités relativement voisines et de faune très-analogue à celles que nous trouvons relatées au Muséum et qui rentrent par conséquent dans l'aire de répartition déjà constatée de ce Stelléride.

"Bien que la description de la *Luidia maculata* de MÜLLER et TROSCHEL s'applique également bien à nos deux échantillons, nous trouvons cependant entre eux quelques différences, à la vérité peu importantes, mais qu'il est utile cependant de signaler. Sur la face ventrale, l'identité est à peu près complète. Même disposition des piquants ambulacraires, même forme, même disposition des grands pédicellaires à trois branches qui les suivent ; même

---

1) Forms marked (A) are those which are distributed throughout the Indian Archipelago and extend into the east coast of Africa.

forme des piquants des plaques ventrales. Mais sur la face dorsale quelques
différences s'accusent : les bras, au nombre de huit, dont un bifurqué au
sommet de l'individu, d'ailleurs monstreux, de Batavia, ont leur surface,
dorsale un peu convexe, tandis qu'elle est plutôt déprimée chez l'individu
de la côte de Coromandel, qui n'a que sept bras. Chez ce dernier les
paxilles sont un peu plus grandes, de forme plus nettement quadrangulaire,
presque carrée, et leur surface libre, plane ou même un peu concave, est
uniformément recouverte de granules hexagonaux tous contigus qui deviennent
un peu plus fins et plus longs sur tout le pourtour de la plaque. Les
granules, étant tous contigus, n'affectent du reste aucune disposition régulière.
Chez l'individu de Batavia, les paxilles ont des angles un peu plus arrondis ;
leur surface libre est légèrement, mais nettement convexe, et les granules
qui la recouvrent sont hémisphériques et non contigus ; il y en a généralement un central et six formant un cercle autour de lui ; assez souvent un
second cercle vient s'ajouter au premier, et sur les grandes paxilles de la
base des bras cette disposition peut devenir plus complexe. Les petits
piquants qui bordent chaque paxille sont plus longs et plus grêles que dans
l'individu de Coromandel ; ils se distinguent bien nettement des granules
centraux de la paxille, tandis que dans l'échantillon africain la différence est
moins sensible. Ce sont là des caractères sans doute peu importants et qui
modifient à peine l'apparence des deux échantillons ; ils ne seraient suffiss-
ants pour motiver une séparation spécifique que s'il venait à être démon-
tré, ce qui est peu probable, qu'ils sont constants pour tous les individus
provenant d'une même localité.

"L'individu de la côte de Coromandel a sept bras normaux.—L'individu
de Manille en présents quatre normalement développés, un bifurqué au sommet
et trois qui ont été brisés à leur base, mais sont en voie de redintégration.
—La longueur de la partie nouvellement formée varie de 1 à 2 centimètres."

WALTER reports this species from Ceylon ['85, p. 368] :

" *Luidia maculata.* M. et TR.

" Ist unter den ceylonischen Echinodermen in 2 prächtigen getrockneten
Exemplaren vorhanden. Die Bestimmung der Art ist durch Prof. E. v.
MARTENS' freundlichen Vergleich mit den Originalexemplaren im Berliner

Museum zweifellos gesichert. Die Ausmasse unserer 2 Exemplare ergaben für einen Arm bei No. 1 215 mm. Länge, für den Scheibendurchmesser 47 mm. No. 2 : Armlänge = 220 mm., Scheibendurchmesser 50 mm. Eine geringe Abweichung von der Originalbeschreibung bei M. et Tr. finde ich in der hier grösseren Zahl der in einer Querreihe fallenden Rückenpaxillen. Zudem findet bei M. et Tr. eine sehr regelmässige und auffallende dunkle Sternzeichnung der Scheibe keine Erwähnung, obgleich sie auch an den Exemplaren des Berliner Museums vorhanden sein soll. Beide Exemplare sind siebenarmig."

Bell simply mentions this species from Ceylon ['87a, p. 648] and the Sea of Bengal ['88, p. 384, 388]. Döderlein ['88, p. 830] reports it from Ceylon.

Sladen reports this species from the Mergui Archipelago and adds the following notes ['89a, p. 327] :

" *Locality.* King Island (native name *Padaw*) ; 24th. Jan. 1882 : sublittoral.

" A single young example having nine rays. The major radial dimension is 65 mm."

Sluiter reports it from Batavia ['89, p. 313] :

" *Luidia maculata* (M. und Tr.). Perrier, ' Revision des Stellérides', pag. 258. Zwei Exemplare (No. 246) aus der Bai von Batavia, von dem Riffe bei der ersten Boje ausserhalb des Hafens von Tandjong Priok. Das eine Exemplar stimmt genau mit dem von der Küste von Coromandel herrührenden, von Perrier beschriebenen. Die Paxillen sind deutlich viereckig. Das Tier ist aber achtarmig. Das andere hat neun Arme und stimmt, was die Form und Paxillen anbelangt, mit dem Perrier'schen Exemplar von Batavia überein."

De Loriol describes an example from the Bay of Amboina as follows ['93, p. 379] : " Un exemplaire à neuf bras inégaux. Couleur brun clair avec de nombreuses macules foncées. Une grande tache brune arrondie au centre du disque, entourée de tache allongées dont une dans l'alignement de chaque bras."

Bell reports this species from Macclesfield Bank and adds ['94, p. 403], " All the specimens collected were of small size."

SLUITER again reports it from the Bay of Batavia ['95, p. 55]. KŒHLER describes a specimen from the Sunda Isles ['95, p. 388] :—" Un échantillon à huit bras inégaux; sur le plus grand R=17. La couleur est gris clair avec de nombreuses taches ' brunes placées de distance en distance sur les bras. Au centre du disque se trouve une tache rayonnante et au niveau de l'insertion des bras une série de taches en forme d'Y."

DÖDERLEIN reports one specimen from Thursday Island ['96, p. 307].

BEDFORD reports it from Singapore [: 00, p. 293] :

" R=7.6 × r.

" *Locality.* A single large 7-armed specimen from between tide-marks, Singapore. The arms had undergone a considerable amount of fracture and regeneration.

" *Distribution.* Extends from Mozambique to Macclesfield Bank and northwards to South Japan.

" R 190　r 25　Arm-breadth 23 " [mm.]

This species is referred to by DÖDERLEIN in his paper on Japanese Asteroidea [: 02, p. 330] :

" *Luidia maculata* MÜLLER u. TROSCHEL.

" Diese schöne schwarz und gelbgefleckte Art, die grösste Asteroiden-Form, die ich in Japan sah, beobachtete ich nur in der Bai von Kagoshima, wo sie in geringer Tiefe (ca. 30 m) ziemlich häufig ist. Exemplare mit 7–9 Armen und einem grossen Radius von 65 bis 350 mm geriethen nicht selten in das bei den dortigen Fischern gebräuchliche Grundnetz; ein grosses Exemplar fieng sich an einer mit einem Fisch beköderten Angel."

It is referred to by HERDMAN in his pearl oyster report [: 03, p. 20, 23, 26, 106], and is again reported from Ceylon by the HERDMANS [HERD-MAN, HERDMAN & BELL, : 04, p. 143] :

" *Luidia maculata*, M. and T.

" Abundant all round Ceylon, and especially in the shallow water of the pearl banks in the Gulf of Manaar, where we obtained specimens of all sizes from a couple of centimetres up to well over a foot across."

HERDMAN [: 06, p. 121, 447] mentions it as an enemy of the pearl oyster. BROWN reports it from the Mergui Archipelago [: 10, p. 30] :

"Localities.—XVIII., off Paway Island, 10 to 21 fathoms, sand and shell; XXV., Gregory Group, 4 to 14 fathoms, sand and shell.

"Several young specimens, in the largest of which R=62 and r=11.

"Found by Dr. ANDERSON at King Island. Also known from Mozambique to Madras, Manilla, and Japan."

It is also reported from Portuguese East Africa by SIMPSON and BROWN [: 10, p. 49] :

"This species is represented by two specimens : the first, from Mtundo Bay, has seven arms ; and the second, which is immature, has five complete arms, and three being regenerated from the disk.

"The diagnostic characters in this species seem to be fairly constant, and our specimens agree very well with those previously described.

"Locality.—Station III., Mtundo Bay (Wamizi Is. to Kifuki Is.) ; Station VI., Kero-Nyuni Bay.

"Previously recorded from—S. Japan ; Philippines ; Singapore ; Malacca; Macclesfield Bank ; Mergui Archipelago ; Tuticorin ; and Mozambique."

This species is also mentioned by KŒHLER from the collection of the Indian Museum [: 10, p. 10] :

"Baie de Balasore. Un très grand échantillon.

"Les bras au nombre de sept, dépassent 35 centimètres de longueur et ils atteignent une largeur maxima de 38 mm. à la base.- Je signale cet exemplaire et j'indique ses principaux caractères dans un travail actuellement sous presse, sur les Échinodermes recueillis par MM. MERTON et ROUX aux îles Kei et Aroe (10).

"Un très jeune exemplaire provenant des îles Andaman, et chez lequel R=10 mm., me paraît devoir être rapporté aussi à la L. maculata : il offre déjà quelques taches sur la face dorsale des bras."

KŒHLER gives a good description and photographic figures of some specimens from the south-eastern Moluccas [: 11, p. 267] :

"Dragage No. 1. 8 Février 1908. Ngaiguli. Profondeur 14 m. Deux échantillons.

"Les deux individus recueillis ne sont pas de très grande taille. Le plus petit a huit bras et R ne dépasse pas 11 cm. ; la plus grande largeur des bras est de 13 à 13.5 mm ; non compris les piquants marginaux ; la face dorsale des bras est fortement convexe. J'ai représenté cet exemplaire, vu par la face dorsale, pl. XV, fig. 1. L'autre individu n'a que six bras, dont l'un est cassé et tous les autres sont en régéneration ; il est plus grand que le précédent, car la plus grande largeur des bras atteint 18 mm. ; la face dorsale des bras n'est pas très convexe. J'en ai représenté la face ventrale pl. XV, fig. 2.

"Ces deux individus n'ont pas atteint tout leur développement ; j'ai pu les comparer à un exemplaire appartenant au Musée de Calcutta dont les bras, au nombre de sept, dépassent 35 cm. de longueur et atteignent une largeur maxima de 28 mm. à la base.

"L'étude de ces exemplaires me permet d'ajouter quelques renseignements complémentaires aux descriptions que les auteurs, notamment MÜLLER et TROSCHEL et PERRIER, ont données de la *L. maculata*. Ainsi, en comparant cette dernière espèce à une nouvelle espèce des îles Hawaï qu'il décrit sous le nom de *L. magnifica*, W. K. FISHER (The Starfishes of the Hawaiian Islands, U. S. Fish Commission Bulletin for 1903, part III, p. 1035) fait remarquer que ni MÜLLER et TROSCHEL, ni PERRIER ne mentionnent de pédicellaires [sur les plaques marginales ventrales de la *L. maculata* et que, vraisemblement, ces pédicellaires ne doivent pas exister. Il se demande aussi si les pédicellaires mentionnés par ces auteurs sur les plaques adambulacraires, sont placés après les piquants adambulacraires ou au milieu d'eux, les descriptions restant muettes sur ce point. Voici ce que j'observe, à cet égard, sur les exemplaires que j'ai sous les yeux.

"Les piquants adambulacraires, au nombre de trois ou de quatre sur chaque plaque, sont suivis d'un nombre variable de pédicellaires à trois valves entremélés de quelques petits piquants très courts ; tantôt il n'y a qu'un seul pédicellaire, tantôt il y en a deux ou même trois sur les deux individus provenant des îles Aru (pl. XVI, fig, 9). Sur l'individu du Musée de Calcutta, il y a généralement trois pédicellaires sur chaque plaque en dehors des piquants adambulacraires (pl. XVII, fig. 8). Ces pédicellaires sont grands et forts ; dans l'exemplaire du Musée de Calcutta, leur longueur peut atteindre 2 mm. ; ils sont assez larges à la base, mais leur longueur dépasse toujours de deux fois leur plus grande largeur.

"Les plaques marginales ventrales portent aussi des pédicellaires qui sont particulièrement abondants dans l'exemplaire de Calcutta : ils se montrent de préférence entre le grand piquant marginal et celui qui le précéde, et généralement il y en a deux à la fois. On en rencontre aussi, mais plus rarement, entre l'avant-dernier piquant et celui qui le précéde. Ces pédicellaires n'ont ordinairement que deux valves : ils sont presque aussi larges que ceux des plaques adambulacraires, mais un peu plus courts.

"Sur les deux individus des îles Aru, les pédicellaires sont moins nombreux et moins développés. Dans l'exemplaire à six bras, j'en observe un d'une manière à peu prés constante entre les deux grands piquants marginaux :, ce pédicellaire est toujour bivalve. Dans l'exemplaire à huit bras, les pédicellaires sont beaucoup plus rares, mais on en trouve cependant. Les pédicellaires bivalves des exemplaires des îles Aru sont moins différenciés que dans le grand individu du Musée de Calcutta et ils ressemblent davantage à deux petits piquants simplement rapprochés.

"J'ai représenté pl. XV, fig. 8, les paxilles de la face dorsale des bras de l'individu provenant du Musée de Calcutta."

I have several specimens of this species of various sizes and from different localities. The radial ratio varies greatly according to the size, as may be seen from the following table.

Specim.	r mm.	R mm.	R : r	Locality
1	7	29.5–32.5	4.2–4.6	Awa.
2	10	88	8.8	?
3	11	92	8.4	?
4	19	164	8.6	Misaki.
5	21	184	8.8	Awa.
6	26	230	8.9	Miyazu.
7	34	263	7.7	Miyazu.

My specimens are either eight or nine armed. The two specimens from Miyazu are both nine-armed. The variegation of the abactinal surface is subject to considerable variation. In some specimens the black portions preponderate so much that the whole abactinal side appears a uniform black at a short distance, while in others they form only small scattered patches in the general orange red surface. The relative distribution of the two patches are entirely irregular, and in some specimens the contrast between the two patches is not so sharp as in others. At the centre of the disk there is generally a black patch, which may be very small or cover nearly the entire disk, and may be of any form. In one specimen in which the patches of the two colours were nearly equally distributed, there was an outline of a star-shaped polygon in black, with as many reëntrant angles as there were arms, each reëntrant angle being interradial in position; a black strand extending inwards from the apex of each projecting angle and all meeting at the centre of the disk, the spaces between the black strands being orange red. In another specimen, in which the orange red portions showed a slight preponderance, there was a black patch of a very irregular shape at the centre of the disk, surrounded at a short distance by a ring of small discontinuous black patches running close to the margin of the disk.

The actinal side is perfectly flat, but the convexity of the abactinal side is somewhat different in different specimens, depending among other things upon the condition in which the specimen was fixed, especially in dry specimens. Generally speaking, alcoholic specimens show a more convex abactinal side than many dried ones, and this appears to me to indicate that the flattened condition of many dried specimens is probably due to pressure before or during dessication.

Superomarginals.—The superomarginals are exactly like the paxillæ, and therefore will not be treated of separately.

Inferomarginals.—The inferomarginals are comparatively large plates forming the lateral margin of the arms and lying between the adambulacral and superomarginal plates. In a cross-section of the arms it occupies nearly two-thirds of the space between the ambulacral furrow and the superomarginals, the remaining one-third being occupied by the adambulacral plates. The armature of the inferomarginals varies slightly in different specimens and on different plates of the same specimen. Most usually there are four large conspicuous spines making a single transverse series among themselves and with the large spines of the adambulacrals (Pl. V, fig. 83). These large spines of the inferomarginals are separated from one another by smaller spines, some of which may become large enough to form a series with the larger ones, so that the number of the latter usually varies between three and five. The smaller spines are mostly grouped along the margin of the infero-marginal plate, but are also present more or less between the larger spines, as stated above. The fasciolar gooves between the infero-marginal plates are well developed especially in the outer part, and are covered with capillary spinelets. MULLER and TROSCHEL state that the larger spines of the inferomarginals form 2–3 longitudinal series on the arms; but in all the specimens ex-amined by me these longitudinal series are too irregular to deserve the name, with perhaps the exception of the outermost spines, which show a tolerably regular longitudinal arrangement. The inferomarginals may bear one or more biscuspid pedicellariæ, as described by KŒHLER.

Adambulacral plates.—The adambulacral plates are tolerably broad, and except for the very occasional plates intercalated between the inferomarginals, there are as many of them as there are inferomarginals, each adambulacral being in a line with the inferomarginal plate just outside it. There is however a hardly noticeable break in the armature between the two series of plates, as is also indicated by MÜLLER and TROSCHEL in their description. This break may be caused by a distinct furrow, or the furrow may be entirely absent. The difference of armature between the inferomarginal and adambulacral plates is also easily noticeable on an actinal view, inasmuch as the spiniferous part of the adambulacrals is more restricted than on the inferomarginals, and the furrowed spaces between two contiguous adambulacral plates are consequently larger and more conspicuous. The armature of the adambulacral plate consists of a transverse series of three large spines on the inner half of the plate, a number (10–15) of small spines on the outer half, and one or more, elongated, pyramidal pedicellariæ with three cusps, situated at the outer end of the series of large spines (Pl. V, fig. 83). The first adambulacral spine is smaller than the other two and are somewhat flattened and curved; the second spine is the largest of the three and is also slightly curved and flattened; the third spine is only slightly smaller than the second and is nearly straight and round. Occasionally the first spine is duplicated and in such a case there are four large spines on one adambulacral plate. Close to the third adambulacral spine, and usually on its proximo-external side there is on most of the adambulacral plates one, or occasionally two, tricuspid large pedicellariæ of the form of a triangular pyramid, only slightly shorter than the spine itself (Pl. V, fig. 84). The rest of the actinal surface of the adambulacral plate is covered with small spines, which may be found also on the proximal side of the third adambulacral spine.

The above may be taken as the typical arrangement of the armature of the adambulacral plate. In one specimen however the third adambulacral spine was wanting and there were two or three pedicellariæ instead of one or two. In this specimen the spines were on the whole less well developed.

The first adambulacral plate usually bears a single series of four or five spines which gradually increase in size outwards, and the second bears usually one less spines of the same form and arrangement.

Mouth-plates.—The mouth-plates are narrow, very slightly curved, and elongated. Each plate has two ridges, the main one on the actinal surface, and a secondary one on the mouth side, parallel with the ambulacral furrow. On the main ridge there are some ten or a few more large spines forming a somewhat irregular series, of which one or two at or near the mouth end may be conspicuously larger than the others (Pl. V, fig. 82). These large spines are usually situated on the inner two-thirds of the plate, while the outer one-third is covered with some twenty or more smaller spines without any regular arrangement. Near the centre of the actinal surface of the coupled plates there are some six or seven spines intermediate in size between the larger and the smaller ones. There are also some small spines on the lateral surface of each mouth-plate. The secondary ridge bears some seven to ten spines similar in form and size to the smaller spines of the main ridge ; one of them at the mouth end may be conspicuously larger than the rest

Ventrolaterals.—The ventrolateral plates are very small and inconspicuous. In each interradius there are some nine or ten plates between the first inferomarginals and the mouth-plates, and a single series of plates extends more than half-way out into the arms between the adambulacral and the inferomarginal plates. Each ventrolateral plate is provided with a prominent, usually more or less flattened projection on the actinal side, which bears some ten to fifteen sharp, slender straight spines, one or two of which are usually larger than the others. The projection becomes less and less conspicuous as it proceeds out into the arms. In the distal part of the arms the ventrolateral plates are very small and are ankylosed with the adambulacrals.

Paxillæ.—The paxillæ are comparatively large and very closely set. As seen from the surface they are more or less roundish-polygonal in the disk and rectangular in the arms. Near the tip of the arms, however, the paxillæ tend to become more or less roundish again. In nearly all cases there is a clear distinction between the central and peripheral coronal armature, the centrals being more or less rounded and granular, while the peripherals are fine and slender or almost capillary in character (Pl. V, fig. 87). The number of these two forms of coronal spinelets varies a great deal according to the size of the paxillæ; on one of the largest there may be as many as seventeen or eighteen central granules and some thirty to thirty-five peripherals, the latter always arranged in a single row, although at times the row may be more or less irregular. On this point, the specimens from Coromandel and Batavia referred to by PERRIER appear to be different from any of the Japanese specimens I have examined. In the specimen from Coromandel the coronal armature consisted of hexagonal granules which were finer and longer along the margin, while in the one from Batavia there was generally one central and one or two rows of peripherals. Occasionally I have observed paxillæ in which the differentiation into the centrals and peripherals was not sharp, and sometimes the spinelets were all fine and slender, exactly like the ordinary peripherals. On the arms the paxillæ are arranged in longitudinal rows, of which I counted eleven at the base of an arm of a specimen with R=29.5—32.5 mm. and sixteen at the same place of a specimen with R=184 mm. In the distal parts of the arms, however, these rows become more or less irregular.

Madreporite.—The madreporite is comparatively small and is situated close to the margin of the disk, between two of the arms. Although naked it is completely hidden from view by the crowns of the surrounding paxillæ. It is very slightly convex and deeply crenate at the margin ; the grooves are mostly radiating (Pl. V, fig. 88).

Terminal plate.—The terminal plate is small and inconspicuous because covered with granules similar to the çoronal granules of the paxillæ. When denuded it is seen to have the form of an incomplete ring.

Locality.—My specimens are from the following localities : Misaki, Awa, Miyazu (Tango). The bathymetrical range appears to be rather wide, there being specimens from the littoral region as well as from a depth of 640 metres.

Specimens in S.C., H.N.S., H.N.S.W.

Remarks.—In small specimens, e.g. in one of R=29.5–32.5 mm., the armature of the mouth-plate is much simpler, and the large pedicellariæ of the adambulacral plates are either absent or imperfectly developed.

Luidia quinaria VON MARTENS.

(Pl. VII, figs. 104–112.)

This species was made known to science by VON MARTENS as a variety of *Luidia maculata* as follows ['65, p. 352] :

" *Luidia maculata* MÜLL. u. TROSCHEL l.c. S. 77.—Herklots fn. jap. tab. inedit.

" Japan, Leidner Museum.

" var. *quinaria* m.

" Nagasaki, SCHOTTMÜLLER.

" Die mir vorliegende Tafel von HERKLOTS stimmt mit der Original-beschreibung der Art. Das von Hrn. SCHOTTMÜLLER von Nangasaki eingesandte Exemplar jedoch unterscheidet sich von demselben dadurch dass :

" 1) nur fünf Arme vorhanden sind ;

" 2) auf der Rückseite der Arme die kleineren unregelmässig gestellten Paxillengruppen eine etwas grössere Breite, die Hälfte der Armbreite über-haupt, einnehmen, (in HERKLOTS Figur kaum mehr als ein Drittel) und demgemäss auch viel zahlreicher sind ;

" 3) keine Flecken an dem in Spiritus aufbewahrten Exemplar zu sehen sind, während doch selbst Jahre lang trocken aufbewahrte andere Exemplare verschiedener Arten *Luidia* noch Flecken zeigen ;

" 4) die Arme verhältnissmässig breiter sind, was wenigstens für ihren Ursprung durch die geringere Zahl bedingt ist. Grosser Radius 135 Mill., kleiner 18, Breite der Arme an ihrem Grunde 23 Mill."

These differences of *quinaria* from the typical *Luidia' maculata* appear to me sufficient to convince any one conversant with Japanese starfishes that v. MARTENS' specimens from Nagasaki belonged to the present species. It must however be remarked that the absence of coloured spots on the back in the alcoholic specimen has hardly any significance in the present question, since they are known to fade away in alcohol much sooner than in dry specimens. But the other characters mentioned taken in conjunction with the fact that *Luidia maculata* always has more than five arms justifies the conclusion that *Luidia maculata* M.T. var. *quinaria* v. MARTENS is a distinct species.

●It was then described very fully by SLADEN under the name of *Luidia limbata* ['89, p. 251]:

"*Luidia limbata*, n. sp. (Pl. XLIV, figs. 3 and 4; Pl. XLV. figs 7 and 8).

"Rays five. R=110 mm., r=15 mm. R=7.3 r. Breadth of a ray near the base, at the broadest part, 18 mm.

"Rays moderately long, very flat, and rather broad, tapering slowly from base to the extremity, which is not very attenuate; frequently with a slight lateral constriction at the base. Abactinal and actinal surfaces subplane, bevelled towards the margin, which is subangular, slightly rounded.

"The paxillæ of the abactinal area are large, compact, closely fitting, and square, except along the median radial line and the centre of the disk. The larger paxillæ bear on the tabulum about a dozen, or even more, low, hemispherical granules, subequal, comparatively large and well-spaced, and round the margin a series of very small, short, cilia-like spinelets, about two dozen or more in number, the series often appearing to be double. There are three to four regular longitudinal series of square paxillæ at the sides of the ray. In the intermediate area along the median radial line the paxillæ are smaller, and have a tendency (more marked in some examples than in others) to become rounded or irregular in form.

"The paxillæ, which represent the aborted supero-marginal plates, are remarkable for bearing on their tabulum a broad, low, valvular pedicellaria resembling the form frequently found in Pentagonasteridæ. These pedicellariæ are most numerous in the neighbourhood of the interbrachial arcs, and may there extend upon the adjacent one or two series of longitudinal paxillæ. Along the ray, however, they are generally confined to the outermost or 'supero-marginal' series. Sometimes on the inner part of the ray there are two pedicellariæ on one tabulum; but towards the extremity of the ray very few are present at all. Rarely a pedicellaria may be composed of three valves, as in some Pentagonasteridæ, the valvate character of the organ being still maintained. Occasionally similar pedicellariæ may be found on the other paxillæ of the abactinal surface, but their occurrence there is rare.

"The infero-marginal plates bear only one lateral spine, placed at the outer end of the plate, directed outward and often more or less appressed to the margin of the ray. This spine is short, not more than 2.75 mm. in length, compressed, though rather robust, tapering, pointed and often subfusiform in outline. On the median keel of the plate is a series of small, subequal, squamiform spinelets, rather more than one-third the length of the lateral spine, compressed, tapering, sometimes pointed, but more frequently obtuse ; more or less appressed to the plate and directed at an angle outward and towards the margin, over the aboral margin of the infero-marginal plate. Additional similar spinelets may be present on the inner side of this series and with the same direction, but this part of the plate is usually occupied by very much smaller spinelets, which are also often directed over the adoral margin, i.e. at an angle adorally and towards the lateral margin. Occasionally the place of one of the larger spinelets is occupied by a two-jawed forficiform pedicellaria, often near the base of the lateral spine. The margins of the ridge are fringed with numerous, delicate, equal, cilia-like spinelets ; and the walls of the deep fasciolar furrow at the outer or upper part of the plates is densely covered with remarkably fine ciliary spinelets.

"The armature of the adambulacral plates normally consists of three spines and one large two-jawed forficiform pedicellaria. The three spines are arranged one behind the other, i.e., in transverse series in relation to the axis of the ray. The innermost or furrow spine, which is the shortest, is delicate, compressed laterally, tapering, and slightly curved ; the second spine is slightly longer, more robust, subtriangular in section, with an edge towards the furrow, sometimes slightly compressed laterally, tapering, pointed, and very faintly geniculate near the base. The outermost spine is subequal in length to the last noticed, but straight, tapering, and obtusely pointed. The pedicellaria stands at the adoral side of the outer spinelet, to which it is subequal in length ; it is large and expanded at the base, with a large luuule, but tapering, attenuate, and pointed at the tips of the jaws. On the outer edge of the plate are usually two or three very delicate, short, cilia-like spinelets.

"The infero-marginal plate is separated from the adambulacral plate throughout the ray by a small well-defined intermediate plate. This on the inner part of the ray may bear a comparatively large two-jawed forficiform pedicellaria, similar to, but rather smaller than, that on the adambulacral plate, and two or three small cilia-like spinelets; but usually along the greater part of the ray, and always on the outer part, only a few small ciliary spinelets are present.

"The mouth-plates are comparatively small. Each plate bears at its innermost point a large forficiform pecicellaria, forming a pair directed horizontally over the actinostome; and sometimes a second pair stand immediately below these, in the place of the first actinal or superficial spines. The marginal spines and actinal spines are subequal to one another in size, and diminish as they recede from the mouth; the actinal spines forming a series parallel to the median suture of the mouth-plates.

"The actinal interradial areas are very small, and do not contain more than two or four small intermediate plates, which bear either a forficiform pedicellaria or a group of small ciliary spinelets.

"The madreporiform body is hidden by paxillæ.

"Colour in alcohol, ranging between light yellowish drab and dirty greenish grey; with a very dark broad band, almost black in some specimens, dark sage green in others, along the median radial line, extending to the centre of the disk, in some specimens fading out gradually at the sides, in others terminating more abruptly. At the extreme tip of the ray the whole area is of this dark colour, whilst the ocular or terminal plate is white, and hence very conspicuous

"*Locality.*—Yokohama. May 6, 1875. Depth 2 to 25 fathoms.

"*Remarks.*—This species is perhaps most nearly related to a *Luidia* from Singapore, preserved in the University Museum at Copenhagen, which bears the name of *Luidia chefuensis*, GRUBE; but the two forms are distinguished by a number of well-defined characters. In *Luidia limbata* the paxillæ are larger and more definitely square, and the lateral spine is of a different shape. The armature of the adambulacral plates is different, wanting altogether the comb of four or more spinelets running parallel to the furrow, which in *Luidia chefuensis*, succeeds the two single curved spinelets. In like manner there is no trace of the second comb, parallel to the furrow, of four to six ciliary spines, probably situated on the intermediate plate between the infero-marginal and adambulacral plates. (For *Luidia chefuensis*, GRUBE, see Jahres-Ber. Schles. Gesellsch. f. vaterl. Cultur, 1876, p. 28.)

" In the University collection at Breslau there is another *Luidia*, which bears the manuscript name of *Luidia singaporensis*, GRUBE, but which seems to me to be the same as VON MARTENS' *Luidia maculata*, var. *quinaria*, and different from *Luidia chefuensis*.

" I am inclined to think that the so-called variety *quinaria* of Dr. VON MARTENS is really a species distinct from *Luidia maculata*."

IVES was the first to recognise clearly the identity of V. MARTENS' var. *quinaria* with the *L. limbata* of SLADEN, and he gives certain figures which leave no doubt that he had the present species before him [IVES, '91, p. 211, Pl. IX, figs. 5–9].

MEISSNER mentions this species under SLADEN's specific name from Yokohama, and after citing the dimensions of SLADEN's orginal specimen gives those of his own as follows ['92, p. 189]: " Die vorliegenden Stücke ergaben: a) R=104; r=16; Armbreite a. d. Basis: 19 mm. b) R=96; r= 13; Armbreite a. d. Basis: 17 mm. (No. 3338.)"

DÖDERLEIN also mentions this species as being very common [: 02, p. 330]:

" *Luidia quinaria* V. MARTENS.

" Syn. *Luidia maculata* var. *quinaria* V. MARTENS;
        Luidia quinaria SLADEN, IVES;
        Luidia limbata SLADEN.

" Sämmtliche 5armige Exemplare von *Luidia*, die ich von der japanischen Küste kenne, gehören zú dieser einen Art. Bei kleineren Exemplaren fehlen die klappenartigen Pedicellarien meist ganz, die an grossen Exem-

plaren oft sehr auffallend sind. Mir liegen Exemplare vor mit einem grossen
Radius von 8 mm bis zu einem solchen von 140 mm.

"Ich fand diese Art überall in geringer Tiefe auf schlammigem Grund,
und zwar in grosser Anzahl : Tokiobai, Enoshima, Kochi, Tango, Kagoshima,
kleine Exemplare traf ich in der Sagamibai noch in einer Tiefe von ca.
200 m."

KŒHLER, using SLADEN's name, describes some specimens from the collection of the Indian Museum [: 10, p. 10] :

"Iles Andaman. Trois échantillons presque intacts.

"Petite Andaman. Profondeur 7–9 brasses. Quelques échantillons.

"Huit miles au W. S. W. de Honawar. Deux échantillons.

"Détroit de Palk. Profondeur 12 brasses. Un bras unique provenant
d'un exemplaire d'assez grande taille at un très petit échantillon incomplet.

"King's Island, Archipel Mergui. Un échantillon.

"Extrémité N. du Golfe Persique. Profondeur 15 brasses. Un échantillon.

"Gopalpore.   Profondeur 25–28 brasses. 25–27 Septembre 1909. Un
échantillon.

"Les deux plus grands individus proviennent de Honawar : dans le
premier les bras sont restés attachés au disque mais deux d'entre eux sont
incomplets, R=76 mm. ; le second est en morceaux. Les autres échantillons
sont plus petits et R varie entre 27 et 47 mm.

"Je remarque sur le grand individu d'Honawar que le grand pédicellaire adoral n'est pas toujours présent et qu'il peut être remplacé par un
simple piquant."

l have only a few particulars to add to the descriptions of
SLADEN and DÖDERLEIN. The radial ratio varies to a considerable
extent, as may be seen from the following :

Specim.	r mm.	R mm.	R : r
1	8	39	4.9
2	8	40	5
3	9	45	5
4	9	46	5
5	10	74	7.4
6	11	62	5.6
7	11	62	5.6
8	12	69	5.7
9	12	60	5
10	13	81	6.2
11	15	75	5
12	15	107	7
13	17.5	102	5.8
14	18	115	6.4
15	19	135	7

The number of regular longitudinal series of large square paxillæ is usually three on either side of the arm (Pl. VII, fig. 104), and the innermost series is usually more or less irregular. In very large specimens there may be four such series, and in this case the third series is more regular. The marginal paxillæ representing the superomarginals bear small, lozenge-shaped, valvate pedicellariæ, one plate sometimes bearing two of them. They are generally more numerous at the interbrachial angle, and here they are usually present not only on the superomarginal, but also on the inferomaginal plates and some of the abactinal paxillæ. In some specimens I have seen similar but very much smaller pedicellariæ scattered all over the abactinal paxillæ, being visible to the naked eye as so many small dots, more lightly coloured than their surroundings. In such cases the pedicellariæ are as a rule larger towards the margin of the ray.

The armature of the adambulacral plates varies somewhat. The first two plates rarely bear pedicellariæ, and the number of large spines is usually four or five; in the following few plates the number of large spines may be three or four, and the pedicellariæ may be absent; but in the greater part of the arms, the typical arrangement of the spines and pedicellariæ described by SLADEN obtains (Pl. VII, figs. 107, 110). In the outer half of the arms the adambulacral pedicellariæ are frequently wanting.

The mouth-plates bear at the mouth end more usually two pairs of forficiform pedicellariæ than one (Pl. VII, fig. 106, 109). Each mouth-plate bears on the actinal ridge a series of six or seven larger spines, and a second series of five or six similar spines parallel to the first; the rest of the plate is covered with a number of smaller spines without any regular arrangement.

The madreporite is completely hidden from view by the surrounding paxillæ, and is situated at the margin of the disk, the outer margins of the disk and plate coinciding with each other. It is perfectly flat and irregular in outline owing to the encroachment of the adjacent paxillæ; it may also bear a paxilla (Pl. VII, fig. 112).

Locality.—This species is the commonest *Luidia* on the coasts of the Main Island but appears to be less common in Hokkaido. I have specimens from Volcano Bay, Misaki, Bay of Tokyo, Gulf of Suruga, Gulf of Isé, the Inland Sea, Gulf of Kagoshima, and Miyazu and Namerikawa on the side of the Japan Sea.

Specimens in S.C., H.N.S., H.N.S.W., S., I.M., F.B.

Luidia moroisoana, n. sp.

(Pl. VI, figs. 95–103.)

I have only one specimen of this species in alcohol. It belongs to the same group of this genus as the *L. aspera* of SLADEN and the *L. hystrix* and *L. magnifica* of FISHER, the first from the Philippines, and the last two from the Hawaiian Islands. The specimen is ten-armed, with a tolerably large disk of the radius of 23 mm. as against the 194 mm. of the arms, the radial ratio consequently being 8.4. The colour is reddish-brown on the abactinal side, the other side being colourless. The paxillæ present in general a rough appearance, and those of the two or three rows bordering on the inferomarginals mostly bear each a conical spine at the centre of the tabulum. The paxillæ of the arms lying between the three rows just mentioned are conspicuously smaller, and do not bear any prominent spines. On the actinal side also the spines are well developed.

Inferomarginals.—Except at the bases of the arms the inferomarginals are very broad, and are the most conspicuous plates on the actinal side, forming the lateral borders of the arms. The fasciolar grooves are very distinct though not deep. The most conspicuous part of the armature of the inferomarginal plates is a transverse series of three or four large conical spines, running along the middle of the plates (Pl. VI, figs. 98, 99). Of these the outermost two may be subequal, but more usually the outermost one is conspicuously larger than the others, and together with the similar spines of the other inferomarginal plates, forms a longitudinal series along the margin of the arms; the next spine lies at the apex of the rounded angle

between the actinal and the lateral side of the arms; the remaining one or two lie entirely on the actinal side and are
mostly less than half as long as the outermost spine. Besides
these large spines, there are many very small, almost ciliary
spines arranged along the margin of the plate. Similar
but slightly larger spines are also present more or less between
the large spines. There are also some 2-jawed forcipiform pedicellariæ of different sizes intermixed with the smaller spines, and
situated mostly along the margin of the plate; on a plate not far
from the disk, I counted in one case as many as ten pedicellariæ,
mostly situated along the adcentral border. The outer part of
the fasciolar grooves are covered with capillary spinelets. On the
abactinal side of the outermost of the large spines, there is a
space sometimes almost naked but mostly covered with small
conical spines somewhat larger than those of the actinal surface
of the plate.

Adambulacral plates.—There are as many adambulacrals as
inferomarginals, and the two are strictly coincident, though
separated from each other by a single series of small ventrolateral plates. Each adambulacral plate presents a rounded surface on the actinal side, and the armature is very characteristic.
There is, namely, a slender forficiform pedicellaria at the furrow
end of the plate (Pl. VI, fig. 98, 99, 101) well down in the
ambulacral furrow; it is slender and elongated, mostly half as
long as the first spine, and often looks, on a cursory view, like
a spine; but when the two jaws are apart, its true nature can
be recognized at once. Close to the pedicellaria lies the first
adambulacral spine, which is sword-shaped, slightly curved and
flattened like the corresponding spine of the species previously

described in this paper; in my specimen it is 2–2.5 mm. long. The second adambulacral spine is mostly about one-third longer than the first, and is also flattened; it is, however, nearly straight and presents only a slightly curved border toward the ambulacral furrow. The third spine is perfectly straight and nearly round, and is as much longer than the second spine as the latter is than the first. On many of the adambulacral plates there are one or two forcipiform pedicellariæ of varying size close to the third adambulacral spine, generally on its adcentral, but sometimes also on its outer side (Pl. VI, fig. 98, 100). There are in addition a few more very small spines on the adambulacral plate, mostly on its outer part, but also more or less on the adcentral side near the pedicellariæ. The first four or five adambulacrals at the month end mostly bear four instead of three spines; but in other respects they are exactly similar to the rest.

Mouth-plates.—The mouth-plates are comparatively small. Each plate bears on its actinal ridge a series of five or six spines, of which the one at the mouth end is stout and long, and with its fellow forms a conspicuous oral armature projecting towards the mouth. The other spines are of similar shape but become smaller towards the abcentral end of the plate (Pl. VI, fig. 97). On the abactinal side of the large oral spine and parallel to the ambulacral furrow is a series of three forcipiform pedicellariæ, of which, however, the middle one is frequently transformed into a flattened spine with truncated end. The pedicellaria nearest the mouth is usually about half as long as the oral spine; the second one is smaller and the third is smallest. When the middle pedicellaria is transformed into a spine, this is

of different lengths, but mostly much longer than the first pedicellaria. On the outer part of the mouth-plate there are some 5–10
small spines.

Ventrolaterals.—There is a single series of small ventrolaterals extending throughout the length of the arms. In the
proximal half of the arms each plate bears a few small spines
and frequently one forcipiform pedicellaria, similar in form and
size to those of the inferomarginals. Further out in the arms the
ventrolaterals are entirely naked, so that the adambulacral and
the inferomarginal plates appear in this part separated by a longitudinal space. The ventrolaterals are strictly coincident with
the adambulacrals and inferomarginals.

Paxillæ.—The paxillæ at the centre of the disk and along
the middle of the arms are smaller than the others (Pl. VI, fig.
95). On either side of the arm, next to the inferomarginals,
there are usually three longitudinal series of paxillæ, almost
square in form and larger than the rest, mostly bearing a conspicuously large conical spine at the centre of the tabulum (Pl.
VI, fig. 103). In the series immediately adjoining the inferomarginals the central spine is particularly large, being nearly one-
third as long as the marginal spine of the inferomarginals, and
is surrounded by some five or seven much smaller, short spines
and 1–3 small pedicellariæ (Pl. VI, fig. 102); the latter may, however,
be entirely absent on some of the plates. The peripheral spinelets are ciliary and are arranged in an irregularly double row
along the margin of the tabulum. The paxillæ of the second
row are smaller than those of the first, there being three or four
of them to every two of the latter; the central spine is much
smaller, and may often be absent, in which case there are some

seven to ten short and thick central spines, and a row of ciliary peripherals; a small pedicellaria may sometimes be present in addition, and the peripheral row may be irregularly double in the larger plates. The paxillæ of the third row present nothing of importance different from those of the second; they are on the whole smaller than the latter. The paxillæ along the middle of the arms do not form any regular longitudinal series, and bear according to their size each 5–10 granuliform centrals and a row of ciliary peripherals; one or two of the centrals may sometimes become larger and assume the form of a conical spine. The paxillæ of the disk are similar in their general character to those of the middle part of the arms; I have not observed any pedicellariæ on them.

Madreporite and Terminal plate.—The madreporite is completely hidden from view. The terminal plates are rather small, horse-shoe shaped and covered with granules; they are colourless and relatively conspicuous.

Locality.—Moroiso (Misaki); littoral.

This species can easily be distinguished from *L. aspera* SLADEN, *L. hystrix* FISHER and *L. magnifica* FISHER by the presence of a slender forficiform pedicellaria at the furrow end of the adambulacral plate, besides some other characters. It appears to be very closely related to *Luidia integra* recently described by KŒHLER [:10, p. 18], but is distinguished from it by the presence of pedicellariæ on the marginal paxillæ.

Specimen in S. C.

Luidia yesoensis, n. sp.

(Pl. V, figs. 89–90; Pl. VI, figs. 91–94.)

This species is nearly allied to *Luidia quinaria*, but distinguished from it by the general facies, shorter arms and the greater distinctness of the lateral faces of the arms. Whereas in *L. quinaria* the abactinal and lateral surfaces of the arms pass into each other without any sharp line of demarcation, they present in this species two distinct faces separated by a rounded angle. In all the specimens in my hands the abactinal side is uniformly dark grey or black, with only the marginal and most distal parts of the arms of a lighter colour. The dark radiating bands present in *L. quinaria* are absent in this species.

The radial ratio may be seen from the following :

Specim.	r mm.	R mm.	R : r
1	8.5	40	4.7
2	9	40	4.4
3	10	47	4.7
4	10	56	5.6

Inferomarginals.—The inferomarginals are broad and short, and coincident with the adambulacrals. At the lateral margin of the plate there is a single, somewhat large, squamiform spine, usually with some irregular fine longitudinal furrows on the actinal surface (Pl. VI, fig. 92). On the surface of the plate there can usually be made out one or two rows of squamiform spines, which are relatively broader near the lateral margin and become relatively larger towards the inner side. In many cases only one of these rows is distinct, and the rest of the surface of the plate is covered with spines, which are subequal to one another. On the abactinal side of the lateral spine, close to its base there are usually three short, squamiform spines, and still further abactinally there are some granules and ciliary spinelets similar to those of the abactinal paxillæ. At and near the interbrachial angle the inferomarginals mostly bear on the abactinal side, each a valvate pedicellaria similar to those of the marginal paxillæ; occasionally there may be two on a plate. Along the margin of the inferomarginal plate is a series of very small spinelets. The fasciolar grooves are covered in the outer part with capillary spinelets. On some plates, especially near the base of the arms, there may be near the inner end of the plate a forficiform pedicellaria similar in shape to those of the adambulacral plates. So far I have not found any in the outer part.

Adambulacrals.—The adambulacral plates are coincident with the inferomarginals and with the ventrolaterals separating the two series and presents a wedge-shaped edge towards the ambulacral furrow. The armature consists typically of three spines and a forficiform pedicellaria (Pl. VI, figs. 92, 93). The first spine is laterally compressed, curved, and sword-shaped, and is attached to the plate close to its inner edge, and therefore lies well in the furrow (Pl. VI, fig. 93a). The second spine is also lightly curved, but is much more stout and is somewhat triangular in cross-section (Pl. VI, fig. 93b). Then follow the third spine and pedicellaria, the two in a line parallel to the furrow, and the pedicellaria always on the adcentral side of the spine. Occasionally the pedicellaria may be found between the third and second spines. The third spine is slightly flattened and stout, and a

notch is present at the base on the inner side i.e. the side facing
the furrow (Pl. VI, fig. 92). On some plates the third spine is
replaced by two spines which are then smaller than when there
is only one (Pl. VI, fig. 93c, d). . The pedicellaria is large and
rather conspicuous, being as long as the third spine. It has two
jaws and a conspicuous lunule near the base (Pl. VI, fig. 93e);
each jaw would be triangular in cross-section, and has usually a
shallow groove on each external face i.e. the face turned away
from the axis of the organ. On the outer side of the third spine
and the pedicellaria there may be two or three smaller spines.

The first and second adambulacrals have a somewhat
different armature from the rest. The pedicellaria is usually
absent, and in its place is found a spine similar in form to the
third spine. There are five or six more spines in the outer part
of the first plate and two or three in the corresponding part of
the second.

Mouth-plates.—The mouth-plates are narrow and long with
an interspace covered over with a membrane. The most characteristic point in the armature · of the mouth-plates–is the presence
of a pair of stout forficiform pedicellariæ at the mouth end of the
plates and of a second pair nearly on the same level with the first
but more removed from the mouth, projecting into the ambulacral
furrow (Pl. VI, fig. 91). These pedicellariæ are similar to those
of *Luidia quinaria* in similar positions, but are relatively thicker
and shorter, and the second pair project in this species more
distinctly into the ambulacral furrow than in *L. quinaria*. On
the actinal surface of each mouth-plate there is a series of six
or seven large, blunt spines along the inner margin (facing
the fellow-plate) and a second series of four or five similar

but slightly smaller spines on the external side parallel to the first series. The most abcentral portion of the mouth-plates is covered with fine, slender, pointed spines, similar to those of the ventrolateral and the outer part of the adambulacral plates.

Ventrolaterals.—There is a series of ventrolateral plates between the adambulacrals and the inferomarginals, extending to the tip of the arms, but becoming somewhat indistinct in the distal part, owing to their fusion with the inferomarginal plates. Those in the interradii bear each five or six slender, pointed spines, and a few of them may be armed with pedicellariæ, similar in form to those of the adambulacral plates but smaller. In the more distal part of the arms, the ventrolaterals bear only two or three spines each.

Paxillæ.—The marginal paxillæ representing the superomarginals are square and mostly bear one or more lozenge-shaped, valvate pedicellariæ. These are especially numerous in and near the interbrachial angles, so that in this part there may be two or even three of them on a single plate. The paxillæ of this row lying near the interbrachial angle usually bear each 10–12 central granules and some 25 peripheral spinelets. The difference between these two sets are generally very distinct, the centrals being short, rounded and granuliform, and the peripherals very fine and almost ciliary in appearance though rather short. In typical paxillæ there is a single row of these peripherals around the margin of the tabulum; but there are others in which this difference is not so sharp, and in such ones there may be one or more rows of spines or granules intermediate in character between the two sets, or these intermediate spines may form a patch among the truly granular ones.

Next to the maginal series of paxillæ there are two more series of them similar in form to those of the first and but very slightly smaller. Except in the interbrachial angle, the paxillæ of these two series are however usually destitute of the valvate pedicellariæ, and where they are present they are much smaller than those of the first series. In the interbrachial angles, however, many of the paxillæ of the second row and some of those of the third row bear pedicellariæ exactly similar to those of the first, or superomarginal, series. The three rows of paxillæ above mentioned form the lateral borders of the arms in the proximal half and are either colourless or more lightly coloured than the rest of the paxillæ of the abactinal side.

The paxillæ of the abactinal side are rather small and closely set, so that on this side the animal appears more smooth, compared with *L. quinaria* and *L. maculata*. On one of the larger paxillæ on the disk or near the base of the arms there may be eight or nine central granules and fifteen or more very fine peripherals. The central granules present rather flattened tops, and in many paxillæ one of them is larger and occupies the very centre of the upper surface of the tabulum; this central granule is, however, sometimes wanting or very imperfectly developed, and in such cases there is an empty space at the centre of the paxillæ, as seen from the abactinal side.

Madreporite and Terminal plate.—The terminal plate is rather broad and bilobed; it is usually colourless and covered with granules. The madreporite is completely hidden from view by the surrounding paxillæ and one or two paxillæ which it bears on its surface (Pl. VI, fig. 94). It is rather small, somewhat triangular in shape, and is situated on the margin of the disk, close to the marginal series of paxillæ. The furrows are mostly radiating, simple and rather shallow.

Locality.—From the vicinity of Otaru and Ishikari, both in Hokkaido.

Remarks.—This species appears to prey upon echinoids, because two of the specimens that have come into my hands contained in their stomach an entire shell of a species probably of the genus *Echinarachnius* in a semi-decalcified state. The echinoid was in one case slightly larger than the disk of the starfish, and the latter was in consequence somewhat disfigured.

This species is perhaps nearly related to *L. chefuensis* GRUBE (Jahres-Ber. schles. Gesellsch. f. vaterl. Cultur, 1876, p. 28), the description of which I have not been able to see.

Specimens in S., some donated to S. C.

PENTAGONASTERIDÆ.

Pentagonaster, Tosia, Mediaster.

I am aware that the genus *Pentagonaster* in the loose sense in which it is used here requires revision. VERRILL ['99, p. 183] has referred *P. japonicus* SLADEN and *P. arcuatus* SLADEN to *Mediaster*; but it seems to me that there is no doubt that the first mentioned species belongs together with such forms as *Pentagonaster patagonicus* SLADEN, *Tosia leptocerama* FISHER and *Pentag. mortenseni* KŒHLER, while *P. arcuatus* SLADEN, *P. misakiensis* n. sp., *P. cuenoti* KŒHLER, *Tosia (Ceramaster) micropelta* FISHER, and *P. ernesti* LUDWIG form another natural group which gradually passes into *Mediaster* through such forms as *M. æqualis* STIMPSON and *M. ornatus* FISHER. There is nearly as gradual a transition between the first group which has been usually called *Tosia* by authors and the second group represented by *P. arcuatus* and others. Under these circumstances the limits between *Tosia* and *Mediaster* must be more or less indefinite and arbitrary, and I have therefore preferred for the time being to describe the species already known under their old names. There is no doubt of the close affinity of *P. arcuatus* to *Mediaster*, but the question is where to draw the line. The name *Mediaster* is retained because the typical species so far described appear to me sufficiently well marked to be placed in a genus by themselves. I look upon *Pentagonaster* as used here as a provisional genus which will undergo further modifications mainly in the way of curtailment. It may also be mentioned that the genus *Tosia* merges into *Hippasteria* through such a form as *T. aurata* Gray, specimens of which I have recently been able to examine.

FISHER [: 11, p. 204] has recently raised VERRILL's subgenus *Ceramaster* to the rank of a genus, and includes in it the following species : *C. japonicus*, *C. leptoceramus*. *C. patagonicus*, *C. clarki* and *C. arcticus*. There is much to be said for this genus, and under any system of classification these species ought to belong together. The materials at my disposal are however not extensive enough to enable me to form a definite opinion on this point, inasmuch as in my opinion examination of intermediate forms is especially needed. FISHER [: 11, p. 166] proposes to restrict the names *Pentagonaster* and *Tosia* to the Australian species, and more recently LUDWIG [: 12] has expressed himself substantially to the same effect. He refers *pulchellus* Gray, *abnormalis* Gray, *crassimanus* MÖBIUS and *dübeni* Gray to *Pentagonaster*, while *australis* Gray, *ornata* MÜLLER & TROSCHEL, *rubra* Gray, *nobilis* MÜLLER & TROSCHEL, *aurata* GRAY and *magnifica* MÜLLER & TROSCEL are referred to *Tosia*.

Pentagonaster japonicus SLADEN.

(Pl. XI, figs. 164–170.)

This species is due to SLADEN who described it as follows ['89, p. 272] :

" *Pentagonaster japonicus*, n. sp. (Pl. XLVI, figs. 1 and 2 ; Pl. XLIX. figs. 1 and 2).

" Rays five. R=68 mm., r=50 mm. R=1.36 r. The minor radius is thus in the proportion of about 73 per cent.

" Body of large size. General form depressed and thin. Abactinal area slightly convex and capable of being more or less inflated, the inflation being greatest in the radial regions, and emphasised by a conspicuous but shallow sulcus which traverses the interradial lines, but terminates at a short distance from the centre. The interradial channels are of uniform breadth, well defined and smooth, their character and regularity suggesting—fancifully, of course—the appearance of a mark produced by the pressure of a heated cylinder. The marginal contour has the form of an almost regular pentagon, with the sides very slightly incurved, the incurvature being produced more by a slight prolongation of the ray than by the regular curve of the side as a whole. Margin thick and vertical, equally rounded actinally and abactinally. Actinal area plane or slightly concave, undulating, more or less flexible, and capable of some inflation ; a slight sulcus usually defined along the median interradial line.

" The whole abactinal area is covered with small, regular, polygonal tabula, which diminish in size as they approach the margin (where they are very small) ; they are also smaller in the central region of the disk and along the edge of the interradial sulcus than in the median radial area and on the actual floor of the sulcus. The larger tabula in the radial areas, which are more or less elevated or paxilliform, are comparatively widely spaced, exposing the papulæ, of which there are about six round each tabulum, separated from one another by the stellate prolongations of the basal portion of the plates. The paxillæ consist of a hexagonal, rhomboid, or polygonal tabulum, slightly raised and faintly convex in the radial regions, where the paxillæ are widely spaced. The tabulum is covered with coarse, low, and almost truncate granules, and the margin is surrounded by a series of thin lamelliform papillæ or flattened granules, which have a striking appearance as compared with other species (see Pl. XLIX. fig. 1). A small excavate pedicellaria with two rather broad jaws and associated pit is present on some of the tabula, and appears to be always placed at the margin of the tabulum, some of the neighbouring granules being scooped away as it were for its reception.

"The supero-marginal plates, which are seventeen in number, counting from the median interradial line to the extremity, form a well-defined and nearly uniformly broad border to the abactinal area. The plates near the interradial line have their length and breadth subequal, the length being perhaps slightly in excess; as they proceed along the ray, however, the length diminishes step by step, until at the extremity the breadth is fully twice as great as the length. The plates are distinctly tumid. The lateral surface of the plates is covered with very small, uniform, crowded granules, but on the abactinal area of the plate there is a large naked quadrangular space which occupies nearly the whole of that surface, being separated from the margin only by two (or rarely three) rows of the small granules. The majority of the plates bear one, or occasionally two, small pedicellariæ placed at the edge of the naked space.

"The infero-marginal plates correspond to the superior series, and are, like them, covered with small crowded granules, excepting, however, a small circular area on the actinal surface of each plate, which is naked. Nearly all the infero-marginal plates bear one of the small excavate pedicellariæ similar to those on the supero-marginal plates; a few plates bear two. The pedicellariæ appear to be invariably placed close to one of the margins of the plate.

"The adambulacral plates are slightly broader than long, and their armature consists of a marginal series of six short, subequal spinelets, excepting the adoral spine of the series, which is smaller. The spinelets are thick and subprismatic or quadrangular in section, and have a roundly truncate tip; their base-line forms a slight curve, trending rather obliquely adorally. At a short distance behind the furrow series are three or four low, prismatic, laterally elongate granules, which form a slightly arched or straight series, traversing the plate slightly obliquely, the aboral end of the series being nearest the marginal or furrow series of spinelets. The remainder of the plate external to these is covered with large, low, subprismatic granules, which may form two or three subregular parallel lines but seldom definitely regular. Normally each adambulacral plate bears a single large two-valved excavate pedicellaria which is usually placed at the adoral extremity of the first series of granules behind the furrow series, although occasionally it is found immediately behind this second series, but always adjacent to the adoral margin of the plate. Rarely near the mouth these pedicellariæ may be rather irregular in construction, and formed of three or more valves.

"The mouth-plates are elongate and triangular, slightly truncate exteriorly, and with the free margin of each plate forming a straight line in continuation of the series of adambulacral plates, the united pair completing exactly the apex of the rectilineal angle of the actinal interradial area, bounded by the two adjacent furrows. The actinal surface of the plates is plane or very faintly convex. The armature of each plate consists of a marginal series of nine or ten short, prismatic, roundly truncate spinelets, exactly similar to those upon the adambulacral plates, but which increase slightly in size as they approach the inner extremity of the mouth-plate. On the actinal surface of the plate a row of about seven or eight large, low, irregular-shaped, prismatic granules runs parallel to the median suture. The innermost two of this series might also be reckoned as parallel to the marginal series, and three or four similar granules continue a line in this direction. Two or three small prismatic granules occupy the angular area between the two main series above described, and along the base-line of this area, which abuts on the first free adambulacral plate, is a straight series of similar granules.

"The actinal interradial areas are paved with a great number of small, normally quadrangular, but occasionally polygonal, intermediate plates, which fit close together and form a compact pavement. The largest are adjacent to the adambulacral plates, and these as well as the next one or two series are a little broader than long, the breadth of each row diminishing as it recedes from the furrow. The remaining intermediate plates have the length and breadth approximately equal, or they may be irregular or trapezoid in shape. All the plates disminish in size as they approach the margin, the plates at the extreme edge of the area adjacent to the infero-marginal plates being very small. The surface of the plates is covered with small, low, uniform granules, which are arranged in straight series along the margins of the plates, but show no definite order within this boundary. In the row adjacent to the adambulacral plates nearly every intermediate plate bears a small excavate pedicellaria with two broad, low, truncate, lamelliform valves, a little broader than high. Occasionally two pedicellariæ are present. Similar pedicellariæ are also present on a number of plates in the neighbourhood of the median interradial line and in an ill-defined region parallel to, and a little removed from, the infero-marginal plates. These pedicellariæ on the intermediate plates are of nearly uniform size throughout, and there is no regularity in their orientation.

"The anal orifice is slightly excentric, and is surrounded by rather larger plates than in the central region generally.

"The madreporiform body, which is rather small and polygonal in form, is situated at about one-third of the distance from the centre to the margin. It is marked with fine, regular, sharply convoluted, centrifugally radiating striations.

"Colour in alcohol, a warm shade of light brown.

"*Locality.*—Station 232. South of Yeddo (Japan). May 12, 1875. Lat. 35° 11′ 0″ N., long. 139° 28′ 0″ E. Depth 345 fathoms. Bottom temperature 41°.1 Fahr.; surface temperature 64°.2 Fahr.

"*Remarks.*—*Pentagonaster japonicus* is distinguished from *Pentagonaster patagonicus*, to which it is most nearly allied, by the more regular pentagonal form, the sides being less curved, and the rays less produced. The general granulation is finer. The structure of the paxillæ and the armature of the adambulacral plates are characteristic, as well as the presence of numerous pedicellariæ."

This species is referred to *Mediaster* by VERRILL ['99, p. 183] :

" *Mediaster japonicus* (SLADEN).

[Reference to SLADEN '89, p. 272 etc.]

" This species has a rather large, sessile, bivalve pedicellariæ with broad valves, on the adambulacral plates ; others of smaller size occur on many of the actinal plates. Some of the pedicellariæ have three valves.

" South of Yeddo, Japan, with the last " [*M. arcuatus*].

FISHER gives a very detailed account of the variation of this species under the name of *Ceramaster japonicus* (*vide supra*) [:11, p. 206, with four figures] :

" *Diagnosis.*—One of the largest if not the largest species in this or any closely allied genus. R=102 mm.; r=65 mm.; R=1.56 r (varying to R=1.36 r). General shape pentagonal or arcuate pentagonal, depressed, the edges of body more or less thickened; abactinal surface swollen on the radial areas; actinal surface nearly plane. Abactinal surface covered with fairly regular, spaced hexagonal to quadrate tabula crowned with a convex group of numerous polygonal granules, the marginal series regular, sometimes compressed, and when dried showing a pit in the top. On the interradial areas tabula are quadrate and smaller. Many of the tabula with a bivalved pedicellaria, having thin, lamelliform jaws which are wider than high. These vary in size, equalling one-third or one-fourth the width of the tabulum. Marginal plates variable, covered with small flat crowded granules, forming an evenly rounded, or a decidedly tumid margin to ray ; a bare space sometimes present on plates of both series, each superomarginal, with one to four small bivalved pedicellariæ, whose thin jaws equal two to four granules in width. These are situated usually on the abactinal surface of the superomarginals near the margins ; inferomarginals with usually only one pedicellaria or none. Actinal intermediate areas very extensive, the plates decreasing in size toward margin ; and covered with crowded low flat polygonal granules, of which the marginal are slightly the largest; individual plates perfectly flat and variously four-sided ; actinal intermediate pedicellariæ in variable numbers. Adambulacral plates usually slightly wider than long, with a slightly opaque furrow series of five or six (four to seven) round-tipped, four-sided, often somewhat flattened, stout spinelets, frequently graduated in length toward the adoral edge of plate, or all but the shorter adoral number subequal. Actinal surface with three to five irregular longitudinal series of low granules, those nearest the furrow spinelets being largest. A bivalved pedicellaria is normally present on most of the plates, the jaws being usually wider than high, but exceptionally higher than wide. The granules of actinal surface of adambulacrals vary greatly in number and those on the outer half are not in very definite series.

"Madreporic body pentagonal, situated one-third distance from centre to inner edge of marginal plates.

"*Anatomical notes.*—Anal opening present, subcentral, surrounded by three to six plates larger than those adjacent. Intestinal cœcum large, with eight long, unequal glove-finger-like radiating divisions. Intestine spacious, stomach small; hepatic cœca long, extending two-thirds to three-fourths length of R; quite a sharp distinction between dorsal and ventral divisions of stomach. Gonads very large, bushy, each with four short ducts and four apertures in a line parallel to and a short distance from the tough, uncalcified interradial septum. Dorsal muscle bands stout, joining at centre of disk. Each band gives off numerous lateral smaller branches obliquely, like the barbs of a feather, on the side toward margin. Polian vesicles large, one ' in each interradius except that of the madreporic canal; ampullæ double; tube feet with large sucking disks. No superambulacral ossicles.

"When the plates which bear the tabula are examined from the inner surface of the stout dorsal integument they are seen to be arranged in very regular rows parrallel with the median radial. The radial row and the two on either side have plates with six short, broad, rounded, or truncate lobes. On either side of these the lobes are very soon lost and the plates become circular. All plates are slightly spaced and there are no internal connecting ossicles as in *Mediaster*. Papulæ are single and distributed all over the abactinal surface, except in a narrow interradial band reaching toward centre of disk as far as the madreporic body; they are absent also at the tip of ray; four to six, or even more, occur around each plate, according to its size and shape.

" *Variations.*—The examination of seventy specimens of this species, including one from off Misaki, Japan, 640 meters, kindly sent by Dr. SEITARO GOTO, of Tokio, reveals a considerable range of individual variation. The differences which the American specimens seem to present are greatly overshadowed by individual differences occurring in examples from the same station. SLADEN apparently had only one specimen, and of course could give no hint of this variation in his excellent description.

" *General form, etc.*—Some specimens are more arcuately pentagonal than SLADEN'S figure and of the same form as his figure of *patagonicus*. The Misaki specimen is about intermediate in form. The Washington specimens are less arcuate than many of those from Bering Sea. The abactinal area varies in the amount of inflation, and corresponding to this the interradial sulcus is more or less evident. There is one four-rayed example.

" *Abactinal plates.*—Two large specimens from stations 3330 and 3331, Bering Sea, in practically the same locality exhibit very nearly the extremes of variation in the size of plates (Pl. 37, figs. 1 and 2). That from station 3330, specimen A (fig. 2) has $R=92$ mm., while B has $R=91$ mm. In A the largest tabula of the median radial region are 3 mm. in diameter, in B 3.5 mm., but in A there are relatively much fewer large tabula and the spaces between them are much wider than in B, as will be seen from the photographs. This difference in the size of the tabula and in the extent of space between is duplicated in other localities. The larger tabula have larger granules, not a correspondingly greater number; eighteen to twenty-four marginal granules, and eighteen to thirty in the central group is the range in large specimens. In dried specimens the marginal granules are concave on the top, the others are flat. Pedicellariæ are usually very numerous, occurring on a majority of plates of the papular areas.

" *Marginal plates.*—The greatest variation of all is in respect to the marginal plates. Typically these form a tumid rounded border, but exceptionally are nearly as thin as in *leptoceramus*. Again, the two specimens mentioned above will serve to illustrate the extremes of size. In A, which is less typical, the plates are entirely covered with close, flat granules and measure in the interradius 4 mm. by 4 mm., being less regular in size and not at all tumid. The thickness of the edge of the disk is 5 mm., which includes the height of the combined plates. In B the plates are very regular, 5 mm. long by 6 wide, and the edge of the disk is 8 mm. thick. The superomarginals are strongly tumid, but entirely covered with granules except for a few distal plates. In A there are forty-two to forty-four superomarginals and about fifty inferomarginals to a side of the pentagon; in B thirty-six superomarginals and forty-two inferomarginals. In typical *japonicus* as described by SLADEN the surperomarginals all had an extensive tumid bare space on the abactinal surface, and the inferomarginals a smaller one on the actinal surface. The size of this spot is variable even when it is present. In my Misaki specimen the spots are smaller than those described by SLADEN. One of the Washington examples has the bare spaces (on superomarginals only) about as in the Misaki specimen; three others lack them. Out of twenty-four specimens from station 3608 only three have a bare spot on the superomarginals and inferomarginals. A small specimen from 3331 has fairly large tumid bare spots, while a large emample ('B') lacks them. The shape of the plates is variable. In specimens with large marginals the width is greater than length, but when the plates are small and not tumid the proximal plates may be square or even longer than wide. The extreme in this respect is a specimen from station 4775 which has proximal plates 4 mm. long by 3.25 wide. Japanese specimens vary in this respect, as the type had longer proximal superomarginals than the Misaki example, which has the plates decidedly wider than long throughout. A remarkable variation is exhibited by a specimen from 4774, which has the marginal plates slightly concave, the transverse sutures being on the summit of a tumidity. The number of pedicellariæ on the superomarginals varies greatly, two or three are commonly present, sometimes three or four on all except the most distally situated. The inferomarginals sometimes lack pedicellariæ, but usually most of the plates have one or even two which are slightly larger than the superomarginal ones.

" *Actinal intermediate plates.*—A characteristic of this species is the very crowded polygonal, flat-topped, smooth actinal intermediate granules. They are not at all globular, and this character seems fairly constant. It is well marked in the Washington specimens and serves admirably in connection with other features to separate *japonicus* from *leptoceramus*. The latter species has spaced hemispherical granules.

"The small bivalved pedicellariæ vary considerably in numbers. They may be restricted to the angle adjacent to mouth plates, or scattered part way or all along the row of plates adjacent to adambulacrals. Many specimens, but not a majority, have them also sparingly elsewhere on the intermediate areas (Misaki, Washington, Bering Sea).

" *Adambulacral plates.*—In proportions the plates vary from being as wide as long to wider than long. The outer edge is frequently oblique or obtusely angular. The furrow spinelets are prismatic or subquadrate in section, blunt, usually but not always with a slightly curved, oblique, base-line. The Washington examples have proximally four or five, rarely six, furrow spinelets, distally three to five; the Misaki specimen proximally six or seven, distally six spinelets, and the base line is more evidently oblique. Large Alaskan examples have proximally five or six spinelets, distally four or five. Young Alaskan specimens (R=30 mm.) have four or five spinelets throughout. The Misaki specimen has rather shorter spinelets than the Alaskan and Washington examples. The first series of actinal granules varies from three to five. Far along the furrow one of these granules is enlarged into a blunt tubercle. The bare space between the furrow spinelets and first series of granules is variable in width, wider in the Japanese specimen than in the Alaskan or Washington examples. The number of granules exclusive of the first series varies from about twelve to twenty-two. The small number occurs when there is a pedicellaria on the plate. In specimen A, before cited, very few of the adambulacrals have pedicellariæ. In the Misaki example nearly all have one. Nearly as great a range of difference is present in the Washington examples, and in Bering Sea specimens, as, for example, station 3608.

" *Young.*—The young bear a great resemblance to *C. granularis* (RETZ.).

In this species, however, the superomarginals are longer than wide, whereas all the young *japonicus* have them wider than long. *Granularis* lacks pedicellariæ altogether, the abactinal tabula are flat-topped with large marginal granules, and the actinal intermediate granules are roundish and not crowded. The adambulacral armature is coarser and the spinelets and granules fewer. In young *japonicus* the papulæ are restricted to broadly ovate areas on the radii, the tabula are lower with fewer granules, and no abactinal pedicellariæ or only few. The marginal plates are sometimes relatively thicker than in the adult.

"*Type.*—In the British Museum.

"*Type-locality.*—Challenger station 232, south of Yokohama, Japan, 345 fathoms, green mud.

"*Distribution.*—Japan (Misaki, and south of Yokohama) to southern Bering Sea, thence south along the American coast to Oregon.

"*Specimens examined.*—Seventy-three, from the following localities:

"Specimens of *Ceramaster japonicus* examined.

Station	Locality	Depth Fathoms	Nature of bottom	No.	Collection
3227	Bering Sea, north of Unalaska. ..	225	Green mud.........	4	U.S. Nat. Mus.
3330	do...............	351	Black sand, mud......	1	do.
3331	do...............	350	Mud	5	do.
3346	Off Tillamook Bay, Oregon	786	Green mud	4	do.
3488	West of Pribilof Islands, Bering Sea........................	106	Green mud, gray sand	1	do.
3489	do...............	184	do.........	2	do.
3502	Bering Sea, south of St. George Island	368	Green mud, dark sand	26	do.
3608	Bering Sea between St. George and Unalaska...............	276	Gray sand...........	24	do.
4768	Bowers Bank, Bering Sea, 54° 12' N.; 179° 07' E.............	771	Green mud.........	1	Albatross, 1906.
4774	Bowers Bank, Bering Sea, 54° 33' N.; 178° 45' E.............	557	Green mud, black speck, foraminifera ..	1	do.
4775	do...............	584	do.........	3	do.
....	Misaki, Japan	320	....	1	Stanford.

" *Remarks.*—In lieu of giving a detailed description of this species, I have listed the chief variations presented by a large series of specimens. The various figures will serve sufficiently, in connection with SLADEN's excellent description of the type-specimen, for purposes of identification. Except in the case of young specimens there is no danger of confusing this species with *C. granularis*, to which it seems to be rather closely related. The salient differences have been already pointed out. The differences between *japonicus* and *leptoceramus* are mentioned below under the latter species."

I have two specimens of this species, both from the neighbourhood of Misaki, and one of them from a depth of about 640 m. The measurements are as follows :

Specim.	r mm.	R mm.	R : r	MS
1	48	74	1.54	20
2	48.5	75	1.54	19

The second specimen shows signs of having regenerated at the tip of three of the arms, and the relatively small number of the marginals is probably to be attributed to this circumstance. In one of the specimens, in which the radii are rather strongly inflated, the interradial sulci are very distinct and presents a smooth appearance, owing to the crowded condition of the paxillæ. In the other specimen, in which the radii are hardly inflated, the interradial sulci are correspondingly inconspicuous ; but the paxillæ are more closely crowded than in the radial areas. The relative conspicuousness of the interradial sulci appear therefore to depend largely on the degree of inflation of the radial areas, which is again due, no doubt, to the condition of the digestive system, the hepatic cœca of which extend to near

the tip of the arms. It may also be remarked that in the interradial sulci the tabula are pronouncedly rhomboidal in form. The
peripheral granules of the paxillæ are usually smaller than the
more central ones (Pl. XI, fig. 169). The pedicellariæ are numerous
on the abactinal surface and are well characterised by SLADEN.

The granules covering the superomarginal plates are so very
small and close set that the plates appear to the naked eye as if they
were covered with a tumid membrane. The presence of a naked area
on the abactinal surface of the superomarginals is a characteristic
which this species shares with some others. The size of this area
varies much even in the same specimen, sometimes occupying, as
SLADEN says, nearly the whole of the abactinal area, but sometimes
being very small. In the latter case the area is close to the inner
margin of the plate and its form tends more to be elliptical ; it may
occupy less than one half of the abactinal surface. On some of
the regenerated plates this area is totally absent. The pedicellariæ may be situated close to the naked area or may be farther
removed from it.

The inferomarginal plates are mostly coincident with the
superomarginals, but as they approach the tip of the arms the
two series become distinctly alternate. In some of the interradii,
also, they are strictly coincident only for one or two plates on
either side of the interradial line. There is a naked area on
each plate on the actinal surface.

The furrow series of the adambulacral armature consists
usually of five, but sometimes six, rather short, stout, truncate
spines of a prismatic form, and is separated from the granules
covering the actinal surface of the plates by a distinct space,
forming a groove on either side of the ambulacral furrow (Pl. XI,

fig. 167). External to this groove there are three or four series of flat, polygonal granules, which are mostly larger than those of the ventrolateral plates. There is a conspicuous, valvate pedicellaria close to the adcentral margin of the adambulacral plate in the same line with the first series of the granules, although it may sometimes be found more externally. Each series of the granules may consist of 3-5. It may also be noted that some of the adambulacral pedicellariæ have three valves. Occasionally there may be a granule in the groove separating the furrow spines from the actinal granules.

The mouth-plates are quite large, and each plate, when denuded of its spines, is distinctly seen to have the form of an isosceles triangle, the two plates being apposed to each other by the base. The number of the spines are somewhat different in my specimens from the description of SLADEN. On the furrow margin there is for each plate a series of seven to nine short, stout, prismatic, truncate spines, the innermost of which is usually conspicuous by its large size. In both of my specimens the mouth is closed, and the marginal spines stand erect. Along the actinal suture line of the plate there is a series of some ten granuliform spines, large and elongate at the mouth end and diminishing in size and height towards the other end. In the area enclosed by the above two series there are some 13-15 granules, which are generally larger and more elongate near the mouth, but smaller and low towards the other end (Pl. XI, fig. 166).

The ventrolaterals are closely set and quadrangular near the ambulacral furrows but irregularly polygonal in the interradial parts. The granules are very small like those of the marginals but perceptibly coarser. The plates next the adambulacrals, with very few exceptions, bear each one valvate excavate pedicellaria placed either parallel or obliquely to the ambulacral furrow, forming a conspicuous series. Many of the other plates also bear similar pedicellariæ, which are larger and more numerous around the mouth.

One of the specimens, which had lain in formalin for some time, and was subsequently transferred to alcohol has a soft pale rosy colour. It was probably red in life.

One specimen in S.C.; one sent to Dr. W. K. FISHER of Stanford University, California.

Pentagonaster arcuatus SLADEN.

(Pl. XI, figs. 171–177.)

This species was first described by SLADEN as follows ['89, p. 277] :

" *Pentagonaster arcuatus,* n. sp. (Pl. LII. figs. 1 and 2 ; Pl. XVIII. figs. 5 and 6).

" Rays five. R=45 mm. ; r=23.5 mm. R=1.93 r. The minor radius is thus in the proportion of 52.2 per cent.

" General form flat, but moderately thick. Marginal contour stellato-pentagonal, with the radial angles produced and tapering to an acute extremity, which is slightly turned upward. Interbrachial arcs widely rounded. Margins equally rounded abactinally and actinally. Abactinal area not elevated above the level of the marginal plates ; slight depressions are present in the interradial areas near the margin, which are probably indicative of a limited capability of inflation. Actinal area subplane, with small well-defined depressions external to the mouth-plates.

" The abactinal area is covered with small, subcircular plates, closely placed, united by short, narrow prolongations, which leave interspaces for comparatively large papulæ in the radial regions. The abactinal plates extend to the tip of the ray, two or more series separating the outermost supero-marginal plates from the corresponding plates on the other side of the ray. Seen from above the abactinal plates have a strikingly paxilliform appearance, when their granulation is intact; the subcircular tabulum is surrounded by a marginal series of small uniform, slightly elongate granules, moderately spaced, and so placed that they appear to radiate slightly apart. Within this ring are several small hemispherical granules, the majority of which are larger than the marginal series, but are in no sense elongate. A small valvate pedicellaria formed by two contingent granules is present on the tabulum of a few of the paxillæ, but these organs are of rare occurrence. The plates on the outer part of the ray and those adjacent to the margin throughout are devoid of stellate prolongations, and appear to have more or less of an imbricating character.

"The supero-marginal plates, which are seventeen or eighteen in number, counting from the median interradial line to the extremity, form a well-defined border to the abactinal area, which diminishes in breadth towards the extremity of the rays. The plates near the interradial line have their breadth rather greater than their length, and the length distinctly increases in a few of the succeeding plates, and then diminishes on the outer half of the ray. The surface of the plate, which is slightly convex in the transverse direction, is covered with rather large, well-spaced, hemispherical granules, those which bound the margin being rather smaller than the others, and regularly disposed in lineal series. The odd terminal plate is very small.

"The infero-marginal plates correspond to the superior series in the neighbourhood of the median interradial line, but alternate with them along the rest of the ray. They are covered with precisely similar granules. I have detected no pedicellariæ on either series of marginal plates.

"The adambulacral plates are slightly broader than long, and their armature consists of three regular series of spinelets. In the furrow series are seven moderately elongate, prismatic spinelets, compressed transversely, with obtusely rounded tips, subequal in length excepting one or two at the adoral end of the series which are rather shorter; these spinelets radiate very slightly apart, but to such a small degree that their disposition is almost that of a compact comb. The second series, which is placed on the actual surface of the plate, consists of five quadrangular prismatic spinelets, which taper slightly, and are shorter but more robust than the furrow series. The adoral spinelet of this series is usually placed rather further back on the plate than the others, a circumstance which causes this series to have the appearance of a slightly oblique position. The third series is close to the external margin of the plate, and consists of four or five low, equal, granuliform spinelets, or spiniform granules, much shorter than the median series just described. The outer spinelet at each end of the series is often accompanied by another placed close behind, and this is sometimes modified into a pedicellaria. Sometimes the whole line may be doubled on the outer part of the ray. The three distinct series of spinelets above described are comparatively widely spaced apart.

" The mouth-plates are small and inconspicuous. Their armature consists of a marginal series on each plate of ten to thirteen powerful representatives of the marginal spinelets on the adambulacral plates, which increase in size as they approach the mouth. On the actinal surface of the plate is a lineal series of about five short, robust, prismatic secondary spinelets, and on the outer part of the plate are a few prismatic granules.

" The actinal interradial areas are large, and extend far along the ray. The intermediate plates have, in consequence of their mode of granulation, a very paxilliform appearance, their armature consisting of a marginal series of slightly elongate subprismatic granules surrounding several larger hemispherical granules, all of which are well spaced.

" The madreporiform body is situated near the centre, its inner margin being less than one-fourth of the distance from the centre to the margin. It is rather large and subcircular in outline, and its surface is marked with fine striations, which are much convoluted in the central region.

" Colour in alcohol, a light shade of umber brown.

" *Locality.*—Station 232. South of Yeddo, Japan. May 12, 1875. Lat. 35° 11′ 0″ N., long. 139° 28′ 0″ E. Depth 345 fathoms. Green mud. Bottom temperature 41°.1 Fahr.; surface temperature 64°.2 Fahr.

"*Remarks*.—This species is distinct from any other *Pentagonaster* with which I am acquainted, and is readily distinguished from all the Pacific forms at present known. Its general character points to some alliance with *Pentagonaster intermedius*, of which *Pentagonaster arcuatus* is perhaps the the Pacific representative. In some of the structural details *Pentagonaster arcuatus* resembles forms grouped in the genus *Nymphaster*, of which I was at first disposed to consider it an aberrant species. I am still in some doubt as to whether this form is correctly referred to the genus *Pentagonaster*, but with only a single example for study, that course appeared to me the most justifiable."

ALCOCK mentions this species from the Indian seas, and makes the following remarks ['93, p. 89]

"*Pentagonaster arcuatus* SLADEN.

"I am not quite certain about this determination, although our species conforms exactly to Mr. SLADEN's description, and although the habitat supports it—the 'Challenger' specimen being dredged in the green mud of the Japanese Sea at 345 fathoms, and our specimen being dredged in the green mud of the Andaman Sea at 271 fathoms."

VERRILL refers this species to *Mediaster*, and there is no doubt that it presents certain resemblances to *Mediaster æqualis* STIMPSON, the type species of the genus, a specimen of which the writer owes to the kindness of Dr. W. K. FISHER. VERRILL adds the following remarks, evidently based on SLADEN's description [VERRILL, '99, p. 183]:

"This species has a few small pedicellariæ on the abactinal plates, similar in size to the granules.

"South of Yeddo, Japan, 345 fathoms."

KŒHLER [: 09a, p. 70] makes some remarks on this species in connection with his *P. cuenoti*, which is no doubt very closely related to it. He says : "M. ALCOCK a signalé un *P. arcuatus* provenant des îles Andaman, par 271 brasses de profondeur. Je n'ai pas vu ce *Pentagonaster*, mais celui que je viens de décrire n'est certainement pas un *P. arcuatus*. Sa forme rapelle bien celle de cette dernière espèce, mais il en diffère par les plaques marginales dorsales et ventrales en grande partie nues, par les plaques dorsales

plus petites et munies de granules peu nombreux, par les pédicellaires
assez abondants et se montrant sur les plaques marginales dorsales et
ventrales, et enfin par l'armature des plaques adambulacraires. Je ne vois
aucune espèce dont on puisse rapprocher le *P. cuenoti*. Il est complétement
different du *P. pulvillus* ALCOCK que l'Investigator a rencontré : il s'en
écarte, en effet par ses plaques marginales plus grandes, moins nombreuses,
et offrant toutes un espace central nu, ainsi que par ses pores très apparents
et très nombreux."

The facts mentioned below tend to lessen the differences between *P.
arcuatus* and *P. cuenoti*, but the latter appears to be a distinct species.

FISHER also refers this species to *Mediaster* and mentions the presence
of long and slender superambulacral plates and of the internal radiating
ossicles of the abactinal side already noted by VERRILL as a characteristic of
Mediaster [FISHER : 11, p. 199, 205].

As this species is sufficiently well characterised by SLADEN's description
my remarks will be confined to a few points of difference, which are of some
systematic importance, presented by my specimens.

I have three specimens, all from the neighbourhood of
Misaki, and from depths ranging between 480 m. and 560 m.
The measurements are as follows :

Specim.	r mm.	R mm.	R : r	MS
1	16	32	2	14
2	27	56	2	21
3	30	58	1.9	19

The number of marginals is thus seen to be somewhat more
variable than in many other starfishes, the specimen with $R=58$
mm. having less of them than the one in which $R=56$ mm.

The interradial areas of the abactinal side is depressed, and
SLADEN thinks that this may indicate a certain capacity of infla-

tion. It can easily be seen that the radiating elevations along the arms are due to the presence of the digestive system which is regularly petaloid in shape.

The pedicellariæ are present both on the actinal and abactinal sides as well as on the marginal plates. Those on the actinal sides are very few in number, there being mostly only one in each interradius and that one is not always present. It is closely similar in form and structure to the pedicellariæ found in similar positions in *Mediaster brachiatus*, but so far as I have observed, it is tricuspid and is mostly situated on one of the ventrolateral plates near the mouth (Pl. XI, fig. 174).

The pedicellariæ of the abactinal side are all valvate, being much more elongated transversely than high. They are borne on the abactinal plates, and are mostly, though not invariably, formed by the homologues of the peripheral granules (Pl. XI, figs. 175, 176). In the specimen with $R=56$ mm. above mentioned I counted as many as 40 pedicellariæ on the whole abactinal surface.

Several of the marginal plates, both inferior and superior, bear pedicellariæ which are closely similar in form and structure to those of the abactinal surface, but they are not very constant in position and number.

The pedicellariæ of this species appear to be subject to much variation in number, so that SLADEN's statement on this point is perfectly intelligible.

As to the adambulacral armature, the spines of the furrow series vary between five and seven, and those of the second series are four or five in number, very rarely six (Pl. XI, fig. 174).

The mouth-plates are also subject to variation in size and general appearance, being tolerably large and conspicuous in the

smaller two of my specimens, and less so in the largest.
A closer examination, however, reveals that this is due more
to the character of their spines than to their actual size, the
former being closely similar to the armature of the ventrolateral plates in the largest of my specimens, while in the other
two the spines are larger and more pointed. There are some ten
or more flattened, robust spines on the furrow margin of each plate,
four or five short, prismatic ones on the border facing the first
adambulacral plate, and some seven spines along the interradial
border. Of the last mentioned series the last one or two spines
next the mouth are especially large and pointed (Pl. XI, fig. 173).

With regard to the abactinal plates, they are, as remarked
by SLADEN, strikingly like paxillæ. The coronal granules are
somewhat elongated and the centrals, when present, are mostly
larger than the peripherals. On one of the larger plates there may
be as many as 15–20 peripherals and 10 centrals (Pl. XI, fig. 175,
176). The latter may, however, be totally absent. The bases of
the abactinal plates are connected by elongated internal ossicles,
between which lie the papular pores, exactly as in *Mediaster*.
The number of these ossicles radiating from a single plate is
mostly six.

Specimens in S.C.

Pentagonaster misakiensis, n. sp.
(Pl. XIII, figs. 194–202.)

The body is regularly stellate, the arms are tolerably long,
bluntly pointed and in my specimen strongly curved upwards in
the distal half (Pl. XIII, fig. 194, 195). This last point must
however depend much on the condition in which the animal was

when killed. The abactinal side is prominently convex and the interradial areas are depressed, so that the dorsal surface is more or less pyramidal in shape with indented bases. The actinal side of the disk is distinctly concave, and there is a radiating depression in each interradius. I have only one specimen of which the measurements are as follows :

r mm.	R mm.	R : r	MS
25	70	2.8	20

' *Superomarginals.*—The superomarginal plates are very well developed and are nearly of the same rectangular form up to near the tip of the arms. At the interradial angle they are twice as broad as long, but become less broad towards the tip of the arms. The surface is completely covered over with coarse, well spaced granules of a polygonal or roundish shape, which are slightly smaller near the inner edge of the plates. Many of the plates bear one or two valvate pedicellariæ similar to the smaller ones found on the paxillæ, and usually surrounded by an empty space, making them the more conspicuous.

Inferomarginals.—The inferomarginal plates are also very well developed and similar in general shape to the superomarginals, but mostly a little longer, so that they are contingent with the superomarginals only for one or two plates on either side of the interradial line and more or less alternate for the rest of the arms. The granules covering the entire surface are similar to those of the superomarginals, and some of the plates bear near their outer border a transversely elongated vice-shaped pedicellariæ, which are usually larger than those of the superomarginals and are like the medium sized ones found on the paxillæ.

Adambulacrals.—The adambulacral plates are nearly square and the armature consists of three rows. On the furrow edge is a series of six or seven, subequal, flattened, long spines, which, when those of both sides are bent towards the furrow, form veritable *chevaux-de-frise* for the latter (Pl. XIII, fig. 195,197). Separated from this furrow series by a groove, is the second series of armature consisting of three or four, or sometimes five, short spines of a four-sided prismatic form, with rather sharp ends. The spines at the two ends of the series are shorter and the one at the adcentral end is usually shorter than the one at the other end. The third row consists mostly of four, sometimes three or five, granules similar to those of the ventrolateral plates. On some of the plates the granules at the abcentral end of this series is duplicated, and towards the tip of the arms the entire row is usually doubled.

Mouth-plates.—The mouth-plates are relatively small and in-conspicuous. The distal end is naked, and there is a quite con-spicuous central space for the two paired plates (Pl. XIII, fig. 196). On the furrow margin of each plate is a series of ten flattened, subequal spines closely set side by side. The mouth being rather tightly closed in my specimen, these spines all stand up erect. On the inner side of this series is a second row of some six, shorter and stouter spines, parallel to the former. Then follows a third series of some four spines, smaller in all respects than those of the second row. Some of the spines of the last two rows may assume such a position that they form a direct continuation of the first series. The third series may also be totally wanting, *e.g.* on one side of fig. 196.

Ventrolaterals.—The ventrolaterals are very numerous, polygonal in shape, and covered with coarse elongated granules, appearing like paxillæ. The plates next the adambulacrals are mostly square and bear each some ten granules. They extend into the arms as far as the twelfth inferomarginals. There are no pedicellariæ on the ventrolateral plates.

Paxillæ.—The paxillæ are comparatively large and closely set, the granules coarse. Those in the interradial depressions are smaller than those on the radial elevations and near the centre of the disk. The pedicellariæ are numerous and many of them are conspicuous by their elongation in the transverse direction (Pl. XIII, fig. 198, 201). A large paxilla without pedicellaria may bear as many as twenty peripheral and ten central granules; the latter are larger and either rounded or polygonal and the former smaller and more or less prismatic (Pl. XIII, fig. 200). The pedicellariæ are of various sizes, and when they are small they are strictly confined to the border of the tabulum, making a series with the peripheral granules (Pl. XIII, fig. 199); but when larger they variously intrude into the central area and may even extend across its whole breadth (Pl. XIII, fig. 201). When their transverse elongation is considerable one or both of the two valves show a trace of fusion somewhere along their length, betraying the formation of a single valve out of more than one. In some cases there are two elongated pedicellariæ placed end to end, but preserving their morphological independence. At the middle of the arms there are five rows (longitudinal) of paxillæ, and each paxilla in this region bears some ten peripheral and two central granules. In the undenuded specimen the papulæ can be seen only here and there.

Madreporite—The madreporite is tolerably large and some-what rhomboidal in my specimen (Pl. XIII, fig. 202) ; it is situat-ed near the centre of the disk, its outer margin being removed from the margin of the disk more than four times as much as its inner margin is from the centre. In the alcoholic specimen the furrows are deep and conspicuous along the margin of the plate but shallow and indistinct in the central area. They are mostly radiating.

Locality.—Misaki ; depth 560 m.

Specimen in S.C.

Remarks.—This species appears to stand very close to *Pentagonaster pulvinus* ALCOCK. In my species the armature of the adambulacral plates consists for the greater part of three rows, and only in the distal part of the arms are there four rows : a furrow comb, two rows of granules on the actinal surface of the plate, and a row of short thick spines between the furrow comb and the rows of granules—a condition characteristic of *P. pulvinus*. ALCOCK, however, does not give the number of granules in the two outer rows, which is of some importance for specific deter-mination in such cases as the present, where two very closely allied forms come in question. I also find that nowhere in my specimen is the adcentral spine of the second row of the adam-bulacral armature replaced by a pedicellaria. The paxillæ are closely set, but their granules are well spaced. What ALCOCK says about the granulation of the marginal plates, the basal in-terradial plates and the madreporite also do not apply to the present species. It may also be mentioned in passing that *P. pulvinus* is an intermediate form between the typical *Tosia* and *Mediaster*. In view of the close relationship of the present species to *P. pulvinus* I reproduce the description of the latter at full length [ALCOCK, '93, p. 90] :

" *Pentagonaster pulvinus*, n. sp.

" Rays 5. R=2.2 r. R=33 mm. in the type specimen.

" Near *Pentagonaster mirabilis* PERRIER (Arch. Zool. expér. vol V., 1876, p. 40).

" Disk pentagonal, much inflated abactinally and hollowed actinally ; rays relatively long and narrow, blunt-pointed, strongly upcurved in the distal half. The strongly convex abactinal surface of disk and rays is covered with hexagonal or polygonal plates, so close-set that, although their boundaries are quite definite, no papulæ are visible on denudation nor any papular pores, and all closely covered with angular granules which show a distinctly paxilliform arrangement ; the basal interradial plates are more than twice the size of any of the other abactinal plates.

" Marginal plates 17 in the upper, 19 in the lower series, measured from mid-interbrachium to tip of ray, all very closely and uniformly covered with granules except the terminal six to eight in the supero-marginal series, which have a central smooth boss.

" Adambulacral plates with a furrow-comb of about seven nearly equal-sized lamellar spinelets, and actinally with two longitudinal rows of granules, and between these and the furrow a row of three coarse spinelets, the adcentral (adoral) of which is often replaced by a pedicellaria with two spathulate valves.

" Actinal surface deeply concave ; the actinal interradial areas are large ; the actinal intermediate plates extend to the fourteenth infero-marginal, they are large and roughly quadrangular, and are so closely covered with granules that their limits are not easily discerned.

" Arms very indistinct. Madreporiform plate small and also very indistinct ; it lies outside a much larger basal interradial plate, and is inconspicuous not merely because it is small, but also because its coarse discontinuous vermicular erosions give it a granular appearance much like that of an ordinary plate.

"Colour in the fresh state salmon-red.

"Laccadive Sea, off Minnikoy, 1200 fathoms, coral and Globigerina-ooze."

Hippasteria imperialis, n. sp.

(Pl. XII, figs. 178–193.)

This is a very large species, unlike any that have been described so far. The body is flat and lightly.pentaradiate, with the marginal border of the disk nearly straight and the arms triangular, rounded and slightly turned upwards at the tip, so that the space that would be enclosed by a line joining the tips of adjacent arms and the margin of the body would form a regular trapezoid. The actinal surface is as a whole nearly plane, but the interradial areas are decidedly hollow. The abactinal side is also plane as a whole, but there is a well marked circular elevation at the centre of the disk, from which a carinal ridge extends to the tip of each arm, while the interradial areas are slightly depressed (Pl. XII, fig. 178, 179). The granulations of the body are coarse, and there are numerous pedicellariæ on the actinal and abactinal sides as well as on the marginal plates, so that the whole body presents a rather rough appearance. I have only a single specimen which measures as follows :

r mm.	R mm.	R : r	MS
71	124	1.74	14

Superomarginals.—The superomarginals are very large and are nearly flush with the inferomarginals. Each plate has a

pronouncedly convex outer surface, and in the specimen before
me the plates are never perfectly coincident with the inferomarginals,
even at the interradial line. Along the larger part of the arms
the two series are strictly alternate. At the interradial line the
plates are nearly as high as long, but less than half as wide
(Pl. XII, fig. 184). In the arms the width increases in comparison
to the height and length, but the general quadrate form is
maintained on a marginal view. Each plate is margined with a
single row of closely arranged, small granules of a cubical form,
and the entire surface bears numerous spherical granules of
various size, which are well separated from one another. When
these granules are detached, which can be done with comparative
ease, there is seen in the place of each a hollow pit in which
the granule was lodged. There are, besides, a few (usually one
to four) valvate pedicellariæ of various sizes, exactly similar
in form to those of the abactinal plates. These pedicellariæ also
lie in excavations of the plates. The fasciolar grooves appear to
be totally absent, the plates being closely contiguous. Relatively
to the size of the body the superomarginals are few, there being
13 or 14 on either side of the interradial line. Each superomarginal plate presents a more or less convex border towards the
paxillar area.

Inferomarginals.—The inferomarginals are closely similar to
the superomarginals in all respects. At the interradial line each
plate is slightly broader than long, and the plates lie well on the
actinal side, so that we can not speak of their height. At the
base of the arms, however, the plates are as high as long and
only half as wide, while near the tip the height is usually greater
than the length and the width is about half as great as the

former. Each inferomarginal plate presents either a more or less convex or a broadly wedge-shaped border towards the ventrolateral areas. The granulations and the pedicellariæ are exactly similar to those of the superomarginals.

Adambulacrals.—In the disk the adambulacral plates are more elevated than the ventrolaterals,[1] but in the arms the two are on a level. Along the greater part of the furrow each adambulacral plate is only very slightly broader than long, so that it is very nearly square in form ; the outer border is, however, frequently either convex or wedge-shaped. The adambulacral armature is subject to some variation. On the furrow border is a series of three or four stout, flattened spines, of which the central one or ones are larger than the others (Pl. XII, fig. 183). Occasionally, however, there are three large and two very small spines. On many of the plates this furrow series leaves the proximal part of the furrow border of the plate free, evidently in relation to the pedicellaria to be mentioned in connection with the second row. This consists of one or two thick conical spines and a pedicellaria at the proximal end. The spines may be as long as those of the furrow series, but are usually shorter and thicker. When there are two spines they may be very unequal in size. The pedicellaria, which is valvate in form, is mostly placed with its length slightly oblique to the ambulacral furrow, and is inequivalve, with the smaller valve facing the furrow (Pl. XII, fig. 190). Next follows a row of two to four large, conico-elliptical granules, and separated from it by a space is the last series of smaller granules placed along the outer margin of the plate, except where this is occupied by the other rows.

---

1) This condition may, however, be altered in inflated specimens.

The number of granules of the last mentioned row is very variable, and may be anywhere between five and ten. In addition to the spines and granules above mentioned there may be one or two others not constant in position.

Mouth-plates.—The mouth-plates are tolerably large, but not very conspicuous, owing to the similarity of the armature of their actinal surface to that of the adambulacral and the adjacent ventrolateral plates. Each plate has the form of a scalene triangle, the two plates being apposed to each other by the longest side, and the shortest side facing the first adambulacral plate (Pl. XII, fig. 182). On the furrow margin is a row of stout, tolerably long spines, six or seven in number, forming a series with the furrow spines of the adambulacral plates. The one at the mouth end is thicker and somewhat longer than the others, but as it is inserted on a lower level, it hardly projects on the actinal side more than the others. On the actinal surface of each mouth-plate there is a row of three or four blunt, conical spines parallel to the furrow series, of which the one at the mouth end is conspicuously larger than the others. On the rest of the actinal surface are some ten granules of various size and form, a few of which may be spinous.

Ventrolaterals.—These are very closely set to one another, and extend into the arms to about the sixth inferomarginals from the tip of the arms. Those lying next the adambulacrals are especially large, but those lying next the inferomarginals are much smaller, while the remaining plates are of intermediate size. Except those next the adambulacral plates, the ventrolaterals do not form any regular rows. Some of them are regularly hexagonal, but the majority are more or less irregularly polygonal. Most of the plates bear each a single pedicellaria, which near the marginals are similar in form to those of the latter, but are considerably larger and of a different form on the inner plates. For a plate with a pedicellaria, the arrangement of the granules is as follows (Pl. XII, fig. 185) : along the border of the plate is a row of small, rather closely set, marginal granules ; in a large excavation in a subcentral position lies a large pedicellaria, the valves of which are of the form shown in figs. 185 and 191 ; and between the pedicellaria and the marginal granules, a variable number of large round or more or less flattened conical granules, with a few additional smaller granules. The pedicellaria is almost always 2-valvate, but I have also observed a 3-valvate one. On the plates without pedicellaria the entire central area is covered with large conical granules, which tend on especially larger plates to be more like the marginal granules both in form and size, though remaining larger than they (Pl. XII, fig. 186). On none of the plates have I observed more than one pedicellaria. The granules when removed leave small excavations on the plates like rain-prints on soft mud, and the pedicellariæ leave rectangular pores.

Abactinal plates.—The abactinal plates are closely set and of very unequal size. The larger one, whose form can be well observed, are more or less round and slightly convex. In the arms a carinal series of somewhat larger plates can be made out, but it is merged in between the other plates in the disk and can not be traced to the central elevation. The plates in the interradial depressions are generally larger than those of the central and radial elevations, but the plates bordering on the supéromarginals are very small. Here and there, where the plates are more separated from one another they are seen to assume a more or less stellate from, owing to the indentations due to the papular pores. The majority of the plates bear each a valvate, transversely elongated pedicellaria, but those without pedicellaria are also very numerous (Pl. XII, fig. 187, 188, 189). Each plate has a marginal series of very small, closely set granules, and in a plate without a pedicellaria the whole central area bears a comparatively small number of nearly spherical granules, which when removed leave circular pits behind them. Even on larger plates the number of these central granules does not appear to exceed fifteen and they therefore leave the larger part of the central area naked. In case there is a pedicellaria this may be situated centrally or excentrically. In the former case it is usually surrounded by a series of conico-elliptical granules usually larger than the marginal granules and arranged concentrically with the latter; but when the pedicellaria is in an excentric position it is usually destitute of such accessories, and the central granules are arranged very much in the same way as where the pedicellaria is absent. In any case the pedicellaria is situated in a large excavation. The papular pores are not very numerous and are found singly between the plates.

Madreporite.—The madreporite is only slightly larger than the largest abactinal plates and, although wholly exposed, is not very conspicuous. Its outer margin is a little less than twice as far removed from the margin of the disk as its inner margin is from the centre. The furrows are very numerous and mostly radiating and the plate is very slightly convex.

Terminal plates.—The terminal plate is very small.

Locality.—Off Misaki, depth 640 m.

I have a very small specimen of a starfish from the deep

water of the same part of the sea where the foregoing specimen
was obtained, which is probably to be referred to this species
(Pl. XII, fig. 180, 181). In its general outline it closely resembles
the young specimens of *Odontaster grayi* figured by LUDWIG [: 05a,
pl. V, fig. 8, 9], but the character of the adambulacral armature,
the form and distribution of the pedicellariæ, the armature of the
mouth-plates and the form and deciduous character of the surface
granules all converge to characterise it as a young example of
the present species, while none of the characteristics of *Odontaster* are apparent.

Specimens in S.C.; a third dried specimen has been subsequently acquired, orange above, light brown below.

Hippasteria nozawai, n. sp.

(Pl. XIII, figs. 203–211.)

This is a well characterised species. The body is flat, the
abactinal surface very slightly convex and the actinal side perfectly level. The disk is relatively large and the arms tolerably
long for a *Hippasteria*, and rounded at the end. I have only one
specimen which measures as follows :

r mm.	R. mm.	R : r	MS
18	37	2	10

Superomarginals.—The superomarginals are quite conspicuous
and are, with the exception of one or two terminal ones, nearly
square in form. They are fairly large on either side of the in-

terradial line but become gradually smaller towards the tip of the arms. Again, in the interbrachial angles the plates gradually slant down towards the margin of the disk, so that one can not conveniently distinguish the lateral and abactinal surfaces; but as the plates approach the tip of the arms the lateral face becomes more distinct. The surface of the plates is decidedly convex. Each plate is bordered along its margin by a single series of small rounded or cubical granules, of which those facing the inferomarginals are frequently larger and irregular in arrangement; the whole remaining surface of the plates is entirely naked and smooth (Pl. XIII, fig. 207).

Inferomarginals.—The inferomarginals are thinner and less convex than the superomarginals, and the plates are on a level with the general actinal surface of the body. On either side of the interradial line the plates are moderately thick; they then become thinner but again thicker toward the tip of the arms. The marginal granules of the outer border of the plates are larger than the others and are frequently arranged in two alternate rows; those of the other borders are similar to one another and to the corresponding granules of the superomarginals (Pl. XIII, fig. 208).

Adambulacrals.—The adambulacral plates are almost square in form and the first one or two plates may bear each a transversely elongated, valvate pedicellaria similar to those of the ventrolateral plates described below. One or two plates further out from the mouth may also bear a pedicellaria. The adambulacral armature consists of two large spines on the furrow margin, and 4 or 5 (exceptionally more) granules on the actinal surface of the plate, the latter mostly in two rows (Pl. XIII, fig. 206). The furrow spines are large and the adcentral one is shorter than the other ; both have blunt ends. The second row consists of two granules, of which the distal one is almost always larger than the other and is sometimes developed into a short spine. The last row consists of 2–4 (exceptionally 5) small granules. In case a pedicellaria is present on an adambulacral plate, it takes the place of the two granules next the furrow spines. The first adambulacrals are not different from the others, except that they more frequently bear pedicellaria.

Mouth-plates.—The mouth-plates are comparatively large and bear relatively few spines. Each plate has the form of a scalene triangle, the two plates being apposed to each other by the longest side, and the shortest side facing the first adambulacral plate. On the furrow margin are four stout spines with blunt ends, of which the one at the mouth end is larger than the others and more pointed (Pl. XIII, fig. 205). On the actinal face of the plate are 5–7 granules either forming a single regular row along the actinal suture line of the plate or more or less deviating from this regular arrangement near the distal end of the plate. It must also be mentioned that the first granule of this row, i.e. the one nearest the mouth, frequently assumes more or less a spiny form.

Ventrolaterals.—The ventrolateral areas are relatively large and the plates reach out into the arms as far as the fifth inferomarginals from the tip. The plates are irregularly arranged, unequal in size, and of various forms. Those forming a row next the adambulacral plates are, however, more uniform in size and shape, and the majority of them bear each a transversely elongated, valvate pedicellaria (Pl. XIII, fig. 204, 206). These pedicellariæ are mostly arranged so that their long axes are directed transversely to the ambulacral furrow, but they may also be more or less obliquely placed or even nearly parallel to the furrow. They are similar in shape to those on the adambulacral and abactinal plates, but mostly larger; rarely they may be 3-valvate (Pl. XIII, fig. 209). They form a row parallel to the ambulacral furrow and are very conspicuous. In my specimen none of the outstanding ventro-laterals bear pedicellariæ. The total number of ventrolateral pedicellariæ in a single interradius is about 17.

The ventrolateral plates with pedicellariæ next the adambulacral plates bear each a single row of granules along the margin, which is, however, usually incomplete on the inner border. All the other ventrolateral plates are covered with granules, which are unequal in size, but of which it is not possible to distinguish the peripherals and centrals (Pl. XIII, fig. 208). The granulation is coarse as a whole, and the granules are generally coarser and more rounded than those of the adambulacral plates.

Abactinal plates.—Of these we may broadly distinguish two classes, the main and the interstitial plates, although it must be distinctly understood that there is no sharp boundary between them, and the distinction is made entirely for the sake of description. The main plates are of various size, but are generally larger than the interstitial plates, and either bear a pedicellaria or present a naked central area (Pl. XIII, fig. 207). They are circular or elliptical in outline, and each plate bears a marginal row of small granules. When there is no pedicellaria, the central area is entirely naked and smooth, and is either level or elevated and mammiform. When a pedicellaria is present it occupies the whole central area. The pedicellariæ are transversely elongated and valvate and are similar in form to those of the actinal side, except that the valves are perhaps slightly more angular (Pl. XIII, figs. 203, 207, 210). The total number of abactinal pedicellariæ is between 80 and 90. I have also observed a 3-valvate pedicellaria on one of the abactinal plates.

The interstitial plates are of various size and shape and are either entirely covered with granules or present a very small central naked area. They fill up the spaces between the main plates (Pl. XIII, fig. 207).

The abactinal plates do not show any regular arrangement in the disk or arms.

The anus is subcentral and is surrounded by some coarse granules of a different appearance from those of the general abactinal surface.

Madreporite.—The madreporite is comparatively large and conspicuous, being much more lightly coloured than the general abactinal surface. It is polygonal in shape and its surface is nearly plain; the furrows are mostly radiating and comparatively simple. It is nearly as far removed from the centre of the disk as from the margin (Pl. XIII, figs. 203, 211).

Terminal plate.—The terminal plate is nearly elliptical in shape and is conspicuous, being as large as or a little larger than the last superomarginal plate.

Locality.—I owe my specimen to my friend, Prof. Nozawa of the Fisheries Department of the Tôhoku Imperial University in Sapporo, who collected it in the vicinity of Volcano Bay, Hokkaido. Colour in alcohol when received umber brown on the abactinal and pale rosy on the actinal side.

Specimen in S., donated to S.C.

This species is closely allied to *H. leiopelta* FISHER, but is distinguished from it by the narrower arms, the mammiform character of many of the main plates, the adambulacral armature and the form of the marginal plates.

Hippasteria spinosa, VERRILL.

This species was first described by VERRILL as follows [: 09, p. 63]

"Very similar in form and size to *H. phrygiana* of the N. Atlantic, but thickly covered with large, tapering, acute spines, usually one to nearly every dorsal plate and 1 to 3 on each marginal. Many of the plates also have large elevated bivalve pedicellariæ, but not so wide as in *phrygiana*.

"Departure Bay, British Columbia, 18 fath. (H. C. YOUNG), Canada Geol. Survey; Puget Sound (Prof. KINCAID)."

It has recently been described minutely by FISHER as follows [:11, p. 224, 6 figs.]:

"*Diagnosis.*—Similar to *H. phrygiana* of the North Atlantic, but primary abactinal and marginal plates with prominent, often long, tapering blunt spines rather than the elongate tubercles of *phrygiana*.

"*Description.*—Rays five. R=110 mm.; r=56 mm.; R=2 r. Breadth of ray at base, 64 mm. General form same as *phrygiana*, but rays a trifle broader. Majority of the larger or primary abactinal plates convex and bearing in centre a stout, upright, rigid, tapering truncate spine, one to two times width of its plate in length; midradial spines slightly the longest, thence decreasing in length to margin of area; same plates in *phrygiana* bearing a very much shorter, stouter tubercle, truncate or rounded at apex. General surface of plates smooth, but rim encircled by a single series of irregular, small, subconical or roundish granules heavily invested with membrane. Many large and smaller plates bear a large central bivalved pedicellaria, the jaws of which are only slightly wider than high; sometimes the two dimensions are equal. These pedicellariæ are usually narrower and higher than those of *phrygiana*, and have thinner blades. Distal edge of jaws is slightly curved, either serrate or smooth. Small intermediate plates intercalated between the larger ones bear one to three or four conical skin-covered granules, or a small central spinelet or pedicellaria, surrounded by irregular granules, depending upon size of plate. As in *phrygiana*, the surface of granules is smooth, and the free edge is rounded, or the whole granule may be subconical and pointed.

"Marginal plates slightly tumid as in *phrygiana*. Superomarginals sixteen or seventeen to ray, corresponding with inferomarginals except on distal part of ray. Plates of both series surrounded by a single row of squarish blunt or flattened skin-covered granules, which in one specimen are more pinched with rounded top. Proximal eight or nine plates of each series with two (less commonly only one, or as many as three) rigid spines, similar to abactinal spines, but usually a little larger; remainder of plates with one such spine. In *phrygiana* marginal plates have only robust tubercles. Occasionally a two-jawed pedicellaria occurs on marginal plates, but widely scattered along the series.

"Actinal intermediate plates arranged as in *phrygiana*, but pedicellariæ with higher jaws which sometimes taper to a narrow tip; these pedicellariæ frequently have serrate jaws, especially in southern examples. Granules pointed, thickly covered with membrane. Tubercles variable, subtruncate, thimble-shaped, or subconical and sharp, especially in southern examples.

"Adambulacral spines arranged practically as in *phrygiana*, but longer and stouter. Actinal spine at least as long as width of plate and subequal to but much stouter than the two furrow spines, which are slightly tapering and compressed at tip. Frequently a pedicellaria stands on the plate.

"Mouth plates with four or five furrow spines slightly larger than those of adambulacral plates.

"*Colour in life.*—A specimen from off Point Pinos, California, was bright scarlet vermilion when taken from the water; this colour disappeared almost at once in alcohol.

"*Anatomical notes.*—No rudimentary superambulacral plates. Intestinal cœcum large, composed of about four radiating branches which immediately subdivide irregularly into numerous slender tubes reaching nearly to margin. Gonad, a thick tuft of branching tubules on either side of the interradial septum, about one-third r from margin. Stomach capacious, the thick hepatic cœca reaching into ray about one-half R.

" *Variations.*—The chief variations concern the length of the abactinal spines and the height of the pedicellariæ. A specimen from station 4292, vicinity of Kadiak Island, and three from station 3080, coast of Washington, have considerably shorter abactinal spines than all the others, but they differ still from typical *phrygiana* in having prominent marginal spines, longer adambulacral spines, and higher pedicellariæ. The actinal pedicellariæ are the most variable, in some cases tending toward the form characteristic of *californica.* The jaws have either smooth or serrate edges.

" In Californian specimens the spines are all very prominent, more so than in the northern examples, and the actinal pedicellariæ are plainly serrate. The Monterey Bay (station 4552) example has more numerous marginal spines than the example from off southern California (station 3664). In both, the rays are notably stouter and shorter than in typical *phrygiana.* The specimen from station 4552 has the following characters : R=71 mm.; r=40 mm.; R=1.75 r. Breadth of ray at base, 48 mm. Height of longest abactinal spines, 6 mm. Disk very rigid, marginal plates massive, the upper with proximally two or three spines as long as those of abactinal plates, distally with one or two; more of the inferomarginals have three spines, occasionally four. Marginal spines of both series stand in line, forming a perpendicular series of six in interradial arc (sometimes seven), then five, four, three, and finally two. Pedicellariæ numerous, the jaws slightly wider than high; actinal a trifle higher than abactinal; none on marginal plates. Actinal intermediate plates with short subconical spines forming a group of four or five (in addition to the granules), or they may surround the pedicellaria when that is present. They are equal to or only slightly higher than pedicellariæ. All granules skin-covered, depressed or conical, but not denticulate or rugose as in *H. californica.* Colour in life, a bright scarlet vermilion.

" *Type-locality.*—Departure Bay, British Columbia, 18 fathoms.

" *Distribution.*—From southern California to Alaska (Kadiak Island) in 27 to 121 fathoms.

" *Specimens examined.*—Eighteen specimens from the following stations :
" Specimens of *Hippasteria spinosa* examined.

Station.	Locality.	Depth. Fathoms.	Nature of bottom.	No.	Collection.
2872	Off Cape Flattery, Washington ..	38	gray sand	1	U. S. Nat. Mus.
2874	do................	27	rocks and shells ..	1	do.
3059	Off Siletz Bay, Oregon..........	77	mud	1	do.
3060	Off Tillamook Rock, Oregon	28	brown mud	1	do.
3080	Off Heceta Bank, Oregon	93	green mud	4	do.
3445	Straits of Fuca, Washington	100	rocky	1	do.
3500	South of St. George Island, Bering Sea	121	fine gray sand .	1	do.
3664	Near Santa Catalina Island, California	80	do..........	1	do.
4233	Near Yes Bay, Behm Canal, Alaska	39–45	soft gray mud	2	Albatross, 1903
4243	Kasaan Bay, Prince of Wales Island, Alaska	42–47	green mud	2	do.
4292	Shelikof Strait, near Kadiak Island, Alaska	66	gray mud	1	do.
4295	Shelikof Strait................	92	soft gray mud	1	do.
4552	Off Point Pinos, Monterey Bay, California	73–66	green mud, rocks ..	1	Albatross, 1904.

A subspecies of this form has been reported by FISHER from the Kurils,
and described as follows [: 11, p. 226, 2 figs.] :

" *Hippasteria spinosa kurilensis*, new subspecies.

" *Diagnosis.*—Similar to *H. spinosa*, but with shorter rays, and relatively larger pedicellariæ; abactinal granules typically pointed; margin massive, spinous; small pincer-shaped marginal pedicellariæ; adambulacral plates with two unequal compressed furrow spines or only one; one prominent subambulacral spine shorter than the longer furrow spine. R=55 mm.; r=32 mm.; R=1.7 r.

" *Description.*—Abactinal surface spiny, the prominent spines being more conical and pointed than in typical *spinosa*, where the spines are usually subcylindrical with a blunt or truncate tip. Granules in a single series surrounding plates and ending in an abrupt usually sharp point, instead of being rounded. Abactinal pedicellariæ, numerous, large, as high or higher than broad, and wider at top than at base; edge of jaw undulating or slightly denticulate. Papulæ numerous, all over abactinal surface except a very narrow interradial band. Superomarginals tumid, ten to twelve to each ray, each plate with two, and inferomarginals with one or two stout conical spines, and in addition a single row of pointed marginal granules. Pedicellariæ, higher than wide and with denticulate jaws smaller than those of abactinal surface, are scattered here and there, usually on the lateral face of the plates. Such pedicellariæ do not occur in typical *spinosa*.

"Actinal surface with numerous pedicellariæ having higher jaws than in *leiopelta*. A prominent series occurs on the row of intermediate plates adjacent to the adambulacrals. The dimensions of these pedicellariæ are variable in typical *spinosa*. The actinal granules are similar in distribution to those of *spinosa*, but often (though not always) have several points, or only one point, rather than a simple low conical form. They are, therefore, somewhat similar to those of *H. californica*. Many of the plates have one or two low thimble-shaped tubercles in the centre, in place of pedicellariæ.

"Usually the adoral furrow spine is the shorter of the two, and both are strongly compressed at the blunt tip. The adoral spinelet is sometimes wanting on the outer part, or along most of the ray. Actinal spine shorter than the longer furrow spine and usually compressed at the base in a plane parallel to furrow. The outer part of plate is covered with granules. Mouth spines four, strongly compressed, chisel-shaped; usually but one suboral standing near margin, and in line with the marginal spines.

" *Type.*—Cat. No. 27885, U. S. N. M.

" *Type-locality.*—Albatross station 4840, off Simushir, Kuril Islands, 229 fathoms, coarse pebbles, black sand.

" *Distribution.*—Known only from the vicinity of Simushir.

" *Specimens examined.*—Eight; from type locality four, and from station 4803, same locality, depth, and bottom, four. (Albatross, 1906.)

" *Remarks.*—This race differs considerably in general appearance from typical *spinosa* of the North American coast, as the photograph of the type-specimen will reveal. The actual differences are not great, however, the most striking being the form of the granules, the marginal pedicellariæ, and adambulacral armature.

"Along with these specimens were taken one small and four large examples of *H. leiopelta armata*. Three specimens listed *kurilensis* are possibly hybrids of this race and the typical form of *kurilensis*.

"The differences between *leiopelta* and *kurilensis* are at once apparent on a comparison of figures. The latter has much higher pedicellariæ, heavily spined abactinals and marginals, marginal pedicellariæ, and compressed, rather spatulate furrow spines, not stumpy conical ones. The naked surface of the marginals in *leiopelta* is also quite different from the condition of *kurilensis*."

Neither the species nor the subspecies of this starfish is represented in the collections mentioned at the outset.

Mediaster brachiatus, n. sp.

(Pl. IX, figs. 142–152 ; Pl. X, figs. 153–155.)

This is a well defined species with long arms, of nearly the same general form as in the already known species of this genus. I have two specimens of different sizes, and the general appearance of the two is somewhat conspicuously different. In the smaller specimen, which was originally preserved in formalin and has R=30 mm., the dermal armature is relatively finer than in the larger specimen, in which R=56 mm; the aboral side is also quite inflated in the smaller specimen (Pl. IX, fig. 144, 145); but I take this to be due to the action of formalin, which is known to cause more or less swelling. But taking all points into consideration there can be no doubt that the two specimens belong to the same species. The area of the ventrolaterals is very spacious, being broad and extending far into the arms. The interradial areas on the abactinal side are depressed.

Radial ratio.—The radial ratios are only slightly different for the two specimens :

Specim.	r mm.	R mm.	R : r	MS	Locality
1	10	30	3.0	24–28	Uraga Channel.
2	18	56	3.1	32	Misaki.

Superomarginals.—These do not appear conspicuous on surface view, owing to the fact that their granulations and the coronal spinelets of the paxillæ bordering on them are similar in form ; but the marginal plates are very well developed. Their number is given in the foregoing table. The superomarginal plates are uniformly covered with coarse granulations of a conico-pyramidal form and rather well spaced. The fasciolar grooves are rather narrow and entirely naked. So far I have not found any pedicellariæ on the superomarginals. No difference worth noting can be detected between the granulations along the margin of the plate and its general surface, except that the former are a shade longer and more pointed.

Inferomarginals.—The inferomarginals are generally coincident with the superomarginals and are similar to them in form. The granulations are also mostly similar to those of the superomarginals, except that those nearer the inner margin assume more a pyramidal shape and form a transition to the armature of the ventrolateral plates. It is also to be noted that there are a few bivalvate pedicellariæ, similar to those found on the paxillæ, on the inferomarginals of the whole body. Their positions, which are evidently not constant, can be more easily seen on a denuded specimen, in which they are marked by a small pore for each.

Adambulacrals.—The first adambulacral plates are not very conspicuously different in form from the rest. The adambulacral armature consists of three rows, and may be shown by the formula 5 (6)–(3) 4–3 (Pl. IX, fig. 148), The first row is very conspicuous and mostly consists of five subequal, almost straight, stout, prismatic spines with blunt ends, slightly flattened in a direction transverse to the ambulacral furrow. This series may consist of six spines on some of the plates. There is a well marked furrow between the first and the second row. The second row usually consists of four spines, similar in form to those of the first, but slightly shorter, and almost as conspicuous as the latter. The third row usually consists of three spines of a pyramidal shape, similar to those of the ventrolateral plates. Occasionally the last row contains four spines ; and on the first adambulacral plate there may be a fourth row of two to four spines.

Mouth-plates.—Each mouth-plate is elongated-triangular and is apposed to its fellow by the longest side. There are eight or nine spines on the furrow border in the larger specimen, but only seven in the smaller. In the larger specimen they are very stout and conspicuous, and flattened transversely to the furrow, the spine at the mouth end being exactly similar in shape and length to the others. In the smaller specimen the corresponding spines are more slender and the one at the mouth end is notice-ably stouter than the others ; they are also nearly cylindrical in form. In both specimens, the spine at the abcentral end of this row is stouter than the others. On the side facing the first adambulacral plate there are four or five spines, of which the one next the ambulacral furrow is largest and similar to the last spine of the first mentioned row, the other spines gradually becoming shorter towards the abcentral end of the mouth-plate. On the actinal surface of the mouth-plate there is a somewhat irregular row of three to five spines along the suture line of the plate, of which the one nearest the mouth is largest. There is an unarmed space left between this spine and the first mentioned row of spines. In the smaller specimen the spines along the suture line form a more regular series (Pl. IX, fig. 147).

Ventrolaterals.—These are very numerous and extend far out into the arms, a single row (the innermost) reaching as far as the fourteenth inferomarginal plate as counted from the apex of the arms. The larger ones bear as many as ten spines, which are tolerably long and distinctly pyramidal in form. Sometimes there are many peripherals and one or two centrals, simulating the appearance of paxillæ. An interesting point is that some of the ventrolaterals bear pedicellariæ, which are different in form from those of the aboral side or the inferomarginals and are mostly 3-valvate, but occasionally 4-valvate (Pl. IX, fig. 149, 150). In denuded parts their positions are marked by a pore in the plate. These pedicellariæ are interesting as having a very primitive form, and can be detected only on careful examination. They appear to be even more primitive than those found on the paxillæ or the inferomarginals.

Paxillæ.—The paxillæ are fairly large, and the top of the tabulum is either circular or elliptical. Those at the centre of the disk are hardly smaller than the others. Those far out in the arms are small, the largest ones being found along the elevated radial zones meeting in the centre of the disk. Thus there is brought about a depressed triangular area in each interradius with smaller paxillæ. Of the coronal spinelets, the peripherals and the centrals are usually well distinguished both by their form and position, the centrals being shorter (Pl. X, fig. 154). On one of the largest paxillæ near the centre of the disk, there may be as many as twenty peripherals and eleven centrals. Some of the paxillæ bear pedicellariæ which are, so far as I have observed, always bivalvate, and are usually formed by the transformation of peripheral spinelets. Each valve is curved and flattened and is apposed to its fellow face to face (Pl. IX, fig. 152). I have never seen a paxilla with more than one pedicellaria. Towards the tip of the arms the paxillar area becomes very narrow, and only a single paxilla is in contact with the terminal plate (Pl. X, fig. 153).

The papular pores are very numerous and are uniformly distributed in the elevated radial bands already mentioned. They are however absent from the apical parts of the arms.

Madreporite.—The madreporite is tolerably large and conspicuous, being elevated to about the level of the surrounding paxillæ (Pl. X, fig. 155). It is nearly twice as far removed from the margin of the disk as from the centre.

Terminal plate.—The terminal plate is fairly large and nearly oval in outline (Pl. X, fig. 153).

Locality.—My specimens are from a depth of 270 m. in Uraga Channel and 550 m. off Misaki.

Two specimens in S.C.

This species stands near to *M. elegans* LUDWIG and *M. tenellus* FISHER.

Johannaster.

The genus *Johannaster*[1] was set up by KŒHLER with the following diagnosis [: 09, p. 7] :

"Le disque est grand et les bras, plutôt minces, sont très allongés. Les plaques marginales dorsales et ventrales sont très nombreuses. La face dorsale du disque et des bras est couverte de petites plaques simplement munies de granules fins et ne formant pas de paxilles ; ces plaques sont disposées sans ordre régulier : entre elles se montrent des papules très nombreuses et très développées. Les plaques marginales dorsales et ventrales ne sont couvertes que de granules et c'est à peine si les plaques marginales ventrales offrent, dans les arcs interradiaux, quelques piquants rudimentaires sur leur bord externe. Les aires interradiales ventrales sont grandes et elles s'étendent sur une grande partie de la longueur des bras ; les plaques y sont disposées en rangées allant des adambulacraires aux marginales ventrales et susceptible de se dédoubler ; elles sont couvertes de granules et chacune porte ordinairement un très petit piquant. Les dents sont peu développées et ne font pas saillie sur la face ventrale. Les tubes ambulacraires sont terminés par une ventouses dont le diamètre est inférieur à celui du tube, mais qui est cependant bien marquée. L'anus est très petit. La plaque madréporique est assez petite, avec des sillons divergents. Il existe de petits pédicellaires alvéolaires sur les plaques marginales et sur les plaques latéroventrales."

KŒHLER thinks that this genus shows in most of its characters an affinity to the *Plutonasteridæ*, the larger specimens presenting an external resemblance to *Plutonaster subinermis* when viewed from the abactinal side. The points of resemblance with that family are the characters of the dorsal plates which are covered with granules and disposed without regularity, the ventrolateral plates which form series, sometimes duplicated, arranged between the adambulacral plates and the

---

1) Dedicated to Mme. JEANNE KŒHLER.

marginals, the presence of an anus, etc. ; the points of difference are the rudimentary condition of the marginal spines, the weak development of the teeth, the presence of suckers on the tube-feet and the pedicellariæ. The author thinks that with these restrictions the genus may be referred to the *Plutonasteridæ*. It appears to me that this genus occupies an intermediate position between the *Archasteridæ* and the *Pentagonasteridæ*, being related to the former by the absence of the superambulacral plates and to the latter by the character of the plates and the pedicellariæ. The total absence of spines from the marginal and ventrolateral plates in the species described below increases its resemblance to the *Pentagonasteridæ*. In fact this genus stands next to *Nymph-aster*, being related to it through such a form as *N. symbolicus* SLADEN, which has been placed by some authors (VERRILL, FISHER) in a distinct genus, *Nereidaster*.

KŒHLER gives a detailed description of one species from the collection of the " Investigator " in the Indian seas [: 09, p. 8]. The species described below, though presenting some notable differences from KŒHLER's, shows a close affinity to it and there is no doubt that both are to be placed in the same genus.

Johannaster differs from *Nymphaster* by the presence of several rows of more or less irregularly arranged plates between the superomarginals of the arms[1] and the papular areas extending into the latter. It may also be added that it will be necessary to modify the generic diagnosis by striking out the phrases bearing on the spines of the inferomarginal and ventrolateral plates. FISHER [:11, p. 162] has recently placed this genus in the *Goniasteridæ* (s. *Pentagonasteridæ*).

---

1) As for *Nereidaster bowersi* FISHER it appears to me that it is more nearly - related to *Pentagonaster arcuatus* and *P. misakiensis* than to *Nymphaster symbolicus*. *Cf.* FISHER [:13, p. 629].

Johannaster giganteus, n. sp.

(Pl. X, figs. 156–163.)

This is a very large species perfectly distinct from the one already described. I have four specimens, all dried, from the same parts of the sea and from depths ranging between 160 m. and 1120 m. One of the specimens is in a half-decomposed state, so the measurements of the other three specimens only will be given.

Specim.	r mm.	R mm.	R : r	MS
1	67	320	4.8	78
2	70	335	4.8	85
3	87	338	3.9	71+

The actinal surface is hard and almost plane, only the parts adjoining the mouth being slightly elevated. The abactinal surface is, in the dried specimens, thrown into light folds, especially in the interradial regions, so that in the fresh state these parts were probably more or less inflated. Another conspicuous feature of the abactinal side is the petaloid papular area extending half-way or more into the arms.

Superomarginals.—The number of superomarginal plates is not strictly proportional to the major radius, as shown in the above table. In the interbrachial arcs as well as sometimes along the arms, the upper and lower marginals may be alternate, but more usually they are coincident. Each superomarginal is rectangular in shape, but triangular plates occur here and there, especially where the arm had been broken off and subsequently

regenerated. In the interbrachial arc the superomarginals slant
down gradually from the abactinal surface towards the inferomarginals, so that here they present only one curved surface towards
the outside ; but as they proceed further out in the arms an edge
gradually becomes apparent, and near the middle of the arms,
there is a distinct outer, as distinguished from the abactinal,
surface to each superomarginal, the angle between the two surfaces
becoming a right angle a little further out than the middle of
the arms. Sometimes however the superomarginals of the interbrachial portions stand erect, presenting only an outer surface.
The superomarginals are uniformly and thickly covered over with
fine granules which are slightly better spaced near the inner margin of each plate. There are numerous small pedicellariæ of some
height on the superomarginals, similar in size and shape to those
of the abactinal plates. In the interbrachial region there may be
as many as three or four pedicellariæ on each plate, but they
decrease further outwards, becoming rare in the distal half of the
arms, although their occurrence is subject to some variation. The
pedicellariæ appear to occur more frequently nearer the margins
of the superomarginal plates rather than near the centre. In
some of the arms which have evidently regenerated, the superomarginals of the two sides are in direct contact with each other
for a short distance near the tip of the arms.

Inferomarginals.—In the interbrachial arcs the inferomarginals
usually project outwards slightly more than the superomarginals,
but become more and more flush with them in the arms, and in
the distal one-third of the latter the two marginals keep strictly
the same front. The inferomarginal plates are mostly coincident
with the superomarginals, but are also frequently alternate,

especially in the distal parts of the arms. They are covered with granulations similar to those of the superomarginals but slightly coarser. They also bear pedicellariæ similar to those of the superior series; they appear to be present more usually near the superomarginals than on the actinal surface, and their distribution along the arms coincides in a general way with that on the superomarginals. The inferomarginals are widest in the inter-brachial arcs, and become narrower along the arms.

Adambulacral plates.—Along the greater part of the furrow the adambulacral plates are, when viewed from the actinal side, nearly square, with a curved side towards the ambulacral furrow, but in the distal parts of the arms they are more or less wedge-shaped. On looking into the furrow, the plates are seen to be tall and have a keeled surface. There is a furrow series of stout, flattened spines strictly parallel to one another, 10–13 in number (Pl. X, fig. 159); in my specimens the longest ones are 3–4 mm. Next this furrow series is a naked space separating it from the spines of the actinal surface of the plate. These spines are very short and even granuliform towards the ventrolaterals; they do not show any definite arrangement and are 20–25 in number for each plate near the base of the arms. In the distal part of the arms, however, they are much more numerous and almost entirely granuliform. Near the mouth end, again, the actinal spines are less numerous, there being only 4–6 forming a single row. On the actinal surface, close to the naked space there are one or two pedicellariæ of large size, usually near the adcentral end of the plate (Pl. X, fig. 159,160). When there are two, one of them is usually smaller than the other; they are both stout and of some length. In the more distal parts there is usually only one pedi-

cellaria to each plate and it may be situated anywhere close to
the naked space, the more usual position being near the centre
of the plate.

Mouth-plates.—The relative size of the mouth-plates is different
in different specimens, being as long as 8 mm. in one with R=
335 mm. and only 5 mm. in another of R=338 mm. On the
furrow margin of each plate there is a series of very stout, flattened spines, 12–15 in number. Those near the mouth end are
usually slightly larger than the others, but not very conspicuously so, so that the spines are more nearly equal to one another (Pl.
X, fig. 158). On the actinal surface of each plate there are some
half a dozen or more spines, either forming a single row along
the suture line of the plate or more irregularly arranged ; one of
them at the mouth end is usually conspicuously larger than the
others, which are usually short and even granuliform. Each
mouth-plate is seen to present towards the ambulacral furrow a
keeled surface similar to those of the adambulacral plates.

Ventrolaterals.—The ventrolaterals are very numerous. There
is a longitudinal row of plates larger than the others immediately outside the adambulacrals running out two-thirds or more into
the arms, and bearing more and larger pedicellariæ than the
other plates. One can also distinguish a second series running
out about half way or slightly more into the arms ; a third series
may sometimes be made out with difficulty, but the plates are
much more irregularly arranged. The ventrolateral plates are,
generally speaking, larger towards the ambulacral furrow and the
mouth, so that the plates near the interbrachial margin are smaller
than the others. The ventrolateral plates are uniformly covered
with granules and are closely set. Many of them bear pedicel-

lariæ, of which those on the plates next the adambulacrals are conspicuously larger than the others (Pl. X, fig. 159,161) ; a plate of this series may bear as many as four pedicellariæ. At the mouth end there is usually a large odd plate which may bear as many as six pedicellariæ. Occasionally there is a small and a large plate instead of a single odd plate. A plate of the next series usually bears only one pedicellaria, which is smaller than those of the first series ; at or near the mouth end, however, there may be two or more pedicellariæ on each plate. Many of the remaining ventrolateral plates also bear pedicellariæ, but they are usually less conspicuous.

Abactinal plates.—The most conspicuous feature on the abactinal side is the papular area of a petaloid shape, radiating from a central patch and extending into the arms to nearly the middle of their length. In this area the plates are better spaced to leave room for the papulæ, while in the interradial areas the plates are closely pressed against one another. In some cases there can be distinguished an indistinct carinal series of plates in each arm running out to the tip, but merged into the general plating at the centre of the disk. Where this series is distinct the constituent plates are frequently lozenge-shaped or more or less hexagonal. Generally speaking the abactinal plates are largest in the papular area at a short distance from the centre, and gradually diminish in size in all directions away from it. The abactinal plates are uniformly covered with small granules, which do not show any differentiation whatever into peripherals and centrals. Many of the plates of the papular area bear small pedicellariæ, there being as many as 3 or 4 on a larger plate (Pl. X, fig. 162,163). The plates of the interradial areas also bear pedicellariæ, but they are far less numerous, and there is rarely more than one pedicellaria on each plate.

Madreporite.—The madreporite is quite large and conspicuous, on the same level with the abactinal plates, polygonal in form and is situated near the centre. The grooves are radiating.

Locality.—All my specimens are from the neighbourhood of Misaki, one from a depth of 160 m., another from a depth of 960 m., and two others from 1120 m. The colour of the dried specimens is light brown for the papular area, and light yellowish brown for the rest.

Specimens in S.C.

Pseudarchaster pretiosus (DÖDERLEIN).

(Pl. VIII, figs. 127–135.)

The original description of DÖDERLEIN is as follows [: 02, p. 326] :

" *Astrogonium pretiosum* nov. sp.

" Die Arme sind kurz, spitz, breit und gleichmässig verjüngt; R : r = 2.7.

" Die dorsalen Paxillen sind klein, am kleinsten am Centrum, am grössten an der Armbasis; mindestens 3 Längsreihen von ihnen grenzen an die Ocellarplatten an der Armspitze; etwa 2 Querreihen von Paxillen entsprechen einer Marginalplatte.

" Die oberen und unteren Marginalplatten sind einander sehr ähnlich, sehr breit und niedrig, die unteren etwas vorstehenden bilden den kantigen Aussenrand der Arme; sie sind fein gekörnelt, die oberen ohne Stacheln, die unteren mit einer Anzahl flacher, spitzer Stachelchen im äusseren Drittel, die nicht über den Aussenrand vorragen.

" Von den 3–4 Längsreihen kleiner Ventrolateralplatten ist die äusserste etwa bis zur Hälfte des Armradius sichtbar.

" Die Adambulacralplatten tragen eine innerste Längsreihe von 4–5 feinen Stachelchen, nach aussen parallel dazu eine Reihe von 3–4 etwas

kürzeren Stachelchen, denen sich ein äusserer kleiner Haufe von noch kürzeren Stachelchen anschliesst, die den Uebergang bilden zu der Granulierung
der anstossenden Ventrolateral- bezw. Marginalplatten.

"Beim kleinsten und grössten der vorliegenden Exemplare beträgt
R=38 u. 60 mm, r=14 u. 22 mm, die Zahl der Marginalplatten 24 u. 28,
die grösste Breite einer Dorsomarginalplatte 5.5 u. 6.5 mm, einer Ventro-
marginalplatte 6.5 u. 8 mm.

"Die Art ist einfarbig, in Alcohol hellgrau.

"Ich fand diese Art in der Tokiobai bei Kadsiyama auf Sandboden in
ca. 20–30 m Tiefe und in der Sagamibai.

"Diese neue Art erinnert am meisten an *Astrogonium aphrodite* PERRIER,
hat aber breitere Marginalplatten.

"Von den ihr einigermassen ähnelnden Arten der Gattung *Astropecten*
unterscheidet sich *Astrogonium* schon durch das Fehlen vorstehender unterer
Randstacheln."

The radial ratios of my specimens are as follows :

Specim.	r mm.	R mm.	R : r	MS	Locality
1	5	12	2.4	12	Uraga Channel.
2	6	13	2.4	13	Sagami Sea.
3	6	14	2.3	—	Tokyo Bay.
4	6	15	2.5	16	Sagami Sea.
5	9	23	2.6	19	,,
6	10	29	2.9	22	,,
7	12	29	2.4	20	,,
8	12	33	2.7	23	,,
9	13	35	2.7	23	,,
10	14	40	2.9	24	(,, ?)
11	14	39	2.8	26	(,, ?)
12	18	49	2.8	25	Sagami Sea.
13	20	53	2.7	24	,,

The testa is very hard, especially on the oral side, where the ventrolateral plates are thoroughly ankylosed with one another. The disk is large and the interbrachial angles are only a little greater than a right angle.

Superomarginals.—The superomarginals are comparatively large and low, and are covered with small polygonal granules, which are only very slightly larger in the outer part of the plates. The number of the superomarginal plates is given in the foregoing table.

Inferomarginals.—The inferomarginals are low like the superomarginals but broader and slightly more projecting and form the lateral margin of the body. They are contingent with the superomarginals through the greater part of the arms, but near the tip the two series become slightly alternate. The inferomarginals are covered with small granules similar to those of the superomarginals for about two-thirds of their breadth, but are replaced in the outer one-third by flattened spines, which are longer in the interbrachial angles and become shorter and more squamiform as they go farther towards the tip of the arms. Even in the interbrachial angles these spines are not very long and are not so very conspicuous as in many species of *Astropecten* (Pl. VIII, fig. 133).

Adambulacral plates.—The adambulacral plates are rather narrow, and there are about nine of them for every seven of the inferomarginal plates. The first adambulacral plate is about twice as broad as long, but from the sixth or seventh on they are about as broad as long; and the inner edge of each plate, except the first few, is wedge-shaped (Pl. VIII, fig. 130). Along this edge there is a curved row of ambulacral spines of some length,

varying from five to seven on each plate; and on the actinal surface of the plate there are, in the basal part of the arms, three
longitudinal rows of shorter spines, each row consisting of from
three to five. The still more external group of shorter spines,
referred to by DÖDERLEIN as forming a transition to the granules
of the inferomarginals or ventrolaterals where present, belongs
undoubtedly to the ventrolateral plates. On the first adambulacral plate, the ambulacral row of spines is straight; and in the
more distal part of the arms the shorter spines on the actinal
surface of the adambulacral plates are frequently more irregular
than in the more proximal part, so that they may form more
than three rows.

Mouth-plates.—The mouth-plate is comparatively narrow and
elongated. and there are distinctly three sides to it, viz. the
longest, almost straight side, by which it is apposed to its fellow,
the second side facing the ambulacral furrow, and the third side
turned towards the first adambulacral plate, the second and third
sides forming a distinct angle between them (Pl. VIII, fig. 132).
The side facing the first adambulacral plate is longer than the
second. The longest side by which the mouth-plates are apposed to
each other are frequently united with that of the fellow plate by a
few transverse trabeculæ. There are about seven rather long spines
on the furrow margin of each mouth-plate, the one next the
mouth being not longer than the others (Pl. VIII, fig. 131). On
the side facing the first adambulacral plate there are one, two,
or sometimes three rather long spines forming the continuation of
the last mentioned row, and the triangular area enclosed by the
third sides of the two mouth-plates is covered with some twenty-
five to thirty short blunt spines, similar to those of the ventro-

lateral plates. Besides, there are some ten or a dozen spines between the rows of the ambulacral sides, of which the more distal are short and blunt, while the more proximal ones (i.e. those nearer the mouth) are longer and just like those of the ambulacral border.

Ventrolaterals.—The ventrolateral plates are mostly lozenge-shaped and are fast ankylosed with one another ; in a specimen of R=40 mm., there are as many as seventy ventrolaterals in one interradius, including those in the arms. A single row of these plates can be followed out as far as two-thirds of the length of each arm. The ventrolaterals are covered with short, truncated spines, of which as many as fifteen may be borne by a single larger plate.

Paxillæ.—The paxillæ are closely set and are smallest around the centre of the disk ; this central area is not, however, so con-spicuous as in *Astropecten*. The coronal spinelets belonging to each paxilla are marked off into polygonal groups by mutual pressure. The centrals and peripherals are very distinct from each other, the latter being considerably smaller. On one of the largest paxillæ, which are usually found about the middle of the base of the arms, there may be as many as fifteen centrals and some twenty-three or more peripherals (Pl. VIII, fig. 134). Both the centrals and peripherals are very short, although the latter are a little longer and more slender. In denuded specimens the paxillar tabula are seen to form regular longitudinal rows in the proximal half of the arms, but are somewhat more irregularly arranged in the distal half (Pl. VIII, fig. 129). The surface of the tabula are mostly circular or slightly elliptical, and their bases are fast ankylosed. The papular pores are uniformly distributed on the aboral surface between the paxillæ, but are absent in the centre of the disk and in the distal half of the arms, where the paxillæ are more irregularly arranged.

Madreporite.—The madreporite is very flat and is very slightly raised above the bases of the paxillæ. It lies about midway between the centre and the margin of the disk, is irregular in outline and notably lobed at the margin. It bears a few paxillæ on its surface, and is completely hidden from view in undenuded specimens (Pl. VIII, fig. 135).

Terminal plate.—The terminal plate is comparatively large and somewhat elliptical in outline when viewed from the aboral side. It is deeply furrowed on the actinal side.

Locality.—My specimens are from the following localities, all near one another : Tokyo Bay, Sagami Sea, Uraga Channel, Misaki. The bathymetrical range of this species appears to be somewhat wide, being from 20 m. to 480 m.

Specimens in S.C., F.B., I.H.S., O.

Pseudarchaster parelii (DÜBEN & KOREN).

According to LUDWIG [: 00, p. 449] this starfish was described for the first time by PARELIUS [1768, p. 350] under the name of *Asterias aranciata* var.[1] It was then redescribed by DÜBEN and KOREN ['46, p. 247] under the name of *Astropecten parelii*, a name adopted by SARS ['50 p. 161], who later changed it to *Archaster parelii* ['61, p. 35]. These citations are made on the authority of LUDWIG.

In discussing the correlation of the form of the ambulacral feet and the presence or absence of an anus, MÜLLER refers to the present species as follows ['54, p. 48] : "*Astropecten Parelii* v. D. et K. hat kürzlich ein Bei-

1) DUJARDIN and HUPÉ write " *Asterias aurantiaca* var."

spiel von der Schärfe dieses Unterschiedes gegeben. Hr. SARS hatte mir mitgeteilt, dass die von mir angegebene Regel hinsichtlich der Füsschen bei diesem *Astropecten* eine Ausnahme erleide ; ich vermutete deshalb, dass dieser Seestern kein *Astropecten* sondern ein *Archaster* sein werde, wovon die europäischen Meere bisher kein Beispiel besassen. Hr. SARS hat mir seitdem ein Exemplar in Weingeist mitgeteilt, woran ich sogleich den After fand, als ich die Paxillen in der Mitte des Rückens von ihrer Krönung bis zum Sichtbarwerden der Haut des Rückens befreite. Jener Seestern wird daher nunmehr *Archaster Parelii* zu nennen sein."

DUJARDIN and HUPÉ give the following description of this species under the name of *Astropecten parelii* DÜBEN et KOBEN, mentioning also PARELIUS in the list of literature ['62, p. 420] :

" Espèce à cinq bras acuminés, séparés par des intervalles arrondis, peu profonds, les bras n'ayant qu'une fois et demie le plus petit rayon du disque. Plaques marginales au nombre de trente sur chaque bras, sans piquants, mais avec de simples granulations : aire paxillaire très-large. La face inférieure ou ventrale est couverte d'un grand nombre de granules, et les plaques marginales, de ce côté, font une saillie comprimée, sublamelleuse.

" Coloration d'un rouge de sang. Dimension : 108 mm.

" Habite les mers du Nord, le Groënland.

" Cette espèce a l'aspect général de l'*Astrop. Andromeda*, mais, outre que ses bras sont moins allongés, elle en diffère, surtout par l'absence compléte de piquants sur le bord de ses plaques marginales."

NORMAN describes this species under the name of *Archaster parelii* as follows ['65, p. 119] :

" Greater to lesser radius is as 3 to 1. Aboral surface entirely covered with closely aggregated paxillæ. Each of these paxillæ consists of a pillar, widening above and supporting about twenty-five (15–30) mamillary spines of different sizes. Madreporiform tubercle nearer to the centre than to the margin of the disk, minute, not so large as one of the paxillæ. Lateral ray-plates thirty, oblong, entirely covered with mamillary spines of the same kind but larger than those of the paxillæ, nearly a hundred on each plate. Oral surface entirely covered with closely packed short papillose spines.

The inferior lateral plates are most beautiful cushions of closely aggregated, appressed papillary spines, each plate having a central row of 3–5 rather larger and more conspicuous spines, which however, like all the rest, are closely appressed to the surface. Indeed there are no spines projecting conspicuously beyond the rest from any part of the body. The spines of the adambulacral plates are so numerous that, spreading from them in all directions, they nearly choke up the ambulacral channels. Greater diameter not quite 4 inches.

"A single specimen of *Archaster Parelii* was dredged by Messrs. JEFFREYS and WALLER, during the past summer, on the Outer Haaf, off Shetland, in 100 fathoms. It is a very interesting addition to our list of British Echinodermata. I have removed this species from the genus *Astropecten*, in which it had been placed by DÜBEN and KOREN, and placed it in *Archaster* on the authority of Professor SARS. I am unable myself to vouch for the correctness of this transfer, as I have been unwilling to injure the only British specimen in order to ascertain the presence of those organs (the anal aperture and pedicellariæ) which separate the genus *Archaster* from *Astropecten*."

LÜTKEN ['71, p. 236] refers to it under the name of *Archaster parelii* in connection with *Astrop. javanicus*. WYV. THOMSON ['72, p. 301] mentions it from the north Atlantic, 300–800 fathoms, with the minimum temperature of 0°—+2°C· It is also mentioned by VERRILL ['74, p. 500][1] under the name of *Archaster parelii*, the form in question being the one later described by SLADEN as a separate species under the name of *Pseudarchaster intermedius*. MÖBIUS and BÜTSCHLI ['75, p. 148] mention it from Bukenfjord and Bömmelfjord, 106 fathoms, on mud or mud with gravel. They found it rare and the distribution is given as "Küste Norwegens und Gross-Britaniens."

PERRIER mentions it in his "Révision" and adds ['76, p. 268] : "Un seul exemplaire dans l'alcool, mais brisé. Donné par le musée de Bergen (Norwége) à l'expédition du prince Napoléon en 1856."

DANIELSSEN and KOREN describe a variety of this species as follows ['77, p. 17 (separate)] :

---

1 This paper is not accessible to me.

"*Archaster Parelii*, Düb. & Koren.   *Varietas longobrachialis*, nob.

"Denne Søstjerne afviger i flere Henseender temmelig meget fra den typiske Form, uden at vi dog have trœt at burde ophøie den til en selvstrændig Art. Forholdet imellem den lille og store Radius er paa Exemplarer af 108–124 mm's Størrelse som 1 : 3⅓–3¼. Randpladerne, der paa Exemplarer af nysnævnte Størrelse ere 38–40 i Antal, ere ved Armens Grund 4.1 mm brede, 2 mm høie og Paxillarfeltet 9 mm bredt; paa Midten af Armen ere de 2 mm brede, 1.5 mm høie, og Paxillarfeltet 2.8 mm bredt. Adambulacralpladerne ere nærmest Ambulcralfuren forsynede med 4, 5 indtil 6 flade, temmelig lange og vidt fra hinanden staaende Papiller; almindeligt er der dog 5. Mundpladerne (Maxillerne) have paa hver Side 8–9 stærke Papiller. Farven cigarbrun paa Rygfladen, graalighvid paa Bugfladen. Sammenholdes nu den typiske *Archaster Parelii* med vor Varietet, saa viser det sig, at denne har en meget brederer Skive og længere Arme, at Randpladerne flere i Antal, smalere og lavere, og at Paxillarfeltet er bredere. Adambulacralpladerne have flere og længere Papiller, hvilket ogsaa er Tilfældet med Mundpladerne, og endelig er Farven ganske forskjellig.

[1]This starfish differs in several respects from the typical form, but we have not thought it right to raise it to an independent species. The proportion between the small and great radius is in examples of 108–124 mm's size as 1 : 3⅓–3¼. The marginal plates, which in examples of the size just stated are 38–40 in number, are at the arm base 4.1 mm broad, 2 mm high and the paxillar area 9 mm broad; in the middle of the arm they are 2 mm broad, 1.5 mm high, and the paxillar area 2.8 mm broad. The adambulacral plates are next the ambulacral furrow armed with 4, 5 to 6 flat, rather long papillæ, placed at some distance from one another; usually however there are 5. The mouth plates (maxillæ) have on each side 8–9 strong papillæ. The colour is chocolate brown on the dorsal side, greyish white on the ventral side. Comparing now the typical *Archaster Parelii* with our variety, it is apparent that this has a much broader disk and longer arms, that the marginal plates are more numerous, narrower and thinner, and that the paxillar area is broader. The adambulacral plates have more and longer papillæ, and this is also the case with the mouth-plates, and finally the colour is quite different.

---

1) In preparing this translation I have consulted Danielssen and Koren's English text published in their full report mentioned further on, but have kept closer to the Norwegian text.

" M. Sars omtaler i sin Over-sigt over Norges Echinodermer en Varietet af *Archaster Parelii*, som nærmer sig vor temmelig meget, men som dog ikke i saa betydelig Grad, som vor, afviger fra den typiske Form. Sars's Varietet har ikke saa bred Skive og heller ikke saa lange Arme, som vor, hvilket Forholdet af den lille og store Radius bedst udviser (Sars's Exemplar var 1 Tomme større end vorst største). Randpladerne ere færre i Antal, og deres Breddeforhold til Paxillarfeltet er næsten omvendt det paa vor va-rietet. Farven er som paa den typiske Art.

" Unegtelig har man her for sig en Form, der synes at ville arbeide sig op igjennem Tidernes Løb til en selvstændig Art, og for saa vidt har den sin Interesse. Sars fandt 1 Ex-emplar i Vestfinmarken (Oxfjord) paa 100–150, senere fandt han den igjen ved Christiansund paa Lerbund, paa 40–70 Favne.

" Vor Varietet fandtes paa den norske Nordhavsexpedition paa følgende Stationer:

10de Stat. 61° 40′ N.B. 3° 20′ L. Ø. f. Gr. 214 Favne. Temp. + 5.8 C. Lerbund blandet med Sand. 6–8 Exemplarer sammen med *Astro-pecten Andromeda* og *Archaster tenui-spinus*.

79de Stat. 64° 50 N. B. 6° 40′ L. Ø. f. Gr. 151 Favne. Temp. + 6.9 C. Lerholdig Sand med større

M. Sars mentions in his Over-sigt over Norges Echinodermer a variety of *Archaster Parelii*, which approaches pretty much to our varie-ty but does not, however, differ from the typical form in such a marked degree as ours does. Sars' variety has not such a broad disk or such long arms as ours, [a fact] which the proportion of the small and great radius shows best (Sars' ex-ample was 1 inch larger than our largest). The marginal plates are fewer in number, and the proportion of their width to that of the paxillar area is almost the reverse of what it is in our variety. The colour is as in the typical species.

We have undoubtedly before us a form which it appears would evolve in course of time into an indepen-dent species, and it is in this that its interest lies. Sars found 1 exam-ple in West Finmark (Oxfjord) at 100–150, later he found it again at Christiansund on clay bottom at 40–70 fathoms.

Our variety was found in the Norwegian North Sea Expedition in the following stations:

10th Stat. 61° 40′ N. 3° 20′ L. E. fr. Gr. 214 fathoms. Temp. + 5.8 C. Clay bottom mixed with sand. 6–8 examples together with *Astro-pecten Andromeda* and *Archaster tenuispinus*.

79th Stat. 64° 50′ N. L. 6° 40′ L. E. fr. Gr. 151 fathoms. Temp. + 6.9 C. Clayey sand with larger

og mindre Stene. Kun et Par Ex-	and smaller stones. Only a couple
emplarer.	of examples.

og mindre Stene. Kun et Par Exemplarer.

92de Stat. 64° N. B. 6° 40' L.
Ø. f. Gr. 173 Favne. Temp. +7.1 C.
Blod, lerholdig Sand. Kun et Ex-
emplar."

and smaller stones. Only a couple
of examples.

92nd Stat. 64° N. L. 6° 40' L.
E. fr. Gr. 178 fathoms. Temp. +7.1
C. Soft, clayey sand. Only one
example.

This species is mentioned by STORM ['78, p. 252 ; '79, p. 19] under the name of *Archaster parelii* and by HOFFMANN ['82, p. 8] under that of *Astropecten parelii*. These citations are made on the authority of LUDWIG. VERRILL ['82, p. 218] mentions *Archaster parelii* as occurring off the New England coast at depths of 225–487 fathoms but scarce ; the form referred to being the one later described by SLADEN under the name of *Pseudarchaster intermedius*.

DANIELSSEN and KOREN in their full report on the Norwegian North Sea Expedition reproduce the description of their variety *longobrachialis* quoted above with a few verbal alterations. The list of the stations where examples were obtained are however extended as follows ['84, p. 89] : Stat. 261, two specimens ; Stat. 262, one specimen ; Stat. 290, one specimen. As to the typical form, they give a list of previous literature and short notes as follows ['84, p. 87] :

" *Archaster Parelii*, (DÜBEN & KOREN) JOH. MÜLLER.

" This rare starfish was found at Station No. 79. A couple of specimens.

" Distribution. Along the Norwegian coast, but, everywhere, sparingly. The Murman coast. The North American coast. The British coast."

VERRILL ['85, p. 543 + 1 fig.] refers to *Pseudarchaster intermedius* SLADEN, a synonym of the present species, under the name of *Archaster parelii* as occurring on the northern coast of the United States : " B. [athymetrical] range, 225 to 1,608 fathoms (547 to 1,608 fathoms, 1883). Not rare, but always in small numbers."

According to LUDWIG this species is mentioned by Jarzinsky ['85, p. 170]. SLADEN refers it to his subgenus *Tethyaster* (under *Plutonaster*) and makes some remarks on it under the name of *Tethyaster parelii* ['89, p. 102] :

" *Localities.*—' Porcupine ' Expedition :

" Station 31. Between the north of Ireland and Rockall. Lat. 56° 15' N., long. 11° 25' W. Depth 1360 fathoms. Bottom temperature 2°.9 C.; surface temperature 13°.8 C.

" In Sir WYVILLE THOMSON's popularly written Depths of the Sea, the occurrence of this species is specially mentioned on at least two occasions (*loc. cit.*, pp. 122,181), but the exact localities are not given, and I have not seen any other specimens in the collections placed in my hands excepting the one under notice.

" *Other localities.*—Finmark, Lofoten, coast of Norway, off Shetland."

According to FISHER the *Pseudarchaster intermedius* of SLADEN is another synonym of this variable species. Its original description is as follows [SLADEN, '89, p. 115] :

" *Pseudarchaster intermedius*, n. sp. (Pl. XIX. figs. 3 and 4 ; Pl. XLII. figs. 5 and 9).

" Rays five. R=35 mm.; r=11 mm. R>3 r.

" Rays moderately long, tapering continuously from the base to a finely pointed extremity ; breadth midway between the centre of the disk and the extremity, 6.5 mm. Interbrachial arcs well-rounded.

" The paxillæ of the abactinal area are rather small, subcircular, and closely placed, surmounted by ten to fifteen short, truncate, polygonal spinelets, two or three central ones usually larger than the rest, but these are irregular in disposition, and smaller ones may appear at the periphery and increase the difficulty of enumerating the spinelets. The paxillæ are disposed in regular longitudinal lines along the ray, a median radial series being clearly distinguishable and slightly larger than the others. The primary embryonic plates are discernible, though not much larger than the neighbouring plates external to them. The paxillæ diminish slightly in size as they approach the margin and proceed along the ray. A considerable number of smaller paxillæ occupy the area within the circle of the primary basal plates, and the dorso-central plate is small and inconspicuous. The madreporiform body is small and sunken, and lies external to its adjacent primary basal plate.

"The marginal plates form a well-rounded lateral wall, the curvature of the inferior series being slightly fuller or more tumid than that of the superior series. The superomarginal plates are thirty-two in number from the median interradial line to the extremity. The height of the plates in proportion to their length is greatest in the interbrachial arc; and the breadth of the marginal border as seen from above is also rather broader in the interbrachial arc. Midway along the ray it is nearly equal to the breadth of the intermediate paxillar area. The superomarginal plates bear no spines, but are covered with a low, truncate, closely packed polygonal granulation. The granules are largest near the summit of the arc of curvature; and the plates are slightly tumid along their median line, transverse to the axis of the ray. The odd terminal plate is of a rounded shield-shape, and subtubercular in appearance.

"The inferomarginal plates correspond to the superior series; their covering, however, is distinctly squamiform, except at the extreme margins, where the granules at the outer end of the plate partake of the character of those of the adjacent superomarginal plates, whilst those at the inner end form a transition to the granules of the actinal intermediate plates. Some of the squamules on each plate are more elongate and spiniform than the rest, but the definite line of small pointed adpressed spines noticed in *Pseudarchaster tessellatus* and *Pseudarchaster discus* is wanting in the present species.

"The armature of the adambulacral plates consists of a furrow series of five spines, their base line forming an acute angle into the furrow. They are moderately long and thickened towards the extremity, which, in the case of the middle spine, is more or less flattened in the direction transverse to the axis of the ray, but in the other spines in the direction of the margin of the plate to which they are attached. External to the furrow series is a line of three short papilliform spinelets parallel to the furrow, rather wide apart, and of which the two outside spines often appear as if they belonged to the furrow series. Behind these is a second and similar longitudinal series of three spines, the middle one being often longer than the others; and these are followed by three or four smaller papillæ, completing the armature of the plate. The furrow series have a decidedly palmo-radiate appearance, and the spines of the external series have a tendency to incline at a slight angle towards the next adjacent adambulacral plate.

" The actinal interradial areas are comparatively small and are ornamented with short papilliform granules, which appear to be arranged rather widely apart round the the margins of the plates; sometimes one or more papillæ are present in the middle of this circlet, and occasionally one is slightly larger than the rest. Consequent on this arrangement the individual plates are more or less defined, but are not distinct; the spinulation is by no means crowded.

" The pair of mouth-plates form a subelliptical or widely fusiform out-line. Each plate bears a straight line of about ten short papilliform spine-lets running parallel to the median suture, and a similar number on the opposite margin of the plate which consequently form a curved series; one or two additional spinelets may be present on the intermediate area of the plate. , The marginal spines are about equal in length to the adambulacral spines, but are rather more robust.

" Colour in alcohol, a yellowish ashy grey.

" *Locality.*—Station 49. Off the coast of the United States, south of Halifax, Nova Scotia. May 20, 1873. Lat. 43° 3′ 0″ N., long. 63° 39′ 0″ W. Depth 85 fathoms. Gravel and stones. Bottom temperature 35°.0 Fahr.; surface temperature 40°.5 Fahr.

" *Remarks.*—This species is allied in many respects to *Pseudarchaster tessellatus.* The rays, however, are more tapering, the abactinal area is less inflated, the paxillæ though smaller are composed of more numerous spine-lets, the marginal border formed by the superomarginal plates is broader, the covering of the inferomarginal plates is more squamiform, and the de-finite single line of pointed spinelets which occurs in *Pseudarchaster tessella-tus* is not present. There is no large prominent spine on the outer part of the adambulacral armature as in the South-African species."

This species is mentioned under the name of *Archaster parelii* by GRIEG [’89, p. 3], BRUNCHORST [’91, p. 30] and HERDMAN [’92, p. 89].

BELL gives the following description of this starfish under the name of *Plutonaster parelii* [’92, p. 63]:

"R = 2 r (nearly).

"Arms short and broad, disk large; marginals very large, extending considerably on to both ventral and dorsal surfaces. Ambulacra rather narrow, bordered by a very large number of spines; these are in five transverse rows, and the inner have again five in each long row; in the outer the spines are much larger and less numerous. The intermediate plates, which extend but a short way beyond the disk, are covered with large flat granules and a few larger flattened spines, of much the same size as the adambulacral spines. The inferomarginals and superomarginals at the rounded angles of the disk very long, wider within than without; further out the plates are wider, and the outer and inner edges are of the same breadth. About twenty-five to thirty plates on either side of each arm; these plates are sometimes broken into two; they are in each row covered by a close pavement of polygonal plates, some of which, in the lower series, become prolonged into distinct spines. The dorsal surface is occupied by a regular pavement of quadrate or polygonal plates, each of which may be resolved into a number of fine spines flattened at their free ends. Among them the madreporite, which is situated not far from the centre, may be easily detected.

"Colour said by DÜBEN and KOREN to be 'intense sanguineus'; in spirit creamy yellow, the marginals lighter than the rest.

"R = 19, r = 8.

"*Distribution.*—Eastern side of North Atlantic north of Great Britain. Very rare in the British seas. 155–1608 fms.

"a. North of Ireland, 1360 fms. 'Porcupine' Exp."

NORMAN ['93, p. 346] mentions this species under the name of *Plutonaster parelii* from the Trondhjem Fiord.

VERRILL describes specimens of this species under the name of *Pseudarchaster intermedius* SLADEN as follows [' 94, p. 249]:

"According to SLADEN, this is distinct from the allied European *parelii*, with which I formerly identified it, but without a direct comparison of specimens.

"Our numerous specimens show considerable variation, especially in the size of the marginal plates as compared with the breadth of the dorsal area of the rays. In some examples the upper marginal plates are so broad that the dorsal area is much reduced in breadth. In others the marginal plates are comparatively narrow, while the dorsal area is wider.

" These differences are not correlated with any others of importance, so that they can hardly be taken as characteristic of permanent varieties.

" The papulæ are confined to the central part of the disk and basomedian part of the rays.

" Distinct fascioles are present in our specimens between the plates next to the adambulacral series, as in P. discus, but SLADEN states that they are wanting in his examples. Moreover, in all our specimens there is a median row of several enlarged spinules decidedly larger than the rest, on each of the inferomarginal plates, which was not the case in SLADEN's specimens. Similar enlarged spinules occur on most of the actinal interradial plates. In consequence of these differences our examples approach much nearer to P. discus SLADEN, from the west coast of S. America, and to P. tessellatus, from off Cape of Good Hope, than is indicated by SLADEN's descriptions.

" It ranges from 110 to 1,608 fathoms, off our coast."

VERRILL ['95, p. 131] again refers to the some form as follows :

" B. [athymetrical] range, 85 to 1608 fath. Most common between 150 and 500 fath. Taken at 33 stations between N. lat. 44° 26' and 37° 59' 30".

" This species is very closely allied to P. discus SLADEN, from off the west coast of S. America in 147 fath., and to P. tessellatus SL., from the Cape of Good Hope.

" Variety, insignis nov.

" A few specimens, much larger than usual and with coarser granules, represent a marked variety or perhaps a distinct species. For the present it may be best to consider it a variety.

" Radii, 75 mm and 23 mm. Upper surface and marginal plates granulated nearly as in the typical form, except that the granules are somewhat larger. Actinal plates covered with unequal, coarse, irregular, angular, fusiform granules, some of those on the middle of each plate longer and larger, spiniform. Lower marginal plates with a median row of small, appressed, fusiform spines much larger than the granules. Adambulacral spines longer than in the type-form, those on the ventral side of the plates, 12–16; the largest, thick and blunt, or clavate. Jaw-spines thick, blunt, angular, longer and more prominent than in the type, those on the actinal surface in two regular rows of about 8 each.

"B. [athymetrical] range, 100 to 1356 fath. Nova Scotia to N. lat. 40° 09′ 30″."

SLUITER ['95, p. 51] mentions a specimen of this species under the name of *Plutonaster* (subg. *Tethyaster*) *parelii* from Bergen (WEBER), and also describes a variety as follows ['95, p. 51]:

"*Pseudarchaster tessellatus* SLADEN, var. *arcticus* (n. var.). Die Art *Ps. tessellatus* wurde bis jetzt nur noch von SLADEN beschrieben und zwar von Simons Bai, bei der Kap der Guten Hoffnung. Unter den Asteriden, welche von der Niederländischen Nord-Polar Expedition mitgebracht sind, finde ich aber 6 Asteriden, welche alle aus der Barents See, 72° 14′ NB. und 22° 30′ OL. Gr., aus einer Tiefe von 150 Faden herstammen, und dem *Ps. tessellatus* von der Kap der Guten Hoffnung überaus nahe verwandt sind. Bei dem grössten Exemplar war R=62 mm., r=18 mm. Die Verteilung der Paxillen auf dem Rücken ist genau als bei *Ps. tessellatus*, nur ist die Reihenstellung auf den Armen öfters recht undeutlich. Die Paxillen selbst sind etwas abweichend, da in der Mitte nur selten ein, gewöhnlich zwei oder drei centrale Höckerchen vorkommen, welche von einem Kranze von 7–11 Dörnchen umgeben sind, ausser noch einigen kleineren. Zahl, Form und Bewaffnung der dorsalen und ventralen Randplatten stimmen genau mit *Ps. tessellatus*. Dasselbe gilt auch von den Bauchplatten, wenn auch die Grenzen der interradialen Platten zuweilen etwas deutlicher zu unterscheiden sind, da die Granula, welche am Rande der Platten stehen, öfters beträchtlich kleiner sind als die grösseren in der Mitte. Die Ambulacralpapillen wie bei der erwähnten Form, 6 in der inneren Reihe, 4 stumpfe papillenförmige in der zweiten Reihe. Ausserhalb dieser noch eine undeutliche dritte Reihe, welche in der Granulation des Bauches übergeht. Auf den Mundplatten kommen vier Reihen von 6 Papillen vor, von welchen aber die inneren kaum grösser sind als die äusseren, wie es beim Typus der Fall ist."

In the copy of this paper which I owe to the kindness of the author a manuscript note is added to the effect that according to Dr. KŒHLER, the form described in the above passage belongs to *Archasterparelii* var. *longobrachialis* DAN. & KOREN, but that it really belongs to the genus *Pseudarchaster (Astrogonium)*.

Further references to this species are : APPELLÖF [*Archaster parelii*, '96, p. 11 ; '97, p. 13] and GRIEG [*Plutonaster parelii*, '96, p. 5, 12 ; '97, p. 37]. The latter says in one place ['96, p. 5] :

" Af denne ved vore kyster sparsomt optrædende art har jeg fundet nogle faa eksemplarer ved Vesø, Skærgehavn (60–80 fv.), Husøen (200 fv.) og Granesund (80 fv.). HANSEN og FRIELE har den fra havet udenfor Sognefjorden og fra Batalden, 250 fv."

Of this species which occurs sparingly on our coasts, I have found a few examples in Vesø, Skærgehavn (60–80 fath.), Husøen (200 fath.) and Granesund (80 fath.). HANSEN and FRIELE have it from the sea outside Sognefjord and from Batalden, 250 fath.

SLADEN ['97, p. 78] mentions it under the name of *Plutonaster (Tethyaster) parelii* from Rockall Island.

VERRILL again describes *Pseudarchaster intermedius* SLADEN as follows ['99, p. 190, 3 figs.] :

" This is the most common species off the eastern coast of the United States and Canada.

" It was taken at about 33 stations by the Albatross and Fishhawk, 1880 to 1887, in 85 to 1608 fathoms, from N. lat. 44° 26' to 37° 59' 30." It has been brought from the fishing banks, off Nova Scotia, by the Gloucester fishermen.

" The variety (*insignis*) named and described by me in 1895, is probably only the full adult form of this species. The largest example has the larger radius, 75 mm ; the lesser, 23 mm. It lacks distinct actinal fascioles. These exist, however, in variable numbers, on other similar specimens, of somewhat smaller size, as well as on quite young examples. Their presence does not depend upon age, for they may be absent or present in specimens of equal size. Most specimens have the odd apical oral spine somewhat larger and longer than those adjacent. The genital pores are opposite and close to the first pair of dorsal marginal plates."

LUDWIG gives a very complete list of previous literature on this species which he calls *Plutonaster parelii*, and adds as follows [: 00, p. 449] :

" Diese nur ostatlantisch bekannte Art kommt der norwegischen Küste entlang von Christianiafjord bis Finmarken (PARELIUS 1768 ; DÜBEN und KOREN 1846 ; M. SARS 1850, 1861 ; MÖBIUS und BÜTSCHLI 1875 ; STORM 1878, 1879 ; DANIELSSEN und KOREN 1846 ; GRIEG 1889, 1896, 1897 ; BRUN-CHORST 1891 ; HERDMAN 1892 ; NORMAN 1893 ; SLUITER 1895 ; APPELLÖF 1896, 1897) von 58° bis 72° n. Br. vor und geht östlich bis zur Murmanschen Küste (JARZYNSKI 1885). Westwärts von Norwegen kennt man sie noch nördlich von den Shetland-Inseln (HOFFMANN 1882) und an diesen Inseln selbst (NORMAN 1865), ferner an Rockall (SLADEN 1889 ; BELL 1892). Von West nach Ost reicht sie von 11° w. L. bis 42° ö. L., von Süd nach Nord von 56° bis 72° n. Br.

" Sie ist nur selten in geringer Tiefe (15 m) angetroffen worden, meist-ens findet sie sich in 75–400 m und geht nach SLADEN (1889) und BELL (1892) auch noch bis in die Tiefe von 2487 m. Gewöhnlich lebt sie auf Lehm oder sandigem Boden, seltener auf Sand oder auf Schlick.

" Das früher behauptete Vorkommen an der Ostküste von Amerika ist nach VERRILL (1875, p. 131) auf eine Verwechselung mit *Pseudarchaster intermedius* SLADEN zurückzuführen."

With regard to SLUITER's *Pseudarchaster tessellatus* var. *arcticus* above referred to, LUDWIG says in a foot-note [: 00, p. 449] : " Ob die von SLUITER als *Pseudarchaster tessellatus* var. *arcticus* bestimmten Exemplare, die zwischen Norwegen und der Bären-Insel aus 274 m erbeutet wurden, wirklich zu dieser südafrikanischen Art zu zählen sind, scheint mir durchaus zweifelhaft, weil seine Angaben über das Verhältnis von r : R, über die Anordnung der Paxillen und namentlich über die Bewaffnung der Ambulacralplatten und den Mundeckplatten dem widersprechen. Dagegen lassen sie sich eher mit der Beschreibung vereinbaren, die DANIELSSEN und KOREN (1876, 1884) von *Plutonaster parelii* var. *longobrachialis* geben."

GRIEG has some remarks on this species under the name of *Plutonaster parelii* [: 02, p. 19]:

" Findested : Sværholt 2 eksemplarer, Foldenfjord 530 m. 1 eksemplar, Svolvær 1 eksemplar og Balstad 150 m., 1 eksemplar.

Skiveradius 10 mm. 7 mm. 5 mm. 5 mm.

Armradius 28 mm. 18.5 mm. 14 mm. 10 mm.

Antal randplader 48 30 26 24.

" Skiven er saaledes forholdsvis større hos de yngere individer, 1 : 2 mod 1 : 2.8 hos det største. DÖDERLEIN paaviser, at den relative længde af armene hos en del arktiske asterider er underkastede store variationer, hvorfor dette merke ikke som hidtil bør benyttes som artsmerke. Denne tilbøielighed til at variere gjælder ikke alene de af DÖDERLEIN nævnte arter, jeg har ogsaa fundet de thos *plutonaster parelii, leptoptychaster articus, psilaster andromeda,* o.s.v. Paa en af de mellemste dorsale randplader hos eksemplaret fra Balstad fandt jeg en større pig, der ganske ligner den, der sidder paa de ventrale randplader.

" Arten er udbredt langs hele vor kyst, men er overalt sjelden. En eiendommelig langarmet varietet fand Nordhavsekspeditionen i Tanafjorden."

Locality : Sværholt 2 examples, Foldenfjord 530 m. 1 example, Svolvær 1 example and Balstad 150 m., 1 example.

Disk radius 10 mm. 7 mm. 5 mm. 5 mm.

Arm radius 28 mm. 18.5 mm. 14 mm. 10 mm.

Number marginal plates 48 30 26 24.

The disk is thus relatively larger in the younger individuals, 1 : 2 against 1 : 2.8 in the largest. DÖDERLEIN indicates, that the relative length of the arms in some arctic starfishes is subject to great variations, on which account this character should not be used as a specific character as heretofore. This tendency to vary concerns not only the species mentioned by DÖDERLEIN, but I have also found it in *Plutonaster parelii, Leptoptychaster arcticus, Psilaster andromeda,* etc. On one of the innermost dorsal marginal plates of the example from Balsted I found a larger spine, quite similar to that which occurs on the ventral marginal plates.

The species is distributed along our whole coast, but is everywhere rare. A peculiar long armed variety was found by the North Sea Expedition in the Tanafjord.

KŒHLER refers to NICHOLS as listing *Plutonaster parelii* var. *longo-brachialis* in his "List of Irish Echinoderms" [: 03], but I can not verify his citation, although I have examined the paper in question. NORMAN mentions it from east Finmark [: 03, p. 407].

MICHAILOVSKIJ reports *Plutonaster parelii* from lat. 71° 20' N., long. 27° 49' E., 413 m., mud, bottom temperature 3.1° C. [: 04, p. 171]:

"Das junge, von Dr. TSCHERNYSCHEV am westlichen Ufer des Murman erbeutete Exemplar gehört augenscheinlich gerade dieser Form an, obgleich die Merkmale dieser letzteren bei ihm durchaus noch nicht alle deutlich ausgesprochen sind. Bei r=5·5 mm. und R=13 mm. sind nur 28 Marginalplatten für den gesammten Armwinkel vorhanden.

"St. 37 (1)[1]."

GRIEG mentions this species under the name of *Plutonaster parelii* from the collections of the 'Michael Sars' and remarks as follows [: 06, p. 7]:

"Fundstätte: 1901 Stat. 52[2](1); 80[3](1).

"1902 Stat. 31[4](1); 51[5](1).

"Diese 4 Exemplare schliessen sich zunächst der von M. SARS ['61, p. 37] vom Kristianiafjord beschriebenen Varietät mit langen, schmalen Armen an, die ich übrigens auch in den Fjorden des westlichen Norwegens gefunden habe. Diese langarmige Varietät ist es, die DANIELSSEN und KOREN ['77, p. 61; '84, p. 68] als *Plutonaster (Archaster) pareli*, var. *longobrachialis* beschrieben haben. Die Varietät kommt mit der typischen Form vor, mit welcher sie Zwischenformen bildet.

	Stat. 31	Stat. 51	Stat. 52	Stat. 80
Scheibenradius	13 mm.	14 mm.	13 mm	15 mm.
Armradius	41 „	41 „	44 „	45 „
r : R	1 : 3.15	1 : 2.93	1 : 3.38	1 : 3
Dorsomarginalplatten	30	34	33	36

---

1) The numeral in parentheses indicates the number of specimens obtained.
2) 71° 0' N. L., 29° 55' E. L., depth 300 m.
3) Ytre Gjæsbaaen, depth 240 m., sand and large stones.
4) Slindingen, Söndmör, depth 75 m.
5) 61° 40' N. L., 3° 11' E. L., depth 400 m., temp. 6.31° C., mud.

" Mitten auf den Armen war das Paxillarfeld ⅔ der dorsalen Marginal-platten. Die ventralen Marginalplatten waren mit 1–4 grösseren konischen Stacheln versehen. Ein einzelner solcher Stachel befand sich auch auf den Ventralplatten, fehlte jedoch an den Exemplaren von Stat. 31 und 52."

KŒHLER describes this species under the name of *Astrogonium longo-brachiale* (DANIELSSEN & KOREN) as follows [: 07, p. 30]:

" Campagne de 1898 : Stn. 960, profondeur 394 m. Un petit échan-tillon.

" Campagne de 1899 : Stn. 1052, profondeur 440 m. Quatre exem-plaires.

" Dans le plus grand échantillon de la Stn. 1052, R = 63 millimètres, r = 18 millimètres.

" Les autres sont plus petits et leur dimensions respectives sont les suivantes :

$$R = 42 \qquad 37 \qquad 31 \qquad 28 \text{ millimètres.}$$
$$r = 13 \qquad 11 \qquad 9.5 \qquad 8 \qquad \text{,,}$$

" Dans les Asteroidea de la Norske Nordhavs Expedition, DANIELSSEN et KOREN ont indiqué, p. 88, une variété de l'*Archaster Parelii* qui diffère du type par les caractères suivants : le disque est beaucoup plus long, les plaques marginales dorsales sont plus nombreuses, plus étroites et plus minces, et l'aire paxillaire est plus large ; enfin les plaques adambulacraires ont des piquants plus longs et plus nombreux. Dans les plus grands individus, R atteint 108 à 124 millimètres et la valeur R/r varie entre 2⅖ et 3⅓. Les auteurs norvégiens ont désigné cette variété sous le nom d'*Archaster parelii*, var. *longobrachialis*.

" Les courtes indications données par DANIELSSEN et KOREN sur cette variété, ne la font pas connaître d'une manière suffisante, tout en apprenant qu'elle se distingue du type de l'espèce à laquelle ils la rapportent par quelques caractères importants. Heureusement, j'ai pu étudier un exemplaire de cette forme provenant des côtes du Finmark et qui m'a été fort aimable-ment communiqués par le Dr. J. GRIEG.

" Dans cet individu, R = 45 millimètres et r = 14.5 millimètres ; il est absolument identique aux échantillons recueillis par la *Princesse-Alice*, et l'étude de ces différents individus et leur comparaison avec des *Archaster* (ou *Plutonaster*) *Parelii* types m'ont convaincu de la nécessité qu'il y avait de les séparer de cette dernière forme et d'élever au rang d'espèce distincte la variété créée par DANIELSSEN et KOREN. Comme ces auteurs n'en ont pas publié de description à proprement parler, il me parait utile d'en faire connaître les caractères d'une façon détaillée.

"Mais auparavant, j'ai une remarque à faire. J'ai donné à l'espèce dont il s'agit le nom d'*Astrogonium longobrachiale* et non pas d'*Archaster* ou de *Plutonaster*. C'est qu'en effet cette Astérie, pas plus que celle qui porte le nom spécifique de *Parelii*, ne peut rester dans le genre *Plutonaster* : c'est un véritable *Astrogonium* ainsi qu'on pourra s'en convaincre par la description ci-dessous. Quant au *Plutonaster Parelii*, j'ai pu m'assurer, par l'étude de plusieurs individus provenant des côtes de Norvège, qu'il devait également rentrer dans le genre *Astrogonium*, tel que l'a défini PERRIER. Je reviendrai sur ce point dans mon mémoire définitif où je publierai des dessins comparatifs d'*Astrogonium Parelii* et *A. longobrachiale*.

"Le disque de l'*A. longobrachiale* est relativement grand, et les bras, qui sont larges à leur origine, s'amincissent brusquement et rapidement pour conserver ensuite une largeur presque constante sur presque toute leur longueur ; du moins ils s'amincissent fort lentement et leur extrémité est arrondie.

"La face dorsale est couverte de paxilles polygonales, assez grandes, très serrés, plus grandes dans la région centrale du disque et devenant plus petites dans les espaces interradiaux où elles se disposent en files radiaires. Chacune d'elles offre un groupe de quatre à sept granules centraux entourés d'un cercle périphérique de granules plus petits. Sur les bras, les paxilles forment une bande longitudinale médiane de trois à cinq rangée longitudinales dans lesquelles la largeur reste à peu près la même, mais, en dehors de cette bande, la largeur des paxilles diminue rapidement tandis que la longueur reste à peu près la même, et les paxilles se disposent en petites rangées transversales perpendiculaires aux plaques marginales.

"La plaque madréporique est petite, située plus près du centre que des bords. L'anus est indistinct.

"Les plaques marginales dorsales sont au nombre de trente-huit à trente-neuf dans le grand exemplaire. Elles sont plutôt petites et elles n'empiétent pas beaucoup sur la face dorsale ; leur longueur est à peu près égale à leur largeur sur une bonne partie de la longueur des bras. Dans les espaces interbrachiaux, elles sont dirigées obliquement en dehors de telle sorte qu' elles paraissent moins larges qu'elles ne sont en réalité ; elles sont un peu plus larges que longues dans cette région. Elles sont recouvertes de granules polygonaux serrés, à peu près aussi gros ou un peu plus gros que les granules qui forment les paxilles dorsales ; ces granules sont disposés sans ordre, mais le long de chaque bord sutural il existe une rangée régulière et constante de granules plus petits.

"Les contours des plaques latéro-ventrales ne sont pas distincts. Ces plaques sont recouvertes de gros granules un peu allongés, à extrémité arrondie et ne formant pas en général de vrai piquants, sauf quelques-uns d'entre eux. On remarque alors que les piquants courts, robustes, terminés en pointe mousse, forment le centre d'un cercle de granules. Sur tout les exemplaires, les plaques de la première rangée contiguë aux adambulacraires offrent sur chacun de leurs bords adossés un alignement régulier de quelques granules formant avec leurs congénères un pedicellaire fasciolaire. Sur le grand exemplaire que je décris, il se trouve que ces pédicellaires sont mal indiqués : il ne sont limités que par trois ou quatre granules de chaque côté et l'on n'en distingue que deux ou trois de chaque côté du pédicellaire interradial. Mais dans les exemplaires plus petits, ils sont certainement mieux marqués : dans l'exemplaire chez lequel R=42 millimètres, on peut en distinguer une demi-douzaine de chaque côté. Même dans le plus petit exemplaire de la Stn. 960, j'en reconnais encore trois dans chaque angle interradial.

"L'exemplaire de Finmark offre aussi une demi-douzaine de ces pédicellaires de chaque côté.

"Les plaques marginales ventrales sont couvertes de granules aplatis, s'allongeant en une petite pointe mousse ; quelques uns de ces granules, au nombre de trois ou quatre, généralement vers le milieu de chaque plaque, s'allongent en un petit piquant aplati, conique, pointu, mais ces piquants ne débordent pas le bord externe des plaques. Le long de chaque bord sutural, il existe une rangée régulière de granules plus petits, mais sans la moindre indication de fascioles.

"Les plaques adambulacraires portent plusieurs rangées de piquants très développés qui forment, de chaque côté du sillon, une bande très large. On peut reconnaître quatre rangées de piquants dont l'interne seule est régulière. Les piquants de cette rangée, au nombre de sept à huit, sont allongés, dressés, aplatis, et ils conservent la même largeur jusqu'à l'extrémité qui est arrondie. En dehors, vient une autre rangée moins régulière de quatre ou cinq piquants plus petits. Les piquants de la troisième rangée au nombre de trois ou quatre, présentent une tendance très nette à s'allonger : généralement un seul de ces piquants s'allonge beaucoup, quelquefois le voisin s'allonge également. Enfin viennent plusieurs piquants plus petits, au nombre de cinq ou six, qui ne sont guère que des granules allongés.

"Les dents portent sur leur bord externe une rangée de grands piquants, aplatis, à extrémité arrondie, ressemblant aux piquants ambulacraires internes, mais plus forts ; le dernier piquant est un peu plus gros. Sur la face ventrale de la dent, on remarque une rangée assez régulière de six ou sept piquants un peu plus courts que les précédents, forts et dressés."

In his full report on the echinoderms of the 'Princesse-Alice' KŒHLER gives a detailed description of *Astrogonium Parelii* var. *longobrachiale* (DANIELSSEN et KOREN) [: 09, p. 75, 10 figs.] :

"LUDWIG a publié une bibliographie compléte relative à l'*Astrogonium Parelii* et sa variété *longobrachiale*.

"Campagne de 1898 : Stn. 690, profondeur 394 m. Un petit échantillon.

"Campagne de 1899 : Stn. 1052, profondeur 440 m. Quatre échantillons.

"Dans le plus grand exemplaire de la Stn. 1052, R=63, r=18 mm.

"Les autres sont plus petits et leur dimensions respectives sont les suivantes :

R=	42	37	31	28 mm.
r=	13	11	9.5	8 „

"Je rapporte également à la même variété un petit exemplaire de la Stn. 922 dans lequel R=12 mm.

"C'est dans les *Asteroidea* de la *Norske Nordhavs Expedition* que

Danielssen et Koren ont indiqué, p. 88, une variété de l'*Archaster Parelii* qui diffère du type par les caractères suivants : le disque est plus grand et les bras sont plus longs, les plaques marginales dorsales sont plus nombreuses, plus étroites et plus minces, et l'aire paxillaire est plus large ; enfin les plaques adambulacraires ont des piquants plus longs et plus nombreux. Dans les plus grands individus, R atteint 108 à 124 mm et la valeur R/r varie entre $2\frac{1}{8}$ et $3\frac{1}{4}$. Les auteurs norvégien avaient désigné cette variété sous le nom d'*Archaster Parelii* var. *longobrachialis*.

" Les exemplaires recueillis par la Princesse-Alice, que j'ai cités plus haut, se rapportent exactement à cette forme, mais, au moment où je les étudiais, je n'avais à ma disposition qu'un très petit nombre d'*Astrogonium Parelii* type, et ceux-ci étaient assez différents de la variété pour que j'aie cru pouvoir élever celle-ci au rang d'espèce dans ma note préliminaire de 1907. Depuis la publication de cette note, j'ai reçu en communication une très belle collection d'*Astrogonium* des mers arctiques qui m'a été confiée par M. le Dr. J. Grieg, et qui renfermait deux exemplaires de la var. *longobrachialis* capturés par l'Expedition norvégienne, ainsi que quelques formes de passage entre le type et la variété. Grieg a, de son côté, indiqué que ces deux formes se reliaient par des états intermédiaires [: 06] Dès lors, la validité de l'*Astrogonium longobrachiale*, en tant qu'espèce distincte, me paraît discutable et je ne crois pas devoir la maintenir. Toutefois il me semble nécessaire de la conserver à titre de variété, car ainsi qu'on le verra plus loin, elle se distingue bien du type quand on compare des individus adultes.

" Je décrirai d'abord les exemplaires recueillis par la Princesse-Alice, puis j'étudierai sommairement quelques autres échantillons de la variété *longobrachiale* que j'ai eus en communication et je comparerai cette variété à l'*Astrogonium Parelii* type.

" Je prendrai comme type le plus grand exemplaire de la Stn. 1052.

" Le disque (pl. XIV, fig. 8 et 9) est relativement grand, et les bras, qui sont assez larges à leur origine, s'amincissent brusquement et rapidement pour conserver ensuite une largeur presque constante sur le reste de leur longueur jusque vers l'extrémité qui est plus ou moins arrondie.

"La face dorsale est couverte de paxilles polygonales, grandes et assez serrées, plus développées dans la région centrale du disque et devenant plus petites dans les espaces interradiaux où elles se disposent en files radiaires. On est surtout frappé par les dimensions de l'aire paxillaire du disque et des bras ; notamment sur le disque, la surface occupée par les paxilles comprend presque la totalité de la face dorsale tellement la bordure que forment les plaques marginales dorsales est mince. Ces paxilles sont très distincts les unes des autres ; elles sont séparées par des intervalles étroits dans lesquels se montrent les papules [pl. XIV, fig. 12]. Chaque paxille offre un groupe de quatre à sept granules centraux entourés d'un cercle périphérique de granules plus petits. Sur les bras, les paxilles forment une bande médiane de trois à cinq rangées longitudinales dans lesquelles la largeur reste à peu près la même, mais en dehors de cette bande, la largeur des paxilles diminue rapidement tandis que la longueur ne varie guère, et les paxilles se disposent en petites rangées transversales perpendiculaires aux plaques marginales.

. " La plaque madréporique est petite, située plus près du centre que des bords. L'anus est indistinct.

" Les plaques marginales dorsales sont au nombre de trente-huit à trente-neuf dans le grand exemplaire. Elles sont plutôt petites et elles n'emptiétent pas beaucoup sur la face dorsale ; leur longueur est à peu près égale à leur largeur sur une bonne partie de la longueur des bras. Dans les espaces interbrachiaux, elles sont dirigées obliquement en dehors de telle sorte qu'elles paraissent moins larges qu'elle ne le sont en réalité ; elles sont un peu plus larges que longues dans cette région. Elles sont recouvertes de granules polygonaux serrés, à peu près aussi gros ou un peu plus gros que les granules qui forment les paxilles dorsales ; ces granules sont disposés sans ordre, mais le long de chaque bord sutural il existe une rangée très régulière et constante de granules plus petits. Les plaques marginales restent beaucoup plus étroites que l'aire paxillaire sur presque toute la longueur des bras et ce n'est que vers l'extrémité de ceux-ci que l'aire paxillaire est aussi large que les plaques marginales correspondantes.

" Les aires ventrales sont de grosseur moyenne (pl. XV, fig. 12). Les contours de plaques latéro-ventrales ne sont pas distincts : on reconnaît cependant que ces plaques ne forment pas de rangées allant des adambula-craires aux marginales, mais, au contraire, des séries plus ou moins accentuées, et toujours peu nombreuses, parallèles aux adambulacraires. Ces plaques sont recouvertes de gros granules un peu allongés, à extrémité ar-rondie et ne formant pas en général de vrais piquants, sauf quelques-uns d'entre eux. On remarque alors que ces piquants courts, robustes, terminés en pointe mousse, forment le centre d'un cercle de granules. Sur tous les exemplaires, les plaques de la première rangée, contiguë aux adambula-craires, offrent sur chacun de leurs bords adossés un alignement régulier de quelques granules formant avec leurs congénères un pédicellaire fasciolaire. Sur le grand exemplaire que je décris, il se trouve que ces pédicellaires sont irréguliers et peu développés ; il ne sont limités que par quelques granules et l'on n'en distingue que trois ou quatre de chaque côté du pédicellaire interradial. Mais dans les exemplaires plus petits, ils sont beaucoup mieux marqués : ainsi dans l'exemplaire chez lequel R=42 mm, on peut en dis-tinguer une demi-douzaine de chaque côté. Même dans le plus petit exem-plaire de la Stn. 960, j'en reconnais encore trois dans chaque angle inter-radial.

"Les plaques marginales ventrales sont couvertes de granules aplatis, s'allongeant en une petite pointe mousse ; quelques-uns de ces granules, au nombre de trois ou quatre et placés généralement vers le milieu de chaque plaque, s'allongent en un petit piquant aplati, conique et pointu, mais ces piquants ne débordent pas le bord externe des plaques. Le long de chaque bord sutural, il existe une rangée régulière de granules plus petits, mais sans la moindre indication de fascioles.

"Les plaques adambulacraires portent plusieurs rangées de piquants très développés qui forment, de chaque côté du sillon, une bande très large. On peut reconnaître jusqu'à quatre rangées de piquants dont l'interne seule est bien régulière. Les piquants de cette rangée, au nombre de sept à huit, sont allongés, dressés, aplatis, et ils conservent la même largeur jusqu'à l'extrémité qui est arrondie. En dehors, vient une autre rangée moins regu-lière, de quatre ou cinq piquants plus petits. Les piquants de la troisième rangée au nombre de trois ou quatre, présentent une tendance très nette à s'allonger : généralement un seul de ces piquants dépasse les autres, quelquefois le voisin s'allonge également. Enfin viennent plusieurs piquants plus petits, au nombre de cinq ou six, qui ne sont guère que des granules allongés et qui passent aux granules des aires triangulaires ventrales.

" Les dents portent sur leur bord externe une rangée de grands piquants, aplatis, à extrémité arrondie, ressemblant aux piquants adambulacraires internes, mais plus forts que ces derniers ; le dernier piquant est un peu plus gros et s'avance vers la bouche. Sur la face ventrale de la dent, on remarque une rangée assez régulière de six ou sept piquants un peu plus courts que les précédents, forts et dréssés, dont le dernier s'avance vers la bouche au-dessous du précédent.

" Les autres exemplaires de la Princesse-Alice, plus petits que celui que je viens de décrire, ont exactement les mêmes caractères que ce dernier : notamment les bras s'amincissent toujours assez rapidement à leur base pour conserver une largeur à peu près constante jusqu' à l'extrémité qui n'est pas pointue. J'ai représenté l'un de ces petits exemplaires (pl. XV, fig. 10).

" Il me paraît utile de mentionner rapidement ici les caractères des autres *Astrogonium Parelii* var. *longobrachiale* que j'ai pu étudier.

" L'un des plus intéressants est celui que SLUITER a décrit sous le nom de *Pseudarchaster tessellatus* var. *arcticus*, rangeant ainsi le premier cette Astérie dans les genre auquelle elle appartient réellement. Il provient de l'Expédition polaire néerlandaise et a été capturé dans la mer de Barents par 150 brasses de profondeur (243 m). A ma demande, M. SLUITER a bien voulu me confier cet exemplaire et je l'ai représenté en vraie grandeur (pl. XIV, fig. 10 et 11). Ses dimensions sont : R=59, r=20 mm. En comparant ces deux figures et les figures 8 et 9 de la même planche qui réprésentent l'un des exemplaires de la Princesse-Alice, on voit que le rapport R/r est légèrement inférieur à celui de ce dernier ; de plus, les bras sont un peu plus larges à la base et s'amincissent plus progressivement jusqu' à leur extrémité : aussi les bords sont-ils un peu moins parallèles que dans ce dernier exemplaire. Les plaques marginales dorsales et ventrales sont plus larges, surtout ces dernières. Les paxilles du disque sont très grandes dans

la région centrale ; elle sont disposées avec une grande regularité et largement séparées par des intervalles dans lesquels se montrent de nombreuses
papules (pl. XIV, fig. 12). Les pédicellaires fasciolaires sont bien distincts :
ils sont au nombre de sept au moins dans chaque espace triangulaire ventral.
Les plaques qui recouvrent cet espace présentent de petits groupes de granules
offrant, au centre, un piquant plus ou moins apparent. Les plaques marginales ventrales sont assez larges ; les plaques marginales dorsales, bien qu'un
peu plus larges que dans l'exemplaire de la Princesse-Alice, restent plus
étroites que l'aire paxillaire des bras et n'arrivent à l'égaler en longueur
que vers l'extrémité des bras.

"Parmi les échantillons provenant des côtes de Norvége que M. le Dr.
GRIEG a bien voulu me communiquer, se trouvaient deux des exemplaires de
l'Expédition norvégienne désignés par DANIELSSEN et KOREN sous le nom de
Pl. Parelii var. *longobrachialis*. Le plus grand individu, qui est le type de
la variété, provient de la Stn. 92 et ressemble beaucoup à celui de SLUITER.
Ses dimensions sont : R=55, r=18 mm. Les plaques marginales dorsales sont
au nombre de trente-six. Les bras sont plus minces que dans l'exemplaire
de SLUITER, tout en se rétrécissant progressivement ; les plaques marginales
dorsales sont étroites et inclinées latéralement comme dans les spécimens de
la Princesse-Alice ; les ventrales sont plus élargies dans les arcs interbrachiaux. Les paxilles sont grandes et bien régulièrement disposées ; elles
sont separées par des espaces où se montrent de nombreuses papules. Ce
n'est que vers le dernier quart du bras que l'aire paxillaire égale la largeur
de la plaque marginale dorsale correspondante.

"L'autre exemplaire (Stn. 79) est très petit : R=45, r=14 mm. Les
plaques marginales dorsales sont au nombre de trente et une (pl. XV, fig. 7
et 8). Bien que le rapport R/r soit voisin du chiffre observé dans l'individu
précédent, les bras sont plus pointus, triangulaire et les plaques marginales
dorsales et ventrales sont plus étroites. Les paxilles sont identiques à celles
du grand exemplaire.

"Cet échantillon était accompagné d'un exemplaire beaucoup plus petit
et dont les caractères ne sont pas encore bien distincts.

"J'ai également eu en main un exemplaire du musée de Bergen qui m'a été communiqué autrefois par M. GRIEG et que j'ai signalé dans ma note de 1907. Dans cet exemplaire, R=45 et r=14.5 mm. Il est absolu-- ment identique aux échantillons de la Princesse-Alice.

"En ce qui concerne les formes de passage entre la variété *longo-brachiale* et le type de l'*Astrogonium Parelii*, je n'en ai trouvé que trois parmi les échantillons qui m'ont été confiés par M. GRIEG. Deux étaient etiquetés 'Lei Kanger, 80 m.' Dans le plus petit, que j'ai représenté pl. XV, fig. 9, R=32, r=12 mm. Les bras sont très largement réunis au disque ; ils sont triangulaires et pointus à l'extrémité. Les paxilles de la face dorsale sont plutôt petites et serrées et la région qu'elles occupent est moins étendue relativement que dans les exemplaires précédents. L'aire paxillaire des bras atteint la largeur de la plaque marginale dorsale correspondante avant le milieu du bras. En revanche, la face ventrale (pl. XV, fig. 11) présente à peu près les mêmes caractères que ceux de l'exemplaire de la Princesse-Alice représenté pl. XV, fig. 10 et dont la taille est à peu près la même : on y observe des pédicellaires fasciolaires bien apparents et au nombre de quatre ou cinq de chaque côté du pédicellaire médian. L'autre exemplaire de la même provenance est plus grands : R=36, r=13.5 mm. Les caractères de la face dorsale sont les mêmes, mais les pédicellaires fasciolaires sont moins nombreux et moins développés : il n'y en a générale- ment que trois de chaque côté du pédicellaire médian. –

"Au contraire, un échantillon de Sondfjord, près de Bergen (pl. XV, fig. 7), qui, par ses bras plus larges à la base et se rétrécissant moins brus- quement, s'éloigne de la var. *longobrachiale*, a les paxilles de la face dorsale grandes et disposées comme dans cette variété, tandis qu'on n'observe dans chaque aire triangulaire ventrale que des traces de trois pédicellaires, un médian et deux latéraux.

"Tous les autres *Astrogonium* des mers du Nord que j'ai étudiés, et qui, pour moi, représentent la forme *Parelii* type, sont caractérisés par des plaques marginales dorsales très grandes et très larges, tandis que l'aire paxillaire du disque et des bras est relativement étroite en raison de l'em- piétement des plaques marginales : d'habitude, les dorsales sont plus larges que les ventrales. Les paxilles sont relativement plus petites et plus serrées ; les pédicellaires sont en général moins développés.

" Voici à titre d'exemple les caractères de quelques exemplaires que j'ai plus particulièrement étudiés et que je représente ici.

" Un exemplaire appartenant au musée de Bergen et provenant d'une profondeur de 120 à 150 m (pl. XV, fig. 5), a comme dimensions, $R = 56$, $r = 20$ mm ; il possède trente-trois plaques marginales dorsales. On voit qu'il est à peu près de la même taille que l'exemplaire de SLUITER ou que le plus grand individu de l'Expédition norvégienne, qui appartiennent tous deux à la variété *longobrachiale*. Or il suffit d'un coup d'œil jeté sur cette figure et sur celle qui représente le premier des deux autres individus (pl. XV, fig. 10 et 11), pour voir combien les aires paxillaires du disque et des bras sont plus étroites, les plaques marginales dorsales plus larges et les paxilles plus petites et plus serrées. Les plaques marginales sont extrêmement larges et dans l'arc interbrachial elles atteignent près de 6 mm de largeur. Elles sont couvertes de gros granules, au moins dans la région centrale ; ces granules diminuent de grosseur vers les bords adjacents qui offrent toujours une rangée assez régulière de granules plus fins (pl. XV, fig. 6). Les plaques marginales ventrales sont moins larges que les dorsales, et, comme les plaques latéro-ventrales, elles n'offrent guère que des granules. En dehors du pédicellaire interradial médian, on en distingue ordinairement quatre de chaque côté. Les bras, larges à la base et se reliant insensiblement au disque, se rétrécissent très progressivement. En aucun point des bras, la largeur de l'aire paxillaire, qui reste toujours très étroite, n'égale celle de la plaque marginale dorsale correspondante.

" J'ai rencontré dans la collection du musée de Bergen que M. GRIEG a bien voulu me communiquer, plusieurs exemplaires analogues au précédent, mais plus petits. J'ai choisi celui-ci parce qu'il était le plus grand.

" Un exemplaire un peu plus petit et desséché m' à été fort obligeamment communiqué par M. le Prof. THEEL : ses plus grandes dimensions sont $R = 51$ et $r = 20$ mm (pl. XV, fig. 1, 2 et 3). Les bras se continuent largement avec le disque à leur base et ils se rétrécissent progressivement jusqu' à l'extrémité qui n'est pas pointue. Les plaques marginales dorsales au nombre de vingt-neuf, sont très larges, et dans l'arc interbrachial elles mesurent 6 à 7 mm. Vers le milieu des bras, elles mesurent encore 6 mm et sont plus larges que l'aire paxillaire qui, à ce niveau, ne mesure que 4 mm de largeur. Les paxilles sont petites, serrées, à contours peu distincts, sans doute parce que l'exemplaire est desséché. Les plaques latéro-ventrales n'offrent pas de contours très distincts, mais chacune d'elles se reconnaît au piquant aplati qu'elle porte en son milieu et qui est entouré d'un cercle de granules gros et obtus. Elles forment, comme d'habitude, quelques séries parallèles aux adambulacraires. Les plaques marginales ventrales, un peu plus courtes que les dorsales, sont couverts de granules aplatis, serrés et imbriqués, parmi lesquels se détachent une ou deux séries de piquants plus grands, allongés et aplatis, mais qui n'atteignent pas le bord externe de la plaque. Les pédicellaires ne sont pas trop développés, mais ils sont cependant assez apparents et l'on en reconnaît trois ou quatre de chaque côté du pedicellaire médian.

"Enfin un dernier exemplaire m'a été communiqué par mon ami M. le Dr. MORTENSEN et provient du musée de Copenhague (pl. XV, fig. 4). Il est plus petit que les précédents : R=38, r=16 mm. Le disque est moins grand et les bras sont comparativement plus courts; ils sont très larges à la base et se rétrécissent très rapidement. Les plaques marginales dorsales sont très larges et l'aire paxillaire devient rapidement très étroite, de telle sorte que la quatrième plaque marginale est à peine plus large que l'aire paxillaire à ce niveau. Dans l'arc interbrachial, les plaques mesurent 4 mm de largeur. J'en compte vingt-deux de chaque côté. Les plaques latéro-ventrales sont couvertes de granules aplatis et arrondis; au milieu de chacune d'elles, se montre un piquant moins développé et moins apparent que dans l'échantillons du musée de Stockholm. Les plaques marginales ventrales, notablement moins larges que les dorsales, sont garnies de simples granules aplatis, dont quelques-uns se relèvent en un piquant conique et plat, moins développé que dans l'autre échantillon; ces piquants sont aussi moins nombreux. Vers les bords adjacents des plaques, on remarque des granules plus petits que les autres et qui forment une bordure assez nette.

"Il est à remarquer que les pédicellaires fasciolaires sont mieux marqués et plus nombreux sur ce petit exemplaire que sur le précédent : de chaque côté du pédicellaire interradial, on peut en compter jusqu' à cinq et même six qui sont formés chacun par deux rangées de quatre ou cinq petits granules se faisant face.

"En résumé, je crois que l'on doit continuer à distinguer deux formes chez l'*Astrogonium Parelii :* l'une, qui est la forme type, a des plaques marginales dorsales remarquablement élargies et empiétant beaucoup sur le disque et sur les bras dont les aires paxillaires se trouvent réduites d'autant ; les paxilles sont petites et serrés et parfois deviennent confluentes au centre du disque. Les aires triangulaires ventrales sont couvertes de granules dont quelques-uns peuvent s'allonger en piquants ; les plaques marginales ventrales sont moins larges que les plaques dorsales. Les bras ont une longueur variable.

"La variété *longobrachialis* est caractérisée, moins par la longueur relativement plus grande des bras que par la réduction de largeur des plaques marginales dorsales, ce qui permet à l'aire paxillaire du disque et des bras de prendre un grand développement ; les bras sont relativement plus minces et mieux séparés du disque. Les paxilles sont grandes, bien séparées, et, dans les intervalles qui les séparent, on observe de nombreuses papules. Les plaques marginales ventrales sont plus grandes que les dorsales. Les pédicellaires fasciolaires sont en général mieux développés que dans le type.

"Il me parait en effet nécessaire de distinguer des formes aussi différentes que celles que j'ai représentées, d'une part pl. XV, fig. 1, 2, 4, 5 et 6 et d'autre part pl. XIV, figs. 8, 9, 10 et 11, et pl. XV, fig. 7 et 8. J'ajouterai que tous les exemplaires adultes que j'ai eu entre les mains se rapportent soit à l'une, soit à l'autre de ces formes, sans qu'il puisse y avoir le moindre doute pour la classification. Chez les jeunes, comme ceux que j'ai représentés pl. XIV, fig. 7 et pl. XV, fig. 9 et 11, il peut y avoir parfois hésitation quand les caractères ne sont pas encore bien établis : il en est d'ailleurs souvent de même pour les individus non adultes qui appartiennent à des espèces considérés comme distinctes par tout le monde. Mais c'est précisément dans les cas où les formes se relient les unes aux

autres par des intermédiaires tandis que les formes extrêmes sont plus ou
moins distinctes, que le terme variété est commode à employer; il indique
précisément, dans une même espèce, la possibilité de variations qu'il est
bon de consacrer par une épithète.

"Ainsi que j'ai déjà eu l'occasion de l'indiquer dans ma note de 1907
l'Asterie qui nous occupe et dont le nom spécifique est *Parelii*, qui a été
successivement placée dans les genres *Archaster*, *Astropecten* et *Plutonaster*,
est en réalité un *Astrogonium*. Elle ne se rapporte nullement au genre
Plutonaster et offre bien les caractères du genre *Astrogonium* : disposition
des plaques latéro-ventrales, présence de pédicellaires fasciolaires entre les
plaques de la première série ou plaques initiales de PERRIER, armature des
plaques latéro-ventrales ainsi que des plaques marginales ventrales, réduction
des dents, etc. La disposition des plaques latéro-ventrales, et la présence
de pédicellaires fasciolaires présentent une importance particulière et ne
permettent pas de laisser notre espèce dans la famille des Archastéridées.

"Cette modification dans la classification avait déjà été présagée par
certains auteurs. Ainsi VERILL avait d'abord donné le nom d'*Archaster
Parelii* à des Astéries qu'il a rangées ensuite dans le genre *Pseudarchaster*
(synonyme d'*Astrogonium*), sous les noms de *Ps. intermedius* et *fallax*. Je
rappelle aussi que SLUITER a décrit l'*Astrogonium Parelii* var. *longobrachiale*
sous le nom de *Pseudarchaster tessellatus* var. *arcticus*."

Under the name of *Tethyaster parelii* (DÜB. & KÓREN) SÜSSBACH and
BRECKNER describe this species from the North Sea [: 10, p. 203] :

"Die Scheibe und die fünf Arme sind flach, Scheibe gross, Arme breit
ansetzend, mässig lang, in den äusseren zwei Dritteln schmal. Die benachbarten
Arme stossen in weit geöffnetem Bogen zusammen. Das Verhältnis des Scheibenradius zum Armradius ist wie 1 : 2 bis 1 : 2¼. Die Ambulakralfurche ist
ziemlich schmal. Die Ambulakralbewaffnung jeder Ambulakralplatte besteht
aus einer inneren, läng der Furche oder etwas schräg dazu gestellten gera-
den oder mässig gekrümmten Reihe von 4-5 und mehreren (2-3) äusseren
Längsreihen von meist je 3 Stacheln. Die der innersten Reihe sind länger
und etwas schlanker als die der anderen. Bei jungen Exemplaren sind die
Stacheln auf der einzelnen Ambulakralplatte nicht in distinkten Reihen,

sondern mehr würfelförmig angeordnet. Der Rand geht gerundet in die Ober- und Unterfläche der Scheibe und Arme über. Die Randplatten setzen sich eine ziemliche Strecke auf die Ober- und Unterfläche fort. In den äusseren Teilen der Arme sind die Platten rechteckig; in den geschwungenen Armwinkeln sind sie trapezförmig, mit der breiteren Grundlinie nach dem Scheibenzentrum. Die Zahl der Randplatten wird von DÜBEN und KOREN mit 30 für jede Armseite, von BELL mit 25–30 angegeben. Wir zählten an unseren grösseren Exemplaren 21, an kleineren 7-8. Ihre Zahl wächst also mit zunehmendem Alter, wie auch von anderen Arten bereits bekannt ist. Sie sind bedeckt mit gleichmässig grossen, flachen, bei jungen Exemplaren rundlichen, später (hauptsächlich auf den oberen Platten) polygonal gegeneinander abgeplatteten, kalkigen Granula; von diesen können einige, namentlich im Armwinkel, zu Stachelchen auswachsen. Zwischen den unteren Armplatten und den Ambulakralia liegen im Bereich der Scheibe und ganz wenig im Anfang der Arme intermediäre Platten; diese sind paxillär ausgebildet, ihre Granula denen der Randplatten gleichend. Die Platten der Oberseite tragen zahlreichere, dichtgestellte Granula, die gruppenweise polygonal gegeneinander abgegrenzt sind. Die Madreporenplatte ist verdeckt.

"Von den Fahrten des 'Poseiden' liegt kein Vertreter dieser Art vor. Wir fanden an zwei Exemplaren aus dem Material von MÖBIUS und BÜTSCHLI von der 'Pomerania'-Expediton:

$$r = 3 \; ; \; 7 \text{ mm}, \qquad R = 6 \; ; \; 18 \text{ mm}.$$

" *Tethyaster parelii* verbreitet sich nur ostatlantisch, von der Murman-Küste entlang westwärts (Finmarken), an der norwegischen Küste südlich bis zum Christianiafjord, ferner kommt er vor bei den Shetlands und nördlich derselben; sein südlichster Fundort ist zwischen Rockall und Nordirland. Sein Ausbreitungsgebiet berührt also eben nur im Norden und Nordosten die Nordsee.

"Die Art findet sich meistens in den Tiefen von 75–400 m, kommt auch noch in grösseren Tiefen, bis 2487 m vor, selten in geringerer Tiefe (15 m), auf Lehm oder sandigem Boden, seltener auf Schlick. LUDWIG, Arkt. Seesterne, p. 444, bringt eine genaue Zusammenstellung der Fundorte, die wir auszugsweise wiedergeben."

FISHER has recently given a very detailed description of this species, *Ps. parelii*, and mentions specimens from the Sea of Japan [: 11, p. 180]:

"*Diagnosis.*—Rays five. R=104 mm. ;[1] r=37 mm. ; R=2.8 r. Breadth of ray at base, 42 mm. Rays well developed and arcuately tapering at base, then very gradually to the blunt extremity; interbrachial arcs wide and rounded; abactinal area subplane, only a trifle inflated on centre of disk. Abactinal paxillæ small, crowded and regular, about two opposite each marginal plate, and with five to seven ploygonal or subprismatic granules surrounded by twelve to fifteen slenderer papilliform spinelets on periphery of tabulm. Abactinal plates with five or six prominent lobes. Marginal plates broad with close-set, flat-topped hexagonal granules. Inferomarginals with a transverse row of small appressed, squamiform, pointed spinules. Adambulacral plates with five or six strongly compressed furrow spinelets, and on actinal surface one or two enlarged spinules surrounded by numerous shorter granuliform spinelets to the number of fifteen to twenty-two, all thick, heavy, and membrane-invested. Mouth plates with median tooth. Actinal interradial areas large; plates covered with rather crowded, swollen, polygonal, papilliform unequal granules, those in centre more robust, clavate, with flaring tips. Many plates with a central enlarged spinule. Fasciolar channel or pectinate pedicellariæ between plates adjacent to adambulacrals, these continued toward margin but less conspicuously. Spinelets forming roof of fascioles slender. Superambulacral plates present.

"*Description.*—Abactinal paxillæ small and crowded fairly regular, largest on proximal radial regions, very crowded and small at ends of rays where only the median radial series attains the terminal plates but the two adradial nearly reach it; one and one-half to two paxillæ correspond to each marginal plate. The larger paxillæ have five to seven polygonal or subprismatic, robust, truncate granules, heavier at tip than at base, and occupying surface of tabulum, while on the periphery are about twelve to fifteen much slenderer papilliform or subprismatic spinelets. The spinelets may be very compactly placed or form more or less open group. Along the border

---

1) " An unusually large specimen for this species, station 3225."

of the area the paxillæ are compressed, the tabulum being elliptical and crowned with two rows of eight to ten granules.

"Abactinal plates with five or six distinct lobes by which the plates touch or overlap; toward margin, plates often very irregularly lobed, or without lobes. They are arranged in series parallel with median radial. Papulæ one to an area, six about each plate; but absent from terminal half of ray, where the plates are without lobes, being irregularly hexagonal or oval.

"Marginal plates broad and short, encroaching conspicuously upon both areas. Superomarginals, fifty-eight in number from interradial line to extremity of ray, vary in width, and are wider in interbrachial arc than elsewhere (8.5 to 9 mm. wide in large examples). They form an even bevel, more or less arched on outer part of ray and are covered with regular, close hexagonal granules in five to seven transverse rows which are coarser at the outer (lateral) end of the plate. The marginal granules are smaller and form very regular series. Granulation has appearance of being very regular, compact, and smooth. Grooves between plates invisible from exterior, and probably not functioning as fascioles. Terminal plate medium-sized, obovoid, covered with granules.

"Inferomarginals correspond to superomarginals in position, though there is usually one additional plate at tip of ray, and they are a trifle wider. Covered with coarse hexagonal granulation, which increases rapidly in coarseness toward margin of ray. In some specimens the granules are slightly squamiform. In the interbrachial arc the plates bear a median transverse series of four to five flattened lanceolate appressed spinules, which are gradually reduced in size and number along ray, being frequently absent from the last few plates, and only one or two beyond the middle.

"Adambulacral plates with an angular furrow margin bearing a palmate series of five or six more or less compressed spinelets, the median (or adoral admedian) the longest and most compressed; tips rounded; lateral spinelets often with flat side uppermost. The furrow series is continued along adoral and aboral margins of plate in three or four spaced, stout, much smaller, papilliform spinelets. On actinal surface one, two, or occasionally three, enlarged very robust bluntly pointed spinules stand in a transverse oblique, or longitudinal series (only one spinule in medium-sized and small specimens). Between them and furrow series is a semicircular row of three or four shorter, blunt, stout, papilliform spinelets or sometimes very strongly flattened spinelets; these sometimes absent; on outer part of plate are several smaller three or four-sided unequal clavate spinelets. Exclusive of furrow series there are about fifteen to twenty-two spinules and spinelets to each plate, the outermost very irregular in distribution and on distal part of ray showing a tendency to group themselves about the two or three larger spinules.

"Mouth plates prominent actinally, with a bristling armature of robust, short, untapered blunt spines disposed in a marginal and two actinal series. Of the latter one stands on the border of the median suture, while the second, an intermediate shorter irregular series, is located between it and a continuation of the marginal series on the edge adjacent to the first adambulacral. These spines are subquadrate, subterete, occasionally spatulate, and are longer than on outer end of plates. They are also variable in number according to age and locality. The true marginal series consists of about seven robust spines in addition to a large median unpaired spine at the inner angle of the combined plates. The three adjacent to inner angle are usually graduated in size, then the next four are stouter and longer. This is the plan of the furrow series; the spine counts vary.

"Actinal interradial areas large, the intermediate plates extending to the tenth to seventeenth inferomarginal, or one-fourth to a little over one-third length of ray measured along side from interradius; plates arranged in rows running from adambulacrals to inferomarginals. From inner side these plates are oval and imbricate with all the surrounding plates. Plates armed with unequal, swollen, more or less crowded granules, those in centre robust and clavate, often with slightly flaring tips bent outward, the peripheral smaller, round-tipped, occasionally subprismatic, very unequal and irregular, and radiating over narrow shallow grooves between the plates. These channels lead in an irregular course from the inferomarginal fascioles to those between the adambulacral plates. In some specimens they are more conspicuous than in others, and all are more conspicuous adjacent to adambulacrals where they are often roofed by pectinate pedicellariæ. Many of the plates bear a central enlarged pointed spinule directed toward margin. The number of plates bearing such spinules is variable.

" Madreporic body small, situated one-third distance from centre to margin; striations very irregular, the ridges between striæ coarse.

" *Anatomical notes.*—Superambulacral plates present, extending along ray as far as do the intermediate plates, and absent from first ambulacral ossicle; they are rather small. Gonads in a single tuft on either side of interradial septum, which is membranous. Anus present. Intestinal cœcum large with ten or eleven radiating, slightly branched divisions. Tube feet with well-developed sucking disks; no deposits.

" *Variations.*—The principal variations have been noticed in the above description. To recapitulate, they concern chiefly the following characters. Rays, which vary in breadth and length, being a little shorter in Japanese specimens, but also vary in Alaskan examples. Paxillæ: more compact in some individuals, with a slight difference in width of paxillar area on rays, due to the variable width of superomarginals. Superomarginals: variable in width and number, being apparently slightly broader in Japanese specimens (but occasional examples, stations 3225 and 3258, from Alaska, also with broad plates); in number varying from thirty-three (Japanese) to forty-three (station 3258) in examples of same size (R, 70 mm.), the Japanese specimens consequently having longer plates. Inferomarginals: variable in width as superomarginals; covering either very compact and polygonal or more open and squamiform. Variations in adambulacral armature, mouth plates and actinal intermediate plates, sufficiently treated above.

" *Young.*—The type described by SLADEN is a small specimen. The smallest in the collection from station 4792 has the following dimensions: R, 19 mm.; r, 8 mm. Superomarginals, twenty. The actinal interradial areas are small and lack enlarged spinules, as do the adambulacral plates on the proximal half of R. Terminal plate large, subglobose; proximal superomarginals with small bare spot; spinules of inferomarginals only a trifle enlarged beyond granules.

" *Type-locality.*—' Christiansund ' (Norway), 30 fathoms.

" *Distribution.*—Off the eastern coast of the United States and Canada (lat. 44° 26' to 37° 59' N.) in 85 to 1,608 fathoms [VERRILL, 1899] : off Norway (Christiania Fjord to Finmark, lat. 58° to 72° N.) eastward to the Murman coast; westward to the Shetland Island, and Ireland [LUDWIG, 1900] ; Bering Sea (Bering Island ; Pribilof Islands) extending at least as far as Kadiak Island in the eastern north Pacific, and south along the coast of Asia to the Sea of Japan, 70 to 351 fathoms.

" *Specimens examined.*—Twenty-seven, from the following stations :

" Specimens of *Pseudarchaster parelii* examined.

Station.	Locality.	Depth. Fathoms.	Nature of bottom.	No.	Collection.
3225	Near Unimak Island, Aleutians..	85	black sand........	1	U. S. Nat. Mus.
3257	do.................	81	grey sand, gravel..	11	do.
3258	do.................	70	black sand, gravel	2	do.
3330	North of Unalaska Island, Aleutians......................	351	mud	1	do.
3487	Bering Sea, west of Pribilof Islands......................	81	green mud, fine sand	1	do.
3490	West of Pribilof Islands	78	do.........	1	do.
3548	North of Unimak Island........	91	black sand........	1	do.
3606	Bering Sea, North of Unalaska Island	87	green mud, fine sand	1	do.
4287	Uyak Bay, Kadiak Island	66–67	grey mud	2	Albatross, 1903
4291	Shelikof Strait, Alaska	65–48	blue mud, sand, gravel	1	do.
4292	do................:	109–94	blue mud, fine sand	1	do.
4784	Near Attu Island, Aleutians	135	coarse pebbles	1	Albatross, 1906
4792	Near Bering Island	72	pebbles	1	do.
4855	Matsushima, Sea of Japan	70–89	green mud........	2	do.

" *Remarks.*—This species is variable in the Atlantic, and the Pacific specimens are certainly no exception to the rule. Greater differences are observable between specimens from station 3258 than between an example from off Newport, Rhode Island, and two from Norway, and a slightly larger one from 3257, near Unimak, Aleutian Islands. The specimens from Kadiak Island and Shelikof Strait are frankly not typical, but appear to be intergrades with *alascensis*. Japanese specimens are variable and not quite typical. The specimens from Matsushima, in proportion to length of ray, have fewer and hence longer marginals, which are wide. All the granulation is low, coarse, and very compact and the adambulacral spinelets are very heavy in proportion to length, those on outer half of plate being granuliform. The wide marginals are duplicated in an Alaskan specimen from station 3258, there being also a more typical specimen from the same dredge haul. The enlarged actinal intermediate spinelets on both Alaskan and Japanese specimens (of which I have examined a number) are quite variable, sometimes being absent, while equal-sized examples have them. Postadambulacral fascioles are not always evident.

"It appears as if this species, spreading south along the Alaskan coast, had changed into a form with narrow superomarginals, having less granuliform armature on the actinal surface, and less compactly placed granules on the abactinals. Along the Asiatic side the development has been toward fewer and broader superomarginals, with more compact abactinal granules, and an accentuation of the granuliform character of the actinal armature. The Japanese form has departed less from the type than has the British Columbian.

"This species is the north Pacific and Atlantic representative of *Ps. discus* SLADEN (from Messier Channel, between Chile and Wellington Island), to which it is closely related. It may be that *Ps. pulcher* LUDWIG (Galapagos to southwest of Acapulco), founded on very small specimens, is the connecting link between the two forms.

"A few words concerning the name adopted may be in order. In the Museum of Comparative Zoölogy are three specimens of '*Astropecten parelii*' from Norway, presented by Professor SARS in 1852. These agree very well with the original description and figures of DÜBEN and KOREN. There are no differences of importance between these specimens and an example of *Pseudarchaster intermedius* from off Rhode Island. Similarly, the Norwegian specimens belong to the same species as the Alaskan. There is far more difference among the various Alaskan examples than exists between these on the one hand and either the Norwegian or Rhode Island specimens on the other. The width of the paxillar area on the arm varies in this species, even in specimens from the same locality. It is not surprising, therefore, that the area is narrower in the specimens from Norway, just mentioned, than in the figure of DÜBEN and KOREN.[1] DÜBEN and KOREN figure also the superomarginal plates and paxillæ, the inferomarginals and adambulacral plates, and fortunately the tube-feet. These have strong sucking disks. The figure is enough to place the animal in *Pseudarchaster*.

"That *Astropecten parelii* and *Pseudarchaster intermedius* are one and the same (somewhat variable) species I have not the slightest doubt. I also doubt if *Pseudarchaster fallax* is anything more than a variety.

"SLADEN placed *parelii* in *Tethyaster*, a subgenus of *Plutonaster*, while BELL, NORMAN, GRIEG, and LUDWIG relegated it to *Plutonaster*. Such a course is untenable, because *Plutonaster* belongs to a different family altogether, and has pointed tube feet, never sucking disks. This character alone is enough to exclude *parelii* from *Plutonaster*. As a matter of fact, *parelii* is so near the type of *Pseudarchaster* [that is, *discus*] that one is obliged to search carefully to find trenchant differences. The following are the considerations which lead one to rank *parelii* in *Pseudarchaster* : tube feet with sucking disks; character of the marginal plates, especially the inferomarginals; the characteristic adambulacral plates and armature, which is totally unlike *Plutonaster*; the actinal intermediate plates with spaced granules and incipient central spinule; the armature of the mouth plates, there being an unpaired median tooth directed over the actinostome; the presence of postadambulacral fascioles very characteristic of *Pseudarchaster* and never found in *Plutonaster* or allied genera.

"In a specimen of *parelii* from the Copenhagen Museum : R$=43$ mm., r$=14$ mm.; superomarginals twenty-three; the superomarginals are considerably arched above the abactinal paxillar area on arms; the latter is narrower than the marginal plates, there being at the middle of ray only

---

1) " Plate 7, fig. 4."

three longitudinal rows of paxillæ, and only a single row reaches terminal plate; about four and one-half to four paxillæ correspond to two superomarginals; furrow spinelets ˉfive or six; one or two actinal spinelets enlarged slightly; first row of actinal intermediate plates has incipient fascioles; a few of the actinal intermediate plates with slightly enlarged spinelets.

"This specimen differs from the Alaskan chiefly in having tumid superomarginals and a narrower paxillar area on rays. There are about five transverse rows of granules on the superomarginals (the same as in Alaskan specimens)."

" FARRAN reports it from the west coast of Ireland as follows [: 13, p. 13] :

" *Helga.*

"S.R. 329—9 v '06. 51° 21' 30" N., 11° 34' W., soundings 215–415 fms. Trawl.—Two.

" S.R. 333—10 v '06· 51° 37' N., 12° 9' W., soundings 557–579 fms., ooze. Trawl.—One.

" S.R. 353—6 vIII '06. 50° 38' 30" N., 11° 32' W., soundings 250–542 fms., muddy sand. Trawl.—Two.

"S.R. 363—10 vIII '06· 51° 22' N., 12° 0' W., soundings 695–720 fms., ooze. Trawl.—One.

"S.R. 400—5 II '07. 51° 18' N., 11° 50' W., soundings 525–600 fms., mud and ooze. Trawl.—One.

" S.R. 440—16 v '07. 51° 45' N., 11° 49' W., soundings 350–389 fms. Trawl.—One.

"S.R. 487—3 IX '07. 51° 36' N., 11° 57' W., soundings 540–660 fms. Trawl.—One.

"S.R. 490—7 IX '07. 51° 57' 30" N., 12° 7' W., soundings 470–491 fms., ooze. Trawl.—One.

"S.R. 491—7 IX '07. 51° 57' 30" N., 12° 13' W., soundings 491–520 fms. Trawl.—Three.

" S.R. 500—11 IX '07. 50° 52' N., 11° 56' W., soundings 625–666 fms. Trawl.—One.

" S.R. 502—11 IX '07. 50° 46' N., 11° 21' W., soundings 447–515 fms. Trawl.—One.

"S.R. 504—12 IX '07. 50° 42' N., 11° 18' W., soundings 627–728 fms., stones and coral. Trawl—Three.

"S.R. 592—6 VIII '08. 50° 39' N., 11° 25' W., soundings 400–510 fms., ooze. Trawl.—Three.

"S.R. 746—14 V '09. 50° 32' N., 12° 13' W., soundings 620–658 fms., ooze. Trawl.—Three.

"As Kœhler (1907) and Fisher (1911) have pointed out, this species was wrongly referred to the genus *Plutonaster*, and properly takes its place in the Family *Pentagonasteridæ* (*Goniasteridæ*) rather than in the *Archasteridæ*, in which it formerly stood. I have for convenience adopted the generic name used by Fisher, without entering into the academic question of whether it or *Astrogonium* is more properly applicable to the genus to which it is here applied.

"Dr. Kœhler, who has had considerable experience both of the type of this species and of the variety *longobranchiale*, has been good enough to examine one of my specimens. He informs me that, although its arms are more pointed than in the specimens he has seen of the type, it cannot be referred to the variety, in which the arms are, relatively, much more slender and longer.

"This species reached a very large size in Irish waters. The two largest specimens obtained by the *Helga* measured R = 192 mm., r = 70 mm. and R = 178 mm., r = 57 mm. respectively.

"Small specimens are rather scarce, for out of nineteen of the *Helga* specimens in which the arms were measured, in five only was R less than 80 mm."

Mortensen reports the long-armed variety of this species from Greenland under the name *Astrogonium parelii* var. *longobrachiale* [: 13, p. 329] :

"Forekomst" (Occurrence). Vest-Grønland : Bredefjord, 225–290 m. (Stephensen, 1912).

"Ikke kendt fra Øst-Grønland" (Not known from East-Greenland).

"Dybde (Depth) : 70–ca. 2500 m.

"Udbredelse : Nordlige-Atlanterhav, varme Area, fra Irland til Murman-Kysten (north Atlantic, warm area, from Ireland to the Murman coast) ; ved Nordamerikas Østkyst ned til ca. 38° N. (on the east coast of North America down to ca. 38° N.). Nordlige-Stillehav, fra Berings-Hav til det Japanske-Hav, og til Kadiak Øen ved den amerikanske Kyst (north Pacific from Bering Sea to the Japan Sea, and to the Kadiak Isl. on the American coast)."

This species is not contained in any of the collections on which this monograph is based, but is represented by numerous examples in the Albatross collection of 1906. It will be treated of in my report on that collection now in preparation.

Stellaster equestris (RETZIUS).

(Pl. XIII, figs. 213–218 ; Pl. XIV, fig. 219–220.)

The genus *Stellaster* is due to GRAY who gives the following diagnosis ['40, p. 277]: "Body depressed, covered with large flat regular six-sided plates, margin with two rows of large tesseræ; the lower rows with a series of compressed mobile spines." The single species described was *St. childreni*, a synonym of *St. equestris* first described by RETZIUS under the name of *Asterias equestris* ['20, p. 12]. GRAY'S description is as follows ['40, p. 278]:

"1. *Stellaster childreni*. Back convex, with 1 or 2 blunt tubercles on the angles of the centre, arms three quarters the length of the width of the body, narrow, attenuated to a blunt recurved tip.

"Inhab. China or Japan?"

MÜLLER and TROSCHEL give the generic diagnosis as follows [42, p. 278]:

"Körper fast pentagonal, auf beiden Seiten platt, mit zwei Reihen grosser granulirter Randplatten, welche beide zur Bildung des hohen Randes beitragen. Jede ventrale Randplatte trägt einen hängenden Stachel. Beide Flächen der Scheibe sind mit granulirten Tafeln bedeckt. After subcentral."

Two species are described in the text, but they are amalgamated into one in an appendicular note ['42, p. 62] :

"Species 1. *Stellaster Childreni* GRAY.

" *Stellaster Childreni* GRAY Ann. IV, p. 278.

" Verhältniss des kleinen Halbmessers zum grossen wie $1 : 2\frac{1}{4}$. Die Arme sind sehr spitz und die Winkel zwischen ihnen abgerundet. Die Furchenpapillen in zwei Reihen, in der Innenreihe sechs auf einer Platte, die mittlere länger ; in der zweiten Reihe platte, breite vereinzelte; zwischen ihnen zangenartige Pedicellarien. Die Bauchseite mit grossen Tafeln besetzt, die Tafeln gleichförmig granulirt und zugleich mit kleinen klappenartigen Pedicellarien. Die Randplatten viel grösser als die Bauch- und Rückenplatten, 16 an jedem Arme, sind eben so granulirt ; jede untere Randplatte trägt einen beweglichen, hängenden, platten Stachel, wie eine Franze, welcher dreimal so lang wie breit ist; sie haben dieselbe Gestalt wie die Papillen in der äussern Reihe am Rande der Armfurchen. Die Rückenplatten sind etwas seltener granulirt.   Die Porenfelder sind klein und beschränken sich auf die Furchen zwischen den Platten, so dass sie nur von einigen Poren besetzt sind. Auf allen Platten, sowohl den ventralen als marginalen und dorsalen, stehen kleine klappenartige Pedicellarien, mehrere oder viele auf jeder Platte.

" Grösse : 4 Zoll.

" Fundort : Japan. Im Museum zu Berlin und Leyden."

" Species 2. *Stellaster equestris* Nob.

" *Asterias equestris* RETZ. Diss. p. 12.

" Verhältniss des kleinen zum grossen Radius wie $1 : 2$. Die Furchenpapillen in zwei Reihen, in der Innenreihe mehrere, bis fünf auf einer Platte, klein, in der Aussenreihe eine auf jeder Platte. Die Randplatten, 13 an jedem Arme, gegen das Ende der Arme an Grösse allmählig abnehmend. Sie sind wie alle übrigen granulirt, ohne Pedicellarien. An jeder ventralen Randplatte ein beweglicher platter Stachel. Von den Rückenplatten liegen die mittleren in einer Reihe bis zum Ende der Arme. Sie sind grösstenteils einfach granulirt. Hier und da zeichnen sich einzelne Platten dadurch aus, dass sich in ihrer Mitte ein kurzer stumpfer, runder Höcker erhebt, nicht höher als breite Poren einzeln.

"Grösse : 3 Zoll.

"Fundort: Ocean. Im Museum zu Lund."

The note in the appendix referred to above on *Stellaster equestris* runs as follows [MÜLLER und TROSBHEL, '42, p. 128] :

"Diese Species ist nach neuern Materialien zu unterdrücken, indem sie mit *Stellaster Childreni* zusammenfällt. Das Exemplar von RETZIUS in Lund hat seine Pedicellarien verloren."

Acoording to the law of priority in vogue we have to adopt the name of *St. equestris* as the older.

In his paper of 1847 [p. 7.] GRAY makes some remarks on this species in connection with the two species described by him. Under *Stellaster incei* he says, "This species is very like *Stellaster Childreni*, GRAY, Ann. and Mag. Nat. Hist. 1840, 278 ; MÜLLER, Aster. 62. 128. t. 4. f. 3 ; *Asterias equestris* RETZIUS, Diss. 12 ; but it is purplish when dry ; the back is tubercular ; the whole surface is minutely granular ; while the Japanese species is always white, the back smooth, and the granules of the surface are so minute and thin that they are very easily eroded, and the lower marginal plates are more convex and the central ones much larger than the others." Again under *Stellaster belcheri* he says, " This species is intermediate between *S. Childreni* and *S. Incei*, having the white colour and the slender arms of the former, and the convex back and tubercles of the latter, but the tubercles are larger and fewer, and the arms are more slender, having only a single series of plates between the marginal ones."

DUJARDIN and HUPÉ describe this species as follows ['62, p. 407] :

"Stellaster de Children. *Stellaster Childreni.*—GRAY.

[References to GRAY, MÜLLER and TROSCHEL and RETZIUS.]

"Espèce à corps pentagonal, dont le plus grand rayon égale deux fois et un tiers le plus petit. Bras très pointus, séparés par un angle rentrant arrondi. Piquants du sillon ventral ou ambulacraire formant deux rangées, ceux de la rangée interne sont disposés par six sur chaque plaque, ceux du milieu étant plus longs ; la rangée externe porte des piquant larges, aplatis et isolés ; des pédicellaires en pinces existent entre ces derniers piquants. La face ventrale est revêtue de grandes plaques uniformément granuleuses. Plaques marginales beaucoup plus grandes que les plaques dorsales et ventrales, au nombre de seize sur chaque bras et aussi granuleuses.

"Chaque plaque marginale inférieure porte un piquant mobile et pendant, de forme aplatie, et trois fois aussi long que large.

"Les plaques dorsales sont un peu moins granuleuses que les ventrales.

"Les pores tentaculaires, peu nombreux, occupent de petits espaces et s'étendent sur les sillons inférieurs entre les plaques.

"Toutes les plaques ventrales, dorsales et marginales, portent de petites pédicellaires bivalves plus ou moins nombreuses.

"Dimension : largeur totale 108 mm.

"Habite les côtes du Japon.

"MM. MÜLLER et TROSCHEL avaient d'abord admis, comme deuxième espèce, sous le nom de *Stellaster equestris*, d'après RETZIUS, une Astérie faisant partie du Musée de Lund, que plus tard ils ont reconnue identique avec la précédente, et qui n'en diffère que par la chute des pédicellaires. MÜLLER et TROSCHEL (Nacht. Syst. der Aster.)."

VON MARTENS mentions *St. childreni* as being represented by Japanese specimens in the Museums of Leyden, London and Berlin ['65, p. 353] and describes *equestris* at some length further on ['65, p. 356]:

" *Goniaster (Stellaster) equestris* RETZ. sp.—*Stellaster Childreni* (GRAY Ann. and Mag. n. h. VI. 1841, p. 278 ?). MÜLL. u. TROSCHEL Syst. Asterid. p. 62 und 128. Taf. 4. Fig. 3.—*St. gracilis* MÖBIUS Abhandl. des naturwiss. Vereins in Hamburg Bd. IV. 1860. S. 12. Taf. 4. Fig. 3. 4.

"Oben und unten mit gekörnten Tafeln bedeckt. Randplatten gleichmässig granulirt, die unteren mit je einen platten, eingelenkten Stachel. Armwinkel ausgerundet, aber Arme lang und gegen die Spitze dünn. An den von mir im chinesischen Meer gesammelten ist der After nicht merklich excentrisch, und es finden sich fünf kleine Höcker, von beiden Seiten etwas zusammengedrückt, ein regelmässiges Fünfeck bildend, das ungefähr gleichweit vom Centrum als vom Rande entfernt ist, und in dessen Umriss auch die Madreporenplatte fällt. Die Pedicellarien (klappenartige) sind auf der

Rückenseite seltener als auf der Bauchseite. Die grösseren Rückentafeln werden reichlich so gross wie der von oben sichtbare Theil der Randplatten, (auch bei den MÜLLER'schen Originalexemplaren). Poren in Gruppen auf dem Rücken der Arme und Scheibe. Drei Reihen von Platten auf dem Armrücken zwischen den Randplatten im mittleren Drittel der Armlänge, weiterhin eine einzige Reihe und zuletzt stossen die Randplatten zusammen.

"In der südchinesischen See und der Formosastrasse aus Schlammgrund, in 40 und 25 Faden Tiefe mit dem Schleppnetz aufgefischt.

"Der Unterschied der Gattung *Stellaster* von *Goniaster* (*Goniodiscus* und *Astrogonium* M. TR.) beruht hauptsächlich auf den Stacheln der unteren Randplatten; wenn man bedenkt, wie wenig constant diese bei Arten von *Oreaster* sind, so erscheint er von wenig Gewicht. *St. equestris* ist in Gestalt wie in Bekleidung der Rücken- und Bauchseite sehr ähnlich dem *Astrogonium capella* MÜLLER u. TROSCHEL, so dass' man erst die Furchenpapillen und die innersten Bauchplatten (vgl. unten) ansehen muss, um *G. capella* nicht für einen kahl gewordenen *Stellaster equestris* zu halten.

"Die kleinen Höcker der Oberseite sowie die Pedicellarien scheinen bei dieser Art mehrfach zu variiren. Das Originalexemplar von *Stellaster Childreni* GRAY im brit. Museum, das Prof. PETERS darauf zu untersuchen die Güte hatte, zeigte gar keine bestimmten Höcker, sondern nur einige leichte Anschwellungen. Das Exemplar des Berliner Museums, nach welchem MÜLLER und TROSCHEL die Art beschrieben, ist an der Oberseite sehr abgerieben und zeigt nur stellenweise Spuren, welche auf früheres Vorhandensein solcher Höcker deuten. *St. Childreni* des Hamburger Museums zeigt die Höcker, aber in unregelmässigen Abständen vom Centrum, von den zwei Exemplaren des *St. gracilis* MÖBIUS ebendaselbt das eine je zwei, das andere 2, 1 oder keinen Höcker in einem Radius. Die Poren stehen auf der Oberseite stets zu 2–5 zusammen, und erstrecken sich bei den von MÖBIUS als *gracilis* bezeichneten Exemplaren weiter gegen die Mitte der Schale als bei den andern. Uebrigens sind bei allen mir vorliegenden Exemplaren von *Stellaster* die Poren auf den Armen zahlreicher als auf der Scheibe und fehlen völlig in der Mitte der Interradialräume, die durch zwei Reihen grösserer

Rückenplatten bezeichnet ist. Die klappenförmigen Pedicellarien sind bei dem Berliner Originalexemplar ebenso zahlreich auf der Oberseite wie auf der Unterseite; bei den von Möbius als *St. Childreni* und den als *gracilis* von ihm unterschiedenen Exemplaren des Hamburger Museums oben weniger zahlreich. Die der Unterseite sind bei dem Hamburger *Childreni* kaum zweimal so breit (scheinbar lang) als dick, bei dem Berliner viele dreimal so breit, doch nicht alle, bei *gracilis* Möbius sogar viermal, bei den von mir gesammelten meist nur doppelt so breit. Doch wechselt dieses Verhältniss zwischen den Pedicellarien desselben Individnums zu bedeutend, als dass ein Artunterschied darauf zu begründen sein dürfte."

As to *Stellaster gracilis* Möbius, which von Martens regards as a synonym of the present species, both Perrier ['76, p. 43] and Sladen ['89, p. 322] agree in referring it to *St. incei* Gray.

Stellaster mülleri, which is regarded by Sladen as a synonym of *Stellaster equestris*, is also described in the same work [von Martens, '65, p. 359]:

" *Goniaster (Stellaster) Mülleri* n. sp.

" Scheibenradius zum Armradius wie 1 : 3. Furchenpapillen der äusseren Reihe nicht grösser als die der inneren, mehrere auf einer Platte. Keine Höcker und keine Pedicellarien auf der Rückenseite. Poren nur einzeln zwischen den Rückenplatten, auf der Scheibe seltener. (Bei allen mir vorliegenden Arten fehlen die Poren in der Mitte der Interbrachialräume, welche zugleich durch zwei Reihen etwas grösserer Platten' ausgezeichnet ist.) Schon von der Mitte der Armlänge an nur eine Reihe Rückenplatten zwischen den Randplatten. Nur 12 Randplatten. Obere Randplatten ohne Höcker, von den unteren tragen bei weitem nicht alle einen Stachel.

" Farbe des trocknen weiss. Grosser Radius 32 Mill., kleiner 11½ Mill.

" Ein Exemplar im Berliner Museum mit der Etikette; *St. Childreni*. Japan. Bachmann.

" Sollte Gray's *Childreni* unser *Mülleri*, sein *Belcheri* unser *equestris* und sein *Incei* unser *tuberculosus* sein? Ich bin desshalb nicht geneigt es zu glauben, weil man nicht wohl annehmen darf, dass Gray die übrigen wesentliche Unterschiede der von mir beschriebenen Arten übersehen hätte.

"Es scheint demnach eine Reihe nah verwandter Arten vom (südlichen) Japan bis in den indischen Archipel sich zu erstrecken."

In his work of 1866 GRAY simply reproduces the descriptions of his earlier papers, but gives figures of this species, and among others makes the following remark ['66, p. 7]: "b. *Stellaster gracilis*, MÖBIUS, Abhandl. IV. 1860, i͘ı. t. 11. f. 3, 4, which is very nearly allied, if not the same in a rather different state. This state has also been described by DUJARDIN under the name of *Astrogonium Souleyetii*.

As to *Astrogonium souleyetii* DUJ. & HUPÉ, it is a synonym of *Iconaster longimanus* (MÖBIUS) [Sladen, '89᾽ p. 262].

VON MARTENS again refers to this species as follows ['67, p. 112]:

"Ein käuflich erworbenes Exemplar, angeblich aus der chinesischen See, zeigt fünf Gruppen von kleinen Höckern auf dem Scheibenrücken, was eine weitere Variation derselben darstellt, und die oben *tuberculosus* genannte Art wieder enger an *equestris* anschliesst."

LÜTKEN combines the three genera, *Stellaster*, *Astrogonium* and *Goniodiscus* under the *Goniaster* of AGASSIZ and mentions or describes six species, *G. equestris* RETZ., *G. incei* GRAY, *G. tuberculosus* v. MART., *G. belcheri* GRAY, *G. mülleri* v. MART., and *G. Dübenii* GRAY. Of these the last named species has since been referred to the genus *Pentagonaster* (*Tosia*). *Goniaster equestris* is described as follows ['71, p. 245]:

"*Goniaster* (*Stellaster*) *equestris* (RETZ.).

"Et Exemplar fra Formosa-Kanalen (R=40 mm., r=16 mm.) stemmer meget godt med Beskrivelsen af *S. equestris* i 'System d. Asteriden,' som er udkastet efter RETZIUS's Original-Exemplar i Lund. Det har ligesom dette 13 Randplader paa hver Side af hver Arm og en Kreds af Knuder (her fire, den femte mangler) paa Ryggen i samme Afstand fra Midtpunktet som Madreporpladen. Disse Knuder omtales ogsaa af v. MARTENS hos hans i det sydlige

A specimen from Formosa Strait (R=40 mm., r=16 mm.) agrees very well with the description of *S. equestris* in the 'System der Asteriden,' which is based on RETZIUS' original specimen in Lund. It has like it 13 marginal plates on each side of each arm and a circle of nodes (in this case four, the fifth is wanting) on the back at the same distance from the centre as the madreporic plate. These nodes are also mentioned by v. MARTENS in his specimens obtained in

Kina-Hav og i Formosa-Strædet fiskede Exemplarer, og paa to foreliggende törre Exemplarer, hvis Opbevaringstilstand er mindre god, kan man endnu, ligesom paa GRAYS citerede Figur, paavise Stedet, hvor de have siddet. Som betegnende for Arten vilde jeg foruden disse fem Knuder endnu anföre (med GRAY) den hvide Farve (eller vel rettere den farvelöse Tilstand af de törrede eller i Spiritus opbevarede Exemplarer), fremdeles den forholdsvis stærke Udvikling af Randpladerne, især i Armvinklerne, de egenlige Rygpladers regelmæsige, i Almindelighed sexkantede Form, de faa Porer, der oftest ere anbragte i Hjörnerne mellem disse Plader og som i Armenes indre Del, hvor Armene stöde til Skiven, danne Grupper paa 5-6, men ellers sidde enkeltvis eller kun ganske faa sammen; de lade sig i övrigt forfölge lige til Stjernens Midtpunkt. Endnu kan anföres, at der ses ikke faa 'Pedicellariæ valvulatæ' paa de almindelige Ryg- og Bugplader; ogsaa paa Randpladerne har jeg iagttaget enkelte. Af Ambulakralpapiller er der gjerne 6 fine i hver Gruppe i den indre Række, og en eller höjst to i den ydre; at de sidstnævnte i Form og Störrelse omtrent stemme med Bugrandpladernes Pigge, er bekjendt.

G. equestris synes at være udbredt fra Japan til Formosa-Strædet og det sydkinesiske Hav.

the Southern China Sea and the Strait of Formosa, and in two dried examples before me, which are in a less good condition of preservation, one can also, as in GRAY's figure above cited, still point out the place, where they were situated. As characteristic for the species I would mention, besides these five nodes, (with GRAY) also the white colour (or perhaps more correctly the colourless condition of the specimens either dried or preserved in alcohol), as also the relatively good development of the marginal plates, especially in the arm angles, the regular, in general hexagonal form of the true dorsal plates, the few pores, which are most frequently situated in the angles between these plates and which form groups of 5-6 in the inner part of the arms, where the latter join the disk, but elsewhere occurring either single or only very few together; they can besides be followed even to the centre of the star. It may, moreover, be mentioned that there are seen not few 'pedicellariæ valvulatæ' on the dorsal and ventral plates in general; I have also observed single ones on the marginal plates. Of the ambulacral papillæ there are usually 6 fine ones in each group of the inner row, and one or at most two in the outer; that the last mentioned ones nearly agree in form and size with the spines of the ventral marginal plates is known.

G. equestris appears to be distributed from Japan to Formosa Strait and the South China Sea.

PERRIER, in his work on pedicellariæ, makes the following remarks ['69, p. 92]:

"*Stellaster Childreni*, GRAY.—Les Pédicellaires en pince de cette espèce ont été figurés par MÜLLER et TROSCHEL: mais la figure qu'ils en donnent est certainement une figure de fantaisie. On ne trouve jamais dans les Pédicellaires des Astéries cette dentelure régulière que représente la planche IV du *System der Asteriden*, la même, du reste, qui porte une figure tout à fait inexacte de Pédicellaires croisés de l'*Asteracanthion gelatinosus*, et qui n'est guère plus heureuse pour les Pédicellaires droite de la même espèce. Les Pédicellaires en pince du *Stellaster Childreni* sont situés un peu en arrière des piquants du sillon ambulacraire, entre ceux-ci et la ligne de piquants aplatis qui suit. Ils sont assez allongés, minces, formés d'un tissu calcaire réticulé, fort irregulier, concave à l'intérieur, et à bords hérissés d'un grand nombre de points calcaires perpendiculaires à la direction de ce bord.

"Les Pédicellaires valvulaires sont situés sur la plupart des plaques ventrales, dorsales et marginales; ils sont assez peu nombreux, et relativement très-petits. La planche IV du *System der Asteriden* de MÜLLER et TROSCHEL donne, dans ses figures 3 et 4, une bonne idée de leur nombre, de leur grandeur et de leur disposition.

"Le *Stellaster Childreni* présente donc dans ses Pédicellaires˙ la même disposition générale que les espèces des genres précédents." Tous ces genres sont du reste intimement unis entre eux."

In his "Révision," PERRIER regards *Stellaster* as a subgenus of *Pentagonaster*, the characteristic of which lies in the "plaques marginales ventrales portant chacune un piquant aplati." After citing the synonyms and literature he makes the following remarks on this species, which he now calls *Pentagonaster (Stellaster) equestris* ['76, p. 42]: "Un seul exemplarie desséché, sans localité, dans la collection du Jardin des Plantes; il provient de la collection MICHELIN. Le type de GRAY est originaire du Japon. On trouve aussi cette espèce, suivant le docteur LÜTKEN, dans la mer de Chine méridionale et le détroit de Formose."

Only two species of *Stellaster* were obtained by the Challenger, *St. incei*

---

1) *Nectria, Goniodiscus*, etc.

(syn. *St. belcheri* GRAY and *St. gracilis* MÖBIUS) and *St. princeps*, the latter a new species. SLADEN, however, makes some general remarks on the *Pentagonasteridæ*, from which the following extract may be made as bearing on this species ['89, p. 261]:

"I have substituted the generic name *Ogmaster* for that of *Dorigona*. The starfish described by GRAY in 1866 under the name of *Dorigona reevesii* is the same species as that previously described by MÜLLER and TROSCHEL in 1842 under the name of *Goniodiscus capella*. In 1865 VON MARTENS placed this form in a subgenus to which he gave the name *Ogmaster*, ranking it under *Goniaster*. The claim of this form to generic recognition has since been admitted, and it follows in my opinion that the name of the starfish in question should therefore be *Ogmaster capella* (M. & T.) VON MARTENS. (Its synonyma are *Dorigona reevesii*, GRAY, and *Goniaster mülleri*, LÜTKEN ; but not *Goniaster* (*Stellaster*) *mülleri* of VON MARTENS). The *Goniaster* (*Stellaster*) *mülleri* of VON MARTENS is a true *Stellaster*, which is so nearly allied to *Stellaster equestris* that I am unable to distinguish it, and I am therefore constrained to consider *Goniaster mülleri* as a synonym of that species. Both LÜTKEN and PERRIER have been in error in regarding VON MARTENS' form as identical with the species described by GRAY as *Dorigona reevesii*." Again, he makes the following remarks on the genus *Stellaster* ['89, p. 321]:

"I consider that this genus well merits independent recognition, and that in any case its structural characters do not justify its being regarded as a mere subdivision of the genus *Pentagonaster*, unless the limits of that genus are made much more extended than has ever yet been proposed by any classifier. To take such a step would be in my opinion to ignore altogether what should be recognised as the characters of a genus, and would almost necessitate a reversion to the old idea of a genus founded on single arbitrary characters rather than on the consideration of the affinities and differences of its morphological structure as a whole.

"*Stellaster* is in many respects structurally related to *Goniodiscus*, as limited by M. PERRIER, and I have placed them in the same sub-family."

Kœhler gives the following description of the specimens studied by him from the Sunda Isles ['95, p. 393]:

"Trois échantillons, dont deux entiers (R=60 mill., r=22 mill.)·et un autre plus grand dont les bras sont cassés (r=30 millim.).

"Von Martens, qui a donné une description très complète de cette espèce, a attiré l'attention sur les variations individuelles qu'elle présente et qui portent en particulier sur les pédicellaires et sur les cinq tubercules qui, dans la plupart des cas, ornent la face dorsale du test. Ces cinq tubercules existent sur les deux échantillons non endommagés; ils manquent chez le troisième, où je ne trouve aucune cicatrice attestant qu'ils s'y trouvaient.

"Les pores de la face dorsale s'étendent presque jusqu'au milieu du disque. Il en existe quatre rangées: une de chaque côté des plaques radiales médianes où ils sont disposés en groupes de deux ou trois, et une autre en dehors de la rangée de plaques suivantes où les groupes sont de quatre à cinq. Ces pores s'étendent jusqu' au milieu des bras; ils font complétement défaut dans les espaces interradiaux. Les pédicellaires sont irrégulièrement répartis sur la face dorsale; sur la face ventrale ils sont un peu plus grands, mais presque exclusivement limités aux plaques qui avoisinent le sillon ambulacraire. L'anus est central; les piquants du sillon ambulacraire sont disposés en deux rangées; l'interne en comprend, sur chaque plaque, huit ou neuf qui sont petits et disposés en éventail; les externes, au nombre de trois, sont grands, larges et inégaux. Le nombre des plaques marginales est de dix-neuf. Les piquants, que portent les plaques marginales ventrales, sont élargis et coupés carrément; ils se continuent jusqu' à l'extrémité des bras."

This species is again mentioned by Kœhler in the collection of the Indian Museum [:10, p. 79]:

"Station 239. 11° 44' 30" Lat. N., 92° 55' Long. E. Profondeur 55 brasses. Nombreux échantillons.

"Tous les exemplaires ont à peu près la même taille et le diamètre ne varie guère que de quelques millimètres: en moyenne, R=27 mm. et r=12.5 mm. Ils sont bien conformes aux dessins que Müller et Troschel ont publiés autrefois sous le nom de *Stellaster Childreni* et ils n'offrent pas de variations. Les plaques dorsales sont constàmment dépourvues de tubercules;

les pédicellaires valvulaires, toujours très petits, se montrent de préférence du côté dorsal sur les plaques voisines des marginales dorsales, et, à la face ventrale, sur les plaques qui font immédiatement suite aux adambulacraires.

" Quelques échantillons portent des Prosobranches parasites dont les uns, plus nombreux, appartiennent au genre *Thyca* et les autres au genre *Eulima*. Les parasites sont presque toujours fixés sur les plaques marginales ventrales ou sur la ligne de séparation des marginales dorsales et ventrales.

" Je me contente de donner ici quelques photographies représentant les parasites en place (Pl. VIII, fig. 7 ; Pl. X, fig. 4 ; Pl. XIII, fig. 5; Pl. XIV, fig. 5 et 6 ; Pl. XV, fig. 9). L'*Eulima* a provoqué sur deux exemplaires certaines déformations des plaques marginales dont les plus marquées s'observent chez celui qui est reproduit Pl. XIV, fig. 6. Je crois que les deux espèces auxquelles ces parasites appartiennent sont nouvelles et je me propose de les étudier en détail dans un autre travail."

I have examined six specimens of this species, all from the same locality. The measurements are as follows :

Specim.	r mm.	R mm.	R : r
1	24	68	2.8
2	26	60	2.3
3	26	62	2.4
4	26	67	2.6
5	27	72	2.7
6	28	68	2.4

In all of my specimens the abactinal surface is plane or very slightly convex, and the actinal is either plane or very slightly concave. The disk is very large and the arms are comparatively long and slender, especially in the distal part. The interbrachial arcs are very open. The plates, both of the actinal and abactinal sides, are uniformly covered with such fine granules that they appear almost perfectly smooth to the naked eye. The papular pores form groups of three to five on the disk, mostly situated in the interspaces between the corners of the abactinal plates, but in the distal half of the arms they occur single. They are absent in the interradial lines.

Superomarginals.—These vary in number from 14 to 19 in my specimens. In the interbrachial arcs they present a lightly convex rounded surface which shows both on the abactinal side and the lateral margin ; further out in the arms the lateral and abactinal surfaces are more distinct from each other, although the edge is well rounded all through (Pl. XIV, fig. 219). In the interbrachial arcs they look almost rectangular in outline both from the abactinal and the lateral side. They are uniformly covered with very fine granules, which are somewhat coarser toward the borders, and with similar granules of the adjacent plates form in some cases an elevated distinct line between the plates. Most of the superomarginal plates bear a few valvate transversely elongated pedicellariæ, which may be very small and inconspicuous and easily overlooked, especially in wet specimens. They are, however, sometimes considerably elongated transversely, and in such cases are rather conspicuous.

Inferomarginals.—The inferomarginals are strictly coincident with the superomarginals, and present a more angular edge, especially in the interbrachial arcs, so that the lateral and actinal faces are very distinct (Pl. XIV, fig. 220). The granules covering them are exactly like those of the superomarginals, but the striking difference between the plates of the two series is that the inferomarginals bear each a movable spatulate spine articulated with the plate by means of a constricted base. These spines are in a line, all borne close to the angular edge of the plates, and may be absent from a few plates near the tip of the arms. On the first inferomarginal plate the spine is but little removed from the middle of the edge, but in all the rest it is attached near the distal end. Occasionally there are two spines on one plate. The inferomarginal plates bear transversely elongated valvate pedicellariæ, exactly similar in form to those of the superomarginals, but usually larger. Many of the more elongated ones show one or a few constrictions, showing clearly that they were formed by the coalescence of two or more pedicellariæ. There are indications that these constituent pedicellariæ are again formed by a linear fusion of two series of granules, each valve consisting in fact of many granuliform portions. In some cases there may be as many as three or four constrictions on a single pedicellaria. There are also some in which the constituent parts stand at a considerable angle to each other. The pedicellariæ of the marginals, both superior and inferior, are exceedingly low and scarcely rise above the general surface of the granules by which they are surrounded.

Adambulacrals.—The adambulacral plates are narrow and comparatively long and their boundary towards the ventrolaterals is not very distinct, so that the first series of the latter are apt to be mistaken for the adambulacrals. The armature is simple and very characteristic : the furrow series consists of 5–7 spines with rounded ends arranged in a palmate form and connected together close to the base by a web-like membrane (Pl. XIII, fig. 213) ; the spines are, however, usually so close together that the connecting membrane is not conspicuous. On the actinal surface at the outer end of the plate on a low tubercle and separated from the furrow series of spines by a furrow is a flattened, spatulate spine, similar in shape to those of the infero-marginal series but shorter, and usually turned away from the furrow. Occasionally this spine is doubled or even trebled (Pl. XIII, fig. 213) and in such cases the individual spines are smaller than when there is only one. Beginning with the second adambulacral plate, there is a forcipiform pedicellaria at the adcentral end of each plate in the longitudinal groove separating the furrow series of spines from the spatulate ones (Pl. XIII, fig. 213, 214). Occasionally it is curved towards one side, but usually it is straight. From near the base of the arms on the adam-bulacral plates are in direct contact with the inferomarginal series.

Mouth-plates.—Viewed from the actinal side each mouth-plate is triangular in form, apposed to its fellow by its longest side and facing the first adambulacral plate by its shortest. On the furrow border of the plate there is a compact series of 7–9 stout, flattened, somewhat wedge-shaped spines (Pl. XIII, fig. 212). On the actinal surface, close to the furrow series, there are on each plate one to three stout spines similar in general form to the outer spines of the adambulacral plates but usually less stout. The rest of the actinal surface of the mouth-plates is covered over with flattened, rounded, polygonal granules, which are much coarser than those of the ventrolaterals.

Ventrolaterals.—The ventrolateral plates are well developed and very conspicuous. There are nearly forty of them in each interradius ; the individual plates are irregularly polygonal and those near the inferomarginal series are smaller than the more ad-central ones, those in contact with the adambulacral plates forming a distinct V-shaped series in each interradius ; a second similarly

shaped series can usually be distinguished, although much more
irregular than the former, and there are some half a dozen
plates between the second series and the inferomarginal
plates. The ventrolaterals are uniformly covered with very fine
granules, and the individual plates are separated only by shallow
grooves. The larger plates bear several valvate pedicellariæ which
are considerably elongated transversely (Pl. XIII, fig. 217). For
instance, on one of the larger plates of the V-shaped series in
contact with the adambulacral plates I counted as many as some
ten pedicellariæ ; it must, however, be remarked that the counting
of the individual pedicellariæ on such a plate is attended with
some difficulties, since several of the elongated pedicellariæ show,
as on the marginal plates, constrictions, and there are all degrees
of coalescence, so that the separation into individual pedicellariæ
is not always distinct (Pl. XIII, fig. 216). The smaller plates in
contact with the inferomarginal series are mostly destitute of
pedicellariæ.

Abactinal plates.—The abactinal plates are rounded-polygonal
and in all my specimens they can be distinguished into two groups
with tolerable precision, viz. radial and interradial plates. The latter
are larger and there are usually 4–6 plates in each interradius,
which are larger than the remaining plates near the superomarginal series, the whole forming a somewhat triangular group. The
radial plates are smaller than the larger interradial plates and
occupy the whole abactinal surface of the arms between the
marginals, and on reaching the disk they are continued on, with
diminishing breadth of the whole group, towards the centre, where
all meet together and form a group of central plates. In this
central group one can mostly, though not always, distinguish a

central plate and five radial plates surrounding it on all sides and forming the apices of the groups of radial plates just mentioned. In the same radial line with these five apical plates, one can usually make out a series of somewhat regularly square plates running through the whole or at least the greater part of the length of the arms. In the basal part of the arms there is one more series on either side of this radial series which terminates towards the centre just before reaching the apical plates. Between this last mentioned series, the superomarginals and the larger interradial plates above mentioned lie a number of small plates of irregular form. The abactinal plates extend mostly to the tip of the arms, and only the last one or two pairs of superomarginal plates are in contact with each other.

The abactinal plates are all uniformly covered with fine granules exactly similar to those of the marginal plates. The majority of them bear pedicellariæ similar in form to those of the superomarginals, but usually so small as to be entirely inconspicuous (Pl. XIII, fig. 215).

The tubercles on the abactinal surface mentioned by several writers as a characteristic of this species are variable in number and position. In what may be looked upon as a typical case, there are five tubercles in all, one for each radius, and all equidistant from the centre of the disk. In one or two of the specimens the tubercles have been abraded, but their position and number can be told, as they leave hollow traces behind. In one specimen there are only three of these tubercles, while in another there are as many as twenty. In the latter case five of the larger tubercles are situated as in the typical case and the remainder lie mostly within the pentagon formed by the former and only one outside it. In a

third specimen with thirteen tubercles the pentagon formed by
the largest five is very regular and the remaining eight all lie
within it.   These conditions bring this species nearer to *St. belcheri*,
but there is no doubt in my mind that the latter is distinct from *St.
equestris*, as the ambulacral armature of *St. belcheri* is according
to LUTKEN's description ['71, p. 23 & Pl. V, fig. 3] very different
from that of the present species.   The individual tubercles are
mostly short-cylindrical in form, but in some cases, especially
where they are more numerous, the larger ones are more or less
elongated transversely, and are sometimes clearly seen to be formed by the coalescence of two tubercles.   The tubercles are to be
regarded either as enlarged granules or as a fused group of them.

The anus is very distinct and is situated in the interradius
next the madreporite, proceeding clockwise, close to the central
plate.   It is surrounded by eight to ten granules larger than those
of the general abactinal surface.

Madreporite.—The madreporite is quite large, though not conspicuous, owing to the fact that its surface is plane and perfectly
level with the surrounding plates.   It is usually more or less
elliptical or rhomboidal in outline and either is covered with
shallow radiating grooves, or the grooves are confined to the
periphery and the more central part is covered with numerous
separate pores.

Terminal plate.—The terminal plates are large and conspicuous,
elliptical or squarish in form, and is covered with fine granules
exactly similar to those of the superomarginals.

Locality.—All my specimens are from Usuki, Prov. Bungo.
Collected by Mr. T. TERAZAKI; no record of depth.   Specimens in
S. C.

Calliaster childreni GRAY.

The genus *Calliaster* was first set up by GRAY with the following diagnosis ['40, p. 280]:

"Body 5-rayed, with flat immersed ossicula armed with flat-based deciduous conical spines, and without any 2-lipped slits on either surface."

Gray then describes a single species ['40, p. 280]:

"*Calliaster childreni*, GRAY. Grey, back slightly convex, with a centre, a ring and 5 radiating lines of small spines; rays slender, tapering, as long as the width of the body; each of the marginal pieces with a central series of 3 distant spines.

"Inhab. ——."

In his "Synopsis" GRAY reproduces *verbatim* the generic diagnosis above quoted, and the specific description with trifling changes in punctuation. But he then adds, "Inhab. Japan," and also gives four figures, which are the only ones of this species that I know ['66, p. 9, 10, pl. 13].

In his "Révision" PERRIER gives a detailed description of the species, which he puts in *Pentagonaster*, regarding *Calliaster* as a subgenus ['76, p. 31]:

"Provient du Japon suivant Gray.—Au Jardin des Plantes, provenance inconnue, achat.

"Il nous paraît inutile de conserver le genre *Calliaster* de GRAY, après l'examen d'un échantillon unique que possède le Jardin des Plantes et qui provient d'un achat. Les *Calliaster* ne diffèrent en réalité des autres *Pentagonaster* que par leurs intervalles brachiaux profondément échancrés, caractère qui se rencontre déjà chez le *Pentagonaster Lamarckii* et qu'on ne peut en conséquence considérer comme générique. Les plaques dorsales sont plus petites que chez la plupart des autres *Pentagonaster* et supportent presque toutes un piquant cylindrique à extrémité arrondie et dont le diamètre est environ la moitié de celui de la plaque, ce qui, joint à sa forme, donne à l'animal une certaine resemblance avec les *Hippasteria*. Il diffère de ces derniers par l'absence complète de pédicellaires valvulaires. Ce qui distingue surtout cette espèce, ce sont les longs piquants mousses et cylindriques que portent la plupart des plaques ventrales et qui forment une double rangée auprès du sillon ambulacraire. Ces épines sont faciles à detacher, caduques, comme cela arrive d'ailleurs aux grosses épines coniques du *Pentagonaster semilunatus*, mais elles ne sont pas mobiles sur la plaque qui les porte. Je considère cette disposition comme d'autant moins propre à caractériser un genre que chez certains individus du *Pentagonaster semilunatus* les tubercules des plaques ventrales présentent une tendence manifeste à se métamorphoser en épines. Voici du reste une description compléte du *Pentagonaster childreni* que j'ai à ma disposition.

"Longueur du plus grand rayon, 80 millimètres ; du plus petit, 35 millimètres ; envergure, 150 millimètres ; à l'état de dessication. Corps aplati ; dos formé de petites plaques irrégulièrement arrondies, entourées chacune d'une rangée unique de granules grossiers, fait qui se reproduit pour les plaques marginales, dorsales et ventrales, comme pour les plaques ventrales. Dans la région centrale du disque, la plupart des plaques se prolongent en une épine allongée légèrement conique, à pointe obtuse ; une rangée de ces épines s'étend le long de la ligne médiane jusqu'à l'extrémité des bras. La plaque madréporique pentagonale est situé au premier tiers du rayon interbrachial, à partir du centre du disque. Les plaques marginales dorsales sont sensiblement rectangulaires, leur petit côté étant dans le sens de la longueur des bras, fortement bombées ; elles diminuent légèrement à mesure qu'on se rapproche de l'extrémité des bras ; chacune d'elles porte cinq ou six gros tubercules de même diamètre que les épines du dos dont ils affectent quelquefois le forme—souvent un certain nombre de tubercules tendent à former une rangée la long de le ligne médiane longitudinale de la plaque. Le nombre de ces plaques dans chaque espace interbrachial est de vingt-deux, sans compter les impaires terminales, soit onze pour chaque bras.

"Sur la face inférieure, les plaques marginales sont plus grandes que sur la face dorsale, rectangulaires au sommet de la concavité de l'angle interbrachial, carrées vers le milieu des bras ; elles diminuent plus rapidement en se rapprochant du sommet de ces derniers, de sorte que leur nombre est de vingt-quatre au lieu de vingt-deux ; elles portent également huit ou dix piquants très irrégulièrement disposés et plus longs que ceux des plaques dorsales.—Les plaques de la face ventrale sont en général polygonales et portent chacune une épine cylindrique trés allongée (3 à 4 millimètres de long sur moins de 1 millimètre de diamètre), facile à briser à sa base.

" Chacune des plaques qui bordent le sillon ambulacraire porte deux de ces longues épines placées l'une derrière l'autre et de plus dans le sillon lui-même cinq épines égales, un peu aplaties et faisant à peine saillie en dehors du sillon.[1)]

" Il n'y a point de pédicellaires.

" La provenance de l'individu de muséum de Paris est, inconnue. Ceux que possède le British Museum viennent du Japon."

VERRILL considers *Calliaster* to be distinct from *Pentagonaster*, *Tosia* and *Hippasteria*. He says [VERRILL, '99· p. 149] :

" *Calliaster* proposed by GRAY, in 1850, for the single type-species, *C. childreni*, is very distinct from the genera already named, not only on account of the spinose plates of both surfaces, but also by reason of its very different adambulacral spines."

I have not seen any specimen of this species.

Calliderma emma GRAY.

The genus *Calliderma* was set up by GRAY with the following diagnosis ['47, p. 197] :

" Body flat, five-sided, rays rather elongated; attenuated end only formed of the marginal plates. Ossicules all minutely granulated; the dorsal ossicules flat-tipped, six-sided, some with a larger, globular, central tubercle-like granule. The marginal ossicules broad, gradually becoming smaller near the tip, short-edged, minutely granular, those of the upper and lower series alternating; the edge of the upper ones with some indistinct spines on the margin, the lower ones with scattered mobile spines on the oral surface. The ossicules of the oral surface three-, four-, or six-sided, granular, with

---

1) " C'est là un fait assez caractéristique que la disproportion entre les épines du sillon ambulacraire et celles des plaques qui les bordent. Dans d'autres espèces de *Pentagonaster* les épines des sillons ambulacraires et les granules des plaques affleurent, au contraire, au même niveau, de sorte qu'il semble y avoir plusieurs rangées d'épines dans le voisinage des sillons ambulacraires."

one (rarely two) central, compressed, acute, mobile spines. The ambulacral spines very small, close, fourteen or sixteen on each ossicule, forming a rounder group, with two or three series of large, scattered, mobile, acute spines on the outer side.

"This genus resembles *Stellaster*, but differs from it in the oral surface being furnished with scattered spines.

"There is a fossil species very like the one here described found in the chalk, and figured in Mr. DIXON's work on the fossils of Worthing, which I propose to call *Calliderma Dixonii*. There are probably several other fossil species from the same locality; they have been referred to the genus *Tosia*, but the ossicules are granular and the oral surface spinose.'

Calliderma emma is then described ['47, p. 193]:

"Flat, pentangular, the sides concave, the arms elongated, produced, tapering to a fine point, about two-thirds the length of the diameter of the disc. The dorsal ossicules six-sided, regular, flat-topped, covered with minute roundish granules; the central granules of the central ossicules and those down the centre of the arms larger, globular, tubercular-like. The margin sharp-edged, concave in the centre; the ossicules of the upper and lower series alternating, minutely granular, with one or two larger subspinose granules on the middle of the upper margin. Marginal ossicules about fifty on each surface on each side, the lower series with scattered, acute, compressed spines on their oral side.

"The ossicules of the oral side four- or six-sided, rather irregular, minutely granular, each armed with a central, compressed, acute, mobile spine.

"Inhab. ——?

"This species most nearly resembles a fossil found in the chalk, which has hitherto been referred to the genus *Tosia*, and figured in Mr. DIXON's forthcoming work on the fossils of Worthing.

"I have named this fine species in compliment to my daughter Mrs. J. P. G. SMITH, who before her marriage commenced a series of plates to illustrate a monograph of this genus."

In his "Synopsis" GRAY reproduces the above description with a few trifling verbal improvements ['66, p. 6]. It is not necessary to quote them here, but it must be noted that in this work he gives two figures of this species [*l. c.* pl. 15], which are the only ones in my knowledge.

PERRIER places this species in the genus *Pentagonaster*, regarding *Calliderma* as a subgenus. He adds some important remarks ['76, p. 41]:

"Il n'existe au British Museum qu'un seul échantillon desséché et en assez médiocre état de cette magnifique espèce. C'est le type de GRAY, qui en a donné une description aussi compléte qui on puisse le désirer dans son *Synopsis* et une bonne figure. Il serait superflu de revenir sur cette espèce, relativement à laquelle nous n'ajouterons que ceci : c'est qu'à l'intérieur de la face dorsale le pavé d'ossicules qui la constitue entièrement est remplacé par un réticulum à mailles hexagonales parfaitement régulières et allongées transversalement. Dans chaque maille se trouve un pore unique peu visible à l'extérieur.

"La provenance de l'individu du British Museum est inconnue ; mais j'ai vu dans la collection de M. COTTEAU, à Auxerre, un individu de taille plus petite qui était arrivé chez un marchand dans une boîte d'insectes du Japon. Il est probable qu'il provenait, lui aussi, de cette contrée."

In his work of 1894, however, the same writer accords *Calliderma* the rank of a genus [PERRIER, '94, p. 337].

This species is not represented in any of the collections studied by me. If it really occurs in Japanese waters it is probably to be looked for in the Ryukyu Islands.

It may be added in passing that a second species of this genus (*C. spectabilis*) has been described [FISHER, : 06, p. 1058] from the Hawaiian seas.

Ogmaster capella (MÜLLER & TROSCHEL).

The occurrence of this species in Japan is questionable. I have brought together its descriptions by previous authors, relying principally on the conclusions arrived at by SLADEN as quoted above under *Stellaster equestris* (*vide svpra*, p. 420) and VERRILL ['99, p. 185]. *Ogmaster capella* appears to have been first described by MÜLLER and TROSCHEL ['42, p. 61]:

" *Goniodiscus capella* Nob. nov. sp.

"Der grosse Durchmesser ist mehr als doppelt so gross wie der kleine. Die Winkel zwischen den Armen sind ausgerundet. Die Arme sehr schlank. Randpapillen 6 auf einer Platte, gleich hoch; nach aussen davon stehen einige sehr niedrige Höckerchen, mehrere auf einer Platte, meist 3. Die Bauchplatten sind platt, polygonal, schwach granulirt; von den Randplatten sind die in den Winkeln breiter als lang, besonders die unteren; gegen die sehr spitzen Enden der Arme nimmt die Breite der Platten so ab, dass sie zuletzt länger als breit sind. Die oberen Randplatten (13 an jedem Arme) sind auch am Ende der Arme nicht länger als breit. Die Randplatten sind überall auch am Rande nackt, aber gegen das Ende der Arme stellt sich auf ihrer Oberfläche eine sehr sparsame Granulation ein. Die Enden der Arme sind übrigens so spitz, dass die 5 letzten dorsalen Randplatten jeder Seite sich berühren. Die Rückenplatten sind glatt, hexagonal, berühren sich dicht, und haben nur einzelne Poren zwischen sich. Sie bilden gegen die Arme hin sehr regelmässige Reihen. Die Rückenplatten sind etwas dichter granulirt, als die Bauchplatten. Die Madreporenplatte liegt in der Mitte zwischen dem Centrum und dem Rande.

"Farbe : röthlich.

"Grösse : 3 Zoll.

"Fundort : China? Aus der Sammlung des Dr. v. D. BUSCH in Bremen durch Dr. A. PHILIPPI mitgetheilt.

"Wir sind nicht sicher, ob wir einen von GRAY beschriebenen flachen Seestern : *Paulia horrida* GRAY Ann. VI, p. 278 hieher ziehen können. Körper gebildet von granulirten stacheltragenden Platten an der Scheibe und am Rande, ohne klappenartige Pedicellarien. Farbe : braun. Fundort : Punta Santa Elena."[1]

DUJARDIN and HUPÉ describe it under the same name as follows ['62, p. 404]:

"Corps pentagonal, à côtés échancrés, et dont les bras, très-étroits, sont deux fois plus longs que le plus petit rayon du disque. Piquants du sillon ambulacraire tous égaux et au nombre de six sur chaque plaque : en dehors de ces piquants se trouvent aussi sur chaque plaque deux ou trois petits tubercules déprimés.

---

1) *Paulia horrida* has been minutely described and figured lately by LUDWIG [:05, p. 143].

"Les plaques ventrales sont aplaties, polygonales, faiblement granuleuses. Les plaques marginales, et particulièrement les inférieures, sont plus larges que longues dans les angles entre les bras, mais leur largeur diminue vers la pointe de ces bras, jusqu'à ce qu'enfin elles soient, au contraire, plus longues que larges.

"Les plaques marginales supérieures, au nombre de treize à chaque bras, ne sont point plus longues que larges à l'extrémité des bras. Toutes les plaques marginales sont nues jusqu'au bord, excepté vers l'extrémité des bras où elles sont couvertes de granules très-clairsemés ; les bras, d'ailleurs, sont tellement pointus que les cinq dernières plaques marginales de chaque côté sont contiguës sur le dos. Les plaques dorsales sont lisses, hexagones, et tellement rapprochées qu'il n'y a que des pores tentaculaires isolés entre elles ; leur granulation est un peu plus serrée qu'à la face ventrale, et elle forme des rangées très régulières vers l'extrémité des bras.

"La plaque madréporique est à égale distance entre le centre et le bord.

"Coloration rougeâtre. Dimension : largeur totale 81 mm.

"Habite les mers de Chine ? "

VON MARTENS sets up a new subgenus *Ogmaster* for this species, characterised by having "die fünf innersten Bauchplatten an ihrer adoralen Seite tief gespalten." The species is then referred to as follows ['65, p. 359] :

"*Goniaster (Ogm.) capella* MÜLL. TROSCH. sp. *Goniodiscus capella* MÜLL. u. TROSCHEL p. 61.

"Die Platten der Rücken- sowie der Bauchseite werden gegen den Rand zu zahlreicher und kleiner. Keine Randstacheln. Gleicht in Form, Farbe und Rückenbekleidung ganz auffallend dem eben beschriebenen *G. Mülleri*. Im Uebrigen siehe die Beschreibung von MÜLLER und TROSCHEL, in der die charakteristische Spalte der innersten Bauchplatten auffallenderweise nicht erwähnt wird.

"China?, v. D. BUSCH."

GRAY mentions it under the name of *Dorigona Reevesii*. The genus *Dorigona* is characterised thus ['66, p. 7] : "*Dorigona*. Body depressed, 5-rayed, smooth ; the dorsal and oral disk covered with many smooth, flat, polygonal squares ; the marginal ossicules without any mobile spine." "*Dori-*

gona Reevesii (T. 7. f. 3). Inhab. China or Japan ; common in boxes of insects brought from China and Japan ['66, p. 7] ." *Dorigona longimana =
Astrogonium longimanum* MÖBIUS ['59, p. 7] is then referred to for comparison. Under *Hosea spinulosa* GRAY refers to *Goniodiscus capella* of MÜLLER and TROSCHEL for comparison ['66, p. 9].

Under the name of *Goniaster Mülleri* v. MARTENS LÜTKEN (syn. *Dorigona Reevesii* GRAY) has some important remarks on this species ['71, p. 248] :

"Godt afbildet hos GRAY ; hvis Fremstillingen er i naturlig störrelse, er Original-Exemplaret ikke lidt större end de foreliggende, der i denne Henseende stemme med Berliner-Museets. Lige saa lidt som GRAY seer jeg Spor til Randpigge hos dem ;[1] i Henseende til Ambulakral-papillerne er Forholdet omtrent som hos *G. Belcheri*. Manglen af Ryg-knuder saavel som af Pedicellarier, de regelmæssige, flade, sexkantede Rygplader, kun adskilte af enkelte Porer i Hjörnerne og Kanterne, og de talrige (13–22) Bugplader inden-for Armvinklerne paa Bugfladen ka-rakterisere i övrigt denne Art, med Hensyn til hvilken jeg vil indskræn-ke mig til at henvise til den en-gelske Forfatters Afbildning og den tydske Zoologs Beskrivelse.

"I Modsætning til de to nærmest foregaaende Arter[2] synes *G. Mülleri* at have Hjem sammen med *G. eques-tris*. GRAY angiver den at være al-mindelig i kinesiske og japanske In-sektensamlinger ; de to foreliggende

Well figured in GRAY ; whose re-presentation is in natural size, the original example is somewhat larger than the one before me, which in this respect agrees with that of the Berlin Museum. As little as GRAY do I see trace of marginal spines in this ;[1] in respect of the ambulacral papillæ the arrangement is nearly as in *G. Belcheri*. The absence of dorsal tubercles as well as of pedicellariæ, the regular, flat, hexagonal dorsal plates, separated only by single pores at the angles and sides, and the numerous (13–22) ventral plates in-side the arm angles on the ventral side, besides other things, characte-rise this species, in consideration of which I will confine myself to referr-ing to the figure of the English author and the description of the German zoologist.

In contradistinction to the two immediately preceding species[2] *G. Mülleri* seems to hail from the same place as *G. equestris*. GRAY states it

---

1) "Derimod bedder det hos v. MARTENS, l.c. S. 359: 'Von den unteren tragen bei weitem nicht alle' (altsaa dog nogle) 'einen Stachel'."

2) *Stellaster tuberculosus* and *Stell. belcheri*.

1) On the contrary it is stated by v. MAR-TENS, l.c. S. 359: 'Von den unteren, tragen bei weitem nicht alle' (therefore some at any rate) 'einen Stachel.'

Exemplarer hidröre netop fra en saadan Kilde, og Berliner-Museets var fra Japan. Ligesom *G. _equestris* er den hvid (farvelös).

" Som anden Art af sin Slægt *Dorigona* (der vel allerhöjst vil kunne gjöre Krav paa at være et kunstigt Afsnit indenfor *Goniaster*) opförer GRAY (l.c.) *D. longimana* c: *Astrogonium longimanum* MÖBIUS (Neue Seesterne, t. I, f. 5–6), og det er vel muligt, at denne Art, der udmærker sig ved, at Rygrandpladerne beröre hinanden i hele de lange Armes Længde, saa at de egenlige Rygplader aldeles ikke komme til Udvikling paa Armene, her vilde faae en ret naturlig Plads.—Der er i övrigt, som jeg tidligere har udtalt,[1] og som ogsaa senere er erkjendt af v. MARTENS,[2] identisk med *A. Souleyeti* DUJ. HUPÉ (Malakka-Strædet), hvilken sidstnævnte Form af GRAY med Uret er opfattet som Afart af *G. equestris*.[3] —Jeg har ligeledes tidlige paavist,[4] at MÖBIUS's *Stellaster sulcatus* (l.c., t. IV, f. 1–2) er en *Archaster*; at den tillige er identisk med den japanske

to be common in insect collections from China and Japan ; the two examples lying before me are from an exactly similar source, and that of the Berlin Museum was from Japan. Like *G. equestris* it is white (colourless).

As the second species of his genus *Dorigona* (which can perhaps at most claim a place as an artificial division inside *Goniaster*) GRAY mentions (l.c.) *D. longimana* = *Astrogonium longimanum* MÖBIUS (Neue Seesterne, t. I, f. 5–6), and it is quite possible that this species, which is distinguished by the fact that the dorsal marginals touch each other through the whole length of the long arms, so that the dorsal plates proper are not developed on the arms at all, would find its most natural place here.—It is besides, as I have previously expressed,[1] and as also later recognised by v. MARTENS,[2] identical with *A. Souleyeti* DUJ. HUPÉ (Malacca Strait), which last mentioned form is erroneously regarded by GRAY as a variety of *G. equestris*.[3]—I have also previously pointed out[4] that MÖBIUS' *Stellaster sulcatus* (l.c., t. IV, f. 1–2) is an *Archaster* ;

---

1) " Videnskab. Medd. 1864, S. 144."

2) " l.c. (1866) S. 86."

3) " Af PERRIERS Arbejde over Söstjernernes Pedicellarier seer man, at VALENCIENNES havde opfattet den som en *Archaster*. Der er virkelig en Del Lighed f. Ex. mellem *Goniaster Mülleri* og *Archaster hesperus*, og en slig Feiltagelse kunde deri finde sin naturlige Forklaring og Undskyldning, hvis· *G. longimanus* er en '*Stellaster*' eller en nærbeslægtet Form."

4) " Videnskab. Medd. 1864, S. 136–38."

---

1) " Videnskab. Medd. 1864, S. 144."

2) " l.c. (1866) S. 86."

3) From PERRIER'S work on the Pedicellariæ of the starfishes it is seen that VALENCIENNES had regarded it as an *Archaster*. There is in fact some resemblance e.g. between *Goniaster Mulleri* and *Archaster hesperus*, and such an error could find a natural explanation and excuse, if *G. longimanus* is a " *Stellaster* " or a nearly allied form.

4) " Videnskab. Medd. 1864, S. 136–38."

A. hesperus M. Tr., derom havde jeg vel en stærk Formodning, som jeg dog ikke fandt det rigtigt at udtale, da jeg savnede Original-Exemplarer af *A. hesperus* til Sammenligning ; da disse have staaet til v. Martens's Raadighed, har han kunnet overbevise sig om deres fuldstændige Identitet[1]"

that it is at the same time identical with *A. hesperus* M. Tr. from Japan, I had a very strong suspicion, although I did not find myself justified to express it, since I did not have original examples of *A. hesperus* in the collection ; as these have stood at v. Martens' disposition, he has been able to convince himself of their perfect identity.[1]

Perrier mentions this species under the name of *Pentagonaster* (subgenus *Dorigona* Gray) *Mülleri* and thinks that it is identical with the *Goniaster* (*Stellaster*) *Mülleri* of von Martens. Sladen, however, thinks this conclusion to be erroneous (*vide supra*, p. 420). Perrier makes the following remarks ['76, p. 44] :

" Cette espèce, facilement reconnaissable d'après la figure que Gray en a donnée, est remarquable en ce qu'elle ne porte ni granules, ni piquants, ni pédicellaires, bien que sa physionomie soit celle des *Stellaster*. Elle se distingue bien nettement du *Pentagonaster longimanus*, Möbius sp., parce que les plaques marginales de ses bras sont séparées en dessus par une rangée de plaques qui manque chez cette dernière espèce, où les plaques marginales constituent à elles seules la face dorsale des bras. Les plaques marginales sont ici au nombre de trente-deux[2] pour chaque côté du corps, presque carrées ; des pores isolés se voient à chacun des sommets des plaques dorsales et la plaque madréporique est située à une distance de $\frac{1}{3}r$ du centre du disque. Les piquants ambulacraires sont disposés sur deux rangées ; chaque plaque interambulacraire porte cinq piquants de la première rangée et deux ou trois de la seconde.

" Mers de Chine ou du Japon.—British Museum, un seul exemplaire desséché."

Leipoldt mentions this species from the " Vettor-Pisani " collection and remarks as follows ['95, p. 649] :

1) " l c. (1865) S. 353."
2) " L'exemplaire décrit par von Martens n'en a que vingt-quatre."

" Drei Exemplare von Assab aus einer Tiefe von 10–15 m. Die Exemplare sind sämmtlich klein (das grösste hat R=25.5 mm, r=9 mm, das kleinste R=17 mm, r=6 mm); die Anzahl der Randplatten beträgt bei allen drei Exemplaren 13 für jede Armseite. *Ogmaster capella* war bisher nur von China[1] und Japan bekannt."

VERRILL makes some passing remarks on this species under *Nymphaster* ['99, p. 185].

KŒHLER mentions it from the collection of the Indian Museum with one figure [:10, p. 79] :

"Station 226. 14° 36' 00" Lat. N. 96° 23' 00" Long. E. Golfe de Martaban. Profondeur 67 brasses. Quelques échantillons.

"Station 239, 11° 49' 30" Lat. N. 92° 55' Long. E. Profondeur 55 brasses.

" Dans les plus grands individus, R_i=50 mm.

" GRAY (66, Pl. VII, fig. 3) a publié une figure suffisante de la face dorsale ; le dessin de la face ventrale est moins satisfaisant. Je donne ici (Pl. III, fig. 12) une photographie de la face ventrale d'un exemplaire de la Station 226."

Anthenea pentagonula (LAMARCK).

This species is stated by GRAY to occur in Japan. It is said to be identical with *Asterias pentaganula* of LAMARCK, which is described as follows ['16, vol. ii, p. 554] :

"Astérie pentagonule. *Asterias pentagonula.*

" *A. inermis, orbiculato-pentagona ; angulis, brevibus, reflexis, emarginatis: paginœ inferioris canaliculis latis, ad margines articulato-plicatis.*

" Mus. n°.

"Habite....Les mers australes ? PÉRON et LE SUEUR. Cette espèce singulière ne tient nullement à l'astérie parquetée par ses rapports, et néanmoins elle est aussi simple, presque discoïde, et n'a que cinq angles courts,

---

1) " Das Vorkommen von *O. capella* bei China ist nach MÜLLER und TROSCHEL sowie nach SLADEN frnglich, wenigstens ist bei Beiden der Fundort mit einem Fragezeichen versehen."

réfléchis en dessus. Son dos est aplati, non parqueté, couvert de papilles courtes. Largeur, huit à dix centimètres."

De BLAINVILLE ['34, p. 237] simply mentions this species, referring to LAMARCK.

The genus *Anthenea* was set up by GRAY and characterised as follows ['40, p. 279] :

"Body 5-rayed, chaffy, with immersed elongated tubercle-bearing ossicula; margin with regular rows of large tesseræ ; both surfaces (especially the under) scattered with large 2-lipped pores."

The species is then described as follows ['40, p. 279] :

"*Anthenea chinensis*, GRAY. *Asterias chinensis*, GRAY, Brit. Mus. Back obscurely netted, rather chaffy, with scattered truncated tubercles in rather diverging lines; marginal plates not tubercled; rays broad, half the length of the width of the body.

"Inhab. China, Japan. *J. REEVES, Esq.*

"See also SEBA, iii. t. 6. f. 5, 6. (*Ast. tessellata*, var. A. LAM.). Similar, but the dorsal tubercles are larger and angular."

MÜLLER and TROSCHEL refer it to their genus *Goniodiscus* (with the synonyms: *Paulia, Randasia, Anthenea, Hosia* GRAY) and describe it as follows ['42, p. 57] :

"*Goniodiscus pentagonulus* Nob.

[Synonyms: LAMARCK, GRAY.]

"Der kleine Halbmesser verhält sich zum grossen wie $1 : 1\frac{1}{2}$—$1\frac{3}{4}$. Zahl der Randplatten 12–13 an jedem Arme. Die Winkel zwischen den Armen flach ausgerundet. Furchenpapillen in drei Reihen : in der innern 5 auf jeder Platte, so hoch wie die der zweiten; in der zweiten 3, von denen die mittlere die grösste; nach aussen von dieser Reihe noch eine niedrige unvollständige. Die Granula der Bauchseite bilden kleine Cylinderchen ; gegen den Rand hin werden sie kleiner und gedrängter. Auf dem Rande selbst sind sie wieder eben so gross, wie auf der Bauchseite. Die Granula der Rückenseite sind sehr klein und nicht gedrängt. Ausserdem auf der Rückenseite kleine cylindrische kurze Tuberkeln, welche nach der Spitze jedes Arms hin einen Zug bilden, in welchem sich meist drei mittlere Reihen unterscheiden lassen, von denen die mittlere unvollständig bleibt, und auch fehlen kann. Die Porenfelder sind sehr unregelmässig. Klappenartige Pedicellarien auf der Bauchseite, auf beiden Reihen der Randplatten und auf der Rückenseite. Diejenigen der Bauchseite sind sehr gross und nehmen die ganze Länge der Platten ein, auf denen sie stehen; auf den unteren Randplatten sind sie schon kleiner, aber doch grösser als auf den obern Randplatten und auf dem Rücken.

"Farbe: auf dem Rücken röthlichblau.

"Grösse: bis 6 Zoll.

"Fundort: China. Im Museum zu Berlin, Paris und Leyden."

PERRIER gives "*Astrogonium articulatum*, VAL.—Coll. Mus. (pars)" as a synonym of this species [PERRIER, '76, p. 91].

DUJARDIN and HUPÉ give the following description ['62, p. 401]:

"Goniodisque pentagonule. *Goniodiscus pentagonulus.*—MÜLLER et TROSCHEL.

[Syn.: LAMARCK, GRAY, MÜLLER and TROSCHEL.]

"Disque pentagonal, à angles très-courts, le plus grand rayon du disque, les bras compris, égalant une fois et demie le plus petit. Plaques marginales au nombre de douze ou treize à chaque bras; les angles, entre ces bras, très-peu échancrés, le sont presque carrément. Les piquants du sillon ambulacraire forment trois rangées, ceux de la rangée interne, au nombre de cinq sur chaque plaque, égalent en hauteur ceux de la dernière rangée, qui sont au nombre de trois seulement, et dont l'intermédiaire est le plus grand. En dehors se trouve encore une rangée plus basse et incompléte. Les granules de la face ventrale forment de petits cylindres qui sont plus petit et plus serrés près du bord, sur lequel se rencontrent d'autres granules plus gros.

"Les granules de la face dorsale sont très-petits et peu serrés, on y voit encore de petits tubercules cylindriques, courts, qui, vers l'extrémité des bras, forment une bande dans laquelle on distingue surtout trois rangées moyennes.

"Les aires des pores tentaculaires sont très-irregulières. Des pédicellaires bivalves se trouvent à la face ventrale, sur les deux rangées de plaques latérales et à la face dorsale ; celles de la face ventrale sont très-grandes et occupent toute la longueur des plaques qui les portent ; celles des plaques marginales inférieures sont plus petites, mais dépassent cependant celles des plaques marginales supérieures ou dorsales.

"Coloration rougeâtre en dessus, plus claire en dessous. Dimension : argeur totale 162 mm.

"Habite les mers de Chine (Mus. Paris)."

PERRIER ['76, p. 91] is of the opinion that the *Goniaster articulatus* (LINN.) of LÜTKEN ['64, p. 147] is a synonym of the present species, but this is contradicted by later authors [DE LORIOL, '84, p. 643 ; SLADEN, '89a, p. 327 ; KŒHLER, '95, p. 392].

In the "Synopsis" GRAY reproduces his description of 1840 above quoted, and gives the following synonyms ['66, p. 9] : "*Asterias chinensis*, GRAY, Brit. Mus. *Asterias pentagonula*, LAMK. ii. 554. *Goniodiscus pentagonula*, MÜLL. TROSCH. *Ast.* 57, t. 4. f. 2. *Goniaster articulatus*, AGASS. Mus. Paris ! " For comparison he refers to " *Goniodiscus scaber* MÖBIUS, Abhandl. iv. 1860, t. 3, f. 4 " and adds "appears to be a nearly allied species."

In his work on pedicellariæ PERRIER describes this species as follows ['69, p. 87] :

" *Goniodiscus articulatus*, ED. P.—*Astrogonium articulatum*, VAL. (Coll. Mus.).—Le rapport du plus petit au plus grand diamètre est comme 4 est à 7 environ. Les bras sont nettement dessinés, et séparés les uns des autres par une échancrure profonde. La face dorsale est couverte de granules très-fins, et présente en même temps un grand nombre de petits tubercules irrégulièrement disséminés, et renflés à leur sommet qui est tronqué. La plaque madréporique est placée un peu avant le quart du plus petit rayon partant du centre du disque ; elle est ovale, et creusée de sillons qui vont en divergeant et en se divisant à partir du centre. Les plaques marginales dorsales sont au nombre de trente pour chaque bord ou quinze pour chaque bras, sans compter les impaires situées au sommet des bras. Elles sont couvertes de granules irreguliers, assez gros et non contigus. Chacune d'elles porte un ou deux petits Pédicellaires valvulaires. Ces plaques vont en diminuant vers l'extrémité des bras ; elles sont plus larges que longues. Il en est de même des plaques marginales ventrales, qui sont au nombre de trente-deux pour chaque bord où seize pour chaque bras, la dernière étant très-petite. Ces plaques sont couvertes de granules plus serrés et plus réguliers que ceux que présentent les plaques dorsales ; elles portent, en outre, deux ou plusieurs Pédicellaires valvulaires un peu plus grands que ceux du dos.

" Toute la face ventrale paraît formée de plaques entourées chacune d'une couronne de gros granules, et portant un Pédicellaire valvulaire qui occupe toute leur longueur. Vers le sommet de chacun des triangles découpés par le sillon ambulacraire, l'une de ces plaques est plus saillante, et a une tendance à former une sorte de corne ou d'arête. Les piquants des sillons ambula-craires sont sur deux groupes : ceux qui forment le rang interne sont disposés par groupe de cinq sur chaque plaque, ceux du milieu étant plus grands et plus forts. Les piquants du second rang sont plus courts, plus gros, tron-qués au sommet, et disposés par groupes de trois sur chaque plaque seule-ment. Entre les piquants du premier et du second rang, à l'angle interne de chaque pièce, on aperçoit en outre, comme chez les *Oreaster*, un Pédicel-laire en pince gros et court.

" Le plus fort diamètre des échantillons desséchés était de 18 centi-mètres.

" Deux individus de la collection du Muséum et deux acquis par lui dans la collection MICHELIN. Des îles Seychelles.

" *Nota.*—Un assez grand nombre de petits Pédicellaires valvulaires se voient sur la face dorsale."

As to the locality see PERRIER's correction in the extract immediately following.

In his " Révision " PERRIER has the following remarks on this species which he calls *Anthenea pentagonula* ['76, p. 91] :

" Cette espèce est-elle bien réellement l'espèce même de LAMARCK ? Il y a lieu d'en douter, si l'on considère que sa forme est bien moins penta-gonale que celle de l'espèce précédente[1] avec laquelle elle se trouvait con-

1) *Anthenea articulata.*

fondue au Muséum sous le nom d'*Astrogonium articulatum* ; mais c'est bien là l'espèce de MÜLLER et TROSCHEL, et, sur un simple doute, il y aurait plus d'inconvénients que d'avantages à changer son nom.

" Nous avons décrit cette espèce très-complétement dans nos *Recherches sur les Pédicellaires*, p. 87. Mais c'est à tort que nous l'avons indiquée comme provenant des îles Seychelles, cette indication se rapporte à l'espèce précédente.

" Trois individus, dont un, de Honkong, donné par le muséum de zoologie comparative de Cambridge (Massachusetts). Le nombre des plaques marginales varie en dessus de vingt-huit à trente pour chaque intervalle interbrachial ; de trente à trente-deux en dessous."

STUDER mentions this species from the " Gazelle " collection ['84, p. 37] : " *Anthenea pentagonula* LAM. Ein Exemplar aus der Meermaidstreet, NW.-Australien aus 5 Faden Tiefe."

It is also mentioned by BELL ['88, p. 384] from the Sea of Bengal.

SLUITER mentions " ein getrocknetes Exemplar von China " ['95, p. 55].

BROWN reports it from the Mergui Archipelago [:10, p. 32] :

" Locality.—XLI., Moskos Islands, 12 to 25 fathoms, rock and sand.

" A single dried specimen in which R=120, r=62 [mm.]. Number of marginal plates, 19. The arms are more acute than in the smaller specimens in the British Museum.

" Known from Hongkong, Madras, and N. W. Australia."

OREASTERIDÆ.

Oreaster modestus (GRAY.)

(Pl. XV, figs. 228–236.)

This species was first described by GRAY, and was placed by him in the same group of the genus with *Pentaceros franklinii* (= *O. nodosus*). The description, though brief, was accompanied by two figures, which place the identity of GRAY's species with the one I have in hand beyond doubt. The description runs as follows ['66, p. 6] :

" *Pentaceros modestus* (T. 9). Arms rather depressed, broad, not quite as long as the diameter of the body, with six or seven large convex tubercle along the middle of the upper surface; the hinder tubercle with a smaller tubercle on each side of it, forming an irregular ring of tubercles exterior to the dorsal ones; back with a circular series of five large ovate blunt spines. Inhab. —— ?"

For comparison GRAY refers to *Oreaster affinis* MÜLL. & TROSCH., *O. chinensis* MÜLL. & TROSCH., *O. tuberculatus* M. T., *Asterias mammillata* AUD. (= *O. mammillatus* M. T.), *O. verrucosus* M. T., *Asterias stellata* Mus. Tessin. (= *O. clavatus* M. T.), *O. carinatus, Asterias obtusata* LAM. (= *O. obtusatus* M. T.), *Asterias obtusangula* Lam. (= *O. obtusangula*), *O. regulus* M. T., *O. orientalis* M. T. and *O. gigas* LÜTKEN.

PERRIER, in his " Revision," regards *P. modestus* as a young form of *P. turritus* ['76, p. 56; *vide infra*, p. 477 under *O. nodosus*], but there is no doubt that it is a distinct species, as may be seen from the description given below.

BELL, in his revision of the genus *Oreaster* ['84], does not recognise this species, probably following PERRIER. SLADEN ['89, p. 346] also regards it as a synonym of *O. nodosus*.

I have only one specimen in alcohol, with r=29 mm. and R=81 mm., giving the radial ratio 2.8. The actinal side is plane, and the ventrolateral plates, though completely covered over with a coarse granulation, are perfectly distinct. The abactinal side of the disk, though elevated in the form of a pentagonal turret, is comparatively low, and the cross section of the arms, though triangular in form, is also much more depressed than in *O. nodosus*, and the marginal plates have a tendency to bulge out on the sides more than in that species (Pl. XV, fig. 228, 229).

Superomarginals.—There are 16–18 superomarginal plates in my specimen; they are comparatively large and project slightly more on the lateral margin than the inferomarginals; another point which is rather notable is the fact that two or three of the superomarginals in the distal part of the arms are larger and more projecting than the others, so as to call forth a somewhat irregularly moniliform appearance of the arms when seen either from above or below. The superomarginal plates are produced toward the abactinal side in the shape of a wedge, and are here separated from one another by pore areas lying between them. The plates are completely covered over with flattened polygonal granules, which are larger and more flattened in the central part of the plates and smaller, more rounded, and comparatively thicker in the more peripheral parts. On some of the plates, especially those nearer the tip of the arms, there are many minute granules between the larger granules. The superomarginals are entirely destitute of pedicellariæ.

Inferomarginals.—The inferomarginals are very distinct and regular. The first two plates next the interradial line are frequently smaller than the others, but the rest are subequal in size with the exception of the last four or five at the tip of the arms. The plates are nearly rectangular in form, with either a rounded or wedge-shaped border towards the ventrolateral plates. The granules are flattened and polygonal, and although those in the central part of a plate are larger than the more peripheral ones, the difference is not so conspicuous as on the superomarginals. Occasionally there are one or more granules on some of the inferomarginal plates, which are larger and more elevated than others ; these may be looked upon as rudimentary spines.

Adambulacral plates.—The adambulacral plates are small, there being about two and a half or three of them corresponding to one inferomarginal plate. The armature consists of two rows of spines, a row of granules and a pedicellaria. The inner row consists mostly of six spines; but since there are on many of the plates one or two very small, rudimentary spines at one end it is possible that in larger specimens there may be more spines. The spines are flattened and arranged like the fingers of the human hand, the one at the middle being longest and those on either side gradually decreasing in length. There is a small elongated pit at the base of each spine (Pl. XV, fig. 230). In the outer row there are mostly three spines, one very short thick and two large flattened ones; but on some plates the larger ones alone may be present. The short spine, when present, always lie at the adcentral end of the plate and may occasionally be represented by two or more smaller spines, or again it may be entirely absent, or, on the contrary, may be nearly as large as the other two spines, which have sharp flattened edges. On the outer side of the outer row of spines, between it and the first series of ventrolateral plates, is a row, sometimes irregular, of somewhat wedge-shaped granules, five or six in number, and separated from the outer row of adambulacral spines by a distinct groove, and from the first series of ventrolateral plates by a narrow groove (Pl. XV, fig. 231). At the adcentral end of each adambulacral plate, between the inner and the outer row of spines is a forcipiform pedicellaria of some size, occasionally with three jaws (Pl. XV, fig. 233, 234). It is sometimes situated exactly between two plates.

Mouth-plates.—The mouth-plates can not be observed from the surface, but the spines are conspicuous and are arranged in two rows. The inner row consists of seven or eight robust, prismatic spines with rounded ends; the spine at the oral end has the form of a triangular prism, and together with its fellow form the apex of the wedge-shaped figure formed by the oral spines; the two or three spines at the distal end are usually smaller than the others. In the outer row there are nearly always two spines, occasionally three for each plate, robust and prismatic in form, in the same line with the outer row of adambulacral spines, and on the outer side of the three most distal spines of the inner row, in close contact with them.

Ventrolaterals.—The ventrolateral plates themselves can not be seen from the surface, since they are completely covered over with flattened polygonal granules exactly similar to those of the inferomarginals; but the individual plates can be easily distinguished, being separated from one another by distinct grooves. They are of different size and shape; but a regular row of them can be distinguished on either side of the ambulacral furrows, reaching out to the tip of the arms, but becoming rather obscure towards the mouth. The plates of this series are rectangular in the middle part of the arms, but more or less roundish or elliptical towards the mouth, and many of them bear one or more, transversely elongated, valvate pedicellariæ. A second series of ventrolateral plates of irregular size and shape can be recognised in the arms, but this stops at some distance from the tip, and becomes merged into the general ventrolateral system at the base of the arms. Valvate pedicellariæ may occasionally be present on some of the plates of this series, but they are exceedingly rare. In the interradial areas two more series, both more or less irregular, are present outside the foregoing; and the rest of the interradial area is filled up with rounded or irregularly polygonal plates without any regular arrangement. A few of these near the mouth are usually larger than the others. Pedicellariæ are present on many of these plates (Pl. XV, fig. 232).

Abactinal skeleton.—The most prominent feature of the dorsal side is the presence of a carinal row of conical or ellipsoidal tubercles in each arm and a conspicuously large, ellipsoidal tubercle in each radius of the disk, the latter together enclosing a rather small central area. These adcentral tubercles have perfectly rounded or even flattened tops and are entirely covered over with flattened, polygonal granules, which are so closely set as to impart a smooth appearance to the tubercles. One or more of the granules at the top are sometimes larger and more elevated than the surrounding ones, hence it is possible that in some specimens there may be one or more spines at the top of these adcentral tubercles. On the outer side of each adcentral tubercle, close to it, there is a pair of smaller, conical tubercles, nearly as large as the last or the penultimate tubercle of the arms, covered over with polygonal granules and with a short, blunt, conical, immovable spine at the tip. In two of the interradii in my specimen, one member of the pair is wanting. The tubercles of the arms are conical in form, and gradually decrease in size towards the tip, there being usually five to seven of them in each arm. They are tipped with a blunt, conical immovable spine, which is, however, usually absent from the small terminal tubercle. Irregularities in the relative size of these tubercles may also be observed. Between these tubercles there are low rounded plates.

The central area enclosed by the five adcentral tubercles is bounded by elongated ossicles stretching between the tubercles and forming the sides of a regular pentagon; this pentagon is again divided into five equal triangular areas by as many depressed ridges radiating from the centre to each tubercle.

On either side of the lophial, or carinal tubercles, there may usually be distinguished two series of ossicles in each arm, the inner of which extends to the last tubercle, but the outer terminates at or before the penultimate one. These ossicles, as well as the depressed ridges of the central pentagon, and the spaces between them are entirely covered over with rounded or polygonal granules; the papular pores are confined to the inter-ossicular spaces. The individual pores are small but very distinct and impart a punctured appearance to the abactinal side. Each pore is usually guarded by a few exceedingly small, short spines.

There are two forms of pedicellariæ on the abactinal side. The more easily observable form is that of small forcipiform pedicellariæ, which are very numerous, especially in the poriferous areas, but are also present on the ossicles (Pl. XV, fig. 235). They project above the general surface and look like so many small prickles. The other form is that of more or less transversely elongated pedicellariæ, similar in form to those of the ventrolateral plates but smaller, which are less numerous than the other form and are confined to the surface of the ossicles (Pl. XV, fig. 236).

Madreporite.—The madreporic plate is tolerably large, nearly elliptical in outline, and is situated at the outer margin of the central pentagon, in the centre of the quadrangular aréa formed by two of the adcentral tubercles and one of their accessory tubercles. The surface of the plate is covered with inconspicuous radiating furrows.

Terminal plate.—The terminal plates are relatively very small, rounded in form, and entirely covered over with flattened granules exactly similar to those of the marginals.

Locality.—Kyamu-Saki, Okinawa Island; littoral.

Specimen in S.C.

Oreaster dœderleini, n sp.

·(Pl. XVI, figs. 244–251.)

This species is closely similar to *O. magnificus* in general form and appearance, but is distinguished from it by a slightly better development of the spines on the abactinal side, the presence of teat-like spines on some of the inferomarginal plates, the larger number of spines in the inner row of adambulacral armature, and the presence of three forms of pedicellariæ on the abactinal side, two of which are totally different from any found in *O. magnificus.*

The body is depressed, flat below and very slightly convex above. The cross-section of the arms is nearly semicircular in outline, and the arm-tips are upturned, as in most other species of the genus. The interbrachial arcs are entirely open. I have only one dried specimen in a tolerable condition, with $r = 73$ mm., $R = 157$ mm., giving the radial ratio 2.15.

Superomarginals.—There are 20–22 of these, and they are comparatively inconspicuous in surface view, although of tolerable size; they can just or hardly be seen from the abactinal side, except near the tip of the arms, where they appear on this side owing to the upturning of the tips. In the interbrachial arcs the plates are more or less rounded-polygonal, and the external, surface is nearly plane; along the sides of the arms the plates are more or less rectangular or pentagonal in outline and the external surface is lightly convex. The plates are entirely covered over with fine granules, and each plate bears some valvate pedicellariæ, which are but slightly elongated transversely or may sometimes be nearly round; there may be as many as ten or a dozen of them on a single plate, but the last four or five plates at the tip of the arms are usually entirely destitute of them.

Inferomarginals.—These are confined to the actinal side of the body and are generally larger than the corresponding superomarginals along the greater part of the arms, where they are pentagonal in outline. The six or seven plates in each interbrachial are are smaller than those on the sides of the arms, more or less mammiform and bear each a teat-like or bluntly conical spine at the top, which is united with the plate by a presumably immovable ball and socket joint. These spiniferous inferomarginals bear some very small valvate pedicellariæ, while the other plates, with the exception of a few at the tip of the arms, bear either round or more or less transversely elongated ones similar to those of the superomarginals. The number of pedicellariæ on one plate may vary from 2 or 3 to about 10. The granules are exactly like those of the superomarginals, except that those on the actinal side of the interbrachial plates are coarser, and are more like the granules of the adjacent ventrolateral plates. In each interbrachial arc there are intercalated between the two marginal series 3–5 ossicles, which are closely invested with granules exactly similar to those of the marginals, and hence are not conspicuous in surface view.

Adambulacral plates. — The adambulacral plates are nearly square in form when seen from the surface. The armature consists of three rows of spines and some pedicellariæ. The inner row is well in the furrow and consists of nine or ten slender, rather fragile spines of very unequal lengths. On the first adambulacral plate the tops of these spines form only a slight curve ; but the curvature becomes pronounced very soon, and from the fourth or fifth plate on these spines form almost a triangle with a rounded apex, the spines at either end being very short. The top of the largest spines of this row usually reaches the base of those of the next row, which consists typically of two or three stout, flattened, blunt spines, like those of the corresponding row in *O. magnificus*. These spines may be subequal or very unequal, and sometimes one or more of them may be so entirely out of the row that the latter becomes duplicated. The third row consists of one to three spines similar in shape to those of the second row, but usually smaller. On the adcentral border of each adambulacral plate there are one or more forcipiform pedicellariæ, and on some plates they may form a row reaching up to the outside of the third row of spines. They may be slightly curved or perfectly straight, and are usually smaller when there are more. There may be in addition one or more forcipiform pedicellariæ either in line with or outside the third row of spines.

Mouth-plates.—As is usual in this genus the mouth-plates themselves can not be seen from the surface, but the spines are very conspicuous, and consists of two rows. In the inner row there are 11–13 spines, of which the four or five at the mouth end are stout and prismatic in form and larger than the rest. The outer row consists of three or four spines, similar in shape to the larger spines of the inner row, and is confined to the abcentral end of the mouth-plates, on the interradial side of the smaller spines of the inner row.

Ventrolaterals.—The ventrolateral plates are as a whole more regular than in *O. magnificus*, and easier to distinguish from the surface. One regular row of plates can be made out on either side of the ambulacral furrow, extending from the tip of the arms to near the mouth, where it is merged in among the plates occupying the mouth corner. A second, somewhat irregular row

can also be made out outside the former, terminating at a
short distance from the tip of the arms. The rest of the ventrolateral area is covered with plates, the boundaries of which are
not so apparent as in the two series just mentioned, and this is
especially the case near the mouth. The ventrolateral plates are
covered with granules, which become coarser or even spiniform
towards the mouth, and bear numerous pedicellariæ, which may
be distinguished into three forms. Close to the mouth and also
more or less on either side of the ambulacral furrow there are
forcipiform pedicellariæ of tolerable thickness (Pl. XVI, fig. 246).
Some of them may be exactly similar in form to those on the
adcentral border of the adambulacral plates, but most of them are
more flattened. There may be some ten of these pedicellariæ at
the mouth corner in each interradius. The majority of the pedicellariæ on the ventrolateral plates are valvate and more or less
elongated transversely (Pl. XVI, fig. 247). The third form is intermediate between the two foregoing and have almost square
valves (Pl. XVI, fig. 248). The number of pedicellariæ on one
ventrolateral plate usually does not exceed three.

Abactinal side.—The abactinal plates are arranged in rows
parallel with the lateral borders of the body, except at the ends
of the arms where the plates are more or less rounded and closely
crowded together, leaving only very small pore areas between.
There is a distinct carinal series of plates in each arm, some of
which are raised into tubercles bearing either a spine or a teat-
like granule at the tip. At the adcentral end of each series is a
large, conspicuous, conical tubercle tipped with a pointed conical
spine, and spreading out at the base into eight trabecular processes, by means of which it is connected with the neighbouring

plates. These adcentral, or apical tubercles enclose a central depression, at the centre of which there are three conical tubercles, exactly similar in form to the apical ones but about one-third as large, and spreading out at the base into four or five trabecular processes, by which they are joined with one another and with the apical tubercles. The plate next the apical tubercle of each series is also raised into a conical tubercle similar in form to the former but somewhat smaller; and in some of the radii there is a small, spine-tipped tubercle between the apical and the next tubercle. Besides the two tubercles just mentioned, which are present in all the radii, there are some more (3–7) tubercles in the carinal series, which are much lower than either of the former, but are tipped each with a teat-like spine. The plates of the carinal series, which are not raised into tubercles, are either cruciform or lozenge-shaped, except those near the tip of the arms, which are generally rounded. On either side of the carinal series there are two more longitudinal series of plates in each arm, which are continued into the disk and are there connected with the corresponding series of the adjacent arm by curved series of plates running more or less parallel to the margin of the disk. There are some more plates without any definite arrangement between the outer of the disk series just mentioned and the superomarginals of the interbrachium. Some of the plates of the dorso-lateral series of the disk are raised into tubercles which are either conical or mammiform and are tipped with a conical or teat-like spine.

Both the abactinal plates and the spaces between them are closely covered over with fine granules, among which there are numerous pedicellariæ, which are of three forms. The most conspicuous of them are the small, vice-shaped ones, which are mostly, though not entirely, confined to the poriferous areas and stick out only like so many prickles above the general surface (Pl. XVI, fig. 249). The second form is that of small, valvate pedicellariæ, which are either round or slightly elongated in a transverse direction, and are entirely confined to the surface of the plates (Pl. XVI, fig. 250). The pedicellariæ of the third form are intermediate between the other two; they are very small, and are nearly round in outline when viewed directly from above (Pl. XVI, fig. 251); they are also confined to the surface of the plates.

The poriferous areas are very irregular in size and shape, corresponding to the irregular form of the plates; the pores are very distinct in the dried specimen, and vary greatly in number according to the size of the areas in which they are found. Near the tip of the arms there are only a few between the rounded plates, but in one of the larger areas near the centre of the disk there may be as many as sixty or more.

Madreporite.—The madreporite is comparatively large but not very conspicuous, owing to the flatness of its surface, which lies in a level with the general surface of the body; it is elliptical in outline, with very fine convoluted furrows on the surface, and is situated well out of the central pentagon formed by the apical tubercles.

Terminal plate.—The terminal plate is small, rounded in outline, and entirely covered over with granules exactly similar to those of the superomarginals.

Locality.—Koniya in Amami-Ôshima, *L*inschoten Islands. Littoral. The dried specimen was light brick-coloured.

This species is named in honour of Prof. L. DÖDERLEIN of

Strassburg, the well known pioneer in the study of Japanese echinoderms.

Specimen in S.C.

Oreaster magnificus, n sp.

(Pl. XV, fig. 237–240.)

I have only one dried specimen of this species in a rather unsatisfactory condition, but the essential characteristics can all be made out, and there is no doubt in my mind that it represents a new species. To judge from the descriptions of preceding authors it appears to present some points of resemblance to *O. australis* LÜTKEN and *O. westermanni* LÜTKEN, but the marginals are entirely destitute of tubercular formations, the ventrolaterals are covered with very coarse granules which have a slight tendency to become spiny, and only a few of the abactinal plates are produced into tubercular structures.

The specimen before me is very large, being the largest specimen of *Oreaster* that has come into my hands so far; r=95, R=220 mm., giving the radial ratio 2.3. The entire body is depressed, being level on the actinal, and lightly inflated on the abactinal side, in the centre of which there is a depression surrounded by five adcentral, radial tubercles. The interbrachial arcs are broadly open; in each there can be seen an almost straight stretch forming the margin of the disk. The arms are fairly long and perfectly rounded on the abactinal side, while on the actinal side it is plain, so that they would present a semicircular outline in cross section. The colour of the dried specimen is greyish brown, being lighter on the actinal side.

Superomarginals.—There are twenty-four of these, and they
are invisible from the abactinal side except in the last third of
the arms. In the interbrachial arcs they lie entirely on the actinal
aspect of the body and thence gradually emerge on to the lateral
margin towards the terminal part of the arms. As seen from
the surface the first superomarginal plate on either side of the
interradial line is larger than the others and has a prolongation toward the abactinal side, the upper end of which articulates
with one of the abactinal plates. The second plate has a much
shorter triangular prolongation·; a few plates at a short distance
from the apex of the arms have also similar prolongations, but
all the rest are destitute of any, and present a decidedly convex
surface towards the outside, especially in the basal part of the
arms. The superomarginals are closely covered over with rather
coarse, flattened granules and mostly bear one or a few transversely elongated, valvate pedicellariæ, which are very conspicuous
to the naked eye. On some of the plates there may be as many
as five of these pedicellariæ, and they may measure as much as
3 mm. in length.

Inferomarginals.—The inferomarginal plates are entirely confined to the actinal aspect of the body. In the interbrachial arcs
they are more or less elliptical or rectangular in form, but in the
arms they are either roundish or irregularly pentagonal. They
are mostly coincident with the superomarginals, but in the more
distal part of the arms they have a tendency to become alternate
with them. They are completely invested with granules exactly
like those of the superomarginals, and bear roundish or more
or less transversely elongated valvate pedicellariæ, which are
usually more numerous than on the superomarginals. They are

mostly simple, but there are also many compound ones, composed of two or three parts. There may be as many as seven pedicellariæ on one plate, counting the compound ones as single.

Adambulacral plates.—The adambulacral plates are short and nearly square in form, but can not be seen from the surface. The armature consists of three rows of spines and a few pedicellariæ, and the rows are well separated from each other, the top of the innermost series nearly coinciding with the base of the second row, and the outermost series being separated from the middle row by a distinct groove. The innermost row consists of eight or nine slender brittle spines arranged like the fingers of the hand, the middle one or ones being longest and the lateral ones shorter, the terminal ones on either side being frequently very short. Near the mouth the difference in length of these spines is less great, and on the first adambulacral plate they may form a straight comb. The second row of spines is separated from the first by a relatively large space, and there are either two or three of them, except on the first plate, which may carry four. The spines of this row are usually very unequal both in size and shape. When there are only two, they may be subequal in size and shape, or one of them may be considerably smaller than the other. When there are three they are, as a rule, of unequal size and shape, one, usually the middle, being considerably larger, and one considerably smaller than the remaining medium sized one. The well developed spines of this series are flattened and spatulate in form, narrower towards the base and with a wedge-shaped, more or less longitudinally grooved truncated end. When there are three the smallest one may be very slender and only half as long as the largest. The spines of the outermost row are in general similar in form to those of the middle row but shorter, and separated from the latter by a spacious groove. There are either two or three of these spines on a single plate ; they may be equal or very different in size and shape, and are generally more irregular in this respect than the spines of the middle row.

At the adcentral end of each adambulacral plate, between the first and second row of spines, there are one to three forcipiform pedicellariæ. They are frequently more or less curved, and when there is only one it is quite large, being about half as long as the well developed spines of the middle row ; but when there are as many as three of them they are usually much smaller, although occasionally there may be three pedicellariæ on one plate, all nearly as large as when there is one. The space between the middle and the outermost row of adambulacral spines is usually destitute of pedicellariæ, but occasionally there may be some, and in such a case their position does not appear to be constant. They are of the same form as those previously described, but their size is variable.

Mouth-plates.—The mouth-plates themselves can not be seen from the surface, but the spines are very conspicuous, and are arranged in two rows. In the inner row there are four thick, prismatic spines with blunt end at the mouth end, and four much smaller and shorter ones at the other, making eight in all. The surface of the! rounded ends of the thick spines is irregularly grooved, somewhat like the irregular markings of some old heart-wood (Pl. XV, fig. 239). The spine at the mouth end has always more or less the form of a triangular prism, and the rest more or less that of a rectangular prism. The outer row is present only at the abcentral end of the mouth-plate, and consists of only three spines, which are nearly as stout as those of the inner row and similar in form, although more irregular.

Ventrolaterals.—The ventrolateral plates are very numerous and small and very irregular both in size and arrangement. In the greater part of the arms there is a somewhat irregular, continuous series of plates between the adambulacral plates and the inferomarginals, besides a few small, intercalated plates which occur here and there; but in the disk the ventrolateral plates are entirely destitute of any regular arrangement. The plates of the arms are covered with granules similar to those of the inferomarginals, but in the disk the granules are very coarse and tend to become conical and spiny towards the corner of the mouth. More or less transversely elongated, valvate pedicellariæ are very numerous on the ventrolateral plates. They are of various sizes; some may be as long as 2 mm. but others may be only half as long or less. The smaller valvate pedicellariæ are particularly numerous in the disk on either side of the ambulacral furrows.

Abactinal plates.—The abactinal side consists of more or less irregularly stellate plates of comparatively small size, except in the most distal part of the arms, where the plates are either round or elliptical and are closely crowded. The processes of the stellate plates may or may not be united with similar processes of the neighbouring plates, and hence the meshes may not be always closed. There are five particularly large, conical tubercular adcentral plates enclosing a central depression, in which there are a few irregular ossicles. Each adcentral plate has some eight radiating processes, which are united with similar processes of the neighbouring plates, and although in the specimen before me the apical portion is abraded, there can hardly be a doubt that the plates were tipped each with a blunt spine. Between the adcentral plates there are as many interradial plates of an elongated, somewhat lozenge-shaped form. In line with the adcentral plates there is, corresponding to each arm, a carinal series of plates, which become gradually smaller towards the tip of the arms, and two to four of which directly following each adcentral plate are produced into spine-tipped tubercles, similar in form to it but smaller. Parallel with the carinal series there are two more rows on either side of the arms, which are continued in the disk across the interradial lines onto the corresponding series of the adjacent arms. The row next the carinal series is continued on to the adcentral plate, with one of the processes of which it is connected. In the interbrachial arc there is a short curved series of some elongated plates and another group of one or two plates between the superomarginal plates and the outer of the interradial rows of plates above mentioned. All the abactinal plates are closely invested with very small granules, and some of them may bear transversely elongated, valvate pedicellariæ.

The meshes between the abactinal plates are also covered over with small granules exactly similar to those of the plates, and bear numerous elliptical or transversely elongated valvate pedicellariæ, which are generally much smaller than those of the marginal or abactinal plates (Pl. XV, fig. 240). There are also in the meshes numerous papular pores, which are generally very distinct in the dried specimen.

Madreporite.—The madreporic plate is tolerably large and elliptical in outline. Its surface is on the same level with the general surface of the surrounding portions, and it is therefore comparatively inconspicuous. It is covered with irregular, discontinuous furrows, which do not show any definite arrangement.

Terminal plate.—The terminal plate is small and circular or elliptical in outline, and is covered with granules exactly similar to those of the superomarginals.

Locality.—Koniya in Amami-Ôshima, Linschoten Islands.

The colour of the dried specimen is dirty brown.

Specimen in S.C.

This species is nearly allied to *O. rouxi* Kœhler recently described from the Moluccas [Kœhler. :11, p. 272].

Oreaster nahensis, n. sp.

(Pl. XVI, figs. 241-243.)

This species appears to be closely related to *O. chinensis, O. decipiens* and especially to *O. reinhardti,* and combines some of the characters of *O. nodosus* and *O. modestus.* It is distinguished from *nodosus* by a lesser development of the abactinal tubercles, the relatively larger size of the ventrolateral plates and the number of spines of the adambulacral armature, and from *modestus* by a greater elevation of the abactinal side, the presence of only one form of pedicellariæ on the abactinal side and the relatively smaller size of the ventrolateral plates. From *O. reinhardti* it is distinguished by the notably greater development of the abactinal tubercles, the total absence of tubercular marginal plates, the presence of at least two series of lateral longitudinal ossicles on the sides of the arms as seen from the surface, the weaker development of the trabeculæ that connect the adcental tubercles, and the presence of two accessory small tubercles on the outer side of each adcentral tubercle.

I have only one specimen of this species, which was ori-ginally in alcohol, but has since been dried after treatment with caustic potash. The dimensions are r=36 mm., R=90 mm., giving the radial ratio 2.5.

Superomarginals.—There are twenty of these, and the plates are nearly triangular in form when viewed from the side, except in the most terminal part of the arms, where they are nearly square. They are completely covered over with flattened poly-gonal granules, which are slightly larger in the central part and smaller towards the periphery of the plates, and are entirely des-titute of pedicellariæ. As seen from the surface the plates are smaller in the interbrachial arcs than in the middle of the arms. When denuded of the superficial granules the plates of the interbrachial arcs are more than twice as broad as long, as meas-ured on well exposed plates. The lateral margin of the body is formed almost entirely by the superomarginal plates, and only in the terminal upturned portion of the arms do the inferomarginal plates show to any extent on the sides.

Inferomarginals.—The inferomarginals are entirely confined to the actinal side, and are rectangular or nearly square in the greater part of the arms. In the interbrachial arcs they are nearly twice as broad as long. They are entirely covered over with flattened granules exactly similar to those of the inferomarginals and are like them entirely destitute of pedicellariæ. When the granules are removed from the marginals, there is seen a series of small intercalary plates between the two series. There are ten or twelve of these in each interradius, but exceedingly small ones may sometimes be seen stretching out to near the tip of the arm.

Adambulacral plates.—The adambulacral plates are narrow and more or less concealed from view ; there are two or two and a half of them corresponding to one inferomarginal plate in the greater part of the arms. The armature consists of two rows of spines, an inner of smaller spines, mostly six in number, arranged like the fingers of the human hand, the middle ones longer than those towards the ends, and an outer of much stouter, prismatic spines, mostly two but sometimes three in number, of subequal length, and with apices either irregularly blunt or in the form of a wedge. When there are only two of these outer spines, there is usually a very short, granuliform spine at the adcentral end of the plate. At the same end of each plate, between the two rows of spines there is a tolerably large, forcipiform pedicellaria.

The foregoing description applies to the adambulacral plates near the mouth, but those which lie some distance away from it have mostly an additional row of five or six granules on the outer side of the outer row of spines, just as in *O. modestus*.

Mouth-plates.—The mouth-plates themselves can not be observed from the outside ; the armature consists of two rows. The inner consists of nine stout spines, of which the last five or six at the abcentral end are slightly smaller than the others, and the one at the mouth end has usually the form of a triangular prism. The outer row consists of four spines, similar in form and subequal in length to the adcentral spines of the inner row. The two rows of oral spines for each pair of mouth-plates have the form of a V telescoped into another V.

Ventrolaterals.—The ventrolateral plates are completely clothed over with coarse, flattened, polygonal granules, and are more distinct than in *O. nodosus*, but less so than in *O. moedstus*. Generally speaking the plates are larger and more regular in the arms than in the disk, where they are of irregular form and size, and being covered over with relatively coarse granules, the individual plates are in this part particularly difficult to make out from the surface. On either side of the ambulacral furrow there is a series of roundish or almost square plates for the entire length of the arms, which becomes merged into the general pavement of the ventral surface of the disk. A second row can also be made out, but this is not so regular as the first one, and terminates at a short distance from the tip of the arms. A third very short row can also be distinguished at the base of the arms.

There are many roundish or transversely elongated, valvate pedicellariæ on the ventrolateral plates, which are most numerous on the plates on either side of the ambulacral furrow and on those forming the interradial areas. In the latter, hardly any are to be observed near the inferomarginal plates. In the longitudinal series on either side of the ambulacral furrow, there may be as many as three or four on one plate, mostly close to the furrow.

Abactinal side.—The lophial tubercles are well developed and conical and armed with a conical spine at the tip. In my specimen there are five of them on each of the four of the arms and four on the fifth. The adcentral tubercles are conspicuously larger than those of the arms and are provided with two conical spines, one at the tip and the other, which is smaller, on the outer side of each tubercle, about half way between the tip and the base. The adcentral tubercles enclose a central area, in which five depressed ridges can be made out radiating from the centre to each tubercle. The adcentral tubercles themselves are connected with one another by similar ridges running between them. The space between the adcentral, BELL's apical, and the first lophial tubercle is larger than that between any two lophial tubercles, and on either side of about the middle of this space lies a tubercle about half as large as the first lophial tubercle, similar to the latter in shape and armed with a terminal spine. The reticulum formed by the ossicles of the abactinal side is tolerably distinct on the surface, and two longitudinal series of them can usually be made out on each side of the arms, the inner of which lies in line with the accessory tubercle between the adcentral and the first lophial tubercles. When the surface granules are removed a third series can be seen extending into the arms about one-third the whole length. The ossicles themselves are covered over with rounded granules, but the meshes between them with rounded or short, conico-cylindrical granules, between which lie the papular pores, of which there are many in a single mesh. On the margin of these pores and between the granules above mentioned there are numerous, minute, short spines. The last form of the armature of the abactinal side is the pedicellariæ, which are all of one form, and are similar to those of the abactinal side of *O. nodosus*, being forcipiform and nearly equal to them in size (Pl. XVI, fig. 243).

Madreporite.—The madreporite is tolerably large and lies just outside the pentagon formed by the adcentral tubercles. It is ovate in form with the broader side turned towards the centre, and is covered with irregular radiating furrows.

Terminal plate.—The terminal plate is relatively small, and is covered over with granules exactly similar to those of the superomarginal plates.

Locality—Naha, Okinawa Island. Collected by Dr. M. MIYAJIMA, May 22, 1900. The colour is greyish white in alcohol. *Littoral.* Specimen in S.C.

Oreaster nodosus (Linné).

(Pl. XIV, figs. 221–227.)

In adopting this name for this species I am simply following the authority of Bell ['84, "Syst. nat. ed. XII. p. 1100 (pars)"]. As certain confusing difficulties will attend the tracing of this name through the literature I will mention them at the outset. There are two species that have been described by Gray under the name of *Pentaceros nodosus*, and Bell in his valuable paper on the genus *Oreaster* ['84, p. 62] gives *clouei* Perrier, *franklini* Gray, *mammosus* Perrier and *turritus* M.T. as synonyms of *nodosus*. Now from what he says under *O. grayi* further on ['84, p. 83] I infer that he uses *nodosus* in the Linnæan sense not in Gray's. In his paper of 1840 Gray mentions only one species of *Pentaceros nodosa*, which he describes as follows ['40, p. 277]: "*Asterias nodosa*, Gmelin (part), Seba, iii. t. 8. f. 11, 12. (t. 5. f. 11, 12. without spines on the margin?). Arms with a double series of hemispherical tubercles; back rather depressed; marginal ossicula unequal, lower one with small blunt conical spines. Inhab. Isle of France. W. E. Leach, M.D." It is placed in the group with the "back formed of irregular flat-topped ossicula, placed in rows so as to appear nearly tessellated; arms elongated, rather narrow." In his paper of 1866, however, Gray mentions two species, apparently distinct, under the same name of *Pentaceros nodosus* on the same page ['66, p. 6], one of which is the species described in the earlier paper and appears as No. 11; but the other is placed in the group with the "back formed of irregular elongated ossicula, apparently reticulated; the spines with elongated bases, interspaces closely punctured." The description is very brief and runs as follows [Gray, '66, p. 6]:

"5. *Pentaceros nodosus*. Arms rather narrow, nearly as long as the width of the body, with a single series of blunt tubercles; back rather depressed, with a large tubercle on each angle of the centre. Gray, Ann. N. H. 1840, p. 276; Linck, t. 26. f. 41. *Ast. nodosa* d, Lamarck; Müll. & Trosch. Ast. 48. Inhab. Isle of France, Dr. W. E. Leach.

"In Linck's figure the spines are rather larger than in our specimens of nearly the same size."

On referring back to GRAY's earlier paper the above species appears as *Pentaceros hiuculus*, and is described in very nearly the same words ['40, p. 276]; it is identical with *O. hiulcus* M.T. according to the joint opinions of GRAY ['66, p. 6], MÜLLER and TROSCHEL ['42, p. 48] and PERRIER ['76, p. 59]. GRAY's *Pentaceros nodosus*, No. 11, (= *O. grayi* BELL) has been more fully redescribed by LÜTKEN ['64, p. 152], BELL ['84, p. 83], DE LORIOL ['85, p. 60] and SLUITER ['89, p. 304], and it is evident from these descriptions and the figures given by DE LORIOL that it is not identical with the *Pentaceros franklinii* of GRAY, a synonym of the present species; the principal differences between the two being the absence of all tubercles from the marginal plates in *P. franklinii* and their presence in *P. nodosus* GRAY's No. 11,[1] the presence here and there of particularly large pedicellariæ on the dorsal side of *P. nodosus* GRAY's No. 11 but of particularly small ones in *P. franklinii*, and the presence of only two spines in the outer row of adambulacral spines in *P. nodosus* GRAY's No. 11 but of three or four in *P. franklinii*.[2] The latter species has been figured by GRAY in his paper of 1866,

1) " .... Randpladerne 17. Af de övre hæver over Halvdelen (7–10) sig op in en bred, Kegle-dannet, kornet, i Toppen nögen Knude; paa de nedre findes der Spor til Smaaknuder, dog kun i Armens inderste og yderste Deel" [LUTKEN, '64, p. 152]. The marginal plates 17. Of the upper over half (7–10) raise themselves in a broad, conical, granulated node, naked at the top; on the lower there are traces of small nodes, but only in the innermost and outermost part of the arm. DE LORIOL gives a similar description ['85, p. 61]: "Les plaques marginales dorsales sont très distinctes, grandes, mais très inégales dans leur développement; les unes, beaucoup plus grandes, se relèvent en tubercules, en forme de mamelon très développé, couvert, comme les autres, de granules polygonaux, serrés, avec une pointe lisse, plus ou moins allongée; les autres, en nombre variable sur chaque bras, ou bien se relèvent moins en formant un mamelon tout en ayant la pointe lisse, ou bien ne se relèvent pas du tout et n'ont point de pointe. Toutes les plaques sont rétrécies et cunéiformes sur leur côté interne, de manière à laisser de la place à de larges aires porifères triangulaires. Le nombre des plaques varie entre quatorze et dix-sept, de chaque côté des bras, soit de vingt-huit à trente-quatre pour chaque arc interbrachial, de l'extré-mité de l'un des bras à celle du voisin: ....."

2) " Klappetænger ere meget sjeldne paa Bugtavlerne (hyppigst i Næhrheden af Fodgangene), mangle ganske paa Randpladerne, men findes hist og her paa Ryggen, hvor Hudskeletts Masker stöde sammen, og have der en betydelig Störrelse. Der er 7 Fodpapiller i den indre, to i den ydre Række; de mellem dem siddende Tænger ere meget smaae." [*O. nodosus*, LÜTKEN, '64, p. 153].

Pedicellariæ are very seldom on the ventral plates (most numerous in the neighbourhood of the ambulacral furrows), wholly wanting on the marginal plates, but present here and there on the back, where the meshes of the cutaneous skeleton meet together, and have there a considerable size. There are 7 foot-papillæ in the inner, two in the outer row; the pedicellariæ which occur between them are very small. [*O. nodosus*, LÜTKEN, '64, p. 153].

and there is no doubt that it is the species we have in hand and the *Pentaceros turritus* of authors.

According to BELL ['84] this species was first described under the binomial system of nomenclature by LINNÉ in the 12th edition of the "Systema naturæ."[1] According to MÜLLER and TROSCHEL ['42] it is figured in "Museum Gottwaldianum testaceorum, stellarum marinarum et coralliorum, 1782," and is identical with the *Asterias nodosa* of LAMARCK, which is described as follows [LAMARCK, '16, p. 557] :

"Astérie couronnée. *Asterias nodosa*.

"A. radiis quinque carinatis, aculeato-muricatis ; margine mutico.

[References omitted.]

"Habite l'océan des Grandes-Indes. Cette belle astérie est fort remarquable par les épines fortes, cuspidiforme ou glandiformes qui couronnent le dos de son disque, et qui régnent le long de ses carênes dorsales. Tantôt ces épines sont toutes très-droites ou verticales, et tantôt elles sont diversement inclinées."

Asteria nodosa L. is simply mentioned by BLAINVILLE ['34, p. 238], with references to GMELIN, LINCK, SEBA and the "Encycl. méthodique."

Pentaceros franklinii is placed by GRAY in the same group of the genus as his *P. hiuculus*, and is described as follows ['40, p. 277] :

"7. *Pentaceros Franklinii*. Rays elongate, as long as the width of the body, with a dorsal series of broad blunt tubercles ; back high, with very large spines at each angle, margin not armed.

"Var. 1. With one or two conical tubercles on each side of the tubercles, near the one at the angle of the central dorsal disk.

"Inhab. Coast of New Holland. G. BENNETT, Esq.

"See also *Pentaceros turritus*. LINCK, t. 22. 23. f. 3. Like the former, but the back is more spinose, and the spines are not so large."

The next description of this species is by MÜLLER and TROSCHEL ['42, p. 47] :

"Species 5. *Oreaster turritus* Nob.

---

1) In the tenth edition of the same work *Asterias nodosa* is described thus : "A. stellata radiis convexis longitudinaliter elevatis muricatis .... Habitat in M. Indico."

[References to Rumphius, Petiver, Linck tab 2 & 3, No. 3, Mus. Gottwald., Linné-Gmelin, Lamarck and Gray.]

"Dem vorigen[1] in allen Punkten ähnlich, ausgenommen in folgendem : Die Furchenpapillen 8–9 auf einer Platte, von denen die mittleren höher sind. Die der äusseren Reihe platt, 3 bis 4 auf einer Platte. Die Bauchseite ist mit sehr ansehnlichen platten, pentagonalen, ungleich grossen Granula besetzt. Sie sind eben so, aber etwas kleiner auf den Randplatten. Auf dem Rücken sind die Granula nur platt da wo sie grossen Tuberkeln bedecken. Der ganze übrige Rücken trägt grössere und kleinere conisch sich erhebende Granula. Die oberen und unteren Randplatten (20 an jedem Arme) sind einfach granulirt und ohne Tuberkeln und Pedicellarien. Die grossen Tuberkeln auf dem Rücken der Arme in einfacher Reihe; auf dem Rücken der Scheibe unregelmässig vertheilt, zuweilen mit ihren Basen zusammenstossend. Die Knoten sind bis zur äussersten Spitze von der Granulation eingehüllt, nur zuweilen ragt eine kleine, glatte Spitze hervor. Die Granulation der Knoten ist platt, pentagonal, am Grunde der Knoten kleiner; nach oben zu werden die Granula länglich. Auf der Bauchseite sieht man mit der Loupe kleine klappenartige Pedicellarien, am deutlichsten und häufigsten in der Nähe der Furchen. Auf dem Rücken fehlen die Pedicellarien, auch auf den tuberkellosen Knoten des Netzes ganz. Die Pedicellarien der Porenfelder sind zangenartig, klein und ziemlich lang. Zwischen je zwei Blättern von Furchenpapillen steht eine zangenartige Pedicellarie.

"Farbe : nach mündlicher Mittheilung des Herrn Salomon Müller aus seinen Manuscripten, roth, die Knoten und die Spitzen der Arme schwarz.

"Grösse : 10 Zoll.

"Fundort : Indischer Ocean. Im Museum zu Berlin aus der Schoenlein-schen Sammlung; im Museum zu Leyden (9 Exemplare); im Museum zu Paris."

This species is also described by Dujardin and Hupé ['62, p. 381], who closely follow Müller and Troschel.

"Oreaster turriclé. *Oreaster turritus.*—Müller et Troschel.

[References omitted.]

---

[1] *O. tuberculatus.*

"Espèce assez semblable à la précédente[1]" par l'ensemble de ses caractères, mais qui en diffère par quelques points : les piquants du sillon ambulacraire sont au nombre de huit ou neuf sur chaque plaque, ceux du milieu étant les plus grands ; ceux de la rangée externe sont plats et au nombre de trois ou quatre sur chaque plaque. La face ventrale est couverte de granules de grosseur notable, pentagones et inégaux, que l'on revoit plus petits sur les plaques marginales. Sur le dos, les granules sont plats là où ils recouvrent de gros tubercules, mais tout le reste du dos porte des granules saillants, coniques, plus ou moins grands. Les plaques marginales, au nombre de vingt à chaque bras, sont simplement granuleuses, sans tubercules ni pédicellaires. Les gros tubercules forment sur les bras une rangée simple, mais sur le reste du dos ils sont irrégulièrement disséminés, et sont quelquefois contigus à leur base. Les nœuds du réseau dorsal sont entourés de granulations jusqu'à leur sommet, d'où s'élève quelquefois une petite pointe lisse ; les granulation sont plates, pentagones, plus petites vers la base et allongées vers le haut. Sur la face ventrale on voit, à l'aide de la loupe, de petites pédicellaires valvulaires qui sont plus distinctes et plus nombreuses auprès du sillon ambulacraire. Les aires des pores tentaculaires ont de petites pédicellaires en pince et assez longues. Entre chaque paire de lames des piquants du sillon ambulacraire, se trouve une pédicellaire en pince.

"Coloration rouge, avec les tubercules noirs, ainsi que la pointe de bras. Dimension : largeur totale 270 mm.

"Habite la mer des Indes (Mus. Paris)."

VON MARTENS ['66, p. 76] refers to this species in the following terms :
"18. *Oreaster turritus* (LINCK) GRAY sp.

[References omitted.]

"Oberseite voll cylindrischer kleiner Papillen und starker spitzer Höcker, welche nicht bis zur Spitze getäfelt sind und auf den Armen eine Reihe bilden, auf der Scheibe gruppenweise vereinigt sind. Unterseite mit flachen, eckigen, ungleichen Körnchen bedeckt. Keine Randstacheln. Zangenförmige Pedicellarien nicht selten zwischen den Papillen der Oberseite. Arme ver-

---

1) *O. tuberculatus.*

hältnissmässig lang und schmal, Armradius 165, Scheibenradius 70, Höhe 39 Mill. Nur an der Spitze der Arme sind die Höcker rund.

" Farbe während des Lebens hellbraungrau, Rand und Armspitzen orange, Höcker schwarz, untere Randplatten in der Mitte schwarz, ringsum orange.

"Amboina auf sandigem Grund, ein Paar Fuss unter Wasser, (am ' Koolenhoofd' dicht bei der Stadt) nicht selten, aber schwer zu trocknen. RUMPH gibt ferner die Nordküste von Ceram und die Insel Bonoa an der Nordwestecke von Ceram als Fundorte an, BLEEKER auch Banda und SAL. MÜLLER Buru."

In his paper of 1866 [p. 6] GRAY reproduces his description of 1840, and adds a plate in which the animal is figured in two aspects, actinal and abactinal.

In his paper on· pedicellariæ, PERRIER gives the following description ['69, p. 73] :

" Oreaster turritus, M. et T.—Il n'existe pas de Pédicellaires à la face dorsale.—Les Pédicellaires indiqués par MÜLLER et TROSCHEL sur les aires des pores tentaculaires, ne sont que de petits 'tubercules entourant l'orifice de chaque pore.

" Sur la face ventrale, nous n'avons vu de Pédicellaires que sur les plaques qui bordent les sillons ambulacraires. Chacune d'elles en porte en général deux formant à peu près deux rangées parallèles aux sillons. Quelquefois on en trouve trois sur la même plaque, dont deux sur la même ligne, très-rapprochés du sillon ambulacraire et un en arrière.—Ces Pédicellaires sont valvulaires et ont à peu près un millimètre de long.—Leur constitution ne présente du reste rien de particulier.

" Sur l'angle le plus rapproché de chacune des pièces calcaires de la bouche qui bordent les sillons ambulacraires, s'insère un Pédicellaires en pince. Il se trouve placé entre le premier et le deuxième rang de piquants et sépare es uns des autres les groupes de piquants appartenant à chaque pièce. Les mâchoires de cette pince, comme nous le trouverons du reste dans presque tous les Oreaster, sont épaisses, presque cylindriques et formées d'une substance calcaire réticulée à mailles très-serrées. Cette substance paraît être constituée par des tiges longitudinales, divergentes, mais très-légèrement à partir de la base et réunies en tous sens par de petites traverses irrégulièrement disposées. Chacune de ces tiges se termine librement à la surface du Pédicellaire par une pointe saillante, et comme ces terminaisons ont lieu à des hauteurs différentes, la surface de l'organe paraît échinulée. Les épines deviennent d'autant plus saillantes qu'on est plus rapproché de l'extremité libre du Pédicellaire. Elles donnent au bord externe du profil de chaque pince l'apparence d'une scie dont toutes les dents seraient dirigées vers le haut.—Le bord interne de la pince présente une denteleure irrégulière composée de dents courtes, arrondies au sommet, très-inégàles, mais sensiblement dirigée perpendiculairement au bord.—Chaque pince présente une face interne sensiblement plane.

"Les tiges primitives, d'ailleurs fort irrégulières, de même diamètre que les traverses et distinctes seulement par l'apparence générale de la substance calcaire, semblent partir du voisinage du bord interne de la pince pour remonter ensuite vers l'extrémité supérieure de l'organe. Cette apparence rayonnée ressort aussi du reste de la disposition uniforme et en ligne des perforations de la substance calcaire et des épines que portent les nœuds du réseau ; épines très-obliques et presque couchées sur la surface de l'organe.

"Longueur double de la largeur."

In the same work PERRIER describes a species under the name of *Oreaster mammosus*, but he subsequently regards it as a variety of *O. turritus*. *O. mammosus* is described as follows ['69, p. 78] :

" *Oreaster mammosus*, VAL. Coll.—L'*Oreaster mammosus* n'a pas encore été décrit. Il se rapproche par sa forme et son ornementation de l'*Oreaster turritus* et de l'*Oreaster hiulcus* ; mais si l'on admet que les caractères spécifiques employés d'ordinaire ont une valeur constante, il s'en distingue néanmoins aisément. Ses formes sont plus trapues. Sur le disque, en face de chaque bras, on remarque un gros tubercle en forme de pain de sucre, et couvert de *granulations aplaties* polygonales. Chacun de ces tubercules est le commencement d'une série de tubercules plus petits qui occupant la ligne médiane des bras. Aucun de ces tubercules, les seuls qui existent sur le corps de l'animal, ne présente de Pédicellaires. On remarque dans les aires tentaculaires un assez grand nombre de Pédicellaires intermédiaires entre les

Pédicellaires valvulaires et les Pédicellaires en pince. Il n'en existe pas d'autres sur le dos. Les plaques marginales, tant dorsales que ventrales, sont aussi dépourvues de Pédicellaires ; elles sont au nombre de quinze pour chaque bras.

" La face ventral est couverte de granulations irrégulièrement polygonales, plus grandes sur le centre des pièces calcaires qui les supportent. Celles de ces pièces qui forment la première rangée après celles des sillons ambulacraires, portent chacune un ou deux Pédicellaires valvulaires. Les plaques du sillon ambulacraire portent chacune de dehors en dedans : 1° deux gros piquants aplatis au sommet ; 2° à leur angle interne, comme d'habitude, un Pédicellaire en pince, analogue à ceux de l'*Oreaster hiulcus* ; 3° sur son bord interne une rangée de trois à cinq piquants, dont les moyens sont de beaucoup les plus allongés. Ces piquants vont en divergeant ; il résulte de là que les piquants des sillons ambulacraires sont sur deux rangs.

" La plaque madréporique est arrondie. Les collines saillantes et ramifiées qui les parcourent naissent les unes du centre, les autres vers le milieu du rayon, s'anastomosent souvent, et semblent passer en sautoir les unes sur les autres. Dans l'*O. hiulcus*, la plaque madréporique est au contraire en forme de losange.

" Zanzibar. (LOUIS ROUSSEAU, 1841)."

The species named *Pentaceros nodulosus* by PERRIER in his " Revision " and regarded by BEDFORD [:00] as identical with the present species is described as follows [PERRIER, '76, p. 53] :

" 149. *Pentaceros nodulosus* (nov. sp.).

" Cinq bras très-pointus, assez grêles, mais pas très-longs, reliés entre eux par un arc interbrachial à grande courbure. Corps élevé en forme de pyramide pentagonale, tronquée au sommet, les arêtes des pyramides étant formées par la ligne médiane des bras et les faces étant représentées par une surface continue légèrement concave. La base supérieure du tronc de pyramide, sensiblement plane. Plaques marginales au nombre de trente-six dorsales et trente-huit ventrales, les dernières diminuant graduellement et la dernière étant très-petite. Toutes ces plaques, parfaitement distinctes les unes des autres et se distinguant du reste aussi très-nettement des faces dorsale

et ventrale, sont complétement inermes et seulement couvertes de petites
plaquettes polygonales aplaties. Ces plaquettes, assez grandes au centre des
ossicules marginaux, vont en diminuant vers leurs bords et finissent par
devenir très-petites et par simuler une bordure de granules. Sur les plaques
marginales dorsales et notamment vers leurs bords dorsal et latéral, on voit un
assez grand nombre de très-petits Pédicellaires valvulaires. Sur la ligne mediane
des bras, formant arête du tronc de pyramide qui représente le corps de
l'animal, onze ou douze des ossicules relativement volumineux, qui constituent
cette arête s'élèvent en tubercules ellipsoïdaux, à grand axe transversal, entièrement couverts de plaquettes polygonales semblables à celles qui recouvrent les
ossicules marginaux. Ces ossicules de la ligne médiane, plus gros et plus
saillants, sont séparés par d'autres ossicules également ellipsoïdaux, ayant un
grand axe de même longueur que le leur, mais de petits axes beaucoup moins
longs, de sorte qu'ils sont beaucoup moins élevés et beaucoup plus courts que
les ossicules entre lesquels ils sont intercalés et paraissent comprimé par eux.
Entre deux grands tubercules on en trouve généralement d'un à trois petits
recouverts de granules arrondis plutôt que de plaquettes polygonales. Cette
disposition reproduit à très-peu près celle qu'on observe chez le *Pentaceros
nodosus* GRAY. Elle a été bien décrite pour cette espèce par le docteur
LÜTKEN, et les figures 11 et 12 de la planche VIII du tome III du *The-
saurus*, de SEBA, la représentent parfaitement. Les grands tubercules de la
ligne médiane des bras vont en décroissant de la base au sommet de ceux-
ci ; les plus volumineux sont placés en conséquence, aux angles du pentagone
qui limite la base supérieure du tronc de pyramide représentant le corps de
l'animal. Ces cinq tubercules ne sont pas contigus, mais on voit entre eux
trois tubercules arrondis plus petits. La base pentagonale qu'ils limitent
ne porte aucun tubercule saillant. Les aires porifères sont nombreuses, légèrement enfoncées, petites et à peu près de même dimension que les ossicules
plus élevés, aplatis et de forme étoileé qui les séparent. Ces ossicules, couverts
par la granulation générale, portent dans leur région centrale un ou deux
petits Pédicellaires valvulaires, enfoncés au milieu des granules, de sorte qu'
on les reconnaît immédiatement à une sorte de trou borgne très-visible au milieu
des ossicules qui les portent. On trouve aussi quelques Pédicellaires identi-

ques parmi les granules des aires porifères. Ces dernières forment, sur les bras proprement dits, trois séries seulement, les inférieures sont plus grandes que les autres et pénètreñt entre les plaques marginales. Les aires porifères du pentagone basilaire supérieur sont semblable à celles des faces de ces pyramides. Sur l'une de ces faces et près de son arête supérieure se trouve la plaque madréporique assez grande et en forme de losange.

"Sur la face ventrale, les ossicules marginaux sont rectangulaires; ils s'élargissent et se rapprochent à mesure qu'on se rapproche du sommet des bras. Les plaques ventrales sont beaucoup plus petites que les marginales, peu distinctes les uns des autres, couvertes de granules polygonaux séparés les uns des autres par un petit intervalle et portant la plupart un ou deux petits Pédicellaires valvulaires. Ces pédicellaires sont surtout constants dans le voisinage de la gouttière ambulacraire. Les piquants de cette gouttière sont disposés sur deux rangs : chaque plaque interambulacraire en porte dans le sillon de sept à neuf prismatiques, tronqués au sommet; et, en dehors, trois plus grands constituent la deuxième rangée. Entre ces deux rangées, sur le bord buccal de chaque plaque, on voit un grand Pédicellaire droit. Les granules qui avoisinent immédiatement la deuxième rangée de piquants sont un peu plus grands que les autres et simulent parfois une troisième rangée. A l'angle buccal, les rangées de piquants ambulacraires sont doubles comme ailleurs.

"$R = (2 + \frac{1}{3})r$. Distance de deux sommets opposés pouvant atteindre 2 décimètres environ.

"Nombreux échantillons desséchés au British Museum. Ils sont originaire d'Australie."

Further on in the same work PERRIER mentions *P. turritus* and discusses at length its identity with *P. mammosus* ['76, p. 56]:

"152. *Pentaceros turritus.*

[References omitted.]

"Deux échantillons dans l'alcool, dont l'un rapporté de l'île de France, par PÉRON et LESUEUR en 1803, correspond exactement au *Pentaceros Franklinii* de GRAY, tandis que l'autre, rapporté de Zanzibar par M. LOUIS ROUSSEAU, et de taille beaucoup plus petite, est le *Pentaceros modestus* de Grya.

Un échantillon desséché est l'*Asterias nodosa* type de LAMARCK. Cette même
espèce se retrouve à la Nouvelle-Guinée (British Museum).

"var :   *Pentaceros mammosus.*

"1869.   *Oreaster mammosus,* E. P.—Péd., p. 78.

"En 1869, dans mon mémoire sur les pédicellaires et les ambulacres des
Astéries et des Oursins, j'ai décrit sous le nom d'*Oreaster mammosus* VAL.,
une Astérie, ainsi nommée par VALENCIENNES dans la collection, mais dont
la description n'avait pas été donnée. L'individu que j'ai décrit provenait
de Zanzibar et avait été recueilli en 1841 par mon regretté collègue M. LOUIS
ROUSSEAU. Un certain nombre d'autres individus, sans indication d'origine,
se trouvaient d'ailleurs sous la dénomination d'*Oreaster turritus*, bien qu'ils
eussent avec l'*Oreaster mammosus* les plus évidentes analogies. Depuis cette
époque, M. BALANSA, M. GERMAIN et M. PETIT ont successivement envoyé de
la Nouvelle-Calédonie, des *Pentaceros* de taille différent et qui présentent
avec l'*Oreaster mammosus* de Zanzibar de telles ressemblances, qu'il me paraît
impossible de les en séparer spécifiquement, dans l'état actuel de nos connais-
sances. D'autre part, tous ces Stéllerides sont eux-mêmes tellement voisins
du *Pentagonaster*[1] *turritus*, qu'il y a lieu de se demander s'ils ne constituent
pas simplement une variété de cette dernière espèce, qui se trouvent également, comme on sait, à Zanzibar, et qui présenterait d'ailleurs à Zanzibar et
à la Nouvelle-Calédonie des variations exactement parallèles. Parmi les six
individus que possède le Musée et qui proviennent authentiquement de la
Nouvelle-Calédonie, il en est un, en effet, le plus grand, à qui l'on peut
appliquer de point en point la description du *Pentagonaster turritus*. Son
plus grand rayon est de 115 millimètres, le plus petit étant de 40 à peu
prés, ce qui donne un rapport de 1 à 3 environ, très-voisin de celui que l'on
constate chez les grands *Pentagonaster turritus*. L'aspect des faces dorsale et
ventrale est exactement le même. Le nombre et la disposition des épines
ambulacraires sont identiques ; il en est encore ainsi des gros piquants de la
carène des bras disposés en une seule rangée et dont le second, à partir du
disque, est flanqué de chaque côté d'un autre piquant aussi gros que lui.
Ces piquants peuvent d'ailleurs exister seuls, le piquant principal faisant

---

1) Evidently a misprint of *Pentaceros.*

défaut. Chez l'individu de la Nouvelle-Calédonie ces gros tubercules sont coniques comme dans le *Pentaceros turritus*, représenté par LINCK ; ils sont hémisphériques chez les *Pentaceros turritus* de l'île de France et de Zanzibar, que nous avons sous les yeux, mais ce sont là des différences individuelles, comparables à celles que l'on trouve entre les divers spécimens de *Pentaceros muricatus.* Le nombre des plaques marginales (vingt-cinq chez l'échantillon de l'île de France, vingt-huit chez celui de la Nouvelle-Calédonie) est un peu différent ; c'est là une question de taille et nous ne voyons pas sur quel autre caractère on pourrait s'appuyer pour distinguer ces deux individus au point de vue spécifique.

D'autre part, si l'on compare le *Pentaceros turritus* de la Nouvelle-Calédonie, du à M. GERMAIN, à deux autres individus de la même provenance recueillis par M. BALANSA on ne trouve d'autre différence qu'une longueur un peu moindre des bras (r=30 millimètres, R=80 millimètres), qui sont en même temps un peu plus larges. De plus, le nombre des piquants ambulacraires de la rangée externe n'atteint pas quatre chez ces individus, mais il peut être de trois, et, comme le nombre quatre n'est qu' accidentel chez le *Pentaceros turritus,* on voit que cette différence n'a rien de fondamental. Le nombre des piquants de la rangée interne varie de quatre à sept, sans cependent atteindre le nombre neuf, qu'on trouve chez les *Pentaceros turritus* de grande taille, mais qui n'est pas atteint non plus chez ceux de petite (individus de PÉRON et LESUEUR). Toutes ces différences s'expliquent facilement par la taille moindre des individus. Reste un dernier caractère négatif, l'absence des gros tubercules en forme de piquants, sauf sur la carène des bras, où ils forment une rangée simple ; mais rien n'est variable chez les *Pentaceros* comme le nombre et la disposition de ces tubercules. La belle série de *Pentaceros muricatus,* rapportée de Zanzibar par M. LOUIS ROUSSEAU, en est un frappant exemple. Chez les *Pentaceros turritus,* la même chose se produit. L'individu de Zanzibar et celui de M. GERMAIN, de la Nouvelle-Calédonie, ont un tubercule au centre du disque, qui manque à l'individu de PÉRON èt LESUEUR. Chez l'individu de M. GERMAIN le second piquant de la carène brachiale, celui qui est flanqué de deux autre piquants, existe pour deux des bras, se trouve placé anormalement pour un troisième et manque aux deux autres. Il manque à tous les bras chez l'individu de Zanzibar et chez celui de Péron et Lesueur ; enfin, chez ce dernier, ses piquants satellites ne se montrent complets que sur un bras ; sur les quatre autres, l'un d'eux est rudimentaire ou manque complétement, de sorte que nous passons ainsi aux formes où il manque d'une manière absolue et pour l'une desquelles M. Valenciennes avait créé le nom d'*Oreaster mammosus*. Cette forme se trouve maintenant représentée au Muséum, outre les deux individus de M. Balansa, par l'individu de Zanzibar qui a servi de type, et six autres individus, dont deux proviennent de la Nouvelle-Calédonie (M. Germain) et quatre ne portent pas d'indication d'origine. Ces neufs individus, dont quatre étaient déjà dans la collection rapportés au *Pentaceros turritus*, forment une série où la taille varie depuis 5 centimètres de diamètre jusqu'à 17, après quoi l'on passe au *Pentaceros turritus* de M. Germain, qui a 21 centimètres de diamètre.

"Cette série continue nous montre que tous ces individus, qu'ils proviennent de la côte d'Afrique ou de la Nouvelle-Calédonie, ne forment qu'une seule espèce. Le nom de *mammosus* ne peut donc être conservé que pour indiquer une variété qui se trouve, du reste, aussi bien à Zanzibar que sur les côtes de la Nouvelle-Calédonie.

"Deux individus dans l'alcool ; onze desséches, dont neuf appartenant à la variété *mammosus*."

Viguier ['78, p. 71] gives a detailed account of the skeletal parts that enter into the formation of the oral parts, with several figures.

Oreaster nodulosus which is, as already stated, regarded by Bedford as a synonym of the present species is described by Bell as follows ['84, p. 66] :

" *Oreaster nodulosus.*

" *Pentaceros nodulosus*, Perrier, Rev. Stell. p. 237.

" R=2.3 r. Disk moderately elevated, arms of moderate width, tapering gradually. The lophial and apical spines absent, and their place taken by the enlargement of the ossicles into convex rounded bodies.

" About 17 marginal plates in either series ; it is only in the more distal regions that the inferomarginals take any share in forming the sides of the arms. Neither series are spinose.

" Adambulacral spinulation diplacanthid, the spines blunt; in the inner row there are ordinarily seven spines, of which the median are the more prominent; in the outer row there are two or three larger spines, one of which is often, when only two are developed, much larger than the other; these spines have a direction a little oblique to the longitudinal axis of the arm. Between the outer and inner rows a well-developed forcipiform pedicellaria is placed. Beyond the outer row there are irregularly shaped separate granules, which appear, at first, to afford indications of a third row of adambulacral spines.

" The ventral ossicles are often distinguishable from one another owing to the larger size of the granules in the centre than at the edge of the ossicle; sessile valvular pedicellariæ are richly developed among the granules. Large and coarse granules are also to be observed on the marginal plates, on which, however, pedicellairæ are only rarely developed.

" The upper surface, both of the disk and of the arms, is delicately reticulated. The pore areas are well separated from one another, and are, in all the more proximal parts of the arm, of some size, and contain more than twenty pores.

" The areas of the two lower series along the sides of the arms sometimes become fused at certain points; the lower series extend into the space between every pair of superomarginal plates. The granulation on the nodal points is rather more delicate than on the ventral surface, and the sessile pedicellariæ are exceedingly small.

" Nearly all the ossicles along the lophial line are elongated; some are more so than the rest, and two or three generally attain to considerable prominence; those which flank the apical region are large and rounded, and are, like the rest, covered with a close-set investment of rather large flat granules. A few pedicellariæ are to be observed among the granules of the apical region, where no spine or protuberance of any kind is developed. The madreporite forms an elongated oval whose longer axis is directed downwards, and is placed just outside the boundary of the apical region.

" Colour (dry) dirty yellow, probably deep yellow in life.

" Measurements :—

    $R=53$; $r=21.5$; breadth of arm at base 18.

    $R=70$; $r=30$; breadth of arm at base 29.

" *Hab.* West Australia (Dick Hartog's Island)."

Further on in the same paper *Oreaster nodosus* is described ['84, p. 70] : " *Oreaster nodosus.*

[References to PERRIER, '76, p. 240 and LINNÉ, Syst. nat. ed. xii. p. 1100 pars.]

" Prof. PERRIER prefers LINCK'S name to that of LINNÆUS, whom, indeed, he abstains from directly quoting, his only reference being to GMELIN'S edition of the Systema Naturæ.

" $R=2.5\,r$ to $3\,r$. Disk considerably elongated ; arms long, rather narrow. Lophial line well marked, with prominent rounded projections ; the apical spines very prominent, and a central one typically developed.

" About 30 superomarginal, and one or two more inferomarginal plates ; both sets obscure, and without any spines, the lower altogether confined to the actinal side.

" Adambulacral spinulation diplacanthid ; ordinarily seven spines in the inner row, of which two or three in the middle are distinctly longer than those at their sides. In the outer row three spines, about twice as stout as those of the inner row ; between the two rows there stands a well-developed forcipiform pedicellaria.

" The separate ventral ossicles are a good deal obscured by the coarse granulation with which they are covered ; the only region in which there can be said to be a distinctly serial disposition of the plates is that which extends along the side of the ambulacral groove. Many of the investing granules are more than a millimetre in length along their longest axis, and the sessile valvular pedicellariæ are very numerously represented. A similar coarse granulation is found on the marginal plates ; but any resemblance to *O. lincki* is opposed by the development of a very large number of pedicellariæ."

" The upper surface might almost be said to be one mass of pedicel-

---

1) " Have we not here another example of the kind of balance between the development of spines and of pedicellariæ ? Cf. the case of *Asterias glacialis*, Zool. Anz. 1882, p. 283."

lariæ, for they not only cover the reticulating bars of the dorsal ossicles, but invade also the poriferous areas; the granules, of ordinary character, are confined to the knobs and spines, the tips of which, however, they do not cover. Along the lophial line the projections are always rather tubercular than spinous, but the five spines at the angle of the apex and the central spine within are exceedingly well developed and rather acutely pointed. A few rounded tubercles, similar in character to those of the lophial line, are developed at the sides of the disk. In dried specimens the disk rises up in an altogether turriform fashion. Madreporite small, rather obscure, on one of the sides of the disk.

"Colour (dried) greyish sandy.

"Hab. Indian Ocean generally.

"Measurements :—

R	165	130	118	100 [mm.]
r	62	44	46	37
Height of disk	58	45	31	27
Length of longest spine ..	15	14	21	18

"It is to be observed that there are some not unimportant variations in the characters of the spines, those of the lophial line are sometimes sharp, are not always blunt; the apical spines appear to be liable to early division into two or three secondary apices, or they may give off a spur or projection, and, lastly, the processes at the sides of the disk may become quite sharp."

BELL ['84 a, p. 128] mentions some "fine specimens from Port Denison, 4 fms."

STUDER refers to this species at length under the name of *P. turritus* LINCK and points out the occurrence of sexual dimorphism ['84, p. 39] :

"Diese Art fand sich im Greetharbour in Neu-Britannien ungemein zahlreich vor. Die tauchenden Eingebornen brachten die Art aus 1 Faden Tiefe in grosser Menge. Es liessen sich unter den zahlreichen Exemplaren zwei Formen unterscheiden. Die eine mit relativ niedriger Scheibe und festem Skelett, mit einer Warze neben dem After im Centrum der Scheibe. Die Farbe der Oberseite ziegelroth, die Höfe um die Warzen schwarzbraun. Sie entspricht der var. *mammosus* PERRIER. Die zweite zeigte eine dorsal stark gewölbte Scheibe, auf der eine centrale Warze fehlt. Das Skelett ist viel nachgiebiger und lockerer. Die Farbe der Dorsalseite war bedeutend heller, mehr fleischroth, die Ränder der Scheibe und der Arme dunkler roth, ebenso die Höfe um die Warzen.

"Beim Oeffnen fand sich, dass die erstere Form männliche, die letztere weibliche Geschlechtsdrüsen hatte, so dass hier ein Geschlechtsdimorphismus vorliegt, der zum Theil dadurch erklärt werden kann, dass die grossen weiblichen Geschlechtsdrüsen den Körper so sehr anschwellen, dass die Skelett-theile auseinander gedrängt werden und daher die Dorsalhaut nachgiebiger und blasser erscheint.

"Ich erhielt die Art ausser in Neu-Britannien auch im Mac Cluergolf in Neu-Guinea. Die dort erhaltenen Exemplare zeigten im Leben eine etwas verschiedene Färbung. Die Grundfarbe war mehr gelb, als roth, die Höfe um die Warzen orange; die Ventralseite braun.

"In Neu-Britannien kamen unter der grossen Zahl von Stücken auch anormal ausgebildete zu Tage.

"So fand sich ein Exemplar mit sechs vollkommen ausgebildeten Radien, zwei mit nur vier Radien.

"Bei dem einen von diesen fehlt der fünfte Strahl vollkommen, bei dem zweiten ist eine fünfte Armfurche vorhanden, ebenso eine fünfte Warzenreihe, nur beschränkt sich beides auf den Scheibentheil, während das freie Arm-stück zurückgeblieben ist. Der *P. turritus* hat einen sehr ausgedehnten Ver-breitungskreis. Das Berliner Museum enthält Stücke aus Port Moresby (Aus-tralien), Celebes, Amboina, Ceram, Banda, Buru; PERRIER citiert die Art aus Neu-Caledonien, Neu-Guinea, Mauritius und Zanzibar." It is again mentioned by STUDER from the collection of the " Gazelle " ['89, p. 254, *fide* Zool. Record].

SLUITER refers to this species as follows ['89, p. 301]: " *P. nodosus* (LINN.) *P. turritus* (LINCK). PERRIER, 'Révision des Stell.,' pag. 56. BELL, 'On the species of Oreaster.' Proc. Zool. Soc. 1884, pag. 70. Ein Exemplar (No. 569) von Billiton. Tandjong Pandan. Den genauen Beschreibungen von PERRIER und namentlich von BELL ist fast nichts hinzuzufügen. Der grösste Armradius meines Exemplars war 158 m. M., der Scheibenradius 57 m. M.

Die Tiere leben in einer Tiefe von 10–16 Faden auf dem sandigen Boden ausserhalb der Reede von Tandjong Pandan. Die Farbe während des Lebens ist ein rötliches Braun. Die Spitzen der Arme sind zwar dunkler braun, so wie auch die Tuberkeln, aber nicht schwarz, wie nach mündlicher Mitteilung SAL. MÜLLER's von MÜLLER und TROSCHEL angegeben wird."

SLADEN, who regards *Pentaceros modestus* GRAY as a synonym of this species, mentions it from the collection of the Challenger under the name of *P. turritus* as follows ['89, p. 346] :

"*Localities.*—Station 212. Off Malanipa Island (Philippine group). January 30, 1875. Lat. 6° 54′ 0″ N., long. 122° 18′ 0″ E. Depth 10 fathoms. Sand. Surface temperature 83.°0 Fahr.

"Off Samboangan (Philippine group). January 29, 1875. Depth 10 fathoms.

"Off Zebu (Philippine group). On the Reefs.

"Station 187. Off Booby Island, Torres Strait. September 9, 1874. Lat. 10° 36′ 0″ S., long. 141° 55′ 0″ E. Depth 6 fathoms. Coral mud. Surface temperature 77.°7 Fahr.

"Station 188. In the Arafura Sea, near the entrance to Torres Strait. September 10, 1874. Lat. 9° 59′ 0″ S., long. 139° 42′ 0″ E. Depth 28 fathoms. Green mud. Surface temperature 78.°5 Fahr."

DE LORIOL describes this species from the Bay of Amboina as follows ['93, p. 380] :

"La taille des exemplaires varie, R=80 mm. à 132 mm. Dans les plus petits il y a cinq tubercules coniques, très elévés, au centre du disque, et une série unique sur la ligne lophiale des bras ; dans les plus grands la face dorsale du disque est très relevée, renflée, et surmontée de six tubercules élevés dont un central. Outre les turbercules très saillants de la ligne lophiale de chaque bras, il s'en trouve encore quatre à cinq semblables dans chaque espace interradial. La face dorsale du disque et des bras est couverte de granules coniques, accompagnés de très petites verrues et de pédicellaires. Des granules aplatis, polygonaux, couvrent les tubercules, parfois entièrement, parfois en laissant lisse la pointe extrême à laquelle une ou deux autres viennent souvent s'ajouter. Point de pédicellaires sur les plaques marginales dont on compte, suivant la taille, 36 à 42 pour chaque arc interbrachial. Chaque plaque adambulacraire porte un faisceau interne de 6 à 7 piquants plats et inégaux, et une série externe de 3 à 4 piquants plus ou moins prismatiques ; ceux qui avoisinent la bouche sont très épais, plus ou moins fortement sillonnés, et même granuleux du côté externe. M. Th. Studer (loc. cit.) s'est assuré que les individus très renflés et turriculés sur la face dorsale du disque sont des femelles ; il leur attribue aussi l'absence d'un tubercule central ; ceci ne paraît pas être constant, j'ai sous les yeux un exemplaire, qui serait un mâle, dépourvu de tubercule central, tandis qu'un autre, turriculé à la façon des femelles, en possède un très saillant."

This species is mentioned by Sluiter ['95, p. 46] as follows : " 35· *Pentaceros turritus* Linck. Vier Exemplare getrocknet von den Molukken und acht in Alcohol von derselben Inselgruppe (Bleeker)." Kœhler also refers to it ['95, p. 395] : " Deux échantillons de grande taille ; chez l'un, les tubercules coniques sont très pointus et très élevés ; chez l'autre ils sont plus courts et en forme de mamelon." Döderlein reports it from Amboina and Thursday Island ['96, p. 309], while Ludwig ['99, p. 540] mentions it as being reported from Zanzibar by Perrier.

Bell ['99, p. 136] has a short but important remark on this species :

" 20.   *Pentaceros lincki,*⎱ see Bell, P. Z. S. 1884, p. 72.
   Loc. Blanche Bay.   ⎰

" 21.   *Pentaceros nodosus,*⎱ id. t. c. p. 70.
   Loc. Blanche Bay.   ⎰

" These two species herd together in the narrow strait which divides the island of Matupi from the mainland. According to Dr. Willey's observations the two species grade into one another both as regards colour and nodosity. He thinks they are either varieties of one species or else that they cross-breed together and produce hybrids."

Bedford has some important remarks on this species [:00, p. 295] :

" 12.   *Pentaceros turritus.*

[References to Müller and Troschel ['42, p. 47], Bell ['84, p. 66] and v. Martens ['66, p. 77].]

" *Locality.* This species is found not uncommonly in about a fathom of water on the shore of the lagoon off the east side of Singapore Island.

" *Distribution.* Indian Ocean, Eastern Archipelago, and N. Australia.

" The systematic classification of the genus *Pentaceros* offers similar difficulties to those found in the Echinoid *Diadema*; in both cases we have exceedingly variable groups of individuals in which the norm of the variations is different in different localities. The four 'species' *Pentaceros turritus* M. & T., *P. hiulcus* M. & T., *P. muricatus* Gray, and *P. alveolatus* PERRIER, which Professor BELL describes as 'apparently distinct,' are based on characters which are admittedly subject to very great variation; and while Prof. BELL instances the presence or absence of marginal spines as a character by which 'we can always safely discriminate between *O. lincki* (= *P. muricatus*) and *O. nodosus* (= *P. turritus*)' (*loc. cit.* p. 59), at the same time Dr. MARTENS had named two varieties of *P. muricatus* (var. *mutica* and var. *intermedia*) in which the marginal spines completely fail: this of course is entirely due to the fact that no two observers are agreed as to the best set of characters to select for specific diagnoses, and in consequence a totally different grouping of the same series of specimens would be resorted to by different systematists. Whether it is advisable in such a case to unite all the forms under one specific name, as has been suggested for the Pacific Diademas by Prof. LOVÉN, I do not feel in the least competent to express an opinion.

" Specimens both with and without a central apical tubercle occur at Singapore as elsewhere, and in the former case the apical tubercles may be much longer and sharper than in the latter, in which they are usually mamilliform; I do not know if individuals with the marginal spines developed occur in the locality.

" Two specimens gave the following measurements :—

" R (act. side)=116 mm., r=46; arm-breadth=54; no. of marg. plates =23; 9 lophial tubercles, 3 or 4 interradials.

" R (act. side)=128 mm., r=52; arm-breadth=58; no. of marg. plates =27; 7 or 8 lophial tubercles, 4 interradials."

PFEFFER simply mentions *Pentaceros turritus* LINCK from Ternate [:00,

p. 83]. HERDMAN refers to *Pen. nodosus* in his pearl oyster report [:03, p. 22, 51, 84].

The HERDMANS mention it as a distinct species from the Gulf of Manaar [HERDMAN, HERDMAN and BELL, :04, p. 144]:

"*Pentaceros nodosus* (GRAY).

"Station III., off Chilaw, 12 fathoms; station XLII., off Barberyn, 40 fathoms; Station LIV., south of Adam's Bridge, 4 to 40 fathoms; Station LXIV., south-east of Modragam Paar, 5 fathoms."

VON MARTENS mentions *P. nodosus* from RUMPF's collection [:04, *fide* Zool. Rec.]. HERDMAN [:06, p. 121, 125, 447] mentions it as an enemy of the pearl oyster.

Under the name of *Oreaster nodosus* CLARK has some important notes on the growth stages of this species [:08, p. 280; references to Syst. nat. ed. 10, p. 661 and BELL, '84]:

"18 specimens, Humboldt Bay, New Guinea.—5 specimens, Sorong, New Guinea.—3 specimens, Ansus, Jappen Island, New Guinea (135° 44′ E. × 1° 47′ S.).—1 specimen, Amboina. BARBOUR collection.

"These specimens range from 80 to 300 mm. in diameter and exhibit the greatest diversity in the development of the great tubercles so characteristic of this species. In the youngest specimen there are present 15 tubercles, one at each radial corner of the disk and two on the ridge of each ray; those on the disk are largest and most nearly pointed, while those nearest the tips of the rays are small and nearly spherical. In specimens a trifle older there are 20 or 25 tubercles, one or two more having developed on each ray. The pair of tubercles which are found in large specimens at the proximal end of the rays, one on each side of the ridge, are first seen in an individual 165 mm. in diameter, but are quite small and rounded, and it is only in much larger specimens that they are fully developed. The tubercle at the centre of the disc is present in only six specimens, and none of these is under 200 mm. in diameter. In the largest individual it is wanting, but there are 72 tubercles, arranged as follows: one large one, with two or even three points, at each radial angle of the disk; one rather small but pointed one in each interradius not far from the margin, and in one interradius there are two such tubercles; eight on the ridge of each ray, with a ninth on two of the rays; the usual pair at the base of each ray; and one, two, or even three extra tubercles on the sides of the rays near the base. No less than 20 of the tubercles terminate in two, three, or even four sharp, bare points.—In life, the colour of this species shows considerable diversity, ranging from clay-colour with the large tubercles muddy brown, or with the large tubercles deep red-brown, becoming vermilion at the base, or with the large tubercles black, with their bases, the tips of the arms, and the centre of the disk claret-red, to a nearly uniform vermilion-red all over. Most of the dried specimens were dirty yellowish, but on being washed with alcohol the vermilion-red colour returned to a greater or less degree in different individuals and has not been lost by subsequent drying. The largest specimen (300 mm.) from Amboina is the most uniform and the brightest vermilion.—This species was found chiefly on bottoms where there was more or less vegetation or in open places about coral reefs."

Under the name of *Pentaceros turritus* LINCK KŒHLER mentions this starfish from the south-eastern Moluccas [:11, p. 277]:

"Dragage No. 6. 28 Mars 1908. Sungi Manumbai. Profondeur 23 m. Deux échantillons.

"17 Avril 1908. Wokam. Profondeur 2 m. Un grand échantillons.

"17 Juin 1908. Nuhu Tawun (îles Kei). Un grand échantillon.

"Dans l'un des individus du dragage No. 6, chez lequel R=95 mm., les tubercules sont coniques et pointus. Dans le second, qui est plus petit (R=75 mm.), ces tubercules sont arrondis au sommet.

"Les deux grands échantillons offrent une différence analogue: celui des îles Kei, dans lequel R mesure 120 mm., a de gros tubercules arrondis, tandis que chez l'autre, qui est plus grand (R=140 mm.), ces tubercules sont coniques et pointus.

"J'observe dans les aires porifères de petits pédicellaires alvéolaires identiques à ceux que j'ai signalés chez *P. rouxi*, et qui se montrent surtout sur les deux petits exemplaires."

I have only one dried specimen of somewhat large size, r= 54 mm., R=144 mm., giving the radial ratio 2.7. The body is

stone hard, the actinal side is concave in the disk and plane in
the arms; the disk is comparatively large and elevated in the
form of an irregular turret on the abactinal side ; the arms are
relatively slender, triangular in cross section and have up-turned
tips; the interbrachial arcs are quite open and well-rounded (Pl.
XIV, fig. 221, 222).

Superomarginals.—There are 23 superomarginal plates, which
are of a triangular shape on the sides of the arms, irregularly
roundish or polygonal in the interradial arcs, and nearly rectangular close to the tip of the arms.   They form the lateral margin
of the arms, and show comparatively little on the actinal side.
The individual plates are comparatively small and are completely
covered over with flattened polygonal granules.   In most of the
interradii there are a few intercalary plates between the inferior
and superior marginal series.   The superomarginals are entirely
destitute of pedicellariæ in my specimen.   This is a difference of
some importance from the descriptions of previous authors, but
I believe this character is subject to a great deal of variation.

Inferomarginals.—The inferomarginals are coincident with the
superomarginals, and are entirely confined to the actinal side.
The individual plates are nearly elliptical in the interbrachial arcs
and at the base of the arms, but they present a straight border
against the superomarginals along the remainder of the arms.
They are completely covered over with granules exactly similar
to those of the superomarginals, and are like them entirely destitute of pedicellariæ.

Adambulacral plates.—The adambulacral plates are comparatively long and narrow, and bear two series of spines, which stand
out erect in the dried specimen.   There are 7–8 spines in the

inner row, arranged somewhat like the fingers of the human hand, those of the middle part being much longer than those at the ends, which are often very short. The individual spines are more or less flattened and prismatic in shape, with rounded apices. The spines of the outer row are much stouter than those of the inner, and there are three or four on each plate. They are more or less flattened and prismatic in shape, and are mostly grooved longitudinally in the distal part; the apices are more or less wedge-shaped. At the adcentral end of each plate between the two rows of spines there is a large forcipiform pedicellaria, usually slightly curved, showing its upper portion above the neighbouring spines of the inner row (Pl. XIV, fig. 223 225).

Mouth-plates.—As seen from the surface the mouth-plates themselves are not very apparent, the space between the rows of oral armature on either side being exactly like the rest of the ventrolateral area. There are two rows of spines corresponding to each mouth-plate; the outer consisting mostly of two, sometimes of three, spines, exactly like those of the outer series of the adambulacral spines in form, but larger, and lying in line with them. The inner row of oral armature consists of nine or ten spines, of which the three or four adjoining the first adambulacral plate are shorter and more slender than the rest, are in all points more like the spines of the inner row of adambulacral armature, and lie in line with them; these spines gradually become longer as they proceed away from the first adambulacral plate. The remaining six spines of the oral armature are subequal in length, but become stouter toward the mouth, and the one at the mouth end is particularly stout and nearly triangular in cross-section. These spines are like those of the outer row of adambulacral spines in form, and are, like them, longitudinally grooved in the terminal parts.

Ventrolaterals.—The ventrolateral plates themselves are completely hidden from view by the coarse granulation that covers them, but the granules are arranged into groups corresponding to the underlying plates, so that their arrangement and general form can be inferred. It appears that the ventrolateral plates are of comparatively small size and have no regular arrangement, except along either side of the ambulacral furrow, where the plates appear to form a single regular row. The plates are covered over with large, polygonal, flattened granules, and the spaces between the plates are covered with smaller granules. Between the large granules covering the plates are seen many transversely elongated valvate pedicellariæ, which are more numerous on either side of the ambulacral furrows and in the angles of the mouth. On some of the plates there may be two or more of these pedicellariæ, but there are also many plates destitute of them (Pl. XIV, fig. 224, 227).

Abactinal plates.—The most conspicuous feature of the abactinal side is the presence of conical tubercles along the lophial line. In my specimen there are eight or nine of these for each arm, of which the most central one, the apical, is especially large, and with their fellows enclose a central space at the summit of the disk, within which there are, in my specimen, two tubercles of nearly the same size as those at the base of the arms. Moreover there is a tubercle on either side of the one next the apical tubercle, and a similar pair for the next lophial tubercle at a greater distance from the latter, so that there are, in each interradius, two pairs of accessory tubercles. In some interradii, however, this arrangement is not strictly adhered to; *e.g.* in one interradius there are five instead of four tubercles, and in another there are five supernumerary smaller tubercles besides the two pairs that are regularly present. All the tubercles are completely covered over with flattened polygonal granules, and tipped by an immovable spine.[1] In the interspaces between the tubercles, there can be seen a reticulated skeleton, with large meshes, in which there are numerous papular pores. The whole is thickly covered over with conico-cylindrical granules of two sizes, larger ones and minute ones scattered between the former. Between these granules there are small pedicellariæ, which are tolerably numerous (Pl. XIV, fig. 226).

Madreporite.—The madreporite is tolerably large, flat, nearly lozenge-shaped in my specimen, and lies just outside the line connecting two of the apical tubercles. The surface is covered with minute groove-like pores.

Terminal plate.—The terminal plates are comparatively small and inconspicuous.

Locality.—Amami-Ôshima, Linschoten Islands. Littoral. Specimen in S.C.

Oreaster lincki (Blainville).

In Rein's work on Japan [:05, p. 291] there is the following passage: " Der gewöhnlichste Hitote, *Oreaster muticus*, wird von den Fischerfrauen als Garnwickler benutzt." I imagine that the writer has here in mind the form

---

1) I take this opportunity of remarking that the words " spines " and " tubercles " have been indiscriminately used by most writers on *Oreaster*. Several of them speak of the " apical naked portion " of the spine. It must however be understood that this "naked portion" is itself a real spine formed either by a single enlarged surface granule or by a fusion of several such and homologous with the movable spines of other parts and other species. Itself being homologous with a granule or granules it is but natural that it should be naked. The remaining part of a tubercle is an enlarged ossicle.

which has been named *Oreaster muricatus* GRAY var. *mutica* by v. MARTENS ['66, p. 80]. If so, it can not be said that it is very common, at least for Japan taken as a whole, for the species of this genus are, so far as well authenticated cases go, confined to the Ryûkyû and Linschoten Archipelagoes and possibly to the southern coasts of the Island of Kyûshû. Neither is the practice of using it as spools at all common. It is true that the name of Itomaki-Hitode *i.e.* spool starfish is widely used, but it is, so far as I know, used mainly for species of *Asterina* and has reference only to its fancied resemblance to certain forms of Japanese spools. I do not know of any actual case where it is used as spools for sewing threads.

A study of the literature of this species leads one to the conclusion that it is probably specifically identical with *O. nodosus*, and it is more in deference to authorities that it is here treated of under a separate heading.

Aside from a single pre-Linnæan reference to this species it is mentioned for the first time by BLAINVILLE in his " Manuel d'actinologie " ['34, p. 238] under the name of *Asteria linkii*, with a reference to LINCK, tab. 7, No. 8. Its first description appears to be that of GRAY, which runs as follows ['40, p. 277] :

" 8. *Pentaceros muricatus*, LINCK, t. 7. f. 8. *Ast. Linckii*, BLAINV. *A. nodosa*, LAM. SEBA, iii. t. 7. f. 3. Arms elongated, nearly as long as the width of the body, with a dorsal series of large, and with 2 or 3 large conical spines near the tips; back rather high, spinose.

"Inhab. ——. Brit. Mus."

It is placed in that section of the genus which has the " back formed of irregular elongated ossicula, apparently reticulated; the spines with elongated bases, interspaces closely punctured."

DUJARDIN and HUPÉ mention this species under the name of *O. muricatus*, giving references to LINCK, t. 7, No. 8, BLAINVILLE ['34, p. 238] and GRAY ['40], and simply adds, " Habite les mers de l'Inde."

LÜTKEN gives a very full description of this species ['64, p. 156] :

" LINCK afbilder t. VII fig. 8 en meget udpræget *Oreaster*- Form, som han i sin Text betegner som ' *Pen-*

LINCK figures in his t. VII fig. 8 a very pronounced *Oreaster*- form, which in his text he names ' *Pentaceros*

taceros B. gibbus muricatus, Castellum montanum dicendus.' Den samme Art findes afbildet hos SEBA t. VII f. 3 som '*Pentaceros gibbus & muricatus seu Castellum montanum.'* De eneste Forfattere, der i disse Figurer have erkjendt en egen Art, ere BLAINVILLE, som citerer LINCKS Figur til sin *Asterias Linckii*, GRAY, som henförer begge Afbildninger til sin '*Pentaceros muricatus*' (der vel egenlig opstilles alene paa dem, thi det antydes ikke, at han selv har seet et Exemplar af den), og DUJARDIN og HUPÉ, som (uden Tvivl efter GRAY) opföre en *Oreaster muricatus* som beboende 'de indiske Have'; heller ikke hos disse sidste Forfattere tyder noget paa, at de mere end deres Forgængere have havt Leilighed til selv at see Arten, hvilken derfor uden Tvivl indtil de senere Tider maa have været meget sjelden i Samlingerne.

"Museet besidder imidlertid, deels fra ældre, deels fra yngre Tid, 4 Exemplarer af noget forskjellig Störrelse af denne *Oreaster Linckii* (BLV.), og jeg er derfor ret godt istand til at oplyse dens Formforhold. Et af dem er ifölge en vedlagt Notits af afdöde Dr. BECK, fra Madagascar; to af de andre fra Zanzibar. For at Aldersforskjellighederne kunne træde desto tydeligere frem, vil jeg beskrive hvert af disse 4 Exemplarer for sig, men dog dvæle meest ved det störste og det mindste.

"Det mindste Exemplar (A) har en större Radius af 55 mm, en mindre

B. gibbus muricatus, Castellum montanum dicendus.' The same species is found figured in SEBA t. VII f. 3 as '*Pentaceros gibbus & muricatus seu Castellum montanum.'* The only authors, who have recognised in these figures an independent species, are BLAINVILLE, who assigns LINCK's figure to his *Asterias Linckii*, GRAY, who refers both figures to his '*Pentaceros muricatus*' (which is properly speaking based on those figures alone, because it is not indicated that he has himself seen an example of it), and DUJARDIN and HUPÉ, who (doubtless after GRAY) cites an *Oreaster muricatus* as inhabiting the Indian seas; nor is anything said by these last authors to show that they had occasion to see the species with their own eyes any more than their predecessors, whence it must have been very rare in the collections, up till recently.

The Museum, however, possesses, partly since long, partly since recently, 4 examples of some different sizes of this *Oreaster Linckii* (BLV.), and I am therefore in good position to shed light on its form relations. One of them is according to an accompanying notice by the late Dr. BECK, from Madagascar; two of the rest from Zanzibar. In order that the differences due to age may come out the more distinctly I will describe each of these 4 examples by itself, but dwell most on the largest and the smallest.

af 21; Forholdet er altsaa omtrent som 1 : 2⅓. Armene ere temmelig smækkre med trekantet Gjennemsnit. Stjernens midterste Deel optages af 5 höie og stærke kegledannede Pigge, som i ⅔ af deres Længde ere beklædte med flade Korn, der opad mod deres övre Grændse blive stedse större og fladere og antage Form af kantede Tavler; ved Grunden af de 3 af disse kegledannede Taarne er der udvendig en lignende tyk, men kort, vandret Pig. Hele dette Centralparti hæver sig op over Armene og sammenlignes ikke upassende med en i Knudepunktet af 5 Bjergrygge opfürt Borg. Rygtavlerne danne kun 3 Rækker paa hver Arm, en bredere i Midten og en smallere paa hver Side. Der er 4 lange Porefelter paa hver Arm, nemlig paa hver Side eet ovenover Randpladerne og eet under den midterste Tavlerække, adskilte ved en sammenhængende Tavlerække. Disse store Porefelter ere selvfölgelig egenlig opstaaede ved en Sammensmeltning af mange mindre, men oplöse sig kun mod Armenes Spidse i saadanne mindre Grupper paa nogle faae Porer. Foruden disse er der kun 5, af Hudskelettets Masker adskilte, mindre Porefelter indeni 'Borgen,' samt eet paa hver af dennes Sider, naar undtages den, hvor Madreporpladen findes. Alle disse Porefelter ere beklædte deels med smaae Pedicellarier, deels med finere Korn end de, som bedække Hudskelttet, af hvilke Randpladerne, Bugtavlerne og

The smallest example (A) has a greater radius of 55 mm, a smaller of 21; the ratio is therefore nearly as 1 : 2⅓. The arms are somewhat slender and triangular in cross section. The central part of the star is occupied by 5 high · and stout conical spines, which are covered ⅔ of their length, with flat grains which become constantly larger and flatter toward the upper limit and assume the form of polygonal tablets; near the base of three of these conical towers there is on the outside a similar thick, but short horizontal spine. The whole of this central part is raised above the arms and may be compared not improperly with a castle raised at the nodal point of 5 mountain crests. The dorsal plates form only 3 rows on each arm, a broader one in the middle and a narrower one on each side. There are 4 long pore-areas on each arm, namely one on each side above the marginal plates and one under the median row of plates, separated by a continuous series of plates. These large pore-areas of course really result from a coalescence of many smaller ones, but resolve themselves into similar smaller groups with a small number of pores only near the apex of the arms. Beside these there are only 5 smaller pore-areas inside the 'castle,' separated by the meshes of the dermal skeleton, with one on each of its sides, with the exception of that in which the madreporic plate is found. All

Armenes midterste Tavlerække have de största i störrelse og Form lignende dem, som beklæde de store Rygpigge. Paa hver Arm findes der dernæst—foruden de ovenfor beskrevne Pigge, som danne 'Borgen' og ikke kunne henregnes til Armene—3 stærke kegledannede Pigge (undtagelsesvis 2 eller 4), de to paa Armens indre Deel, den tredie tæt ved dens Spidse, alle hvilende paa den midterste Tavlerække ; samt paa hver Side 2 (undtagelsesvis 1 eller 3), udgaaende i en nogen skraa Retning fra to (1–3) af de övre Randplader, som ligge henimod Armens Spidse. Randpladernes Antal c. $\frac{14}{9}$; de nedre frembyde intet mærkeligt ; Pedicellarier savnes paa dem begge. Derimod findes en enkelt stor Klappetang hist og her paa Bugtavlerne, hvis Beklædning forresten bestaaer af Korn af meget forskjellig Störrelse, 2–6 store tavleformige Gryn, omgivne af mange smaae, men kantede og flade ligesom de större ; som hos andre Oreastre fortsætter et smalt Bælte af disse Bugtavler sig mellem Armfurerne og Randpladerne lige til Armspidserne. De indre Fodpapiller ere meget fine, 6–7 i hver Gruppe, men af den hos Slægten sædvanlige Form ; de ydre ere brede, flade og butte, to for hver Gruppe af hine. En langgrenet Tang er anbragt mellem hvert Sæt.

" Det næstmindste Explr. (B) (R = 70 mm, r = 26) afviger væsenlig kun fra det mindre ved en mere sammensat Bygning af 'Borgen,' som bedre

these pore-areas are covered partly with small pedicellariæ, partly with finer granules than those which cover the dermal skeleton, of which the marginal plates, the ventral plates and the median row of plates of the arms have the largest, and similar in form and size to those that clothe the large dorsal spines. On each arm there are found next—beside the above described spines, which form the 'castle' and can not be referred to the arms —3 stout conical spines (exceptionally 2 or 4), the two on the inner part of the arm, the third close to its apex, all resting on the median row of plates ; also two (exceptionally 1 or 3) on each side projecting in a somewhat oblique direction from two (1–3) of the superior marginal plates, which lie near the apex of the arm. Number of marginal plates c. $\frac{14}{9}$; the inferior ones present nothing worth noting ; pedicellariæ are absent from both. There is found, however, here and there single large pedicellaria on the ventral plates, the covering of which consists among others of grains of very different size, 2–6 large tablet-shaped grains, surrounded by several small, but angular and flat ones similar to the larger ; as in other Oreaster species a narrow belt of these ventral plates is continued on to the apices of the arms between the ambulacral furrows and the marginal plates. The inner foot papillæ are very fine, 6–7 in each group, but of the form usual in this genus ; the outer are broad, flat and blunt, two for each group of

beskrives under det fölgende Explr.
I Fodgangenes indre Deel er der 3
ydre Fodpapiller i hvert Sæt.

"Hos det næststörste (C) (R=84
mm, r=31 mm) bestaaer 'Borgen'
foruden af de under A omtalte 5
'Taarne' normalt[1] af to for hver af
hine, anbragte længere nede, umid-
delbart over Armryggens Begyndelse
og forbundne indbyrdes og med hine
ved et stærkt Net af Rygtavler, som
omslutte særskilte Porefelter. Deres
Stilling nærmer sig snart til det lodrette, snart til det vandrette. At de
smaae Porefelter, hvoraf de ovenfor
(ved A) beskrevne 4 större sammen-
sættes, hos dette og det fölgende
Exemplar i det hele vise sig mere
sondrede, er maaskee tildeels en Fölge
af Indtörringen, men kunde maaskee
ogsaa tildeels være en Fölge af, at
Hudskelettet med Alderen udvikledes
noget stærkere, og at Grændsen mellem de enkelte Porefelter derved blev
noget tydeligere.—Selv i Armens indre Deel findes der hos dette Exemplar kun 2 ydre Fodpapiller i
hver Gruppe.

"Det störste Exemplar (D) har
et Forhold mellem Stjernens Radier
=omtrent 1 : 3 (r=35, R=100 mm);
Randpladernes Antal er $\frac{17}{17}$. Ved
Grunden af hver af de fem store
Taarnpigge findes der 3 i en Tver-
række, rettede skraat udad, altsaa 20

the former. Apedicellaria with long
jaws is placed between each set.

"The next smallest Example (B)
(R=70 mm, r=26) deviates essentially from the smaller only by a more
complicated structure of the 'castle,'
which may better be described under
the following example. In the inner
part of the ambulacral furrow there
are three outer foot papillæ in each
set.

"In the next largest (C) (R=84
mm, r=31 mm) the 'castle' consists,
in addition to the 5 towers enumerated under A, normally[1] of two for
each of them, situated more inferi-
orly, immediately over the beginning
of the back of the arms and joined
with one another and with the 'towers'
by a stout net of dorsal plates, which
enclose the separated pore-areas.
Their attitude is sometimes more
nearly vertical, sometimes more nearly
horizontal. That the small pore-
areas, of which the above (in A)
described 4 larger ones are composed,
present themselves on the whole better
separated in this and the following
example, is perhaps partly a result
of dessication, but may perhaps also
be partly a result of the fact that
the dermal skeleton is developed more
strongly with age, and that the boundary between the several pore-areas
became thereby somewhat more distinct.—Even in the inner part of the
arm there are in this example only
2 outer foot papillæ in each group.

"The largest example (D) has the
ratio between the radii of the star=
nearly 1 : 3 (r=35, R=100 mm); the

---

[1] "De mangle nemlig undtagelsesvis (ere ikke
komne frem endnu) paa eet af de fem Steder,
hvor de skulde findes."

[1] They are namely absent exceptionally
(they have not yet appeared) from one of the
five places where they should be found.

i alt.[1] Midt ud ad hver Arm er der 5 (undtagelsesvis 6) höie, kegledannede, lodrette Pigge; desuden er der, paa hver Side af den · inderste af disse, en lavere, skraat ud'ad til Siden rettet Pig, som hviler paa en af de Tavler, der adskille Armens övre og nedre Porefelter. Derimod er Randpiggenes Antal ikke forøget med Alderen, det er ogsaa her kun 1 eller 2 paa hver Side, henimod Spidsen. Bugtavlernes Klappetænger ere talrigere end paa yngre Exemplarer; der er i Regeln 1, ofte 2 eller 3 paa hver Tavle, og navnlig langs med Fodgangene synes deres Forekomst at være constant. Fodpapillernes Antal er $\frac{3}{2}$ i den indre Deel af Fodgangen, de indre Mundpapiller c. 12, hvoraf dog kun de 4 inderste sees, da de andre ere betydelig finere og lavere; disse fire suppleres imidlertid af den ydre Række (svarende til de ydre Fodpapiller), saa at der omkring hvert Mundhjörne er en Række af 16 tykke, butte, rynkede Papiller af lignende Form som nærmeste ydre Fodpapiller. Kornbeklædningen ledsager her de store Rygpigge lige til deres Top, men Grændsen mellem de enkelte Korn eller Smaatavler er ofte saa utydelig, at de synes at danne et sammenhængende Lag."

number of marginal plates is $\frac{11}{11}$. At the base of each of the five large tower-like spines there are 3 in one transverse row, directed obliquely outwards, therefore 20 in all.[1] Along the middle of each arm there are 5 (exceptionally 6) high, conical, vertical spines; there is beside, on each side of the innermost of these, a lower spine directed obliquely outwards towards the side, resting on one of the plates that separate the upper and lower pore-areas of the arm. On the contrary the number of marginal spines does not increase with age, there is in this case also only 1 or 2 on each side near the apex. Pedicellariæ of the ventral plates are more numerous than in younger examples; there is as a rule 1, often 2 or 3 on each plate, and especially along the ambulacral furrow their occurrence appears to be constant. Number of foot papillæ is $\frac{3}{2}$ in the inner part of the furrow, the inner oral papillæ c. 12, of which, however only the four innermost can be seen, as the others are much finer and lower; these four are however reinforced by the outer row (corresponding to the outer foot papillæ), so that around each mouth angle there is a row of 16 thick, blunt, fluted papillæ similar in form to the nearest outer foot papillæ. The granular covering accompanies in this case the large dorsal spines nearly to their top, but the boundary between the several grains or tablets is often so indistinct that they appear to form a continuous sheet.

---

1) " Lige i Stjernens Midte udvikles undertiden en lille Pig saaledes som paa det af SEBA afbildede Exemplar."

---

1) Sometimes there is developed exactly in the centre of the star a small spine similarly as in the example figured by SEBA.

GRAY in his "Synopsis" reproduces his previous description but gives some additional synonyms ['66, p. 6]:

" 8. *Pentaceros muricatus.* Arms elongated, nearly as long as the width of the body, with a dorsal series of large, and with two or three large conical spines near the tips; back rather high, spinous. GRAY, Ann. N. H. 1840, p. 277; LINCK, t. 7. f. 8. *Ast. Linckii*, BLAINV. *A. nodosa*, LAM. SEBA, iii. t. 7. f. 3. *Oreaster turritus*, MÜLL. & TROSCH. Ast. 47. Inhab.........(Brit. Mus.)."

VON MARTENS gives a detailed description of this species and distinguishes three varieties ('66 p. 77]:

" 19. *Oreaster muricatus* (LINCK) GRAY. *Pentaceros muricatus* LINCK de stellis marinis p. 23. Taf. 7. Fig. 8, kopirt in der Encyclopaedie 106, 1.—SEBA Band III. Taf. 7. Fig. 3.—*Asterias nodosa* LINNÉ zum Theil.— *Ast. nodosa* var. 3 LAM.—*Pentaceros muricatus* GRAY Ann. and Mag. n. h. VI. 1841, p. 277.—? *Oreaster tuberculatus* MÜLLER und TROSCHEL Syst. Asterid. p. 47.—*Oreaster castellum* GRUBE Breslauer Zeitung vom 7. Febr. 1865.

" Rückenseite gekörnt mit grössern Höckern, die in ihrer unteren Hälfte stets mit flachen Körnchen bedeckt sind und auf den Armen eine mittlere Reihe bilden; Unterseite mit gekörnten Täfelchen und einzelnen klappenartigen Pedicellarien bekleidet, Furchenpapillen in 2 Reihen, in der inneren je sieben bis zehn, schlank, die mittleren länger, in der äusseren 2-3 flache, stumpfe, gleichlange; Randstacheln sehr variabel, nie an allen Randplatten vorhanden.

" Es liegt mir eine grössere Anzahl von Exemplaren sowohl desselben als verschiedener Fundorte vor, und erlaubt mir, die fast masslose Schwankung aller Artkennzeichen etwas näher zu verfolgen. Auf der Rückenseite tritt das Balkennetz, dessen Maschen die Porenfelder darstellen, bald mehr bald weniger hervor, indem die Granulation entweder gröber und flacher ist als die der Porenfelder und ziemlich genau derjenigen der Höcker gleicht, bald hierin kaum ein anderer Unterschied als der durch das Zwischentreten der Poren selbst bedingte stattfindet; Uebergänge von dem einen zum andern kommen an demselben Individuum vor. Wo die Körnelung ganz gleichmässig wird (Exemplare von Larentuka), ist das Balkennetz als solches nicht mehr ins Auge fallend. Ueberhaupt bietet es ein netzartiges Ansehn mit gradlinig begränzten Maschen nur auf der Scheibe. An den Armen treten die Verbindungslinien zurück und man sieht nur Reihen von rhombischen etwas erhabenen gekörnten Platten, getrennt durch die Porenfelder, und auf jedem Arm anfänglich fünf, gegen die Spitze zu nur drei, indem die äussere bald früher bald später in den Randplatten endigt. Die Höcker sind eigentlich nichts anderes, als solche gekörnte Platten, welche hoch konisch oder kugelig sich erheben und in der Mitte (Spitze) fast immer der Körnelung entbehren; diese nackte Spitze ist bald mehr bald weniger von dem mehr gewölbten gekörnten unteren Theile abgesetzt. Solche Höcker finden sich stets in der Mittellinie der Arme, mehr oder weniger regelmässig abwechselnd mit gewöhnlichen Platten, meistens 8 in einem Radius; die fünf innersten bilden ein Fünfeck auf der Scheibe und ihnen gesellen sich bei grösseren Exemplaren fast immer noch einige ähnliche ausserhalb dieser Reihen in den Armwinkeln liegende bei, nur bei einem grossen Exemplar von Mossambique entwickelt auch die nächstobere Plattenreihe des Armrückens einige solche Höcker, so dass dieselben hier (abgesehen von den Randplatten) in drei Reihen stehen. Ein centraler Höcker in der Mitte der Scheibe ist meist vorhanden, aber kleiner als die fünf umgebenden; zuweilen fehlt er völlig und zwar ist sein Vorhandensein und seine Grösse nicht im Verhältnisse zur Deutlichkeit des Balkennetzes, welches hier in der Mitte der Scheibe eine regelmässige fünfstrahlige Sternfigur bildet. Diese Höcker sind mit flacher Granulation besetzt und zeigen in der Regel eine abgesetzte Spitze; an denselben Exemplaren finden sich aber auch solche die oben abgerundet und vollständig von Granulation bedeckt sind.

"Die grösste Variation zeigen die Stacheln der Randplatten; bald sind einzelne an Grösse und Form vollständig den Höckern auf dem Armrücken gleich wie auch in den angeführten Abbildungen, so namentlich an der Armspitze, bald fehlen sie, und zwar beides ebensowohl bei kleineren (jüngeren) als bei grösseren Exemplaren. In den Armwinkeln fehlen sie den oberen Randplatten fast immer, von der Mitte bis zur Spitze der Arme zeigen sie Neigung zum Abwechseln, indem auf eine Randplatte mit grossem Stachel (Höcker) eine andere mit kleinem oder gar keinem folgt. Hier kann man

besonders dentlich sehen, dass der Höcker nichts anderes ist, als die ange-
schwollene gekörnte Platte selbst plus dem aufsitzenden nackten Stachel.
Die Stacheln der unteren Randplatten sind meistens kleiner, mehr gleichmässig und mit wenigen Ausnahmen auf allen Platten vorhanden, namentlich auch in den Armwinkeln; nur nahe der Spitze der Arme werden sie bei
einzelnen Exemplaren grösser, selbst so gross wie die grössten Randplatten.
An manchen Exemplaren ermangeln alle untern Randplatten der Stacheln,
so an den grössern Exemplaren von Mossambique, während an den kleineren
von ebenda einzelne vorhanden sind.

"Auch die Körnelung der Platten der Bauchseite zeigt Schwankungen;
im Allgemeinen sind es eckige, flache Körnchen, meist gröber und immer
unregelmässiger als die Körnelung der Randplatten und der Rückenhöcker,
zuweilen aber auch wenig davon unterschieden (Exemplare von Mossambique);
manchmal erhebt sich nahe dem Munde auf fast jeder Platte ein höheres,
fast stachelartiges Körnchen über die anderen, so bei einigen Exemplaren
von Larentuka.

"Endlich zeigen die Furchenpapillen Verschiedenheiten, indem in der
äusseren Reihe bald drei bald zwei auf je eine Platte kommen, bei kleineren
Exemplaren gegen die Armspitze sehr oft auch nur eine, in der inneren aber auf
einer Platte bald eine grade, bald eine ungrade Zahl steht, so dass bald eine
bald zwei in der Mitte die längsten sind und hierdurch wie durch Hinweg-
fallen der kleinsten äussersten der auf einer Platte vorhandenen ihre Anzahl
von 7 bis 10 schwankt.

"Alle diese Variationen kreuzen sich so sehr durcheinander, dass man
darnach keine irgendwie bestimmbaren Lokalvarietäten aufstellen kann;
dagegen gränzen sich einige der mir vorliegenden Exemplare durch die
gleiche Combination der an anderen zerstreut vorkommenden Abweichungen
zu bestimmbaren Abarten ab:

"19 a. var. *multispina*. SEBA III. 6. 1, 2, 11, 12. Klein. Höcker in einer
Reihe, die meisten, wie die meisten Randplatten mit einem Stachel.
Balkennetz nicht vortretend.

"19 b.' *mutica*. JOH. MÜLLER's und TROSCHEL's *Oreaster hiulcus* S. 48
und wahrscheinlich auch GRAY's *Pentaceros hiulcus* l. c. S. 276, aber

nicht der von beiden citirte *Pentaceros hiulcus* LINCK. tab. 26. fig. 41, welcher übrigens irgend eine andere Form dieser vielgestaltigen Art sein mag.

" Nicht nur die erste, sondern auch die zweite Plattenreihe zwischen Armrücken und Randplatten bis nahe zur Spitze der Arme fortlaufend ; keine Höcker oder Stacheln weder an den oberen noch unteren Randplatten ; die Höcker des Armrückens und der Scheibe ungewöhnlich gross und kugelartig, meist ohne, selten mit kurzer nackter Spitze, die fünf mittleren noch höher als die übrigen, dagegen nie ein unpaarer Höcker im Mittelpunkt der Scheibe. Bezüglich des Unterschieds der Granulation auf Balkennetz und Porenfeldern, sowie des Fehlens oder Vorhandenseins sekundärer Höcker auf der Scheibe ausserhalb der fünf grossen ist diese Varietät denselben Variationen unterworfen.

" 19 c. var. *intermedia.*

" In der Rückenbedeckung und dem Mangel aller Randstacheln mit der vorigen übereinstimmend ; keine einzelnen Höcker ausserhalb der fünf Reihen ; einige Höcker endigen bei der Mehrzahl der Exemplare in einer längeren nackten Stachel wie es bei *muricatus* Regel ist, bei anderen sind sie stumpf abgerundet ; die Arme sind verhältnissmässig breiter und kürzer als bei der gleich grossen *mutica* aus Timor.

" Farbe während des Lebens grau oder röthlich, der Rand lebhaft roth, die Höcker schwärzlich ; je nach der Lebhaftigkeit der Färbung kann man zwei Abänderungen unterscheiden : a) die *rothe*: oben und unten purpurroth, die Höcker schwarz mit intensiver rothem Hof, der Rand tief karminroth : b) die *graue*: oben und unten braungrau, die Armspitzen und die Höcker schwarzbraun, der Rand orangefarbig.

" An einzelnen Individuen geht die graue Färbung gegen die Spitze der Arme zu plötzlich in die rothe über. Die schwarze Farbe erblasst in Spiritus meist völlig oder wenigstens schieferblau, die rothe wird erdbraun.

" *Oreaster tuberculatus* MÜLLER und TROSCHEL S. 47 möchte, nach dem einzigen im Berliner zoologischen Museum vorhandenen Exemplar, einem verkrüppelten und verzerrten, zu urtheilen, auch noch in den Kreis dieser Art gehören ; es zeichnet sich aus durch kleine, aber zahlreiche spitze Höcker,

Stacheln auf fast allen unteren, aber auf keiner der oberen Randplatten,
ferner durch vereinzeltes Vorkommen von klappenartigen Pedicellarien auf
den oberen Randplatten und der Rückenfläche, was ich an allen anderen
Exemplaren vermisse.

" *Oreaster mammillatus* AUDOUIN Descr. Eg. pl. 5 ; MÜLL. TROSCH. S.58
aus dem rothen Meer ist nur durch die schlankere Arme und die Grösse von
meinen Exemplaren aus Flores zu unterscheiden.

"Dimensionen in Millimetern.

	Armradius	Scheiben-radius.	Breite der Arme an ihrer Wurzel.	In der Mitte.	Höhe der Scheibe.
muricatus von Mossambique . .	116	44	49	45	29
,, ,, ,,	25	10	11½	8	9
,, ,, ,,	60	24	27	15	20
,, ,, ,,	35	14	18	10	11½
,, ,, ,,	30	12	14	8	10
Var. *multispina* von Larentuka	75	30	38	19	27½
Var. *mutica* von Timor	94	38	46	34	28
,, ,, ,, ,,	51	20	22	21	15
,, ,, (*hiulcus* M. Tr. Originalexemplar). . .	86	34	32	37	24
Var. *intermedia* von Amboina	34	17	18	13	11
turritus von Amboina	166	99	58	34	34½

"Die Hauptform habe ich nicht selbst im indischen Ocean gesehen
(A).[1] Var. *multispina* auf Larentuka, Insel Flores.

"Var. *mutica.* m. (*hiulcus* MÜLL. TROSCH.) auf der Insel Timor bei
Kupang und Atapupu.

"Var. *intermedia* m. auf Batjan (eigentliche Molukken) ; ROSENBERG
sandte dieselbe aus Amboina dem Kgl. Museum zu."

PERRIER gives the following description of the pedicellariæ of this
species ['69, p. 74] :

---

1) Occurring throughout the Indian Archipelago and extending to the east coast of Africa.

" *Oreaster muricatus*, Duj. et Hupé."[1] Syn.—*Pentaceros muricatus*, Linck. *Asterias Linckii*, de Blainv. *Oreaster Linckii*, Val. (Coll. Muséum).

"Sur la face dorsale, dans les aires tentaculaires, on aperçoit entremêlés avec les granulations de la peau, un nombre considérable de petits Pédicellaires valvulaires, mais à valves très-peu allongées transversalement, de telle façon que le Pédicellaire présente l'aspect d'un petit bouton arrondi, fendu suivant l'un de ses diamètres. Ces Pédicellaires sont à peine plus gros que les granulations de la peau.

"À la face ventrale, on distingue deux sortes de Pédicellaires, les uns en pince, les autres valvulaires. Chacune des pièces calcaires qui bordent les sillons ambulacraires porte à son angle interne le plus rapproché de la bouche un Pédicellaire en pince. Ces Pédicellaires se trouvent donc placés comme dans l'espèce précédente.[2] Ces Pédicellaires sont très-allongés (trois fois environ aussi longs que l'ensemble de la largeur des deux mâchoires).— La pièce calcaire qui les constitue est moins dense que dans l'espèce précédente ; on n'y aperçoit pas non plus de pointes aussi nombreuses, elles sont réduites à de simple petites saillies irrégulières et arrondies.

"Les pièces de la seconde rangée portent aussi le plus souvent un et quelquefois deux Pédicellaires valvulaires, très-allongés quand ils sont seuls. Ces Pédicellaires forment une rangée assez régulière le long des sillons ambulacraires. Les pièces suivantes peuvent également porter ces Pédicellaires valvulaires, mais la distribution de ces organes se fait sans régularité. Ils ne sont abondants que vers le sommet du triangle compris entre deux sillons ambulacraires consécutifs. Dans cet espace, chaque pièce calcaire en porte au moins un. Au delà du milieu de la distance qui sépare le sommet de ce triangle du sommet de l'arc rentrant qui limite les bras, les pièces calcaires qui portent des Pédicellaires valvulaires sont l'exception.

"Les plaques marginales des bras sont constamment dépourvues de Pédicellaires valvulaires."

Lütken in his paper of 1871 [p. 259] has the following remarks with reference to the species of *Oreaster* :

---

1) " Pl. 2, fig. 3 *a* et *b*."
2) *Oreaster reticulatus* M. T.

" En Reduktion, hvorved *O. Linckii*, *O. dosratus* L. (SEBA t. VI, f. 1–2), *O. nodosus* GR. (SEBA t. VI. f. 11–12),[1] *O. hiulcus* M. TR. og *O. mamillatus* AUD. reduceres til kun at udgjöre een Art, er mig saa ufattelig, at jeg skal afholde mig fra enhver Kritik af den og kun tillade mig at udtale min fuldstændige Uenighed med den ærede Forfatter med Hensyn til det Resultat, hvortil han mener at være kommet."

A reduction, by which *O. Linckii*, *O. dorsatus* L. (SEBA t. VI, f. 1–2), *O nodosus* GR. (SEBA t. VI, f. 11–12)[1], *O. hiulcus* M. TR. and *O. mamillatus* AUD. are reduced to form only a single species, is so incomprehensible to me, that I shall abstain from any criticism of it and permit myself to express my complete disagreement with the honoured author [v. MARTENS] in regard to the result to which he believes to have come."

HOFFMANN ['74] mentions and figures this species under the name of *O. muricatus* (LINCK) and var. *mutica* (*fide* Zool. Record).

PERRIER in his "Révision" gives some interesting informations as to the locality of this species, which he calls *Pentaceros muricatus* ['76, p. 55] :

" En 1864, au moment où elle a été décrite en détail par le docteur LÜTKEN, cette espèce n'était encore que très-peu connue et considérée comme très-rare dans les collections. Le muséum de Paris n'en possédait pas moins à cette époque une magnifique série composée de vingt-neuf exemplaires, dont vingt-quatre avaient été recueillis par M. LOUIS ROUSSEAU, à Zanzibar et aux îles Seychelles. Cette belle espèce est parfaitement distincte et présente de nombreuses variations dans le nombre et la disposition de ses piquants, remarquables d'ailleurs par leur développement.

" La plupart de ces variétés ont été décrites par VON MARTENS dans

---

1) " Ogsaa i Tillæget til V. D. DECKENS Rejse (l.c. S. 130) forenes *O. mamillatus* med *O. muricatus*; derimod opföres *O. nodosus* GRAY særskilt, hvilket synes at stride mod Henförelsen af SEBAS t. VI, f. 11–12 til ' *var. multispina*' af ' *O. muricatus*.' At den af LINCK (l. c. t. VII, fig. 8) og SEBA (t. VII, fig. 3) afbildede Form ikke kan benævnes *Oreaster muricatus*, men maa benævnes *O. ·Linckii* BLAINV., er udviklet i mine ' Kritiske Bemærkninger om forskjellige Söstjerner' o.s.v. (1864), S. 156 (34), hvor jeg atter fremdrog denne mere end halvt forglemte Art og beskrev den udförligt."

1) Also in the appendix to v. D. DECKEN's travel (l.c. p. 130) *O. mamillatus* is united with *O muricatus*; but *O. nodosus* GRAY is cited separately, which appears to be inconsistent with the reference of SEBA's t. VI, f. 11–12 to ' *var. multispina* ' of ' *O. muricatus*.' That the form figured by LINCK (l c. t. VII, fig. 8) and SEBA (t. VII, fig. 3) can not be called *Oreaster muricatus*, but must be called *O. Linckii* BLAINV., is explained in my ' Critical Remarks on Different Starfishes' etc. (1864), p. 156 (34), where I brought to light this more than half forgotten species anew and described it in detail.

son travail sur les Échinodermes de l'Asie oriental : nous n'y reviendrons pas ; mais nous devons protester, comme le docteur LÜTKEN, contre la réuinon proposée par VON MARTENS de cette espèce avec les *Pentagonaster*[1] *mamillatus, hiulcus* et *turritus*, M. T., qui sont des espèces parfaitement distinctes.

"Les individus que possède le Muséum se répartissent ainsi :

"Deux exemplaires desséchés, de l'île de France, PÉRON et LESUEUR, 1803.

"Deux exemplaire desséchés de l'île Bourbon, M. LANZ, 1865.

"Un exemplaire desséché des Seychelles, M. LOUIS ROUSSEAU, 1841.

"Vingt-trois exemplaires (trois desséchés, vingt dans l'alcool), Zanzibar, M. LOUIS ROUSSEAU, 1841.

"Un exemplaire desséchés, Zanzibar (Mus. Comp. Zool.), 1864.

"Il existe au British Museum deux exemplaires de cette espèce provenant de Ceylan."

VIGUIER gives a detailed description of the skeletal system of this species ['78, p. 197] :

"Cette dernière espèce [*P. muricatus*], qui habite la côte orientale d'Afrique, diffère au premier coup d'œil du *Pentaceros reticulatus*. Ici, en effet, les piquants ne sont pas simplement implantés sur les pièces squelettiques, et caducs ; mais ce sont les ossicules dorsaux eux-mêmes qui prennent un énorme développement et donnent á l'animal sa physionomie particulière.

"La figure 8 (pl. xii) a été faite de manière à bien mettre en évidence les dimensions des piquants dorsaux. La figure 10 représente le centre du disque, qu'on n'aperçoit que de profil sur la figure 8. Enfin la figure 9 montre la face inférieure d'un bras. Toutes les trois sont de grandeur naturelle.

"Les bras, fortement carénés, portent sur la ligne médiane une série de pièces en forme d'hexagones allongés dans le sens transversal, et légèrement convexes. Les pièces 1, 7, 11, 14 et 16 en partant de l'extrémité du bras sont surélévées en forme de gros piquants arrondis à leur sommet, et près de 1 centimètre de haut. Toutes les pièces de cette série médiane sont à peu près exactement contiguës. Après le cinquième piquant viennent deux ossicules à peu près cubiques, puis un dernier très-allongé dans le sens longitudinal et qui aboutit à une des cinq grandes pièces *apicales* du disque.

---

1) Evidently a misprint of *Pentaceros.*

"De chaque côté de cette série médiane, on en voit une autre composée aussi de plaques hexagonales vers l'extrémité du bras, où elles sont exactement contiguës à la série médiane. À peu près vers le milieu du bras, elles s'écartent de cette série qu'elles ne touchent plus que par un prolongement qui part de leur côté interne, limitant ainsi des aires quadrangulaires. Leur bord externe est devenu droit à ce niveau. Enfin, à partir du cinquième piquant, elles perdent tout rapport avec la série médiane, et se terminent par une grosse pierre allongée qui aboutit au piquant *apical*, en limitant une aire porifère très-allongée.

"Nous avons ainsi tout le long du bras trois séries longitudinales d'ossicules, dont les deux latérales sont toujours composées de pièces à peu près planes. Vers le milieu du bras quelques pièces irrégulières viennent encore s'interposer entre les séries externes et les plaques marginales supérieures ; enfin l'espace interbrachial est occupé par des pièces irrégulières dont la forme paraît dériver de l'hexagone, et dont on voit toujours une paire située de part et d'autre de la ligne interbrachiale, immédiatement en dehors du disque. C'est sur une de ces paires que se trouve la plaque madréporique *m* (fig. 10), qui présente à peu près la même forme et les mêmes caractères que chez le *Pentaceros reticulatus*. Ici également, les deux pièces de support sont échancrées en forme de rein à leurs faces contiguës, pour livrer passage au canal hydrophore.

"Les cinq grosses pièces apicales dont nous avons parlé forment un pentagone à peu près régulier sur le centre de la face dorsale. On voit très-bien sur les figures 8 et 10 (pl. xii) leur forme assez difficile à décrire.

"De la base de chacun de ces piquants partent six pièces, dont trois, que nous avons déjà vues, sont le commencement des séries longitudinales des bras ; deux se portent aux piquants apicaux voisins ; enfin la sixième se porte vers le centre du pentagone. C'est là, un peu à gauche de la ligne médiane, en supposant la plaque madréporique en arrière, que se trouve l'anus entouré, comme chez l'autre type de *Pentaceros*, de plaques très-petites.

"Les côtés du pentagone sont formés par de grosses pièces arquées, allant d'un piquant à l'autre ; parfois il y en a deux bout à bout. Du milieu des côtés de ce pentagone, en face de la paire de plaques décrite plus haut, une pièce irrégulièrement branchue se porte aussi vers le centre. Les aires porifères sont donc fort irrégulières sur le disque, ainsi que sur les espaces interbrachiaux.

"Les plaques marginales dorsales diffèrent aussi très-notablement de ce que nous avons vu dans l'autre *Pentaceros*.[1] Ici, en effet, au lieu d'être allongées dans le sens du bras, on les voit, dans l'angle formé par deux bras, très-allongées dans le sens de la ligne interbrachiale, tandis que leur largeur atteint tout au plus de tiers de leur longueur. Du reste, elles ne sont point contiguës dans l'angle des bras, et affectent une disposition rayonnante.

"En s'écartant de la ligne interbrachiale les plaques marginales se raccourcissent assez vite, en même temps qu'elles s'élargissent un peu et se pressent les unes contre les autres. Elles n'ont jamais de rapports bien intimes avec les séries latéro-dorsales, auxquelles les relient seules de petites pièces, irrégulières de forme et de disposition, et qui manquent assez souvent. Elles décroissent assez régulièrement de volume jusqu'à la plaque ocellaire, qui est ici encore très-réduite, *oc*. La cinquième ou sixième plaque marginale avant la plaque ocellaire, est surélevée en un gros piquant arrondi, semblable à ceux de la série médiane des bras ; au-dessus d'elle vient une plaque ordinare, puis une autre plaque transformée en piquant d'un volume encore plus considérable.

"Il y a donc, à l'extrémité de chaque bras, quatre piquants marginaux, deux de chaque côté, et à des hauteurs à peu près correspondantes. Le reste des plaques marginales a la surface lisse, légèrement convexe.

"Les plaques des séries margino-ventrales ont leur plus grandes dimensions vers le milieu du bras ; de là elles décroissent régulièrement, d'une part vers la ligne interbrachiale, de l'autre vers l'extrémité du bras. Elles correspondent du reste exactement aux plaques margino-dorsales, sauf tout à fait au bout du bras, où trois plaques ventrales répondent à deux dorsales. Le relèvement de la pointe du bras, qui permet cette concordance, laisse voir cette disposition sur notre figure 8. Ces plaques margino-ventrales ont leur surface libre à peu près circulaire, très-légèrement convexe, et quelquefois marquée

---

1) *P. reticulatus.*

de très-petits alvéoles à pédicellaires. Ainsi qu'on le voit sur notre figure 9, c'est le très-petit nombre qui en possède; toutefois il est, sur la gauche du dessin, une plaque qui en porte une dizaine à elle seule; c'est donc à tort que M. PERRIER a dit qu'il n'existait jamais, chez cette espèce, de pédicellaires valvulaires sur les plaques marginales.

"Le bord interne des plaques marginales ventrales est recouvert par les ossicules ventraux; ceux-ci, très-variables de forme et de dimension, étroitement pressés les uns contre les autres, forment une sorte de mosaïque, où il est impossible de reconnaître des rangées régulières, sauf tout à fait au contact de la série adambulacraire; encore cette rangée, composée de pièces assez grosses, et distincte sur toute la longueur du bras, devient-elle, près de la bouche, assez difficile à suivre par la position et l'irrégularité des ossicules qui la constituent. Toutes les pièces de ces rangées portent chacune au moins un, le plus souvent plusieurs alvéoles, où se logent de très-petits pédicellaires valvulaires.

"Dans l'angle interbrachial, où les plaques ventrales sont fort petites et s'avancent jusque sur les dents, chacune de ces plaques porte aussi au moins un alvéole à pédicellaire, comme l'avait très-bien vu M. PERRIER; le long des bras elles en sont presque toujours dépourvues.

"Toutes ces plaques ventrales ont leur surface très-légèrement convexes, ce qui donne à la face inférieure de l'animal une apparence pavimenteuse; leur épaisseur est assez faible.

"Les plaques adambulacraires sont presque cubiques, un peu moins pressées les unes contre les autres que chez le *Pentaceros reticulatus*. Chacune d'elles porte, le long du sillon, à l'angle tourné du côté de la bouche, un petit alvéole très-fin où se loge un pédicellaire en pince. Ces petits alvéoles disparaissent vers l'extrémité du bras. Il n'y a rien de particulier à dire des pièces ambulacraires, qui ressemblent à celles de l'autre *Pentaceros*, bien qu'un peu moins fortes. Ainsi qu'on peut le voir, elles ne présentent rien d'extraordinaire dans leur disposition. La coupe de bras qu'a donnée M. GAUDRY de cette espèce, qu'il nomme *Oreaster linckii*, a sans doute porté sur un échantillon très-déformé, car elle est absolument méconnaissable; et l'on serait bien embarrassé pour distinguer sur cette figure une plaque marginale d'une pièce dorsale, ventrale, ou même ambulacraire, autrement que par la situation.

" La bouche du *Pentaceros turritus*, qui est tout à fait semblable au *Pentaceros muricatus*, nous ayant servi de type pour notre description générale et notre schéma A (p. 70), on n'aura qu'à se reporter là pour en avoir une description détaillée. La figure 11 (pl. xii) représente l'odontophore, dont on remarquera la grande similitude avec celui du *Pentaceros reticulatus*.

" Le système interbrachial[1] rappelle tout à fait ce que nous avons vu chez la Culcite, et diffère complètement de la muraille d'ossicules de l'autre type de *Pentaceros*. En se reportant au schéma A, on verra la disposition des muscles dorso-ventraux."

BELL gives a full description of this species under the name of *O. lincki* ['84, p. 72] :

[References to DE BLAINVILLE and PERRIER ['76, p. 239].[2]]

" R=3 r. Disk moderately high ; arm moderately wide, not at all acutely pointed. Lophial spines well developed, the apical very prominent ; a spine or two sometimes developed within the apical region.

" About 18 marginal plates ; the superomarginals alone form the sides of the arms, and are alone provided with spines ; these are confined to the distal end, and vary considerably ; from one to four may be developed, and in some specimens they are twice as long as they are in others.

" Adambulacral spinulation diplacanthid ; in the inner row eight poorly developed spines, in the outer two, which are much stouter, for each plate ; the tips of the latter are often marked by several shallow grooves ; as so frequently happens, a forcipiform pedicellaria is developed between each inner group of adambulacral spines.

" The separate ventral ossicles are hardly, if at all, to be made out under the exceedingly coarse granulation by which they are covered ; the separate granules vary considerably in size, and a few valvular pedicellariæ are scattered among them. The granules on the marginal plates are hardly

---

1) Interbrachial septum.

2) " M. PERRIER here adopts the name of LINCK ; a course in which, I regret, I cannot follow him."

less coarse. The dorsal surface is rendered markedly reticulate by the great size and close approximation of the poriferous areas, two of which pass along each side of every arm ; in the middle ofthe arm the second of these may equal in length as much as half the whole height of the arm ; sometimes the connecting processes of the ossicles become very delicate, when the whole side of the arm appears to form a huge poriferous area. Spines are very irregularly developed at the angle of the areas ; sometimes they are distributed so regularly that one may almost speak of a regular row of spines running on either side of the lophial series ; in other cases they are completely absent. This happens sometimes also to the spines of the lophial ridge itself, but they are ordinarily very well developed, as are, too, the apical spines and the spines that stand below them on the sloping sides of the disk. The granulation on the dorsal spines and ossicles is very coarse and extends sometimes quite to the tips of the spines. Madreporic plate rather small, not conspicuous.

"Colour (when dry)—lower surface reddish, upper reddish where the granules are developed, with grey poriferous areas ; in some cases the dried specimens are almost white, but this may be due to the mode of drying.

"The above description has been drawn up from a set of five specimens, which were collected at the same time and place (between tidemarks, at the Mozambique, in May 1882) by Dr. COPPINGER, H.M.S. , Alert,' and illustrate the exactness of the statement of Dr. VON MARTENS :— 'Alle diese Variationen kreuzen sich so sehr durcheinander, dass man darnach keine irgendwie bestimmbaren Lokalvarietäten aufstellen kann.' The variations are so marked that it seems to be impossible to follow Dr. VON MARTENS in establishing definite 'varieties.' The exact state of the case is, I think, this. The strength of the marginal and ventral plates, with their coarse granulation, is sufficient for the safety of the Starfish ; the spines are additional defences that are not constantly needed, and are developed more according to the conditions of individual environment than in obedience to the necessities of the species. They are organs which have begun to disappear, and their importance to their possessor may be judged of by the extent to which they vary in number and size on the different

arms of one and the same individual. The species stands midway between
O. alveolatus, in which inferomarginal spines are also developed, and *O.
nodosus,* in which there are no marginal spines at all.

" *Hab.* Indian Ocean (Mauritius, Timor).

" R 110   95   80   64

" r 40   34   28   26

" Greatest breadth of arm 35   34   31   26."

BELL mentions *O. lincki* from Mosambique ['84a, p. 510]. STUDER
mentions it from New Britain as follows ['84, p. 40]: "Neben den oben
genannten Arten[1]) ist im Neu-Britanischen Archipel noch *P. muricatus* GRAY
häufig, wie zahlreiche von FINSCH in Neu-Britanien gesammelte Exemplare
des Berliner Museums zeigen. Ich erhielt dort diese Art nicht." BELL
mentions it from Ceylon and adds ['87a, p. 647], "I think it is probable
that further search will be rewarded by the discovery of other species of
Oreaster on the shores of Ceylon." The same writer again mentions it
from the Sea of Bengal ['88, p. 384, 388]. MEISSNER ['92, p. 187] mentions it under the name of *Pentaceros muricatus* (GRAY) from Zanzibar, "1
getrocknetes Exemplar." RUSSO mentions it from the Red Sea ['94, p. 162,
fide Zool. Record].

SLUITER mentions this species in his paper on the Asteroidea of the
Amsterdam Museum ['95, p. 56]:

" 29. *Pentaceros muricatus* LINCK. Ein getrocknetes Exemplar von
Neu-Ierland, zwei getrocknete von den Molukken (V. D. HUCHT), und ein in
Alkohol von Zanzibar."

It appears to be mentioned also by PFEFFER from Zanzibar ['96].
DÖDERLEIN mentions the presence of the "Krystallkörper" in this species
['98, p. 493].

It is also mentioned by LUDWIG ['99, p. 540] from Zanzibar as
follows:

" *Pentaceros muricatus* LINCK.

" 1 Exemplar dieser weitverbreiteten indopacifischen Art von Sansibar,
von wo sie schon durch LÜTKEN (1875), PERRIER (1875), MEISSNER (1892)

---

1) *O. hiulcus* and *O. turritus.*

und PFEFFER (1896) bekannt war ; von Mozambique wird sie von v. MAR-
TENS (1866) und BELL (1884) erwähnt, von Madagaskar von LÜTKEN (1864)
und HOFFMANN (1874)."

BELL mentions it as either specifically identical with *O. nodosus* or else
closely interbreeding with it (see under *O. nodosus*, p. 486, BELL, '99).

It is again mentioned from Zanzibar by BELL [:03, p. 244] :

" 5. *Pentaceros Lincki*, DE BL.

"This is the form catalogued by Prof. LUDWIG as *P. muricatus* of
LINCK, but that zoologist was not a binomialist.

"Zanzibar shore."

HERDMAN [: 03, p. 107] mentions this species under the name of
Pentaceros lincki as occurring in very large numbers on some of the pearl-
oyster banks of the Gulf of Manaar and destroying the oysters. He also
reproduces a figure of a specimen lying on a large pearl-oyster.

The HERDMANS mention it as a distinct species [HERDMAN, HERDMAN
& BELL, : 04, p. 144] :

" *Pentaceros lincki*, DE BL.

" From lagoon inside reef, Galle. Common on the pearl banks in the
Gulf of Manaar, and of importance as an enemy of the pearl oyster (see
figure on p. 147)."

HERDMAN [: 06, p. 121, 125, 447, pl. 1, fig. 1] again mentions *Pentaceros lincki* as an enemy of the pearl oyster.

BROWN reports *P. lincki* from the Mergui Archipelago [: 10, p. 32] :

"Localities.—XIV., Bushby Island, 15 to 23 fathoms, sand, shell, and
rock ; XVII., Sir John Malcolm Island, 14 fathoms, sand and rock ; XXV.,
Gregory Group, 4 to 14 fathoms, sand and shell.

" Very frequent on the pearl banks, where it is reputed by the divers
to work havoc among the mother-of-pearl oysters. The collection includes
a series of nine dried specimens of this variable species.

"In some specimens the development of spines is very luxuriant and
in these cases the distal supero-marginals bear conspicuous spines : in other
cases, however, all the spines are more poorly developed, and those of the
supero-marginals are not prominent. Two specimens have no central apical

spine. The number of pedicellariæ which develop varies considerably. In some cases they are numerous on the reticulating bars of the dorsal ossicles up to the base of the lophial spines; in other cases they are rare even on the superomarginals.

"There is great variation in the colour of this species when alive. Most individuals are bright red or carmine except for the poriferous areas which are brown or grey, but many examples were noticed of a bright yellow or even orange colour.

"Distributed from Mozambique and Zanzibar to Ceylon."

It is also reported from Portuguese East Africa by SIMPSON and BROWN [: 10, p. 51]:

"Locality.—Station I. to X., bottom—sand, or sand and rock.

"Previously recorded from—Mergui; Tuticorin; Ceylon; Mozambique; and Zanzibar."

In a foot-note SIMPSON says [: 10, p. 51], "These three well-known species [*P. lincki*, *superbus* and *gracilis*] are extremely abundant over almost the whole coast, and are a distinct menace to pearl-oyster beds. During the period over which my work extended on the coast, more than five thousand of these were brought up in the dredge, while on the shallow reefs thousands may be seen daily at low tide. The colour patterns on all these species, but especially on *P. lincki*, are worthy of attention. The general tone is in most cases blue, but the following variations in the colour of the spines were observed—(1) central spines orange, the others creamy-white; (2) all the spines vermilion red; (3) all the spines creamy white. Another type had bright yellow as a ground-work, while the spines were orange-coloured.

"These few observations demonstrate the futility of basing any specific character on colouration in brightly-coloured asteroids."

Culcita.

The following synoptical key to the species of this genus,

modified from Döderlein ['96, p. 315], will be found useful for identification.

I.   Body pentagonal or roundish.

    A.   Poriferous areas separate or more or less continuous; a marginal zone of variable width free from papulæ. Non-poriferous areas forming a network between the poriferous areas, or may form patches of variable size in the midst of poriferous areas; spines or tubercles may occur in them as far as the ventral margin of the body. Ventral surface finely granulated and with coarser pearl-shaped, flattened or rod-shaped granules. Inner adambulacral spines mostly 5 (at most 7).

        a)   No spines in poriferous areas. The coarse granules of the ventral side forming separate groups corresponding to the ventral plates, with fine granulation between :

                        1. *C. schmideliana* Retzius.

           α)   Poriferous areas small, roundish, separated by non-poriferous areas; dorsal tubercles rather small, numerous, spiniform :

                        var. *ceylonica* (Ceylon).

           β)   Poriferous areas large, more or less continuous; dorsal tubercles very large, wart-like, seldom spiniform, numerous or very few :

                      var. *africana* (East Africa,

                            Mauritius, Seychelles).

        b)   Dorsal tubercles all spiniform. Poriferous areas with spines, which are mostly smaller than those of the non-poriferous areas. The coarse granules of the ventral side may be segregated into groups by intervening fine granulation :

2. *C. novæ-guineæ* M.T.

α) Poriferous areas small, roundish, separated by well developed network of non-poriferous areas :

var. *plana* (Sumatra to Samoa).

β) Poriferous areas large, 3–6-sided, continued into one another, separated by rows of larger spines and non-poriferous patches :

var. *typica* (Amboina to Samoa).

γ) Poriferous areas indistinctly separated into patches, uniformly covered with numerous small spines, with smaller non-poriferous patches between with few coarse spines :

var. *arenosa* (Amboina to Sandwich Islands).

δ) Poriferous areas continuous, with small non-poriferous patches between ; dorsal side uniformly covered with numerous coarse spines :

var. *acutispinosa* (New Hebrides, Viti Islands).

c) Poriferous areas covered with small spines, small, round, entirely separated from one another by a continuous network of wide non-poriferous areas free from spines. Coarse granulation of the ventral side very weakly developed:

3. *C. grex* M.T.

B. Papulæ uniformly distributed on the whole dorsal side
    as far as close to the ventral edge, with very small
    non-poriferous patches. Dorsal side uniformly covered
    with very small, isolated tubercles (rarely spines). Ven-
    tral side finely granulated all over and with pearl-
    shaped coarser granules, which may be segregated into
    groups corresponding to the ventral plates. Coarse
    granules of the ventral side more or less larger than
    those of the dorsal side. Inner adambulacral spines
    mostly 6 (at most 8) :   4. *C. coriacea* M.T.

C. Papulæ uniformly distributed on the whole dorsal side
    as far as close to the ventral edge. All the granules
    and spines of both the dorsal and ventral sides covered
    over by continuations of the soft skin of the body.
    Dorsal side with numerous sharp spines under the skin.
    Inner adambulacral spines 2 (rarely 3):

                                   5. *C. veneris* E. PERR.

II. Body pentagonal or stellato-pentagonal.

A. Body stellato-pentagonal. Papulæ in patches on the dorsal
    side, which is entirely destitute of tubercles. Inner
    adambulacral spines 7 :   6. *C. niassensis* SLUITER.

B. Body pentagonal or slightly stellate. Papulæ on the
    dorsal side in eleven patches, one circumanal and two
    triangular ones in each interradius ; or in five petaloid
    radial patches, each patch being divided into two by a
    naked radial line. Outer adambulacral spines in two
    or three rows:[1]   7. *C. borealis* Süssbach & Breckner
                                        (North Atlantic).

---

1) Neither Süssbach and Breckner nor Farran give the number of the inner adambulacral
spines.

Culcita novæ-guineæ Müll. & Troschel.

(Pl. XVII, figs. 252–262.)

This starfish appears for the first time in post-Linnæan literature under the name of *Goniaster sebæ* Gray. The author contrasts it with *G. cuspidatus* and simply says [Gray, '40, p. 280], " 2. *Goniaster Sebæ*, Seba, iii. t. 8. f. 2. differs in the sides of the rays being angularly inflexed." It will be seen further on that *Goniaster sebæ* is a young form of *C. novæ-guineæ*.

The first description of *C. novæ-guineæ* is due to Müller and Troschel and runs as follows ['42, p. 38]:

" Species 3. *Culcita novæ guineæ* Nob. nov. sp. .

" Körper fünfeckig, selten sechseckig. Gestalt und Verhältnisse wie beim vorigen.[1] Furchenpapillen gross, vorstehend, fünf auf jeder Platte, eine etwas schräge Reihe bildend, die mittlere etwas höher. Dicht neben diesen Furchenpapillen eine Reihe kürzerer, dickerer, von denen je zwei auf eine Platte kommen. Grössere niedrige Knötchen auf der Bauchseite zwischen der feinern Granulation stehen sehr dicht, werden jedoch an den Seiten des Körpers seltener und höher. Mitten auf den Seitenflächen beginnen plötzlich sehr grosse Porenfelder mit vielen Poren und kleinen stachelartigen Tuberkeln. Auf den Räumen zwischen den Porenfeldern stehen einzelne etwas grössere stachelartige Granula zerstreut. Porenfelder wie Zwischenräume überall granulirt. Pedicellarien sind nicht beobachtet.

" Grösse : bis 10 Zoll.

"Fundort : Neu Guinea. Im Museum zu Leyden durch Salomon Müller."

Goniodiscus sebæ which has recently been proved to be a young form of *C. novæ-guineæ* is thus described by Müller and Troschel ['42, p. 58]:

" Species 2. *Goniodiscus sebæ* Nob.

" Artocreas Altera Seba thes. tab. 6, fig. 7, 8.[2]

" Körper pentagonal mit äusserst wenig eingebogenen Seiten. Der

---

1) *Culcita coriacea* ; r : R :: 1 : 1⅓.

2) Lutken ['64, p. 147], Perrier ['76, p. 46] and de Loriol ['84, p. 643] have called attention to the fact that the starfish represented in Seba's figures 7–8, tab. vi, is a different species from the one described by Müller and Troschel under the name of *Goniodiscus sebæ*.

kleine Halbmesser verhält sich zum grossen wie 1 : 1¼. 6 Randplatten an
jedem Arme. Furchenpapillen alle gleich hoch, fünf bis sechs auf einer
Platte in der innern Reihe; nach aussen von diesen dickere, 2–3 auf
jeder Platte, die übrigen gehen in die Granulation der Bauchseite über.
Zunächst den Furchenpapillen verläuft eine Reihe sehr kleiner Bauchplättchen. Die übrigen Bauchplatten sind doppelt so breit, hexagonal, und
nehmen nach dem Rande hin an Grösse etwas ab. Die Granulation auf
den Bauchplatten ist dicht, aber nicht gleichförmig, sondern auf der Höhe
einer jeden Platte stehen einige dickere Granula. Die Randplatten sind
dicht granulirt und tragen ausserdem in einer meist verticalen Reihe
drei bis vier grössere platte Körner, in der Art der grösseren Körner auf
den Bauchplatten. Die Täfelchen des Rückens stossen nicht dicht an
einander, sondern sind durch Balken mit einander netzartig verbunden.
Die Maschen des Netzes sind Porenfelder, von denen immer je sechs eine
Platte umgeben. Die Platten, Balken und Porenfelder sind granulirt, die
Platten tragen jedoch ausserdem noch ein oder mehrere grössere Granula,
wie die Randplatten und Bauchplatten. Auf jedem Porenfelde finden sich
20–24 Poren. Mit der Loupe erkennt man auf der Bauchseite klappenartige
Pedicellarien von der Grösse der grösseren Granula; auf den Randplatten
keine; auf der Rückenseite sind sie sehr klein und nähern sich der
zangenartigen Gestalt. Madreporenplatte ein Drittel vom Centrum.

"Farbe : trocken gelbbraun.

"Grösse : 2–3 Zoll.

"Fundort: Rothes Meer, Neu-Guinea. In den Museen zu Berlin
und Leyden; in letzterem durch MACKLOT und SALOMON MÜLLER, in
ersterem durch HEMPRICH und EHRENBERG."

Culcita pentangularis, which is a synonym of *C. novæ-guineæ*, was
first described by GRAY ['47, p. 195] :

"Body pentangular; back flat when dry, convex beneath, minutely and
closely granulated; back with obscure reticulations, the reticulations armed
with small conical tubercles; the interspaces closely and minutely porous.
The oral surface protected with distinct well-defined ossicules, defining the
lower edge of the margin, covered with close and minute granules and larger

round-topped tubercles, those near the ambulacra and the oral angles being largest and highest.

"Inhab. Reef of Oomaga.

"This species is very distinct from the former[1], and forms the passage to the genus *Randasia*, but there is a series of concave, minutely porous spaces in place of the upper marginal plates."

Randasia granulata first described by GRAY is a young form of this species (HARTLAUB ['92, p. 77] and PERRIER ['76, p. 83]). The genus is characterised by GRAY ['47, p. 195] as follows:

" *Randasia*, GRAY.

"Body pentagonal, depressed, minutely granular; back nearly flat, minutely granular, reticulated, reticulations rather tubercular, interspaces sunken (when dry) and covered with very minute close perforations. Dorsal tubercles roundish, single, subcentral. Margins furnished with an upper and lower series of oblong ossicules, the upper one narrower internally, with a central series of tubercles, the lower ones oblong, close together and convex. The oral surface protected by close, regular, squarish, convex ossicules, covered with short crowded granules. The ambulacral spines in rounded groups; the series of tubercles nearest the ambulacra larger, crowded, and placed in groups of three or five, and those in the oral angles largest and flat-topped.

"The genus differs from *Pentaceros* in the back being flat, elevated, and not angular; it is in several respects intermediate between *Culcita* and *Pentaceros*."

Two species are described, *R. granulata* and *R. spinulosa*. The former is the species now under consideration, and is described as follows ['47, p. 196]:

" *Randasia granulata*, n. s.

"Body five-sided; back minutely granular, with roundish convex subconical tubercles in the reticulations; the marginal plates fourteen on each side, the upper ones with a central series of tubercles.

"Inhab. Reefs of Attagor, Torres Straits.

"There are two specimens of this species in the British Museum, one in a very bad state."

---

1) *Culcita schmideliana*.

Hosia spinulosa which has also been proved recently to be a young form of *C. novæ-guineæ* was first described by GRAY as follows ['47, p. 199]

"Body flat, pentagonal, sides concave; arms not half the length of the diameter of the body; ossicules large, subequal, six-sided, very minutely granular. Marginal ossicules $\frac{1}{1}\frac{0}{0}$ on each side, convex, deeply separated from each other, with a series of two or three small, acute, spine-like tubercles in the centre of each. The ossicules of the oral surface flat, minutely granular, with small two-lipped pores.

"Inhab. Indian Ocean: Philippines.

"This species nearly resembles the shape of *Tosia australis*, but is at once known from that species by the granular ossicules, the spines on the margin, and the two-lipped pores beneath; it differs from *Hosia flavescens* in its being five-sided instead of five-armed, and in having no spines on the middle of the back."

According to PERRIER ['76] *G. sebæ* is mentioned by PETERS ['52, p. 178] from Mosambique.

Culcita novæ-guineæ is described by DUJARDIN and HUPÉ as follows ['62, p. 372] :

"Corps pentagone, rarement hexagone, de même forme que les espèces précédentes. Piquants du sillon ambulacraire gros, saillants, au nombre de cinq sur chaque plaque et formant une rangée un peu oblique. Tout auprès, plus en dehors, se trouve une autre rangée de piquants plus courts, plus épais; il y en a deux sur chaque plaque, sur la face ventrale il existe, entre les fines granulations, de gros tubercules déprimés devenant plus rares et plus hauts sur les côtés du corps. Au milieu des faces latérales brusquement limitées, les pores tentaculaires occupent de grandes aires avec des pores nombreux et des petits tubercules épineux. Les intervalles de ces aires portent quelques granules plus gros, épineux et disséminés.

"Corolation jaunâtre foncé. Dimension: largeur totale 200 à 270 mm.

"Habite les côtes de la Nouvelle-Guinée."

DUJARDIN and HUPÉ's description of *Goniodiscus sebæ* is based on that of MÜLLER and TROSCHEL [DUJARDIN et HUPÉ, '62, p. 402[1]] :

---

1) References to SEBA, GRAY ['40, p. 280] and MULLER and TROSCHEL omitted.

" Corps pentagonal, à côtés très-peu échancrés, et dont le plus grand rayon égale une fois et un quart le plus petit. Piquants du sillon ambulacraire longs et égaux, au nombre de cinq ou six sur chaque plaque pour la rangée interne ; il n'y en a que deux ou trois pour les autres, mais ils sont plus épais et passent par des transitions peu sensibles, aux granulations de la face ventrale. Près de ces piquants, plus en dehors, se trouve une rangée de très-petites plaquettes. Les autres plaques de la face ventrale sont deux fois plus larges, hexagones, et vont en diminuant un peu vers le bord. La granulation de cette face ventrale est compacte, mais non uniforme ; quelques granules plus épais occupent le sommet de chaque plaque. Les plaques marginales, au nombre de six à chaque bras, sont couvertes de granules serrés et portent, en outre, trois ou quatre gros granules aplatis, formant ordinairement une rangée verticale. Les plaques de la face dorsale ne sont pas contiguës entre elles, mais sont réunies par des barres, de manière à former une sorte de réseau dont les mailles, occupées par les pores tentaculaires, entourent chaque plaque. Toutes ces parties sont granuleuses, et chaque plaque porte, en outre, un ou plusieurs gros granules, comme les plaques marginales et ventrales. Chaque aire contient de 20 à 24 pores. Les plaques marginales manquent de pédicellaires, mais à l'aide d'une loupe, on distingue à la face ventrale des pédicellaires bivalves aussi grandes que les gros granules, et sur la face dorsale, on en voit de très-petites qui se rapprochent de la forme en pince. La plaque madréporiforme est environ au tiers de la distance entre le centre et le bord.

" Coloration brun-jaune, à l'état sec. Dimension : largeur 50 à 80 mm.

" Habite la mer Rouge, les côtes des îles Molluques et de la Nouvelle-Guinée."

In his " Synopsis " GRAY reproduces his former descriptions of *Culcita pentangularis, Randasia granulata, Hosea*[1] *spinulosa* and *Goniaster sebæ* and adds two figures of the first, two of the second and three of the third ['66, p. 5, Tab. 2 ; p. 8, Tab. 2 ; p. 9. Tab. 4 ; p. 10].

VON MARTENS ['66, p. 85] simply mentions *C. novæ-guineæ* as being represented in the Leyden Museum and also refers ['66, p. 86] to *Goniaster*

---

1) He now spells " *Hosea* " instead of " *Hosia*."

(= *Goniodiscus*) *sebæ* and *Goniaster* (= *Hosia*) *spinulosus* as appearing in the literature and represented in museums; the former from the Moluccas through MACLOT and SAL. MÜLLER, from New Guinea in the Leyden Museum, again from the Moluccas in the Museum of Amsterdam and from Amboina according to BLEEKER; the latter from the Philppines. According to PERRIER ['76] *C. novæ-guineæ* is mentioned by v. MARTENS from the collection of v. D. DECKEN in east Africa ['69].

C. novæ-guineæ is mentioned by PERRIER in his work on pedicellariæ. He says ['69, p. 66], "un seul échantillon dans la collection du Muséum, tellement détérioré qu'il est impossible de rien dire de ses Pédicellaires."

The starfish described by PERRIER as *Culcita grex* is probably referable to the present species, as PERRIER himself suggests in his later work ['76, p. 78]. Here is its description [PERRIER, '69, p. 64]:

"*Culcita grex* (?), M. et T.—La Culcite en question ne me paraît être en aucune façon l'animal désigné sous ce nom par MÜLLER et TROSCHEL. Je crois donc devoir, sans lui donner un nom nouveau, accompagner ma description des Pédicellaires d'une description complète de l'animal.

"Le corps est un pentagone régulier à côtés légèrement concaves. La face ventrale est plane, entièrement couverte de granules fins et partagée par des lignes enfoncées en aréoles plus ou moins pentagonales dont l'aire est légèrement convexe. Toutes ces aréoles sont contiguës absolument comme les alvéoles d'un gâteau de cire. Le long des sillons ambulacraires le nombre de ces aréoles, qui sont bien distinctes, est de treize ou quatorze, dont une impaire dans l'angle et faisant partie de deux séries à la fois. Dans l'intérieur de chaque aréole on voit le plus souvent un certain nombre de tubercules peu considérables et quelques Pédicellaires valvulaires très-allongées. Tantôt les Pédicellaires peuvent manquer, tantôt les tubercules. —Ils sont en général en nombre inverse les uns des autres. Quand il n'existe qu'un seul Pédicellaire, les tubercules ont une tendance à se ranger en cercle autour de lui. Les uns et les autres peuvent manquer à la fois.

"Dans le sillon ambulacraire, chaque plaque porte quatre piquants dépassant fort peu le bord, à peu près de même grandeur; ces piquants deviennent beaucoup plus gros en se rapprochant de la bouche; ils forment alors avec ceux du sillon voisin un angle circonscrivant un espace au sommet duquel sont accumulés et pressés les uns contre les autres un grand nombre de tubercules assez gros.—Derrière les piquants du sillon ambulacraire, on voit une autre rangée de tubercules au nombre de deux derrière chaque plaque.—Un peu en arrière entre chaque couple de ces tubercules, on voit un Pédicellaire dont les valves sont à peu près aussi larges que hautes et ne sont à proprement parler ni des Pédicellaires en pince, ni des Pédicellaires valvulaires, bien qu'ils se rapprochent plutôt de ceux-ci.

"Sur la face dorsale, les aires tentaculaires sont déprimées entre le réseau qui les circonscrit et qui porte un certain nombre de courts aiguillons à base large, mais à pointe très-acérée.—Dans l'intérieur de ces aires, on voit aussi un ou deux piquants ; elles sont d'ailleurs couvertes de fines granulations, parmi lesquelles on distingue les Pédicellaires qui sont un peu plus gros, un peu plus longs que larges et à mâchoire dentée sur les bords. Leur tissu paraît fait d'une série de bâtons calcaires parallèles réunis par des traverses très-serrés.

"La plaque madréporique est située dans l'intervalle de deux angles à un tiers environ du bord ; elle est légèrement saillante.—Le diamètre de l'animal est de 8 à 9 centimètre. Il provient de l'expédition d'URVILLE."

C. arenosa, which is put down as a synonym of the present species by DÖDERLEIN, is described for the first time by PERRIER, as follows ['69, p. 66]:

"*Culcita arenosa*, VAL.—Corps formant un pentagone régulier, mais à côtés légèrement concaves.—Dessus du corps très-finement granuleux et parsemé d'un grand nombre d'épines, courtes, pointues, paraissant couchées sur le dos.—Les trous tentaculaires sont nombreux, serrés, petits, également distribués sur la face dorsale.—On remarque aussi vers cette région de nombreux Pédicellaires en pinces couchés sur le dos de l'une des valves. Chacune de ces dernières est légèrement courbée sur sa partie dorsale ; en même temps, elle se relève légèrement vers le milieu de sa longueur. Ces valves ne présentent de dents bien nettes que sur le pourtour du cuilleron terminal ; aussi ces dents irrégulières et espacées rassemblent-elles plutôt à de simples lacunes de la substance calcaire.

"La plaque madréporique est large, aplatie, irrégulière, assez saillante ; au lieu de présenter, comme dans la plupart des cas, une série de sillons plusieurs fois bifurqués, rayonnant d'un centre commun, elle est couverte de sillons formant de très-nombreuses circonvolutions à sa surface sans que l'on puisse distinguer aucune disposition radiée.—Sa forme, au lieu d'être ronde, se rapproche de la forme de la croix qu'on obtiendrait en décrivant des demi-circonférences sur les côtés d'un carré comme diamètres.

"Sur la face ventrale, on distingue deux sortes de granules, les uns rares relativement et plus gros, les autres très-nombreux et contigus. On ne retrouve pas ici la disposition alvéolaire que nous avons fait remarquer dans la *Culcita grex* et la *Culcita discoidea.*—Parmi ces granules se trouvent les Pédicellaires, dont les valves sont tantôt plus longues que larges, tantôt au contraire plus larges que longues, de telle façon qu'on trouve à peu près tous les intermédiaires entre les Pédicellaires en pince et les Pédicellaires valvulaires, sans toutefois avoir ni les uns ni les autres.

"Tout le long du sillon ambulacraire, les plaques portent 4, 5, 6 ou même 7 piquants dont les rangées obliques s'imbriquent de manière que les deux ou trois piquants les plus rapprochés du centre, et qui sont les plus petits, sont souvent en partie masqués. Derrière cette rangée de piquants, on voit une rangée de gros tubercules géminés de manière á simuler deux rangées.

"Au sommet des aires angulaires limitées par les sillons, les piquants marginaux atteignent presque la grosseur de ces tubercules qui eux mêmes deviennent plus robustes, de telle façon qu'il semble y avoir en ces points une accumulation de gros tubercules.

"Entre deux paires de tubercules consécutives, on aperçoit deux piquants réunis, plus petits, souvent inégaux, et qui rappellent par leur position les Pédicellaires en pince qu'on trouve à leur place dans quelques espèces.

"L'individu, conservé dans l'alcool, d'après lequel est faite cette description, mesure 18 centimètres de diamètre, cette mesure étant prise entre deux sommets sépares par un troisième.

"Cette Culcite a été rapportée des îles Sandwich par la *Bonite.* (EYDOUX et SOULEYET, 1837.)"

Culcita pulverulenta was originally described by PERRIER as a distinct species, but subsequently withdrawn by him and referred to *C. novæ-guineæ*. The original description is as follows ['69, p. 68] :

"*Culcita pulverulenta*, Val.—Le Muséum ne possède qu'un seul échantillon de cet animal. Il est conservé dans l'alcool, mais fortement endommagé.

"Dessus du corps très-finement granuleux présentant les aires tentaculaires larges et irrégulières, séparées par des espaces très-finement granulés, presque nus ou portant seulement de petites épines peu nombreuses.

Les Pédicellaires dorsaux se trouvent seulement dans les aires tentaculaires. Ils sont petits, nombreux, en forme de pinces, à mâchoires assez délicates, dentées largement sur leur bord, plus élargis vers le milieu de leur longueur et légèrement recourbés à leur sommet. La lame calcaire qui constitue ces Pédicellaires est perforée de trous nombreux et présente même quelques petites épines.

"Je n'ai pu voir la plaque madréporique; elle se trouvait dans une portion de la face dorsale qui a été détruite.

"Sur la face ventrale, on remarque de nombreux tubercules aplatis, serrés les uns contre les autres et entre lesquels se voient de petites granulations et des Pédicellaires. Les tubercules sont d'autant plus gros qu'ils se rapprochent davantage des sillons ambulacraires, auxquels ils forment une double bordure irrégulière. Les tubercules situés à l'angle des sillons sont, comme d'ordinaire, le plus gros.

"Les Pédicellaires situés entre les tubercules sont petits, peu distincts, assez nombreux, affectant sensiblement la forme valvulaire. La hauteur de chacune des valves est à peu près égale à sa largeur.

"Dans les sillons ambulacraires, chaque plaque porte cinq ou six piquants, serrés, à peu près cylindriques, tous à peu près de même taille, sur chaque plaque. Les deux dernières plaques les plus rapprochées de la bouche portent des piquants beaucoup plus gros et en nombre moindre.

"Diamètre : 14 centimètres environ.

"Individu rapporté en 1829 du détroit de la Sonde par M. Raynaud, expédition de la *Chevrette*."

In his paper of 1872 Gray mentions *Goniaster sebœ* from the Red Sea and refers to *Goniodiscus sebœ* Müll. et Troschel as its synonym [Gray, '72, p. 118].

Pentagonaster spinulosus, Perrier's name for *Hosia spinulosa* Gray, is thus described in detail [Perrier, '76, p. 34] :

"Diamètre 25 à 40 millimètres. Corps de forme pentagonale, et rappelant beaucoup, sauf une épaisseur légèrement plus grande, le *Pentagonaster australis*, GRAY, mais entièrement couvert d'une fine granulation, s'étendant même dans les intervalles des plaques, où les pores tentaculaires sont groupés par trois ou quatre. Plaques marginales dorsales au nombre de quatorze, les dernières graduellement plus petites, unies par la granulation générale, qui n'est modifiée en rien sur leurs bords ; chacune d'elles porte dans son aire médiane de six à huit petits tubercules disposés peu régulièrement. Ces plaques sont plus larges que longues. Les plaques formant la surface dorsale sont arrondies, granuleuses ; celles qui forment la double rangée interbrachiale sont un peu plus grandes que les autres ; sur beaucoup d'entre elles, tant sur les plaques ordinaires que sur les interbrachiales, on voit un petit pédicellaire valvulaire, mince et assez allongé. Les plaques marginales ventrales sont au nombre de seize, dont les dernières sont très-petites ; presque toutes portent quatre ou cinq petits tubercules. Les plaques occupant le voisinage du sommet de l'aire interbrachiale, en sont dépourvus. Plaques ventrales granuleuses, comme les dorsales, mais portant à leur centre quelques granules plus gros que les autres, entourant ordinairement un pédicellaire valvulaire, de forme allongée.—Piquants ambulacraires formant deux rangées très-serrées, assez distinctes des granules de la face ventrale. Chaque plaque porte sur le bord même du sillon quatre ou cinq piquants, suivis en arrière de deux ou trois piquants formant la deuxième rangée et un peu plus gros, mais affleurant au même niveau. Les granules de la face ventrale qui suivent immédiatement sont un peu plus gros que les autres et ils sont eux-mêmes suivis assez fréquemment d'un petit pédicellaire valvulaire. La plaque madréporique est petite et compris entre la première interbrachiale et les deux suivantes.

" Deux échantillons desséchés au British Museum, l'un des Philippines (type de GRAY), l'autre, plus grand, des îles Fidji."

In the same work PERRIER ['76, p. 46] also describes *Goniodiscus sebæ*, but as pointed out by DE LORIOL ['84, p. 643] he must have had *Goniaster articulatus* LÜTKEN before him.

PERRIER mentions under *C. novæ-guineæ Culcita pulverulenta* as being due to VALENCIENNES, whose manuscript is preserved in the Muséum d'histoire naturelle, Paris, the date being unknown. The specimens studied by him are described as follows ['76, p. 78]:

"Deux exemplaires, l'un desséché, l'autre dans l'alcool, mais en mauvais état, portent dans la collection du Muséum le nom de *Culcita pulverulenta*, je les ai décrits en 1869 dans mes *Recherches sur les Pédicellaires des Asteries et des Oursins*. Ils ont été recueillis, en 1829, au détroit de la Sonde par M. REYNAUD, naturaliste de l'expédition de *la Chevrette*, et ne semblent pas avoir été vus en 1840 par TROSCHEL. Cependant une comparaison minutieuse de la description de la *Culcita Novæ Guineæ* des auteurs du *System der Asteriden* avec les deux échantillons que nous avons sous les yeux ne permet pas de douter qu'il s'agisse du même animal. Ces échantillons proviennent d'ailleurs d'une region océanique très-voisine de la Nouvelle-Guinée, d'où le type est originaire.

"Bien que le nom de VALENCIENNES soit probablement antérieur à celui de MÜLLER et TROSCHEL, comme il n'a été fixé que bien plus tard par une description il convient de rendre à ces échantillons le nom de *Culcita Novæ Guineæ* qui a la priorité, comme publication."

Culcita pentangularis, of which *Randasia granulata* of GRAY and *C. grex* of PERRIER ['69, p. 64] are regarded as synonyms, is referred to at length in the same work [PERRIER, '76, p. 78]:

"Je regarde comme identique à la *Culcita pentangularis* de GRAY l'espèce à laquelle j'ai laissé avec doute, dans mes *Recherches sur les Pédicellaires*, le nom de *Culcita grex*, M.T., qu'elle portait dans la collection. J'ai indiqué toutefois dans mon travail le peu de confiance que m'inspirait cette détermination. Il suffira de reproduire ici la description que MÜLLER et TROSCHEL donnent de leur *Culcita grex* et celle que nous avons donnée en 1869 de l'animal qui porte ce nom dans la collection du Muséum pour se convaincre qu'il s'agit bien ici de deux espèces distinctes, et que l'individu en question ne saurait avoir reçu ce nom de TROSCHEL."

He then reproduces side by side the principal points in the descriptions of his *C. grex* and of MÜLLER and TROSCHEL, and concludes as follows:

"Cette espèce se distingue nettement des *Culcita grex* et *Novæ Guineæ* de Müller et Troschel par les aréoles de sa face ventrale, la finesse de sa granulation, l'allongement de ses pédicellaires, le nombre des aires porifères de sa face dorsale, le nombre et la dimension des petits piquants qu'elle porte. La *Randasia granulata* de Gray n'en est que le jeune âge.

"Hombron et Jacquinot. Expédition d'Urville. 1841.

"M. Filhol vient de rapporter au Muséum un bel échantillon de cette espèce recuelli aux îles de Fidji. Le type de Gray vient des récifs d'Attagor, dans le détroit de Torres.—Un individu de Mozambique au musée de Cambridge (Massachusetts)."

Culcita arenosa, briefly described in Perrier's former publication, is referred to in the "Revision" in the following terms ['76, p. 80]:

"Cet échantillon que Valenciennes avait réuni à ceux que nous rapportons à la *Culcita grex* de Müller et Troschel sous le nom de *Culcita arenosa*, mais qu'il semble en avoir séparé plus tard, est réellement une espèce distincte, que nous avons decrite en 1869 dans nos *Recherches sur les Pédicellaires*.

"La *Culcita arenosa* se rapproche de la *Culcita grex* par sa forme plus aplatie qui la distingue nettement de la *Culcita Novæ Guineæ* Les caractères qui l'eloignent de la *Culcita grex* sont les suivants : les côtés du disque pentagonal sont légèrement concaves, les aires porifères indistinctes, les pores étant uniformément répartis sur toute la surface dorsale de l'animal qui est couverte de petits granules très-fins parsemés de pédicellaires en pince. Un assez grand nombre d'aiguillons isolés, courts, très-pointus, disséminés sur toute la surface du disque, assez régulièrement espacés, s'élèvent de la très-fine granulation générale. Ces aiguillons sont plus grêles, plus aigus, plus régulièrement espacés, et moins nombreux que ceux de la *Culcita grex*.

"La face ventrale est entièrement couverte d'une granulation moins fine que celle du dos, d'où s'élèvent de nombreux granules arrondis, plus gros, souvent presque contigus, mais irrégulièrement disséminés. Parmi ces granules on voit d'assez nombreux pédicellaires à contour circulaire, ressemblant par conséquent à des granules, mais plus gros que les plus petites granulations et plus petits que les gros granules tuberculiformes.

"Les piquants des sillons ambulacraires sont disposés sur deux rangées obliques, sauf dans le voisinage de la bouche. Ceux de la première rangée sont au nombre de cinq dans cette dernière région, puis de quatre ou cinq sur chaque plaque ; le plus extérieur en regardant vers la bouche est dans chaque groupe plus petit et plus grêle que les autres. Tous sont reliés ensemble jusque vers le dernier tiers de leur hauteur, de manière à former une lame continue, de simples sillons indiquant sur cette lame les limites de chaque piquant ; tous les piquants sont libres d'ailleurs, mais contigus dans leur dernier tiers. Les piquants de la seconde rangée sont très-gros, courts, cylindriques, d'abord au nombre de trois, puis de deux sur chaque plaque ; ils sont placés obliquement, de manière à simuler une double rangée. Entre eux et les piquants de la rangée interne, deux piquants, très-rapprochés, en général inégaux, simulent le pedicellaire en pince qu'on trouve d'ordinaire à leur place sur chaque plaque. En arrière de la rangée de piquants externes, il existe vis-à-vis de chaque plaque un ou plusieurs pédicellaires en pince formant, à 1 ou 2 millimètres en dehors du canal ambulacraire, une ou deux rangées dans laquelle les pédicellaires sont distants les uns des autres d'au moins 3 millimètres.

"La couleur générale de l'individu unique, conservé dans l'alcool, que possède le Muséum, est le blanc sale.

"Des îles Sandwich. Expédition d'EYDOUX et SOULEYET sur *la Bonite*, 1837.

"*Nota.*—Nous ne pouvons comparer l'échantillon bien conservé dans l'alcool de la *Culcita arenosa* que nous venons de decrire, qu'à des échantillons dessechés et très-détériorés de la *Culcita grex*. Les caractères differenciels que nous venons d'indiquer sont par conséquent susceptibles d'être un peu modifiés. Quoi qu'il en soit, il ne saurait être douteux que les deux espèces que nous avons en vue sont bien distinctes."

On the genus *Randasia* PERRIER makes the following remarks ['76, p. 82] :

"Les espèces que GRAY a placés dans son genre *Randasia* ne nous paraissent être que de jeunes *Culcita* : elles en ont toute la physionomie et jusqu'à l'ornementation générale. Les seules différences résident dans leur moins grande épaisseur et dans la plus grande netteté de leurs plaques

marginales; mais ce sont là des caractères que l'âge modifie considérablement et sur lesquels on ne saurait fonder un genre.

"Les deux formes suivantes," que nous décrivons complètement d'après
les types de GRAY, se laissent rapprocher, l'une de la *Culcita grex*, MÜLLER
et TROSCHEL, l'autre de la *Culcita pentangularis*, GRAY; nous pensons donc
qu'elles ne devront pas être conservées comme espèces distinctes. Malheureusement nous n'avons pas vu suffissamment d'echantillons de grandeur
intermédiaire pour faire d'une manière absolue la preuve du fait. C'est pour-
quoi nous conservons provisoirement les noms specifiques de GRAY."

"*Randasia granulata*.

"Fome générale de la *Randasia spinulosa*, GRAY. Plaques marginales
bien distinctes; seize dorsales et vingt-deux ventrales. Plaques dorso-
marginales s'écartant l'une de l'autre vers le haut et laissant entre elles un
espace anguleux occupé par une aire porifère: ces plaques portent, outre la
fine granulation générale, quelques petits tubercules saillants. Le système
squelettique dorsal est fort simple; il se compose de deux séries d'ossicules
occupant, l'une la ligne médiane des bras, l'autre, celle qui va du centre du
disque au sommet de l'arc interbrachial. Entre ces deux séries on voit
encore deux ossicules isolés et l'ensemble de toutes ces pièces est réuni
par des systèmes de six trabécules calcaires partant de chacune d'elles pour
aller aboutir aux arêtes et former ainsi un réseau hexagonal. De petits
tubercules hérissent les parties saillantes de ce réseau, comme chez la *Culcita
pentagularis*, GRAY, qui habite la même localité et dont cette espèce n'est
certainement que le jeune. Les mailles de ce réseau sont occupées par les
aires porifères contenant une vingtaine de pores environ. Parmi les granules
qui recouvrent ces aires on voit de petits pédicellaires valvulaires arrondis.—
La face ventrale est semblable à celle de la *Randasia spinulosa*.—Les piquants ambulacraires sont sur deux rangs; chaque plaque interambulacraire
en porte quatre de la première rangée, tous égaux entre eux, et deux de la
seconde, affleurant au même niveau et suivi d'une rangée de granules plus
gros que ceux du reste de la face ventrale.—Détroit de Torrés (British
Museum)."

---

1) *Randasia spinulosa* and *R. granulata*.

Goniodiscus sebæ of Viguier ['78, p. 186] is evidently identical with that of Perrier ['76, p. 46] and therefore does not refer to the species before us (*vide supra*, p. 528).

Goniodiscus sebæ is reported by Möbius from Mauritius ['80, p. 50, *fide* de Loriol].

Culcita acutispinosa, another synonym of the present species, was originally described by Bell ['83, p. 334] as a distinct species:

"Resembling *C. coriacea*, and distinguished from all other species of the genus by the fact that the apices of the upturned ambulacra are below the level of the dorsal or abactinal surface. The body is almost completely discoidal in shape, the angles of the rays being very nearly altogether rounded off; the sides of the disk are very deep; in the dry specimens, at any rate, the actinal surface gradually slopes downwards, so that the animal is very much deeper along a line drawn dorso-ventrally through the actinostome than it is at the margin of the disk (62-40 millim.).

"The adambulacral spines are in two rows; in the inner there are ordinarily four on each plate, and they are not so well developed as in some allied species: about five plates out from the actinostome they measure about 5 millim. in length. In the outer row there are generally two spines, one of which is very strong and blunt, while the other is much smaller. The spines on the intermediate plates sometimes lie quite close to the outer interambulacral series, and occasionally appear to invade it.

"The actinal surface is not marked out into areolæ, and is richly invested by a number of short, blunt, stout processes, hardly to be called spines, amidst which a coarse granular covering is to be observed. The poriferous region begins quite suddenly, at about the second fourth of the length of the side of the disk (counting from the actinal margin); while this is the point at which the pores begin at the middle of each side of the disk, the line of demarcation gradually curves upwards, so that along the radial axes the pores begin just above the apex of the ambulacral groove. The greater part of the sides and the whole of the abactinal surface of the disk are covered with short sharp spines, which are scattered over them with considerable profusion, though in no definite order; dotted

among the spines are pores of moderate size, which are very indistinctly
grouped into pore-areas; surrounding and separating the pores are fine
granules, and these are, at apparently irregularly disposed points, closely
aggregated into distinct patches; such a patch is found outside the spines
which fringe the arms. On the abactinal surface the pedicellariæ are small
and scarce; on the actinal they are, though small, much more numerous.
The madreporic plate is large and raised above the surface. (It has unfortunately been injured in the specimen under description.)

"The single dried specimen has the upper surface of a pinkish and
the lower of a yellowish-white hue.

" *Hab.* Aneiteum, New Hebrides.

"As already stated, the greatest height is 62 millim.; its diameter is
120 mm."

STUDER mentions this species from the collection of the " Gazelle "
['84, p. 37] :

"Von Neu-Hannover aus der Rifflagune im Gazellehafen.

" Die Färbung dieses Seesterns war im Leben folgendermassen : Die
polygonalen Felder, in welche die Oberfläche durch die dorsalen Wärzchen
zerfiel, waren schwarz, am meisten im centralen Theil der Scheibe, wo sie
einen Kranz um das Centrum bilden.   Die Wärzchen, wie der übrige Theil
des Rückens olivengrün, die Bauchseite orange." .

DE LORIOL reports *Goniodiscus sebæ* from Mauritius and makes some important remarks. He also gives very good figures ['85, p. 48; Pl. XV, figs. 6–6e]:

" Diamètre total 61 mm. à 65 mm. r=27 mm. R=35 mm.

" Les individus envoyés de Maurice correspondent fort exactement à la
bonne description de MÜLLER et TROSCHEL, et je n'ai aucun doute sur leur
identité. Ainsi que je l'ai exposé ailleurs (*Recueil de zoologie suisse*, t. I,
p. 638, pl. XXXV, fig. 1), c'est à tort que MÜLLER et TROSCHEL ont envi-
sagé leur espèce comme se rapportant à la figure de SEBA (t. III, pl. VI, fig.
7–8) qui représente *Goniodiscus articulatus*, LINNÉ ; cette espèce diffère, entre
autres, par un nombre de plaques marginales dorsales presque double, des
bras bien plus détachés, etc. J'ai comparé les individus de Maurice avec
un exemplaire des îles Viti, et avec un autre de Macassar, et j'ai constaté

une identité complète. La description de MÜLLER et TROSCHEL convient en tous points á tous mes échantillons. Je n'ai que quelques observations à ajouter. MÜLLER et TROSCHEL comptent six plaques marginales dorsales pour chaque bras, mais, en réalité, il y en a sept, soit quatorze pour chaque arc interbrachial de l'une des extremités á l'autre. Cette différence peut provenir de ce que la dernière plaque de chaque côté, à l'extrémité des bras, est petite et comme soudée à la voisine, dont elle n'est pas séparée par des pores, elles pourraient donc n'avoir été comptées que pour une seule. Entre chacune des plaques marginales dorsales se trouve une aire porifère cunéiforme, plus ou moins large, et contenant plus ou moins de pores. La granulation de la face dorsale est bien plus fine que celle de la face ventrale, et les granules plus gros, qui apparaissent çà et là, deviennent souvent de petits tubercules coniques et pointus, il en est de même sur les plaques marginales dorsales. Un ou plusieurs petits piquants semblables se voient ordinairement près de la plaque madréporique, et près de l'orifice anal; ce dernier est entouré de plusieurs gros granules allongés et obliques qui le recouvrent tout à fait. Les pédicellaires paraissent être des pédicellaires valvulaires, ils varient considérablement en nombre, tantôt il n'y en a qu'un seul sur une plaque, tantôt trois ou quatre; il y a bien plus sur la face ventrale que sur la face dorsale. Les piquants ambulacraires externes sont toujours très développés à l'extrémité des bras.

"La couleur des individus secs est orange en dessus, jaune en dessous.

"Cette espèce n'a encore jamais été figurée."

The same writer also describes *Pentagonaster spinulosus* further on in the same work and gives eight figures of it [DE LORIOL, '85, p. 52; Pl. XVI, figs. 1–1 g] :

"Diamètre total 32 mm. r=12 mm. R=17 mm. Epaisseur 5.

"Jusqu'ici un seul exemplaire appartenant à cette espèce m'a été envoyé de Maurice. Il est parfaitement conservé et correspond très exactement, soit à la description et à la figure de GRAY, soit à la description donnée plus tard par M. PERRIER d'un autre individu du British Museum. Sa dimension est exactement intermédiaire entre celle du type de GRAY et celle de l'individu de M. PERRIER (26 mm. et 40 mm.); il en est de même du nombre de ses plaques marginales, qui est, pour chaque angle interbrachial, de douze dorsales contre quatorze ventrales, dont les dernières sont très petites, tandis que, dans le type plus petit de GRAY, il y a dix dorsales et dix ventrales, et dans l'exemplaire plus grand, décrit par M. PERRIER, quatorze dorsales et seize ventrales. Les plaques de la face dorsale sont, les unes régulièrement hexagones, les autres hexagones tendant à s'arrondir ; elles sont couvertes de cette granulation d'une extrême finesse mentionée par GRAY ; plusieurs d'entre elles, surtout vers le pourtour, portent un pédicellaire valvulaire très étroit et très allongé. Les pores sont les plus souvent isolés, rarement deux ensemble, et, encore plus rarement trois ensemble. Les plaques adambulacraires portent, sur le bord du sillon, une rangée de quatre ou cinq piquants internes, cylindriques, presque égaux et peu saillants en dehors ; serrée contre celli-ci vient une rangée externe de piquants moins nombreux, mais plus robustes ; le reste de la plaque porte de nombreux granules et, très souvent, un pédicellaire valvulaire de faible dimension.

"L'orifice anal, presque central, est fermé par plusieurs valves fort étroites, et accompagné de deux petits tubercules coniques. La plaque madréporique, jaunâtre, est un peu renflée, avec des sillons assez écartés.

"La couleur, à l'état sec, est un vert olivâtre sur la face dorsale, avec cinq taches jaunes dans les espaces interbrachiaux, comprenant les deux plaques marginales médianes et trois autres plaques dont une de la rangée la plus externe et deux de la voisine. La face ventrale est entièrement jaune.

"Cette espèce n'était encore connue que des îles Philippines et des îles Fidji."

Under *C. schmideliana* DE LORIOL observes as follows ['85, p. 66] :

"Quelques individus jeunes permettent de suivre le développement de l'espèce et de constater que les *Randasia* de GRAY sont certainement des jeunes Culcites. Au diamètre total de 58 mm., le *C. Schmideliana* a toute l'apparence d'un *Goniodiscus*, les côtés sont un peu échancrés, les plaques marginales très distinctes, les dorsales étroites, au nombre de treize pour un arc interbrachial, les ventrales, au nombre de dix-sept, plus larges au milieu de l'arc, plus petites vers l'extrémités, formant exclusivement le bord sans participer à la face ventrale. Les trabécules de la face dorsale sont très distincts, et ils portent déjà les trabécules coniques caractéristiques. Dans un individu de 62 mm., les trabécules sont déjà un peu moins distincts; dans un autre de 73 mm. l'apparence de la Culcite s'accentue déjà tout á fait, les côtés se redressent, les trabécules et les plaques marginales deviennent indistinctes: Dans un individu de 80 mm. de diamètre la forme est élevée, les côtés très droits, on ne voit plus de plaques marginales, c'est l'individu le plus adulte en diminutif. Par contre, dans un individu de 90 mm. de diamètre, la forme reprend du rapport avec celle des jeunes, et les plaques marginales sont encore assez distinctes. Tous ces individus jeunes auraient constitué une espèce de *Randasia* pour GRAY, mais elle ne me paraît pas comprise parmi celles qu'il a décrites. Dans tous, les plaques ventrales sont exactement disposées comme dans les adultes, seulement elles sont plus exactement hexagones, et les gros granules de second ordre ne se développent que graduellement au centre de chaque plaque. Dans quelques-uns les pédicellaires valvulaires sont bien plus abondants que dans les adultes, où ils sont rares; dans un individu on trouve même un pédicellaire valvulaire fort étroit et assez long au milieu d'un grand nombre des plaques ventrales. Dans le sillon ambulacraire de ces jeunes individus, il y a généralement quatre piquants dans la série interne et un ou deux dans la série externe, pour chaque plaque adambulacraire."

In his paper of 1887 BELL says under the heading *Culcita* sp. as follows ['87, p. 142]: "There are two specimens of what would, a short time ago, have been set down as *Randasia granulata*. As however, M. DE LORIOL has lately shown, the form so called by the late Dr. GRAY is really a young stage of *Culcita*. It will be remembered that Prof. PERRIER has expressed himself in a similar sense. Further series are required before the several stages of each species can be accurately defined."

The next year BELL ['88, p. 388] mentions *C. novæ-guineæ. C. grex* and *C. schmideliana* from the Sea of Bengal and adds, "*Randasia granulata* may be the young of one of these species of *Culcita* or of an unknown species; it has been taken at the Andamans." DÖDERLEIN ['88, p. 827] reports *Goniodiscus sebæ* from Ceylon and adds, "Diese Art scheint sehr häufig zu sein bei Trincomali."

SLADEN ['89, p. 352], after mentioning *Culcita pulverulenta* as a synonym
of this species, makes the following remarks :

"*Localities.*—Off Zebu, Phillipine Group. On the Reefs.

"Station 212. Off Malampa Island, Philippine group. January 30,
1875. Lat. 6° 54 ′0 ″ N., long. 122° 18′ 0″ E. Depth 10 fathoms. Sand.
Surface temperature 83°. 0 Fahr.

"*Remarks.*—Examples of this form may be selected which correspond
exactly to the type of GRAY's *Culcita pentangularis.* . After a careful study,
however, of the various examples referable to *Culcita novæ-guineæ* which I
have been able to examine, I am unable to indicate any characters by which
these two forms can be separated specifically, and I have therefore referred
the material now in my hands for description to the first described species."

The *C. schmideliana* of SLUITER ['89, p. 305] is admitted by the author
himself (*vide infra*) to be in reality *C. novæ-guineæ*. He makes some interesting remarks on its habit ['89, p. 305] :

"*C. Schmideliana* (GRAY) *C. discoidea* (AG.) PERRIER, 'Révision des
Stellérides,' pag. 74. P. DE LORIOL, 'Catalogue raisonné,' pag. 64.—Fünf
Exemplare (No. 511, 607) aus der Bai von Batavia. Wie von PERRIER und
DE LORIOL schon hervorgehoben worden ist, zeigt die *C. Schmideliana* ziemlich viele Varietäten, und ist wahrscheinlich auch die *C. coriacea* (M. u.
TROSCHEL) als dieselbe Art anzusehen. Auch die Tiere aus der Bai von
Batavia weichen mehr oder weniger von einander ab. Die Zahl der Dornen
auf den Ambulacralplatten variiert zwischen 4 und 7, gewöhnlich kommen
aber 6 vor. Die Tuberkeln auf den Bauchplatten sind fast alle gleich gross
und nicht nur auf der Mitte der Platten sondern auch auf den Rändern
entwickelt. Die Porenfelder auf dem Rücken sind deutlich zu unterscheiden.
Die Farbe der Tiere ist beim Leben auch ziemlich verschieden. Der Rücken
ist heller und dunkler gelblich braun gefleckt, zuweilen aber auch mehr
graubraun. Die kleinen konischen Rückentuberkeln sind hellrot. Der Bauch
ist mehr rötlich orange, zuweilen mehr rotbraun, die Ambulacralfurchen
gewöhnlich dunkler rot.

"Bei einem Exemplare waren die Porenfelder des Rückens sehr weit
von einander entfernt und die kleinen konischen Rückenpapillen nur auf den

Porenfeldern entwickelt. Das breite Netz zwischen diesen war fast ganz glatt; nur an den Rändern der Scheibe kommen einige platte Tuberkeln vor.

" Die Tiere sind auf den Korallenriffen in der Bai von Batavia nicht selten. Sie kriechen gewöhnlich langsam auf dem Sande, zwischen den Korallenstücken, im seichten Wasser umher. Im Aquarium erhalten sie sich ganz gut und bleiben öfters 8–10 Monate am Leben. Gewöhnlich kriechen sie an der senkrechten Wand des Bassins hinauf, oder bewegen sich langsam auf dem Sandboden mittelst der grossen Ambulacralfüsschen fort. Sehr merkwürdig ist es die Formveränderungen, deren diese Tiere fähig sind, zu beobachten. Indem der Körper gewöhnlich so starr und hart aussieht und sich auch so anfühlt, ist er dennoch viel schmiegsamer als man erwarten sollte. Wenn die Tiere vom Boden an der senkrechten Wand hinauf klettern wollen, so folgen sie mit der Bauchfläche gänzlich dem rechten Winkel der beiden Flächen, und der Körper nimmt fast die Gestalt eines Kugelsektors an. Noch viel mehr ändert sich die Gestalt, wenn die Tiere, nachdem man sie auf den Rücken umgekehrt hat, versuchen, sich wieder auf die Bauchfläche zu legen. Sie schwellen alsdann die eine Hälfte des Körpers kugelig an, damit sie mit einer der sich eine Strecke weit auf den Rücken fortsetzenden Bauchfurchen, dem Boden nahe genug kommen, um einigen der Saugfüsschen einen Anhaltspunkt zu geben. Indem nun allmählich die folgenden Füsschen sich anheften, schwillt die ganze Bauchfläche kugelig an und zieht der Rücken sich mehr zusammen, so dass der ganze Körper zuweilen fast wie eine Gurke aussieht. Zum Umwenden braucht das Tier etwa zehn Minuten.

" Als Futter dienen ihm verschiedene Echiniden und andere Tiere. Im Aquarium hat er öfters den *Echinometra lucunter* gefressen, was allerdings sehr auffällig ist, da man doch von diesem Tiere, der starken Stacheln wegen, nicht vermuten sollte, dass es leicht der so harmlos aussehenden *Culcita* zum Opfer fallen würde.

" Einmal sah ich, wie eine *Culcita* eine *Echinometra*, welche gegen die Glasscheibe sass, bewältigte. Sie legte sich mit der ganzen Bauchfläche über ihre Beute, welche sie mit den grossen Saugfüsschen festhielt. Indem sie mit dem starken Mundgerüst arbeitete, wurde die *Echinometra* allmählich umgekehrt, so dass die Bauchfläche der letzteren ihr zugekehrt war.   Dann zerbrach sie den Mundraum der *Echinometra* und stülpte den Munddarm in das Innere der *Echinometra* hinein und sog die Schale fast völlig aus."

HARTLAUB[1] in his critical revision of the genus, makes several important remarks, particularly in regard to the original type specimen ['92, p. 77] :

" *C. novæ guineæ* M. T. 1842, l.c.

[Here is reproduced the original description of MÜLLER and TROSCHEL.]

"Unser Göttinger Museum besitzt ein Exemplar von Amboina, 1864 durch BLEEKER gesammelt, welches als *C. novæ guineæ* M. T. bestimmt war. Wie ich jedoch dasselbe mit der Beschreibung sorgfältig verglich, fiel mir auf, dass es in Reihen stehende Gruppen gröberer Granula auf der Bauchseite besass, und dass seine Porenfelder nicht gross, sondern im Gegenteil relativ klein waren. Als ich kurze Zeit darauf nach Hamburg kam, fand ich ein ganz ähnliches Stück von Samoa mit derselben Bestimmung, und als ich später Leyden besuchte, fand ich auch hier ein von Neu Guinea stammendes durch SALOMON MÜLLER gesammeltes und in Spiritus vortrefflich conservirtes Exemplar, welches als *C. novæ guineæ* M. T. bezeichnet war und den beiden Stücken des Göttinger und Hamburger Museums vollkommen glich ; offenbar hatte es für die Bestimmung dieser als Vorbild gedient. Da es aber, wie sich bald herausstellte, augenscheinlich nicht das Exemplar ist, welches den Autoren der Species zum Typus diente, sondern vielmehr zweifelsohne einer andern Art angehört, so fragt es sich, ob von Seiten späterer Autoren das erwähnte Spiritus Exemplar in Leyden als Typus aufgefasst würde oder aber das richtige Original Stück, welches trocken conservirt und in einem Auszuge aufbewahrt gewesen ist. Ich möchte auf Grund des Hamburger und Göttinger Stückes das erstere für wahrscheinlicher halten. Dann würde es mir erklärlich sein, warum SLADEN in seinem Challenger Report sagt, er sei nach sorgfältiger Prüfung nicht im Stande *Culcita pentangularis* GRAY von *C. novæ guineæ* zu unterscheiden.   Mir

---

1) In the section on the skeletal system, in which *C. schmideliana* serves as the principal material, HARTLAUB makes several interesting remarks on the skeleton of *C. novæ-guineæ*, which should be consulted by those who would make a detailed study of the skeleton of this genus.

scheint in der That auch das fälschlich als *C. novæ guineæ* bestimmte
Spiritus Exemplar in Leyden grosse ähnlichkeit mit der GRAY'schen Species
zu haben, obwohl ihr die bei letzterer vorhandene sechseckige Felderung
fehlt, doch wage ich ohne den GRAY'schen Typus gesehen zu haben nicht
den Pseudotypus von *C. novæ guineæ* mit ihr zu identificiren; ich ziehe es vor
letztere Art auf die Gefahr hin sie später wieder einziehen zu müssen,
unter neuem Namen zu beschreiben. Leider ist in der spärlichen Literatur
über die Gattung ein sicheres Urteil über die von einem Autor besprochene
Art in den seltensten Fällen zu gewinnen, wenn man von der einzigen durch
gute Abbildungen bekannt gewordenen *C. schmideliana* absieht. Ganz unsicher scheint z. B. PERRIER gewesen zu sein. Der Mangel von Abbildungen
macht sich ausserordentlich fühlbar. Die Bestimmungen scheinen überall
auf gut Glück und nach den Fundorten gemacht zu sein, jedoch selten auf
Kenntniss der Originale oder auf Grund guter Beschreibungen. Letztere
fehlten eben bislang gänzlich.

"Die Gründe, die mich bewegen, das trockne *C. novæ guineæ* Exemplar als allein gültigen Typus der Art anzusprechen, sind folgende: das
Spiritus Exemplar hat einen Durchmesser von 117 mm., ist also um Vieles
kleiner als die von den Autoren angegebene Maximalgrösse; es ist ferner
kaum anzunehmen, dass die Autoren die Gruppenständigkeit der gröberen
Granula auf der Bauchseite sowie die besondere Kleinheit der dorsalen
Porenfelder nicht von ihm erwähnt haben sollten; es ist schliesslich sehr
wahrscheinlich, dass die Autoren die von SALOMON MÜLLER in Spiritus
conservirten Stücke nicht gesehen haben, weil sie sonst den Fundort von
C. grex M. T. nicht als unbekannt angegeben haben würden, denn das
Leydener Museum besitzt ein von diesem Reisenden gesammeltes ausgezeich-
netes Exemplar dieser Art von den Moluccen, in Spiritus conservirt. Das
trockne Exemplar von *C. novæ guineæ*, welches in seinem Habitus mit dem
in Spiritus garkeine Aehnlichkeit hat, misst im Durchmesser 158 mm. (R + r),
ist also, zwar grösser wie jenes, doch auch viel kleiner als 10 Zoll. Dennoch müssen wir allein dieses als Original auffassen, denn nur so finden die
Worte der Autoren eine Erklärung, wenn sie die Porenfelder sehr gross
nennen, und wenn sie von der ventralen Granulirung sagen 'grössere nie-

drige Knötchen auf der Bauchseite zwischen der feineren Granulation stehen
sehr dicht.' Einige Stellen freilich bleiben dennoch dunkel in ihrer Beschreibung; die Worte 'selten sechseckig' passen weder auf das eine noch
auf das andre Exemplar. Dies und die Grössenangabe beruhen vielleicht auf
mündlichen Aussagen des Reisenden. Viel unerklärlicher ist, dass es von
der Gestalt heisst 'wie beim vorigen.' Die hier gemeinte C. *coriacea* ist nach
einem allerdings ausgetrockneten Originale des Berliner Museums und
anderen Exemplaren zu urteilen, ganz flach, während der Typus von *C.*
novæ guineæ entschieden gewölbt ist, doch ist vielleicht in Betracht zu
ziehen, dass es auch in der Beschreibung von *C. coriacea* 'Gestalt und
Verhältnisse der vorigen' heisst, und dass hier *C. discoidea* AGASS. gemeint
ist, von welcher das Leydener Museum ein Original Exemplar von ganz
ungewöhnlich hoher Form besitzt. Dieser Art können allerdings schlechter
conservirte, gequetschte Stücke von *C. novæ guineæ* in der Gestalt so
ähnlich sehen, dass wie erwähnt wurde, ein Hamburger von Java stammendes Stück offenbar nur seiner Form wegen als *C. discoidea* AGASS. bestimmt war. (vergl. pag. 67.)

    "Eine eingehende Beschreibung des von mir als Typus der Art auf-
gefassten Exemplares anzufertigen ermangelte mir leider bei meinem
Leydener Aufenthalte die Zeit. Ich musste mich darauf beschränken
dasselbe von der Bauch und Rückenseite zu photographiren und möchte
nach den so erhaltenen Bildern und nach etwa einem Dutzend Exemplaren,
die ich genau untersuchte, folgendes zur näheren Kenntniss der Species
anführen.

    "Diagnose.

    "Gestalt hoch gewölbt, mit convexen Seiten und abgerundeten Ecken.
Porenfelder des Rückens rosettenständig, oft gross und polygonal, seltener
mittelgross und dann weitläufiger stehend und abgerundet, an den Seiten
sehr gross und meist bis an die Ventralkante reichend. Rücken und Seiten
bedornt. Dornen der Porenfelder feiner oder fehlend. Bauchseite von einer
sehr groben, dichtstehenden, selten gruppenständigen Granulation bedeckt,
deren Charakter von einem polygonal plattenförmigen bis zu einem dornförmi-
gen variirt. Dazwischen eine feine Grundgranulirung. Innere Furchen-

papillen kräftig, in Gruppen von 3–6, meist 5. Aeussere Bewaffnung, wenn deutlich differenzirt, eine einfache oder Doppelreihe grober, manchmal dornförmiger Tuberkel. Kleine Pedicellarien auf den Porenfeldern und der Ventralseite. Madreporenplatte nicht sehr gross, oft von einem Dornenkranze umgeben.

"Färbung in Spiritus: meist hell graugelb, seltener hell bräunlich oder weiss.

"Grösse: bis 151 mm. Dm. (R+r).

"Fundorte: Viti, Marshall Inseln, Neu Guinea, Neu Hanover, Amboina, Java, W. Küste v. Sumatra (Padang). Mascarenen?

"Ich kann in Anschluss an die Diagnose nur empfehlen der weitgehenden Variation der Culciten bei der Bestimmung Rechnung zu tragen. Die geringe von mir gesehene Anzahl von Exemplaren beweist dieselbe in hohem Maasse. Ich kann unter diesem Materiale drei Varietäten unterscheiden:

"Die erste ist die Ausbildungsform des trocknen M. T. Exemplares in Leyden, welcher ein von J. BROOK auf Amboina gesammeltes schön erhaltenes Spiritus Exemplar unserer Sammlung vollkommen gleicht. Ihnen eigentümlich sind in der Regel grosse drei, bis sechseckige Porenfelder, die durch schmale, ein Netzwerk bildende, porenfreie Züge getrennt und um grössere porenfreie Stellen rosettenartig gruppirt sind. Auf diesen porenfreien Räumen stehen in kleinen oder mässigen Zwischenräumen ziemlich kräftige Dornen, die aber höchstens ein Drittel so stark sind wie die dicken Rückendornen von *C. schmideliana* RETZ. Auf den Porenfeldern stehen in ziemlich derselben Dichtigkeit bedeutend feinere Dornen, die jedoch aus der Granulation derselben deutlich hervorragen. Die ventrale Fläche ist ausser ihrer feinen Grundgranulirung mit einer dichtstehenden sehr groben rundlichen Granulation bedeckt, die keine deutliche Gruppenständigkeit oder gar Felderung zeigt.

"Die zweite Varietät ist durch eine besonders dichte Bedornung ausgezeichnet, die namentlich auf den Seiten des Körpers sehr kräftig wird und hier ohne Unterbrechung in die grobe Granulirung der Bauchseite übergeht. Die Tendenz zur Dornenbildung ist so gross, dass an dem Hamburger Exemplar von den Viti Inseln sogar die ganze grobe Ventralgranulation

dornigen Charakter hat, incl. der äusseren Ambulacralbewaffnung. Für die
Bedornung des Rückens und der Seiten ist sodann eigentümlich, dass der
Grössenunterschied zwischen den Dornen der Porenfelder und denen der
Zwischenräume sehr gering ist, und dass namentlich die grossen Porenfelder der Seitenflächen starke Dornen tragen. Die Porenfelder sind weniger gross als bei der ersten Varietät und neigen sehr zur Verschmelzung.
Die grobe Granulirung der Bauchseite ist gleichmässig, aber manchmal
weniger dicht verteilt, ohne Andeutung von Gruppenbildung oder von
Zügen. Von dieser Varietät sah ich ausser dem Hamburger Exemplare
nur eins im Bremer Museum aus dem 'Mare indicum.' Beide Stücke
zeigen ziemlich zahlreiche kleine Klappenpedicellarien.

   "Für einen dritten von den beiden bis jetzt beschriebenen ziemlich
verschiedenen Habitus sind als Beispiele ein Exemplar von Pulo Edam
)Java) und das HUBRECHT'sche von Padang zu nennen, beide in Alcohol
tadellos erhalten. Ihnen ist eigentümlich eine spärlichere, aber kräftigere
Bedornung des Rückens und der Seiten, die fast ausschliesslich auf die porenfreien Räume beschränkt ist. Die feinere Bedornung der Porenfelder fehlt
fast ganz. Die Porenfelder sind von mässiger Grösse und mehr abgerundet.
Die Bauchseite ist entsprechend dem Verlauf der subcutanen Plattenreihen
mehr oder minder deutlich gefurcht und die grobe Granulation mit Ausnahme
der oralen Umgebung in deutlicheren Gruppen gesondert. Diese groben
Granula haben bei dem Pulo Edam Exemplare die Form kleiner polygonaler
Tafeln, dei dem andern sind sie etwas weniger dicht gestellt und mehr
perlartig.

   "Dass zwischen diesen drei Ausbildungsformen allemöglichen Uebergänge vorkommen, unterliegt kaum einem Zweifel. Schon das HUBRECHT'-
sche Exemplar nähert sich durch den perlartigen Charakter der groben
Granulation, und dadurch, dass die groben Granula wohl in scharf getrennten Zügen aber nicht in vollständig gesonderten Gruppen stehen, den beiden
andern Varietäten. Ein Hamburger Exemplar von Java dagegen hat ganz
die feine polygonale Täfelung und Gruppenständigkeit unsres von Pulo
Edam stammenden Stückes, aber auf der Rückenseite viel dichter stehende
Dornen und auch Dornen auf den Porenfeldern.

"Ein sehr merkwürdiges mittelgrosses Exemplar von Pulo Edam besitzt die Göttinger Sammlung, welches zu *C. novæ guineæ* zu stellen etwas gewagt sein dürfte. Ziehen wir indessen in Erwägung, dass es von einem Fundorte stammt, wo diese Art erwiesener Maassen vorkommt, dass es ferner ein jüngeres Exemplar ist, und dass es fast nur durch die höchst eigentümliche Felderung der Bauchseite abweicht, so werden unsre Bedenken, wenn auch nicht ganz schwinden, so doch wesentlich verringert. Es hat die charakteristische hochgewölbte abgerundete Form unsrer Art. Die Bedornung des Rückens ist die typische, indem sie der des Leydener Originales gleicht, obwohl sie im allgemeinen etwas schwächer ist. Die Madreporenplatte ist von einem Kranze kräftiger Dornen umgeben. Die Porenfelder sind rundlich und auffallend klein für die Art. Sie erreichen auf den Seitenflächen bei weitem nicht die Ventralkante (was sich indessen auch an ganz zweifellosen Individuen der Species gelegentlich wiederholt). Die grobe Granulation ist in Gruppen gesondert, die ihrer Lage nach den Tafeln des unter der Haut liegenden Skelettes entsprechen. Die Gruppen sind von beträchtlicher Grösse, enthalten etwa 20–25 Granula von der Form kleiner polygonaler Täfelchen und sind im ganzen schwach vorgewölbt. Sie sind von einander durch etwa 1 mm. breite Züge einer mit zahlreichen groben Körnern untermischten Granulation getrennt, welche im Interradius zu einem Doppelstrang zusammentreten. Die die äussere Ambulacralbewaffnung vertretende grobe Granulation der Ambulacralplatten ist perlartig, und sind dadurch die Gruppen von der Ambulacralrinne durch ein ziemlich breites Band von sehr differenten Aussehen getrennt. Da das Exemplar die Bedornungsart des Rückens unsrer ersten Varietät hat, welche auf den Porenfeldern bedeutend feiner ist als auf den Zwischenräumen, dagegen auf der Bauchseite gruppenständige polygonale Täfelchen besitzt wie das Pulo Edam Exemplar unsrer dritten Varietät, so kann man es als eine Art Bindeglied zwischen beiden auffassen.

"Als Hauptmerkmale der Art seien nochmals die grobe Granulation der Bauchseite und für gute Spiritus Exemplare die hoch gewölbte Körperform und die abgerundeten Ecken hervorgehoben.

"Die als dritte Varietät beschriebenen Formen mit gruppenständiger grober Bauchgranulirung und mangelnder Bedornung auf den Porenfeldern würden sich bei anatomischer Untersuchung vielleicht als sexuell differenzirt erweisen, sie als eigene Art aufzufassen scheint mir einstweilen nicht geboten."

Culcita plana, which is regarded by DÖDERLEIN as a synonym of *C. novæ-guineæ* is described in the same work as a new species [HARTLAUB, '92 p. 84]:

" *Culcita plana*, n. sp.

"Körper flach scheibenförmig (an jüngeren Exemplaren gewölbter), mit schwach eingebogenen Seiten und vortretenden Ecken. Porenfelder rosettenständig, klein, zahlreich; auch auf den Seiten klein. Bedornung des Rückens fein. Dornen der Porenfelder kleiner als die der Zwischenräume. Bauchseite von feiner Granulation bedeckt, zwischen welcher schwach vortretende Gruppen etwas gröberer Granula stehen, die jedoch nicht auf scharf begrenzten Feldern liegen. Die Gruppen bilden Reihen und entsprechen ihrer Lage nach den subcutanen Skelettteilen. Innere Furchenpapillen in Gruppen von 5–6 ziemlich gleichmässigen Stäbchen. Aeussere Furchenpapillen in zuweilen ähnlichen, der Rinne parallel liegenden Gruppen von meist drei Tuberkeln, deren Stärke gering ist. Bisweilen eine dritte undeutlich differenzirte Reihe. Manchmal kleine Pedicellarien von der Grösse und Form kleinerer Granula in Menge auf der Bauchseite. Madreporenplatte von einem Dornenkranze umgeben.

"Färbung in Spiritus: meist weisslich, seltener graugelb.

"Grösse : bis 160 mm. Dm.

"Fundorte : Samoa, Viti, Neu Guinea, Philippinen, Amboina, Mauritius.

"Zu dieser Species gehört das erwähnte fälschlich als *C. novæ guineæ* M. T. bestimmte Exemplar des Leydener Museums, welches von SALOMON MÜLLER auf Neu Guinea gesammelt wurde. Sie ist möglicher Weise mit *C. pentangularis* identisch; um den Lesern darüber das Urteil zu erleichtern, citire ich hier wörtlich die GRAY'sche Beschreibung seiner Art :

[This is omitted, because reproduced before ; *vide supra*, p. 520.]

" *C. pentangularis* GRAY scheint sich also von unsrer Art dadurch zu unterscheiden, dass ihre ventralen Granulationsgruppen scharf begrenzt sind. Ich halte es aber auch für möglich, dass beide zu einander in einem ähnlichen Verhältntiss stehen wie unsre dritte Varietät von *C. novæ guineæ* zu unsrer ersten, dass sie also nur eine Art bilden.

"Mit *C. novæ guineæ* hat unsre Species nur in der dorsalen Bedornungsart eine gewisse Aehnlichkeit. Der Bestimmer des Leydener Spiritus-Exemplares hielt dasselbe möglicherweise für ein junges Individuum dieser Art, da seine Grösse, wie die fast aller übrigen Exemplare, die ich sah, ziemlich gering ist (R + r 117 mm.). Dass diese Annahme jedoch durchaus verkehrt gewesen wäre, zeigt ein grösseres Exemplar von den Philippinen im Göttinger Museum, welches in jeder Hinsicht die typischen Merkmale besitzt.

"In Bezug auf die ventralen Gruppen etwas gröberer Granula variirt die Art sehr. Diese Gruppen können ganz klein sein und nur aus wenigen, kaum merklich grösseren und lose vereinigten Körnern bestehen, oder aber ziemlich gross und körnerreich sein und sich in letzterem Falle so nähern, dass ihre gegenseitige Abgrenzung verwischter wird. Ohne dass ich derartige Exemplare gesehen habe, halte ich es für wahrscheinlich, dass bei weitergehender Ausbildung im letzteren Sinne auch Stücke vorkommen, die auf der Bauchseite mit einer mässig groben Granulation gleichmässig bedeckt sind. Für ein solches Exemplar halte ich z. B. ein als *C. coriacea* M. T. bestimmtes Stück von Mauritius (ROBILLARD), von dem mir Herr Prof. v. MARTENS in Berlin auf meine Bitte Photographieen anfertigen liess. Dasselbe gleicht von der Rückenseite durchaus unsrer Art und unterscheidet sich von *C. coriacea* M. T. sehr wesentlich, durch den Besitz getrennter Porenfelder. Andrerseits würde das Extrem mangelhafter Gruppenbildung eine ganz gleichmässige sehr feine Granulirung der Bauchseite sein. Das Hamburger Museum besitzt ein Stück von Samoa, bei welchem ein derartiges Verhalten fast erreicht ist.

"Sehr charakteristisch für die Species ist ferner die äussere Ambulacralbewaffnung, welche der inneren durch ihre in der Richtung der Rinne stehenden Tuberkelgruppen manchmal (Philippinen-Exempl. in Göttingen) in auffallender Weise gleicht.

"Ein junges Exemplar von den Viti Inseln im Lübecker Museum unterscheidet sich durch einen gewölbteren Rücken. Die Seiten aber sind eingebogen und die Ecken stark vortretend. Die Gruppen der Bauchseite sind sehr deutlich und vorgewölbt, aber nirgends durch scharfe Furchen begrenzt. Das Exemplar hat einen Durchmesser von 80 mm. (R+r). Die Lage der unteren wie oberen Marginalplatten ist äusserlich noch erkennbar.

"Die Porenfelder, auf deren Kleinheit nochmals hingewiesen sei, scheinen ventralwärts niemals über die dorsale Grenze der oberen Marginalplatten hinauszureichen. Die Granulation der Bauchseite kann der von *C. coriacea* sehr ähnlich sehen, bei welchen die gröberen Körner auch gelegentlich in undeutlichen Gruppen stehen (trocknes Exempl. in Stuttgarter Museum vom Roten Meer); in solchen Fällen dürfte jedoch nicht nur die so verschiedene Verteilung der Poren sondern auch der für *C. coriacea* eigentümliche Charakter der dorsalen Granulirung sofort entscheidend sein."

C. arenosa is also described in the same work as a distinct species [HARTLAUB, '92, p. 92] :

"*C. arenosa* PERRIER, 1869, l. c.

"Gestalt flach scheibenförmig, Ecken nicht abgerundet, Porenfelder mehr oder minder rosettenständig, zur Verschmelzung neigend, von mittlerer Grösse, auch die Seitenflächen bis an die ventrale Kante bedeckend. Rücken und Seiten mit nur einer Sorte kleiner schlanker spitzer Dornen bedeckt, die ziemlich zerstreut stehen aber gleichmässig verteilt sind. Bauchseite mässig grob granulirt. Die gröbere Granulation dicht, nicht gruppenständig, perlartig. Verlauf der ventralen Plattenreihen nur stellenweise durch seichte Furchen schwach angedeutet. Innere Furchenpapillen ziemlich kräftig, in Gruppen von 5-6. Aeussere in Gruppen von 2-3 groben, oft cylindrischen oder conischen Tuberkeln. Zuweilen noch eine dritte Reihe weniger grosser Tuberkel. Kleine zangenförmige Pedicellarien in der Nähe der Rinnen, sowie zahlreiche kleinere, von der Grösse und Form gröberer Granula, auf der Bauchseite. Madreporenplatte ohne Dornenkranz.

"Färbung in Spiritus: schmutzig weiss, oder dunkel olive mit schmutzig violetten Hautkiemen.

"Grösse : bis 200 mm. Dm. (R+r).

"Fundorte : Sandwich Inseln (Typus), Amboina, Ceram Laut.

"Die Bestimmung dieser zweifellos guten Art war insofern bisher mit Schwierigkeiten verbunden, als PERRIER die Porenfelder als undeutlich und die Poren als gleichmässig über den ganzen Rücken verteilt beschrieben hatte. Diese Aussage beruhte, wie der Autor die Güte hatte mir brieflich mitzuteilen, auf einem Irrtum. Die Art hat also, wie die meisten andern, getrennte Porenfelder, obgleich diese entschieden zur Verschmelzung neigen. An einem Göttinger Exemplare von Amboina, dessen Photographie ich Prof. PERRIER zur Begutachtung meiner Bestimmung schickte, sind z. B. die einzelnen Felder eines Rosettenringes in der Regel miteinander etwas verwachsen. Da PERRIER die Poren für gleichmässig verteilt hielt und die Porenfelder als 'indistinctes' beschrieb, so hätte nach unserm Dafürhalten ein Vergleich mit *C. coriacea* M. T. nahe gelegen, mit welcher die Species ohne Frage. viel Aehnlichkeit hat. Dagegen scheint der Autor eine grössere Verwandtschaft mit *C. grex* angenommen zu haben, mit welcher die Art eingehender verglichen wird. Mit dieser hat jedoch *C. arenosa* kaum etwas gemein. Die Form von *C. grex* ist nicht abgeplattet, wie PERRIER glaubt, und abgesehen davon ist der Charakter der Granulation und die Verteilung der dorsalen kleinen Dornen bei *C. grex* ein so eigentümlicher, dass sie mit keiner andern bekannten Art verwechselt werden kann. Mit *C. coriacea* M. T. aber teilt unsre Species die Körperform, ferner einen ganz ähnlichen Habitus der Bauchseite und schliesslich die Einförmigkeit der dorsalen Dornen. Letztere stehen jedoch bei jener dicht, bei letzterer zerstreut. Diese Einförmigkeit der dorsalen Dornen unterscheidet *C. arenosa* leicht von *C. novæ guineæ* und Verwandten, bei denen die Dornen der Porenfelder viel kleiner sind als die der Zwischenräume.

"Von den Exemplaren dieser Art, die mir zu Gebote standen, waren zwei von den Sandwich Inseln, eins dem Stuttgarter, eins dem Göttinger Museum gehörig. Beide sind ausgezeichnet durch sehr kräftige schlank conische Form ihrer äusseren Ambulacralpapillen. Dieselben scheinen an dem PERRIER'schen Originale ähnlich zu sein, insofern sie hier als 'cylindrisch' bezeichnet wurden; anders aber erscheinen sie an den Stücken von Amboina und Ceram Laut, wo sie kurz, dick und abgerundet sind. Sie. stehen selten einzeln, sondern kommen meist als Zwillingstuberkel vor und an dem Amboina Exemplare bilden sie stellenweise sogar Gruppen von drei mit. einander verwachsenen. Von den Zwillingstuberkeln ist gewöhnlich der eine viel grösser wie der andre. Die Stellung dieser Gruppen ist häufig eine zur Rinne quere, braucht es aber nicht zu sein. Das Amboina Exemplar besitzt eine fast überall deutliche dritte Bewaffnungsreihe in Gestalt einfacher dicker rundlicher Tuberkel.

"Der ziemlich schmale Raum zwischen der ventralen Grenze der Porenfelder und der Bauchkante ist mit stärkeren Dornen besetzt, die in die perlartige gröbere Granulation der Ventralfläche allmälig übergehen.

"Die von PERRIER beschriebenen zangenartigen Pedicellarien des Rückens habe ich nicht gefunden, ebenso ist der von ihm erwähnte eigentümliche Verlauf der Furchen der Madreporenplatte kein constantes Merkmal. An dem Amboina Exemplare z. B. haben dieselben einen entschieden radiären Verlauf.

"Die Art ist aufs nächst verwandt, wenn nicht gar identisch, mit der folgenden."

Immediately following is described *C. acutispinosa*, also a synonym of *C. novæ-guineæ* according to DÖDERLEIN [HARTLAUB, '92, p. 95] :

"*Culcita acutispinosa* BELL, 1883, l.c. p. 334.

"Diese Art soll von allen anderen dadurch unterschieden sein, dass die Enden ihrer Ambulacralrinnen nicht auf den Rücken übergreifen. Es ist dies jedoch unsres Erachtens ein etwas trügerisches Merkmal, da ich Exemplare von *C. novæ guineæ* sah, bei denen dies ebenso wenig der Fall war, und bei denen die Länge der einzelnen Rinnen ganz erheblich variirte (vergl. pag. 72). Da aber die Art auf nur ein Individuum begründet wurde, so scheint mir das Verhalten der Ambulacralrinnen noch kein genügender Grund für die Annahme der Species zu sein. Die übrigen von BELL angeführten Merkmale weisen aber zum grossen Teil auf Uebereinstimmung mit *C. arenosa* PERR. hin. Zur Bestätigung dessen führe ich folgende Citate an :

"'Resembling *C. coriacea*'.—'The body is almost completely discoidal in shape.'—'The adambulacral spines are in two rows'; 'in the outer row there are generally two spines, one of which is much smaller'. 'The actinal surface is not marked out into areolæ.' 'The greater part of the side and the whole of the abactinal surface of the disk are covered with short sharp spines, which are scattered over them with considerable profusion, though in no definite order; dotted among the spines are pores of moderate size, which are very indistinctly grouped into pore-areas.'

"Ich kann nicht läugnen, dass mich diese Stellen mit einigem Zweifel erfüllen; einige andre freilich könnten vielleicht doch für die Species sprechen, so z. B. 'the angles of the rays being very nearly altogether rounded off.' Aber dann versteht man kaum, wie die Art *C. coriacea* ähneln soll, bei der die Ecken keineswegs abgerundet sind. Bemerkenswert ist ferner ('the actinal surface') 'is richly invested by a number of short, blunt, stout processes, hardly to be called spines, amidst which a coarse granular covering is to be observed.' Dies Verhalten wäre vielleicht dasjenige, welches noch am meisten für die Berechtigung der Species ins Gewicht fällt.

"Der Fundort der Art sind die neuen Hebriden (Aneityum Insel)."

BELL ['94, p. 394] mentions *C. pentangularis* GRAY from north-west Australia.

SLUITER mentions *Goniodiscus sebæ* in the collection of the Museum of Amsterdam ['95, p. 55]: "*Goniodiscus sebæ* M. u. TR. Sechs Exemplare, und zwar drei in Alkohol von Ambon (BLEEKER), und drei getrocknet von den Molukken (V. D. HUCHT)." In the same paper [p. 57] *C. novæ guineæ* is described as follows:

"39. *Culcita novæ guineæ* M. u. TR. Zwei getrocknete Exemplare von den Molukken (V. D. HUCHT), neun in Alkohol von den Molukken (BLEEKER) und zwei in Alkohol von der Bai von Batavia (SLUITER). Es ist nur mit einem gewissen Zweifel, dass ich alle diese Exemplare zu derselben Species *C. novæ guineæ* rechne. Die sorgfältige und ausführliche Besprechung der Culcita-Arten von HARTLAUB zeigt aufs deutlichste, wie schwer es ist die verschiedenen Culcita-Species scharf aus einander zu halten. Unter den dreizehn Exemplaren, welche ich vor mir habe, besitzen sechs die typische grobe nicht gruppenständige Granulation der Bauchseite. Die Porenfelder des Rückens sind rosettenständig, gross und mit einer Neigung zur Verschmelzung. Die Rückentuberkeln zwischen den Porenfeldern sind grösser als die auf den Porenfeldern. Bei drei weiteren Exemplaren von den Molukken ist die Granulation des Bauches etwa die gleiche, nur sind die

Granula noch etwas grösser. Die Porenfelder des Rückens sind aber überall durch ziemlich breite Zwischenräume scharf von einander getrennt. Diese Zwischenräume sind sehr fein granulirt, während aus der feinen Granulation mehrere vereinzelt stehende grosse Tuberkeln hervorragen. Ein von mir in der Bai von Batavia gesammeltes Exemplar stimmt mit diesen drei zuletzt erwähnten überein, nur dass die grösseren Tuberkeln auf dem Rücken gänzlich fehlen. Die ganze Rückenseite gleicht also genau der photographischen Abbildung, welche HARTLAUB von *C. grex* giebt, und wie ich dieselbe auch mit dem Leidner Exemplar habe vergleichen können. Ich möchte denn auch das Tier zur *C. grex* rechnen, wenn es nicht durch die grobe Granulation des Bauches und die hohe gewölbte Form doch wieder der *C. novæ guineæ* näher stand. Ein zweites Exemplar aus der Bai von Batavia zeigt wieder, in Betreff der Verteilung der Porenfelder auf dem Rücken, ein Zwischenstadium zwischen diesen beiden Extremen. Bei einem jüngeren Exemplar von den Molukken (BLEEKER) stehen die Granula des Bauches deutlich in Gruppen, während die Porenfelder und Tuberkel des Rückens sich verhielten wie bei den sechs zuerst erwähnten Tieren von den Molukken.

"Indem ich die Tiere längere Zeit lebend beobachtet habe, ist es mir nicht möglich, der allgemeinen Körpergestalt den Wert beizulegen, wie HARTLAUB es thut. Dasselbe Tier kann seine Gestalt derartig ändern, dass es zuweilen hochgewölbt, zuweilen flach scheibenförmig erscheinen kann. Kränkliche Tiere sind immer scheibenförmig, wodurch auch alle Exemplare, welche nicht unmittelbar, nachdem sie gefangen sind, in Alkohol versetzt wurden, scheibenförmig erscheinen, was bei vielen Museumstücken wohl der Fall gewesen sein wird.

"Die früher von mir[1] als *C. schmideliana* erwähnten Tiere aus der Bai von Batavia sind, wenigstens der geografischen Verbreitung nach, auch wohl zu *C. novæ guineæ* zu zählen. Ob aber die beiden Arten, wenn das Vergleich-Material sich mehrt, als gesonderte Arten aus einander zu halten sind, scheint mir sehr zweifelhaft. Von DE LORIOL (l.c. p. 182) wurde ein von PICTET und BEDOT gesammeltes Exemplar als *C. grex* bestimmt, aber noch kein Rücksicht genommen von der ausführlichen Arbeit von HARTLAUB."

---

1) SLUITER, '89' p. 305.

LEIPOLDT mentions this species under the name of *C. plana* HARTLAUB ['95, p. 637] :

"Zwei Exemplare von San Jacintho (Philippinen), 'Stretto di San Bernardino,' die ich zu dieser von HARTLAUB neu aufgestellten Art gestellt habe. Es sind beides augenscheinlich noch junge Exemplare, doch zeigt das grössere schon den Habitus der erwachsenen Tiere, während das kleinere noch die Kennzeichen der jungen Tiere an sich trägt.

"Das erstere hat einen Durchmesser (=R+r) von 90 mm. Die Gestalt ist flach, jedoch mit etwas gewölbter Rückenfläche, schwach eingebogenen Seiten und vortretenden Ecken. Die aktinale Seite zeigt die von HARTLAUB für *C. plana* angegebene Granulation und Gruppirung der Granula, d. h. die aktinale Seite ist mit feinkörnigen Granula bedeckt, zwischen denen sich, den darunterliegenden, ventrolateralen Platten entsprechend, Gruppen gröberer Granula befinden. Die letzteren sind hier relativ gross und zahlreich. Die Ambulacralbewaffnung besteht aus einer inneren Längsreihe von fünf Stacheln, welche in der Mitte etwas, aber nur wenig, höher sind, als an den Seiten und aus einer äusseren Reihe, die sich aus je zwei—nur dicht in der Nähe des Mundes trifft man ihrer drei an, von denen der mittlere am stärksten ist,—dickeren und kräftigeren Tuberkeln zusammensetzt. Diese Tuberkel sind mit einander verwachsen und der adorale ist kräftiger als der aborale. Diese äussere Reihe von Paxillen liegt parallel der Ambulacralrinne.

"In der Art ihres Auftretens weicht die äussere Reihe der Ambulacralbewaffnung also etwas von der HARTLAUB'schen Beschreibung ab, und nähert sich dem bei *Culcita arenosa* beschriebenen Verhalten. Eine dritte, noch weiter nach aussen liegende Reihe fehlt.

"Die rosettenständigen Porenfelder der abaktinalen Seite gruppiren sich bei unserem Exemplare um Felder, welche grösser oder kleiner als die Porenfelder selbst sein können. Letztere sind nur durch sehr schmale Zwischenräume, die schwinden können, so dass die einzelnen Porenfelder mit einander verschmelzen, von einander getrennt. An den Seiten scheinen mir die Porenfelder etwas tiefer zu reichen als die dorsale Grenze der oberen Randstücke geht; nach der Armspitze zu zeigen sie auch hier die Neigung unter einander zu verschmelzen. Aus der feinen Granulation der abaktinalen Seite ragen feine Dornen hervor, die feiner auf den Porenfeldern sind als auf deren Zwischenräumen. Die Dornen auf den Porenfeldern der Seite waren jedoch stärker als die auf dem Rücken.

"Pedicellarien waren auf der Bauchseite in Gestalt von Klappenpedicellarien vorhanden, die sich in Grösse und Umriss von den umgebenden Körnern wenig unterschieden. Eben so standen kleine, winzige Pedicellarien auf den Porenfeldern und ferner fanden sich endlich noch Pedicellarien zwischen den einzelnen Gruppen der Stacheln der äusseren Reihe der Ambulacralbewaffnung.

"Die Madreporenplatte ist von einem, jedoch nur an einer Stelle erhalten gebliebenen Dornenkranz umgeben. Die Farbe dieses älteren Exemplares ist gelblich-weiss; nur an einer Stelle der abacktinalen Seite zeigt sich ein grösseres Fleck von etwas dunklerer Farbe.

"Bei dem kleineren Exemplare [mit Durchmesser $(R+r)=82$ mm], das sich durch etwas mehr eingebogene Seiten und etwas stärker hervortretende Ecken vor dem ersteren auszeichnet, sowie es auch HARTLAUB von einem jüngeren Exemplare angiebt, sind obere und untere Randstücke noch deutlich durch die Haut hindurch zu erkennen und die Platten beider Reihen sind mit einem oder mehreren grösseren, halbrunden Tuberkeln besetzt. Die Porenfelder zeigen nur geringe Neigung zu verschmelzen; sie reichen an den Seiten nur bis an die dorsale Grenze der oberen Randstücke. Die Gruppen gröberer Granula auf der Bauchseite sind deutlich und vorgewölbt und durch seichte Furchen getrennt. Die Pedicellarien stimmen in ihrem Auftreten mit denen des grösseren Exemplares durchaus überein.

"Die Madreporenplatte besitzt einen Dornenkranz. Die Farbe dieses Exemplares ist weisslich, an einer Stelle des Rückens mehr graugelblich, doch geht letztere Farbe allmählich in die weissliche über."

KŒHLER describes two specimens from the Sunda Islands and adds some important critical remarks ['95, p. 388]:

"Deux échantillons de 16 centimètres de diamètre; $R=80$ millim. Le corps est en forme de pentagone à côtés à peu près rectilignes et à angle peu arrondis. La face dorsale est bombée sur l'un des échantillons, mais elle est plane et même concave sur l'autre. Elle est découpée, par un réseau irregulier, en champs porifères polygonaux ayant plus d'un centimètre de largeur, qui ne sont pas déprimés. Les trabécules de ce réseau sont plus étroits dans le milieu que vers les bords du disque. Toute la face dorsale est recouverte d'une granulation très fine, de laquelle s'élèvent de nombreux tubercules pointus, qui sont développés aussi bien sur les trabécules du réseau que sur les aires porifères. Ces aires deviennent plus larges sur les bords du disque et passent sur les faces latérales, mais sans atteindre la face ventrale. Leur limite est indiquée par une ligne courbe, en forme d'un oméga très allongé, et se trouve toujours à 12 ou 15 millim. de la face ventrale. La portion des faces latérales qui n'est pas occupée par les aires porifères est recouverte d'une granulation très fine, de laquelle s'élèvent des granules un peu plus gros, mais qui n'atteignent pas les dimensions des tubercules dorsaux avec lesquels ils se continuent, ainsi qu'avec ceux de la face ventrale. Celle-ci porte des granules assez gros, peu saillants, les uns arrondis, les autres en forme de tablettes polygonales contiguës et d'autres plus fins, disséminés entre les précédents. Les gros granules ne forment pas de groupes distincts, sauf une première rangée parallèle au sillon ambulacraire : encore les limites de chaque groupe sont elles indécises. Sur le reste de la face ventrale, ces gros granules sont réunis par petits amas qui se distinguent très mal des granules plus fins qui les séparent. D'assez nombreux pédicellaires sont parsemés sur la face ventrale ; leur taille est à peine supérieure à celle des petits granules. Ils sont surtout nombreux au voisinage du sillon ambulacraire.

"Ce sillon est limité par deux rangées de piquants. Les piquants internes, au nombre de cinq, et parfois de six sur chaque plaque, sont parallèles, très serrés et assez réguliers : c'est le piquant adoral qui est le plus petits. La rangée externe comprend deux gros piquants arrondis, souvent inégaux. En dehors de ceux-ci on distingue, mais sur l'un des échantillons seulement, quelques gros granules qui passent à la granulation générale de la face ventrale et qu'on ne peut pas considérer comme formant une troisième rangée. Entre chaque groupe de piquants, se trouve un pédicellaire en pince. Dans l'angle buccal, les piquants s'élargissent et forment une douzaine de très gros tubercules.

"La plaque madréporique est allongée, ovale, assez saillante et entourée d'un cercle de granules. Elle est située à un tiers environ de la
distance qui sépare le centre du bord du disque.

"La détermination des Culcites n'est pas chose facile, car les espèces,
peut-être trop nombreuses, qui ont été créées, sont reliés les unes aux autres
par de nombreuses formes intermédiaires. HARTLAUB, qui a fait récemment
une révision des espèces de ce genre, accorde une grande importance à la
forme du corps qui est haut et bombé chez les *C. grex* et *Novæ-Guineæ*,
aplati et discoïde chez les *C. acutispinosa, arenosa, pentangularis, plana* et
schmideliana. Je crois que c'est donner trop d'importance à un caractère
qui n'est pas toujours facilement appréciable, surtout quand il s'agit d'échantillons desséchés ou même conservés dans l'alcool ; l'état des deux échantillons que j'ai étudiés, dont l'un est bombé et l'autre discoïde, est bien fait
pour confirmer cette manière de voir. STUDER,[1] qui a eu fréquemment
l'occasion d'observer à Java des Culcites vivantes, se refuse à accorder à la
forme du corps l'importance que lui attribue HARTLAUB. Ces animaux, dit-il,
peuvent changer de forme et tantôt se montrent très bombés, tantôt au contraire tout-à-faits plats. C'est cette dernière forme qu'ils prennent toujours
lorsqu'ils cessent d'être vigoureux, et les échantillons qui ne sont pas mis
immédiatement dans l'alcool restent constamment discoïdes, ce qui est le cas
de beaucoup d'échantillons que l'on trouve dans les musées.

"Les caractères de la *C. Novæ-Guineæ* me paraissent loin d'être nettement établis et l'espèce est incontestablement très polymorphe. HARTLAUB,
qui a eu entre les mains de nombreux échantillons, les a divisés en trois
variétés d'après la grandeur des aires porifères, le développement des tubercules dorsaux et la granulation de la face ventrale. Mes échantillons ne
se rapportent exactement à aucune de ses variétés, et ils sont surtout
caractérisés par la faible différence que présente, dans leur taille, les tubercules des aires porifères et ceux du réseau qui les sépare, et par la forme
polygonale des gros granules de la face ventrale, qui sont très rapprochés.
En étudiant les échantillons du Musée d'Amsterdam, dont les uns proviennent des Moluques, et les autres, moins nombreux, de Batavia, STUDER a pu

---

1) Evidently a misprint for "SLUITER."

observer des différences plus importantes encore que celles signalées par HARTLAUB. Il cite même un échantillon de Batavia dont la face dorsale offre un aspect identique à celui de la *C. grex* photographiée par HARTLAUB, dont STUDER a étudié l'original au Musée de Leyden. Il aurait même déterminé ce spécimen comme un *C. grex*, si le disque eût été moins haut et la granulation de la face ventrale moins grossière.

"HARTLAUB, dans sa caractéristique de la *C. Novæ-Guineæ*, dit que les côtés sont *convexes* et les angles *arrondis·* Ces caractères ne s'appliquent pas exactement à mes échantillons, qui sont franchement polygonaux. Ici encore, je crois que ces différences de *forme* sont tout à fait secondaires. Je possède dans ma collection un exemplaire de *C. Novæ-Guineæ* provenant des Iles Samoa, acheté au Muséum Godeffroy, qui est tout à fait conforme à la seconde variété décrite par HARTLAUB. La forme en est bien polygonale, mais les côtés sont légèrement convexes, et, sous ce rapport, cet échantillon est intermédiaire entre le type de HARTLAUB et les échantillons recueillis par M. KOROTNEV.

"Les espèces avec lesquelles la *C. Novæ-Guineæ* pourrait être confondue sont les *C. pentangularis, grex, plana* et *arenosa*.

"La *C. pentangularis* GRAY n'est pas distinguée par SLADEN de la *C. Novæ-guineæ*. HARTLAUB n'en donne pas de description, mais il l'admet comme espèce distinct. D'après PERRIER (*Revision des Stellérides*), cette espèce a les granules de la face ventrale réunis en groupes à contours très nets et séparés par des lignes enfoncées. Les aires porifères de la face dorsale constituent des aréoles plus ou moins triangulaires de 3 ou 4 millim. de longueur, déprimées; la plaque madréporique est peu saillante. Aucun de ses caractères ne se retrouve sur mes échantillons.

"Chez la *C. grex*, le corps est presque arrondi, et les granules de la face ventrale, qui ne sont jamais réunis en groupe, sont peu développés. Les aires porifères, arrondies, sont très distinctes et forment saillie; les tubercules de la face dorsale sont exclusivement limités à ces aires. Tous ces caractères séparent nettement cette espèce de la *C. Novæ-Guineæ*; néanmoins SLUITER a trouvé des formes intermédiaires, ainsi qu'on l'a vu plus haut.

"Parmi les espèces qui, pour HARTLAUB, sont caractérisées par leur corps aplati, la *C. plana*, avec ses aires porifères petites, les granules de la face ventrale réunis en groupes et la disposition des piquants externes du sillon ambulacraire, se distinguent facilement de la *C. Novæ-Guineæ*. Il n'en est pas de même de la *C. arenosa*. Dans la description originale de PERRIER, cette espèce était caractérisée par des aires porifères indistinctes, les pores étant uniformément répartis sur toute la surface dorsale qui est couverte de petits granules très fins parsemés de pédicellaires en pince. Or, d'après HARTLAUB, les aires porifères sont distinctes tout en présentant une tendence à la fusion et PERRIER a lui-même reconnu qu'il en était réellement ainsi. Ces aires s'étendent jusqu'à la face ventrale, et elles sont petites. Les granules de la face ventrale ne sont pas disposés par groupes. Tous ces caractères sont plus ou moins communs à cette espèce et à la *C. Novæ-Guineæ*, et les seuls caractères que l'on puisse invoquer pour séparer les deux espèces, consistent dans une forme différente des piquants adambulacraires externes et dans l'absence du cercle de granules autour de la plaque madréporique. Je ne suis pas convaincu de la validité de la *C. arenosa*, mais si cette espèce devait être confondue avec la *C. Novæ-Guineæ*, c'est ce dernier nom qui devrait être conservé comme étant le plus ancien ; la détermination que j'ai faite de mes deux échantillons n'en resterait donc pas moins correcte."

DÖDERLEIN, in his report on SEMON's collection of Asteroidea, mentions two specimens of *Goniodiscus sebæ* from Amboina ['96, p. 309] and also gives a critical review of the species of *Culcita*, mainly with reference to the species before us, and makes several valuable remarks of systematic importance. The part having direct reference to the present species will be cited ['96, p. 310] :

"*Culcita novæ-guineæ* MÜLLER u. TROSCHEL. [References omitted.]

"Vier Exemplare von Amboina.

	a	b	c	d
R + r	190 mm	157 mm	145 mm	91 mm
Höhe		83 „	67 „	25 „

"Von den vorliegenden Stücken sind drei, und zwar die grossen Exemplare (a, b, c) zweifellose Vertreter von *C. novæ-guineæ* mit stark gewölbter Körperform (Fig. 1), die sich von HARTLAUB's erster Varietät dieser Art unterscheiden durch die Neigung der Porenfelder, abgerundete, nicht polygonale Umrisse anzunehmen; sie verschmelzen wohl stellenweise mit einander, sind aber grösstenteils weit von einander getrennt durch ein wohl entwickeltes Netz von breiten porenfreien Zügen, das nicht sehr zahlreiche, aber beträchtlich grössere Dornen trägt als die Porenfelder. Die Unterseite ist ziemlich gleichmässig perlartig gekörnelt, die äusseren Furchenpapillen werden nicht sonderlich gross und zeigen bei einem Exemplar verschiedene Gruppen von je drei etwa gleich grossen Papillen.

"Das vierte, kleinere Exemplar (d) von 91 mm Durchmesser (Fig. 2) muss dagegen jedenfalls zu *C. plana* HARTLAUB gestellt werden; es ist ziemlich flach (25 mm hoch), besitzt scharf vorspringende Ecken und concave Seiten; die rundlichen Porenfelder sind 'klein' zu nennen, die Bedornung des Rückens ist die der typischen Exemplare von *C. novœ-guineœ*; auf der Unterseite sind die den Bauchplatten entsprechenden Gruppen von gröberen Granula und auch die Felderung sehr deutlich (Fig. 2a); zwischen diesen Gruppen sind aber zahlreiche gröbere Granula noch überall zerstreut. Die äusseren Furchenpapillen bilden mit den inneren einigermassen parallele Reihen; neben und durch einander finden sich hier entweder drei Papillen von gleicher Grösse, oder drei, deren äussere etwas oder viel kleiner sind als die mittleren, oder drei, von denen die aborale oder adorale viel kleiner ist als die beiden gleich grossen anderen oder drei unter einander verschiedene, oder nur zwei Papillen von gleicher Grösse oder zwei von ungleicher Grösse, also wohl alle Formen von Ausbildung, die innerhalb der Gattung schon beschrieben wurden; es überwiegen die mit zwei grossen Papillen, während drei gleich grosse selten vorkommen. Eine dritte äusserste Reihe von Furchenpapillen ist angedeutet. Auf Grund von HARTLAUB'schen Ausführungen dürfte dieses Exemplar nicht zu *C. novœ-guineœ* gestellt werden, während ich es unbedenklich für den Jugendzustand der durch die drei grossen Exemplare vertretenen Form von *C. novœ-guineœ* halten möchte.

"Zur Vergleichung liegen mir nun eine Reihe von Exemplaren der Gattung *Culcita* aus der Strassburger Sammlung vor (Taf. XIX und XX).

"Eines davon (e), aus dem Godeffroy Museum stammend, nach dessen

Catalog als ' *C. novæ-guineæ* von den Viti oder Samoa-Inseln Nr. 1196 '
bezeichnet, hat einen Durchmesser von 95 mm und gehört ohne Zweifel
auch der angegebenen Art an; es besitzt einen hochgewölbten Körper (59
mm hoch), convexe Seiten, abgerundete Ecken und die gewöhnliche Rückenbedornung dieser Art; die Porenfelder sind aber kaum grösser als die der
oben erwähnten *C. plana* (Ex. d), nämlich ca. 4 mm in tangentialer Richtung gemessen, die gröbere Granulirung der Bauchseite gleicht ziemlich
genau der von jenem Exemplar, indem zwischen den in Reihen stehenden
Gruppen von gröberen Granula noch einzelne grobe Granula überall zerstreut auftreten; die äussere Reihe von Furchenpapillen (eine äusserste ist
schwach entwickelt) besteht meist aus je zwei ungleich grossen Papillen,
zwischen denen aber in ziemlicher Anzahl Gruppen von je drei Papillen
zu sehen sind, die allerdings niemals gleich gross werden. Als einzigen
wesentlichen Unterschied zwischen diesem Exemplar von. *C. novæ-guineæ*
und jenem von *C. plana* (d) kann ich nur die hohe Wölbung der Scheibe
und die dadurch veranlasste etwas grössere Länge der Porenfelder in
radialer Richtung bei ersterem Exemplar bezeichnen.

"Diesem Charakter vermag ich aber keinen specifischen Wert zuzuer-
kennen, und ich befinde mich damit in Uebereinstimmung mit SLUITER und
KŒHLER, welche auf Grund von Beobachtungen an *C. novæ-guineæ* derselben
Ansicht sind. Meine Beobachtungen sind an zahlreichen Exemplaren von
C. schmideliana gemacht, von denen ich vollkommen flache Scheiben besitze
(Fig. 10, 11 u. 14), die fast die Gestalt von *Goniodiscus sebæ* zeigen, neben
solchen, die in einer Weise aufgebläht sind, wie es von einer typischen
C. novæ-guineæ nicht übertroffen werden kann (Fig. 12); von dieser Aufbläh-
ung bängt die Abrundung der Ecken, die Convexität der Seiten, die Verlän-
gerung der Porenfelder in radialer Richtung direct ab; die flachen wie die
aufgeblähten Exemplare dieser Art, die ich vor mir habe, sind trefflich in
Alkohol conservirt.

"Ein anderes, ebenfalls aus dem Godeffroy Museum stammendes Exemplar (f) von 75 mm Durchmesser, dort als ' *C. pentangularis* von den
Samoa oder Viti-Inseln Nr. 3457 ' bezeichnet, zeigt einen vollkommen verschiedenen Habitus (Fig. 8 u. 8a.) Es ist ganz flach, 20 mm hoch, fünfeckig

mit concaven Seiten und vorspringenden Ecken, auf dem Rücken und den
Seiten mit zahlreichen groben Dornen, die keinen Grössenunterschied zeigen,
ganz gleichmässig bedeckt; wohlabgegrenzte Porenfelder sind nicht vorhanden, sie fliessen alle in einander, und zwischen ihnen lassen sich un-
zusammenhängende porenfreie Strecken von geringer Ausdehnung erkennen,
die Reste des bei anderen *Culcita*-Formen die Porenfelder trennenden Netzes.
Die Unterseite zeigt im Wesentlichen das oben bei *C. plana* (d) geschilderte Bild, in Reihen gestellte Gruppen grösserer Granula, zwischen ihnen
aber noch überall einzelne grobe Granula zerstreut; sie finden sich jedoch in
etwas geringerer Anzahl als bei der besprochenen ' *C. plana* '. Eine Felderung tritt deutlich hervor. Die äussere Ambulacralbewaffnung zeigt Gruppen
von je 2, seltener auch von je 3 Papillen in ähnlicher Ausbildnng wie bei
dem kleinen Exemplar (e) von *C. novæ-guineæ*. Dieses Exemplar könnte
man zu *C. arenosa* PERRIER ziehen, obwohl es dieser mit seiner groben
Rückenbedornung nicht recht entspricht. Wäre es hoch gewölbt, so könnte
es zur zweiten Varietät von *C. novæ-guineæ* nach HARTLAUB gestellt werden;
vielleicht ist es auf *C. acutispinosa* BELL zu beziehen

"Ein grosses Exemplar (g) aus der Südsee mit 160 mm Durchmesser
(Fig 7) zeigt den hochgewölbten Rücken von *C. novæ-guineæ* (in getrocknet-
em Zustande noch 70 mm hoch); die Rückenbedornung erinnert sehr an die
des eben erwähnten Exemplars (f) und besteht aus zahlreichen, dicht stehenden, gleichmässig verteilten, grossen Dornen von nahezu gleicher Grösse;
die der Porenfelder sind nur wenig kleiner als die anderen; die Porenfelder
fliessen überall zusammen und nur Reste des sie trennenden porenfreien
Netzes sind vorhanden; es ist meist nur durch die Reihen der etwas grösseren Dornen gekennzeichnet. Die Unterseite zeigt die perlartige Granulirung von typischen *C. novæ-guineæ*. Die äussere Ambulacralbewaffnung zeigt
meist je 2 Papillen, die eine sehr gross und dick, die andere sehr klein;
öfter sind zwei gleich grosse vorhanden, hier und da zwei kleine und eine
mittlere grosse. Dies Exemplar könnte man vielleicht für ein grosses
Stück von *C. arenosa* ansehen; es steht in der Tat dem eben geschilderten
Exemplar (f) sehr nahe. Da es aber hoch gewölbt ist und auch die perlartige Granulirung der Bauchseite ohne deutliche Gruppenbildung der

groben Granula zeigt, muss es als *C. novæ-guineæ* gelten und entspricht
ziemlich gut HARTLAUB's zweiter Varietät dieser Art.   Ich teile KŒHLER's
Ansicht, dass *C. arenosa* sich nicht von *C. novæ-guineæ* trennen lässt; es
liegt hier wieder ein Beispiel für die Hinfälligkeit der Unterscheidung nach
der äusseren allgemeinen Körperform vor.

" Ein Exemplar (h) von den Samoa-Inseln (Godeffroy-Museum Nr.
1196) mit 140 mm Durchmesser und hohem Körper (Fig. 3 u. 3a) zeigt auf
der Oberseite grosse, dreieckige Porenfelder mit kleinen Dornen besetzt, getrennt durch ein Netz von in einfacher Reihe stehenden grösseren Dornen.
Die porenfreien Felder sind inselartig und hängen nicht zusammen.   Die
Dornen stehen viel spärlicher als bei dem vorigen Exemplar und zeigen
sehr auffallende Grössenunterschiede.   Die groben Granula der Unterseite
sind sämmtlich als spitze Dornen entwickelt, die äusseren Ambulacralpapillen bilden hohe, spitze Kegel; sie stehen meist zu zweien, einer davon
bleibt gewöhnlich winzig klein.

" Ein anderes Stück (i) von Gunong (Bandasee) mit 180 mm Durchmesser ähnelt dem vorigen vollständig auf der Oberseite, während die
Unterseite die runden perlartigen Granula der typischen *C. novæ-guineæ*
zeigt.   Diese beiden Exemplare entsprechen ziemlich gut HARTLAUB's erster
Varietät von *C. novæ-guineæ*.

" Ein Exemplar (k) von den Sandwich-Inseln mit 120 mm Durchmesser
(ursprünglich hochgewölbt, nun aber in trockenem Zustande mit tief einge-
sunkenem Rücken) zeigt eine feine, aber ziemlich gleichmässige Bedornung
über den ganzen Rücken und die Seiten.   Die Porenfelder fliessen alle in
einander und lassen zwischen sich nur noch ganz vereinzelte porenfreie
Inseln, auf denen sich dann sehr wenige, etwas grössere Dornen erheben.
Die Unterseite zeigt eine perlartige Granulirung mit ungleich grossen groben
Granula (Fig. 6).   Die den Ambulacralplatten benachbarte Reihe von Ventrolateralplatten trägt je einen stabförmig hervorragenden Stachel; die
äusseren Ambulacralpapillen stehen in Reihen von je zwei oder drei, von
denen meist einer dick und stachelartig verlängert ist, die anderen sehr klein
bleiben.   Diese Form steht wohl der typischen *C. arenosa* am nächsten.

" Wenn ich diese meine Beobachtungen nun zusammenhalte mit den von HARTLAUB, SLUITER, KŒHLER und LEIPOLDT publicirten, so komme ich zu dem Schluss, dass die Variationsbreite von *Culcita novæ-guineæ* noch viel grösser anzunehmen ist, als bisher geschah, so gross, dass die unter dem Namen *Culcita acutispinosa* BELL, *arenosa* PERRIER, *pentangularis* GRAY, *plana* HARTLAUB aufgestellten Formen in diese Art einzubeziehen sind. Sämmtliche Charaktere, nach welchen diese sogenannten Arten unterschieden werden sollen, erweisen sich also so ausserordentlich variabel, wie dies schon HARTLAUB in seiner sehr lehrreichen Abhandlung überzeugend nachgewiesen hat, dass die vorgeschlagenen Namen allenfalls zur Bezeichnung von Varietäten, die jedoch wenig Constanz zeigen, aber nicht als Bezeichnung verschiedener Art zu verwenden sind.

"Von den zur Unterscheidung der Arten benutzten Charaktere hängen einige meines Erachtens vom Alter der Individuen ab. Dies gilt vor allem für die Felderung der Unterseite. Von den mir vorliegenden Exemplaren zeigen die kleineren Stücke (d, e, und f) unter 100 mm Durchmesser, den 'Arten' *C. novæ-guineæ*, *plana* und *arenosa* (?) angehörg, eine Felderung der Unterseite mehr oder weniger deutlich, während die grossen Exemplare höchstens noch Spuren davon zeigen, die durch radiäre Furchen angedeutet sind. Dies hängt mit der Entwickelung der gröberen Granula zusammen. Bei den jüngsten Exemplaren von *C. schmideliana*, die mir vorliegen (Fig. 11), ist die Unterseite mit einer gleichartigen feinen Granulirung bedeckt, in der sich die darunter liegenden, die Felderung bedingenden Bauchplatten deutlich hervorheben. An der den Adambulacralplatten benachbarten Plattenreihe, und zwar zuerst an den adoral gelegenen Flatten beginnen sich einige der über der Mitte der einzelnen Flatten gelegenen Granula zu verbreitern, und nach und nach zeigt sich über jeder der Bauchplatten eine Gruppe gröberer Granula. Diese vergrössern sich mit zunehmenden Alter und ragen perl-, platten- oder stachelartig aus der über den Plattenrändern unverändert gebliebenen feinen Körnelung hervor (Fig. 14a u. 15). So heben sich bei *C. schmideliana* auch im Alter die Plattenreihen sehr deutlich ab, da bei dieser Art stets die mittleren Granula allein in grösserer oder geringerer Anzahl sich vergrössern. Anders ist es bei den zu *C. novæ-guineæ* gehörigen Formen; hier vergrössern sich nicht nur über der Mitte jeder

Platte Gruppen von Granula, sondern auch unter den die Plattenränder
bedeckenden Granula nimmt eine mehr oder weniger grosse Zahl an der
Vergrösserung Anteil, während die übrigen klein bleiben und die Grundgranulirung bilden (Fig. 2a. u. 8a). Sobald nun diese zu Perlen, Stacheln
oder Plättchen sich entwickelnden grossen Granula eine gewisse Grösse
erreicht haben, wird die Felderung der Unterseite durch sie verdeckt, und
nur ausnahmsweise lassen sich den Plattengrenzen entsprechende, meist
radiär verlaufende Furchen noch erkennen. Nur wenn die Grösse der gröberen Granula bei erwachsenen Exemplaren eine unbedeutende ist, also in
dieser Hinsicht einen jugendlichen Charakter bewahrt haben, wie es bei
einigen von HARTLAUB beschriebenen Stücken der Fall zu sein scheint, tritt
die Reihenbildung der Granulagruppen noch deutlich hervor. Umgekehrt
mag auch einmal in seltenen Fällen bei *C. schmideliana* die Reihenbildung
undeutlich werden, wie HARTLAUB einen solchen erwähnt; mir selbst ist noch
kein Stück vor Augen gekommen, das die Unterdrückung der Felderung in
einer Weise zeigt, dass es mit einer *C. novæ-guineæ* verwechselt werden könnte.

"Auch die Ausbildung der äusseren Furchenpapillen wird durch das
Wachstum stark beeinflusst. Bei jungen Exemplaren sind sie noch klein.
Der Unterschied unter den zu einer Gruppe gehörigen ist noch nicht so
übermässig ausgeprägt wie später, und sie bilden leicht sehr regelmässige
Reihen, welche denen der inneren Furchenpapillen, die ihnen auch an
Grösse noch nicht allzusehr nachstehen, parallel laufen; sehr bald aber
überwiegt eine, seltener zwei der äusseren Furchenpapillen derart an Grösse,
dass die neben ihnen stehenden fast verschwinden und die Regelmässigkeit
der Reihe gestört wird. Selten bleiben auch bei grossen Exemplaren die
äusseren Furchenpapillen verhältnissmässig klein und regelmässig, von
jugendlichem Charakter, wie das wohl bei HARTLAUB's grossem Exemplar von
C. plana, auch bei meinen grossen Exemplaren von Amboina einigermassen
der Fall ist.

"Was den Zusammenhang der von mir nur als Varietäten von *Culcita
novæ-guineæ* betrachteten Formen anbetrifft, so können wir dabei von der
mir vorliegenden Amboina-Form ausgehen. Bei ihr sind die Porenfelder von
mässiger Grösse, ziemlich selbständig und meist von abgerundeter Form

(Fig. 1 u. 2), von einander getrennt sind sie durch ein zusammenhängendes Netz von breiten porenfreien Zügen; darauf stehen ziemlich sparsam Dornen, welche beträchtlich grösser sind als die meisten Dornen der Porenfelder. Die Unterseite ist bei Erwachsenen perlartig granulirt. Zu dieser Form gehört *C. plana* HARTLAUB, die auf Exemplare von zum Teil jugendlichem Charakter gegründet zu sein scheint; in der Tat sind auch die verschiedenen von HARTLAUB zu dieser Form gestellten Exemplare mit einer Ausnahme (160 mm) von geringer Grösse. Durch Unterdrückung der Dornen auf den Porenfeldern mag HARTLAUB's dritte Varietät von *C. novæ-guineæ* daraus entstehen (Fig. 4). Vielleicht ist auch *C. grex* von solchen Formen abzuleiten, die u. a. durch die kleinen runden Porenfelder und das Fehlen von ausserhalb der Porenfelder stehenden Dornen ausgezeichnet ist. Auf der anderen Seite dürfte aus jener Amboina-Form durch Vergrösserung der Porenfelder, welche allmählich eine polygonale, meist dreiseitige Gestalt annehmen, eine Form entstanden sein, welche HARTLAUB's erster Varietät von *C. novæ-guineæ* entspricht. Bei typischer Ausbildung dieser Form (Fig. 3) ist das Netz porenfreier Strecken schon vielfach unterbrochen und grösstenteils nur noch durch Reihen grösserer Dornen vertreten, welche die Porenfelder von einander trennen, ohne ihr Zusammenfliessen zu verhindern (Exemplare h und i); die auf den Porenfeldern befindlichen Dornen sind hier viel kleiner als die anderen; ist der Unterschied in der Grösse der Dornen nicht mehr so auffallend, so entsteht eine Form, bei welcher der ganze Rücken und die Seiten ziemlich gleichmässig von Poren bedeckt ist mit Ausnahme einiger porenfreier Stellen, die wie Inseln, in dem zusammenhängenden Porenareal liegen, das nicht mehr in Felder geteilt erscheint. Eine solche Form dürfte als *C. arenosa* PERRIER angesprochen werden; ihr Zusammenhang mit der vorigen Form wird durch Exemplare vermittelt, bei welchen (wie bei Exemplar k) nur die porenfreien Inseln noch einzelne grössere Dornen tragen (Fig. 5). Bei der typischen *C. arenosa* sind alle Dornen klein; sind sie dagegen alle grob, aber gleichmässig dicht über den ganzen Rücken verteilt (Fig. 8), so entsteht eine Form, die ich als Varietät von *C. arenosa* (Exemplar f) oben geschildert habe (möglicherweise ist *C. acutispinosa* auf ein solches Exemplar gegrundet); eine Uebergangsform

zwischen dieser und der typischen *C. novæ-guineæ* bildet mein Exemplar g
(Fig. 7), wo noch ein geringer Unterschied in der Grösse der Dornen
erkennbar ist, durch welche eine Felderung des Porenareals angedeutet ist.
Von einer Form, wie sie als typische *C. arenosa* bezeichnet wird, ist
vielleicht *C. coriacea* abzuleiten, indem die porenfreien Inseln fast völlig
verschwinden und die ganze Rückenseite von dichtstehenden feinen Dörnchen oder Tuberkeln gleichmässig bedeckt erscheint (Taf. XXI, Fig. 1);
von einzelnen Porenfeldern ist hier nicht mehr die Rede. Die Entwick-
elungsrichtung, die sich innerhalb der sehr variablen *C. novæ-guineæ* kund
gab, hat in dieser Art ihr Extrem erreicht.

"Innerhalb der Art *Culcita schmideliana* lässt sich eine Parallelentwick-
elung in der Ausbildung der Porenfelder verfolgen, wie sie oben bei *C.
novæ-guineæ* angedeutet wurde. Exemplare von Ceylon zeigen auffallend
kleine, weit von einander getrennte Porenfelder (Fig. 10, der *C. plana* HART-
LAUB entsprechend), während an der Ostküste von Afrika Formen mit
grossen polygonalen, aber durch ein Netz von schmalen porenfreien Zügen
getrennten Porenfeldern (wie bei der typischen *C. novæ-guineæ*) dominiren,
die bei anderen Exemplaren grosse Neigung zeigen, mit einander zu verschmelzen (Fig. 13, wie bei *C. arenosa*). Durch die Ausbildung scharf abge-
grenzter Granulagruppen auf der Unterseite, durch das constante Fehlen
von Dornen auf den Porenfeldern und durch die meist sehr plumpen, oft
kugeligen Warzen (Fig. 14) auf der Rückenseite ist *C. schmideliana* von
C. novæ-guineæ und ihren Abkömmlingen wohl unterschieden.

"Uebrigens stehen diese beiden vielgestaltigen Arten einander doch
nicht ganz unvermittelt gegenüber. Bei der Ceylon-Form von *C. schmideliana* (Fig. 10) finden sich kleine, wohlgetrennte Porenfelder, und die
Tuberkeln der Rückenseite sind verhältnissmässig klein, meist spitz und
ziemlich zahlreich, während ich die grossen abgerundeten Warzen nur an
Exemplaren von der Ostküste Afrikas, von Mauritius und den Seychellen
kenne; bei Sumatra und Java dagegen ist eine Form von *C. novæ-guineæ*
entwickelt (Fig. 4, HARTLAUB's dritte Varietät), ebenfalls mit kleinen, wohl-
getrennten und fast dornenfreien Porenfeldern, sowie mit einer spärlichen,
aber kräftigen Bedornung des Rückens und der Seite; dazu ist die Unter-

seite ausgezeichnet durch die deutlichen Gruppen, welche die grobe Granulirung bildet. Zwischen beiderlei Formen dürfte nur noch ein sehr geringer
Unterschied bestehen.

"*Culcita veneris* PERRIER, die ich selbst nicht gesehen habe, scheint
ziemlich isolirt zu stehen.

"Die verschiedenen Arten von *Culcita* haben ihre wohlabgegrenzten
Verbreitungsgebiete. *C. schmideliana* scheint auf den westlichen Teil des
Indischen Oceans beschränkt zu sein; sie ist von Ceylon, den Andamanen,
der Ostküste von Afrika und von Mauritius und den Seychellen sicher
bekannt. *C. novæ-guineæ* nimmt dagegen den östlichen Indischen Ocean
und das pacifische Gebiet ein; ihr Verbreitungsgebiet ist begrenzt durch
folgende sichere Fundorte: Sumatra, Java, Philippinen, Sandwich-Inseln,
Samoa-Inseln, Torresstrasse, West-Australien. Wo die Verbreitungsgrenze
beider Arten aneinanderstösst, finden sich Varietäten, die einander auffallend ähneln, in Ceylon einerseits, bei Sumatra und Java andererseits. Mitten
im Verbreitungsgebiet von *C. novæ-guineæ*, bei den Molukken und Amboina,
hat sich *C. grex* ausgebildet; ein local getrennter Ausläufer ist vielleicht *C.
coriacea* aus dem Rothen Meer. Ausserhalb der Tropen, bei St. Paul, hat
sich die eigentümliche *C. veneris* entwickelt.

"Nachdem das Vorhergehende bereits druckfertig niedergeschrieben
war, erhielt ich durch das freundliche Entgegenkommen von Herrn Dr.
HARTLAUB, dem ich auch an dieser Stelle meinen verbindlichsten Dank
dafür aussprechen möchte, eine grössere Anzahl von Photographien, welche
die wichtigeren der von ihm in den 'Notes from the Leyden Museum Vol.
XIV' besprochenen Formen von *Culcita* in so ausgezeichneter Weise darstell-
en, dass sie die Originale fast ersetzen können. Ich bin um so mehr über
diese unerwartete Vervollständigung des mir zur Verfügung stehenden
Materials erfreut, als diese Photographien meine oben niedergelegte Ansicht
über die gegenseitigen Beziehungen der bisher unterschiedenen 'Arten' von
Culcita in einer Weise bestätigen, wie ich es nicht besser hätte erwarten
können. Auch Herrn Prof. LUDWIG in Bonn bin ich für einige wohlerhal-
tene Spiritus-Exemplare von *C. coriacea* sehr zu Dank verpflichtet.

"Besonders wichtig unter dem neuen Material war mir die Abbildung

des von HARTLAUB auf p. 83 besprochenen Exemplars von *C. novæ-guineæ*
von Pulo Edam (Java) mit auffallend gefelderter Unterseite (Fig. 9); diese
Felderung erwies sich absolut verschieden von der bei *C. schmideliana* auftretenden, indem hier, genau wie bei allen *C. novæ-guineæ*, die ich kenne,
auch von den über den Plattenrändern liegenden Granula eine grosse
Anzahl verbreitet ist, in derselben Weise, wie die über der Mitte der
Flatten gelegenen; es hat den Anschein, als sei die eigentümliche, sonst
bei erwachsenen Stücken dieser Art nicht auffallende Felderung nur durch
eine Methode der Conservirung entstanden, bei der die Bauchseite überklei-
dende Haut in einer Weise contrahirt wurde, dass die Grenzen der darunter
liegenden Bauchplatten deutlich hervortreten.   Einen Uebergang von dieser
auffallenden Felderung zu der öfter bei *C. novæ-guincæ* zu beobachtenden
radiären Furchung der Unterseite (Fig. 3a u. 6) zeigt das grössere, von
HARTLAUB öfters erwähnte Exemplar seiner ' *C. plana*,' welches SEMPER auf
den Philippinen sammelte.

"Ein Exemplar von Amboina, nach HARTLAUB'S Mitteilung von PERRIER
selbst als *C. arenosa* bestimmt (Fig. 5), zeigt noch Reste porenfreier Inseln
auf der Rückenseite, welche mit einzelnen groben Dornen bestanden sind,
und vermittelt direkt zwischen der typischen *C. novæ-guineæ* und der typisch-
en *C. arenosa*, bei welcher solche Inseln und grossen Dornen gar nicht
erwähnt werden.

"Zwischen meinem Exemplar 'g' aus der Südsee, bei welchem nur
noch ein geringer Unterschied in der Grösse der Dornen auf der Rückenseite besteht, und der typischen *C. novæ-guineæ* mit grossen dreiseitigen
Pornfeldern vermittelt ein dem Bremer Museum gehöriges Exemplar aus
dem 'Indischen Meere,' das HARTLAUB als Typus seiner zweiten Varietät
von *C. novæ-guineæ* anführt."

In his paper of 1898 DÖDERLEIN has a section headed "Ueber die
Wachsthumsverhältnisse von *Goniodiscus sebæ*," in which he makes many
remarks of much taxonomic importance.   The entire section is therefore reproduced below ['98, p. 494]:

"Als ich gelegentlich der Untersuchung einiger Skeletplatten von
Goniodiscus sebæ das Dorsalskelet bei einem grossen Exemplar dieser Art

von Mauritius frei präparirte, konnte ich die Beobachtung machen, dass die das Dorsalskelet zusammensetzenden sternförmigen Flatten nicht unmittelbar zusammenstossen, sondern durch Reticularia ('Trabeculæ' oder Connectivplatten), welche die Fortsätze der Sternplatten von aussen her bedecken, mit einander verbunden sind, wodurch das Rückenskelet einen ausgesprochen netzartigen Charakter erhält. Es ist das ein Befund, wie ihn MÜLLER und TROSCHEL in der Originalbeschreibung von *Goniodiscus sebæ* angegeben haben, und das mir vorliegende Exemplar entsprach auch in jeder anderen Beziehung dieser Beschreibung. Aus den trefflichen Abbildungen, welche DE LORIOL von *Goniodiscus sebæ* giebt, ist deutlich zu erkennen, dass auch ihm dieselbe von MÜLLER und TROSCHEL beschriebene Form vorlag, die sich durch das Vorhandensein der Reticularia auszeichnet. Derartige Exemplare mit einem Durchmesser von 61–66 mm habe ich vor mir von Mauritius, den Seychellen und Ceylon (Taf. XL, Fig. 6 u. 7).

"Um so auffallender war es mir, in den zwei neueren Werken von PERRIER und von SLADEN, wo die Systematic der Asteroidea eingehendèr behandelt wird, übereinstimmend als wesentlichen Charakter der Familie, zu der die Gattung *Goniodiscus* gestellt wird—der *Pentagonasteridæ* — den mosaikartigen Bau des Dorsalskelets aufgestellt zu finden im Gegensatz zu dem netzartigen Bau des Rückenskelets, durch den die Pentacerotiden unterschieden werden.

"Nach PERRIER sind die Familie der *Pentagonasteridæ* und *Gymnasteriidae* charakterisirt durch ein Dorsalskelet, das aus mosaikartig an einander stossenden Tafeln besteht, gegenüber den Familien der *Antheneidæ* und *Pentacerotidæ*, deren Dorsalskelet netzartig ist. Die Unterfamilie der *Goniodiscinæ* besitzt nach ihm sternförmige flache Rückenplatten, die bei der Gattung *Goniodiscus* selbst nackt sein sollen.

"Nach SLADEN sind die *Pentagonasteridæ* gleichfalls durch ein mosaikartig getäfeltes (tessellata) Rückenskelet ausgezeichnet, dessen Flatten rund, polygonal oder sternförmig sind, während es bei den *Pentacerotidæ* netzförmig ist. Die *Goniodiscinæ* sind bei ihm ebenfalls mit sternförmigen, flachen Rückenplatten versehen, aber von einer gleichförmigen, körnigen Membran bedeckt, während die *Pentagonasterinæ* durch runde oder polygonale Platten ausgezeichnet sind. *Goniodiscus* selbst ist nach SLADEN charakterisirt durch ein Rückenskelet mit sternartigen Flatten, die weite Zwischenräume lassen für die Papulae.

"In der That finde ich eine Anzahl Seesterne von Ceylon, Amboina und den Tonga-Inseln, die bei einem Durchmesser von 39–45 mm in ihrem ganzen Habitus sonst durchaus den Charakter von *Goniodiscus sebæ* zeigen, deren Dorsalskelet aber nicht netzartig ist, wie in der MÜLLER und TROSCHEL'-schen Beschreibung angegeben, sondern ganz den von PERRIER und SLADEN der Unterfamilie der *Goniodiscinæ* zugeschriebenen Charakter zeigt; es besteht aus sternförmigen Flatten, die mit ihren Fortsätzen einander direkt berühren, aber dazwischen weite Lücken lassen zum Durchtritt der Papulae (Taf. XL, Fig. 4 u. 5). Die Rückenplatten scheinen bei diesen Exemplaren, solange sie in Alkohol liegen, nackt zu sein; bei genauerer Untersuchung aber, oder wenn man ein derartiges Exemplar trocknet, überzeugt man sich leicht von dem Vorhandensein einer feinen, in einer dünnen Membran gelegenen Körnelung; dadurch erklärt sich wohl der scheinbare Widerspruch zwischen der PERRIER'schen und SLADEN'schen Angabe.

"Endlich liegen mir noch aus Ceylon eine Anzahl kleiner Seesterne mit einem Durchmesser von 15–29 mm vor, die ebenfalls ganz den Habitus von *Goniodiscus sebæ* haben, und die ohne Frage mit *Hosea spinulosa* GRAY identisch sind. Nach dem Vorgang von PERRIER wird von DE LORIOL diese Form bei *Pentagonaster* untergebracht und sehr deutlich beschrieben und abgebildet, während SLADEN sie, allerdings mit Vorbehalt, zu *Anthenea* stellt. Ihr Rückenskelet besteht aus polygonalen, meist sechseckigen Tafeln, die mosaikartig an einander stossen, ohne grössere Lücken zwischen sich zu lassen (Taf. XI, Fig. 1–3). Diese Formen haben ganz den Charakter der *Pentagonasterinæ*, wie sie von PERRIER und SLADEN charakterisirt sind, und zwar den der Gattung *Calliaster*.

"Nach dem mir vorliegenden Material ist es aber nicht mehr zu bezweifeln, dass diese verschiedenen Formen von Seesternen nur die verschiedenen Alterszustände einer einzigen Art, und zwar von *Goniodiscus sebæ* M.T., darstellen. Die jüngsten der zur Beobachtung gekommenen Exemplare zeigen den Charakter der *Pentagonasterinæ*, die mittelgrossen den der *Goniodiscinæ*

	Gonioliscus sebæ											Culcita schmideliana.				
	Ceylon					Amboina		Cey-lon	Sey-ch.	Maur.	Cey-lon	Cey-lon	Mauritius			Zanzibar
	*a	b	c	*d	*e	f	*g	*h	*i	*k	*l	*a	*b	e	d	*e
Durchmesser in mm	15	21.5	23	25.5	29	39	44	45	62	65	66	60	75	74	88	109
R in mm	9	12	13	15	17	23	27	26	36	37	39	35	41	40	47	61
r in mm	6.5	9	10	11	12	16	17	19	26	28	28	25	34	34	42	48
Zahl der { Radialia	5	6	6	6	6	6	7	7	8-9	8	7	8	8	8		
Adradialia	5	5-6	5-6	6	6	6	7	7	8	8	7	7-8	8	8	8	
Dorsolateralia	2	3	3	3	3	3	3	3	3	3	3	3	3	3		
Ventrolateralreihen	3	4	4	4	4	5	5	6	7	7	6	7	7-8	7	8	9
Ventrolateralia der I. Reihe	7	10	10	11	11	12	14	14	16	18	15	16	18	16	17	21
Dorsomarginalia	4	5	5	6	6	6	7	6	7	7	7	7	7	8	9	9
Ventromarginalia	4	5-6	5-6	6	6	7	7-8	8	9	9-10	9	9	10	11	11	13
inn. Amb. Papillen	4	4	4	4	4	5	5	5	5	5-6	6	6	5-6	6	5	5
Papulæ in Porenf.	1	1-2	2	2-3	3-4	8	9-12	15	20	15-30	25-30	20-25				
Breite der Dorsomarginalia in mm	2.3	2.8	2.9	3	3.6	4.4	4.5	4.5	5.5	6.2	5	4.8		6.2		

Die mit * bezeichneten Exemplare sind auf Taf. XXXIX u. XL abgebildet.

und die ältesten den der Familie der *Pentacerotidœ* nach der SLADEN'schen und PERRIER'schen Auffassung.

"Einige der auffallenderen beim Wachsthum von *Goniodiscus sebœ* auftretenden Veränderungen sollen auf den folgenden Seiten geschildert werden. Einige der bemerkenswerthesten Aenderungen während des Wachsthums sind auf einer Tabelle übersichtlich zusammengestellt [p. 571].

"Die äussere Gestalt ändert sich im Verlauf der Entwickelung bei Exemplaren von 15 mm an bis zu solchen von 66 mm Durchmesser nicht bemerkenswerth; es beruht auf individueller Variabilität, dass bei manchen Exemplaren die Seiten des etwa pentagonalen Seesterns fast gerade sind, bei anderen sehr stark concav, sowie dass die Arme bei einigen Exemplaren auffallend spitz enden, bei anderen stark abgerundet sind (Taf. XL); auch die Höhe des Seesterns am Rande ist variabel, insofern manche Stücke auffallend niedrig sind, andere ziemlich hoch. Einigermassen constant ist die Wölbung der Randplatten, die meist auffallend geschwollen erscheinen und durch tiefe Furchen von einander getrennt sind; gewöhnlich ist auch an den den oberen Randplatten angrenzenden Dorsalplatten eine kugelige Aufblähung bemerkbar; alle Exemplare aber zeigen ferner an den Winkeln zwischen den oberen und unteren Randplatten trichterförmige bis punktförmige Vertiefungen oder förmliche Löcher. Alle zeigen auch die Dorsalseite verhältnissmässig spärlich mit spaltförmigen Pedicellarien versehen, deren Alveolen in den Platten des Dorsalskelets selbst liegen, während die Ventrolateralplatten reichlicher damit besetzt sind. Den Randplatten fehlen Pedicellarien stets ganz. Stets findet sich auch eine Anzahl von Höckern oder Warzen auf den Marginalplatten in wechselnder Menge (2–10, Charakter der Gattung *Calliaster*); im Durchschnitt scheinen diese bei den grossen Exemplaren etwas weniger zahlreich zu sein als bei den kleinen; bei den meisten Exemplaren sind sie spitz, öfter aber auch stumpf. Die Flatten des Dorsalskelets zeigen stets zahlreiche Krystallkörper.

"Die die Dorsalseite bedeckende Membran ist bei den kleineren Exemplaren zart und äusserst fein granulirt, so dass die Rückenplatten leicht für nackt gehalten werden können; bei den grösseren Exemplaren tritt allmählich eine gröbere Körnelung auf, und die grössten Exemplare zeigen eine auffallend gekörnelte Rückenfläche und eine dicke Membran, unter der die Flatten des Dorsalskelets und zugleich auch ein Theil der oberen Randplatten völlig verborgen sind (Taf. XL, Fig. 2 u. 6).

"Höcker oder gröbere Warzen sind bei den kleinsten Exemplaren kaum warhzunehmen auf den Rückenplatten; beim Grösserwerden werden solche allmählich deutlicher; die älteren Exemplare zeigen stets eine grössere Anzahl stumpfer oder spitzer, niedriger Warzen, sie sind aber immer spärlich vorhanden und zeigen wenig Constanz in ihrem Auftreten; gern erscheinen sie auf einigen der den Randplatten benachbarten aufgeschwollenen Dorsalplatten, ohne sich darauf zu beschränken; mitunter trägt die Mehrzahl der dorsalen Skeletplatten einen oder zwei dieser Höcker.

"Die Ventrolateralplatten sind schon bei den jüngsten Exemplaren deutlich gekörnelt, und stets findet sich über der Mitte jeder Platte eine Anhäufung etwas gröberer Körner.

"Die Zahl der Ambulacralpapillen nimmt mit dem Grösserwerden nur wenig zu; die 10. Ambulacralplatte zeigt bei den jüngsten Exemplaren etwa 4 innere Furchenpapillen, bei den grössten kann diese Zahl auf 6 steigen.

"Die Marginalplatten nehmen mit dem Alter an Zahl zu; die Zahl der oberen Marginalplatten ist bei den kleinsten Exemplaren von 15 mm Durchmesser nur 4 jederseits; sie steigt allmählich mit dem Wachsthum des Seesterns, bei den grösseren Exemplaren auf 7 jederseits; die äussersten sind natürlich von sehr geringer Grösse; die Zahl der unteren Marginalplatten ist bei den kleinsten Exemplaren die gleiche wie die der oberen oder höchstens um eine voraus, solange die Ambulacralfurche auf die Ventralseite beschränkt bleibt. Bei Exemplaren von ca. 50 mm aber beginnt die Furche sich an der Armspitze aufwärts zu ziehen, so dass sie auf der Dorsalfläche endet, und bei den grössten Exemplaren wendet sie sich auf der Dorsalseite wieder nach der mitte der Scheibe, so dass die Terminalplatte nicht mehr die Spitze bildet. An dieser Aufwärtsbiegung der Ambulacralfurche nehmen die unteren Randplatten in erheblichem Maasse Theil, während die oberen Randplatten weniger davon berührt werden. Infolge dessen tritt eine stärkere Vermehrung in der Zahl der letzten unteren Randplatten ein, so dass, während die ersten 5, selbst 6 Randplatten sich oben

und unten fast genau entsprechen, auf die letzte obere Randplatte plötzlich
eine grössere Zahl unterer Randplatten kommt (vergl. Tabelle).

"Auch die Grösse der Marginalplatten, verglichen mit der Grösse des
ganzen Seesterns, zeigt auffallende Veränderungen beim Wachsthum; bei
den kleinsten Exemplaren von 15 mm ist die Entfernung des Innenrandes
der oberen Marginalplatten vom Centrum der Scheibe nur 1⅔mal so gross
wie die Breite der mittleren Marginalplatten jeder Seite; dieses Verhältniss
wird beim Wachsthum immer grösser, bis bei den grössten Exemplaren
diese Entfernung etwa das Vierfache von der Breite der Randplatten beträgt; das Breitenwachsthum der oberen Randplatten hält eben durchaus
nicht Schritt mit der Grössenzunahme der Scheibe.   Noch auffallender
scheint dieses Zurückbleiben des Breitenwachsthums an den unteren Randplatten; während die unteren Randplatten bei den jüngsten Exemplaren
eine breite Zone um den Seestern bilden (Taf. XL, fig. 1a u. 3a), sind sie
förmlich verschwunden von der Unterseite bei den grössten Exemplaren;
doch ist dieses Verschwinden nicht nur auf das schwächere Breitenwachsthum der Randplatten, sondern zum Theil auf Rechnung der Ventrolateralplatten zu setzen, welche sich ganz allmählich über die unteren Randplatten schieben und deren Unterseite schliesslich grösstentheils verdecken
(Taf. XXXIX, fig. 5 u. 5a).

"Sehr auffallend ist die beim Wachsthum allmählich stattfindende
Vermehrung der Ventrolateralplatten; diese bilden stets sehr regelmässige
Längsreihen, die den Ambulacralfurchen ungefähr parallel laufen, und
gleichzeitig regelmässige Querreihen, die von den Ambulacralfurchen zu den
Randplatten verlaufen; während bei den kleinsten Exemplaren von 15 mm
Durchmesser aber auf jeder Seite der Furche nur 3 Längsreihen von Ven-
trolateralplatten vorhanden sind, von denen die den Adambulacralplatten
benachbarte aus 7 Platten besteht, finden sich bei den grössten Exemplaren
von Ceylon (66 mm) 6 Längsreihen, deren erste aus 15 Flatten besteht; bei
Exemplaren von Mauritius steigt die Zahl der Längsreihen auf 7, die der
Platten in der ersten Reihe auf 16 und 18.

"Nur den 3 ersten Längsreihen von Ventrolateralplatten jederseits in
jedem Interradialraum entspricht je eine unpaare im Interradius gelegene

Platte als Scheitelplatte. Die erste unpaare Platte, die Scheitelplatte der ersten Längsreihe, ist jedoch bei den kleinsten Exemplaren noch kaum sichtbar, während die Scheitelplatte des 2. und 3. Paares von Längsreihen mindestens die Grösse der ersten paarigen Platte der entsprechenden Längsreihe hat. Diese erste Scheitelplatte nimmt aber allmählich an Grösse zu und wird schliesslich die grösste von allen Ventrolateralplatten. Dem vierten Paare von Längsreihen, wie den folgenden, entspricht keine unpaare Ventrolateralplatte.

" Dass die äussersten, kleinsten Ventrolateralplatten sich allmählich mehr und mehr über die unteren Randplatten schieben und zuletzt diese Randplatten fast ganz verdecken, ist bereits oben erwähnt.

" Die auffallendsten Veränderungen beim Wachsthum zeigt aber das Dorsal-Skelet, bei dessen Darstellung ich mich der von LUDWIG angewandten Terminologie bedienen werde. Bei den jüngsten Exemplaren (15 mm Durchm.) bilden die Platten des Rückenskelets ein mosaikartiges Pflaster, bestehend aus flachen, polygonalen, meist 6-eckigen Platten, unter denen 10 merklich grössere einen geschlossenen Ring um die Mitte der Scheibe bilden; sie lassen sich leicht als die 5 primären Radialplatten und die 5 primären Interradialplatten erkennen. Von den primären Radialplatten aus verläuft nach der Spitze jedes Armes zu eine Reihe von weiteren 4 Radialplatten, deren äusserste durch das zusammenstossende letzte Paar von Randplatten von der Terminalplatte getrennt ist. Etwa parallel zu der Reihe der Radialia verläuft beiderseits eine Reihe von Adradialplatten, und nach aussen von diesen eine Reihe von Dorsolateralplatten. Die erste Adradialplatte und die erste Dorsolateralplatte berühren einander und grenzen längs des Interradius an die entsprechenden Platten des benachbarten Armes. Nach der Scheibenmitte zu grenzen diese 2 ersten Adradialplatten an die primäre Interradialplatte, die als unpaare Scheitelplatte der Adradialplatte erscheint; an das erste Paar von Dorsolateralplatten grenzt nach aussen eine unpaare Platte, die als unpaare Scheitelplatte einer nicht weiter entwickelten zweiten Reihe von Dorsolateralplatten angesehen werden kann, und nie direct an das mittlere Paar von Randplatten stösst; sie ist als 3. unpaare Interradialplatte zu bezeichnen.

"Wo je 3 von den Rückenplatten an einander stossen, findet sich ein Porus zum Austritt von einer Papula (Taf. XL, Fig. 1). Nur längs des Interradius, und zwar zwischen den beiden Interradialplatten und den von ihnen eingeschlossenen ersten Paaren von Adradialia und Dorsolateralia, sind keine Oeffnungen für Papulæ vorhanden.

"Die Weiterentwickelung des Dorsalskelets hängt direct mit der Ausbildung der Papulæ zusammen; bei weiterem Wachsthum vermehrt sich die Zahl der in den Plattenwinkeln austretenden Papulæ so, dass bei Exemplaren von etwa 20 mm Durchmesser etwa je 2, bei solchen von 30 mm etwa 4, bei 40 mm etwa 8, bei den grössten Exemplaren (60-65 mm) 15-30 Papulæ neben einander auftreten, Porenfelder (Papularien) bildend, die zuletzt Neigung zeigen, mit einander zu verschmelzen. Infolge dessen müssen die zwischen den Flatten vorhandenen Lücken zum Austritt der Papulæ allmählich grösser werden, doch so, dass der Zusammenhang der Flatten bewahrt bleibt. Zunächst geschieht das durch sternförmige Ausbildung der Flatten, indem jede Platte nach jeder der angrenzenden Dorsalplatten hin einen Ausläufer bildet, und da die meisten Flatten mit je 6 anderen in Berührung stehen, stellen sie meist sechseckige Sterne vor (Taf. XL, Fig. 4). Mittelst dieser Ausläufer wird eine Zeitlang die gegenseitige Verbindung der Dorsalplatten erhalten. Zuletzt, wenn die grösser werdenden Porenfelder die Flatten allzu weit auseinanderdrängen, verlängern sich die Ausläufer nicht weiter, sondern es beginnen sich besondere Verbindungsstücke anzulegen, die eine Brücke bilden zwischen je 2 benachbarten Flatten: dieselben legen sich oberflächlich an, und zwar jedesmal über je 2 an einander stossenden Ausläufern von 2 benachbarten Sternplatten, die sie von oben her bedecken.

"Diese Verbindungsstücke, Reticularia oder Connectivplatten, werden zunächst als ganz kleine Körnchen gebildet, die von den der Rückenhaut allenthalben eingelagerten Granulæ nur dadurch zu unterscheiden sind, dass sie etwas mehr in die Tiefe eingesenkt gefunden werden; sie sind aber offenbar auf solche Granulæ zurückzuführen; sie wachsen nun allmählich zu balkenartig verlängerten Gebilden aus in dem Maasse, als die Sternplatten weiter und weiter auseinandergeschoben werden. Bei der ersten Anlage von

Reticularia finden sich oft mehrere Körner neben einander über den Ausläufern von zwei an einander stossenden Sternplatten; nur eines davon wird
aber zum definitiven balkenartigen Reticulare. Erst spät legen sich augenscheinlich Reticularia an zwischen den Scheitelplatten in der Mitte der
Scheibe, noch später zwischen äussersten (admarginalen) Flatten des Dorsalskelets und zuletzt zwischen diesen und den Marginalplatten selbst; an der
Armspitze bilden sich zwischen den jüngsten Flatten keine Reticularia.

"Während dieser Vorgänge vermehrt sich langsam die Zahl der
Radialia, die an jedem Arm von 5 auf 6–8 steigen kann. Die jüngste Radial-
platte ist stets durch ein Paar, mitunter durch zwei Paare in der Mittellinie zusammenstossender Marginalplatten von der Terminalplatte getrennt.

"Ebenso vermehrt sich die Zahl der Adradialplatten und zwar von 5
nach und nach bis auf 8; die der Dorsolateralia erreicht sehr früh das
Maximum von 3 Flatten jederseits; sehr bald tritt auch ferner für die
Dorsolateralreihe eine unpaare Scheitelplatte, die 2. oder mittlere Interradialplatte, auf, welche sich interradiär zwischen die ersten Paare der Adradialia
und der Dorsolateralia einschiebt und sie beim Grösserwerden allmählich
auseinanderdrängt; doch bleibt diese zweite Interradialplatte immer klein.
Nie legen sich paarige Platten einer zweiten Reihe von Dorsolateralia an;
diese Reihe bleibt stets auf die einzige dritte Interradialplatte beschränkt.

"Zwischen den an den Interradius grenzenden, zwischen dem ersten
und dritten Interradiale gelegenen Flatten legen sich auch später nie Papulæ
an, die sonst überall, wo je 3 Platten auf der Dorsalseite zusammenstossen,
sich finden einschliesslich der Winkel, welche die Marginalplatten mit den
benachbarten Dorsalplatten bilden. Auch Reticularia bilden sich längs des
Interradius nur zwischen dem primären Interradiale und dem ersten Paare
von Adradialia, sowie zwischen dem dritten Interradiale und dem ersten
Paare von Dorsolateralplatten, um die Vergrösserung der seitlichen Porenfelder zu ermöglichen. Nie zeigen sich aber bei *Goniodiscus sebæ* Reticularia zwischen dem ersten Paare von Adradialia oder von Dorsolateralia oder
zwischen ihnen und dem von ihnen eingeschlossenen mittleren Interradiale,
so dass diese 5 Flatten in jedem Interradialraum ein charakteristisches Bild
geben, da es die einzigen Platten im Dorsalskelet sind, welche stets ihre

Seiten dicht an einander gepresst halten und nie Ausläufer oder Reticularia zur gegenseitigen Verbindung besitzen.

"Das späte Erscheinen der ersten unpaaren Ventrolateralplatte auf der Bauchseite und der mittleren Interradialplatte auf der Rückseite des Seesterns hängt offenbar mit der Ausbildung des interradialen Septenpfeilers zusammen, der eine innere Verbindung zwischen Rücken und Bauchskelet darstellt. Das erste unpaare Ventrolaterale dient hauptsächlich als Fussplatte für diesen Pfeiler; durch Vermittelung mindestens von einem Paar von Zwischenstücken erreicht der Pfeiler das zweite Interradiale des Dorsalskelets, das ventralwärts stark verlängert ist; erst wenn sich der Pfeiler bei weiterem Wachsthum kräftigt, schiebt sich das erste unpaare Ventrolaterale an die Oberfläche des Bauchskelets und seine Gipfelplatte, das zweite Interradiale an die Oberfläche des Rückenskelets. Der Fuss des Pfeilers wird nun durch Theilnahme der benachbarten Platten beträchtlich verstärkt, desgleichen durch das Auftreten einer grösseren Zahl von Zwischenstücken, die nach und nach dazutreten; die Verbindung mit dem Rückenskelet wird aber ausschliesslich durch das dritte Interradiale hergestellt.

"Die ursprünglich tafelförmige Gestalt der Dorsalplatten, die bei jüngeren *Goniodiscus sebæ* eine verhältnissmässig geringe Dicke und ungefähr parallele Aussen- und Innenfläche besassen, macht im Laufe des Wachsthums allmählich einer mehr oder weniger kugelförmigen Gestalt Platz, die dadurch entsteht, dass bei geringem Flächenwachsthum ein verhältnissmässig bedeutendes Dickenwachsthum eintritt, und dass die nicht mehr aneinanderstossenden Seiten sich abrunden; an diesen kugeligen Skeletplatten treten die sternartigen Ausläufer als kurze Warzen hervor."

In the next section of the same work, headed "Verwandtschaftsbeziehungen von *Goniodiscus sebæ*," the author discusses the taxonomic relationships of this starfish. Only the parts having direct bearing on the present species will be reproduced here [DÖDERLEIN, '98, p. 499] :

"Das Bild, welches das Dorsalskelet eines erwachsenen *Goniodiscus sebæ* darbietet, erinnert mm in ganz überraschender Weise an das, welches das Dorsalskelet einer kleinen *Culcita schmideliana* etwa von der Grösse der älteren Exemplare von *Goniodiscus sebæ* (Taf. XL, Fig. 8) zeigt. Der einzige nennenswerthe Unterschied ist der, dass bei *Culcita* zwischen den 5 Platten in jedem Interradius, zwischen denen bei *Goniodiscus sebæ* nie Reticularia sich zeigten, auch solche angelegt werden; es sind aber die kleinsten und offenbar am spätesten angelegten; auch stellen sich dabei keine neuen Porenfelder ein, sondern längs des Interradius fehlen die Porenfelder auch hier, genau wie bei *Goniodiscus sebæ.* Das Auftreten von weiteren Reticularia wird hier wohl nothwendig durch das Grösserwerden der benachbarten Porenfelder, durch das schliesslich auch die Platten auseinandergedrängt werden, die keine Porenfelder zwischen sich selbst dulden. Die Zahl der Papulæ in einem Porenfeld ist bei solchen *Culcita* ca. 20–25; wenn wir annehmen, dass sich *Goniodiscus sebæ* in der gleichen Richtung weiter entwickeln würde, wie es sich während seines individuellen Wachsthums entwickelt hat, müsste es genau den Zustand des Dorsalskelets zeigen, den die junge *Culcita schmideliana* darbietet.

 "Noch überraschender wo möglich ist die Aehnlichkeit von *Culcita* mit *Goniodiscus sebæ*, wenn wir das Ventralskelet und die Marginalplatten betrachten. Bis ins Einzelnste stimmen hier die beiden Formen überein, welche man in 2 verschiedenen Familien untergebracht hat. Die Beschreibung eines erwachsenen *Goniodiscus sebæ* ist fast Wort für Wort die Beschreibung der jungen *Culcita* von etwa gleicher Grösse (vergl. die Tabelle auf Seite [571]).

 "Selbst die Oberflächengebilde bieten zum Verwechseln ähnliche Verhältnisse dar. Die inneren wie die äusseren Furchenpapillen, die Granulæ der Ventrolateralplatten, die in der Mitte einer jeden Platte etwas gröber erscheinen, die gleichmässige Körnelung der Dorsalfläche, unter der nur vereinzelte grobe, spitzige Warzen auftreten, die Anordnung der groben, spitzigen Warzen auf oberen wie unteren Marginalplatten, die Vertheilung und Gestalt der Pedicellarien zeigen durchaus den Typus von *Goniodiscus sebæ.* Auch die Krystallkörper sind vorhanden wie bei *Goniodiscus sebæ*, erreichen aber die Oberfläche der Platte nicht mehr. (Vergl. Taf. XXXIX, Fig. 5 u. 6 und Taf. XL, Fig. 6–8).

 "Ueberhaupt ist es gar nicht einfach, die beiden Formen auseinanderzuhalten. *Culcita* zeigt die Aufblähung der Marginalplatten nicht, die für

Goniodiscus sebæ charakteristisch ist, sie treten trotz ihrer Grösse etwas
mehr zurück, vor allem fehlen in Folge dessen der *Culcita* die lochartigen
Vertiefungen in der Mitte zwischen den beiden Reihen von Marginalplatten,
und die Körperhaut lässt die Grenzen der beiden Randplattenreihen kaum
erkennen; ohne Berücksichtigung der letzteren, doch recht unbedeutenden
Unterschiede würde es sehr misslich sein, eine junge *Culcita* von einem
Goniodiscus sebæ zu unterscheiden. Die Hauptmerkmale der Gattung *Culcita*
treten erst später auf, vor allem die kolossale Verdickung der Körperhaut
auf der Rückenseite, infolge deren es so ausserordentlich erschwert ist,
einen Einblick in den Aufbau ihres Dorsalskelets zu thun.

"Ich stelle als Resultat dieser Darlegung die Ansicht auf, dass wir in der
Gattung *Culcita* nichts anderes vor uns haben als eine weiter entwickelte
Form vom Charakter des *Goniodiscus sebæ*, und dass wir in dem lebenden
Goniodiscus sebæ eine Form haben, die der Stammform von *Culcita schmideliana* und überhaupt der Stammform der Gattung *Culcita* ausserordentlich
nahe steht. Die Verbreitung von *Goniodiscus sebæ* scheint etwa dieselbe zu
sein wie die der Gattung *Culcita*.

"Wenn wir die Weiterentwickelung von *Culcita schmideliana* verfolgen,
so finden wir darin in den meisten Punkten eine Fortsetzung der Entwicke-
lungsrichtung, die wir bei *Goniodiscus sebæ* verfolgen konnten. Die Körperhaut wird dicker, bis zuletzt das ganze Skelet schliesslich völlig dadurch verdeckt wird; die Granulirung der Bauchseite wird immer gröber; die Pedicellarien werden spärlicher; die Warzen an den Randplatten verschwinden;
die Ambulacralfurchen erstrecken sich mehr oder weniger weit auf die
Dorsalfläche; die oberen Randplatten nehmen nur langsam an Zahl
zu, die unteren viel rascher infolge der verstärkten Aufwärtsbiegung
der Ambulacralfurchen; die unteren Randplatten werden fast völlig
von den immer zahlreicher werdenden Ventrolateralplatten überwallt. Die
Papulæ treten in immer grösserer Zahl auf, die Porenfelder werden
immer grösser und fliessen in einander; das Dorsalskelet bleibt aber in
Zusammenhang; eine Vermehrung der bei *Goniodiscus sebæ* vorhandenen
Dorsalplatten scheint aber wie bei *Goniodiscus* nur in ganz unbedeutendem Maasse stattzufinden, während die Reticularia stark in die Länge

wachsen. Die bereits bei *Goniodiscus* mehr oder weniger aufgeblähten admarginalen Flatten des Dorsalskelets sind auch bei *Culcita* noch von auffallender Grösse und Dicke, so dass sie auf HARTLAUB den Eindruck machten, wie wenn eine dritte Marginalplattenreihen entwickelt wäre (Taf. XXXIX, Fig. 66).

"Einen Charakter noch finden wir bei *Culcita* sehr viel weiter entwickelt als bei *Goniodiscus*: die interradiären Septenpfeiler sind bei *Goniodiscus* noch verhältnissmässig einfach und von unbedeutender Beweglichkeit. Bei *Culcita* finden wir schon bei den kleineren Exemplaren eine beträchtlichere Anzahl von Zwischenstücken entwickelt, gegenüber *Goniodiscus sebœ*; davon sind die den Fuss bildenden sehr zahlreich, aber klein, während sie dorsalwärts viel spärlicher werden, aber beträchtliche Grösse erreichen (Taf. XXXIX, Fig. 7). Ferner sind nicht nur am Bauchskelet eine Anzahl von benachbarten Platten zur Verstärkung der Fussplatte beigezogen, sondern auch hier sind die benachbarten Platten mit dem Septenpfeiler verwachsen, der sich marginalwärts bis zum dritten Interradiale fortsetzt; die meisten der den Pfeiler zusammensetzenden Stücke sind leicht gegen einander beweglich, so dass er nunmehr den mannigfaltigsten Formveränderungen des Seesterns von Kugel- bis Plattengestalt zu folgen vermag (Taf. XXXIX, Fig. 66 u. 7). In dieser Beziehung hat sich die Gattung *Culcita* hoch über den Zustand erhoben, an den die Gattung *Goniodiscus* und ihre näheren Verwandten noch gefesselt sind, die ihre plattenförmige Körpergestalt nicht zu ändern vermögen."

In subsequent paragraphs the author discusses the mutual relationships of *Pentagonasteridœ*, *Pentacerotidœ*, *Antheneidœ* and *Gymnasteriidœ*, laying special weight on the character of the dorsal skeleton, the interradial septa and the distribution of alveolar pedicellariæ on the abactinal side.

Under the genus *Hosia* VERRILL expresses his opinion as follows ['99, p. 149]: "When GRAY established this genus, in 1840, he referred to it only *H. flavescens*. PERRIER (1876) has redescribed the types of this species and refers them to two distinct species of true *Anthenea* (GRAY, 1849). Therefore *Hosia* becomes a synonym of the latter. In 1847 and 1866, GRAY added another species *H. spinulosa* to *Hosia*, but according to PERRIER (1876), who reëxamined GRAY's type, this species belongs to a different genus. He referred it to his section C of *Pentagonaster*. It has spinulose marginal plates, and also valvular pedicellariæ. It is probably an immature species of *Tosia* or of some closely related genus."

LUDWIG ['99, p. 539] mentions *Goniodiscus sebæ* from the Zanzibar region, "von Ibo (Querimba) durch PETERS (1852), von den Seychellen durch DÖDERLEIN (1898)." He also mentions " *Culcita pentangularis* GRAY (vielleicht identisch mit *C. plana* HARTLAUB, von Mosambique durch PERRIER (1875) " ['99, p. 540].

BEDFORD describes *C. novæ-guineæ*, var. *arenosa* from Singapore [:00, p. 296] :

"*Locality*. This species is fairly common on Pulo Rengkam, Singapore, between tide-marks on the reef; it lives in company with, and appears to have a mode of life similar to, the large tropical species of *Muelleria*.

" *Distribution*. This variety has previously been recorded from Amboina, Ceram Laut, and the Sandwich Islands.

" Unfortunately I have only examined one specimen with care, so that I do not know whether the other recorded varieties of the species also occur in the district. This specimen must certainly be regarded as the *C. arenosa* of PERRIER ; the pore-areas are not distinct from each other but tend to unite over the whole abactinal surface, leaving small, distinct islands which are devoid of pores and granular, the larger of these spaces being about the same size as the madreporite; they are devoid of large spines, thus differing from typical *arenosa* and approaching *C. coriacea* M. & T. The madreporite is an irregular oval structure, and is quite prominent, being bordered with a few large blunt spines; it measures 13 mm. × 6.5 mm.

" Dr. DÖDERLEIN has shown that *C. novæ-guineæ* differs from the Western species *C. schmideliana* in the character of granulation of the ventro-laterals, the larger granules not being divided up into distinct groups corresponding to the subjacent plates in the former. In the specimen now described this holds good, and parallel to the ambulacral furrow up to within a distance of about 7 mm. from it the granules enlarge gradually up to the furrow, where they form a conspicuous edging; in the adambulacral armature there are 3 to 6 furrow-spines (usually 5) on each plate, the outer row being extremely irregular and poorly developed, over a large part of the arms it seems to disappear completely.

"Dr. SLUITER has pointed out how impossible it is in this genus to rely on shape for specific diagnosis; and in this species, at any rate, this character depends almost entirely on the mode of preservation. During life, especially when left dry on the reef at low-tide, the outline may alter considerably, and, if placed on a flat surface with the actinal side downwards, the upper side would flatten out over the edge, so that the ambulacral furrow would appear confined to the lower side (cf. *C. acutispinosa* BELL); also when specimens are placed in a basin and spirit is poured over them, the lower surface (abactinal or actinal) will often accommodate itself to the shape of the bottom of the vessel. R + r = 195 mm."

As late as 1903 this species is mentioned by BELL [:03, p, 245] from Zanzibar under the name of "*Culcita pentangularis*, GRAY."

LUDWIG [:05, p. 156] mentions specimens of this species from the following localities: one specimen from the reef near Papeete, Society Islands, September 26, 1899; one specimen from the reef near Neiafu, Vavau Island, Tongo Island, December 5, 1899; two specimens from the reef near Ponape, Caroline Islands, February 12, 1900. He then adds the following remark: "Die stark verdrückten Exemplare haben einen Durchmesser (R + r) von 145–185 mm. und eine Höhe von 33–55 mm."

FISHER has the following remark on *C. arenosa* [:06, p. 1076]. "This species was not taken by the *Albatross* expedition. The specimen which PERRIER described was collected by the expedition of EYDOUX and SOULEYET in the *Bonite* in 1837, 'des îles Sandwich.'" In the same work he proposes a new generic name for *Goniodiscus sebæ* and describes it at length as follows [:06, p. 1070]:

"Genus *Goniodiscides*,' new name.

"*Goniodiscus* MÜLLER and TROSCHEL, System der Asteriden, 1842, p. 57. Emended by PERRIER, Révision des Stellérides, 1875, p. 229. Type, sens. nov., *Goniodiscides sebæ*.

"This genus is equivalent to that long known as *Goniodiscus*. Under existing rules of nomenclature *Goniodiscus* is untenable because it was proposed by MÜLLER and TROSCHEL to include previously described genera of GRAY (*Anthenea*, *Nectria*, *Tosia*) as well as species unknown to GRAY. If this group had really constituted a genus the oldest name, *Anthenea*, should have been used. Likewise the type (first species) of *Goniodiscus* is the same species, under a different name, as the type of GRAY's *Anthenea*. MÜLLER and TROSCHEL included the following species under their *Goniodiscus*: *Pentago-nulus* [*Anthenea*], *sebæ*, *placenta* [*Tosia*], *regularis* [unknown], *pleyadella*, *ocel-liferus* [*Nectria*], *cuspidatus*, *mammillatus* [*Tosia*], *capella* [since made the type of *Ogmaster* v MARTENS]. This left *sebæ*, *regularis*(?), *pleyadella*, and *cuspidatus*. *Goniodiscus* has subsequently been used for these species, but since the name was originally applied to a composite group and was a synonym as soon as made, it should be discarded for all time. As there appears to be no subsequent name[1] available I propose *Goniodiscides*, with *Goniodiscides sebæ* as type.

" *Goniodiscides sebæ* (MÜLLER and TROSCHEL). Pl. XIX, fig. 3.

" *Goniodiscus sebæ* MÜLLER and TROSCHEL, System der Asteriden, 1842, p. 58.

" An example of this curious species, the first from the Hawaiian Islands, was taken by Mr. H. W. HENSHAW at Hilo, on the windward side of the Island of Hawaii. (Accession no. 42800, U. S. Nat. Mus.) No specimens were secured by the expedition of 1902.

" Concerning the capture of this specimen, Mr. HENSHAW has sent me the following notes: ' They are by no means rare in a small inlet some three miles south of Cocoanut Island. If I remember rightly, all I found were under stones in shallow water, two or three feet deep. In other words, so far as I have observed it, it was a littoral, shallow water species.' This is the only species of starfish, so far as I am aware, that may be collected along shore in the islands, unless *Ophidiaster lorioli* be excepted.

---

1) " *Metopaster* SLADEN (Monog. on Brit. Fossil Echinod. from Cretaceous, II, Asteroidea, pt. II.(Pal. Soc. Monog. 1893, p. 13) is near this genus, but is hardly identical, as has been claimed. (VALETTE, Note sur quelques Stellérides de la Craie Senonienne du Dép. de l'Yonne. Bull Soc. l'Yonne, lvi, 1902, p. 7.) "

"This specimen agrees very well with the original description of MÜLLER and TROSCHEL, and with the notes given by DE LORIOL (Mém. de Société Phys. et d'Hist. Nat. Genève, XXIV, 1885, p. 48). DE LORIOL also gives a good figure (op. cit., pl. XV, fig. 6) with which our example shows a few unimportant points of difference.

"Form pentagonal, the sides of disk only very slightly curved inward. R = 29 mm.; r = 23.5 mm. R = 1.23 r. As noted by DE LORIOL there are 14 superomarginals to a side, or 7 to the 'ray', instead of 6, as stated by MÜLLER and TROSCHEL. The ultimate plate of each series is very small, and is wedged between the penultimate and ocular plates. There are 9 inferomarginals to the ray, the last plate being very small indeed.

"The superomarginals are somewhat tumid, and are broader than high. Besides the even, fine granulation, each bears from 2 to 10 conspicuous, low, hemispherical, or subconical, tubercular granules, unevenly disposed. Inferomarginals are similarly armed. There are no pedicellariæ on any of the marginal plates. Between the two series of marginal plates, on the lateral wall of the body, is a row of 5 to 7 pits, each of which occurs at the junction of a dorsoventral with the horizontal suture.

"Abactinal surface is covered with a much finer granulation than the actinal, and each plate is surrounded by 6 to 8 papular areas which appear in many cases practically confluent. These areas contain 8 to 20 pores, and a cuneiform area containing about 15 to 18 pores occurs between the dorsal ends of the superomarginal plates (excepting between 5 and 6 and 6 and 7). Each abactinal plate bears near the centre 1 to 3 of the tubercular granules, a few of which occur also, here and there, over the papular areas. A number of plates toward centre of disk bear small bivalved pedicellariæ similar to those of *Pentaceros* and flush with the general level of the granulation. These are not very numerous and are irregularly scattered, never more than 2 to a plate. The madreporic body is raised above the general surface and is situated about one-third the distance from the centre to margin. About its border are several tubercular granules. The plates toward the end of ray and adjacent to superomarginals appear to be a trifle convex, the tumidity being surmounted by the granule, or granules.

"The actual intermediate plates are arranged in chevrons and decrease
in size toward the margin. They are polygonal, and covered with a
coarser granulation than the dorsal plates—a granulation which increases
somewhat in coarseness toward the centre of each slightly convex plate,
which is surrounded by 1 to 5 enlarged granules, usually of unequal size.
Scattered here and there are bivalved pedicellariæ, 0.25 to 0.75 mm. in
length.

"The furrow spinelets are 4 to 5 in number, robust, short, truncate,
slightly flattened, the adoral spinelet shorter than the others. On the
actinal surface of the plate stands a longitudinal series of 2 or 3 shorter,
thicker, granuliferous spinelets, with often 1 to 2 smaller granules standing
in line at either end of the series. Occasionally a very small bivalved
pedicellaria stands at the adoral end of the series out of line. Behind the
actinal series the fine granulation of the general surface begins, decreasing
in size toward outer end of plate. The furrow spinelets appear a trifle
shorter and heavier than those figured by DE LORIOL. A few of the proximal adambulacral plates have 6 furrow spinelets.

"This species has a wide range, being found in following localities :
Red Sea, Moluccas, New Guinea [MÜLLER and TROSCHEL], Mauritius,
Macassar (Celebes), Fiji Islands [DE LORIOL], Ceylon, Madagascar, ʻEast-
ern Archipelago' [SLADEN]. Its capture in the Hawaiian Islands considerably extends its known range."

The following notes on this species by CLARK is important as pointing
out clearly for the first time that the starfish *Goniodiscus sebæ* is a young
form of *C. novæ-guineæ* [CLARK, : 08, p. 281][1] :

"3 specimens, 80–130 mm. in diameter. Sorong, New Guinea.—1 specimen, 75 mm. in diameter. Amboina. BARBOUR collection.

"The small series of Culcitas brought home by Mr. BARBOUR is of
great interest because they prove that the starfish hitherto known as
Goniodiscus sebæ is the young of *Culcita novæ-guineæ* and not a distinct
species related to the ancestral stock from which Culcita has sprung, as
DÖDERLEIN has so ably argued (SEMON's Zool. Forsch. Aust., 5, lf. 4, p. 489

---

1) References to MULLER and TROSCHEL, Syst. p. 38 and 58 omitted.

-504). The specimen from Amboina is clearly *Goniodiscus sebæ*, agreeing not only with MÜLLER and TROSCHEL's description, but with DE LORIOL's (1885. Mém. Soc. Phys., Genève, 29, p. 48 ; Plate 15, figs. 6–6e) description and figures, and with specimens in the Museum of Comparative Zoölogy collection from the Gilbert and Marshall Islands. It cannot, however, be separated in any way from the slightly larger young Culcita from Sorong, which is certainly identical with the other two specimens. On the actinal side the latter are exactly like DÖDERLEIN's (1896. SEMON's Zool. Forch. Aust., 5, lf. 3, p. 301–322) figures (Plate 20, fig. 9) of *C. novæ-guineæ*, but abactinally one is like *C. n. plana* (Plate 19, fig. 1), while the other (the largest of all) is like *C. n. arenosa* (Plate 19, fig. 5). Judging from the 54 Culcitas accessible to me, it seems doubtful whether the varieties (or subspecies) of *C. novæ-guineæ*, so carefully worked out by DÖDERLEIN, are really sufficiently distinct to warrant their recognition.—Mr. BARBOUR's specimens were collected about the reefs and were of a yellowish-brown colour, with something of an olive tint when alive. They were all flat and more or less discoidal in life and showed no tendency to the spherical form characteristic of many adult Culcitas.''

KŒHLER in his recently published monograph on Indian starfishes makes several remarks of interest on this species and points out that *Hosea spinulosa* GRAY is a very young form of it [KŒHLER, : 10, p. 119] :

" Iles Andaman. Quelques échantillons.

" Tous les individus que je rapporte à la *Culcita novæ-guineæ* sont très jeunes, et, à part le plus grand dans lequel R=43 mm., ils offrent les caractères du *Goniodiscus sebæ*, forme qui ne représente, comme on le sait, que de jeunes Culcites. Je ne crois pas me tromper en rapportant ces échantillons à la *Culcita novæ-guineæ*, en raison de la présence de tubercules au milieu des aires porifères et des caractères des plaques latéro-ventrales ; mais il est déjà difficile de séparer les espèces du genre Culcite lorsqu'il s'agit d'adultes, à plus forte raison les différences sont-elles délicates à apprécier lorsqu'on est en présence de formes jeunes.

" Les caractères des jeunes Culcites ont déjà été étudiés depuis long-temps par plusieurs auteurs. Sans parler des Asteries que GRAY a représentées sous les noms de *Randasia granulata* et de *R. spinulosa* (66, Pl. II, fig. 1 et Pl. XII, fig. 3), je rappellerai que DE LORIOL (85, p. 64 et suivantes) a décrit différents stades jeunes de la *C. schmideliana* de l'île Maurice, et il a même fait remarquer qu'une Culcite n'ayant que 58 mm. de diamètre avait tout à fait l'apparence d'un *Goniodiscus*; les dessins que ce savant a publié sont très intéressants (85, Pl. XX, fig. 1 à 6). Par une coïncidence très curieuse, DE LORIOL a étudié, dans le même travail, à la fois le *Goniodiscus sebæ* et un autre *Goniodiscus* auquel il a donné le nom de *G. studeri* (Pl. XV, fig. 6 et 7), et l'on peut être surpris qu'un observateur aussi sagace n'ait pas songé à rapporter au *Goniodiscus sebæ* les jeunes Culcites qu'il décrivait d'autre part. Il fait remarquer cependant que la petite Culcite représentée Pl. XX, fig. 3 de son mémoire, a tous les caractères d'un *Goniodiscus.*

"Les jeunes Culcites représentées par DE LORIOL ont toutes des tubercules bien développés et assez nombreux sur les trabécules du réseau calcaire de la face dorsale du corps, et, chez des exemplaires mesurant 58 mm. de diamètre, les granules des plaques latéro-ventrales offrent bien les caractères de la *C. schmideliana*, c'est-à-dire qu'ils laissent distinguer un amas central de gros granules entourés de granules péripheriques très fins.

"Dans son important travail sur la croissance et les affinités du *G. sebæ* (98, p. 404, Pl. XXXIX et XL), DÖDERLEIN a montré les ressemblances que cette forme présente avec la *C. schmideliana* et il insiste sur les importants changements que subit le *G. sebæ* au cours de son évolution. C'est grâce à ces changements qu'il se présente d'abord sous la forme de *Pentagonaster spinulosus* rangée dans les Pentagonasteridés, puis sous celle qui l'avait fait classer parmi les Goniasteridés sous le nom de *G. sebæ*, pour prendre enfin, à l'état adulte, les caractères des Pentacerotidés lorsqu'il est devenu une Culcite.

"Cette manière de voir a été confirmée en 1908 par L. Clark (08, p. 281), qui a pu s'assurer, en étudiant une serie considérable de Culcites, que le *G. sebæ* était le jeune de la *Culcita novæ-guineæ.*

"DÖDERLEIN a décrit et photographié une série très intéressante de *G. sebæ* de différentes tailles dont le diamètre est compris entre 15 et 66 mm.

Dans le plus petit échantillon du Musée de Calcutta, R=23 mm. et le diamètre est de 35 à 37 mm.; dans le plus grand, R=43 mm. et le diamètre est de 65 mm. Ces échantillons diffèrent quelque peu de ceux qui ont été étudiés par DÖDERLEIN et il me paraît utile de les examiner en détail.

"Les deux plus petits exemplaires portent le no. 2235 ; leurs dimensions respectives sont les suivantes : R=23 et 25 mm., r=17 mm. Dans le plus petit (Pl. IX, fig. 3), le corps est pentagonal avec les côtés légèrement excavés, tandis que dans l'autre les différences entre R et r sont plus accentuées, les côtés sont un peu plus excavés et l'on commence à apercevoir une indication des bras. Le plus petit est intermédiaire comme taille entre les jeunes Culcites représentées par DÖDERLEIN (98, Pl. LX, fig. 3 et 4), dont le diamètre était compris entre 29 et 44 mm. Les plaques marginales dorsales sont remarquablement saillantes dans leur région médiane et elles sont séparées par des sillons très profonds. Elles sont au nombre de sept de chaque côté et constituent une bordure très large et très apparente ; elles sont séparées les unes des autre sur leur tiers interne par les aires porifères externes petites et triangulaires ; la première plaque de chaque côté est notablement plus large que les suivantes. Leur surface est couverte de granules très fins comme on en voit sur les autres plaques de la face dorsale ; mais, de plus, chacune d'elles offre sur la ligne médiane une rangée de trois à cinq petits tubercules. Les plaques latéro-dorsales sont tout à fait planes ; elles offrent une forme hexagonale souvent très régulière, avec des côtés concaves entre lesquels prennent place les petites aires porifères. Indépendamment de la granulation générale très fine, chacune d'elles porte, dans sa région centrale, soit un petit tubercule unique, soit un groupe de deux ou trois petits tubercules toujours plus fins que ceux des plaques marginales. Les aires porifères, arrondies, sont bien distinctes et beaucoup plus petites que les plaques elles-mêmes ; elles renferment le plus souvent un ou deux petits granules chacune. Des pédicellaires valvulaires se montrent sur un certain nombre de plaques et les plus grands se trouvent sur la rangée située immédiatement en dedans des marginales, ou sur les plaques qui précèdent cette rangée. La disposition générale des plaques est la suivante. Il n'existe ni radiales ni interradiales primaires ; la région centrale est occupée par quelques plaques irrégulièrement disposées, dont quatre entourent l'anus, puis on observe, dans chaque radius, une rangée carinale d'une demi-douzaine de plaques hexagonales, de chaque côté de laquelle se montre une rangée latérale de plaques, ayant à peu près la même taille et la même forme. Les autres plaques sont un peu plus petites et leur forme est moins régulière ou simplement arrondie. La plaque madréporique est ovalaire et allongée suivant l'interradius. Les plaques latéro-dorsales placées immédiatement en dedans des marginales forment une rangée bien distincte ; elles sont arrondies, saillantes et elles portent généralement chacune un petit tubercule en leur milieu. Les cinq rangées ainsi formées sont parallèles aux cinq côtés du corps et chacune d'elles comprend une dizaine de plaques ; les trois plaques moyennes, qui correspondent au fond des arcs interradiaux, sont moins saillantes que les autres et elles ne dépassent guère le niveau des autres plaques latéro-dorsales ; leur tubercule central est aussi plus petit ou même fait complètement défaut ; la plaque médiane correspond toujours à l'intervalle qui sépare la première plaque marginale de chaque série de sa congénère de l'autre série.

"Les plaques latéro-ventrales forment des rangées très regulières, les unes longitudinales et parallèles aux adambulacraires, et les autres transversales allant des adambulacraires aux marginales ventrales. On peut distinguer trois rangées longitudinales, la première allant jusqu'à la cinquième marginale ventrale, et la troisième allant jusqu'à la limite de séparation de la troisième et de la quatrième marginales ventrales ; quelques autres plaques occupent le reste des aires interradiaires. Ces plaques ont une forme très régulière, carrée ou hexagonale, et elles offrent des granules un peu plus gros que sur la face dorsale ; on distingue, au centre, un petit groupe de trois ou quatre granules plus forts, dont l'un est en général plus développé encore que les autres. De plus, chacune de ces plaques porte un pédicellaire valvulaire.

"Les plaques marginales ventrales, au nombre de six, correspondent aux dorsales, sauf la sixième qui est placée en face des deux dernières marginales dorsales. Ces plaques ont une face externe très saillante et elles sont séparées par des sillons profonds, mais la face ventrale est moins convexe; elles sont beaucoup moins larges que les marginales dorsales. Comme ces dernières, elles possèdent un recouvrement général de granules très fins, et de plus, sur la ligne médiane, quelques granules plus gros, mais moins développés cependant que sur les plaques dorsales; les pédicellaires font également défaut.

"Les plaques adambulacraires offrent une rangée interne de cinq piquants assez forts, obtus, généralement placés un peu obliquement; la rangée externe est ordinairement formée de deux piquants plus forts, puis le reste de la plaque est occupé par quelques rangées de granules assez fins, au milieu desquels on remarque ordinairement un petit pédicellaire valvulaire.

"Dans le deuxième individu, la bordure des plaques marginales dorsales est un peu moins large et la rangée de plaques latéro-dorsales qui lui est parallèle est un peu moins distincte et moins saillante.

"Comme on le voit, ces deux échantillons répondent absolument au *Pentagonaster spinulosus*.

"Deux individus un peu plus grands portent le no. 2217: les diamètres sont respectivement de 50 et 53 mm.; R=29 et 31 mm. Ils sont très intéressants à comparer l'un à l'autre parce que, malgré leurs dimensions très voisines, ils offrent, dans les caractères de la face dorsale, des différences très marquées. Dans le plus petit (Pl. IX, fig. 4), les côtés sont assez excavés et les plaques marginales dorsales, au nombre de sept, forment une bordure bien apparente. Les premières plaques de chaque série sont très saillantes, séparées par des sillons très profonds et la première est toujours plus large que les autres; toutes sont séparées les unes des autres, sur la moitié de leur longueur au moins, par de grandes aires porifères, et la rangée médiane de tubercules que porte chaque plaque est bien développée. La rangée de plaques latéro-dorsales qui vient immédiatement en dedans des marginales est toujours très apparente, et les plaques distales font une saillie bien marquée; les autres plaques latéro-dorsales sont distinctes les unes des autres et elles offrent le même arrangement que dans les deux échantillons précédents, mais leur forme hexagonale s'exagère et on constate qu'elles se relient les unes aux autres par des travées étroites limitant des aires porifères comparativement plus developpées. Sur la ligne médiane interradiale, les plaques sont un peu plus grands que les autres et leurs limites sont moins nettes ; les tubercules qu'elles portent sont peu nombreux et relativement petits. Les pédicellaires valvulaires sont de petite taille, mais assez nombreux. Chaque aire porifère présente un ou deux petits granules distincts. La plaque madréporique est piriforme, saillante et assez petite ; la plaque apicale, très petite et triangulaire, est terminée par deux tubercules : elle est entièrement située sur la face dorsale du corps.

"Les plaques latéro-ventrales forment quatre rangées parallèles aux adambulacraires. Leur granulation est assez uniforme et des granules centraux plus gros que les autres ne se montrent guère que sur les plaques péripheriques. En revanche toutes les plaques portent des pédicellaires assez grands et atteignant souvent le nombre de trois par plaque ; ces pédicellaires sont dirigés en tous sens, mais ils sont toujours rapprochés du centre de la plaque. Les plaques marginales ventrales sont au nombre de huit ; les dernières sont beaucoup plus petites que les dorsales et les cinq premières seules correspondent aux marginales dorsales. Ces plaques sont peu développées sur la face ventrale, mais leur faces latérales sont convexes et elles sont séparées des plaques marginales dorsales par un sillon très profond qui présente une dépression très marquée au niveau de chaque point d'intersection. Elles portent, dans leur région médiane, des tubercules comme ceux des plaques dorsales, mais ceux-ci sont au nombre de deux ou même d'un seul sur les premières plaques et toujours d'un seul sur les suivantes. On remarque, en revanche, que ce tubercule unique se développe davantage sur les dernières plaques marginales ventrales et il arrive à former un petit piquant épais, court et émoussé. Les piquants adambulacraires sont disposés comme dans les échantillons du no. 2235, mais les pédicellaires valvulaires sont plus rares et plus petits.

"Dans le deuxième individu, les côtés sont un peu moins excavés ; les plaques marginales dorsales, au nombre de sept, sont moins développées et la bordure qu'elles constituent est moins apparente et moins large : la première plaque de chaque série est encore plus large et plus saillante que les suivantes. Toutes portent une rangée médiane de tubercules qui deviennent plus apparents et plus pointus sur les dernières plaques ; elles sont séparées par les aires porifères sur les deux tiers au moins de leur largeur. On ne peut plus reconnaître, en dedans des marginales, que quelques plaques de la rangée qui était bien distincte dans les exemplaires précédents. Toutes les plaques latéro-dorsales forment ici un réseau très apparent dans lequel il est encore possible de distinguer les limites des plaques hexagonales, mais celles-ci se relient les unes aux autres par des travées plus larges et les aires porifères sont devenues presque aussi grandes que les plaques elles-mêmes. Chaque plaque porte ordinairement en son centre, un ou quelquefois deux tubercules arrondis et bien développés, mais, en revanche, les granules des aires porifères sont rares et petits ; sur la ligne interradiale médiane, les plaques restent plus larges et elles se soudent ensemble de manière à former une bande plus ou moins apparente.

" Les plaques marginales ventrales sont au nombre de neuf : les cinq premières cörrespondent aux dorsales, mais les suivantes alternent de plus en plus avec ces dernières. La ligne de séparation des deux rangées est moins profonde et elle constitue une ligne en zig-zag n'offrant de dépressions aux points de rencontre des plaques que dans la première moitié des bras ; dans la seconde moitié, la séparation des deux rangées dorsale et ventrale est à peine indiquée. Les marginales ventrales sont très peu développées sur leur côté ventral ; elles offrent toujours quelques tubercules médians qui se développent davantage sur les dernières.

" L'échantillon portant le no. 2218 a un diamètre de 48 à 49 mm. ; R = 28 mm. Il correspond assez exactement à la figure de *Randasia granulata* donnée par GRAY (66, Pl. II, fig. 1). Cet individu est peu différent du précédent, mais les tubercules des plaques sont plus développés ; d'autre part, les contours des plaques sont plus distincts et chacune porte en son centre un petit tubercule. En dedans des plaques marginales dorsales, il existe encore une rangée distincte de plaques latéro-dorsales assez saillantes. Les aires porifères offrent de petits granules. Les plaques marginales dorsales sont très saillantes et forment une large bordure ; elles sont à peine séparées les unes des autres sur la moitié de leur largeur par des aires porifères très amincies. Les plaques marginales ventrales sont assez développées sur leur face ventrale. La plupart des plaques latéro-ventrales portent, dans leur région centrale, un et rarement deux ou trois petits tubercules qui deviennent plus gros sur les plaques périphériques et qui sont accompagnés d'un ou de deux pédicellaires, rarement davantage. Cet exemplaire est aussi voisin de celui que FISHER a représenté (06, Pl. XXIX, fig. 3), mais sur celui-ci les plaques marginales dorsales sont moins saillantes, la granulation de la face dorsale est plus uniforme et la rangée de plaques latéro-dorsales en dedans des marginales est moins distincte.

"Dans l'exemplaire no. 8728, $R=27$ mm. et le diamètre est de 47 mm. Bien qu'à peine plus petit que le précédent, il en diffère par les plaques dorsales formant déjà un réseau plus marqué ; les marginales dorsales sont moins larges et moins hautes, et la rangée de plaques dorsales en dedans des marginales est à peine reconnaissable. Les plaques latéro-dorsales n' offrent pour ainsi dire pas de tubercules ; les marginales ventrales sont un peu moins développées sur leur côté ventral. L'exemplaire n'est d'ailleurs pas bien conservé et toute une moitié est plus ou moins endommagée.

"Dans l'exemplaire no. 8339, $R=34$ mm. et le diamètre est de 57 mm. ; les côtés sont assez excavés et les bras sont distincts et pointus (Pl. IX, fig. 5). Cet individu rappelle beaucoup le deuxième exemplaire du no. 2217, seulement la disposition en réseau des plaques latéro-dorsales est moins apparente ; celles-ci offrent d'assez nombreux petits tubercules arrondis, et les aires porifères présentent quelques petits granules. Les plaques marginales dorsales, au nombre de sept, ne sont pas très larges, mais elles sont très saillantes et séparées sur les deux tiers de leur longueur par des aires porifères assez larges ; chacune d'elles porte une rangée médiane de quatre ou cinq gros tubercules. La rangée qui vient immédiatement en dedans des plaques marginales dorsales est encore indiquée.

"Les plaques marginales ventrales sont au nombre de neuf ; elles correspondent aux marginales dorsales dans la première moitié des bras, ensuite elles alternent quelque peu avec ces dernières. La ligne de séparation entre les deux rangées est assez profonds avec des dépressions bien marquées. Les premières plaques offrent sur leur ligne médiane deux ou trois granules chacune : la quatrième et la cinquième en ont quatre ou cinq, puis le nombre diminue et tombe à un ou deux sur les dernières ; la face ventrale de ces plaques est peu développée, mais la face externe est assez saillante. Les plaques latéro-ventrales offrent ordinairement deux ou trois granules centraux plus gros que les autres, mais les pédicellaires valvulaires sont peu abondants. Les piquants adambulacraires de la rangée externe sont souvent au nombre de trois.

" Les trois individus portant le no. 2213 sont sensiblement plus grands que les précédents ; deux d'entre eux ont à peu près la même taille et R varie entre 40 et 42 mm., le diamètre est de 70 mm. ; dans le troisième, R=42 à 44 mm. et le diamètre arrive à 75–76 mm. L'intérêt de ces échantillons provient surtout des différences qu'ils présentent, malgré leurs dimensions très voisines : les deux plus petits ont encore des caractères de *Goniodiscus*, tandis que le troisième est devenu une vraie Culcite ; tous trois sont d'ailleurs remarquables par le faible développement des tubercules sur la face dorsale du corps.

" Dans les deux plus petits, les côtés sont légèrement excavés. Sur l'un d'eux, on peut distinguer certaines plaques de la face dorsale, principalement dans les interradius ; un très petit tubercule se montre çà et là, surtout vers l'extrémité des bras et les aires porifères sont plus grandes que les plaques. Les plaques marginales dorsales, au nombre de sept de chaque côté, sont encore distinctes : elles sont légèrement renflées, mais courtes et séparées sur toute la longueur de leur face dorsale par des aires porifères triangulaires ; l'on n'observe un petit tubercule que sur la dernière ou sur les deux dernières. Les marginales ventrales, au nombre de neuf, sont séparées des dorsales par un sillon qui n'est bien apparent qu'au milieu des arcs, et qui n'offre plus de fossettes aux points de séparation ; elles sont encore assez saillantes sur les côtés, mais peu développées sur la face ventrale ; les quatre ou cinq dernières plaques portent un petit tubercule central assez gros. Les plaques latéro-ventrales présentent le plus souvent un petit groupe d'un à trois granules centraux plus ou moins développés.

" Dans le deuxième individu, le réseau dorsal est constitué à peu près comme chez le premier avec des pores un peu plus grands, mais les plaques marginales dorsales sont beaucoup moins apparentes : elles ne sont pas du

tout saillantes et elles sont séparées les unes des autres ¡par des aires porifères plus larges ; cependant elles offrent chacune deux ou trois tubercules
qui se développent davantage sur les dernières. Les tubercules de la face
dorsale sont un peu moins rares que sur l'échantillon précédent et l'on en
retrouve un ou deux plus marqués vers l'extrémité des bras. Les plaques
marginales ventrales sont à peine séparées des dorsales par une ligne en
zig-zag. Chacune d'elles porte un ou deux très petits tubercules qui deviennent plus accusés vers l'extrémité des bras ; les granules centraux des
plaques latéro-ventrales sont un peu plus gros que les autres.

"Le troisième individu est une vraie Culcite. Le corps est exactement
pentagonal et les aires porifères sont considérablement développées par rapport au réseau calcaire qui est formé de trabécules étroites. Les plaques
marginales dorsales sont complètement indistinctes ; elles sont rejetées sur le
côté du corps et sont confondues avec les plaques marginales ventrales. On
observe quelques rares tubercules vers l'extrémité des bras ; d'autres se montrent
çà et là et ils sont un peu moins rares que dan sles deux échantillons précédents. Les plaques marginales ventrales ont aussi les contours indistincts : on
retrouve encore cependant, vers l'extrémité des bras, une rangée comprenant
une demi-douzaine de petits tubercules qui indiquent l'emplacement d'un
nombre correspondant de marginales ventrales. Les plaques latéro-ventrales
offrent, en leur milieu, un groupe de quelques tubercules un peu plus gros
que les autres et parfois un pédicellaire.

"Les caractères spécifiques de la *Culcita novæ-guineæ* ne sont pas
encore bien apparents sur ces trois exemplaires. Le plus grands individu
n'offre qu'un petit nombre de tubercules fort peu développés sur le réseau dorsal, et il rappelle bien à ce point de vue la *C. novæ-guineæ* ; mais
les aires porifères sont uniformément recouvertes de fins granules et il est
rare d'y rencontrer quelques tubercules plus gros. La face ventrale est bien
identique à celle de l'individu que DÖDERLEIN a représenté (98, Pl. XIX,
fig. 3). Dans les deux autres individus, les aires porifères n'offrent guère
de tubercules plus gros que les granules voisins. Je ne crois pas cependant
me tromper en rapportant ces trois exemplaires à la *C. novæ-guineæ* car s'ils
appartenaient à la *C. schmideliana*, les tubercules du réseau dorsal seraient

plus développés et les plaques latéro-ventrales devraient offrir un groupe central de granules beaucoup plus gros que les autres. Cette dernière remarque s'applique d'ailleurs à tous les autres exemplaires dont j'ai parlé ci-dessus chez lesquels ces granules centraux sont, d'une manière générale, toujours peu développés. Dans la jeune *C. schmideliana* ayant 58 mm. de diamètre que DE LORIOL a représentée (86, Pl. XX, fig. 3), on peut remarquer que les plaques latéro-ventrales offrent déjà un groupe central de granules relativement gros.

"L'étude des échantillons du Musée de Calcutta confirme donc les remarques de DÖDERLEIN et de LYMAN CLARK, et on peut considérer comme un fait bien certain maintenant que les Culcites passent successivement par les stades de *Pentagonaster spinulosus* et de *Goniodiscus sebæ*. On voit de plus que les modifications et l'évolution des différents caractères ne sont pas toujours en rapport avec la taille des individus."

I have examined eight specimens of this species, of which two are dried and in such a bad state of preservation that they must be left out of consideration. They are all from the Ryûkyû Archipelago, but from different localities, the most southern being Yaeyama. Of the alcoholic specimens three are of the same general appearance, being quite thick and coarsely granulated on both sides; the ventral side is either slightly concave or plane, the dorsal side decidedly convex, and the lateral margin either forming a rather distinct edge or a bounding plane forming nearly a right angle with the ventral surface. The coarse granules of the dorsal side are more numerous in the non-poriferous areas but are also present in the poriferous areas, where they are smaller; they show a certain tendency, in the specimens now under consideration, to become conical and spinous. The poriferous areas extend to very near the ventral margin of the body; they are more or less circular or elliptical in outline, and may fuse together to form larger areas of irregular form. In one of the speci-

mens this tendency is carried so far and the poriferous areas are
so close together that comparatively little space is left between
them, the adjacent poriferous areas being in many cases separated from each other by a single row of coarse granules, which
again may entirely disappear. In none of these three specimens
can I see distinctly the arrangement of these poriferous areas in
the form of rosettes around a number of central non-poriferous
areas mentioned by preceding authors; the poriferous areas do
not show any regular arrangement. In the three specimens under
consideration, the total poriferous areas far outweigh the non-
poriferous. The latter also bear small forcipiform pedicellariæ,
which appear to vary considerably in number in different specimens, there being sometimes several of them in one poriferous
area, while in other cases one has to look for them for some
time to find one (Pl. XVII, fig. 259, 260).

In the three specimens in question the ventral side is entirely covered over with coarse granules, which are not spinous.
The more or less spiniform coarse granules of the dorsal side
become rapidly rounded towards the edge of the ventral side,
and pass on without any sharp contrast to the coarse granules
of the latter. The demarcation into polygonal areas referred
to by preceding authors is seen in only one of these specimens.
The granules of the ventral side are, generally speaking, decidedly coarser near the ambulacral furrows and towards the mouth,
at the angles of which there are always some particularly well
developed granules. The spaces between the coarse granules of
the ventral side are covered over with finer granules, and the
areal ratio between the fine-grained and coarse-grained portions
varies largely in the different specimens. In the fine-grained areas

near the ambulacral furrows and the mouth there are small valvate pedicellariæ (Pl. XVII, fig. 257, 258); they vary considerably in number in the different specimens, but appear to be in general more numerous in the specimens with more extensive fine-grained areas. They also vary in form to a certain extent, some having nearly square valves, while others are transversely elongated.

The inner ambulacral spines are with very few exceptions five in number and in one of the specimens the spines of the successive plates are imbricating, the adcentral spine or spines of one plate being covered over on the furrow side by the abcentral spines of the next adcentral plate. The spines of the mouth-plates are either seven or eight for each plate, of which the one at the mouth end is wedge-shaped when viewed from the actinal side; this spine may be subequal in size to its fellow of the other side, or it may so far surpass it as to bring about the appearance of an unpaired spine at the mouth end. The outer ambulacral spines are two in number and one of them is considerably smaller than the other; both are rounded conical in form. The outer spines of the mouth plates are the direct continuations of those of the adambulacral plates, and may vary from five to eight for one interradius.

The dimensions of these three specimens are as follows:

Specim.	r mm.	R mm.	Height	Locality
1	52–58 mm.	62–74 mm.	40 mm.	Naha Harbour.
2	55–57 mm.	61–69 mm.	35 mm.	,,
3	52–55 mm.	55–62 mm.		Yaeyama.

A fourth specimen, also full-grown, is in a very good state of preservation in alcohol. It is subglobular in shape, being dis-

tinctly convex on the ventral, and much more so on the dorsal
side, and is very different in general appearance from the specimens described above, so much so that one is tempted to make
another species of it. But from what we know about the
variability of this species, so ably set forth by DÖDERLEIN, it appears to me that it is merely a varietal form of this polymorphic
species. What strikes one at the first sight of this specimen is
the smoothness of the dorsal side, which is entirely destitute of
any coarse granules, except around the madreporite and the anus,
and a few other places at intervals. The poriferous areas are less
well developed than in the specimens previously described, the
intervening non-poriferous areas are relatively larger, and the
rosette-like arrangement of the pore areas is more or less evident.
The dorsal pedicellariæ appear to be exceedingly few, if any.
Spiniform granules of different sizes begin to appear towards the
margin of the dorsal side, mostly in the poriferous, but also more
or less in the non-poriferous, areas, and these gradually pass on
to the rounded coarse granules of the ventral side, which are
less well developed than in the specimens described above. The
ventral valvate pedicellariæ are also far less numerous in this
specimen. The inner adambulacral spines are only apparently four
in number, and a small fifth spine is always found covered over
by the abcentral spines of the next adcentral plate. The outer
spines are two and subequal and are more spiniform than in the
specimens 1–3. The oral spines are eight or nine for each plate.
Another point which may be noted is the presence of an irregular
row of coarse granules directly outside the outer adambulacral
spines, separated from the general granulation of the ventral side by
a smooth space. The dimensions of this specimen are as follows :

Specim.	r mm.	R mm.	Height	Locality
4	55–62 mm.	58–68 mm.	75 mm.	Kyamu Saki, Okinawa Isl.

My specimen No. 5 is so notably different from all the others and like *C. schmideliana* in some points that it will be described in detail. Its dimensions are given below :

Specim.	r mm.	R mm.	Height maxim.	Locality
5	62–68 mm.	57–83 mm.	40 mm.	Okinawa Isl.

As may be seen from the table, this specimen is very irregular in shape, evidently owing to distortion (Pl. XVII, fig. 261, 262). The flatness of the body is also characteristic, the height being only 25–30 mm. for the greater part of the body. The actinal side is very nearly plane, and the dorsal is slightly convex. The apex of one of the ambulacral furrows has evidently been injured and is regenerating in an abnormal way (left side of fig. 261). The outline of the body is pentangular, the side being almost straight except where they have evidently undergone distortion. The poriferous areas are generally rounded in outline and of various sizes ; they appear to have a rather strong tendency to coalesce with the neighbouring ones, and in many places their arrangement in rosette-like groups is very distinct. The poriferous areas extend to the outer borders of the actinal side, and they mostly bear each a few (1–5) coarse mammiform granules which are smaller than those of the non-poriferous areas.

There are also a tolerably large number of small, short, forcipiform pedicellariæ in the poriferous areas, which stand out like so many prickles when the specimen is drained of the alcohol.   The non-poriferous areas are very irregular in form and distribution, and in the specimen now under consideration, I have estimated them to be somewhat greater on the whole than the poriferous areas put together.   The general granulation of the non-poriferous areas is exceedingly fine and looks smooth to the naked eye ; but there rise from this general granulation some coarse granules which are distinctly mammiform, there being a teat-like portion on the top of each.   In some parts they are more elongated and conical in form, but they are distinctly mammiform, except in a very few cases, mostly towards the ventral side.   This mammiform charac-ter of the coarse granules may also be observed in specimen 4, but is not so conspicuous there.   The madreporite is irregularly elongated in outline and surrounded by a number of especially large mammiform granules.

On the actinal side a tolerably distinct row of groups of 4–7 coarse granules can be made out on either side of the ambula-cral furrows, which apparently correspond each to a subjacent plate ; but they may be totally absent from some places.   The general granulation of the actinal side is distinctly coarser than that of the abactinal side, and there stand out in it a large number of particularly coarse granules, which are either flattened and circular or at most hemispherical in form and not mammi-form, as on the abactinal side.   There are small valvate pedicel-lariæ of varying height on the actinal side, which are more numerous just outside the adambulacral plates but may be found in smaller numbers all over this side.

The inner adambulacral spines are slightly palmate in arrangement and mostly six for each plate, but sometimes only five, of which the one at the adcentral end is especially short and usually overlapped by the abcentral spine or spines of the next adcentral plate. The outer row consists of only two short stout spines, of which the abcentral one is usually very much smaller. At the adcentral end of each adambulacral plate, between the two rows of spines there is an elongated forcipiform pedicellaria, but this may be wanting on some of the plates.

The inner oral spines are either seven or eight for each plate, and one of the pair at the mouth end may so far surpass its fellow in size as to bring about the appearance of the presence of an unpaired spine. The outer spines are the direct continuations of the outer spines of the adambulacral plates and may vary from five to seven for one interradius.

Specimen 6, belonging to the Fisheries Bureau, is intermediate in character between specimens 1–3 on the one hand and specimen 4 on the other. It is very irregular in form. The actinal side is slightly and the abactinal very decidedly convex. What is rather conspicuous on the actinal side is the presence of a row of groups of coarse granules directly external to the outer adambulacral spines, separated from the general coarse granules of the actinal side by a groove-like space. Each group contains 1–3 spines and is in line with an adambulacral plate. Groups of coarse granules similar to those of specimen 5 may be found in some places on the actinal side.

The abactinal surface is relatively smooth in appearance, the spines being comparatively far distanced and few in number for each non-poriferous area. The poriferous areas also bear spines, but they are few in the more central part. There are also some sharply pointed conical spines on the dorsal side, which are more numerous towards the margin, where they gradually pass on to the coarse granules of the actinal side. The dimensions of this specimen are as follows :

Specim.	r mm	R mm.	Locality
6	60 mm.	68 mm.	Ryûkyû Archip.

In all the specimens the adambulacral plate bears a forcipiform pedicellaria mostly at the adcentral end (Pl. XVII, fig. 256), but it may be absent from many of the plates.

One alcoholic specimen in F.B. ; one alcoholic specimen in I. M. ; three alcolholic and two dried specimens in S.C.

My specimens present such a medley of characters that it is difficult to refer them to any of the varieties distinguished by DÖDERLEIN.

Choriaster granulatus LÜTKEN.

(Pl. XVII, fig. 263 ; Pl. XVIII, figs. 264—269.)

This species was first described by LÜTKEN as follows ['69] :

" Seestern aus der Gruppe derjenigen mit deutlichem After und zweizeiligen Saugfüsschen mit deutlicher Saugscheibe ; mit hohem Körper und fünf kurzen, dicken, beinahe cylindrischen, am Ende abgerundeten Armen, deren Länge nicht dem Durchmesser der Scheibe gleichkommt. Körper und Arme oben und unten völlig glatt, ohne Stacheln irgend einer Art, und ohne sichtbare Hautplatten, nur mit einer weichen, lederartigen, dichtgekörnten Haut überzogen. After gross, mitten auf dem Rücken. Porenfelder mit zahlreichen Poren, unregelmässig rundlich, scharf geschieden, auf dem Rücken

des Körpers und auf den Seiten und Rücken der Arme, mit Ausnahme des letzten Drittels der Arme, wo sie, wie auf der Unterseite, ganz fehlen, sonst acht Reihen der Länge nach auf jedem Arme bildend. Ambulacralpapillen zweireihig in handförmigen Gruppen, die innern zu 6–7, die äussern gröbern gewöhnlich zu 4. Grosser Radius c. 105 mm., kleiner Radius c. 50 mm.

" Fundort. Pelew-Inseln von Capt. TETENS und Viti-Inseln von Dr. GRAEFFE."

In his paper of 1871, LÜTKEN reproduces the preceding description without substantial alterations, and in the summary appended to the same paper he gives a French translation of the text. These together with a short introductory note are reproduced below ['71, p. 243, summ. p. 21]:

" *Choriaster* LTK. (n. g.)

" Den följgende korte Karakteristik af en ny Söstjerne-Form fra Pelew- og Viti-Öerne har jeg allerede tidligere meddelt til 'Museum GODEEF-ROY''s 4de Katalog, efter Opfordring af dettes Udgiver, Hr. SCHMELTZ. Museet har desværre ikke havt Evne til at erhverve sig denne Söstjerne, af hvilken der kun forelaa to Exemplarer.

The following short characterisation of a new form of starfish from the Pelew- and Viti-Islands I have already previously communicated to the 4th. catalogue of 'GODEFFROY'S Museum', at the invitation of its editor Mr. SCHMELTZ. The Museum [Copenhagen] has unfortunately not been able to procure this starfish, of which only two example lay before me.

" *Choriaster granulatus* LTK.

" Denne nye Söstjerne - Slægt hörer til dem, der have et tydeligt Gat og Sugefödderne stillede i to Rækker samt forsynede med tydelig Sugeskive ; dens höje Legeme og fem korte og tykke, næsten cylindriske, i Enden afrundede Arme, der ikke engang ere saa lange som Skivens Tvermaal, stiller den i Nærheden af Slægten *Oreaster*. Fra denne og nærbeslægtede Former (*Goniaster* o.

" Ce nouveau genre d'Astérie appartient à ceux qui ont un anus évident et les pieds ambulacraires disposés en deux rangées et munis d'un disque terminal distinct. Son haut corps et ses cinq bras courts et epais, presque cylindriques, arrondis vers l'extrémité, et qui ne sont pas même aussi longs que le diamètre du disque, le rapprochent du genre *Oreaster*. De celui-ci et des genres voisins (tels que le *Goniaster* etc.), il se distingue en ce que le

s.v.) udmærker den sig derved, at Krop og Arme paa begge Sider, saavel paa Ryggen som paa Bugen, ere fuldkommen glatte uden Pigge af nogensomhelst Art og uden synlige Hudplader, kun overtrukne med en blöd og læderagtig, om end med fine Korn tæt beklædt Hud. Det forholdsvis store Gat ligger midt paa Ryggen. Porefelterne ere skarpt sondrede fra hinanden, uregelmæssigt afrundede og indeholde hver et större Antal Porer; de danne otte Længderækker paa hver Arm og findes overhovedet paa hele det egenlige Legemes (Skivens) Rygside saavel som paa Armenes Sider og Ryg, med Udtagelse af Armenes yderste Trediedel, hvor de aldeles mangle; ligeledes mangle de aldeles paa Bugsiden. Fodpapillerne ere ordnede i haandformige Grupper; langs med hver Side af hver Fodgang er der to Rækker af saadanne Grupper, en indre af mindre Papiller, 6–7 i hver Gruppe, en ydre af större, i Almindelighed 4 i hver. Pedicellarier ere ikke iagttagne.——Store Radius (R) c. 105 mm. mindre Radius (r) c. 50 mm.

corps et les bras des deux côtés, tant au dos qu'au ventre, sont complètement lisses, sans épines d'aucune espèce et sans plaques dermiques visibles, seulement recouverts d'une peau tendre et coriacée, quoique bien garnie de grains fins. L'anus, relativement grand, se trouve au milieu du dos. Les aires porifères sont distinctement séparées les unes des autres, irrégulièrement arrondies, et contiennent, chacune, un assez grand nombre de pores; elles constituent huit rangées longitudinales sur chaque bras; en somme, elles se trouvent sur le côté dorsal du corps proprement dit (du disque), ainsi que sur les côtés et le dos des bras, excepté le tiers extrême des bras, où ils manquent entièrement; ils manquent aussi tout à fait sur le côté ventral. Les papilles ambulacraires sont disposées en groupes maniformes; le long de chaque côté de chaque ambulacre sont deux rangées de tels groupes; une rangée intérieure de papilles plus petites, chaque groupe en contenant 6 ou 7; et une extérieure de papilles plus grandes, chaque groupe en contenant ordinairement 4. Des pédicellaires n'ont pas été observés. Le grand rayon est d'environ 105 mm; le petit d'environ 50 mm.''

In his "Révision," PERRIER makes the following remarks ['76, p. 72]:

"Un bel exemplaire dans l'alcool de ce remarquable genre des îles Pelew et Fidji, récemment décrit par LÜTKEN, existe au British Museum.

Il provient du Musée Godeffroy. Ce genre est évidemment voisin des *Oreaster* et des Culcites et on peut le considérer comme intermédiaire entre ces deux sortes de Stellérides, bien qu'il présente des caractères speciaux le distinguant nettement de l'un et de l'autre. Le docteur LÜTKEN a donné une description en français de cette espèce dans les Videnskabelige Meddelelser de 1871, page 21 des résumés français insérés, dans ce volume."

VIGUIER refers to this species as follows ['78, p. 192]:

"Une espèce seulement de ce genre, le *Choriaster granulatus*, a été décrite par M. le docteur LÜTKEN; elle vient de Fidji, et n'est pas représentée au Muséum de Paris. La forme de son corps en fait un intermédiaire entre les *Culcita* et les *Pentaceros*. Le caractère saillant de la physionomie est l'aspect absolument lisse de l'animal, qui est recouvert, tant en dessus qu'en dessous, d'une peau coriace. La description de M. LÜTKEN ne nous apprend rien sur la constitution du squelette, qui doit être, en effet, comme chez les Culcites, à peu près impossible à voir sans préparation.

"On ne rencontre pas de pédicellaires."

SLADEN mentions it from the collection of the Challenger as follows ['89, p. 354]: "Localities.—Station 212. Off Samboangan, Philippine group. Depth 18 fathoms. Off Kandavu, Fiji Islands. Depth and conditions not recorded."

This species has never been figured in my knowledge.

I have examined three specimens, of which only one is in alcohol, the other two being dried. The following description is based on the alcoholic specimen, which is in a very good condition.

The body is very thick, the disk is large and the arms are short, so that the latter are almost cylindrical in form. The abactinal side is slightly convex and the actinal nearly plane. The average of the smaller radii is 41 mm. and that of four of the larger radii is 82 mm., giving the radial ratio of 2; one of the arms being bent considerably was not taken into account. The dried specimens are all dark brown in colour, but the alcoholic one is almost colourless with a very slight tinge of brownish yellow. The surface of the body is entirely covered over with fine granules, which are coarser along either side of the ambulacral furrows as well as on, and in the vicinity of, the mouth-plates. Beneath the surface granules is a thick leathery skin, which becomes hard on drying. On the abactinal side of the disk, close to the centre is the anus, which can be detected at once by the presence of coarse granules around it. The papular pores are very small and form patches, which are present on the abactinal and lateral sides of the whole disk and the basal half of the arms. These papular patches are generally roundish in shape, and two or three of them may unite and form compound ones. The number of pores in a single patch varies considerably according to its size, from half a dozen to sixty or more being present in each. On the actinal side the papular pores are entirely absent, but there are many not very distinct grooves running parallel to one another and transversely to the ambulacral furrow and reaching out to the papular areas of the sides of the arms (Pl. XVIII, fig. 265). I have counted as many as eleven of these grooves on one side of an arm. They appear to have no definite relation to the adambulacral plates or any other structures visible on the surface. In the interradial line the grooves of the neighbouring arms abutt against each other, and leave a triangular space corresponding to each pair of mouth-plates.

The adambulacral plates are not visible from the surface, but the armature is very distinct and characteristic. As described by LÜTKEN the spines of these plates are arranged in two series, but they also carry pedicellariæ, a fact which was overlooked by that observer. The furrow series consists of 7–9 spines, arranged in a palmate form and connected together by a web-like membrane which is not very distinct and is present only between the basal halves of the spines (Pl. XVIII, fig. 267). The individual spines have somewhat the form of a rectangular prism, with the end shaped like a wedge. Where the spines are more numerous *i.e.* in the proximal half of the arms, the abcentral spines of each plate overlap the adcentral ones of the next following, in such a way that the former lie on the inner side of the latter. The outer series of adambulacral spines is separated from the inner by a distinct groove running through the whole length of the arms; and each plate bears mostly three, sometimes two, and rarely four spines, which are much thicker and shorter than those of the furrow series. When there are three the middle one is usually the largest; and each spine is more or less prismatic in shape, with rounded end. At the adcentral end of each adambulacral plate, nearly in line with the outer series of spines, but a little to the inner side, there is a forcipiform pedicellaria about half as tall as the longest spine of the inner series (Pl. XVII, fig. 263; Pl. XVIII, fig. 268). This was probably mistaken by LÜTKEN for an ordinary spine, because he says that the outer series consists usually of four spines.

The mouth-plates are entirely hidden from view, and bear a series of spines on the furrow margin in the same line with the inner series of adambulacral spines. There are 10–12 spines to each mouth-plate, the two series belonging to a pair meeting each other at an acute angle; the spines at the mouth end are not particularly large. The individual spines are more or less flattened and prismatic, and they generally decrease gradually in size away from the mouth (Pl. XVIII, fig. 266). When this series is removed there may be found a few spines nearly similar in form imbedded in the leathery skin covering the mouth-plates on the actinal side, corresponding to the outer series of adambulacral spines.

The madreporite is plane or concave and is exposed on the surface ; it is irregularly elliptical in outline, and is covered with irregularly radiating fine grooves. It is situated about mid-way between the centre of the disk and the margin (Pl. XVIII, fig. 269).

My alcoholic specimen has r=39–43 mm., R=79–85 mm. Obtained at Kyamu Saki, Okinawa Island. Alcoholic specimen in S. C. ; two dried specimens in I. M.

GYMNASTERIIDÆ.

Gymnasteria carinifera (LAMARCK).

(Pl. XVIII, figs. 270–271.)

According to FISHER [: 08a, p. 90] the oldest valid name for this star-fish is Asterope carinifera auct. MÜLLER et TROSCHEL [MÜLLER, '40, p. 140, April], which should therefore be used by those who would uphold the law of priority at any cost. The next name in point of publication was Asterop-sis carinifera auct. MÜLLER et TROSCHEL ['40a, p. 322, about September], and GRAY's Gymnasteria was published only in December of the same year (fide FISHER). I have however used the name which is by far the best known.

Notwithstanding the doubt expressed by PERRIER on the point, as given below, the first description of this species is generally attributed by authors to LAMARCK, whose Asterias carinifera is identified with the present species. It is the thirteenth species of Asterias in LAMARCK's work ['16, p. 556] and is described as follows :

"Astérie carinifère. *Asterias carinifera*. "A. pentagona, angulis porrectis; margine aculeato; dorso carinis quinque aculeatis muricato.

"Mus. n°.

"Habite......Elle provient du voyage de *Péron* et *le Sueur*. Cette astérie ressemble tellement à la précédente[1] par son aspect, qu'on pourrait présumer qu'elle n'en est qu'une variété. Cependant, au lieu de papilles digitiformes sur ses scutelles marginales, elle offre une série de piquants simples, et sur son dos on voit cinq côtes tranchantes et spinifères."

PERRIER, as will be seen below, thinks that the phrase "dorso carinis quinque aculeatis muricato" does not well fit the present species, but this difficulty is obviated if one assumes that it refers to the animal as a whole and not to each arm.

BLAINVILLE ['34, p. 238] simply mentions this species, referring to LAMARCK. It is also mentioned by MÜLLER and TROSCHEL in their paper of 1840 (*vide supra*).

The generic name *Gymnasteria* is due to GRAY, who mentions two species, which, however, are regarded by subsequent writers as one ['40, p. 278]:

"1. *Gymnasteria spinosa*. Rays triangular, tapering, about one quarter longer than the width of the body, with a dorsal series of conical cylindrical tubercles. Young with a few spines on the margin and back of the arms. Allied to *Porania*.

"Inhab. Panama, fine sand 16 fathoms. H. CUMING Esq.

"2. *Gymnasteria inermis*. Rays rapidly tapering, convex above without any spine.

"Inhab. Panama, fine sand, 10 fathoms. Half the size of the young, spined specimens of the former species."

MÜLLER and TROSCHEL give the following description in the "System" ['42, p. 63]:

"Species 1. *Asteropsis carinifera* Nob.

"*Asterias carinifera* LAM. II, p. 556.

"Fünf Arme. Verhältniss des kleinen zum grossen Radius wie $1 : 2\frac{1}{2}$.

---

1) *Asterias equestris = Hippasteria plana = Astrogonium phrygianum.*

Arme anderthalbmal so lang wie breit. Furchenpapillen in zwei Reihen;
in der innern Reihe fünf auf jeder Platte, von denen die mittlere die
längste; in der äussern eine dicke Papille auf jeder Platte. Die Plättchen
der Bauchseite sind äusserst fein granulirt, alle Zwischenräume häutig und
ohne Granulation. Der Rand ist gekielt und mit einer Reihe von 16 etwas
stärkeren, sonst aber ähnlichen Plättchen besetzt, welche sich in von der
nackten Haut grösstenteils überzogene Stacheln erheben. Neben diesem
Kiele verläuft auf der Bauchseite eine zweite Reihe grösserer Flatten, die
aber nur selten etwas kleinere Stacheln trägt. Die Plättchen der Rückseite
stehen in regelmässigen Reihen, besitzen dieselbe feine Granulation wie auf
der Bauchseite mit nackter Zwischenhaut und Porenfeldern. Die Mitte
des Rückens der Arme ist gekielt. Die Erhebung des Rückens beginnt
erst abgesetzt vom scharfen abgeplatteten Rande der Arme und der Scheibe.
Die gekielte Mitte des Rückens der Arme trägt eine Reihe ähnlicher
Stacheln wie am Rande und grösstenteils von der nackten Haut bedeckt.
Zuweilen sind die Stacheln der mittlern Reihe wenig ausgebildet. Die
Pedicellarien sind sehr lang und zangenartig und stehen bloss auf der
Rückseite nahe dem Rande. Bei trocknen Exemplaren findet man sie in der
Regel bis zur Spitze zurückgezogen.

" Farbe : nach LESUEUR braun, ins Rötlichgelbe.

" Grösse : 4½ Zoll.

" Fundort : Indischer Ocean ; Rothes Meer. Im Museum zu Berlin durch
HEMPRICH und EHRENBERG ; im Museum zu Paris durch PÉRON und LESUEUR ;
im Museum zu Leyden durch SALOMON MÜLLER."

According to PERRIER ['76, p. 99] this species is reported from Mozambique by PETERS ['52, p. 178].

DUJARDIN and HUPÉ describe this species under the name of *Asteropsis
carinifera* as follows ['62, p. 409]

" Espèce pentagone, le plus grand rayon égalant deux fois et demie le
plus petit. Bras une fois et demie aussi longs que larges. Piquants du
sillon ambulacraire en deux rangées : ceux de la rangée interne au nombre
de cinq sur chaque plaque, dont les intermédiaires plus longs ; ceux de la
rangée externe sont plus épais et isolés sur chaque plaque. Les plaques de

la face ventrale sont très-finement granuleuses et séparées par une peau nue. Bord caréné, portant une rangée de seize plaques un peu plus épaisses, prolongées chacune en un piquant, que la peau nue recouvre en grands partie ; à côtés de cette carène, sur la face ventrale, se trouve une deuxième rangée de plaques plus grandes, portant quelquefois de petites épines.

" Les plaques de la face dorsale forment des rangées régulières et sont, comme celles de la face ventrale, finement granuleuses et séparées par des places nues.

" La face dorsale des bras est carénée, et porte au milieu une rangée de piquants analogues à ceux du bord, et en grande partie recouverts par la peau nue, mais quelquefois peu développés. Pedicellaires très-longues et en pince, situées seulement près du bord, à la face ventrale.

" Coloration brune, passant au jaune rougeâtre. Dimension : largeur totale 120 mm.

" Habite la mer des Indes, la mer Rouge (Musée de Paris)."

GRAY's descriptions of 1866 ['66, p. 8] are simple reproductions of his former ones already quoted, with references to GRAY ['40], LAMARCK ['16] and MÜLLER and TROSCHEL ['42] under *Gym. spinosa* and to *Asteropsis ctenacantha* of MÜLLER and TROSCHEL ['42, p. 63] under *Gym. inermis*.

VON MARTENS mentions this species and also describes a new one, which is however regarded by DE LORIOL and SLUITER as a young form of the former. VON MARTENS says ['66, p. 74]:

" 16. *Gymnasterias carinifera* LAM. sp. *Asteropsis c.* MÜLL. TROSCHEL S. 43.

" Amboina, ein ungewöhnlich grosses Exemplar, (Armradius 80, Scheibenradius 30 Mill.), v. ROSENBERG (A.)[1]

" 17. *Gymnasterias liserrata* n. sp.

" Fünf Arme. Verhältniss des Scheibenradius zum Armradius wie 1 : 2. Arme mehr als anderthalbmal so lang wie breit. Furchenpapillen in 2 Reihen, in der inneren 3-4 auf jeder Platte, die eine oder zwei mittleren

---

1) The (A) indicates that the species thus marked occur not only throughout the Indian Archipelagoes but also on the each coast of Africa.

etwas länger, in der äussern eine dicke Papille auf jeder Platte. Die
Plättchen der Bauchseite fein granulirt, meist sich berührend, die seltenen
Zwischenräume zwischen ihnen einfach häutig, sie bilden eine regelmässige
Reihe längs der Armfurche und eine regelmässige Reihe unterer Randplatten; diese beiden Reihen liegen von der halben Länge der Arme an bis
zu deren Spitze unmittelbar aneinander und alle ihre Platten sind stets
viereckig; der Zwischenraum beider Reihen in den Armwinkeln wird von
mehr regellos gestellten ungleich grossen, oft fünfeckigen Platten ausgefüllt.
Die unteren Randplatten sind grösser als die übrigen Platten der Bauchseite, und jede trägt an ihrer äusseren und der Armspitze zugewandten
Ecke einen kurzen Stachel. Sie bilden nicht selbst den Rand, sondern
werden überragt durch die oberen Randplatten, welche eine mehr abgerundete Gestalt haben, ebenfalls fein gekörnt sind und von denen ebenfalls
jede an ihrem äusseren Rande einen Stachel, wie die untere trägt. Die
Stacheln der oberen wie der unteren Randplatten sind dick konisch, ohne
Körnelung, und schief nach der Armspitze zu gerichtet, wie Zähne einer
Säge; die der oberen noch einmal so gross als die der unteren. Obere
und untere Randplatten an jeder Armseite 12–13. Die Plättchen der Rückseite sind ziemlich kreisförmig, alle nahezu gleich gross, etwas kleiner als
die oberen Randplatten, fein gekörnt und lassen regelmässig nackthäutige
Zwischenräume zwischen einander; sie ordnen sich in radiale Reihen, von
denen namentlich eine mittlere auf jedem Armrücken deutlich hervortritt,
aber sich nicht als Kiel über die andere erhebt und so wenig als die anderen Rückenplatten Stacheln trägt. Pedicellarien über den Rücken zerstreut,
häufiger nahe am Rande, auf der Bauchseite keine sichtbar.

"Farbe während des Lebens dunkelroth, unten orange. Armradius
24 Mill.

"Larentuka auf der Insel Flores.

"Nahe verwandt *G. carinifera*, unterscheidet sich durch die Stacheln
der unteren Randplatten, den Mangel der Armkiele u. a. Bei *G. carinifera*
trägt je eine der oberen Randplatten einen grösseren und die nächste einen
kleineren oder gar keinen Stachel; bei *G. biserrata* tritt ein solcher Wech-
sel nur zunächst der Armspitze auf, während die grosse Mehrzahl der oberen

Randplatten gleich grosse Stacheln trägt. Angesichts der grossen Variabilität der Randstacheln bei *Oreaster* würde es nicht sehr zu verwundern sein, wenn Zwischenformen gefunden würden, welche *G. biserrata* eng an *carinifera* anknüpften.

"Da die Gattungsnamen *Gymnasterias* (GRAY schreibt minder gut ohne s *Gymnasteria*) und *Asteropsis* ziemlich gleichzeitig gegeben zu sein scheinen, dürfte der weit mehr bezeichnende dem fast bedeutungslosen vorzuziehen sein. Ausnahmsweise rührt in diesem Falle jener von GRAY, dieser von den deutschen Forschern her."

VON MARTENS also found this species (*carinifera*) in Zamboanga, Island of Mindanao ['67, p. 111].

Gymnasteria spinosa GRAY and *Gymnasteria inermis* GRAY are mentioned by VERRILL from the Bay of Panama ['67, p. 330, 343].

PERRIER gives an account of the pedicellariæ of *Asteropsis* (= *Gymnasteria*), from which the following is reproduced as referring particularly to *carinifera* ['69, p. .92]:

"Dans le genre *Asteropsis*, la variété est bien plus grande que dans les précédents[1] sous le rapport des Pédicellaires. Nous trouvons, en effet, une espèce pourvue de Pédicellaires en pince, une espèce pourvue de Pédicellaires valvulaires, et enfin des espèces complètement dépourvues de ces organes[2]. Il n'y a donc rien de général à dire sur les Pédicellaires des *Asteropsis*, si ce n'est pourtant qu'ils ne nous semblent pas, comme dans les genres qui précédent, logés dans des alvéoles spéciaux pratiqués dans l'épaisseur des pièces calcaires du squelette. Cela est de toute évidence pour l'*Asteropsis carinifera*, qui porte des Pédicellaires en pince distribués tout le long des bords de la région dorsale des bras   Ces Pédicellaires se trouvent au milieu des aires tentaculaires de cette région; ils sont petits, peu allongés, et presentent quelque chose d'analogue à ce que nous avons rencontré chez les *Asteracanthion*.

"Ici aussi le Pédicellaire est composé de trois pièces, une basilaire et

---

1) *Stellaster, Nectria*.
2) PERRIER, in this paper, includes in this genus also *A. vernicina, A. ctenacantha, A. pulvillus* and *A. imbricata*, now referred to other genera.

deux formant la pince. La pièce basilaire présente une cavité légère, dans laquelle sont implantées les deux branches de la pince. Celles-ci sont courtes et irrégulières ; elles sont reliées à la pièce basilaire par des fibres musculaires verticales qui écartent les deux branches l'une de l'autre. A leur base, ces deux branches s'entrecroisent légèrement en se servant mutuellement de point d'appui, et des fibres musculaires horizontales peuvent déterminer leur rapprochement. Il semblerait au premier abord que ces Pédicellaires remarquables fissent exception au type général que nous avons décrit pour les Pédicellaires des Astérides à deux rangées d'ambulacres ; mais il est facile de se convaincre que ces organes s'éloignent plus de ceux des *Asteracanthion* que de ceux des autres Étoiles de mer. En effet, la pièce qui leur sert de support n'est pas, comme dans les *Asteracanthion*, dans le même plan que les branches des pinces et portée par un pédicule mou ; elle se développe au contraire dans la peau elle-même, et dans un plan perpendiculaire à celui des pinces. Elle occupe par conséquent, relativement à celles-ci, la même position que les pièces calcaires empruntées au squelette pour servir de support au Pédicellaire.

"Dans l'*Asteropsis carinifera*, la peau est bourrée de petits spicules calcaires, quelquefois un peu bifurqués, le plus souvent, au contraire, arrondis aux deux bouts."

VERRILL mentions this species under the name of *Gymnasteria spinosa* from Lower California ['71, p. 574] :

" *Gymnasteria spinosa* GRAY[1]).

"A starfish collected at La Paz by Capt. PEDERSEN, seems to be identical with this species, originally obtained at Panama by Mr. H. CUMING. There are three specimens in the collection.

"Form pentagonal, with rather broad, tapering, somewhat depressed, triangular rays. Radii as 1 : 2.2. The skeleton consists of moderately large, rounded and polygonal plates, joined by their edges, so as to leave variously shaped spaces between, with their surface roughened by minute, granule-like prominences and covered with a thin membranous skin, which allows the roughness of the plates to show through it. The median dorsal

---

1) References to previous literature omitted.

plates on each ray are stout, rather rhomboidal, with the angles produced and rounded and the centre tubercular ; they bear a row of eight to twelve, stout, elevated, blunt spines, one to each plate. The sides of the rays near the base are formed by about four series of plates ; in the two intermediate rows rounded ; in the upper and lower ones with lateral prolongations, which articulate with the dorsal and marginal plates in such a way as to leave rather large, transverse, oblong openings between ; toward the end of the rays the plates become more regular and uniform, mostly polygonal, and more closely united, except that there are still larger openings next to the marginal plates, forming a regular series. Marginal plates stout, prominent, projecting laterally, and rounded on the outer side, much broader than high, forming a single row, with the plates placed alternately a little above and below the median line, about 12 to 16 on each side of the ray, each one bearing a stout, elongated, conical spine. Plates of the lower side rounded and subpolygonal, unequal, some of them bearing a very small central tubercle, mostly closely united, so as to leave only small pores between. Each interambulacral plate bears an outer, stout, oblong, blunt spine, compressed or wedge-shaped at the tip, and an inner group of four or five slender ones, of which the lateral are very short and the two middle ones considerably longest, all connected together by a thin web. On each margin of the mouth there is a group of five to eight, rather slender, subequal, obtuse spines, connected together by a web. Near the margin of the disk and rays, above and below, there are many rather large pedicellariæ, oblong or subcylindrical in form, obtuse at the tips. The dried specimens are light red above, yellowish below.

"The largest specimen is 1.50 inches from centre to edge of disk; 2.75 to tip of rays; breadth of rays at base 1 to 1.25; length of largest spines .20 to .22; diameter at base, about .08.

"A smaller one has the radius of disk .68 of an inch; of rays 1.50; length of dorsal and marginal spines .10 to .12; diameter .05 or .06; diameter of upper and lower plates .05 to .10, mostly about .08."

·Further on in the same paper [VERRILL, '71, p. 594] *Gymnasteria spinosa* GRAY is mentioned as occurring in La Paz, Lower California, according to the author's own observation.

PERRIER in his "Révision" gives some valuable informations about the type specimen and the variability of this species ['76, p. 101]:

"Les indications données par LAMARCK relativement à son *Asterias carinifera*, la ressemblance qu'il indique entre elle et l'*Asterias equestris* (*Hippasteria plana*), les 'cinq côtes spinifères' qu'il décrit sur son dos se rapportent assez mal à l'*Asteropsis carinifera* de MÜLLER et TROSCHEL qui n'offre qu'une seule carène médiane spinifère. Ne retrouvant pas dans la collection l'échantillon de LAMARCK, il nous est impossible de vérifier cette synonymie, que tous les auteurs ont du reste acceptée. Quant à la synonymie de GRAY, bien que les individus décrits par l'auteur anglais proviennent de localités très-éloignées (Panama et mer Rouge), nous avons pu nous assurer par l'examen direct des échantillons qu'elle était parfaitement exacte.

"Le Muséum de Paris possède quatre individus desséchés, de la mer Rouge (M. CLOT-BEY, 1850); deux individus également desséchés dans la collection MICHELIN, l'un sans indication de provenance, l'autre des îles Sandwich; trois individus dans l'alcool, des îles Fidji (M. FILHOL, 1875). Le British Museum possède, de cette même espèce, de nombreux individus desséchés, provenant de la mer Rouge, de l'île de France, des îles de Sandwich; et de Panama. Un individu à six bras provient d'Oomaga. Malgré l'étendue de cette aire de répartition, il est à remarquer que, dans ces localités si diverses, on observe exactement les mêmes variations du type. Ces variations sont au nombre de deux principales et elles portent sur la disposition des plaques du squelette. Chez tous les jeunes individus que j'ai pu observer et chez un grand nombre d'adultes, le squelette est formé d'ossicules plus ou moins rapprochés les uns des autres, et qui présentent dans le jeune âge exactement la disposition et la structure que nous avons précédemment décrites chez les jeunes *Anthenea* et chez les *Anthenea* de petite taille. Mais chez d'autres, la disposition du squelette devient tout autre. Les ossicules formant la carène dorsale s'allongent beaucoup, s'imbriquent réciproquement et développent latéralement près de leur extremité supérieure deux courtes apophyses. Sur ces dernières s'appuient les plaques des deux rangées voisines, à droite et à gauche, lesquelles s'allongent aussi beaucoup, de manière à figurer avec la rangée longitudinale comme une série de côtes courtes et larges s'appuyant sur une sorte de colonne vertébrale. Ces côtes sont séparées l'une de l'autre par un intervalle au moins aussi grand que leur largeur. Elles servent, en quelque sorte, d'amorce aux autres parties du squelette formé de pièces étoilées irrégulières, dont la disposition échappe à toute description. Les plaques marginales sont seulement un peu plus robustes que d'ordinaire, et les plaques ventrales étoilées, au lieu d'être irregulièrement arrondies. Entre cette forme et celle où tous les ossicules sont à peu près arrondis, on trouve des intermédiaires, les plaques prennent des formes moins régulières, des pièces accessoires plus ou moins volumineuses se développant entre la rachis médian et les rangées de plaques voisines indiquent les pièces costiformes dont nous avons parlé. Cette circonstance et le fait que les mêmes formes se rencontrent dans toutes les localités où l'une d'elles a été signalée, nous conduisent à ne voir en elles qu'une seule et même forme spécifique.

"En ce qui concerne la *Gymnasteria inermis* de GRAY, l'examen que nous avons fait du type du savant anglais, et sa comparaison avec les jeunes *Gymnasteria carinifera* des autres localités ne nous laisse aucun doute sur leur identité. Voici, du reste, la description de cet échantillon typique :

"Face dorsale plane, ou à peu près, formée de plaques arrondies, sans piquants, granules ou pédicellaires, mais finement ponctuées. Chaque point saillant portant une soie courte et fine visible seulement à la loupe. Plaques marginales dorsales, minces, aplaties, au nombre de vingt-quatre, en carré arrondi sur ces bords, portant sur leur angle externe apical un très-court piquant plat et pointu au sommet. Plaque madréporique petite, arrondie, mais bien visible.—Plaques marginales ventrales carrées, inermes près du sommet de l'arc interbrachial et plus loin portant près de leur angle extérieur apicial un piquant encore plus petit que celui de la plaque marginale dorsale correspondante. Toutes les plaques ventrales arrondies et d'ailleurs semblables aux plaques dorsales. Piquants ambulacraires bisériés, chaque plaque portant dans le sillon trois petits piquants cylindriques divergents, et en arrière, sur sa surface ventrale, un piquant isolé plus gros et conique. La forme est du reste, sauf l'aplatissement du corps, la même que dans les individus de la *Gymnasteria carinifera*. Un individu de cette dernière espèce, provenant de la mer Rouge et deux fois plus grand que le type de la *Gymnasteria inermis* de GRAY, n'en diffère que parce que les cinq plaques apiciales du deuxième cercle dorsal (cercle des dix plaques) portent un petit piquant arrondi, et établit nettement la liaison des deux formes. Il n'est donc pas possible de douter que la *Gymnasteria carinifera* (LINCK. sp.) et la *Gymnasteria inermis* (GRAY) ne sont qu'une même espèce ; et l'on peut s'étonner que les naturalistes américains, mieux placés que personne pour résoudre la question, continuent à distinguer les deux espèces. Cela ne s'explique guère que si quelque autre espèce a été prise en Amérique pour la *Gymnasteria inermis* de GRAY. Peut-être est-ce celle que nous avons décrite plus haut sous le nom de *Dermasterias inermis* et qui a été envoyée au Jardin des Plantes par le muséum de zoologie comparative de Cambridge."

VIGUIER gives a detailed description of the skeletal system of this starfish which he places in the tribe *Goniasterinæ* ['78, p. 202] :

"La *Gymnasteria carinifera* habite les localités fort éloignées les unes des autres : la mer Rouge, l'île de France, les Fidji, les Sandwich, Panama, etc., et présente, toutefois, une remarquable uniformité.

"Tout le squelette est composé de plaques minces, noyées dans une peau assez épaisse, qui les masque à peu près complètement sur les sujets conservés dans l'alcool. Sur les sujets desséchés, au contraire, l'arrangement des plaques se voit assez distinctement. Les bras sont très-marqués, comme on le voit sur la figure 5, pl. xiii, et fortement carénés comme le montre la figure 8. La ligne médiane du bras est occupée par une rangée d'ossicules, irrégulièrement circulaire vers la pointe du bras, et qui s'allongent un peu en pointe vers sa base. Ces ossicules sont assez régulièrement imbriqués, et portent, de deux en deux, un petit tubercule perforé pour l'insertion d'un piquant ; mais tout ceci est sujet à des variations assez grandes.

"De chacune des pièces de la série médiane partent des lignes d'ossicules à peu près perpendiculaires à la direction du bras. Ces ossicules sont elliptiques et légèrement déprimés à leur centre. Ils forment, sur les côtés de la ligne médiane, trois séries parallèles qui arrivent jusqu'à l'extrémité du bras, où toutes les séries se confondent en une accumulation de pièces sans disposition ni imbrication régulière. La série latérale la plus interne est reliée à la série médiane par de petites pièces connectives très-courtes.

"Notre sujet étant assez déformé, la disposition des séries dorsales présente sur notre dessin une assez grande irrégularité ; leur imbrication dans le sens latéral se voit bien sur la figure 8. Dans les angles interbrachiaux, qui sont très-arrondis, on voit des séries supplémentaires d'ossicules imbriqués, à surface libre à peu près circulaire, venir combler l'espace laissé libre entre les séries dorsales.

"Au centre du dos se voit un pentagone dont les sommets sont formés par les premières pièces des séries médianes des bras, et dans l'intérieur duquel l'arrangement des ossicules ne présente plus assez de régularité pour se prêter à une description.

"C'est sur le bord de ce pentagone central que se trouve la plaque madréporique m, qui est petite, ovalaire, assez épaisse, et marquée de sillons divergeants extrêmement fins. Au milieu du pentagone, un peu à gauche du centre, se voit l'anus, an.

"Les plaques marginales dorsales sont assez régulièrement imbriquées, mais en sens inverse des ossicules de la série médiane du bras. On trouve, en effet, au sommet de l'angle interbrachial, une plaque impaire qui recouvre latéralement ses deux voisines. A partir de là, l'imbrication continue dans ce sens, chaque plaque recouvrant celle qui lui est voisine du côté de la pointe du bras ; mais cet arrangement devient irrégulier en approchant de l'extrémité. A ce niveau, les plaques marginales deviennent très-petites, ne se differencient plus des autres pièces dorsales, et se terminent à une plaque ocellaire de fort petites dimensions.

"Les plaques marginales inférieures, qui sont toujours débordées par les supérieures, sont assez irrégulièrement circulaires, à contour sinueux. Elles sont assez exactement juxtaposées, sans imbrication bien apparente, et en même nombre que les plaques margino-dorsales.

"Tout à fait vers l'extrémité du bras, elles viennent au contact des séries adambulacraires, mais dans presque toute la longueur du bras il s'interpose entre elles une ligne de petites plaques irrégulièrement circulaires à imbrication très-variable. Près de la base du bras et dans les angles interbrachiaux, il vient s'en intercaler d'autres encore, d'un diamètre un peu plus grand, de façon que tout le triangle compris entre les deux séries adambulacraires et la série marginale est à peu près exactement rempli par des pièces imbriquées de la base vers le sommet du triangle.

"Les pièces adambulacraires sont petites, quadrangulaire, peu serrées les uns contre les autres. Les pièces ambulacraires sont assez hautes, mais n'atteignent point cependant le squelette dorsal. La première porte une apophyse en aile très-développée, et la dent, qu'elle supporte est forte et pointue comme chez toutes les *Goniasterianœ*. L'odontophore présente, ici encore, la forme typique de la tribu, et ses apophyses sont bien développées.

"Le système interbrachial[1] est composé de deux lignes d'ossicules, et forme en conséquence un arc plus resistant que chez le *Pentaceros muricatus* ou la *Culcita* ; toutefois la disposition est bien évidemment la même.

"Si l'on joint à ces caractères, tirés de la constitution de la bouche, la présence de spicules dans les ambulacres, on verra que ce type doit bien se ranger dans les *Goniasterinœ*. Toutefois, outre les autres particularités que nous a présentées l'étude de son squelette, il faut dire que la disposition des spicules dans les ambulacres est assez singulière, et qu'ils semblent plutôt garnir l'extrémité du tube que former une couronne plate à la ventouse. En outre, les pédicellaires présentent une disposition tout à fait remarquable, etant ici composés de trois pièces, comme chez les Astéries ambulacraires. Il faut en lire la description détaillée dans le mémoire de M. PERRIER sur les pédicellaires.

"Les *Gymnasteria* présentent donc des particularités assez singulières ; mais leur place doit bien être ici, et nous allons voir, en étudiant les *Porania*, quelles différences séparent ces deux genres autrefois reunis sous un même nom."

BELL mentions this species from the collection of the "Alert" as

---

1) Interbrachial septum.

occurring in Darros Island [BELL, '84a, p. 510]. STUDER mentions it from the collection of the "Gazelle" ['84, p. 41]: "Zahlreich von Mauritius. Ein Exemplar fand sich in 25 Faden Tiefe vor dem östlichen Riff der Insel auf Corallinengrund."

DE LORIOL gives some interesting informations on the young stages of this starfish with figures of a series of small specimens ['85, p. 67]:

"L'espèce n'est pas rare à l'île Maurice, dans les plus petit individus envoyés R=13 mm., dans les plus grands, R=135 mm.

"L'envoi de nombreux individus jeunes permet de suivre le développement de l'espèce, qui est intéressant. Au diamètre total de 25 mm. le corps est moins étoilé, $R=1\frac{9}{7}$ r, au lieu de $R=3\frac{3}{8}$ r, dans les grands exemplaires. Toute la face dorsale est formée de plaques octagones, formant des séries très regulières, presque tout à fait contiguës, avec de petits pores aux angles, couvertes d'une granulation extrêmement fine. Le disque est très plats, les bras larges, courts et coniques, nullement carénés sur leur ligne médiane formée par une série de plaques entièrement semblables aux autres et ne portant aucun piquant. Quatre piquants très petits protègent l'orifice anal central. Les plaques ventrales sont contiguës, bien moins régulières de forme que les dorsales, pentagones, hexagones ou irrégulières et inégales. Plaques marginales, dorsales et ventrales, concourant également à former le bord, presque deux fois aussi grandes que les plaques dorsales, carrées, egales dans les deux rangées ; chacune d'entre elles porte, sur le bord externe, une série de petits piquants qui sont plus nombreux et plus petits sur les plaques ventrales et dont un ou deux, surtout dans la série dorsale, sont plus longs que les autres. Au diamètre total de 35 mm., l'aspect est encore identique, seulement les plaques de la face dorsale commencent un peu à s'écarter, les bras ne sont nullement carénés au milieu, mais un piquant assez saillant se montre sur la première plaque de la série médiane dans trois d'entre eux ; sur les plaques marginales dorsales le piquant qui prédominait s'allonge beaucoup, au détriment des autres, et ceux des plaques marginales ventrales diminuent de nombre. Au diamètre de 52 mm. le corps est bien plus étoilé $R=2\frac{7}{11}$ r, les plaques dorsales sont un peu écartées, trois ou quatre plaques de la série médiane portent un piquant, sur chaque bras, mais elles sont encore en tout semblables aux autres ; chaque plaque marginale dorsale ne porte plus qu'un piquant unique déjà assez long, mais leurs dimensions relatives ont diminué, les plaques marginales ventrales ne contribuent presque plus à la formation du bord, mais elles portent encore de petits piquants, dont l'un situé à l'angle externe de chaque plaque est assez long, surtout vers l'extrémité des bras. Je n'ai pas, malheureusement, les intermédiaires conduisant à un individu dans lequel R=60 mm. ; à cette dimension les plaques octagones de la face dorsale sont encore très distinctes et assez régulières, mais écartées, celles qui avoisinent la série médiane tendent déjà à s'allonger, celles de la série médiane elle-même, pronfondément modifiées par la formation de la carène, sont devenues étroites, et s'imbriquent, plusieurs portant de forts piquants ; les plaques marginales dorsales forment le bord et chacune est armée d'un long piquant, par contre les ventrales sont tout à fait relégnées à la face inférieure, quoique encore bien distinctes, mais sans traces de piquants, la face ventrale est encore très régulièrement parquetée.

"Le *Gymnasteria biserrata*, de M. DE MARTENS, me paraît correspondre à une de ces phases de développement du *Gym. carinifera*. Les exemplaires dans lesquels les bras sont relativement courts, ressemblent assez à l'*Hippasteria plana*, et LAMARCK avait probablement un de ces exemplaires en vue lorsqu'il comparait à cette dernière espèce son *Asterias carinifera*, et lorsqu'il dit, en le décrivant, *dorso carinis quinque aculeatis muricato*, il entend évidemment par là les cinq carènes formées par la carène médiane de *chacun des bras* et il ne veut pas dire qu'il y a cinq côtes spinifères *sur chaque bras* comme semble le comprendre M. PERRIER (loc. cit) ; je ne vois rien qui puisse faire douter que l'*Asterias carinifera* de LAMARCK, dont le type est perdu, ne soit pas exactement cette espèce. D'après les identifications de M. PERRIER au British Museum, le *Gymn. spinosa*, et le *Gymn. inermis* de GRAY doivent être réunis au *Gymnasteria carinifera*."

CUÉNOT, in his work of 1887, makes some remarks on the pedicellariæ of this species, which are of direct systematic importance. He says ['87, p. 18]: "La *Gymnasteria carinifera*, espèce voisine de la Culcite et du

Pentaceros, va nous fournir un exemple analogue[1]; j'en ai étudié deux échantillons conservés dans l'alcool, provenant de la mer Rouge (env. d' OBOCK). M. PERRIER décrit chez cette espèce des pédicellaires assez singuliers, situés sur le bord inférieur des bras, parmi les pores des branchies lymphatiques. Rien de pareil n'existe sur mes échantillons : les bras ne portent aucun pédicellaire ; mais dans la rainure ambulacraire, attachés aux pièces qui la bordent, se voient de nombreux pédicellaires didactyles (pl. I, fig. 14). La figure remplacera avec avantage une longue description ; les branches dentées s'appuient l'une sur l'autre sur une partie de leur hauteur ; par la base, elles se rejoignent également. Un muscle transversal (a), détermine le rapprochement des valves ; des muscles externes (b), s'attachant sur le calcaire qui sert de support, président à l'écartement ; enfin un fort faisceau des fibrilles conjonctives (f) rattache le pédicellaire à l'échancrure dans laquelle il est enchâssé.

"M. PERRIER parle également de spicules qui existeraient dans le derme ; il n'y en a certainement pas sur mes échantillons, qui sont pourtant bien des *Gymnasteria carinifera*. Je considère toutes ces différences comme simplement individuelles."

DÖDERLEIN ['88, p. 828] simply mentions this species from Ceylon.

SLUITER mentions this species from Batavia ['89, p. 307] :

" *G. carinifera* (LAM.). PERRIER, 'Révision des Stellérides,' p. 101. P. DE LORIOL, 'Catalogue raisonné,' p. 67. Drei Exemplare (No. 249) aus der Bai von Batavia. Die Tiere sind auf den hiesigen Riffen nicht selten, und leben zwischen den lebendigen und toten Korallenstücken. Im Aquarium erhalten sie sich längere Zeit ganz gut. Sie sind für Seesterne sehr schnell in ihren Bewegungen, da sie 14 c. M. in der Minute zurücklegen können. Die Farbe ist am Rücken grünlich grau, der Bauch ist etwas heller, die Ambulacralfurchen orangerot.

"Dass die *G. biserrata* von v. MARTENS nur ein Jugendzustand der *G. carinifera* sei, wie LORIOL vermutet, scheint mir nicht wahrscheinlich, da das von MARTENS beschriebene Tier von Flores am Rücken dunkelrot, am Bauche orange war, eine Farbe, welche weder bei den jungen noch bei

---

1) Variability of pedicellariæ.

den alten Exemplaren von *G. carinifera* jemals vorkommt, wenigstens nicht bei den zahlreichen, von mir hier gesehenen Exemplaren."

It may be remarked that the colour differences are frequently of no specific importance in the starfish. We may recall in this connection the cases of *Asterina pectinifera* and *Asterias rollestoni*.

This species was also represented in the Challenger collection and is referred to as follows [SLADEN, '89, p. 357], figures of a young specimen being given :

" *Locality.*—Off Kandavu, Fiji Islands. On the Reefs.

" *Remarks.*—Two adult examples (presenting the well-known characters of this form), and one young one, were collected by the Challenger at the above-named locality. I have given drawings of the young specimen (see Pl. LII. figs. 5–8), in order to show the presence of spinelets on the infero-marginal plates, and their absence along the median line of the ray : a state of things exactly the reverse of what occurs in the adult condition. This circumstance leads me to believe that the form described by VON MARTENS as specifically distinct under the name of *Gymnasteria biserrata*, is only the young stage of *Gymnasteria carinifera*, or at most only a variety in which the juvenile characters are maintained throughout life.

" Valuable observations on the growth stages of this species, as presented by a series of examples from Mauritius, have been made by DE LORIOL, who has also given figures, with which it will be interesting to compare those on Pl. LII. of the young example from Kandavu.

" It may be remarked that some variation occurs in the spinulation of the infero-marginal plates of adult specimens from different localities. Normally in fully grown examples these plates bear no trace of spines, but in large specimens from the Red Sea (which are amongst MÜLLER and TROSCHEL's types in the Berlin Collection), traces of a few incipient spines are present on the outer part of the ray. Also in examples from Mozambique and from Timor the occasional spinulation of the infero-marginal plates is more pronounced. I do not, however, look upon these as anything more than local, and perhaps even individual variations."

LEIPOLDT describes this species from the Vettor-Pisani Expedition ['95, p. 649], with figures of the adambulacral armature and dermal spicules:

"Zwei Exemplare von Massauah aus einer Tiefe von 18 m (24. April 1884).

"Die Maasse der Exemplare sind R=78 mm, r=26 mm, also R=3 r und R=57 mm, r=21 mm, also R=2.8 r. Das grössere Exemplar trägt die bei den verschiedenen Autoren als charakteristisch für die Art angegebenen Kennzeichen, nur fehlen sowohl ihm als auch dem kleineren Exemplare die Stacheln auf den unteren Randstücken. Auch finden sich bei ihm die Pedicellarien in weiterer Verbreitung, als dies gewöhnlich angegeben wird. MÜLLER und TROSCHEL (1842 p. 63) bemerken, dass Pedicellarien bei 'Asteropsis carinifera' nur auf der 'Rückseite nahe dem Rande' stünden. Nach PERRIER (1869 p. 285) sind die Pedicellarien in ihrem Vorkommen ebenfalls auf die Ränder der abaktinalen Seite beschränkt und er giebt richtig an, dass sie sich immer in der Mitte der Porenfeldern befinden. CUÉNOT (Contrib. à l'étude anatomique des Astérides. Arch. zool. exp. 2. série, T. V. [Suppl.] 1888 p. 18 u. f. Taf. I, Fig. 14) konnte dagegen bei seinen Exemplaren die Pedicellarien an den von PERRIER angegebenen Stellen nicht finden, er entdeckte sie vielmehr nur 'dans la rainure ambulacraire, attachés aux pièces qui la bordent.' Bei unserem grösseren Exemplare kamen sie dagegen nicht nur an den von MÜLLER und TROSCHEL, PERRIER und CUÉNOT angegebenen Stellen, sondern auch auf der ganzen abaktinalen Seite vor und zwar, wie diess PERRIER angiebt, ausschliesslich auf den Porenfeldern. Auf der Bauchseite stehen sie immer isolirt an dem adoralen Rande der Adambulacralplatten und zwar nach innen von der Basis des äusseren Stachels der Adambulacralbewaffnung, so dass sie sich also in dem Zwischenraume zwischen dem letzteren und der inneren Längsreihe von Adambulacralstacheln befinden (s. Fig. a).

"Das kleinere Exemplar, weches in seinem äusseren Habitus der oben citirten SLADEN'schen Abbildung der Jugendform von G. carinifera sehr ähnlich ist und nur schon zahlreichere Poren in den einzelnen Porenfeldern und einige Stacheln mehr in der Medianlinie der abaktinalen Seite der Arme besitzt, liess mich auf der abaktinalen Seite die Pedicellarien nur in einer den Angaben MÜLLER und TROSCHEL's, sowie PERRIER's entsprechenden

Verbreitung auf den am Rande der Seite befindlichen Porenfeldern erkennen. Auf der aktinalen Seite standen Pedicellarien auch auf den Adambulacral-platten, aber nur an einzelnen Stellen.

"Die von PERRIER für *G. carinifera* angegebene Zusammensetzung der Pedicellarien aus drei Stücken, nämlich zwei Zangen und einem Basalstück, fand sich bei den Pedicellarien unserer beiden Exemplare nicht. Ich kann vielmehr den Angaben CUÉNOT's zustimmen, welcher die Pedicellarien der *G. carinifera* als 'didactyles' bezeichnet und abbildet. Nur stimmt die Gestalt der Pedicellarien, wenigstens derjenigen des grösseren Exemplares, in so fern nicht ganz mit der CUÉNOT'schen Figur überein, als sie im Verhältnis zu ihrer Breite viel länger[1] waren, als jene dies zeigt. Besser stimmten die Pedicellarien des kleineren Exemplars mit jener Abbildung überein, und daher möchte ich vermuthen, dass auch CUÉNOT kleinere Exemplare vor sich hatte. CUÉNOT erwähnt auch noch, dass er die von PERRIER[2] beschriebenen Kalkspicula, welche in der Haut der *G. carinifera* vorkommen sollen, bei seinen Exemplaren nicht habe entdecken können. Es gelang mir, sie bei den beiden Exemplaren und zwar sowohl in der Rücken- als auch der Bauchhaut zu finden. Sie treten bei dem kleinen Exemplare dicht neben einander und in mehreren Schichten über einander liegend und dabei zahlreicher auf, als bei dem grösseren, wo sie in ziemlich weiten Zwischenräumen von einander liegen. Bei dem letzteren Exemplare nehmen die Spicula die Form kleiner, in Bildung begriffener oder rudimentär entwickelter Stachelchen an (Fig. *b*); bei dem kleineren Exemplare haben sie hingegen meist die Form eines einfachen, bald etwas kürzeren, bald etwas längeren Stäbchens, das sich in letzterem Falle an beiden Enden etwas verjüngt (Fig. *c*). Dadurch, dass sich dann etwa in der Mitte der Länge eines solchen Stäbchens ein kleiner Seitenast ausbildet, scheint mir der Uebergang zu den etwas komplicirteren Formen gegeben zu sein, wie sie auf derselben Figur abgebildet sind."

SLUITER in his paper of 1895 refers to this species as follows ['95, p. 59] :

---

1) "Die Länge der Pedicellarien des grösseren Exemplars konnte bis zu 2 mm steigen"
2) *Vide supra*, p. 616.

" *Gymnasterias carinifera* (LAM.) v. MARTENS. Drei getrocknete Exemplare (zwei jüngere und ein erwachsenes) von den Molukken (V. D. HUCHT), ein junges Exemplar in Alkohol von Ambon (BLEEKER) und ein erwachsenes Exemplar in Alkohol von der Bai von Batavia (SLUITER)."

DÖDERLEIN mentions it from Amboina ['96, p. 316], and refers to the occurrence of the " Krystallkörper " in this species ['98, p. 493].

LUDWIG ['99, p. 540] mentions it from the Zanzibar region, although it was not represented in the collection of VOELTZKOW which he studied. He simply writes, " *Gymnasteria carinifera* (LAMARCK), von Querimba durch PETERS (1852), von den Amiranten durch BELL (1884)."

This species is mentioned by BELL from WILLEY's collection [BELL, '99, p. 137] :

" *Gymnasteria carinifera*.

[References to LAMARCK, GRAY and SLADEN omitted.]

" Loc. Lifu.

" This is a species which appears to extend from the Red Sea to Panama, and is often found in collections from the Pacific."

Under *Gymnasteria valvulata* PERRIER, CLARK says [:02, p. 524], " A species based on a young individual which may be a young *Nidorellia* or possibly an immature *Gymnasterias carinifera*, a well-known Panama species."

This species is also described by LUDWIG from the tropical Pacific [:05, p. 156] :

" Ein junges Exemplar von der Küste von Papeete, Gesellschafts-Inseln, 9. November, 1889. Ein zweites noch jüngeres Exemplar ebendaher von dem Riff, 13. November, 1899.

" Maasse : a. R=32 mm., r=12 mm., r=1 : 2.67. b. R=20 mm., r=9 mm., r=R=1 : 2.22.

" Das grössere Exemplar (Taf. XII, Fig. 61, 62) steht in einem Maassen zwischen den von LORIOL l.c. Fig. 9 und 10 abgebildeten und bietet in der Bestachelung seiner Randplatten bereits das Verhalten des erwachsenen Tieres dar. Das kleinere hat ungefähr die Grösse des von LORIOL in seiner Fig. 8 abgebildeten ; es besitzt auf dem Scheitel noch gar keine Stacheln (mit Ausnahme der winzigen, die den After umstellen) und seine unteren Randplatten sind fast alle mit einem (selten zwei) kleinen Stachel ausgerüstet, entsprechend der SLADEN'schen Fig. 6.

"Der Ansicht LORIOL's und SLADEN's, dass *G. biserrata* v. MARTENS nur eine Jugendform der *G. carinifera* ist, kann ich mich nur anschliessen.

"An der Madreporenplatte liess sich feststellen, dass sie ein selbständige Skelettplatte ist, die nach aussen von der betreffenden primären Interradialplatte liegt. *Gymnasteria* gehört demnach wie wahrscheinlich alle Gymnasteriiden zu den euplacoten Seesternen.

"In Betreff des Baues der Pedicellarien, die bei dem vorliegenden grösseren Exemplare nur zwischen den inneren und äusseren Furchenstacheln vorkommen, stimmen meine Beobachtungen mit denen von LEIPOLDT überein."

FISHER simply mentions this species [:06, p. 1087] remarking, " This species is recorded from the Sandwich Islands by PERRIER (Révision des Stellérides, p. 286), the specimens being in the British Museum. SLADEN likewise records it in the report on the starfishes collected by the *Challenger* expedition (p. 357), very likely from the same specimens. No examples were secured by the *Albatross* expedition." It must be remarked that the specimens which PERRIER described from the British Museum were the original types of GRAY, while those that SLADEN described were collected by the Challenger. Needless to say that SLADEN must have had the original examples of GRAY for comparison (*cf. supra*, p. 626).

CLARK [:08, p. 281] mentions one specimen from Sorong, New Guinea, " 130 mm. in diameter, yellowish brown (dried). BARBOUR collection."

I have examined two specimens. One has R=30 mm., r= 12 mm., R : r=2.5 ; the other has R=64 mm., r=23 mm., and R : r=2.8. The smaller specimen is from Naha, Okinawa Island. For the larger specimen the locality is not recorded, but there is no reasonable doubt that it is somewhere from the Ryûkyû Islands.

The body is regularly star-shaped, the disk comparatively large and the arms are very distinct. The whole surface of the body is covered with a humid leathery membrane, but the plates are very distinct. The granulations of the plates mentioned by previous authors can be seen only indistinctly in alcoholic specimens, but in dried specimens they ought to come out more clearly. Of the two specimens in my hands, both of which are in alcohol, the smaller one is in a better state of preservation in all respects, so that it will be described first in detail.

On the dorsal side, the anus is very distinct, subcentral in position and surrounded by some nine or ten small granules, which are themselves surrounded by five plates, one of which bears in this specimen an indistinct tubercle. From this circle of perianal plates there radiates a series of circular carinal plates to the tip of each arm, being terminated by the ocular plate which bears two short spines. The carinal plates are smaller than those of the adjacent rows on either side, and some of them bear each a blunt conical spine. The number of carinal plates is some 20, and the carinal spines are 4 or 5 for each arm. On either side of the carinal plates there is a row of similar but slightly larger plates reaching to the tip of the arm, and outside it another row of similar plates terminating at a short distance from the arm tip, followed by a third row extending about half way out into the arm. Around the anus the arrangement of the plates is more or less irregular.

The spaces between the dorsal plates are occupied by a thick humid membrane, which in alcoholic specimens is more elevated than the plates, and bear the papular pores which are very small but distinct and form groups of 2–5, all over the dorsal surface.

The superomarginal plates occupy the very margin of the

body, are very stout and more or less triangular in shape near the interradial line but square outwards. Each plate bears on the outer border a stout, short conical spine projecting obliquely outwards, so that the sides of the arms are coarsely serrated.

On the actinal side, the comparatively large inferomarginal plates are most conspicuous. They are nearly square in outline and the last few plates of the arms bear each a short but distinct spine at the disto-external corner of the plate. The inferomarginal plates are throughout coincident with the superomarginals. Just external to the adambulacral plates there is a series of square or rectangular plates much smaller than (about one-fourth of) the inferomarginals and extending to the tip of the arms. The triangular space between this row and the inferomarginals is occupied by some roundish or square plates, which are generally speaking arranged in rows parallel to the first mentioned series, and extending into the arm only about one-third of its length.

The adambulacral plates bear each a furrow series of three or four, rarely five, comparatively thick, short spines arranged in a palmate form, followed by a single stout spine on the actinal face. Previous descriptions lead one to infer that there must be considerable variation in the number and position of the pedicellariæ, and in fact, in the specimen now under consideration these are present only on a few plates in the whole body, and are borne at the adcentral end of the adambulacral plate, between the furrow series and the actinal spines, which latter form a regular series along the ambulacral furrow.

The mouth-plates bear on the furrow border each 7–8 spines similar to the furrow spines of the adambulacral plates, and the one at the mouth end is usually larger than the rest. On the actinal face of each mouth-plate there may be one or two short, thick spines forming the continuation of the outer spines of the adambulacral plates.

The nearly circular madreporic plate is very distinct and is covered with indistinct fine grooves.

In the larger specimen (Pl. XVIII, fig. 270, 271) the carinal spines are very large and distinct, but the plates themselves are less distinct than in the smaller specimen, the number of carinal spines varying between seven and nine for each arm; and in addition there are two similar spines close to the anus. Of the lateral series of plates there are four on either side in each arm, two of which extend to the tip of the arm, the third about one-half and the last about one-fourth as far. The madreporic plate is very conspicuous and elliptical in outline. The dorsal plates are relatively small, the membranous interspaces larger than in the smaller specimen, and the papular pores are much more numerous, there being as many as twenty in the larger groups. The superomarginal spines are longer and more finger-shaped than in the smaller specimen.

On the ventral side there is nothing noteworthy, except that none of the inferomarginals bear spines. The pedicellariæ are very few if not entirely absent.

One small specimen in S. C.; one larger specimen in F. B.

ASTERINIDÆ

Asterina pectinifera (MÜLLER & TROSCHEL).

(Pl. XVIII, figs. 272-273; Pl. XIX, fig. 274.)

This species was described for the first time by MÜLLER and TROSCHEL under the name of *Asteriscus*[1] *pectinifer* ['42, p. 40]:

"*Asteriscus pectinifer* Nob. nov. sp.

"Körper fast pentagonal mit ausgerundeten Seiten. Verhältniss des kleinen zum grossen Radius wie 1:1½. Bauchseite platt, Rückenseite wenig gewölbt. Die Furchenpapillen in einer Reihe, vier bis fünf auf einer Platte. Neben ihnen verläuft eine Reihe Flatten, deren jede drei bis vier platte breite Stachelchen trägt, von denen der dem Ende der Arme zunächststehende der grösste ist, von wo aus sie abnehmen. Uebrigens ist die Bauchfläche mit Flatten bedeckt, die in der Nähe des Centrums sehr gross sind, nach dem Rande zu sehr klein werden. Jede trägt einen Kamm von vier bis sieben kurzen, stumpfen, platten Stachelchen, deren Spitzen nur von einander getrennt sind. Der Rücken ist ganz mit kleinen granulaartigen Körperchen (Stachelchen) besetzt. Zwischen ihnen erheben sich aber quer ovale Haufen von grösseren (12-15) Granula, die sehr regelmässige Reihen auf dem Rücken der Arme und parallel mit ihnen bilden. Diese Haufen ordnen sich auch regelmässig um das Centrum des Rückens.

"Grösse: 5 Zoll.

"Fundort: Japan. Im Museum zu Leyden durch v. SIEBOLD."

---

1) According to MÜLLER and TROSCHEL ['42, p. 39] "*Asteriscus*" used by LUIDIUS is the oldest name (1703) for a starfish of this genus. On this point PERRIER ['76, p. 214] has the following remarks: "Le nom d'*Astericus* que MÜLLER et TROSCHEL font remonter à LUIDIUS était pour cet auteur un simple nom spécifique, que l'on trouve remplacé dans le *Gazophylacium* de PETIVER par celui de *Stellula*, se rapportant à la figure 8 de la planche XVI. Il est donc impossible de l'admettre comme dénomination générique dans la nomenclature actuelle, ainsi que nous l'avons fait pour les noms de LINCK qui se rattachent à la nomenclature binaire. La premier nom générique des Etoiles de mer qui nous occupent est celui de NARDO, adopté en 1835 par AGASSIZ, en 1840 par GRAY et qui prime par consequent celui de MÜLLER et TROSCHEL qui ne date que de 1842. On doit donc nommer *Asterina* tous les Stellérides nommés *Asteriscus* d'apres MÜLLER et TROSCHEL, sauf, bien entendu, les espèces qui doivent en être distinguées pour former le genre *Palmipes*."

DUJARDIN and HUPÉ describe this species in almost exactly the same words as MÜLLER and TROSCHEL ['62, p. 374].

VON MARTENS refers to it in his paper on Japanese starfishes ['65, p. 352]:

"*Asterina pectinifera* MÜLL. u. TROSCHEL (*Asteriscus*) S. 40.

"Farbe während des Lebens oben schwärzlich grün mit grossen, blutrothen Flecken, unten pomeranzengelb.

"Yokohama nicht selten, ganz oberflächlich an Steinen, wie *A. gibbosa* PENN. (*verruculata* RETZ.) in Neapel. In den ersten Tagen meines Aufenthalts daselbst fand ich einmal mehrere dieser Seesterne unter dem Küchenabfall mit Schalenfragmenten von Krabben neben einem Bauernhause, aber da sie unversehrt waren, konnte ich daraus nicht schliessen, dass sie als Speise dienen; vielleicht wurde der Inhalt eines Grundnetzes erst zu Hause von den Leuten sortirt."

It need not be remarked that this or any other starfish is not eaten in this country.

GRAY ['66, p. 15] mentions *Asteriscus pectinifer* MÜLLER & TROSCHEL under *Palmipes Stokesii*, but there is nothing in the context to lead us to infer that he regarded it as a synonym of that species.

PERRIER mentions this species under the name of *Asterina pectinifer* and refers to it as follows ['76, p. 234]:

"Il existe au British Museum deux échantillons de cette belle espèce qui appartient très-nettement au groupe de *Patiria* de GRAY et présente les plus grandes affinités, tant de forme que d'ornementation, avec la *Patiria coccinea* GRAY, du cap de Bonne-Espérance. Elle en diffère cependant par la plus grande finesse et le plus grand nombre des piquants qui forment le peigne de ses plaques ventrales. Ce nombre varie de sept à quatre, le premier étant plus fréquent au voisinage de la bouche. Les nombres cinq et six sont les plus ordinaires.

'L'un de ces échantillons desséchés provient de Hakodadi (Japon), l'autre de Chee-foo; ce dernier, qui a près de 1 décimètre de diamètre, est conservé dans l'alcool et a été donné par M. SWINHOE."

SLADEN mentions this species from Yokohama (5–25 fathoms) in his Challenger report ['89, p. 393].

IVES ['91, p. 212] also mentions it from Japan and adds the following note :

" Nine specimens of this species were sent by Mr. STEARNS, indicating that it is found in abundance. It agrees in general character with *Asterina miniata*, but differs from that species in the more heap-like arrangement of the spinelets of the paxillæ, in the absence of a well marked series of paxillæ along the middle line of the dorsal surface of each arm, and in the colour."

MEISSNER mentions this species from Yokohama as follows ['92, p. 187] : " 4+2 Exempl.—Yokohama, 10. VII. 84 und 3. X. 84. (No. 3335.) "

SLUITER mentions this species in his paper on the Asteroid collection of the Museum of Amsterdam ['95, p. 90] : " 52. *Asterina pectinifera* M. Tr. Ein Exemplar in Alkohol von Japan (v. SIEBOLD, Reichs Museum)."

DÖDERLEIN refers to this species as follows [:02, p. 330] :

" Dieser schöne Seestern ist sehr häufig bei Japan. Ich erhielt ihn in grosser Menge aus der Tokio- und Sagamibai, auch bei Kochi (Shikoku), und beobachtete grosse Herden von dieser Art in einer Tiefe von etwa 5 m auf sandigem und steinigem Boden in der Bucht von Miyadzu (Provinz Tango). 4armige, sowie 6–8armige Exemplare sind bei dieser Art nicht sehr selten. Sie ist im Leben von dunkelblauer Farbe mit zinnoberrothen Flecken, die Unterseite ist orange. Die grössten Exemplare erreichen gegen 90 mm Durchmesser."

CLARK mentions it from Japan [:08, p. 282] : " 2 specimens, Misaki, Sagami Bay, Japan.—3 specimens, Tokyo, Japan. OWSTON collection.

" These specimens are 68–90 mm. in diameter, and the colour in alcohol is a more or less indistinct orange-red, which becomes paler on drying."

In his description of *Asterina miniata* (BRANDT) FISHER contrasts it with the present species as follows [:11, p. 257] :

" This species [*A. miniata*] is closely related to *A. pectinifera* (MÜLLER and TROSCHEL) from Japan. The general type of armature is strikingly alike, and the differences are mainly in details of the skeleton and in colour.

A. pectinifera seems to vary more widely than does *miniata*, judging by the specimens in the Stanford University collection. The principal differences are as follows:

" *miniata*

" General form stellate with broad rays and curved interbrachial arcs ; R=1.7 to 2 r.

" Papular areas large, with about fifteen to twenty pores on proximal radial regions.

" Primary abactinal plates rather well spaced with fifteen to twenty-five secondary plates on the intervening meshes of proximal radial region.

" Medioradial (carinal) series of abactinal plates and armature usually fairly regular and conspicuous proximally, the crescents being much larger than adradial series.

" Primary groups of abactinal granules of a wide crescent form, except toward margins, and with many granules (about forty on proximal carinals).

" Spinelets of actinal combs longer than base line of a comb of four spinelets (much longer than one of three) ; usually three or four spinelets to a comb.

" Serrate scars on actinal plates larger (difference apparent on comparison).

" *pectinifera*

" General form stellato-pentagonal to pentagonal, with nearly straight sides (Mororan, Hokkaido) ; R=1.4 r to 1.5 r.

" Papular areas small ; less than fifteen papulæ on proximal radial regions.

" Primary abactinal plates more closely placed with fewer (five to fifteen) and relatively larger secondary plates on the papular areas of proximal radial region.

" Carinal series of abactinal plates and their granules not conspicuously larger than adradials, and not forming a noticeable series (viewed internally they do, however).

" The primary groups of abactinal granules only exceptionally of the wide crescent form (then disk is pentagonal) but generally more elliptical, or bowed elliptical, with fewer granules (about twelve to twenty on proximal carinals).

" Spinelets of actinal combs not longer than base line of comb of four. Difference in size of actinal spinelets in the two species more evident on comparison of specimens. Usually five or four,[1] rarely three spinelets to a comb.

" Serrate scars with fine teeth or bas-relief serrations.

---

1) " These numbers refer to specimens 40 mm. or more major radius."

"Three furrow and two or three actinal adambulacral spinelets, longest furrow and actinal spinelets equal to or exceeding two adambulacral plates with suture."

"Frequently four furrow and four or five[1] actinal adambulacral spinelets; longest actinal and furrow spinelets less than two adambulacral plates with suture."

I have examined many specimens of this species, among which there are some 4-rayed and 6-rayed individuals, but I have yet to see specimens with more than 6 rays. The radial ratio appears to be subject to a good deal of variation, as may be seen from the following table, in which the specimens have been taken at random from among large numbers in my hands.

Specim.	r mm.	R mm.	R : r	Locality
1	5	8	1.6	Inubôsaki.
2	7	11	1.5	
3	11	17	1.5	
4	17	26	1.5	?
5	19	35	1.8	Isoko (Tokyo Bay).
6	20	36	1.8	Fukuoka.
7	20	39	2.0	Isohama.
8	22	35	1.6	Otaru.
9	22	36	1.6	Miyazu.
(6-rayed, one normal and one abnormal larger madreporite in the same interradius.)				
10	22	39	1.8	Misaki.
11	22	40	1.8	Isohama.
12	24	40	1.7	Otaru.
13	24	43	1.8	Fukuoka.
14	25	40	1.6	Otaru.
15	25	39	1.6	,,
16	25	45	1.8	Fukuoka.

1) "These numbers refer to specimens 40 mm. or more major radius."

Specim.	r	R	R : r	Locality
17	25	47	1.9	Fukuoka.
18	27	43	1.6	Otaru.
19	27	44	1.6	Fukuoka.
20	28	49	1.8	Miyazu.
21	28	47	1.7	Shikuzushi (Hokkaido).
22	28	47	1.7	? (6-rayed.)
23	28	50	1.8	Otaru.
24	29	49	1.1	Isoko.
25	29	43	1.5	Miyazu.
26	29	54	1.8	Fukuoka.
27	31	49	1.6	Miyazu (6-rayed).
28	31	45	1.4	Otaru.
29	31	49	1.6	Shikuzushi.
30	31	52	1.7	Fukuoka.
31	32	49	1.5	Misaki.
32	34	55	1.6	?
33	34	56	1.6	Misaki.
34	35	55	1.6	?
35	39	61	1.6	?
36	39	57	1.5	?
37	41	59	1.4	Shikuzushi.

Unless the specimens have been subjected to distortion before or during preservation, the actinal surface is perfectly plane, being usually firmly applied to the substratum in life. The abactinal side is notably convex, but the degree of the convexity varies a good deal, and there is sometimes an elongated radiating groove in each interradius. Although the whole body is very flexible in life and can be adapted almost to any unevenness of the surface of the rocks, on which this specimen commonly lives, the skeleton becomes very hard in alcohol and the body is tolerably stiff. In the 5-rayed specimens the sides are lightly arcuate, but in specimens with supernumerary rays the interradial indentations are deeper and more angular.

Adambulacral plates.—The adambulacral armature consists of two series of spines. The inner spines are borne well down in the ambulacral furrow and are three or four in number for each plate, except on the first adambulacral plate, which may bear only two. When there are three spines the one at the adcentral end of the plate is smaller than the others, which are subequal in size. The outer spines are borne on the actinal surface of the adambulacral plates and are also three or four in number for each plate, the one at the adcentral end being usually shorter than the rest. When denuded of the spines the adambulacral plates are seen to present two convexities separated by a groove, one facing the ambulacral furrow and the other facing the actinal side and lying on a level with the ventrolateral plates. The two series of spines are borne on the two convexities.

Mouth-plates.—The mouth-plates are tolerably large and conspicuous and bear as a rule each two series of somewhat flattened spines. One of the series is borne on the furrow margin of the plate and consists of five or six spines, of which the one at the mouth end is conspicuously large, being sometimes more than twice as broad as the next one, and with its fellow of the other side, forms the oral spines. One or both of the pair may however be sometimes no larger than the next ones. The other series is borne on the actinal surface of the mouth-plate on its abcentral part, is parallel to the first, and consists of five or six somewhat flattened spines with rounded or truncated ends.

Besides these two series there may be a few very short but comparatively thick spines on the actinal surface of the mouth-plates, between the second row of spines of either side (Pl. XIX, fig. 274). The mouth-plates are usually separated from the adambulacral and ventrolateral plates by a groove.

Ventrolaterals.—When denuded of the spines the ventrolateral plates are irregularly polygonal in outline, slightly imbricating, quite firmly ankylosed with one another, and presenting each a convex surface on the actinal side. The plates are largest near the mouth and ambulacral furrows, and those of the arms and vicinity of the disk margin are exceedingly small. Each ventrolateral plate bears a comb-like row of somewhat flattened, pointed spines, of which the number may vary from nine or ten on a very large plate to about four on the smallest plates and those of moderate size. On the smaller plates these spines are regularly arranged in one row, but on the larger ones some of the spines may depart from the row, so that two rows may occasionally be found on some plates (Pl. XIX, fig. 274).

Dorsal plates.—Along the margin of the disk and arms there is a band-like portion on the dorsal side, in which the plates are all of the same form and nearly equal size, each plate being either rounded or lozenge-shaped in outline, and bearing each from six to a dozen granuliform spinelets. In the remaining parts of the dorsal side there are two orders of plates according to their size, which may be called, for the sake of description, principal and accessory. The principal plates are larger and are either elongated or crescent-shaped in outline. Around the centre of the disk the principal plates are arranged somewhat concentrically, with the concavities of the crescent-shaped plates turned towards the centre. The principal plates bear granuliform spinelets, which are larger than those of the accessory plates, and may vary anywhere between 40 and 20 or less according to the size of the plate. The accessory plates are very much smaller than the principal plates, are simply round or at most elliptical in outline, and fill up the spaces between the principal plates. They bear granuliform spinelets which may vary in number from half a dozen to fifteen or more according to the size of the plate, and which are sufficiently smaller than those of the principal plates to make the distinction of the two orders of plates at once apparent without denuding the plates of the paxillar spinelets. It may be remarked that there are also some plates which are intermediate in character between the two.

Madreporite.—The madreporite is large and conspicuous and round or slightly elliptical in outline. The surface is perfectly flat or very slightly concave and bears a large number of generally radiating grooves.

The colour of this species has been accurately described by DÖDERLEIN, and it may be added that the relative area of the dark blue and cinnabar red portions may vary to any extent between the two extremes of complete rufism and cyanism, so that in some cases the species may be distinguished into two colour varieties. The colour of the actinal side appears to be nearly constant.

The papulæ are found singly all over the dorsal side of the arms to their tip and, in the interradial areas, as far out as the distinction between the two orders of plates above mentioned is apparent.

This species is very widely distributed in Japan, being found in Hokkaido, both on the Pacific coast and the Japan Sea side of the Main Island, the Inland Sea, and Kyûshû. I have also specimens from Cheefoo, Port Arthur, Cape Kolokoltsof (northern Corea) and Saghalin.

Specimens in S. C., H. N. S., H. N. S. W., I. H. S., I. M., F. B,. O.

Asterina novæ-zelandiæ PERRIER.

(Pl. XIX, figs. 279–281.)

This species appears to have been first recognised as new by PERRIER, who describes it as follows ['76, p. 228]:

" *Asterina novæ-zelandiæ* (nov. sp.).

" Parmi les nombreux individus de la variété pentagonale de l'*Asterina Gunnii* recueillis à la Nouvelle-Zélande, j'ai trouvé un individu unique qui en diffère beaucoup et qui constitue évidemment une espèce distincte, dont voici la description.

" Corps de forme pentagonale, à côtés légèrement échancrés. R=20 mil., r=13 ; R=⅔ r environ.

" Plaques de la face dorsale entièrement couvertes de petits tubercules mousses, serrés [les uns] contre les autres et formant pour chaque plaque un groupe tantôt distinct circulaire, tantôt plus ou moins allongé. Des groupes allongés sont généralement concaves vers le centre du disque et séparés les uns des autres par un ou plusieurs groupes arrondis. Dans les groupes allongés les petits piquants mousses ou tubercules forment au moins trois ou quatre rangées. Ces groupes de tubercules ne dessinent pas sur le disque de figure de forme particulière. Plaques marginales imbriquées et formant autour du disque une sorte de bordure où la partie apparente constitue une sorte de pavage à éléments rhomboidaux. La plaque madréporique arrondie est située à 3 millimètres du centre du disque et entourée de groupes de granules plus allongés que les autres. Les sillons qu'elle porte ne sont pas rayonnants autour d'un centre. En somme, sauf le nombre et la disposition des piquants, l'ornementation de la partie dorsale de cette espèce ressemble beaucoup à celle de l'*Asterina Gunnii*.

"La face ventrale est au contraire toute différent. Les plaques du sillon ambulacraire portent chacune dans le sillon trois piquants divergents, réunis en une lame un peu oblique relativement à la direction du sillon par un repli des téguments qui n'atteint pas le sommet des piquants. En dehors ces plaques portent en outre une rangée oblique de trois piquants, un peu plus grands que ceux qui existent sur les autres plaques de la face ventrale et qui sont aussi au nombre de trois ou même quatre, surtout au voisinage de la bouche sur chaque plaque. Ces piquants sont d'autant plus petits et d'autant plus serrés les uns contre les autres qu'on se rapproche davantage des bords du disque, où ils sont fréquemment aussi au nombre de quatre sur chaque plaque.

"Un seul individu de la Nouvelle-Zélande, en très-bon état et conservé dans l'alcool."

FARQUHAR ['97, p. 196] has expressed the opinion that PERRIER's *A. novæ-zelandiæ* is probably identical with *Asterina regularis* VERRILL; but the latter species is, according to PERRIER, identical with the *Asterina cabbalistica* of LÜTKEN, and an examination of the description and figures of the latter author leaves no doubt in my mind that the specimens I have in hand, which agree in all essential points with the description of *A. novæ-zelandiæ* as given by PERRIER, can not be referred to *Asterina cabbalistica* = *Asterina regularis*. In HUTTON's "Index faunæ Novæ-zealandiæ" *Asterina regularis* and *A. novæ-zelandiæ* are listed as separate species [:04, p. 291].[1]

─────────────────

1) For the sake of comparison and to bring out more clearly the relationships of the two species in question I reproduce the descriptions of VERRILL, LÜTKEN, and FARQUHAR. The first is the original description of *Asterina regularis* [VERRILL, '67, p. 250]:

"*Asterina (Asteriscus) regularis* VERRILL, sp. nov.

"Pentagonal, depressed, with the interradial spaces evenly concave, and the rays short, broad and acute; greatest radius to least as 15 : 10. Ambulacral pores large; interambulacral plates each with two slender acute spines, forming a single row. Those near the mouth larger, obtuse and flattened. Ventral plates of the first row stout and prominent, each bearing a conical, acute spine, twice as large as the preceding. Exterior to these the ventral or interradial plates are flattened and imbricated, diminishing in size as they recede from the centre, each bearing an acute conical spine; these diminish in size like the plates, the larger ones being about as thick as the interambulacral spines, but shorter; near the margin these spines become very small and crowded, many of the plates bearing two. Plates of the upper surface rather

FARQUHAR ['98, p. 312] mentions this species from New Zealand on the authority of PERRIER and SLADEN. Under the name *Asterina neozelanica* PERRIER, FARQUHAR writes in part as follows [:09, p. 127] : " I have col-

large, increasing towards the centre, regularly imbricated, the free margin evenly rounded and thin, bearing near the end a cluster of five to nine very small, nearly equal spines ; toward the centre the plates become less regular in form and unequal in size, the larger ones often bearing twelve or fourteen spines in a transverse cluster. Madreporic plate large and prominent, at about one-third of the distance from the centre to the margin. The large dorsal pores are in groups on the sides and within the bases of the rays, arranged in about four rows, which run parallel with the median line of the rays, with from six to twelve pores in a row. A few irregularly arranged pores between adjacent rays connect these groups.

" Colour, when dried, dark olive-green above, yellow below.

" From centre to end of ray 1.5 inches, to edge of disk, 8.

" Aukland, New Zealand.—H. EDWARDS."

LÜTKEN's description of *Asterina cabbalistica* is very clear and is accompanied by two figures showing the two sides of the animal ['71, p. 242] :

" *Asterina cabbalistica* LTK. (Taf. IV, Fig. 1.)

" *Diagnosis. Asterina crassa, convexa, brachiis brevibus ; spinæ ventrales singulæ conicæ, fortes ; ambulacrarum lineæ ; pori dorsales vulgo singuli ; paxillæ dorsles brachiorum mediocres, lunatæ, e spinis minutissimis 8-16 compositæ ; interbrachiales minores, spinas 4-8 gerentes ; in parte centrali disci in lineas coalescunt, triangulum duplicem vel potius stellam pentagonam, corpus madreporiforme quoque circumvallantem, formantes.*

" Et Exemplar af denne nye Art er hjembragt fra Sydhavet af Skibslæge MÖLLER ; en bestemtere Lokalitetsangivelse foreligger desværre ikke.—*Den dobbelte Stjernefigur midt paa Ryggen* antager jeg vil vise sig konstant og karateristisk for Arten.[1]

" Denne hører til de tykkere og mere kortarmede Asteriner ; Forholdet mellem Radierne er omtrent som 19 : 13 ; der gives altsaa paa den ene Side Asterina, der ere forholdsvis mere kortarmede end denne, men ogsaa Arter med forholdsvis længere Arme. De Dele af Rygsiden, der ligge nærmest ved Armvinklerne, ere som sædvanlig frie for Porer og besatte med vel adskilte Grupper af fine Smaapigge eller korte Börster, 4–8 i hver Gruppe ; de ere ordnede i Rækker, der løbe parallelt med Armenes Radier eller Axer ; herfra undtages dog to Rækker, som nærmest Stjernens Omkreds løbe parallelt med denne og uden Tvivl modsvare de f. Ex. hos

One example of this new species has been brought home from the South Sea by the ship-physician MÖLLER ; a more definite statement of the locality is however not given.— *The double star-shaped figure at the middle of the bark* I assume to be constant and characteristic for the species[1].

This belongs to the thicker bodied and shorter armed Asterinæ ; the proportion between the radii is nearly as 19 : 13 ; there are therefore on the one side Asterinæ with relatively shorter arms than this, but also those with relatively longer arms. The parts of the dorsal side which lie next the arm angles are as usual free from pores and are covered with well separated groups of fine spinelets or short bristles, 4–8 in each group ; they are arranged in rows, which run parallel with the radii or axes of the arms ; with the exception however of two rows, which are

1) Den omtales ikke i PERRIERS Beskrivelse af *A. calcarata* VAL. (Valparaiso) ; Bugpladernes Pigge ere desuden dobbelte i Bugfelternes ydre Halvdel hos denne Art.

1) This is not mentioned in PERRIER's description of *A. calcarata* VAL. (Valparaiso) ; the spines of the ventral plates are moreover double in the outer half of the ventral areas in this species.

lected a considerable quantity of *A. regularis* in the neighbourhood of Wellington, but I have not found a specimen of *A. neozelanica*. (I write the
name '*neozelanica*' in accordance with the rule suggested by VON MARTENS,

Goniasterne og de fleste Kamstjerner foie-
kommende Randplader, men ikke, som hos
disse, i andre Henseender ere forskjellige fra
dem, der bedække Interradialfelterne. I de
övrige Rækker tiltage Paxillerne i Störrelsa
fra Omkredsen til Armryggen, hvor de gaae
jevnt over i dem, der ligge mellem Rygfods-
porerne. Hvor de porebærende Partier ere
bredest, vil man kunne tælle 13 Rækker af
Porer; disse sidde i Regeln enkeltvis, sjeld-
nere to eller tre sammen; paa Hudpladerne
mellem disse Porer kan Antallet af Smaapigge stige til c. 16; de ere da ordnede i dobbelt
eller tredobbelt Række til skrant stillede,
krumme, halvmaanedannede Smaagrupper. Paa
Stjernens midterste Del flyde disse Piggrupper
tildels sammen til længere sammenhængende Linier, som danne to hinanden krydsende
Triangler og tilsammen en femdelt Stjerne,
hvis Spidser vende mod Armvinklerne; det
större femkantede Rum i Midten og de fire
af de smaae trekantede Rum, der dannes
udenom dette ved hine Liniers Krydsning,
udfyldes af Porer og af Piggrupper af samme
Beskaffenhed som i Nærheden af Armvinklerne; det femte optages af Madreporpladen.
Bugpladerne bære derimod ikke Piggrupper,
men kun enkelte Figge, som ere temmelig
stærke, kegledannede og butte; deres Ordning
i Rækker vil ses af Figuren; de tiltage i
Störrelse fra Stjernens Omkreds ind imod
Fodgangene, saaledes at den med disse parallele inderste Række er den störste af dem alle
Ambulakralpapillerne ere stillede i Grupper
paa to (sjelden tre); nærmest Munden optræde de som stærkere, butte Mundpigge —
Störrelse : R = 19 mm., r = 13 mm. (maalte paa
Rygsiden; tagne paa Bugsiden vilde de samme Maal falde lavere ud)."

nearest to the margin of the star and run
parallel with it and doubtless correspond to
the marginal plates which are present e. g. in
Goniasters and in most Astropectens, but not
different in other respects from those that cover
the interradial areas, as is the case in these
starfishes. In the remaining rows the paxillæ
increase in size from the margin towards the
back of the arms, where they smoothly
pass into those that lie between the papular
pores. Where the poriferous areas are broadest, 13 rows of pores can be counted; these
lie as a rule singly, more rarely two or three
together; on the dermal plates between these
pores the number of spinelets may rise to
some 16; they are then arranged in double
or triple rows in obliquely placed, curved,
crescent-shaped grouplets. In the middle part
of the star these groups of spines partly flow
together into elongated continuous lines, which
form two intersecting triangles and together
a five-rayed star, whose apices are turned
towards the arm angles ; the larger pentagonal space in the middle and four of the
small triangular spaces, which are formed
around it by the intersection of those lines,
are filled with pores and groups of spines of
the same character as those in the neighbourhood of the arm angles; the fifth is taken up
by the madreporic plate. The ventral plates
on the contrary do not bear groups of spines
but only single spines, which are tolerably
stout, conical and blunt ; their arrangement in
rows will be seen from the figures; they increase in size from the margin of the star
inwards to the ambulacral furrows, so that the
innermost row parallel with the furrow is the
largest of all. The ambulacral papillæ are
placed in groups of two (seldom three: ; next
the mouth are somewhat stouter, blunt oral
spines.—Size : R = 19 mm , r = 13 mm. (measured on the dorsal side; taken on the ventral side the measurements would be less).

FARQUHAR ['97, p. 196] refers to *A. regularis* as follows:

"A number of specimens of this species which I have collected near Wellington shows that
it often belies its name. Two of them have seven rays each, five have six rays each, several
have more than one madreporic plate; and there is a specimen in the Colonial Museum with
eight rays. Several of these are so distinct from normal specimens of *A. regularis*, that if a
series were found in a separate locality, a new species might safely be established for them

and adopted by New Zealand naturalists—see Trans. N. Z. Inst., vol. xxi, p. 238—that specific names should be one word only)." He then reproduces PERRIER's description.

It appears from a comparison of the descriptions of *Asterina regularis* and *Asterina novæ-zelandiæ* that the two are closely related. *Asterina novæ-zelandiæ* is again nearly allied to *Asterina batheri* and more remotely to the Mediterranean species *Asterina gibbosa*. So far as my specimens show *A. novæ-zelandiæ* seems to be confined in Japanese waters to the southern parts, being found from Kyûshû southwards, *Asterina batheri* taking its place in the more northern parts. These two species may be regarded as the Japanese representatives of the Mediterranean *A. gibbosa*. *Asterina cephea* also seems to come near these species; in fact *A. batheri*, *A. regularis*, *A. novæ-zelandiæ*, *A. gibbosa* and *A. cephea* appear to form a series of nearly related forms, and a more minute comparative study of these forms on a sufficient number of specimens will probably bring out interesting facts in regard to their geographical and taxonomic relationships.

The body is regularly star-shaped, perfectly flat on the under side and convex on the upper, with a circular area in the middle of the abactinal surface, distinct in some specimens but obscured in others, enclosed by larger, roughly crescent shaped plates, from

---

Prof. PERRIER has described another New Zealand species of this genus, *A. novæ-zealandiæ*; but unless it be very different from *A. regularis*, or has been described from a good series showing a constant difference, it may be one of these abnormal forms, which are not uncommon. I stated (Trans. N. Z. Inst. vol. xxvii, p. 199) that this species occurs in Australia. My authority was the 'Alert' Report. Mr. WHITELEGGE states, however, that it is not found at Port Jackson (Proc. Roy. Soc. N. S. W. vol. xxiii, p. 202, 1889); and the good 'series' of the 'Alert' Report probably belongs to some other form. I have a series of very fine specimens which were sent me by Mr. LUKINS from Nelson. Amongst these are several with two spines on each of the interradial plates on the actinal surface; others have one spine on the large plates near the mouth and two on the smaller ones near the margin. Not unfrequently the four or five large plates immediately outside the mouth-plates are without spines. The adambulacral plates bear two or three spines. forming a single row in the furrow."

which the convex ridges of the arms radiate, separated from each
other by tolerably distinct interradial furrows (Pl. XIX, fig, 279.)
The upper surface appears more or less rough, owing to the presence of plates which are larger and more prominent than the
surrounding ones and bear larger spines.   In some specimens these
larger plates are very numerous, but in others they are few.
The interradial margin is well scooped out but not angular.   The
papular pores are single and step about two plates from the
superomarginals on the sides of the arms and four or five plates
from the same in the interradial lines.   The radial ratios are
given below :

Specim.	r mm.	R mm.	R : r	Locality
1	6	13	2.1	Shimokoshiki Shima.
2	7	13	1.5	Shimadaira, Satsuma.
3	9	17	1.9	Shimokoshiki Shima.
4	9	17	1.9	
5	9	21	2.3	
6	10	20	2.0	,,
7	10	21	2.1	Natsui, Hyûga.
8	12	23	1.9	Tanoura, Satsuma.
9	12	25	2.0	Shifushi, Hyûga.
10	12	26	2.2	Tanoura, Satsuma.

The superomarginal plates are more distinct than in *A.
batheri*, being larger than the adjoining plates, and bearing at the
interbrachial angle each 6–10 spines frequently arranged in two
rows along the length of the plate.   The inferomarginals occupy
the very margin of the body and can be seen both from the
actinal and abactinal sides.   The spines of the abactinal side of
the plates are like those of the superomarginals, but slightly

smaller and 5–8 on each plate at the interbrachial angle ; on the actinal side the spines are longer and shaped like those of the ventrolateral plates but smaller, six or seven on each plate and arranged parallel to one another in the form of a brush.

Adambulacral plates.—The adambulacral plates bear two rows of spines. The furrow series consists of four or five spines united together by a web-like membrane ; of these however only three are usually large and face the ambulacral furrow, while the other one or two are exceedingly small and generally covered over entirely by the membrane and turned more or less away from the furrow, so that they are apt to be overlooked, especially in wet alcoholic specimens. When the spines of a group are well separated the palmate form of the group and its slight obliqueness with respect to the furrow can be distinctly seen, but when they are close together only the three large spines can be seen ; and it is on this account that I refer my specimens to PERRIER's *A. novæ-zelandiæ* in spite of his discordant statement on this point. On the actinal face of the adambulacral plates there is a second row of spines which are stouter than those of the inner, and are also connected together by a web-like membrane. There are in this row likewise four or five spines on each plate, but generally only three of them are stout ; and when they are pointed away from the ambulacral furrow only the stout ones can be seen (Pl. XIX, fig. 281).

Mouth-plates.—The mouth-plates are comparatively large, and each bears on the furrow margin 5 or 6 straight spines united together by a membrane in the basal half, and becoming gradually shorter away from the mouth. On the actinal surface of each mouth-plate there are three or four spines forming a curved row roughly parallel with the ambulacral furrow, of which the one or two next the mouth are particularly stouter. The two rows of spines of the mouth-plates are separated by quite a distinct space.

Ventrolaterals.—The ventrolateral plates are arranged regularly in V-shaped rows parallel to the ambulacral furrows, except at the interradial lines near the margin of the disk, where they become more or less irregular and very small. By far the larger number of these plates bear two spines which are comparatively stout and blunt, but there are also quite a number of them with three spines, and only occasionally there may be four spines on one plate. Where there are three spines on one plate, they are arranged in a row, but when there are four, they appear to have no regular arrangement. The spines of each plate are separated from those of the neighbouring plates by distinct spaces. The ventrolateral spines regularly decrease in size away from the ambulacral furrows.

Abactinal plates.—The general appearance of the abactinal surface has been described above. The arrangement of the dorsal plates appear to be destitute of regularity along the ridge of the arms, where the plates are also of various sizes. The larger plates project more prominently than the others and give rise to the rough appearance mentioned above. The number and position of these larger plates are very variable, but a general idea can be obtained from the photographic picture reproduced in fig. 279, Pl. XIX. On the sides of the arms there are 4–6 longitudinal rows of small plates, all looking similar to one another and different from those of the abactinal ridge, running parallel to the long axis of the arms and forming acute angles with the superomarginal series. The number of spines on these plates may vary from 4 to 7. The more prominent plates near the dorsal ridge may bear as many as 25 spines, while the smaller plates may bear any number less than that down to two or three.

Madreporite.—The madreporite is tolerably large but not very conspicuous, owing to the presence of some prominent plates around it, whose spines tend to overhang the margin of the madreporic plate. It is more or less elliptical in outline and is covered with short, irregular, discontinuous small furrows.

Terminal plates.—The terminal plates are relatively large but not conspicuous in undenuded specimens, owing to its being covered over with rough granules.

Locality.—All my specimens are from the southern parts of Kyûshû and the adjacent islands. This species appears to be the southern representative of *Asterina batheri*, but the distributional limits of the two species can not be determined with the data at hand. The exact localities of my specimens are given above.

Specimens in S. C.

Asterina batheri, n. sp.

(Pl. XIX, figs. 275–278.)

In the Challenger Report, SLADEN ['89, p. 393] mentions *Asterina penicillaris* from Kobé; but as there was a grave doubt in my mind as to the occurrence of this species in Japanese waters I asked my friend, Dr. F. A. BATHER of the Natural History Department of the British Museum to examine the original specimens of the Challenger collection, and he was so kind as to furnish me with the following notes, dated April 3, 1905.

" The Challenger specimens referred by SLADEN to *Asterina penicillaris* are in the British Museum, and are two in number, having R. 14 and 19 mm. respectively.

" The ambulacral spines are distinctly arranged in palmate form, being from 4 to 6 in a group. They are very short, so that the whole arrangement looks like a stumpy human hand. The plates bearing these spines are succeeded by another row of plates almost similar in structure, and then by other rows gradually merging into the ordinary plates of the oral surface.

" The number of spines on the dorsal plates in the larger specimen, varies from 10 to 12, according to the size of the plates. On the majority of the plates they are little more than tubercles, but on the 3 plates surrounding the madreporite, they are more digitate in form. On the smaller specimen, these spines on the dorsal plates are mere tubercles or pustules, and some plates appear to have more than 12.

" The specimens are preserved in alcohol, and are rather stiff, so that it is not easy to separate the plates distinctly, so as to count every single spine. They would be more easily counted on a dried specimen."

The points mentioned in these notes make it clear that the Challenger specimens are to be referred to the same species as those that I have before me, which can not be identified with *Asterina penicillaris*; the principal differences being the arrangement of the adambulacral spines, and the number of spines on the actinal and abactinal plates[1].

---

1) For the sake of comparison MULLER and TROSCHEL's description of *Asterina penicillaris* is here reproduced ['42, p. 42] :

" Die Arme sind mehr verlängert als bei den vorhergehenden Arten. Der kleine Radius verhält sich zum grossen wie 1 : 2⅔. Die Einschnitte zwischen den Armen sind winklig. Die Furchenpapillen stehn in einem Bündel auf jeder Platte und von diesen kommen drei an den Rand der Furche. Jedes Bauchplättchen trägt einen dichten Haufen von vielen (8–15) cylindrischen Stacheln. Die Plättchen des Rückens sind noch viel dichter mit kürzeren Papillen (30–60) besetzt, so dass sie einer Bürste gleichen. Die Poren in vielen Reihen kommen dem Rande sehr nahe, mehr als bei den übrigen Arten. Häufig stehn mehrere Poren nahe an einander, so dass sie fast das Ansehen eines Porenfeldes geben, indessen finden sich auch meist zwischen diesen kleinere Stachelhäufchen.

Of the genus *Asterina* the present species appears to be next to *Asterina pectinifera* the commonest species on the coasts of the Main Island. The arms are not long, and the interradial angles are rather deep and mostly rounded at the tip, though sometimes sharply pointed. The body is perfectly flat on the actinal, but prominently convex on the abactinal side, and presents a pentaradiate, regularly stellate appearance in normal specimens. An indistinct radiating groove is mostly present in each interradius. Measurements of twenty-three specimens gave the following results:

Specim.	r mm.	R mm.	R : r	Locality.
1	6	15	2.5	Misaki.
2	7	16	2.3	,,
3	9	20	2.2	,,
4	9	20	2.2	,,
5	10	20	2.0	,,
6	10	20	2.0	Miyazu.
7	10	20	2.0	Misaki.
8	10	24	2.4	,,
9	11	24	2.2	,,
10	11	25	2.3	,,
11	11	24	2.2	,,
12	11	25	2.3	,,
13	11	28	2.5	,,
14	11	26	2.4	Miyazu.
15	12	22	1.8	Misaki.
16	12	25	2.1	,,

"Farbe: oben mit blaulichem Grunde und rothbraunen Stacheln, unten blau nach den Zeichnungen von LESUEUR.

"Grösse: 2½ Zoll.

"Fundort: Rothes Meer und Indischer Ocean. Im Museum zu Berlin durch HEMPRICH und EHRENBERG."

Specim.	r mm.	R mm.	R : r	Locality.
17	12	26	2.2	Miyazu.
18	12	27	2.3	Misaki.
19	12	27	2.2	Miyazu.
20	13	28	2.2	Misaki.
21	14	30	2.1	,,
22	14	32	2.3	,,
23	14	34	2.4	,,

Along the margin of the arms there is a series of plates probably corresponding to the inferomarginals, which are larger than the adjoining plates of both the actinal and abactinal sides and bear a larger number of minute spines (20 or more).

Adambulacral plates.—The adambulacral plates bear two series of spines arranged in a curved fan-like shape and each series is connected together in the basal parts by a web-like membrane, so that the individual spines are rather difficult to count in alcoholic specimens. In the first series which directly faces the furrow, there are 6 or 7 spines, of which the middle two or three are larger than the others and either directly bounds or project into the ambulacral furrow, the rest gradually diminishing in size towards either side and borne more on the abcentral and adcentral margins of the plates. On the actinal face of the adambulacral plates there is a similar row of shorter spines, separated from the first series by a groove (Pl. XIX, fig. 278). It may be remarked here that the plates bearing this second row of spines are regarded by Dr. BATHER as distinct from those of the first series, hence the apparent discrepancy between his description and mine. The spines of the second series are also connected together by a web-like membrane, and in alcoholic specimens are more difficult to count than those of the first series.

Mouth-plates.—Each mouth-plate bears on the furrow margin 6–7 spines, of which the first two or three are conspicuously larger than the adambulacral spines and the rest nearly equal to the latter, the length gradually decreasing away from the mouth. On the actinal face of the plates there is a group of spines, which are sometimes arranged in a curved line parallel to the first series of spines, but sometimes more like those of the ventrolateral plates. When they are arranged parallel to the first series, there are usually 4 or 5 larger and a few smaller spines for each mouth-plate; in the other case the number of spines remain nearly the same, only the arrangement being different (Pl. XIX, fig. 277).

Ventrolaterals.—The ventrolateral plates are all small, but are largest along the ambulacral furrow and gradually decrease towards the margin, where they are bounded by the series of larger plates already mentioned. The plates are regularly arranged in rows parallel to the ambulacral furrow, so that those belonging to any two adjacent arms together form a V-shape; about ten of these V-shaped series can be counted in each interradius. It is very difficult to count the spines on each plate in alcoholic specimens, as they are covered over for the most part by a tumid membrane, but in dried specimens they are very distinct though small, and each plate bears 6–10 of them, except on the very small plates near the margin which may bear less.

Dorsal plates.—On the sides of the arms the dorsal plates are distinctly arranged in longitudinal rows parallel to the length of the arms, so that the rows of the adjacent arms form about

five V-shaped series within the papular area ; but on the ridge
of the arms the plates do not show any regular arrangement. Except for the very small plates close to the margin of the
body and those that are found elsewhere among other larger
plates, each plate bears some 7–12 short pointed spines, which are
mostly covered over by a tumid membrane in alcoholic specimens.
A few plates bordering on the madreporite are usually larger than
others and may bear more than 12 spines, which are also slightly longer than those of other plates. The papular pores are strictly
isolated and are absent for a certain stretch along the margin of the
arms, where the arrangement of the plates in longitudinal rows is
especially distinct.

The madreporite is conspicuous and round and is covered
with meandering grooves.

I have the pleasure of naming this species in honour of my
friend, Dr. F. A. BATHER of the Natural History Department of
the British Museum.

I have specimens of this species from Tokyo Bay, Misaki,
Miyazu, Kashiwajima (Tosa) and Shimabara (Hizen).

Specimens in S. C. and I. H. S.

Palmipes[1] *tenuis*, n. sp.

(Pl. XIX, figs. 282–283.)

The body is very thin and translucent, the margin is very
lightly scooped out, so that the outline is nearly pentagonal. The
under side is perfectly flat except around the mouth, where there

---

1) According to BELL and FISHER the valid post-Linnæan name for this genus is *Anseropoda*. The latter author says [:06, p. 1088]: "This genus has usually been called *Palmipes*.
Palmipes, however, was not used by a Linnæan writer until a year after *Anseropoda* NARDO. It
seems somewhat discouraging in the first years of the twentieth century to be obliged to call
attention to the fact that LINCK (1733), who first used *Palmipes* and who was quoted in the

is a slight bulging out, owing to the presence of the circular stomach, which shows well through the body wall when the specimen is held up against light. There is also a noticeable out-bulging in the corresponding portion on the dorsal side. On this side there is also a pentaradiate papular area consisting of a central pentagonal portion and five radiating strips extending to the very tip of the arms. The papulæ are not very numerous, there being some twenty-five in the central pentagon and some twenty-five or thirty on each arm, the latter being arranged in two somewhat irregular longitudinal rows along the margin of each radiating strip. In my specimen the papulæ are very unequal in size, but this may be due to different degrees of expansion and contraction. In the papular area the dorsal plates are of unequal sizes, more widely spaced, destitute of any regular arrangement, and covered with minute granules, which may vary in number from one to ten or so. In the interradial triangular areas bounded by the radiating strips of the adjacent arms and the margin of the body, the dorsal plates are of nearly uniform size, except near the margin of the body where they are slightly smaller, are regularly arranged in rows parallel to the radiating strips of the arms and also oblique to these and are more thickly set than in the papular area. The plates, which are not very distinct in surface view, bear each some twelve or fifteen minute spines which are hardly more than pointed granules. The supero-

last extensive paper on European starfishes as the authority for the name, was in no way a binomial writer, and that consequently his so-called genera (!) have no place even if 1758 had not been agreed upon as the starting point of zoological nomenclature. For a statement of the facts in the present case see Prof. F. JEFFREY BELL in Annals of Natural History, ser. 6, Vol. VII, 1891, p. 233. The Rev. Canon A. M. NORMAN (op. cit., p. 382) admits that *Anseropoda* has priority, but objects to its use because it happens to be etymologically a hybrid. What owuld happen if all generic names which confess this fault were thrown out for the same reason?"

marginals are hardly different in appearance from the adjacent dorsal plates.

The madreporite is very small and can hardly be seen with the naked eye. In the single specimen before me (R=22 mm.) it lies about 2 mm. from the centre of the disk, which is marked by a conical prominence.

The *adambulacral armature* consists of a single series of ten or eleven slender delicate spines arranged in a deep horse-shoe shape, four or five of them usually facing the ambulacral furrow and being longer than the others, which are more or less turned away from the furrow. The latter spines are sometimes disposed in such a way as to simulate a second outer row of adambulacral spines, but close examination shows the horse-shoe arrangement to be the true one. The spines are easily flexible and do not offer any perceptible resistance when touched with a metal point. Each *mouth-plate* bears on its free margin a regular row of seven or eight spines which gradually increase in size towards the mouth, and of which the last two or three at the central end are particularly stouter ; they are less flexible than the adambulacral spines. On the actinal ridge of each mouth-plate there is a somewhat irregular row of spines nearly equal in size to those of the furrow series of the mouth-plate, in close apposition with those of the fellow plate, so as to form a single group. The two series of a plate are separated by a distinct space.

The adambulacral plates are followed by a series of ventrolateral plates, which are distinctly smaller than the other ventrolaterals and bear very short pointed spines, varying from two to seven on each plate. All the remaining ventrolaterals are similar and only decrease in size towards the margin of the body. The larger plates near the centre of the body bear each 5–7 sharply pointed, slender spines nearly always arranged in a single row, but sometimes with two or three spines outside the row. The spines at the centre of each plate are longer than the others, so that in the more peripheral plates with only three spines the middle one is longer than the other two. The ventrolateral plates are arranged in regular rows parallel to the ambulacral furrows, except near the mouth where this arrangement is more or less obscured. The plates occupying the very margin of the body do not present any distinctive features.

I have a single specimen of this species from Misaki, of which the dimensions are $r = 16$ mm., $R = 22$, $R : r = 1.4$. The specimen which is in alcohol is colourless, except where the stomach shows through.

Type specimen in S. C.

Palmipes petaloides, n. sp.

(Pl. XIX, figs. 284–285.)

The general outline of the body is like that of *Palmipes-membranaceus*. The body is very thin and when it is entirely flattened out, the arms are very broad and the interradial indentations very shallow; but when it is curved towards the actinal side, as appears to happen often in alcoholic specimens, there is formed a more or less distinct notch in the interradial line, as is also suggested to occur in *P. membranaceus* in LUDWIG's figures ['97, Pl. 5]. Most of my specimens were originally preserved in formalin and they were then light brown on the dorsal side and colourless on the ventral. On the dorsal side a central pentagonal area can be distinguished, as in *Palmipes tenuis*, in which the

plates are arranged without any definite order, and along the
borders of which the relatively large papulæ are situated. These
are comparatively few in number and extend in two irregular
rows along the carina of each arm and almost reach the apex. In
a specimen whose R=27 mm. there were 15–17 papulæ to each
arm and one or two to each interradius, while in another with
R=32 mm. there were about 27 papulæ to each arm. Each papula
is surrounded by small plates different from the rest of the abactinal surface and stands out conspicuously. The anus may also be
tolerably distinct, but the madreporite is very small, circular in outline and is covered with a small number of radiating furrows meeting
in the centre. In alcoholic specimens it appears as an opaque white
body near the anus, while the rest of the body is transparent. The
stomach shows through the body wall as a roundish sac, and is
larger than the central pentagonal area referred to above.

In well preserved specimens the plates of the central pentagon
of the abactinal side are marked off from one another by narrow
but distinct smooth grooves. The plates are of different size and
shape and are, as already mentioned, without any definite arrangement, although in some places, traces of their arrangement in
concentric rings around the anus may be made out. Each plate bears
on its surface a number of groups of short paxilla-like spinelets, the
number of these groups varying from about a dozen on the larger
plates to about half as many on the smaller ones. The plates
immediately around the anus is very small. The plates of the
papular areas of the arms are much smaller than those of the
central pentagon, are also without any regular arrangement and
mostly bear each a single tuft of short spinelets similar to those
of the central pentagon.

Outside the papular areas the abactinal plates are very regularly arranged in rows parallel to the carinæ of the arms and also in rows oblique to them, and are marked off from one another by distinct grooves. The plates regularly decrease in size towards the periphery of the body. The plates immediately outside the central pentagon are only slightly smaller than those of the latter, regularly logenze-shaped and bears each about half a dozen tufts of short, paxilla-like spinelets. The number of these tufts on a single plate decreases outwards and from about half way out from the centre of the disk each plate bears only a single tuft of delicate spinelets.

The *adambulacral plates* bear two groups of tolerably long, slender spines. On the furrow border there is a row of long spines arranged in a curved palmate form. Each plate bears six (sometimes seven) of these spines, of which the middle ones are longer than those towards either end; on the actinal surface of the plate there are 4–6 similar spines, arranged roughly in a straight line, but sometimes very irregularly, and of unequal length. The latter spines may be more or less incomplete in young specimens.

The *ventrolateral* plates are regularly arranged in rows exactly similar to those of the dorsal plates outside the central pentagon. Each plate bears according to its size 3–10 spines arranged in a light curve, whose convexity is turned toward the centre of the body.

The *mouth-plates* are comparatively large, and each bears on the furrow margin 7–10 subequal spines which are considerably longer than the furrow spines of the adambulacral plates, excepting the last one or two at the abcentral end of the plate. On the actinal face of each mouth-plate there is a group of smaller spines similar to those of the ventrolateral plates, some ten in number, and sometimes separated from the same group of the fellow plate by a groove. Between the two groups of spines of each mouth-plate there is a distinct space.

I have examined five specimens of this species, all from Misaki. Of these one is too much injured to be measured; the rest are as follows:

1) $r=22$ mm. $R=32$ mm. $R:r=1.5$
2) $r=16$ „ $R=27$ „ $R:r=1.7$
3) $r=14$ „ $R=22$ „ $R:r=1.6$
4) $r=13$ „ $R=16$ „ $R:r=1.2$

Specimens in S. C.

Palmipes rosaceus (LAMARCK).

The occurrence of this species in Japanese waters appears to me not well authenticated. It appears to have been first described by LAMARCK as a variety of *Asterias rosacea*, although previously figured. It is the nineteenth species of LAMARCK's *Asterias* and is characterised as follows ['16, p. 558]:

" Astérie rosacée. *Asterias rosacea.*

" A. complanata, submembranacea, utrinque tuberculis minimis et subhispidis granulosa; lobis obtusis brevissimis; disco dorsali nudo.

" Encycl. pl. 99, f. 2-3.

" 2. var. lobis senis. Mus. n°.

" 3. var. lobis quindenis. Mus. n°.

" Habite......Quelque voisine que soit cette astérie de la précédente[1] par ses rapports, elle me paraît s'en distinguer constamment par la forme de ses lobes, et par le défaut d'écailles au centre et sur les côtés de son disque dorsal. Effectivement, la surface supérieure ou dorsale de l'astérie rosacée n'offre partout que de petits tubercules, tous semblables, qui lui donnent l'aspect d'une peau de chagrin.

---

1) *Asterias membranacea = Palmipes membranaceus.*

"La variété 3 est fort grande et singulièrement remarquable, ayant quinze lobe courts, qui la font ressembler à une *rose des vents*."

BLAINVILLE ['34, p. 237] simply mentions this species, with references to LAMARCK and the Encycl. méthodique.

GRAY mentions it under the name of *Palmipes Stokesii* ['40, p. 288], "*Palmipes Stokesii.* Rays 15, acute. Mus. Mr. STOKES." He then refers for comparison to the following: *Asterias pulvillus*[1] MÜLLER, Zool. Dan. t. 19, f. 1,2; *Ast. equestris* and *Ast. militaris*[2] MÜLLER, of the North Sea; and *Ast. Luna* LINNÆUS, from India, all not seen by him.

MÜLLER and TROSCHEL describe it under the name of *Asteriscus rosaceus*, giving as synonyms *Asterias rosacea* LAM. Var. lobis quindenis and *Palmipes Stokesii* GRAY ['42, p. 40]:

"Fünfzehn Arme. Verhältniss des kleinen Radius zum grossen wie 1:1¼. Die Furchenpapillen fünf auf jeder Platte, dicht neben ihnen läuft ein Zug von je zwei längeren Stacheln auf jeder Platte. Der Körper ist so flach und membranartig wie beim vorigen[3]; auch die Bedeckung des Rückens ist ähnlich, doch sind die Platten des Rückens mit zahlreicheren äusserst feinen Stachelchen bedeckt. Auf den Bauchplatten stehn nur 3-5 Stacheln, von denen die mittleren die längsten sind.

"Farbe: oben graulich mit dunkelbraunen Punkten und dunkelbraunem Rande; unten hell mit vielen blauen Punkten und einigen rothbraunen Flecken und eben solchem Rande. Nach der Originalzeichnung LESUEURS.

"Grösse: 7 Zoll.

"Fundort: unbekannt. Im Museum zu Paris."

This species is again described by DUJARDIN and HUPÉ, who place it in *Palmipes*. The description is as follows ['62, p. 373]:

"2. Palmipes rosacé. *Palmipes rosaceus——*Nobis.

[References to LAMARCK, GRAY and MÜLLER and TROSCHEL.]

"Espèce à quinze bras, dont la longueur dépasse d'un septième le plus petit rayon du disque. Les piquants du sillon ambulacraire sont au

---

1) *Ast. pulvillus* MÜLLER = *Porania pulvillus*.
2) *Ast. militaris* MÜLLER = *Pteraster militaris*.
3) *Palmipes membranaceus*.

nombre de sept sur chaque plaque ; tout d'auprès d'eux se trouve une bande formée de deux piquants plus longs sur chaque plaque. Le corps est plat et membraneux, comme chez l'espèce précédente. L'ornementation du dos est également semblable, seulement les plaques dorsales sont couvertes de petits piquants plus nombreux et extrêmement fins. Les plaques ventrales portent seulement trois à cinq piquants, dont les intermédiaires sont les plus longs.

"Coloration grise en dessus, avec des point et le bord brun foncé. Le dessous est plus clair, avec beaucoup de points blancs et quelques taches brun-rouge, ainsi que le bord. Dimension : 190 mm."

In the "Synopsis" GRAY has the following ['66, p. 15] :

"*Palmipes Stokesii*. Rays 15, acute. GRAY, Ann. N. H. 1840, p. 288. *Asterias rosacea*, var., LAMK. ii. 558. *Asteriscus rosaceus*, MÜLL. and TROSCH. Ast. 40 (Mus. Mr. STOKES). Inhab. Japan."

PERRIER refers to this species as follows ['76, p. 210] :

"L'étiquette de l'échantillon unique du Muséum porte comme seule indication de localité : mers australes. Voyage de PÉRON et LESUEUR. Expédition du capitaine BAUDIN, 1803. GRAY indique le Japon comme lieu d'origine de cette magnifique espèce."

According to SLADEN this species occurs in the Bay of Bengal (Berlin Museum) ['89, p. 394].

BEDFORD mentions this species from Singapore, as follows [:00, p. 297] :

"*Distribution*. Japan and Bay of Bengal.

"*Locality*. One large and rather damaged specimen in 5 fathoms, Singapore.

"This species, being the only 15-armed Asterinid, is easy to recognise it appears to be rare.

R=121.5 mm., r=82 mm."

It is reported from the Mergui Archipelago by BROWN [:10, p. 34] :

"Localities.—XVII., between Sir John Malcolm and Charlotte Islands, in 18 fathoms, coarse sand ; XXV., Gregory Group, 4 to 14 fathoms, sand and shell. Two damaged specimens, in the largest of which R=105 and r=80.

"Recorded from the Bay of Bengal and Japan."

KŒHLER mentions it from the collection of the Indian Museum [:10, p. 127], with two figures.

" *Palmipes rosaceus* (LAMARCK).

" 6° 0' Lat. N. 80° 16' Long. E. Profondeur 34 brasses. Un échantillon.

" L'exemplaire n'est pas en très bon état de conservation et une portion du corps au moins est fortement endommagée. R=75 mm., r=55 à 60 mm.

" Cette espèce a été décrite d'une manière très suffisante, mais comme elle n'a jamais été figurée, j'ai cru devoir donner ici deux photographies représentant la face dorsale et la face ventrale.

" Vers l'extrémité de l'un des bras, et sur la face ventrale, se trouvent fixées deux *Mucronalia*, l'une petite et à coquille cassée, l'autre plus grande à coquille entière et mesurant 6.5 mm. de longueur. Ces Prosobranches me paraissent appartenir à une espèce nouvelle et je les étudierai dans un travail ultérieur, en même temps que les Prosobranches parasites du *Stellaster equestris* que j'ai signalés plus haut."

BIBLIOGRAPHY.

In this bibliography are enumerated only those papers which are referred to in the text, and in the case of those that I have been able to consult personally, I have brought together under each title the species treated of in it, with the principal items in the lists of synonyms, where such are given. On this last point the method adopted was this : of the whole list of references under each specific name, I have picked out only those names, the identity of which with the species in question is not apparent at once. To take an example, MICHAILOVSKIJ [: 02, p. 487] gives the following references under *Ctenodiscus corniculatus* (LINCK) :

1733. *Astropecten corniculatus* LINCK, &c. 1842. *Ctenodiscus polaris* MÜLLER & TROSCHEL, &c. 1842. *Ct. pygmæus* MÜLLER & TROSCHEL, &c. 1846. *Ct.· crispatus* DÜBEN & KOREN, &c. 1857. Ditto LÜTKEN, &c. 1861. Ditto M. SARS, &c. 1862. Ditto DUJARDIN & HUPÉ, &c. 1869. Ditto PERRIER, &c. 1876. *Ct. corniculatus* PERRIER, &c. 1878. Ditto VIGUIER, &c. 1881. Ditto DUNCAN & SLADEN, &c. 1886. *Ct. krausei* LUDWIG, &c. 1886. *Ct. crispatus* LEVINSEN, &c, 1892. Ditto BELL, &c. 1899. *Ct. corniculatus* DÖDERLEIN, &c. 1900. Ditto DÖDERLEIN, &c. 1900. *Ct. crispatus* LUDWIG, &c.

In such a case as this only the following are reproduced in this bibliography : *Ct. corniculatus* (LINCK) = *Astropecten corniculatus* LINCK 1733 = *Ct. polaris* M. T. '42 = *Ct. pygmæus* M. T. '42 = *Ct. crispatus* DÜBEN & KOREN '46 = *Ct. krausei* LUDWIG '86·

Pentaceros and *Asteriscus* are regarded as strictly synonymous for the purposes of this bibliography with *Oreaster* and *Asterina* respectively, so that if *e.g.* the name *Asteriscus burtoni* should occur after *Asterina burtoni* the former is not mentioned in the bibliographical notices above explained.

1800–1899 are abbreviated thus : '00 —'99 ; 1900—1913 thus : : 00—: 13.

Publications marked with an asterisk were not accessible to the writer. It may be added that in the case of those works which deal also with other groups than the Asteroidea, only the contents of the part treating of the latter are reproduced.

ALCOCK, A. (See also under WOOD-MASON.)

'93. Natural History Notes from H. M. Indian Marine Survey Steamer 'Investigator,' Commander C. F. OLDHAM, R.N., commanding.—Series II, No. 7. An Account of the Collection of Deep-sea Asteroidea. Ann. Mag. Nat. Hist., 6th. ser., vol. xi, p. 73–121, pls. iv, v, vi.

ARCHASTERIDÆ. PARARCHASTER SLADEN. *P. semisquamatus* SLADEN. *P. huddlestoni,* n. sp. *P. violaceus,* n. sp. PONTASTER SLADEN. *P. mimicus* SLADEN. *P. cribellum,* n. sp. *P. hispidus* ALCOCK & WOOD-MASON. *P. pilosus,* n. sp. DYTASTER SLADEN. *D. exilis* SLADEN. *D. anacanthus* ALCOCK & WOOD-MASON. PERSEPHONASTER ALCOCK & WOOD-MASON. *P. croceus* ALCOCK & WOOD-MASON, 2 figs. *P. rhodopeplus*

ALCOCK & WOOD-MASON. *P. cœlochiles*, n. sp. PSEUDARCHASTER SLADEN. *P.* *mosaicus* ALCOCK & WOOD-MASON. PORCELLANASTERIDÆ. PORCELLANASTER WYV. THOMSON. *P. cœruleus* WYV. THOMSON. STYRACASTER SLADEN. *St. horridus* SLADEN. *St. armatus* SLADEN. *St. clavipes* ALCOCK & WOOD-MASON. HYPHALASTER SLADEN. *H. tara* ALCOCK & WOOD-MASON. ASTROPECTINIDÆ. ASTROPECTEN LINCK. *A. sp.* DIPSACASTER, n. g. *D. sladeni*, n. sp., 2 figs. PENTAGONASTERIDÆ. PENTAGONASTER LINCK, SLADEN. *P. investigatoris*, n. sp. *P. arcuatus* SLADEN. *P. intermedius* PERRIER. *P. pulvinus*, n. sp. MILTELIPHASTER n. g. *M. wood-masoni*, n. sp., 3 figs. DORIGONA GRAY. *D. pentaphylla*, n. sp. NYMPHASTER SLADEN. *N. florifer*, n. sp. *N. protentus* SLADEN. *N. basilicus* SLADEN. *N. nora*, n. sp. PARAGONASTER SLADEN. *P. tenui-* *radiis*, n. sp. *P. sp.* MEDIASTER STIMPSON. *M. roseus*, n. sp. ANTHENOIDES PERRIER. *A. sarissa*, n. sp. ASTERINIDÆ. PALMIPES LINCK. *P. pellucidus*, n. sp. LINCKIIDÆ. CHÆTASTER M. T. *Ch. sp.* ZOROASTERIDÆ. ZOROASTER WYV. THOMSON. *Z. alfredi*, n. sp. *Z. barathri*, n. sp. *Z. planus*, n. sp. *Z. angulatus*, n. sp. *Z. carinatus*, n. sp. *Z. gilesii*, n. sp. *Z. squameus*, n. sp. *Z. zea*, n. sp. PTERASTERIDÆ, MARSIPASTER SLADEN. *M. hirsutus* SLADEN. HYMENASTER WYV. THOMSON. *H. nobilis* WYV. THOMSON. ECHINASTERIDÆ. CRIBRELLA AGASSIZ, FORBES. *Cr. præstans* SLADEN. DICTYASTER ALCOCK & WOOD-MASON. *D. xenophilus* ALCOCK & WOOD-MASON, 2 figs. PEDICELLASTERIDÆ. PEDICELLASTER SARS. *P. atratus*, n. sp. ASTERIIDAE. ASTERIAS LINN. (subg. *Stolasterias*.) *A. mazophorus* ALCOCK & WOOD-MASON. BRISINGIDÆ, BRISINGA ASBJØRNSEN. *B. insularum* ALCOCK & WOOD-MASON. *B. andamanica* ALCOCK & WOOD-MASON. *B. bengalensis* ALCOCK & WOOD-MASON. *B. gunnii*, n. sp. FREYELLA PERRIER. *F. tuberculata* SLADEN. *F. benthophila* SLADEN.

'93a. Natural History Notes from H. M. Indian Marine Survey Steamer, 'Investigator,' Commander C. F. OLDHAM, R. N., Commanding. Series II., No. 9. An Account of the Deep Sea Collection made during the Season of 1892-93. Journ. Asiat. Soc. Bengal, vol. lxii, pt. ii, no. iv, p. 169-184, pls. viii, ix. Asteroidea, p. 171-173.

ARCHASTERIDÆ. PSEUDARCHASTER. *Ps. mosaicus* ALCOCK & WOOD-MASON. ASTROPECTINIDÆ. DIPSACASTER, n. g. (no diagnosis.) *D. pentagonalis*, n. sp. PENTAGONASTERIDÆ. CALLIASTER. *C. mammillifer*, n. sp., 2 figs. ZOROASTERIDÆ. ZOROASTER. *Z. alfredi*. *Z. barathri*.

APPELLÖF, A.

'96· Faunistiske undersøgelser i Herløfjorden. Bergens Mus. Aarbog 1894-'95, no. 11. 11 pp.

Archaster parelii. Astropecten irregularis. Pentagonaster granularis. Stichaster roseus. Solaster endeca. Asterias glacialis. Asterias mülleri. Asterias rubens.

'97· Faunistiske undersøgelsær i Osterfjorden. Bergans Mus. Aarbog 1896, no. 13. 13 pp.

Archaster parelii. Astropecten irregularis. Astropecten andromeda. Pentagonaster granularis. Stichaster roseus. Asterias rubens. Brisinga endecacnemos.

AURIVILLIUS, C. W. S.

*'86. Hafsevertebrater från nordligaste Tromsöamt och Vestfinmarken. Bihang kongl. Svenska Vetensk.-Akad. Handlinger, XI. Cited after LUDWIG.

BARRETT, L.

*'59. Descriptions of four Species of Echinodermata. Ann. Mag. Nat. Hist., 2. ser., vol. xx, p. 46–48.

BEDFORD, F. P.

: 00. On Echinoderms from Singapore and Malacca. Proc. Zool. Soc. London, p. 271–299, pl. xxi–xxiv.

Archaster typicus M. T. '40· Craspidaster glauconotus, n. sp., 2 figs. Astropecten javanicus LÜTKEN var. malaccanus n. = Astrop. andersoni SLADEN '89· Astropecten pleiacanthus, n. sp., 3 figs. Luidia longispina SLADEN '89· Luidia penangensis DE LORIOL '91· Luidia maculata M. T. '42· Iconaster longimanus MÖBIUS '59 = Astrogonium longimanum MÖBIUS '59 = Dorigona longimanum BELL '84· Goniodiscus articulatus (LINN.) DE LORIOL '84 = Pentaceros granulosus GRAY '66 = Goniodiscus scaber MÖBIUS '59· Stellaster incei GRAY '66 = St. gracilis MÖBIUS '59 = St. belcheri GRAY '66 = St. squamulosus STUDER '84· Anthenea flavescens PERRIER '76 = Hosea flavescens part. GRAY '66· Pentaceros turritus M. T. '42 = Oreaster nodulosus BELL '84· Culcita novæ-guineæ M. T. '42 var. arenosa = C. arenosa PERRIER '76· Palmipes rosaceus M. T. '42 = Asteriscus rosaceus M. T. '42· Retaster cribrosus = Pteraster cribrosus v. MARTENS '67 = Retaster insignis SLADEN '89·

BELL, F. J.

'83· Descriptions of Two New Species of Asteroidea in the Collection of the British Museum. Ann. Mag. Nat. Hist., 5th. ser., vol. 12, p. 333–335.

Asterias nautarum, n. sp. Culcita acutispinosa, n. sp.

'84· Contributions to the Systematic Arrangement of the Asteroidea, II. The Species of Oreaster. Proc. Zool. Soc. London, p. 57–87.

O. australis LÜTKFN '71· *O. forcipulosus* LUTKEN '64· *O. nodulosus* PERRIER '76·
O. mammillatus M. T. '42· *O. valvulatus* M. T. '42· *O. granulosus* GRAY '47· *O.
chinensis* GRAY '40· *O. decipiens*, n. sp. *O. nodosus* = *Pentaceros turritus* PERRIER
'76 = *Asterias nodosa* LINNÉ xii part. *O. hiulcus* M. T. '42· *O. hedemanni* LÜTKEN
'71· *O. lincki* = *Asterias lincki* DE BLAINVILLE Dic. sc. nat. lx = *Pentaceros muricatus*
PERRIER '76· *O. alveolatus* PERRIER '76· *O. reinhardti* LÜTKEN '64· *O. productus*, n.
sp. (?) *O. lütkeni*, n. sp. *O. occidentalis* VERRILL '67· *O. dorsatus* (PERRIER '75) =
Asterias dorsata LINNÉ 1753 = *Asterias nodosa* LINNÉ 1766 (part.) = *O. clavatus* M. T.
'42 = *O. dorsatus* LÜTKEN (PERRIER's synonymy inexact). *O. reticulatus* = *Asterias
reticulata* LINNÉ 1766 = *Pentaceros grandis, gibbus, reticulatus. aculeatus* GRAY '40 = non
Oreaster gigas GRUBE '57 = *O. tuberosus* MÖBIUS. *O. armatus* = *Pentaceros (Nidorellia)
armata* GRAY '40 = *Nidorellia armata* PERRIER '76 = *Nidorellia michelini* PERRIER '76·
O. westermanni LÜTKEN '71· *O. superbus* MÖBIUS '59· *O. gracilis* LÜTKEN '71· *O.
grayi*, n. sp. = *Pentaceros nodosa* GRAY '40· *O. orientalis* M. T. '42· *O. troscheli*, n.
sp. *O. mülleri*, n. sp. *O. carinatus* M. T. '42· Sp. incertæ : Sp. α. Sp. β· List of
apparently distinct species: *affinis* M. T., *alveolatus* PERRIER, *armatus* PERRIER, *australis* LÜTKEN, *carinatus* M. T., *chinensis* M. T., *dorsatus* M. T., *forcipulosus* LUTKEN,
gracilis LÜTK., *granulosus* PERRIER, *hedemanni* LÜTK., *hiulcus* M. T., *lincki* LUTK.,
mammillatus M. T., *nodosus* M. T., *nodosus* (GRAY), *nodulosus* PERRIER, *occidentalis*
VERRILL, *orientalis* M. T., *regulus* M. T., *reinhardti* LÜTK., *reticulatus* M. T., *superbus*
MÖBIUS, *tuberculatus* M. T., *valvulatus* M. T., *verrucosus* M. T., *westermanni* LUTK.
List of the more important synonyms: *aculeatus* M. T. = *reticulatus, clavatus* M. T. =
dorsatus, clouei PERRIER = *nodosus, francklini* GRAY = *nodosus, gigas* LINNÉ = *reticulatus,
lapidarius* GRUBE = *reticulatus, mammosus* PERRIER = *nodosus, michelini* ¦PERRIER =
armatus, muricatus GRAY = *lincki, nodosus* GRAY = *grayi* (BELL), *obtusangulus* M. T. =
Goniaster obtusangulus, sebæ DE BLAINVILLE = *reticulatus, turritus* = *nodosus.*

'84a. **Echinodermata.** Report on the Zoological Collections made in the
Indo-Pacific Ocean during the Voyage of H. M. S. 'Alert,' 1881-
2, p. 117–177 + p. 509–512.

Asterias calamaria. Asterias polyplax. Echinaster purpureus = *Othilia purpurea* GRAY
'40 = *Echin. fallax* M. T. '42· *Metrodira subulata. Linckia lævigata. Linck. nodosa.
Linck. marmorata* = *Ophidiaster marmoratus* MICHELIN '45· *Linck. pauciforis. Linck.
megaloplax*, n. sp. *Linck. sp. Anthenea flavescens. Oreaster gracilis. Oreast. nodosus*
= *Asterias nodosus* LINNÉ = *Pentaceros turritus* PERRIER '75· *Oreast. sp. Stellaster
belcheri. Stellaster incei. Pentagonaster coppingeri*, n. sp. *Pentagonaster validus*, n.
sp. *Dorigona longimana. Asterina belcheri. Asterina calcar. Asterina cepheus. Asterina gunnii. Asterina regularis. Asterina (Nepanthia) brevis*, 2 figs. *Patiria crassa*
GRAY. *Luidia sp. Astropecten coppingeri*, n. sp. *Astropecten polyacanthus. Archaster*

*typicus.   Retaster insignis.   Echinaster purpureus.   Linckia lævigata.   Linck. multiforis.
Linck. diplax.   Scytaster variolatus.   Oreaster lincki.   Culcita schmideliana.   Gymnasteria carinifera.   Archaster typicus.   Astropecten polyacanthus.   Astrop. hemprichii (?).*

'87.   Report on a Collection of Echinodermata from the Andaman Islands.   Proc. Zool. Soc. London, p. 139–145.   (January.)

Acanthaster echinites.   Fromia indica.   Culcita schmideliana.   Culcita grex.   Culcita sp.
Listed are the following species: *Acanthaster echinites* E. & S., *Fromia indica*
PERRIER, *Fromia tumida* BELL, *Linckia lævigata* L , *Linckia pacifica* GRAY, *Scytaster
novæ-caledoniæ* PERRIER, *Culcita grex* M. T., *Culc. schmideliana* RETZIUS, *Culc. sp.*
(" *Randasia granulata* "), *Astropecten polyacanthus* M. T., *Astropecten sp.*, *Archaster
typicus* M. T.

'87a.   Echinoderm Fauna of the Island of Ceylon.   Trans. Roy. Dublin
Soc., iii, ser. 2, p. 643–658 + 2 pls.

Linckia lævigata GMEL. = *Ophidiaster miliaris* M. T.   *Linckia sp.*   *Scytaster novæ-caledoniæ* PERRIER.   *Scyt. variolatus* (RETZIUS) = *Asterias variolata* RETZIUS.   *Fromia milleporella* GRAY = *Asterias milleporella* LAMARCK.   *Fromia tumida* BELL.   *Oreaster lincki*
(DE BLAINVILLE) = *Asterias lincki* DE BLAINVILLE = *Pentaceros muricatus* PERRIER.
Astropecten polyacanthus M. T. = *Astrop. hystrix* M. T. = *Astrop. armatus* M. T.   *Luidia
maculata* M. T.

'88.   Report on a Collection of Echinoderms made at Tuticorin, Madras,
by Mr. EDGAR THURSTON, C. M. Z. S., Superintendent, Government
Central Museum, Madras.   Proc. Zool. Soc. London, p. 383–389.

Oreaster thurstoni, n. sp.   Listed are the following species.   (a) From Mr. E. THURS-
TON's collection: *Echinaster purpureus* GRAY, *Linckia lævigata* GMELIN, *Anthenea
pentagonula* LAMK., *Oreaster lincki* DE BLAINVILLE, *Oreaster superbus* MÖBIUS, *Oreaster
thurstoni*, n. sp., *Asterina cepheus* M. T., *Luidia hardwickii* GRAY, *Luid. maculata* M.
T., *Luid. sp.* (juv.), *Astropecten hemprichii* M. T., *Astropecten sp.* (juv.).   (b) From
the Sea of Bengal: *Acanthaster echinites* E. & S., *Echinaster purpureus* GRAY,
Fromia indica PERRIER, *Fromia milleporella* LAME., *Fromia tumida* BELL, *Linckia
lævigata* GM., *Linckia pacifica* GRAY, *Scytaster ægypticus* GRAY, *Scyt. galatheæ* LÜTKEN,
Scyt. variolatus RETZIUS, *Oreaster lincki* DE BLAINVILLE, *Oreas. superbus* MÖBIUS,
Oreas. reinhardti LÜTKEN, *Oreas. thurstoni* BELL, *Oreas. westermanni* LÜTKEN, *Culcita
grex* M. T., *Culc. novæ-guineæ* M. T., *Culc. schmideliana* RETZIUS, *Asterina burtonii*
GRAY, *Asterina cepheus* VALENCIENNES, *Luidia hardwickii* GRAY, *Luid. maculata* M.
T., *Astropecten hemprichii* M. T., *Astrop. polyacanthus* M. T., *Astrop. euryacanthus*
LUTKEN, *Archaster typicus*.

'92. Catalogue of the British Echinoderms in the British Museum. 202 pp. + 16 pls.

ARCHASTERIDÆ. *Pararchasterinæ.* PONTASTER SLADEN '85· *P. tenuispinis* = *Astropecten tenuispinus* DÜB. & KOREN '46 = *Archaster tenuispinus* SARS '61 = *P. tenuispinus* SLA-DEN '89 = *P. tenuispinus* var. *platynota* SLADEN '89 = *P. hebitus* SLADEN '89 = *P. limbatus* SLADEN '89· *Plutonasterinæ.* PLUTONASTER SLADEN '85· *P. bifrons* SLADEN '89 = *Archaster bifrons* WYV. THOMSON '73· *P. pareli* SLADEN '89 = *Astropecten parelii* DÜB. & KOREN '46 = *Archaster parelii* Sars '61· PORCELLANASTERIDÆ. *Ctenodiscinæ.* CTENODISCUS. *Ct. crispatus* DÜBEN & KOREN = *Asterias crispata* RETZ. '05 = *Asterias polaris* SABINE '24 = *Astropecten polaris* GRAY '40 = *Ct. polaris* M. & TR. '42 = *Ct. pygmæus* M. & TR. '42 = *Ct. corniculatus* PERRIER, 3 figs. ASTROPECTINIDÆ. *Astropectininæ.* LEPTOPTYCHASTER = LEPTYCHASTER SMITH '76 = LEPTOPTYCHASTER '79· *L. arcticus* SLADEN '89 = *Astropecten arcticus* SARS '51 = *Archaster lütkeni* BARRETT '57 = *Archaster arcticus* VERRILL '78· ASTROPECTEN C. F. SCHULTZE 1760 = STELLARIA NARDO '34 = ASTERIAS AG. '36 = CRENASTER D'ORB. '49 *A. irregularis* GRAY '40 = *Asterias irregularis* PENN. 1777 = *Asterias aranciaca* O. F. MÜLLER 1776 = *Asterias aurantiaca* FORBES '39 = *A. mülleri* M. & Tr. '44 = *A. echinulata* M. & Tr. '44 = *A. acicularis* NORM. '65· *A. sphenoplax* BELL '92· 3 figs. PSILASTER SLADEN '85· *Ps. andromeda* SLADEN '89 = *Astropecten andromeda* M. & Tr. '42 = *A. christii* DÜB. & KOREN '44· BATHYBIASTER DAN. & KOREN '83 = ASTROPECTEN (pars) DAN. & KOBEN. *B. vexillifer* SLADEN '89 = *Archaster vexillifer* WYV. THOMSON '73 = *Astropecten vexillifer* VIGUIER '78· *Luidiinæ.* LUIDIA FORBES '39 = HEMICNEMIS M. & TR. '40 = PETALASTER GRAY '40· *L. ciliaris* DUJ. & HUPÉ '62 = *Asterias rubens* JOHNSTON '36 = *Asterias ciliaris* PHILIPPI '37 = *L. fragilissima* FORBES '39 = *Asterias pectinata* COUCH '40 = *L. ? ciliaris* GRAY '40 = *Asterias imperati* DELLE CHIA. '41 = *Luidia savignii* M. & Tr. '42· *L. sarsi* DÜB. & KOREN '45 = *L. fragilissima* FORBES '39 = *L. savignii* DUB. & KOREN '46· PENTAGONASTERIDÆ. *Pentagonasterinæ.* PENTAGONASTER C. F. SCHUL-TZE 1760 = GONIASTER (pars) AG. '36 = ASTROGONIUM (pars) M. & TR. '42· *P granularis* KOREN & DAN. '84 = *Asterias granularis* RETZ. 1783 = *Asterias tessellata* var. *A.* LAM. '16 = *Astrogonium granulare* M. & Tr. '42 = *Astrogonium boreale* BAR-RET '57 = *Goniaster granularis* LÜTK. = *P. balteatus* SLADEN '91 = *P. concinnus* SLADEN '91, 3 figs. *P. greeni* BELL '89, 1 fig. NYMPHASTER SLADEN '85· *N. subspinosus* NORM. in BOURNE '90 = *Pentagonaster subspinosus* PERR. = *Nymphaster protentus* SLADEN '89· HIPPASTERIAS (incertæ sedis) = HIPPASTERIA GRAY '40 = GONIASTER (pars) AG. '36 = ASTROGONIUM M. & TR. '42· *H. phrygiana* = *Asterias equestris* (pars) L. 1766 = *Asterias phrygiana* PARELIUS '68 = *Goniaster phrygianus* '36 = *Asterias johnstoni* GRAY '36 = *Hippasterias plana, H. europœa, H. johnstoni* GRAY '41 = *Goniaster equestris* FORBES '40 = *Asterias (Goniaster) equestris* GOULD = *Astrogonium phrygianum* M. & Tr. '42 = *Goniaster abbensis* FORBES '43 = *Astrogonium aculeatum* BARRETT = *Hippasteria phry-*

giana VERRILL '85· *Mimasterina.* MIMASTER SLADEN '82· *M. tizardi* SLADEN '82· GYMNASTERIIDÆ. PORANIA GRAY '40 = GONIASTER (pars) AG. '36 = ASTEROPSIS (pars) M. & TR. '42· *P. pulvillus* GRAY '48 = *Asterias pulvillus* O. F. MÜLLER 1788 = *Goniaster templetoni* FORBES '39 = *Porania gibbosa* GRAY '40 = *Asteropsis pulvillus* M. & TR. '42· 2 figs. RHEGASTER SLADEN '83· *R. murrayi* SLADEN '83· CHEILASTER = MARGINASTER PERRIER '81· *Ch. fimbriatus* = *Marginaster fimbriatus* SLADEN '89· LASIASTER SLADEN '89· *L. villosus* SLADEN '89· ASTERINIDÆ. ASTERINA NARDO '34· *A. gibbosa* FORBES '39 = *Asterias gibbosa* PENN. 1777 = *Asterias verruculata* RETZ. '05 = *Asterias exigua* DELLE CHIAJE '25 = *Asterias pulchella* BL. '34 = *A. minuta* NARDO '34 = *Asterias membranacea* GRUBE '40 = *Asteriscus verruculatus* M. & TR. '42 = *Asteriscus gibbosus* FISCHER '72· 2 figs. PALMIPES AG. '36 = ANSEROPODA NARDO '34 = ASTERISCUS (pars) M. & TR. '42· *P. placenta* NORM. = *Asterias placenta* PENN. 1777 = *Asterias membranaceus* RETZ. 1783 = *Asterias cartilaginea* FLEM. '28 = *Anseropoda membranacea* NARDO '34 = *P. membranaceus* AG. '36 = *Asteriscus palmipes* M. & TR. '42· STICHASTERIDÆ. STICHASTER M. & TR. '40 = TONIA GRAY '40 = ASTERACANTHION M. & TR. = CŒLASTERIAS (pars) VERR. '71 = *Stephanasterias* VERR. '72· *St. roseus* SARS '61 = *Asterias rosea* O. F. MULLER 1788 = *Linckia rosea* THOMPSON '40 = *Cribrella rosea* FORBES '40 = *Asteracanthion roseus* DUB. & KOREN '46 = *Henricia rosea* GRAY '48 = *Asteracanthion roseus* (pars) M. & TR. '42· NEOMORPHASTER SLADEN '89 = GLYPTASTER SLADEN '85· *N. eustichus* SLADEN '89· ZOROASTER WYV. THOMS. '73· *Z. fulgens* WYV. THOMSON '73· SOLASTERIDÆ. SOLASTER FORBES '39 = CROSSASTER M. & TR. '40 = LOPHASTER VERR. '78· *S. papposus* M. & TR. '42 = *Asterias helianthoides* (?) PENN. 1777 = *Asterias papposa* FABR. 1780 = *Stellonia papposa* AG. '36 = *S. papposa* FORBES '39 = *Solaster (Polyaster) papposa* GRAY '40 = *Crossaster papposus* M. & TR. '40· Ditto var. *septentrionalis* = *Crossaster papposus* var. *septentrionalis* SLADEN '84· *S. endeca* FORBES '39 = *Asterias endeca* L. 1771 = *Asterias aspera* O. F. MULLER 1776 = *Asterias endica* FLEM. '28 = *Stellonia endeca* AG. '36 = *Solaster (Endeca) endeca* GRAY '40· *S. furcifer* DÜB. & KOREN '44 = *Lophaster furcifer* VERR. '78 = *Chœtaster borealis* DÜB. '45· CORETHRASTERIDÆ. CORETHRASTER MARENZ. '78 = KORETHRASTER WYV. THOMSON '73· *C. hispidus* WYV. THOMSON '73· PTERASTERIDÆ. PTERASTER M. & TR. '42. *Pt. militaris* M. & TR. = *Asterias militaris* O. F. MÜLLER 1776 = *Asteriscus militaris* M. & TR. '42· Ditto var. *prolata* = *Pt. militaris* var. *prolata* SLADEN '84· *Pt. personatus* SLADEN '91· HYMENASTER WYV. THOMSON '73· *H. pellucidus* WYV. THOMSON '73· *H. giganteus* SLADEN '92· ECHINASTERIDÆ. HENRICIA GRAY '40 = LINCKIA FORBES '39 = CRIBELLA FORBES '40 = CRIBRELLA LÜTKEN '57 = ECHINASTER M. & TR. '42· *H. sanguinolenta* = *Asterias sanguinolenta* O. F. MÜLLER 1776 = *Echinaster sanguinolentus* SARS '44 = *Cribrella sanguinolenta* LÜTK. '57 = *Asterias pertusa* O. F. MÜLL. 1776 = *Asterias oculata* PENN. 1777 = *Asterias spongiosa* FABR. 1780 = *Asterias seposita* RETZ. '05 = *Linckia oculata* FORBES '39 = *H. oculata* GRAY '40 = *Cribella oculata* FORBES '40 = *Cribrella oculata* PERRIER '75

= *Echinaster oculatus* M. & Tr. '42 = *Echinaster eschrichtii* M. & Tr. '12 = *Echinaster sarsi* M. & Tr. '44· Ditto var. *abyssicola* = *Cribrella sanguinolenta* var. *abyssicola* Norman '69 = *Cribrella oculata* var. *cylindrella* Sladen '84· Ditto var. *curta* = *Cribrella sanguinolenta* var. *curta* Norman '68· Asteriidæ. Asterias Linn. 1758 = Stellonia Nardo '34 = Uraster Forbes '40 = Asteracanthion M. & Tr. '40 = Diplasterias Perrier '91· *A. glacialis* Linn. 1758 = *Uraster glacialis* Forbes '40 = *A. spinosa* Penn. 1777 = *A. angulosa* O. F. Müller 1788 = *A. echinophora* Dell. Chiaje '25 = *Stellonia angulosa* Ag. '36 = *Stellonia webbiana* D'Orb. '39 = *Asteracanthion webbianus* Duj. & Hupé '62 = *A. madeirensis* Stimpson '62 = *Marthasterias foliacea* Jullien '78 = *Asterias (Stolasterias) glacialis* Sladen '89 = *Stellonia glacialis* Forbes '39· *A. mulleri* Norman '65 = *Leptasterias mülleri* Verr. '66 = *Asterias (Leptasterias) mulleri* Sladen '89· *A. rubens* Linn. 1758 = *A. clathrata* Penn. 1777 = *A. glacialis* Penn. '28 = *A. holsatica* Retz. '05 = *A. minuta* Retz. '05 = *Uraster rubens* Forbes '40 = *A. violacea* O. F. Müller 1788 = *Uraster violacea* Forbes '40 =? *A. helgolandica* Ehrenberg '37· Ditto var. *attenuata* = *A. rubens* var. *attenuata* Hodge '72· *A. murrayi* Bell '91· 2 figs. *A. hispida* Penn. 1777 = *Stellonia hispida* Forbes '39 = *Uraster hispida* Forbes '40 = *A. rubens* var. *hispida* Hodge '72, 2 figs. *A. tenuispinis* = *A. tenuispina* Lam. '16· Brisingidæ. Brisinga Asbjørnsen '56· *B. endecacnemos* Asbj. '56· *B. coronata* G. O. Sars '71· Odinia Perr. '85 = Brisinga auct. (pars). *O. pandina* Sladen '89 = *Brisinga coronata* Wyv. Thomson (pars) '73.

'93. On the Names or Existence of three Exotic Starfishes. Ann. Mag. Nat. Hist., 6. ser., vol. 12, p. 25–29.

Asterina stellifer = *Asterina marginata* (Val.) Perrier = *Asteriscus minutus* M. T = *Asteriscus brasiliensis* Lütken '59 = *Asteriscus marginatus* Val. MS., Perrier '69 = *Asterina stellifera* Lütken '71· *Goniodiscus articulatus* = *Goniaster articulatus* Lütken '64 = *Asterias aranciaca* Linné 1758 = *Goniodiscus sebæ* M. T. '42 = *Goniodiscus articulatus* de Loriol '84· *Pentaceros clavatus* = *Asterias nodosa* Linné 1758 = *Oreaster dorsatus* Lütken. *Goniodiscus gracilis* Gray (phantom sp). *Gymnasteria valvulata*. (I can not trace the synonymy of *Goniaster articulatus* Lutken '64 with *Goniodiscus sebæ* Mull. & Troschel '42).

'94. On the Echinoderms collected during the Voyage of H. M. S. "Penguin" and by H. M. S. "Egeria," when surveying Macclesfield Bank. Proc. Zool. Soc. London, p. 392–413.

Archaster typicus M. T. *Archaster tenuis*, n. sp., 3 figs. *Luidia maculata* M. T. *Luid. longispinis* ? *Luid. forficifer. Luid. sp. Goniodiscus sp. Culcita sp.*, 1 fig. *Patiria briareus*, n. sp., 3 figs. *Chætaster moorei*, n. sp. *Ophidiaster helicosuchus* Sladen. *Rhipidaster vannipes* Sladen. *Asterias volsellata* Sladen. Listed are the following species. (a) From North-west Australia : *Astropecten polyacanthus* M. T., *Astrop. schœnleini* M. T., *Astrop. zebra* SLADEN, *Luidia hardwickii* GRAY, *Luid. aspera ?* SLADEN, *Iconaster longimanus* MÖBIUS, *Stellaster incei* GRAY, *Pentaceros nodulosus* PERRIER, *Culcita pentangularis* GRAY, *Ophidiaster helicostichus* SLADEN, *Linckia marmorata* MICHELIN, *Linck. megaloplax* BELL, *Nardoa tuberculata* GRAY, *Metrodira subulata* GRAY, *Echinaster purpureus* GRAY. (b) From Macclesfield Bank: *Archaster typicus* M. T., *Archaster tenuis*, n. sp., *Astropecten polyacanthus* M. T., *Luidia ? aspera* (juv.) SLADEN, *Luid. forficifer* SLADEN, *Luid. hardwickii* GRAY, *Luid. longispinis* SLADEN, *Luid. maculata* (juv.) M. T., *Goniodiscus rugosus* (juv.) PERRIER, *Culcita* (juv.), n. sp , *Patiria briareus*, n. sp., *Chœtaster moorei*, n. sp., *Asterina cepheus* M. T., *Fromia milleporella* LAMARCK, *Leiaster leachi* (juv.) GRAY, *Leiaster speciosus* (juv.), MARTIN, *Nardoa tuberculata* GRAY, *Rhipidaster ? vannipes* SLADEN, *Mithrodia clavigera* LAMARCK, *Echinaster purpureus* GRAY, *Asterias volsellata* SLADEN.

'99. Report on the Echinoderms (other than Holothurians) collected by Dr. WILLEY. With Figures on Pl. XVII, and One Figure in the Text. Zoological Results based on Material from New Britain, New Guinea, Loyalty Islands and elsewhere, collected during the Years 1895, 1896 and 1897, by ARTHUR WILLEY. Part II, p. 133–140.

Astropecten monacanthus. Pentaceros lincki + Pentaceros nodosus. Pentaceropsis obtusata. Culcita, 2 figs. *Gymnasteria carinifera. Asterina exigua. Fromia milleporella. Linckia multiforis. Nardoa tuberculata. Acanthaster echinites. Mithrodia clavigera. Echinaster purpureus. Echinaster eridanella.* 2 figures of an unknown starfish, perhaps allied to the *Pythonasterina* of SLADEN.

: 03. Report on a Collection of Echinoderms from the Neighbourhood of Zanzibar. Part I. Ann. Mag. Nat. Hist., 7. ser., vol. xii, p. 244–248.

ASTROPECTINIDÆ. *Astropecten Hemprichii* M. T. *Astropecten polyacanthus* M. T. *Luidia Savignii* AUDOUIN. PENTAGONASTERIDÆ. *Goniodiscus* yg. *Pentaceros Lincki* DE BL. *Pentaceros hiulcus* M. T. *Pentaceros* yg , ? *hiulcus. Culcita pentangularis* GRAY. ASTERINIDÆ. *Asterina cepheus* VAL. LINCKIIDÆ. *Ophidiaster fuscus* GRAY. *Leiaster coriaceus* PETERS. *Linckia Ehrenbergi* M. T. *Linckia marmorata* MICHELIN. *Nardoa variolata* RETZIUS. PTERASTERIDÆ. *Retaster cribrosus* v. MARTENS. ECHINASTERIDÆ. *Mithrodia clavigera* LAMK. *Echinaster purpureus* GRAY.

BIDENKAP, O.

'99. Undersøgelser over Lyngenfjordens evertebratfauna. Tromsø Mus. Aarsheft., xx (1897), p. 80–103.

Asterias glacialis LINN. *Asterias mulleri* M. SARS. *Asterias rubens* LINN. *Asterias stellionura* E. PERRIER. *Solaster endeca* FORBES. *Pteraster militaris* MÜLL. *Henricia sanguinolenta* MÜLL. *Ctenodiscus crispatus* RETZ.

'99a. Tromsøsundets Echinodermer. Tromsø Mus. Aarsheft., xx (1897), p. 104-112.

Astropecten andromeda MÜLL. & TROSCH. *Solaster papposus* FABR. *Solaster endeca* FORBES. *Pteraster militaris* O. F. MÜLL. *Pteraster pulvillus* M. SARS. *Henricia sanguinolenta* O. F. MÜLL. *Asterias rubens* LINN. *Asterias mulleri* M. SARS. *Asterias glacialis* LINN. *Asterias sp. Ctenodiscus crispatus* RETZ.

BLAINVILLE, H. M. D. de.

'34· Manuel d'actinologie ou de zoophytologie. viii + 695 pp. + 100 pls. Stelleridea, p. 233-242 + pl. xxii-xxiii. I have been enabled to consult this work through the kindness of my friend, Dr. K. KINOSHITA of this Institute.

Asteria luna L. *A. discoidea* LAMK., 1 fig. *A. granularis* L. *A. pentagonula* LAMK. *A. membranacea* L., 1 fig. *A. rosacea* LAMK. *A. calcar* LAMK. *A. pulvillus* L. *A. minuta* L. *A. gibbosa* PENN. *A. pulchella* BLAINV., 1 fig. *A. tessellata* LAMK., 1 fig. *A. equestris* L. *A. carinifera* LAMK. *A. nobilis* L. *A pleyadella* LAMK. *A. oculata* LINCK. *A. ocellifera* LAMK. *A. punctata* LAMK. *A. cuspidata* LAMK. *A. vernicina* LAMK. *A. obtusangula* LAMK. *A. reticulata* L. *A. sebæ* BLAINV. *A. nodosa* L. *A. Linkii* BLAINV. *A. auranciaca* L. *A. calcitrapa* LAMK. *A. irregularis* LINCK. *A. regularis* LINCK. *A. fimbriata* LINK. *A. bispinosa* OTTO. *A. rubens* L., 4 figs. *A. violacea* L. *A. spongiosa* FAB. *A. acuminata* LAMK. *A. striata* LAMK. *A. glacialis* L. *A. angulosa* MÜLLER. *A. clathrata* PENN. *A. milleporella* PENN. *A. multifora* PENN. *A. variolata* PENN. *A. granifera* LAMK. *A. spinosa* LINCK. *A. lævigata* L. *A. cometa* BLAINV. *A. bicolor* LAMK. *A. reticulata* LINCK. *A. phrygiana* L. *A. pertusa* L. *A. clavigera* LAMK. *A. seposita* L. *A. cylindrica* LAMK. *A. subulata* LAMK. *A. ophidiana* LAMK. *A. tenuispina* LAMK. *A. arenata* LAMK. *A. senegalensis* LAMK.⁻ *A. endeca* L. *A. papposa* L. *A. echinitis* LAMK. *A. helianthus* LAMK., 1 fig.

BROWN, R. N. R.

: 10. Echinoidea and Asteroidea from the Mergui Archipelago and Moskos Islands, Lower Burma. Proc. Roy. Phys. Soc. Edinburgh, vol. 18, p. 21-35. Feb. 17.

Archaster typicus M. & T. *Craspidaster hesperus* (M. & T.) SLADEN. *Astropecten mauritianus* GRAY. *Astropecten zebra* SLADEN. *Astropecten polyacanthus* M. & T.

Luidia forficifer SLADEN. *Luidia maculata* M. & T. *Goniodiscus articulatus* (LINN.) LÜTKEN. *Stellaster incei* GRAY. *Anthenea flavescens* (GRAY) PERR. *Anthenea penta-gonula* (LAM.) PERRIER = *Goniodiscus articulatus* PERR. '69· *Pentaceros granulosus* GRAY. *Pentaceros lincki* (DE BLAINV.) = *P. muricatus* PERR. *Pentaceros superbus* MÖBIUS. *Pentaceros gracilis* (LÜTKEN). *Palmipes rosaceus* (LAM.) DUJ. & HUPÉ. *Fromia milleporella* GRAY. *Retaster cribrosus* (v. MART.). *Echinaster purpureus* (GRAY) BELL.

BRUNCHORST, J.

'91· Die biologische Meeresstation in Bergen. Bergens Mus. Aarsberet. for 1890. 31 pp. + 5 pls.

Asteracanthion rubens, glacialis, mulleri. Stichaster roseus. Echinaster sanguinolentus. Solaster furcifer, papposus, endeca. Pteraster pulvillus, militaris. Astrogonium granu-lare, phrygianum. Asteropsis pulvillus. Archaster parelii. Brisinga endecacnemos. Astropecten mülleri. Luidia sarsii. All simply listed.

BRØGGER, W. C.

*:01. Om de senglaciale og postglaciale nivåforandringer i Kristianiafeltet (Molluskenfaunan). Norges geol. Undersøg., No. 31. Cited after the Zoological Record.

CLARK, H. L.

:02. Papers from the Hopkins Stanford Galapagos Expedition, 1898–1899, xii. Echinodermata. Proc. Washington Ac. Sci., vol. iv, p. 521–531.

Luidia bellonæ LÜTKEN. *Luidia columbiæ* (GRAY). *Pentaceros occidentalis* (VERRILL). *Nidorellia armata* (GRAY). *Paulia horrida* GRAY. *Pharia pyramidata* (GRAY). *Phata-ria unifascialis* (GRAY). *Heliaster cumingi* (GRAY). *Heliaster multiradiata. Culcita schmideliana* (RETZIUS). *Gymnasterias valvulata* PERRIER. *Acanthaster ellisii* (GRAY).

*:05. Fauna of New England. 4. List of the Echinodermata. Pap. Boston Soc. vii. 16 pp. Cited after the Zoological Record.

:08. Some Japanese and East Indian Echinoderms. Bull. Mus. Comp. Zoöl. Harvard Coll., vol. li, No. 11, p. 279–311.

Archaster typicus. Oreaster nodosus. Culcita novæ-guineæ. Gymnasteria carinifera. Asterina cepheus. Asterina exigua. Asterina pectinifera. Linckia lævigata. Nardoa

tuberculata. Pteraster obesus, n. sp. *Pteraster multiporus,* n. sp. List of and key to
Pteraster. *Echinaster eridanella. Asterias rollestoni. Asterias similispinus.*

: 13. Echinoderms from Lower California, with Descriptions of New
Species. Scient. Res. Exped. to Gulf of Cal. in Charge of C. H.
Townsend, by the U. S. Fish. S. S. Albatross in 1911. Commander
C. H. Burrage, U. S. N., commanding. Bull. Amer. Mus. Nat.
Hist. xxxii, Art. viii, p. 185–236. Iss. July 9.

Eremicaster tenebrarius Fisher : 05. *Erem. pacificus* (Ludwig : 05). *Ctenodiscus
crispatus* (Retzius '05). *Leptychaster inermis* (Ludwig : 05). *Astropecten erinaceus*
Gray. *Psilaster pectinatus* (Fisher : 05). *Thrissacanthias penicillatus* (Fisher : 05).
Luidia phragma H. L. Clark : 10. *Pectinaster agassizii* (Ludwig : 05). *Nearchaster
aciculosus* (Fisher : 10). *Pseudarchaster pectinifer* Ludwig : 05. *Pseudarch. pusillus*
Fisher : 05. *Ceramaster leptoceramus* (Fisher : 05). *Ceram. patagonicus* (Sladen
'89). *Hippasteria californica* Fisher : 05. *Hippast. spinosa* Verrill : 09. *Oreaster
occidentalis* Verrill '66· *Amphiaster insignis* Verrill '68· *Linckia columbiæ*
Gray '40· *Phataria unifascialis* (Gray '40). *Echinaster tenuispinus* Verrill '71·
Henricia clarki Fisher : 10. *Henri. læviuscula annectens* Fisher : 10. *Solaster paxillatus* Sladen '89· *Sol. borealis* Fisher : 06. *Heterozonias alternatus* (Fisher : 06).
Lophaster furcilliger Fisher : 05. *Peribolaster biserialis* Fisher : 05. *Pteraster
jordani* Fisher : 05. *Hymenaster perissonotus* Fisher : 10. *Hymen. quadrispinosus* Fisher
: 05. *Zoroaster evermanni* Fisher : 05. *Zoro. ophiurus* Fisher : 05. *Zoro. platyacanthus,* n. sp. *Heliaster kubiniji* Xantus '60· *Pedicellaster hyperoncus,* n. sp. *Pedicell.
improvisus* Ludwig : 05. *Pisaster ochraceus* (Brandt '35). *Asterias forreri* de Loriol
'87· *Brisinga panamensis* Ludwig : 05.

Cuénot, L.

'87. Contribution a l'étude anatomique des Astérides. Arch. zool. expér.
gén., 2. sér., t. v, suppl. 144 pp.+9 pls.

In the section on pedicellaria, which is the part of direct taxonomic importance, the
following species are dealt with : *Luidia ciliaris, Gymnasteria carinifera, Asterina
gibbosa, Asterias glacialis.*

Danielssen, D. C., & Koren, J.

'77· Fra den norske Nordhavsexpedition, I. Nyt. Mag. Naturvidens.,
Bd. xxiii. I have only separate, on the cover of which the year of
publication is printed as above. Ludwig gives it as '76·

Solaster affinis Brandt *Archaster tenuispinus* Düb. & Koren. *Archaster parelii*

Düb. & Koren, var. *longobrachialis* nob. *Astropecten pallidus* nob. *Hymenaster pellucidus* W. Thomson.

Derjugin, K. M.

: 06.	Murmansche Biologische Station, 1899–1905. Trav. soc. imp. nat. St.-Pétersbourg, zool. physiol. xxxv, livr. 4. 10 pls. + 5 charts + 4 plans.

Ctenodiscus crispatus (Retzius , forma *brevimana* Kal *Rhegaster tumidus* (Stuxberg) var. *tuberculatus* (Sladen). *Crossaster papposus* (L.). *Solaster endeca* (Retzius). *Pteraster militaris* (O. F. Müller). *Pter. pulvillus* M. Sars. *Cribrella sanguinolenta* (O. F. Müller). *Asterias linckii* (Müll. & Troschel). *Ast. rubens* L., var. *violacea* Perr.

Dewhurst.

* '34·	The Natural History of the Order Cetacea. Cited after Perrier and Duncan and Sladen.

Döderlein, L.

'88·	Echinodermen von Ceylon. Bericht über die von Herren Dres. Sarasin gesammelten Asteroidea, Ophiuroidea und Echinoidea. Zool. Jahrb., System., Bd. 3, p. 821–846.

Acanthaster echinites Ellis = *Echinaster solaris* M. T. '42· *Pteraster cribrosus* v. Martens '67· *Asterina cephea* Valenciennes-Perrier '76· *Disasterina ceylanica*, n. sp , 6 figs. *Linckia multiforis* Lamarck. *Fromia milleporella* Lam. *Fromia tumida* Bell '82· *Scytaster ægypticus* Gray = *Scy.* zodiacalis M. T. '42· *Scy. novæ-caledoniæ* Perrier '75. *Goniodiscus sebæ* M. T. *Pentaceros sp.* *Culcita schmideliana* Retzius = *C. discoidea* M. T. '42· *Gymnasteria carinifera* Lamarck. *Astropecten tamilicus*, n. sp., 4 figs. *Astropecten polyacanthus* M. T. *Luidia maculata* M. T.

'96.	Bericht über die von Herrn Professor Semon bei Amboina und Thursday Island gesammelten Asteroidea. Zool. Forschungsreisen in Australien u. d. Malayischen Archipel, Bd. v, p. 301–326, Taf. xviii–xxii.

Archaster typicus M. T. '42· *Astropecten granulatus* M. T. '42· 2 figs. *Astropecten zebra* Sladen '83 = *Astrop. coppingeri* Bell '84· 2 figs. *Astropecten velitaris* Martens '65· 2 figs. *Luidia maculata* M. T. '42· *Iconaster longimanus* (Möbius '59) = *Astrogonium longimanum* Mobius '59 = *Pentagonaster longimanus* Perrier '76· 1 fig. *Stellaster incei* Gray '47 = *Stell. belcheri* Gray '47 & '66 = *Goniaster belcheri* Lütken '71 = *Pentagonaster incei* & *belcheri* Perrier '76· *Goniodiscus pleyadella* (Lamarck '16) = *Asterias pleyadella* LAMARCK '16, 7 figs. *Goniodiscus sebæ* M. T. '42, *Anthenea tuberculosa* GRAY '47, *Pentaceros turritus* (M. T. '42). *Pentaceropsis obtusatus* (BORY ST. VINCENT '27) = *Asterias obtusata* BORY ST. VINCENT '27, *Culcita novæ-guineæ* (M. T. '42) = *C. plana* HARTLAUB '92, 13 figs. *Gymnasteria carinifera* (LAMARCK '16) = *Asterias carinifera* LAMARCK '16, *Nepanthia brevis* PERRIER '76 = *Asterina (Nepanthia) brevis* PERRIER '76, *Asterina cepheus* (M. T. '42). *Ophidiaster pustulatus* (MARTENS '66) = *Linckia pustulata* VON MARTENS '66 = *Ophidiaster purpureus* PERRIER '69, *Ophidiaster tuberifer* SLADEN '89, *Bunaster ritteri*, n. g. n. sp., 7 figs. *Linckia miliaris* MARTENS '66, *Linckia multifora* (LAMARCK '16) = *Asterias multifora* LAMARCK '16 = *Ophidiaster multiforis* M. T. '42, *Nardoa tuberculata* GRAY '40, *Retaster cribrosus* (MARTENS '67) = *Pteraster cribrosus* VON MARTENS '67 = *Retaster insignis* SLADEN '82, *Acanthaster echinites* (ELLIS & SOLANDER 1786) = *Asterias echinites* ELLIS & SOLANDER 1786, 6 figs. *Mithrodia clavigera* (LAMARCK '16) = *Asterias clavigera* LAMARCK '16, *Echinaster eridanella* M. T. '42, Also 6 figs of *Culcita schmideliana* and 1 fig. of *Cul. coriacea.*

'98. Ueber " Krystallkörper " bei Seesternen und über die Wachsthums-erscheinungen und Verwandtschaftsbeziehungen von Goniodiscus sebæ. Zool. Forschungsreisen in Australien u. d. Malay. Archip., Bd. v, p. 491–503, Taf. xxxviii–xl.

'99. Einige Beobachtungen an arktischen Seesternen. Zool. Anz., Bd. 22, p. 337–339. Errata, p. 432.

Pteraster hexactis. Solaster syrtensis. Ctenodiscus krausei = Ct. corniculatus. Solaster (Crossaster) papposus. Solaster affinis. Solaster (Crossaster) helianthus. Solaster endeca. Cribrella oculata. Asterina lincki.

: 00. Die Echinodermen. Zoologisch. Ergebn. einer Untersuchungsfahrt d. deutsch. Seefisch.-Ver. n. d. Böreninsel u. Westspitzbergen, aus-geführt i. Sommer 1898 auf S. M. S. "Olga." Wissenschaftl. Meeresunters. herausg. v. d. Kommiss. z. Untersuch. d. deutsch. Meere i. Kiel u. d. Biolog. Anst. a. Helgoland, N. F., Bd. iv, Abtl. Helgoland, Hft. 2, p. 195–248, Taf. iv–x.

Asterias rubens O. F. MÜLLER, 1776, 1 fig *Asterias hyperborea* DANIELS. & KOREN '82, 3 figs. *Asterias lincki* (MÜLL. & TROSCH.) = *Asteracanthion lincki* MÜLL. & TROSCH. '42 = *Asterac. stellionura* PERRIER '69 = *Asterias stellionura* DANIELS. & KOREN '84 = *Asterias gunneri* PFEFFER '94, *Asterias grœnlandica* (LÜTKEN) = *Asteracanthion grœnlandicum* LÜTKEN '57, 3 figs. *Asterias panopla* STUXBERG '79, 3 figs. *Solaster papposus* (FABRICIUS) = *Asterias papposa* FABRICIUS 1780 = *Solaster affinis* BRANDT '35 =

Crossaster papposus MÜLL. & TROSCH. '40 = *Crossaster affinis* PFEFFER '94· 5 figs.
Solaster papposus var. *anglica*. *Solaster papposus* var. *squamata*. *Solaster endeca* (L.)
= *Asterias endeca* LINNÉ 1771, 2 figs. *Solaster syrtensis* VERRILL '94· 2 figs. *Solaster
furcifer* DÜBEN & KOREN '46 = *Lophaster furcifer* DUNCAN & SLADEN '81· *Cribrella
sanguinolenta* (O. F. MÜLLER) = *Asterias sanguinolenta* O. F. MÜLLER 1776 = *Asterias
oculata* PENNANT 1777 = *Cribrella oculata* DUNCAN & SLADEN '81· *Pteraster obscurus*
(E. PERRIER) = *Pteraster pulvillus* M. SARS, p.p. '61 = *Hexaster obscurus* E. PERRIER '91
= *Pteraster (Temnaster) hexactis* VERRILL '94 = *Temnaster hexactis* VERRILL '95· 9 figs.
Pteraster pulvillus M. SARS '61· 2 figs. *Pteraster militaris* (O. F. MÜLLER) = *Asterias
militaris* O. F. MÜLLER 1776 = *Pt. militaris* MÜLL. & TROSCH. '42· *Hymenaster pel-
lucidus* W. THOMSON '73· *Hippasteria phrygiana* (PARELIUS) = *Asterias phrygiana*
PARELIUS 1770 = *Hippasteria plana* GRAY '41· *Rhegaster tumidus* (STUXBERG) = *Solaster
tumidus* STUXBERG '79 = *Asterina tumida* KOREN & DANIELSSEN '80 = *Solaster tumidus*
STUXBERG '83 = *Rhegaster tumidus* PFEFFER '94· 2 figs. *Ctenodiscus crispatus* (RETZIUS)
= *Asterias crispata* RETZ. '05 = *Asterias polaris* SABINE '24 = *Ct. polaris* MÜLL. &
TROSH. '42 = *Ct. corniculatus* DUNCAN & SLADEN '81 = *Ct. krausei* LUDWIG '86· 2 figs.
Pontaster tenuispinus (DÜBEN & KOREN) = *Astropecten tenuispinus* DÜBEN & KOREN
'46 = *Archaster tenuispinus* M. SARS '61 = *Pontaster tenuispinus* SLADEN '89· *Leptoptych-
aster arcticus* (SARS) = *Astropecten arcticus* SARS '51 = *Archaster arcticus* VERRILL '79·

: 02.   Japanische Seesterne.   Zool. Anz., Bd. 25, p. 326–335.

Astrogonium pretiosum, n. sp. ASTROPECTEN. *Astropecten ludwigi* DE LORIOL '99 =
Astrop. japonicus IVES. *Astropecten scoparius* M. T. = *Astrop. japonicus* (M. T.) SLA-
DEN. *Astropecten polyacanthus* M. T. = *Astrop. armatus* M. T. *Astropecten kagoshi-
mensis* DE LORIOL '99· *Astropecten kagoshimensis* var. *kochiana*, n. var. *Astropecten
latespinosus* MEISSNER '92· *Luidia quinaria* v. MARTENS = *Luidia maculata* var. *quina-
ria* v. MARTENS = *Luid. quinaria* SLADEN, IVES = *Luidia limbata* SLADEN. *Luidia maculata*
M. T. *Asterina pectinifera* M. T. *Nardoa semiregularis* M. T. var. *japonica* v.
MARTENS. *Cribrella sanguinolenta* O. F. MÜLLER = *Cribrella densispina* SLADEN '78·
ASTERIAS. *Asterias calamaria* var. *japonica*, n. var. *Asterias volsatella* SLADEN var.
sakurana, n. var. *Asterias rollestoni* BELL '81 = *Asterias amurensis* SLADEN, IVES,
MEISSNER = ? *Asterias versicolor* SLADEN. *Asterias nipon*, n. sp. *Asterias satsumana*,
n. sp. *Asterias japonica* STIMPSON MS. BELL '81 = *Asterias torquata* SLADEN.

DÜBEN, M. W. von, & KOREN, J.

* '46.   Öfversigt af Skandinaviens Echinodermer.   K. Vet. Akad. Handl.
   Stockholm (1844), p. 229–328.   Cited after DUJARDIN and HUPÉ
   and LUDWIG.

DUJARDIN, F., & HUPÉ, H.

'62. Histoire naturelle des zoophytes échinodermes comprenant la description des crinoïdes, des ophiurides, des astérides, des échinides et des holothurides. 627 pp. + atlas, 7 pp. + 10 pls.

1. Tribus. ASTERACANTHION M. & TR. = STELLONIA NARDO AG. = ASTERIAS LAM. GRAY = URASTER AG. FORBES. *A. glacialis* M. & TR. = *Sol echinatus cancellatus* LINCK = *A. spinosa* PENNANT = *Pentadactylosaster spinosus* LINCK = *Stella hibernica echinata* PETIV. = *Asterias glacialis* O. F. MÜLL. = *Stellonia glacialis* NARDO = *Asterias angulosa* O. F. MÜLL. = *Stellonia angulosa* AG. = *Uraster glacialis* = *Asterias echinophora* DELLE CHIAJE. *A. gelatinosa* M. & TR. = *Asterias rustica* GRAY. *A. rubens* M. & TR. = *Tetractis petaloides* LINCK = *Stella coriacea vulgaris Luidii* LINCK = *Stellonia rubens* AG. = *Uraster rubens* FORBES = *A. clathrata* PENNANT = *Asterac. polaris* M. & TR., 2 figs. *A. violaceus* LIN. = *Sol coriaceus planus* LINCK = *Asterias rubens violacea* MÜLL. = *Uraster violaceus* FORBES = *Asterias holsatica* RETZIUS = *Asterias minuta* RETZIUS. *A. roseus* M. & TR. = *Cribrella rosea* FORBES = *Henricia rosea* GRAY. *A. africanus* M. & TR. *A. tenuispinus* M. & TR. = *Stella marina echinata* COLUMNA = *Asterias saveresii* D. CHIAJE = *A. glacialis* GRUBE. *A. bootes* M. & TR. *A. linckii* M. & TR. = *Pentadactylosaster reticulatus* LINCK = *Mithrodia spinulosa* GRAY. *A. striatus* M. & TR. *A. polyplax* M. & TR. *A. aster* M. & TR. *A. graniferus* M. & TR. = *Asterias serrulata* Ency. méth. *A. janthina* (BRANDT). *A. margaritifer* M. & TR. *A. ochraceus* (BRANDT). *A. aurantiacus* M. & TR. = *Stichaster striatus* M. & TR. = *Tonia atlantica* GRAY. *A. globiferus* = *Uniophora globifera* GRAY. *A. germanii* PHIL. *A. luridum* PHIL. *A. echinata* M. & TR. = *Asterias echinata* GRAY. *A. katherinæ* M. & TR. = *Asterias katharinæ* GRAY. *A. wilkinsonii* M. & TR. = *Asterias wilkinsonii* GRAY. *A. calamaria* M. & TR. = *Asterias calamaria* GRAY. *A. webbianus* DUJ. & HUPÉ = *Stellonia webbiana* D'ORBIGNY. *A. grœnlandicus* STEENST. *A. mülleri* SARS. *A. polaris* M. & TR. = *Asterias rubens* FABRIC. *A. problema* STEENST. = *A. albulus* STIMPS. *A. epichlora* (BRANDT). *A. helianthoides* = *Asterias ochotense* BRANDT. *A. camschaticum* (BRANDT). *A. pectinatum* (BRANDT). *A. forbesi* DESOR. *A. hispidum* DUJ. & HUPÉ = *Stella coriacea acutangula hispida* LINCK = *Asterias hispida* PENNANT = *Asterias distichum* BRANDT = *Stellonia hispida* FORBES. *A. miniatum* (BRANDT). Fossil species : *A. tenuiradiatus* (HALL) = *Cœlaster tenuiradiatus* D'ORBIGNY. *A. matutina* (HALL) = *Cœl. matutina* D'ORBIGNY. *A. americanus* (GRAHAM) = *Cœl. americanus* D'ORBIGNY. *A. constellatus* (THORENT) = *Cœl. constellata* D'ORBIGNY. *A. gaveyi* (FORBES) = *Uraster gaveyi* FORBES. *A. lumbricalis* (SCHLOTH. GOLDFUSS). *A. lanceolata* (GOLDF.). *A. yvaryana* (D'ORP.) *A. desmoulinsii* (D'ARCH.). *A. rubens* M. T. HELIASTER GRAY = ASTERIAS auct. = ASTERACANTHION M. & TR. *H. helianthus* (LAMK.) = *Asterias helianthus* LAMK. = *Stellonia helianthus* AGASSIZ = *Asterias cumingii* GRAY = *Asterias multiradiata* GRAY = *Asteracanthion helianthus* M. & TR. 2. Tribus. ECHINASTER M. & TR. = ASTERIAS LAMARCK = RHOPIA, OTHILIA, HENRICIA GRAY. *E. spinosus* M. & TR.

= *Echinaster* seu *Stella coriacea, Pentadactyla echinata* PETIVER = *Pentadactylosaster
spinosus* LINCK = *Asterias echinophora* LAMK. = *Stellonia spinosa* NARDO = *Othilia spinosa*
GRAY = *O. aculeata* GRAY. *E. crassus* M. & TR. *E. gracilis* M. & TR. *E. eri-
danella* M. & TR. *E. serpentarius* M. & TR. *E. deplanatus* GRUBE. *E. rigidus*
GRUBE. *E. lacunosus* GRUBE. (*Ferdinia flavescens* GRAY. *F. cumingii.*) CRIBRELLA
AGASSIZ = OTHILIA GRAY = ECHINASTER (pars) M. & TR. *C. sanguinolenta* SARS = *Pentadactylosaster oculatus* LINCK = *Asterias sanguinolenta* MÜLLER = *Asterias pertusa* MÜLL.
= *Asterias oculata* PENNANT = *Asterias spongiosa* FABR. = *Asterias seposita* RETZIUS =
Linckia oculata FORBES = *Crib. oculata* FORBES = *Echinaster oculatus* M. & TR. = *Echin.
eschrichtii* BRANDT = *Echin. sanguinolentus* SARS = *Echin. sarsii* M. & TR. = *Linckia
pertusa* STIMPSON. *C. eschrichtii* (M. & TR.) = *Echinaster eschrichtii* M. & TR. *C.
fallax* (M. & TR.) = *Othilia purpurea* GRAY = *Oth. luzonica* GRAY = *Echinaster fallax*
M. & TR. *C. brasiliensis* DUJ. & HUPÉ = *Echinaster brasiliensis* M. & TR. =
Othilia multispina GRAY. *C. seposita* DUJ. & HUPÉ = *Asterias seposita* RETZIUS =
Asterias sanguinolenta (pars) O. F. MÜLL. = *Stellonia seposita* NARDO = *Rhopia seposita*
GRAY = *Rh. mediterranea* GRAY = *Echinaster sepositus* M. & TR. (*Asterias acuminata*
LAMARCK.) ACANTHASTER GERVAIS = ECHINASTER GRAY (non M. & TR.) = ECHINITES
M. & TR. *A. solaris* DUJ. & HUPÉ = *Stella pentekai de kactis* COLUMNA = *Asterias
echinus* SOLANDER & ELLIS = *Asterias solaris* Naturforscher = *Asterias echinites* LAMK. =
Stellonia echinites AGASSIZ = *Echinaster ellisii* GRAY = *Echin. solaris* GRAY = *Acanthaster
echinus* GERVAIS = *Echinites solaris* M. & TR. = *Echinaster solaris* M. & TR. SOLAS-
TER FORBES = SOLASTERIES BLAINV. = STELLONIA (pars) NARDO AG. = CROSSASTER M. &
TR. = POLYASTER GRAY. *S. papposus* FORBES = *Stella rubra coriacea, 12 radiis* LUIDIUS
= *Asterias stellata* RETZIUS ? = *Asterias papposa* FABR. = *Stellonia papposa* AG. = *Asterias
helianthosus* PENNANT. *S. endeca* FORBES = *Asterias endeca* LINCK = *Asterias aspera*
MÜLL. = *Solasterias endeca* RETZIUS = *Stellonia endeca* AG. *S. decanus* (M. & TR.) =
Echinaster decanus M. & TR. *S. furcifer* DÜBEN & KOREN. *S. affinis* = *Asterias
affinis* BRANDT. *S. alboverrucosa* = *Asterias alboverrucosa* BRANDT. Fossil species : *S.
moretonis* FORBES. CHŒTASTER M. & TR. = ASTERIAS auctorum. *Ch. longipes* (RETZ.)
= *Asterias longipes* RETZIUS = *Asterias subulata* LAMK. = *Chœtaster subulata* M. & TR.
Ch. hermanni M. & TR. *Ch. troschelii* VAL. *Ch. tessellatus* M. & TR. = *Nepanthia
tessellata* GRAY. *Ch. maculatus* M. & TR. = *Nepanthia maculata* GRAY. OPHIDIASTER
AG. = LINCKIA (pars) NARDO GRAY. *O. ophidianus* AG. = *Asterias ophidiana* LAMK. =
? *O. aurantius* GRAY. *O. hemprichii* M. & TR. *O. cylindricus* M. & TR. = *Asterias
cylindrica* LAMK. = *O. leachii* GRAY. *O. attenuatus* M. & TR. = *Asterias coriacea*
GRUBE = *A. attenuata* GRAY. *O. pusillus* M. & TR. *O. miliaris* M. & TR. = *Dacty-
losaster miliaris* LINCK = *Asterias lævigata* L. = *Linckia typus* NARDO = *L. brownii* GRAY.
O. suturalis M. & TR. = ? *Linckia unifascialis* GRAY = *L. bifascialis* GRAY. *O. diplax*
M. & TR. *O. ornithopus* M. & TR. *O. multiforis* M. & TR. = *Asterias multifora*

LAMK. = *O. ehrenbergii* M. & TR. *O. tub rculatus* M. & TR. *O. echinulatus* M. & TR. *O. pyramidatus* GRAY. *O. colombiæ* M. & TR. = *Linckia colombiæ* GRAY. *O. pacifica* M. & TR. = *Linckia pacifica* GRAY. *O. guildingii* M. & TR. = *Linckia guildingii* GRAY. *O. bicolor* M. & TR. = *Asterias bicolor* LAMK. *O. arenatus* M. & TR. = *Ophiura arenata* LAMK. DACTYLOSASTER *cylindricus* GRAY. *D. gracilis* GRAY. TAMARIA *fusca* GRAY. CISTINA *colombiæ* GRAY. LINCKIA *crassa* GRAY. *L. leachii* GRAY. MEDIASTER *æqualis* STIMPSON. SCYTASTER M. & TR. = LINCKIA (pars) NARDO AG. GRAY = METRODIRA GRAY = FROMIA GRAY = NARDOA GRAY = GOMOPHIA GRAY (?) = MITHRODIA GRAY (?) = NARCISSIA GRAY (?). *S. variolatus* M. & TR. = *Pentadactylosaster variolatus* LINCK = *Asterias variolata* RETZ. = *Linckia variolata* NARDO = *Nardoa agassizii* GRAY. *S. milleporellus* M. & TR. = *Asterias milleporella* LAMK. *S. zodiacalis* M. & TR. *S. pistorius* M. & TR. = *Fromia milleporella* GRAY. *S. semiregularis* M. & TR. *S. kuhlii* M. & TR. *S. subulatus* M. & TR. = *Metrodira subulata* GRAY. *S. canariensis* DUJ. & HUPÉ = *Asterias canariensis* D'ORB. *S. cancellatus* GRUBE. MITHRODIA. *M. clavigera* GRAY = *Asterias clavigera* LAMK. = *Asterias reticulata* BLAINVILLE. GOMOPHIA GRAY. *G. ægyptica* GRAY. LINCKIA GRAY. *L. pulchella* GRAY. *L. intermedia* GRAY. *L. erythræa* GRAY. NARCISSIA *teneriffæ* GRAY. CULCITA AG. = ASTERIAS LIN. LAMK. = les oreillers BLAINVILLE. *C. discoidea* AG. = *Asterias discoidea* LAMK. = *C. schmideliana* GRAY. *C. coriacea* M. & TR. = *Asterias coriacea* Encyc. méth. *C. novæ-guineæ* M. & TR. *C. grex* M. & TR. PALMIPES LINCK = ASTERISCUS (spec.) M. & TR. *P. membranaceus* AG. = *Asterias placenta* PENNANT = *Asterias membranacea* L. = *Asteriscus palmipes* M. & TR. *P. rosaceus* DUJ. & HUPÉ = *Asterias rosacea* LAMK. = *P. stokesii* GRAY = *Asteriscus rosaceus* M. & TR. ASTERISCUS M. & TR. = ASTERINA et ANSEROPODA NARDO AG. = ASTERINA GRAY. *A. pectinifer* M. & TR. *A. verruculatus* M. & TR. = *Asterias verruculata* RETZIUS = *A. gibbosa* PENNANT = *A. membranacea* GRUBE = *Asterina gibbosa* GRAY = *Asterina burtoni* GRAY = *Asterias pulchella* Bv. *A. minutus* M. & TR. = *Asterina minuta* GRAY = *Asterias exigua* LAME. *A. cephea* M. & TR. *A. krausii* M. & TR. = *Asterina krausii* GRAY. *A. pentagonus* M. & TR. *A. penicillaris* M. & TR. = *Asterias penicillaris* LAMK. *A. australis* M. & TR. = *Asterina calcar* GRAY = *A. gunnii* GRAY. *A. calcar* (LAMK.) = *Asterias calcar* LAME. *A. diesingii* M. & TR. *A. coccinus* M. & TR. *A. setaceus* VAL. *A. trochiscus* M. & TR. = *A. trochiscus* RETZ. *A. antiquus* FORBES = *Asterias antiqua* HISINGER = *Palm pes antiquus* FORBES. OREASTER M. & TR. = PENTACEROS LINCK GRAY = GONIASTER (pars) AG. = NIDORELLIA GRAY. *O. reticulatus* M. & TR. = *Stella reticulata sive cancellata* RONDELET = *Pentaceros gibbus, reticulatus et lentiginosus* LINCK = *Asterias reticulata* LINNÉ = *Asterias pentascyphus* RETZ. = *Pentaceros grandis* GRAY = *P. gibbus* GRAY. *O. affinis* M. & TR. *O. chinensis* M. & TR. *O. tuberculatus* M. & TR. *O. turritus* M. & TR. = *Asterias nodosa* L. = *P. franklinii* GRAY. *O. muricatus* DUJ. & HUPÉ = *Asterias linckii* BLAINV. *O. hiulcus* M. & TR. = *Pentaceros gibbus* LINCK.

O. mamillatus M. & Tr. = *Asterias mamillata* Audouin, 2 figs. *O. verrucosus* M. &
Tr. *O. clavatus* M. & Tr. *O. carinatus* M. & Tr. *O. aculeatus* M. & Tr. *O.
obtusatus* M. & Tr. = *Asterias obtusata* Lamk. *O. obtusangulus* M. & Tr. = *Asterias
obtusangula* Lamk. *O. regulus* M. & Tr. *O. nodosus* M. & Tr. *O. armatus* M.
& Tr. *O. orientalis* M. & Tr. *O. valvulatus* M. & Tr. *O. lapidarius* Grube.
O. desjardinsii Michelin. Fossil species : *O. coronatus* Forbes. *O. boysii* Forbes. *O.
bulbiferus* Forbes. *O. obtusus* Forbes. *O. ocellatus* Forbes. *O. pistilliformis* Forbes.
O. squamatus Forbes. Astrogonium M. & Tr. = Goniaster Ag. = Hippasteria,
Goniaster, Pentagonaster, Tosia Gray. *A. phrygianum* M. & Tr. = *Pentaceros
planus seu Oxyceros* Linck = *Pentaceros macroceros* Linck = *Asterias phrygiana* O. F.
Müll. = *Asterias equestris* Gmelin ? = *Goniaster equestris* Ag. = *Hippasteria europœa*
Gray = *Asterias johnstonii* Gray = *Hippasteria johnstonii* Gray = *Hippasteria equestris*
Gray. *A. magnificum* M. & Tr. *A. astrologorum* M. & Tr. *A. geometricum* M.
& Tr., 2 figs. *A. pulchellum* M. & Tr. = *Pentagonaster pulchellus* Gray = *Stephan-
aster elegans* Ayres. *A. australe* M. & Tr. = *Tosia australis* Gray. *A. ornatum* M.
& Tr. *A. lamarckii* M. & Tr. *A. cuspidatum* M. & Tr. = *Pentagonaster semi-
lunatus* Link = *Asterias tessellata* (pars) Lamk. = *Goniaster cuspidatus* Gray. *A.
granulare* M. & Tr. = *Asterias granularis* O. F. Müller = *Asterias tessellata* (pars)
Lamk. *A. nobile* M. & Tr. *A. fonki* Philippi. *A. aculeatum* Barrett. *A. boreale*
Barrett. *A. punctatum* (Lamk.) = *Asterias punctata* Lamk. *A. souleyetii*, n. sp., 1
fig *A. abbensis* = *Goniaster abbensis* Forbes = *Hippasteria abbensis* Gray. Fossil
species : *A. jurense* M. & Tr. = *Asterias jurensis* Munster Goldf. = *Goniaster ? jurensis*
Ag. = *Pentetagonaster jurensis* D'Orb. *A. tabulatum* (Goldf.) = *Asterias tabulata* Goldf.
= *Pentetagonaster tabulatus* D'Orb. *A. scutatum* (Goldf.) = *Asterias scutata* Goldf. =
Pentetagonaster scutatus D'Orb. *A. stelliferum* (Goldf.) = *Asterias stellifera* Goldf. =
Pentetagonaster stellifer D'Orb. *A. fleuriausum* (D'Orb.) = *Pentetagonaster fleuriausa*
D'Orb. Cretaceous species : *A. variabile* (Koch.) = *Cidaris variabilis* Koch = *Asterias
dunkeri* Rœmer = *Pentetagonaster variabilis* D'Orb. *A. malbosii* (D'Orb.) = *Pentetagonas-
ter malbosii* D'Orb. *A. porosum* (Ag.) = *Goniaster porosus* Ag. = *Pentagonaster porosus*
Pictet. *A. couloni* (Ag.) = *Goniaster couloni* Ag. = *Pentagonaster couloni* Pictet. *A.
dutempleanum* (D'Orb.) = *Pentetagonaster dutempleanus* D'Orb. *A. parkinsoni* (Forbes)
= *Asterias regularis* Parkinson = *Goniaster parkinsoni* Forbes = *Pentagonaster parkinsoni* Pictet. *A. bowerbankii* (Forbes) = *Goniaster bowerbankii* Forbes = *Pentagonaster
bowerbankii* Pictet. *A. compactum* (Forbes) = *Goniaster compactus* Forbes = *Pentagonaster compactus* Pictet. *A. mosaicum* Forbes. *A. combii* Forbes. *A. latum*
Forbes. *A. smithii* Forbes. *A. angustatum* Forbes. *A. hunteri* (Forbes) = *Goniaster
hunteri* Forbes = *Pentagonaster hunteri* Pictet. *A. mantellii* (Forbes) = *Goniaster mantellii* Forbes = *Pentagonaster mantellii* Pictet. *A. rugatum* (Forbes) = *Goniaster rugatus*
Forbes = *Pentagonaster rugatus* Pictet. *A. sublunatum* (Forbes) = *Goniaster sublunatus*

FORBES = *Pentagonaster sublunatus* PICTET. *A. uncatum* (FORBES) = *Goniaster uncatus* FORBES = *Pentagonaster uncatus* PICTET. *A. lunatum* FORBES. *A. rectilineum* McCoy = *Goniaster rectilineus* PICTET. *A. quinquelobum* (GOLDF.) = *Asterias quinqueloba* GOLDF. = *Pentetagonaster quinqueloba* D'ORB. *A. chilopora* (DESM.) = *Asterias chilopora* DES-MOULINS = *Pentetagonaster chilopora* D'ORB. = *Pentagonaster chilopora* PICTET. *A. strati-ferum* (DESM.) = *Asterias stratifera* DESMOULINS = *Pentetagonaster stratifera* D'ORB = *Pentagonaster stratifera* PICTET. *A. costatum* (D'ORB.) = *Pentetagonaster costata* D'ORB. *A. dutemplei* (D'ORB.) = *Pentetagonaster dutemplei* D'ORB. *A. moulinsii* (D'ORB.) = *Pente-tagonaster moulinsii* D'ORB. Species from the Tertiary : *A. marginatum* FORBES. *A. stokesii* FORBES. *A. tuberculatum* FORBES. GONIODISCUS M. & TR. = ASTERIAS auct. = PAULIA, RANDASIA, ANTHENEA, HOSIA et NECTRIA GRAY. *G. pentagonulus* M. & TR. = *Asterias pentagonula* LAMK. = *Anthenea chinensis* GRAY. *G. sebæ* M. & TR. = *Artocerus altera* SEBA = *Goniaster sebæ* GRAY. *G. placenta* M. & TR. *G. regularis* M. & TR. = *Pentagonaster regularis* LINCK = *Asterias tessellata* LK. = *Goniaster regularis* GRAY. *G. pleyadella* M. & TR. = *Asterias pleyadella* LAMK. *G. cuspidatus* M. & TR. = *Asterias cuspidata* LAMK. *G. mamillatus* M. & TR. *G. capella* M. & TR. *G. singularis* M. & TR. *G. seriatus* M. & TR. *G. verrucosus* PHILIPPI. *G. horridus* M. & TR. = *Paulia horrida* GRAY (?). *G. luzonicus* M. & TR. = *Randasia luzonica* GRAY. *G. flavescens* M. & TR. = *Hosia flavescens* GRAY. NECTRIA GRAY = ASTERIAS sp. LAMK. = GONIODISCUS (pars) M. & TR. *N. ocellifera* GRAY = *Asterias ocellifera* LAMK. = *Goniodiscus ocelliferus* M. & TR. STELLASTER GRAY. *St. childreni* GRAY = *St. equestris* M. & TR. = *Asterias equestris* RETZ. Fossil species: *St. comptoni* FORBES. *St. schultzii* PICTET = *Asterias schultzii* colta ROEMER = *Pente-tagonaster schultzii* D'ORB. COMPTONIA GRAY. *C. elegans* GRAY = *Stellaster elegans* PICTET. ASTEROPSIS M. & TR. = GYMNASTERIA, PORANIA GRAY. *A. carinifera* M. & TR. *A. ctenacantha* M. & TR. *A. pulvillus* M. & TR. = *Asterias pulvillus* O. F MÜLL. = *Asterias gibbosa* LEACH = *Goniaster templetoni* FORBES = *Porania gibbosa* GRAY. *A. vernicina* M. and TR. = *Asterias vernicina* LAMK. *A. imbricata* GRUBE. (*Gymasteria spinosa* GRAY. *Gymn. inermis* GRAY. *Porania gibbosa* GRAY = *Asterias gibbosa* LEACH = *Asterias equestris?* THOMPSON = *Goniaster templetoni* FORBES.) ARCHASTER M. & TR. = ASTERIAS (pars) auctorum. *A. typicus* M. & TR. = *Astropecten stellaris* GRAY. *A. hesperus* M. & TR. *A. angulosus* M. & TR. = *Astropecten mauritianus?* GRAY. ASTROPECTEN LINCK = ASTERIAS LAMK. AG. = CRENASTER LUID = STELLARIA NARDO M. & TR. *A. crenaster* (LUIDIUS) = *Crenaster luidius* 1699 = *A. echinatus major* LINCK = *Asterias aurantiaca* (pars) LAMK. = *Astropecten aurantiacus* M. & TR., 6 figs. *A. irregularis* LINCK = *Asterias aurantiaca* MÜLLER = *Stella marina minor* JOHNSTON = *A. mülleri* M. & TR. *A. brasiliensis* M. & TR. = *A. duplicatus* GRAY?. *A. valenciennii* M. & TR. = *Astropecten stellatus* GRAY?. *A. tiedemanni* M. & TR. *A. bispinosus*

M. & Tr. = *A. echinatus minor* Linck = *Asterias aurantiaca* (pars) Lamk. = *Asterias bispinosa* Otto. *A. polyacanthus* M. & Tr. *A. platyacanthus* M. & Tr. = *Asterias platyacanthus* Philippi. *A. hystrix* M. & Tr. *A. armatus* M. & Tr. *A. scoparius* M. & Tr. *A. hemprichii* M. & Tr. *A. articulatus* M. & Tr. = *Asterias articulata* Say. *A. tenuispinus* Düben & Koren. *A. andromeda* M. & Tr. = *A. christi* Düben & Koren. *A. parelii* Düben & Koren = *Asterias aurantiaca* var. *parelius*. *A. rappa* M. & Tr. *A. buschii* Phil. *A. triseriatus* Philippi. *A. echinulatus* M. & Tr. *A. johnstoni* M. & Tr. = *Asterias aurantiaca* var. 2 Lamk. = *Asterias johnstoni* Dell Chiaje. *A. serratus* M. & Tr. *A. spinulosus* M. & Tr. = *Asterias spinulosa* Phil. *A. japonicus* M. & Tr. *A. hispidus* M. & Tr. *A. longispinus* M. & Tr. *A. pentacanthus* M. & Tr. = *Asterias pentacantha* Delle Chiaje = *Asterias aranciaca* Johnston. *A. subinermis* M. & Tr. = *Asterias subinermis* Phil. *A. marginatus* M. & Tr. *A. schœnleinii* M. & Tr. *A. granulatus* M. & Tr. *A. pressii* M. & Tr. *A. squamatus* M. & Tr. *A. ciliatus* Grube. *A. arcticus* Sabs. *A. lutkeni* Barrett. *A. calcitrapa* = *Asterias calcitrapa* Lamk. *A. corniculatus* Linck. *A. fimbriatus* Linck = *Asterias fimbriata* Blainville. (*A. longipes* Gray. *A. dubius* Gray. *A. gracilis* Gray. *A. mesodiscus* Gray. *A. erinaceus* Gray. *A. regularis* Gray. Nauricia *pulchella* Gray.) Fossil species : *A. priscus* (Goldf.) = *Asterias prisca* Goldf. = *Crenaster prisca* D'Orb. *A. hastingiœ* Forbes = *Crenaster hastingiœ* Pictet. *A. orion* Forbes = *Crenaster orion* Pictet. *A. mandelslohi* (Munster) = *Asterias mandelslohi* Munster = *Cœlaster mandelslohi* D'Orb. = *Crenaster mandelslohi* Pictet. *A. arenicolus* (Charlesworth) = *Asterias arenicola* Charlesworth = *Crenaster arenicolus* Pictet. *A. cotteswoldiœ* Bucken = *Crenaster cotteswoldiœ* Pictet. *A. philipsii* Forbes = *Crenaster philipsii* D'Orb. *A. cottaldinus* D'Orb. = *Crenaster cottaldina* D'Orb. *A. nodotianus* D'Orb. = *Crenaster nodotiana* D'Orb. *A. rupellensis* D'Orb. = *Crenaster rupellensis* D'Orb. *A. rectus* (M'Coy) = *Asterias rectus* M'Coy = *Crenaster rectus* Pictet. *A. propinquus* (Munster) = *Asterias propinquus* Munster = *Crenaster propinquus* Forbes. *A. castellanensis* (D'Orb.) = *Crenaster castellanensis* D'Orb. *A. armatus* Forbes = *Crenaster armatus* Pictet. *A. colei* Forbes. *A. crispatus* Forbes. *A. poritoides* (Desmoulins) = *Asterias poritoides* Desmoulins = *Crenaster poritoides* D'Orb. *A. lævis* (Desmoulins) = *Asterias lævis* Desmoulins = *Crenaster lævis*. *A. adriaticus* (Desmoulins) = *A. adriatica* Desmoulins = *Crenaster adriatica* D'Orb. Ctenodiscus M. & Tr. Ct. *crispatus* Lütken = *Asterias crispata* Retzius = *Asterias polaris* Sabine = *Asterias aurantiaca* Dewurst = *Astropecten polaris* Gray = Ct. *polaris* M. & Tr. = Ct. *pygmaus* M. & Tr. Luidia Forbes = *Asterias* Lamk. = Hemicnemis M. & Tr. *L. savignii* (Aud) = *Asterias savignii* Aud. *L. ciliaris* (Phil.) = *Asterias ciliaris* Philippi = *L. fragilissima* Forbes = *L. savignii* M. & Tr. = *Asterias* n. sp. Sars = *L. sarsii* Düben & Koren. *L. maculata* M. & Tr. *L. senegalensis* M. & Tr. = *Asterias senegalensis* Lamk.

Fossil species: *L. williamsoni* (FORBES) = *Asterias williamsoni* FORBES = *L. murchisoni* PICTET. (PETALASTER *hardwickia* GRAY. *P. columbia* GRAY) PTERASTER M. & TR. *Pt. militaris* M. & TR. = *Asteriscus militaris* M. & TR. *Pt. capensis* GRAY. Fossil genera: PALÆASTER HALL = URASTER sp. FORBES. *P. niagarensis* HALL. *P. obtusus* SALTER = *Uraster obtusus* FORBES. *P. ruthveni* SALTER = *Uraster ruthveni* FORBES. *P. hirudo* SALTER = *Uraster hirudo* FORBES. *P. coronella* SALTER. *P. asperrimus* SALTER. PALÆASTRINA MACCOY. *P. primæva* SALTER = *Uraster primæva*. TROPIDASTER FORBES. *T. pectinatus* FORBES. LEPIDASTER FORBES. *L. grayi* FORBES. CŒLASTER AG. *C. couloni* AG. (*Asterias matutina* HALL. *A. tenuiradiata* HALL. *A. constellata* THOR. *A. americana* GRAHAM. *A. mandelslohi* MUNSTER.) PLEURASTER AG. *P. obtusa* AG. = *Asterias obtusa* GOLDF. *P. arenicola* AG. = *Asterias arenicola* GOLDF. ARTHRASTER FORBES. *A. dixoni* FORBES. BDELLACOMA SALTER = PALÆOCOMA SALTER (non D'ORB.). *Bd. marstoni* SALTER = *Palæocoma marstoni* SALTER. *Bd. colvini* SALTER = *Palæocoma colvini* SALTER. *Bd. cygnipus* SALTER = *Palæocoma cygnipus* SALTER. *Bd. vermiformis* SALTER = *Palæocoma vermiformis* SALTER. *Bd. pyrotechnicum* SALTER = *Palæocoma pyrotechnicum* SALTER. App. BRISINGA ASBJØRNSEN. *B. endecacnemos* ASBJ.

DUNCAN, P. M., & SLADEN, W. P.

'81. A Memoir on the Echinodermata of the Arctic Sea to the West of Greenland. 82 pp. + 6 pls. Asteroidea, p. 23–53 + pl. ii–iii.

Asteracanthion polare M. T. '42 = *Aster. rubens* FABRICIUS 1780 (non LINNÉ) = *Aster. minuta* FABRICIUS 1780 = ? *Aster. violacea* SABINE '21 = ? *Aster. ochotense* BRANDT '51 = *Asteracanthion rubens* (pars) DUJARDIN & HUPÉ '62, 5 figs. *Asteracanthion grœnlandicum* (STEENSTRUP) LÜTKEN '57 = ? *Uraster violacea* FORBES '52 = *Aster. mülleri* STIMPSON '53, 4 figs. *Stichaster albulus* (STIMP.) VERRILL '66 = *Asteracanthion roseus* (pars) M. T. '42 = *Asterac. albulus* STIMPSON '53 = *Asterac. problema* STEENSTRUP '55, 5 figs. *Cribrella oculata* (LINCK) FORBES '39 = *Pentadactylosaster oculatus* LINCK 1733 = *Aster. sanguinolenta* O. F. MÜLLER 1776 = *Aster. pertusa* O. F. MÜLLER 1776 = *Aster. oculata* PENNANT 1777 = *Aster. spongiosa* FABRICIUS 1780 = *Aster. seposita* RETZ. '05 = *Henricia oculata* GRAY '40 = *Echinaster oculatus* M. T. '42 = *Echin. eschrichtii* M. T. '42 = *Echin. sanguinolentus* SARS '44 = *Echin. sarsii* M. T. '44 = *Linckia oculata* FORB. '39 = *Linck. pertusa* STIMPSON '53 = *Cribrella sanguinolenta* LÜTKEN '57 = *Crib. eschrichtii* DUJARDIN & HUPÉ '62, 4 figs. *Pedicellaster palæocrystallus* SLADEN '80 = *Asteracanthion palæocrystallus* SLADEN '77, 5 figs. *Crossaster papposus* (LINCK) M. T. '40 = *Triskaidecactis papposa* LINCK 1733 = *Dodecactis reticulata in dorso* LINCK 1733 = *Asterias helianthemoides* PENNANT 1777 = *Aster. papposa* FABRICIUS 1780 = *Aster.* (Solasterias) *papposa* BLAINVILLE '34 = *Aster. affinis* BRANDT '34 = ? *Aster. alboverrucosa* BRANDT '34 = ? *Stellonia papposa* AGASSIZ '35 = *Solaster papposa* FORBES '39 = *Solaster* (Polyaster) *papposa* GRAY '40, 4 figs. *Solaster* endeca (GMEL.) FORBES '39 = Aster. aspera O. F. Müller 1776 = Asterias endeca GMELIN 1788 = Aster. endica FLEMING '28 = Aster. (Solasterias) endeca BLAINVILLE '34 = Stellonia endeca AGASSIZ '35 = Solaster (Endeca) endeca GRAY '40· 4 figs. Lophaster furcifer (DÜBEN & KOREN) VERRILL '78 = Chataster borealis DÜBEN '44 = Solaster furcifer DÜBEN & KOREN '44· 4 figs. Pteraster militaris (O. F. Müller) M. T. '42 = Aster. militaris O. F. Müller 1776 = Asteriscus militaris M. T. '42· 4 figs. Ctenodiscus corniculatus (LINCK) PERRIER '75 = Astropecten corniculatus LNK. 1733 = Asterias crispata RETZIUS '05 = Aster. polaris SABINE '21 = Aster. auranciaca DEWHURST '34 = Astrop. polaris GRAY '40 = Ctenodiscus polaris M. T. '42 = Cten. pygmaus M. T. '42 = Cten. crispatus DÜBEN & KOREN '44 = Anodiscus (= Ctenodiscus) crispatus PERRIER '69· 4 figs.

D'URBAN, W. S. M.

'80· The Zoology of Barents Sea. Ann. Mag. Nat. Hist., 5. ser., vol. vi, p. 253–277.

Species listed: Asteracanthion grœnlandicus ST., Crossaster papposus (LINCK) var. affinis BRANDT, Ctenodiscus crispatus RETZ. = corniculatus LINCK, Archaster tenuispinus DÜB. & KOREN, Archaster bifrons WY. THOMSON, Asterias stellionura PERRIER.

FARQUHAR, H.

'97· A Contribution to the History of New Zealand Echinoderms. Journ. Linn. Soc. London, Zool., vol. 26, p. 186–198, pl. 13–14. Asteroidea, p. 192–197.

Asterias rodolphi PERRIER. Asteropsis imperialis, n. sp., 2 figs. Gnathaster rugosus HUTTON, 1 fig. Astrogonium sp. Ophidiaster sp. Asterina regularis VERRILL. Stichaster polyplax MÜLL. & TROSCHEL. Stichaster suteri LORIOL.

'98. On the Echinoderm Fauna of New Zealand. Proc. Linn. Soc. N. S. Wales for 1898, vol. 23, p. 300–327.

ASTROPECTINIDÆ. Astrop. edwardsii VERR. '67· A. polyacanthus M. T. '42 = A. vappa M. T. '43 = A. hestrix M. T. '42 = A. armatus M. T. = A. chinensis & A. ensifer GRUBE '65· Psilaster acuminatus SLADEN '89· PENTAGONASTERIDÆ. Astrogonium pulchellum SLADEN '89 = Pentagonaster pulchellus GRAY '40 = (Dorigona) pulchellus TENISON-WOODS '79 = Stephanaster elegans AYRES '54 = Astrogonium crassimanum MÖBIUS '59· A. abnormale GRAY '66 = A. pulchellum var. B. HUTTON '72 = A. sp. FARQUHAR '97· Gnathaster miliaris SLADEN '89 = Astrogonium miliare GRAY '47 = Pentagonaster (Astrogonium) miliaris PERRIER '76· Gnathaster rugosus FARQ. '97 = Astrog. rugosum HUTTON '72· Gnath. dilatatus SLADEN '89 = Pentagonaster (Astrogonium) dilatatus PERRIER '76· PEN-

TACEROTIDÆ. *Choriaster granulatus* LÜTKEN '69· ASTERINDÆ. *Nepanthia maculata* GRAY '40 = *Chœtaster maculatus* M. T. '42 = *Chœt. cylindratus* MÖBIUS '59· *Asterina regularis* VERR. '67 = *Asterina gunnii* var. GRAY '40 = *Asteriscus australis* (pars) M. T. '42 = *Asterina cabbalistica* LÜTKEN '71· *Asterina novæ-zealandiæ* PERRIER '76· *Stegnaster inflatus* SLADEN '89 = *Pteraster inflatus* HUTTON '72 = *Palmipes inflatus* HUTTON '76· LINCKIIDÆ. *Metrodira subulata* GRAY '40 = *Scytaster subulatus* M. T. '42· STICHASTE-RIDÆ. *Stichaster australis* SLADEN '89 = *Cœlasterias australis* VERRILL '67· *Stichaster polyplax* SLADEN '89 = *Asteracanthion polyplax* M. T. '44 = *Tarsaster neozealanicus* FARQ. '95· *Stichaster suteri* LORIOL '94 = *Asterias rupicola* HUTTON (not VERRILL) '78 = *St. littoralis* FARQ. '95· *St. suteri* var. *lævigatus* FARQ. '98 = *Asterias rupicola* var. *lævigatus* HUTTON '78· *St. insignis* FARQ. '95· ECHINASTERIDÆ. *Cribrella compacta* SLADEN '89 = *Henricia occulata* HUTTON '72. *Crib. ornata* PERRIER '75 = *Echinaster* (*Cribrella*) *ornata* PERRIER '63· *Crib. lutkinsii* FARQ. '98· *Echinaster purpureus* BELL '84 = *Othilia purpurea* GRAY '40 = *Cribrella fallax* DUJARDIN & HUPÉ '62 = *Echin. fallax* M. T. '42· ASTERIIDÆ. *Asterias calamaria* GRAY '40 = *Asterac. calamaria* M. T. '42· *Asterac. tenuispinus* MICHELIN '45 = *Coscinasterias muricata* VERRILL '67 = *Asteracanthion australis* PERRIER '69 = *Asterias jehennesei* PERRIER '75. *Asterias scabra* SLADEN '89 = *Margaraster* (?) *scaber* HUTTON '72· *Asterias mollis* HUTTON '72. *Asterias fragilis* STUDER '84· *Uniophora granifera* BELL '81 = *Asterias granifera* LAMARCK '16 = *Asteracanthion graniferus* M. T. '42 = *Margaraster graniferus* GRAY '66·

: 09. Further Notes on New Zealand Starfishes. Trans. N. Zeal. Inst. for 1908, vol. 41, p. 126–129, pl. xii. Issued June, 1909.

Stegnaster inflatus HUTTON, 2 figs. *Asterina neozelanica* PERRIER. *Asterias fragilis* STUDER. *Asterias calamaria* GRAY. *Asterias calamaria* var. *reischkei*, var. n. *Stichaster polyplax*.

FARRAN, G. P.

: 13. The Deep-water Asteroidea, Ophiuroidea and Echinoidea of the West Coast of Ireland. Sci. Investigations, 1912, No. vi. Fish. Branch Dept. Agric. and Techn. Instruction Ireland. 66 pp. + 2 pls.

BENTHOPECTINIDÆ. BENTHOPECTEN *armatus* (SLADEN) = PARARCHASTER *armatus* SLAD. '89· 1 fig. CHEIRASTERIDÆ. *Pontaster tenuispinus* (DÜB. & KOR.). PLUTONASTERIDÆ. PLUTONASTER *bifrons* (WYV. THOMSON). ASTROPECTINIDÆ. ASTROPECTEN *irregularis* PENN. *Psilaster andromeda* M. TR. *Psilasteropsis patagiatus* (SLAD.) = *Psilaster patagiatus* SLADEN '89 = *Psilaster andromeda* (pars) PERRIER '94 & '96· *Luidinium.* LUIDIA *ciliaris* PHILIP. *L. sarsi* DÜB. & KOR. PENTAGONASTERIDÆ. PENTAGONAS-TER *balteatus* SLAD. '91 = *P. granularis* (pars) BELL '92 = *P. gosselini* PERRIER '94· *P.*

dentatus PERRIER '81 = *P. grandis* PERR. '85· *P. perrieri* SLAD. '89 = *P. concinnus* SLAD. '91· *P. granularis* (pars) BELL '92· NYMPHASTER *arenatus* (PERRIER) = *P. arenatus* PERRIER '84 = *N. protentus* SLAD. '89 = *N. subspinosus* BOURNE '90 = *Doriqona arenata* PERRIER '94· PSEUDARCHASTER *pareli* (DÜB. & KOR) = *Astropecten parelii* DÜB. & KOR. '44 = *Archaster parelii* WYV. THOMS. '73 = *Plutonaster (Tethyaster) parelii* SLAD. '89 = *Astrogonium pareli* KŒHLER : 07 & : 09. *Mimasterbur.* MIMASTER *tizardi* SLAD. PENTACEROTIDÆ. *Culcita borealis* SUSSBACH & BRECKNER : 11, 2 figs. GYMNASTERIDE. *Porania pulvillus* O. F. MULLER. *Poraniomorpha villosa* (SLAD.) = *Lasiaster villosus* SLAD. '89· *Asterinæ. Palmipedinæ.* PALMIPES *placenta* (PENN.). STICHASTERIDÆ. STICHASTER *roseus* O. F. MÜLLER. *Stich. roseus* var. *ambiguus*, nov., 1 fig. *Zoroaster fulgens* WYV. THOMS. *Z. diomedeæ* VERR. '84 = *Z. trispinosus* KŒHLER '96· 1 fig. SOLASTERIDÆ. *Solasterinæ.* SOLASTER *papposus* (FABR.) = *S. affinis* KEMP : 05. *Korethrasterinæ.* KORETHRASTER sp. PTERASTERIDÆ. *Pterasterinæ.* PTERASTER *personatus* SLAD. *Hymenaster giganteus* SLAD. ECHINASTERIDÆ. HENRICIA *abyssicola* NORMAN = *Cribrella sanguinolenta* var. *abyssicola* NORM. '69 = *Henricia sanguinolenta* var. *abyssicola* BELL '92· 1 fig. ASTERIIDÆ. ASTERIAS *rubens* L. *A. glacialis* L. BRISINGIDÆ. BRISINGA *endecacnemas* ASBJ.

FEWKES, J. W.

'91.   An Aid to a Collector of the Coelenterata and Echinodermata of New England. Bull. Essex Inst. Salem, vol. xxiii, p. 1–92.

Asterias vulgaris. A. forbesii. A. polaris. Leptasterias tenera. Cribrella sanguinolenta. Solaster endeca. Crossaster papposa. Ctenodiscus crispatus. Asterina borealis. Pteraster militaris. Hippasterias phrygiana.

FISHER, W. K.

: 05.   New Starfishes from Deep Water off California and Alaska. Bull. Bur. Fisheries for 1904, vol. xxiv, p. 291–320. Issued June 10.

PORCELLANASTERIDÆ. PORCELLANASTER WYV. THOMSON. EREMICASTER, n. subg. *P. (E.) tenebrarius*, n. sp. ASTROPECTINIDÆ. BATHYBIASTER DANIELSSEN & KOBEN. *B. pectinatus*, n. sp. DIPSACASTER ALCOCK. *D. eximius*, n. sp. PERSEPHONASTER WOOD-MASON & ALCOCK. *P. penicillatus*, n. sp. BENTHOPECTINIDÆ. BENTHOPECTEN VERRILL. *B. acanthonotus*, n. sp. PLUTONASTERIDÆ. DYTASTER SLADEN. *D. gilberti*, n. sp. Subf. *Mimasterinæ.* MIMASTER SLADEN. *M. swifti*, n. sp. ODONTASTERIDÆ. ODONTASTER VERRILL. *O crassus*, n. sp. PSEUDARCHASTERIDÆ. PSEUDARCHASTER SLADEN. *Ps. alascensis*, n. sp. *Ps. pusillus*, n. sp. GONIASTERIDÆ. Subf. *Goniasterinæ.* TOSIA. *T. leptocerama*, n. sp. Subf. *Mediasterinæ.* MEDIASTER STIMPSON. *M. tenellus*, n. sp. Subf. *Hippasterinæ.* HIPPASTERIA GRAY. *H. heathi*, n. sp. *H. californica*, n. sp. CRYPTOPELTASTER, n. g. *C. lepidonotus*, n. sp. SOLASTERIDÆ.

LOPHASTER VERRILL. *L. furcilliger*, n. sp. KORETHRASTERIDÆ. PERIBOLASTER SLADEN. *P. biserialis*, n. sp. PTERASTERIDÆ. PTERASTER M. T. *Pt. jordani*, n. sp. HYMENAS_ TER WYV. THOMSON. *H. quadrispinosus*, n. sp. ZOROASTERIDÆ. ZOROASTER WYV. THOMSON. *Z. ophiurus*, n. sp. MYXODERMA, n. subg. *Z. (M.) sacculatus*, n. sp. *Z. (M.) evermanni*, n. sp. BRISINGIDÆ. BRISINGA ASBJØRNSEN. *B. exilis*, n. sp. FREYELLA PERRIER. *F. fecunda*, n. sp.

: 06. The Starfishes of the Hawaiian Islands. U. S. Fish Commission Bulletin for 1903, pt. iii, p. 987–1130, pl. i–xlix. Issued June 30, 1906.

ASTROPECTINIDÆ. ASTROPECTEN SCHULZE 1760. *A. polyacanthus* M. T. '42, 4 figs. *A. velitaris* v. MARTENS '65, 3 figs. *A. ctenophorus*, n. sp., 7 figs. *A. pusillulus*, n. sp., 4 figs. *A. productus*, n. sp., 10 figs. *A. callistus*, n. sp., 7 figs. CTENOPHORASTER, n. g. *Ct. hawaiiensis*, n. sp., 10 figs. TRITONASTER, n. g. *T. craspedotus*, n. sp., 14 figs. PSILASTER SLADEN '85. *P. attenuatus*, n. sp., 7 figs. PSILASTEROPSIS n. g. *Ps. cingulata*, n. sp., 7 figs. DIPSACASTER ALCOCK '93. *D. nesiotes*, n. sp., 8 figs. PATAGIASTER, n. g. *P. nuttingi*, n. sp., 4 figs. LUIDIIDÆ. LUIDIA FORBES '39. *L. hystrix*, n. sp., 6 figs. *L. magnifica*, n. sp., 6 figs. *L. sp.* *L. brevispina* LÜTKEN '71. PSEUDARCHASTERIDÆ. PSEUDARCHASTER SLADEN '89 = ASTROGONIUM PERRIER '94. *Ps. myobrachius*, n. sp., 6 figs. *Ps. jordani*, n. sp, 4 figs. BENTHOPECTINIDÆ. Subf. *Pontasterinæ*. CHEIRASTER STUDER '83 = PONTASTER part. SLADEN. *Ch. snyderi*, n. sp., 3 figs. *Ch. horridus*, n. sp., 4 figs. *Ch. inops*, n. sp., 2 figs. ARCHASTERIDÆ. ARCHASTER M. T. '40. *A. typicus* M. T. '40. GONIASTERIDÆ. Subf. *Mediasterinæ*. MEDIASTER STIMPSON '57. *M. ornatus*, n. sp., 5 figs. NEREIDASTER VERRILL '99. *N. bowersi*, n. sp., 4 figs. Subf. *Goniasterinæ*. PENTAGONASTER GRAY '40. *P. ammophilus*, n. sp., 4 figs. TOSIA GRAY '40. Subg. PLINTHASTER. *T. (P.) ceramoidea*, n. sp, 4 figs. Subg. CERAMASTER VERRILL '99. *T. micropelta*, n. sp., 3 figs. ASTROCERAMUS, n. g. *A. callimorphus*, n. sp., 4 figs. CALLIDERMA GRAY '47. *C. spectabilis*, n. sp., 6 figs. CALLIASTER GRAY '40. *C. pedicellaris*, n. sp., 4 figs. GILBERTASTER, n. g. *G. anacanthus*, n. sp., 4 figs. Subf. *Hippasterinæ*. EVOPLOSOMA, n. g. *E. forcipifera*, n. sp., 8 figs. Subf. *Leptogonasterinæ*. ANTHENIASTER VERRILL '89. *A. epixanthus*, n. sp, 8 figs. Subf. *Goniodiscidinæ*, new name. GONIODISCIDES, n. name. *G. sebæ* M. T. '42, 1 fig. PENTACEROTIDÆ. PENTACEROS SCHULZE 1760. *P. hawaiiensis*, n. sp., 5 figs. NIDORELLIA GRAY '40. *N. armata = Pentaceros (Nidorellia) armatus* GRAY '40. ASTERODISCUS GRAY '47. *A. tuberculosus*, n. sp., 8 figs. CULCITA AGASSIZ '35. *C. arenosa = C. novæ guineæ* var. *arenosa* DÜDERLEIN '96. LINKIIDÆ. OPHIDIASTER AGASSIZ '35. *O. lorioli*, n. sp, 6 figs. *O. squameus*, n. sp, 4 figs. *O. triseriatus*, n. sp., 3 figs. *O. sclerodermus*, n. sp., 4 figs. *O. tenellus*, n. sp., 2 figs. *O. rhabdotus*, n. sp, 2 figs. LEIASTER PETERS '52. *L. callipeplus*, n. sp, 3 figs. LINCKIA NARDO '34. *L.*

diplax M. T = *Ophid. diplax* M. T. '42 = *Linckia pacifica* var. *diplax* SLADEN '89· 2 figs. *L. multifora* LAMARCK = *Asterias multifora* LAM. '16 =? *L. typus* GRAY '40 =? *L. leachii* GRAY '40 = *Ophid. multiforis* M. T. '42 = *L. multiforas* GRAY '66· NARDOA GRAY '40 = LINCKIA part. M. T. = GOMOPHIA GRAY '40 = SCYTASTER part. M. T. '42 = OPHIDIASTER part. M. T. *N. œgyptica* GRAY = *Gomophia œgyptica* GRAY '40 = *Scytaster œgyptiacus* PERRIER '75· GYMNASTERIDÆ. GYMNASTERIA GRAY '40· *G. carinifera* LAMARCK = *Asterias carinifera* LAM. '16· ASTERINIDÆ. ASTERINA NARDO '34· *A. granulosa* PERRIER '75· ANSEROPODIDÆ, n. f. ANSEROPODA NARDO '34 = PALMIPES AGASSIZ '35· *A. insignis*, n. sp., 2 figs. ECHINASTERIDÆ. Subf. *Echinasterinæ*. HENRICIA GRAY '40 = LINCKIA FORBES '39 non NARDO = CRIBRELLA FORBES '41 non AGASSIZ = ECHINASTER M. T. (part.) '42· *H. robusta*, n. sp., 4 figs. *H. pauperrima*, n. sp., 5 figs. ECHINASTER M. T. '40· *E. sp.* Subf. *Valvasterinæ*. VALVASTER PERRIER '75· *V. striatus* LAMARCK = *Asterias striata* LAM. '16· 2 figs. MITHRODIIDÆ. MITHRODIA GRAY '40· *M. bradleyi* VERRILL '69· 5 figs. Ditto, peculiar specimen, 2 figs. MYXASTERIDÆ. ASTHENACTIS, n. g. *A. papyraceus*, n. sp., 2 figs. PTERASTERIDÆ. PTERASTER M. T. '42· *Pt. reticulatus*, n. sp.. 4 figs. HYMENASTER WYV. THOMSON '73· *H. pentagonalis*, n. sp., 5 figs. BENTHASTER SLADEN '82· *B. eritimus*, n. sp., 4 figs. ZOROASTERIDÆ. ZOROASTER WYV. THOMSON '73· *Z. spinulosus*, n. sp., 4 figs. HELIASTERIDÆ. HELIASTER GRAY '40· *H. multiradiata* GRAY '40· ASTERIIDÆ. COSCINASTERIAS VERRILL '69· *C.* (Subg. *Distolasterias*) *euplecta*, n. sp., 8 figs. HYDRASTERIAS SLADEN '89· *H. verrilli*, n. sp. BRISINGIDÆ. ODINIA PERRIER '85· *O. pacifica*, n. sp., 6 figs. BRISINGA ASBJØRNSEN '56· *B. panopla*, n. sp., 10 figs. *B. alberti*, n. sp., 10 figs. *B. evermanni*, n. sp., 7 figs. *B. fragilis*, n. sp., 5 figs.

: 06a. New Starfishes from the Pacific Coast of North America. Proc. Washington Acad. Sc., vol. viii, p. 111–139. Issued Aug. 14.

ASTROPECTINIDÆ. LEPTYCHASTER. *L. pacificus*, n. sp. *L. anomalus*, n. sp. ASTROPECTEN. *A. ornatissimus*, n. sp. LUIDIIDÆ. LUIDIA. *L. ludwigi*, n. sp. *L. asthenosoma*, n. sp. ECHINASTERIDÆ. HENRICIA. *H. aspera*, n. sp. *H. polyacantha*, n. sp. SOLASTERIDÆ. CROSSASTER. *C. alternatus*, n sp. *C. borealis*, n. sp PYCNOPODIIDÆ. RATHBUNASTER. *R. californicus*, n. sp.

: 08. Necessary Changes in the Nomenclature of Starfishes. Smithsonian Miscellaneous Collections (Quarterly Issue), vol. 52, p. 87–93. Publ. May 27.

ANASTERIAS LUDWIG : 03 = LYSASTERIAS nom. nov. = non ANASTERIAS PERRIER '75· ASTEROPSIS M. T.? Sept., '40 = GYMNASTERIA. CRENASTER PERRIER '85 = DYTASTER SLADEN. CTENASTER PERRIER '81 = LÆTMASTER, nom. nov. DIPLASTERIAS PERRIER '88· nom. nud. = DIPLASTERIAS PERRIER 'J1 = COSMASTERIAS SLADEN = PISASTER M. T.

'40· GONIODON PERRIER '9⊢=DIPLODONTIAS nom. nov. GYMNASTERIA GRAY December, '40=ASTEROPSIS M. T. Sept., '40=ASTEROPE M. T. Apr., '40· therefore ASTEROPE M. T. the correct name. PATIRIA GRAY '40=ASTERINA non PATIRIA PERRIER '75· PATIRIA PERRIER '75=PARASTERINA, nom. nov. PARARCHASTER SLADEN '85=BENTHOPECTEN VERRILL '84· PENTACEROS SCHULZE 1760 non-binomial=OREASTER M. T. '42· ASTRO-PECTEN SCHULZE 1760 non-binomial=ASTROPECTEN GRAY '40 PENTAGONASTER SCHULZE 1760 non-binomial =GONIASTER AGASSIZ '35=PENTAGONASTER GRAY '40=STEPHANASTER AYERS '52=PHANERASTER PERRIER '94·

: 08a. Some Necessary Changes in the Generic Names of Starfishes. Zool. Anz., Bd. 33, p. 356-359. Publ. Aug. 18.

CRIBRELLA AGASSIZ '35=LINCKIA NARDO '34· CRIBRELLA FORBES '41 (non AGASSIZ)= HENRICIA GRAY '40· PALMIPES AGASSIZ '35=ANSEROPODA NARDO '34· The other genera dealt with are contained in the preceding paper : 08.

: 10. New Starfishes from the North Pacific. I. Phanerozonia. II. Spinu-losa. Zool. Anz., Bd. 35, Nr. 18, March 29. I, p. 545-553. II, p. 568-574.

I. ASTROPECTINIDÆ. Leptychaster propinquus, n. sp. DIPSACASTER. D. eximius FISHER. D. borealis, n. sp. D. laetmophilus, n. sp. D. anoplus, n. sp. BENTHOPECTINIDÆ. BENTHOPECTEN. B. acanthonotus FISHER. B. claviger, n. sp. B. mutabilis, n. sp. ACANTHARCHASTER VERRILL=MARCELLASTER KŒHLER : 07. A. dawsoni VERRILL. A. aciculosus, n. sp. A. variabilis, n. sp. A. variabilis pedicellaris, n. subsp. A. inter-medius, n. sp. CHEIRASTER agassizi evoplus, n. subsp. GONIASTERIDÆ. PSEUDARCHASTER dissonus, n. sp. CERAMASTER clarki, n. sp. CLADASTER validus, n. sp. HIPPASTERIA leiopelta, n. sp.

II. ECHINASTERIDÆ. PORANIOPSIS inflata flexilis, n. subsp. HENRICIA. H. sanguino-lenta (MÜLLER). H. leviuscula (STIMPSON). H. annectens, n. subsp. H. multispina, n. subsp. H. aspera FISHER. H. asthenactis, n. sp. H. leviuscula annectens, n. subsp. H. longispina, n. sp. H. polyacantha FISHER. H. clarki, n. sp. SOLASTERIDÆ. SOLAS-TER exiguus, n. sp. S. hypothrissus, n. sp. LOPHASTER furcilliger vexator, n. subsp. N. B. The name annectens n. subsp. occurs twice in the synopsis of the genus Henricia.

: 11. Asteroidea of the North Pacific and Adjacent Waters. Part. 1. Phanerozonia and Spinulosa. U. S. Nat. Museum, Bull. 76. vi+ 419 pp. + 122 pls.

Classification of Asteroidea History of systematic works on North Pacific Asteroidea. Distribution and relationships of North Pacific Asteroidea.

Phanerozonia. PORCELLANASTERIDÆ. EREMICASTER. *E. tenebrarius* (FISHER) = *Porcellanaster* (*Erem.*) *tenebrarius* FISHER : 04 = *Porcell. waltharii* LUDWIG : 05 = *Erem. tenebrarius* LUDWIG : 07 = *Eremicaster waltharii* LUDWIG : 07, 7 figs. *E. pacificus* (LUDWIG) = *Porcell. pacificus* LUDWIG : 05 = *E. pacificus* FISHER : 07, 3 figs. CTENODISCUS. *Ct. crispatus* (RETZIUS) = *Asterias crispata* RETZIUS '05 = *Ct. australis* LÜTKEN '71 = *Ct. krausei* LUDWIG '86 = *Ct. procurator* SLADEN '89, 10 figs. ASTROPECTINIDÆ. LEPTYCHASTER. *L. arcticus* (SARS) = *Astropecten arcticus* M. SARS '51 = *Astropecten lütkeni* BARRETT '57 = *Archaster arcticus* VERRILL = *Leptoptychaster arcticus* SLADEN '89, 2 figs. *L. pacificus* FISHER : 06, 4 figs. *L. anomalus* FISHER : 06 = *Glyphaster anomalus* VER. : 09, 4 figs. *L. propinquus* FISHER : 10, 1 fig. ASTROPECTEN. *A. armatus* GRAY '40 (non MÜLLER & TROSCHEL '42) = *A. erinaceus* GRAY '40 = *A. örstedii* LÜTKEN, 6 figs. *A. californicus* FISHER : 06, 6 figs. *A. ornatissimus* FISHER : 06, 7 figs. PSILASTER. *P. pectinatus* (FISHER) = *Bathybiaster pectinatus* FISHER : 04 = *Plutonaster abyssicola* LUDWIG : 05, 9 figs. THRISSACANTHIAS. *T. penicillatus* (FISHER) = *Persephonaster penicillatus* FISHER : 05, 12 figs. DIPSACASTER. *D. eximius* FISHER : 04, 9 figs. *D. borealis* FISHER : 10, 7 figs. *D. laetmophilus* FISHER : 10, 7 figs. *D. anoplus* FISHER : 10, 4 figs. DYTASTER. *D. gilberti* FISHER : 05, 5 figs. LUIDIA. *L. foliolata* GRUBE = *L. foliata* SLADEN '89, 7 figs. *L. ludwigi* FISHER : 06, 5 figs. *L. asthenosoma* FISHER : 06, 3 figs. BENTHOPECTINIDÆ. PECTINASTER. *P. agassizi evoplus* (FISHER) = *Cheiraster agassizi evoplus* FISHER : 10. LUIDIASTER. *L. dawsoni* (VERRILL) = *Archaster dawsoni* VER. '80 = *Acantharchaster dawsoni* VER. '94, 9 figs. NEARCHASTER. *N. aciculosus* (FISHER) = *Acantharchaster aciculosus* FISHER : 10. *N. variabilis* (FISHER) = *Acantharchaster variabilis* FISHER : 10, 6 figs. *N. pedicellaris* (FISHER) = *Acantharchaster variabilis pedicellaris* FISHER : 10, 4 figs. MYONOTUS. *M. intermedius* (FISHER) = *Acantharchaster intermedius* FISHER : 10, 9 figs. BENTHOPECTEN. *B. acanthonotus* FISHER : 04, 7 figs. *B. mutabilis* FISHER :10, 4 figs. *B. claviger* FISHER : 10, 5 figs. ODONTASTERIDÆ. ODONTASTER. *O. crassus* FISHER :04, 5 figs. GONIASTERIDÆ. Notes on certain genera. Synopsis of the family. GEPHYREASTER. *G. swifti* (FISHER) = *Mimaster swifti* FISHER : 04, 10 figs. PSEUDARCHASTER. *Ps. parelii* (DÜBEN & KOREN) = *Astropecten parelii* DÜBEN & KOREN '44 = *Archaster parelii* M. SARS '61 = *Archaster parelii* var. *longobrachialis* DANIELSSEN & KOREN '76 = *Plutonaster* (*Tethyaster*) *parelii* SLADEN '89 = *Plutonaster parelii* BELL '92 = *Pseudarchaster intermedius* SLADEN '89 = *Pseudarch. tessellatus* var. *arcticus* SLUITER '95 = *Astrogonium parelii* KŒHLER : 07, 4 figs. *Ps. parelii alascensis* FISHER = *Ps. alascensis* FISHER : 04, 6 figs. *Ps. pusillus* FISHER : 04, 9 figs. *Ps. dissonus* FISHER : 10, 5 figs. MEDIASTER. *M. æqualis* STIMPSON '57, 7 figs. *M. tenellus* FISHER : 04, 8 figs. CERAMASTER. *C. japonicus* (SLADEN) = *Pentagonaster japonicus* SLADEN '89 = *Mediaster japonicus* VERRILL '99, 4 figs. *C. leptoceramus* (FISHER) = *Tosia leptocerama* FISHER : 04, 6 figs. *C. patagonicus* (SLADEN) = *Pentagonaster patagonicus* SLADEN '89 = *Astrogonium granulare* WHITEAVES ? '86 = *Mediaster patagonicus* VER. '99, 4 figs. *C. clarki*

FISHER : 10, 7 figs. *C. arcticus* (VERRILL) = *Tosia arctica* VER. : 09, 1 figs. CLADAS-
TER. *C. validus* FISHER : 10, 2 figs. HIPPASTERIA. *H. spinosa* VERRILL : 09, 6 figs.
H. spinosa kurilensis, n. subsp., 2 figs. *H. leiopelta* FISHER : 10, 5 figs. *H. leiopelta
armata*, n. subsp , 3 figs. *H. heathi* FISHER : 04, 5 figs. *H. californica* FISHER : 04,
6 figs. CRYPTOPELTASTER. *C. lepidonotus* FISHER : 04, 5 figs. LINCKIIDÆ. LINCKIA.
L. columbiæ GRAY '40 = *L. ornithopus* VER. '67 = *L. diplax* PERRIER '75 = *L. pacifica* var.
diplax SLADEN '89 = *Phataria* (*L.*) *fascialis* MONKS : 03 = *Phataria* (*L.*) *unifascialis*
GRAY var. *bifascialis* MONKS : 04, 7 figs. ASTEROPIDÆ. DERMASTERIAS. *D. imbricata*
(GRUBE) = *Asteropsis imbricata* GRUBE '57 = *D. inermis* PERRIER '75 = "*Dermaster*"
imbricatus WHITEAVES '86, 3 figs.

Spinulosa. ASTERINIDÆ. ASTERINA. *A. miniata* (BRANDT) = *Asterias miniata* BRANDT
'35 = *Asteriscus miniatus* STIMPSON '57, 8 figs. ECHINASTERIDÆ. PORANIOPSIS. *P. inflata*
(FISHER) = *Alexandraster inflatus* FISHER : 06, 5 figs. *P. inflata flexilis* FISHER : 10, 3 figs.
HENRICIA. *H. sanguinolenta* (O. F. MULLER) = *Asterias sanguinolenta* O. F. MÜLLER 1776
= *Asterias pertusa* O. F. MÜLL. 1776 = *Asterias oculata* PENNANT 1777 = *Asterias spongiosa*
FABRICIUS 1780 = *Asterias seposita* RETZIUS 1783 = *Linckia oculata* FORBES '39 = *Henricia
oculata* GRAY '40 = *Cribrella oculata* FORBES '41 = *Echinaster oculatus* MÜLL. & TROSCH. '42
= *Echinaster sanguinolentus* M. SARS '44 = *Echinaster sarsii* MÜLL. & TROSCH. '44 = *Lin-
ckia pertusa* STIMPSON '53 = *Cribrella sanguinolenta* LUTKEN '57, 8 figs. *H. sanguinolenta
eschrichtii* (MÜLL. & TROSCHEL) = *Echinaster eschrichtii* MULL. & TROSCHEL '42 = *Cribrella
eschrichtii* DUJ. & HUPÉ = *Cr. sanguinolenta* MURDOCH '85 = *Cribrella oculata* LUDWIG '86
= *Henricia tumida* VER. : 09, 5 figs. *H. leviuscula* (STIMPSON) = *Linckia leviuscula*
STIMPSON '57 = *Chætaster californicus* GRUBE '65 = *Cribrella laviuscula* WHITEAVES '78 =
H. leviuscula crassa CLARK : 01 = *H. leviuscula attenuata* CLARK : 01, 7 figs. *H. levius-
cula multispina* FISHER : 10, 6 figs. *H. leviuscula dyscrita*, n. subsp., 5 figs. *H.
leviuscula annectens* FISHER : 10, 3 figs. *H. aspera* FISHER : 06, 5 figs. *H. spiculifera*
(CLARK) = *Cribrella spiculifera* CLARK : 01. *H. asthenactis* FISHER : 10, 3 figs. *H.
longispina* FISHER : 10, 4 figs. *H. longispina aleutica* n. subsp., 1 fig. *H. polyacantha*
FISHER : 06, 5 figs. *H. clarki* FISHER : 10, 4 figs. SOLASTERIDÆ. SOLASTER. *S. endeca*
(LIN.) = *Asterias endeca* LIN. 1771 = *Asterias aspera* O. F. MULLER 1776 = *Asterias
alboverrucosa* BRANDT '35 = *Stellonia endeca* AGASSIZ '35 = *Solaster* (*Endeca*) *endeca* GRAY
'40 = *Solaster galaxides* VER. : 09, 4 figs *S. stimpsoni* VER. '80 = *Asterias endeca* var.
decemradiata BRANDT (?) '35 = *S. decemradiatus* STIMPSON (?) '57 = *Crossaster vancouver-
ensis* DE LORIOL '97 = *S. endeca decemradiata* CLARK : 01, 6 figs. *S. dawsoni* VER. '80 =
S. endeca MURDOCH '85, 7 figs. *S. paxillatus* SLADEN '89, 6 figs. *S. exiguus* FISHER
: 10, 5 figs. *S. borealis* (FISHER) = *Crossaster borealis* FISHER : 06, 7 figs. *S. hypothrissus* FISHER : 10, 4 figs. *S. papposus* (LIN.) = *Asterias papposa* LIN. 1767 = *Asterias
helianthemoides* PENNANT 1777 = *Asterias affinis* BRANDT '35 = *Stellonia papposa* AGASSIZ
'35 = *Solaster* (*Polyaster*) *papposa* GRAY '40 = *Crossaster papposus* MÜLL. & TROSCH.

'40· 6 figs. *S. japonicus*, n. sp., 1 fig HETEROZONIAS. *H. alternatus* (FISHER) = *Crossaster alternatus* FISHER : 06, 6 figs. LOPHASTER. *L. furcilliger* FISHER : 04, 11 figs. *L. furcilliger vexator* FISHER : 10, 4 figs. KORETHRASTERIDÆ. PERIDOLASTER. *P. biserialis* FISHER : 05, 7 figs. PTERASTERIDÆ. PTERASTER. *P. militaris* (O. F. MÜLLER) = *Asterias militaris* O. F. MÜLLER 1776 = *Asteriscus militaris* MÜLL. & TROSOH. '42 = *Pt. aporus* LUDWIG '86· 2 figs. *P. trigonalon* FISHER : 10, 5 figs. *P. gracilis* (CLARK) = *Retaster gracilis* CLARK : 01. *P. jordani* FISHER : 04, 5 figs. *P. marsippus* FISHER : 10, 4 figs. *P. coscinopeplus* FISHER : 10, 3 figs. *P. temnochiton* FISHER : 10, 5 figs. *P. pulvillus* SARS '61· 2 figs. *P. multispinus* CLARK : 01. *P. tesselatus* IVES '88· 5 figs. *P. tesselatus arcuatus*, n. subsp., 1 fig. *P. obscurus* (PERRIER) = *Hexaster obscurus* PERRIER '91 = *P. (Temnaster) hexactis* VER. '94 = *P. hexactis* DÖDERLEIN '89 = *Temnaster hexactis* VER. '95 = *P. octaster* VER. : 09, 6 figs. DIPLOPTERASTER. *D. multipes* (SARS) = *Pteraster multipes* SARS '65 = *Retaster multipes* SLADEN '89· 2 figs. HYMENASTER. *H. kœhleri* FISHER : 10, 2 figs *H. perissonotus* FISHER : 10, 10 figs. *H. quadrispinosus* FISHER : 05, 5 figs.

: 13. Four New Genera and Fifty-eight New Species of Starfishes from the Philippine Islands, Celebes, and the Moluccas. Proc. U. S. Nat. Mus. 43, p. 599–648. Publ. Feb. 5.

PORCELLANASTERIDÆ. ¡SIDONASTER *psilonotus*, n. sp. *Ctenodiscus orientalis*, n. sp. GONIOPECTINIDÆ. *Goniopecten asiaticus*, n. sp. PRIONASTER VERRILL. *P. analogus*, n. sp. *P. gracilis*, n. sp. *P. megaloplax*, n. sp. ASTROPECTINIDÆ. *Astrop. acanthifer phragmorus*, n. subsp. *Astrop. eremicus*, n. sp. *Astrop. luzonicus*, n. sp. *Astrop. tenellus*, n. sp. *Astrop. pedicellaris*, n. sp. CTENOPLEURA, n. g. *Ct. astropectinides*, n. sp. CTENOPHORASTER *diploctenius*, n. sp. PSILASTER *gotoi*, n. sp. *Psil. robustus*, n. sp. ASTROMESITES, n. g. *A. compactus*, n. sp. PERSEPHONASTER ALCOCK. *P. euryactis*, n. sp. *P. anchistus*, n. sp. *P. luzonicus*, n. sp. *P. tenuis*, n. sp. *P. multicinctus*, n. sp. *P. suluensis*, n. sp. *P. œdiplax*, n. sp. *P. habrogenys*, n. sp. *P. monostœchus*, n. sp. TRITONASTER *evorus*, n. sp. DIPSACASTER *diaphorus*, n. sp. PATAGIASTER *sphærioplax*, n. sp. DYTASTER SLADEN. KOREMASTER n. subg. *D. (K.) evaulus*, n. sp. GONIASTERIDÆ. MIMASTER *notabilis*, n. sp. PSEUDARCHASTER *oligoporus*, n. sp. APHRODITASTER *microceramus*, n. sp. PARAGONASTER SLADEN. *P. ctenipes hypacanthus*, n. subsp. *P. stenostichus*, n. sp. PERISSOGONASTER, n. g. *P. insignis*, n. sp. ROSASTER PERRIER. *R. nannus*, n. sp. *R. mimicus*, n. sp. *R. mamillatus*, n. sp. NYMPHASTER SLADEN. *N. euryplax*, n. sp. *N. dyscritus*, n. sp. *N. mucronatus*, n. sp. *N. leptodomus*, n. sp. *N. moluccanus*, n. sp. *N. arthrocnemis*, n. sp. *N. meseres*, n. sp. *N. labrotatus*, n. sp. *N. atopus*, n. sp. CERAMASTER *smithi*, n. sp. PELTASTER *cycloplax*, n. sp. SPHÆRIODISCUS *scrotocryptus*, n. sp. ICONASTER *perierctus*, n. sp. ASTROCERAMUS *lionotus*, n sp. *Astroc. sphæriostictus*, n. sp. CALLIASTER *corynetes*, n. sp. ASTROTHAUMA,

n. g. *A. euphylacteum*, n. sp. Anthenoides Perrier. *A. granulosus*, n. sp. *A. lithosorus*, n. sp. *A. rugulosus*, n. sp.

Forbes, E.

   * '52· Sutherland's Journal of a Voyage, vol. ii, App. p. ccxiv. Cited after Duncan and Sladen.

Giebel, C. G.

   * '62· Bemerkungen über einige Astropecten-Arten. Zeitschr. ges. Natur- wiss., Bd. 20, p. 324–326. Cited after Ludwig.

Gray, J. E.

   '40· A Synopsis of the Genera and Species of the Class Hypostoma (Asterias, Linnæus). Ann. Mag. Nat. Hist., vol. vi, p. 175–184 + p. 275–290.

Comparison of Linck's genera and those of later authors : Pentagonaster = Goniaster Agassiz = Scutasteries Blainv. Pentaceros = Goniaster Agassiz = Asterina Nardo = Plattasterias Blainv. Astropecten = Stellaria Nardo = Asterias Agassiz. Palmipes = Anseropoda Nardo = Palmasterias Blainv. Stella coriacea = Stellonia Forbes = Stellonia part Nardo = Pentasterias + Solasterias Blainv. Pentadactylosaster = Cribrella Agassiz non Edwards = Linckia Nardo non Agassiz. Octactis + Enneactis + Decactis + Dodecactis + Triskaidecactis = Solaster Forbes = Stellonia part Nardo & Agassiz.

Comparison of Nardo's genera ('33) with those of Linck : Stellaria = Astropecten Linck. Stellonia = Stella coriacea Linck and his other genera enumerated above. Asterina = Pentaceros Linck part. Anseropoda = Palmipes Linck. Linckia = Pentadactylosaster Linck.

Comparison of Agassiz's genera and those of others : Asterias = Astropecten Linck = Stellaria Nardo. Cœlaster (fossil). Goniaster = Pentagonaster + Pentaceros Linck. Ophidiaster, a new sp. Linckia = Cribrella = Pentadactylosaster Linck. Stellonia Nardo = Stella coriacea Linck etc. asa bove. Asterina Nardo. Palmipes Linck = Anseropoda Nardo. Culcita Agassiz for *Ast. discoidea* Lam.

Asteriadæ. Asterias. *Ast. aster*, n. sp. *Ast. calamaria*, n. sp. *Ast. glacialis* Linck = *Ast. spinosa* Pennant. *Ast. rustica*, n. sp. *Ast. echinata*, n. sp. *Ast. holsatica* Retzius = *Ast. violacea* O. F. Müller = *Ast. glacialis* John. *Ast. rubens* Linné. *Ast. katherinæ*, n. sp. *Ast. wilkinsonii*, n. sp. *Ast. helianthus* Lam. *Ast. cumingii*, n. sp. *Ast. multiradiata*, n. sp. Tonia, n. g. *T. atlantica*, n. sp. Astropectinidæ. Nauricia, n. g. *N. pulchella*, n. sp. Astropecten. *Astrop. corniculatus* Linck. *Astrop. polaris* = *Asterias polaris*. *Astrop. stellaris*. *Astrop. duplicatus*, n. sp. *Astrop. auran-*

tiacus = *Asterias aurantiaca* LINN. *Astrop. stellatus*, n. sp. *Astrop. armatus*, n. sp. *Astrop. echinatus* LINCK. *Astrop. marginatus*, n. sp. = ? *Astrop. fimbriatus* LINCK. *Astrop. regalis*, n. sp. *Astrop. erinaceus*, n. sp. *Astrop. mauritianus*, n. sp. *Astrop. mesodiscus* LINCK. *Astrop. gracilis*, n. sp. *Astrop. irregularis* LINCK. *Astrop. dubius*. *Astrop. regularis* LINCK. *Astrop. longipes*, n. sp. LUIDIA. *Luid. fragillisima* FORBES '39 = *Asterias rubens* JOHNSTON. *Luid. savignii* = *Asterias savignii* AUDOUIN. *Luid.? ciliaris* = *Asterias ciliaris*. PETALASTER, n. g. *Petal. hardwickii*, n. sp. *Petal. columbia*. SOLASTER. *Sol. endeca* FORBES = *Asterias endeca* LINN. = *Ast. aspersa* O. F. MÜLLER. *Sol. papposa* FORBES = *Asterias papposa* LINN. = *Ast. stellata* RETZIUS. HENRICIA (FORBES non NARDO), n. g. *Hen. oculata* GRAY = *Asterias oculata* PENNANT = ? *Asterias seposita* PENNANT. PENTACEROTIDÆ. CULCITA, n. g. *C. schmideliana* = *Asterias schmideliana* RETZIUS = *Ast. placenta* = *Ast. discoidea* LAMARCK. PENTACEROS, n. g. *Pentac. grandis* SEBA. *Pentac. reticulatus* = *Asterias reticulata* LINN. *Pentac. gibbus* LINCK. *Pentac. cumingii*, n. sp. *Pentac. hivlcus* LINCK = *Asterias nodosa a* LAMARCK. *Pentac. chinensis*. *Pentac. franklinii*. *Pentac. muricatus* LINCK = *Asterias linckii* BLAINVILLE = *Ast. nodosa* LAMARCK. *Pentac. nodosa* = *Asterias nodosa* GMELIN (part). *Pentac. aculeatus* SEBA. *Pentac. armatus*, n. sp. STELLASTER, n. g. *St. childreni*. COMPTONIA, n. g. *Comp. elegans*, n. sp. (fossil). GYMNASTERIA, n. g. *Gymn. spinosa*. *Gymn. inermis*. PAULIA, n. g. *P. horrida*. RANDASIA, n. g. *R. luzonica*. ANTHENEA, n. g. *An. chinensis* = *Asterias chinensis* GRAY. HOSEA, n. g. *H. flavescens*. HIPPASTERIA, n. g. *Hipp. europaea* = *Asterias equestris* PENNANT. *Hipp. johnstoni* = *Asterias johnstoni* GRAY '36. CALLIASTER. *Call. childreni*, n. sp. GONIASTER. *Gon. cuspidatus* = *Pentagonaster semilunatus cuspidatus* LINCK. *Gon. sebæ* SEBA. *Goniaster regularis* SEBA = *Pentagonaster regularis* LINCK. PENTAGONASTER, n. g. *Pent. pulchellus* = *Asterias pulchella* GRAY. TOSIA, n. g. *T. australis*. ECHINASTER, n. g. *Echin. ellisii* GRAY = *Asterias echinus* SOLANDER & ELLIS = *Ast. echinites* LAMARCK. *Echin. solaris* = *Asterias solaris*. OTHILIA, n. g. *Oth. spinosa* = *Asterias spinosa* RETZIUS = *Pentadactylosaster spinosus* LINCK = *Ast. echinophora* LAMARCK. *Oth. aculeata*. *Oth. multispina*. *Oth. purpurea*. *Oth. luzonica*. METRODIRA, n. g. *Met. subulata*. RHOPIA, n. g = STELLONIA AGASSIZ part. *Rh. seposita* = *Asterias seposita* RETZIUS 1783 = *Pentadactylosaster reticulatus* LINCK = *Stellonia seposita* AGASSIZ. *Rh. mediterranea*. FERDINA, n. g. *F. flavescens*. *F. cumingii*. DACTYLOSASTER, n. g. *Duct. cylindricus* = *Asterias cylindrica* LAMARCK. *Dact. gracilis*. TAMARIA, n. g. *T. fusca*. CISTINA, n. g. *Cist. columbiæ*. OPHIDIASTER. *Oph. aurantius*. *Oph. leachii*. *Oph. guildingii*, n. sp. *Oph. attenuatus*. *Oph. pyramidatus*. LINCKIA (non MICHELI) NARDO & AGASSIZ non PERSOON nec CAV. *L. typus* NARDO = *Pentadactylosaster miliaris* LINCK = *Asterias lævigata* LINN. LAM. *L. crassa*. *L. brownii* RUMPH. *L. leachii*. *L. guildingii*. *L. pacifica*. *L. columbiæ*. *L. unifascialis*. *L. bifascialis*. *L. pulchella*. *L. intermedia*. *L. erythraea*. FROMIA, n. g. *Fr. milleporella* = *Asterias sebæ* BLAINV

= *Ast. millepora* LAMARCK. GOMOPHIA, n. g. *G. egyptiaca.* NARDOA, n. g. *N. variolata* = *Asterias variolatus* LAMARCK = *Pentadactylosaster variolatus* LINCK = *Linckia variolosa* NARDO. *N. agassizii,* n. sp. *N. tuberculata,* n. sp. NARCISSIA, n. g. *N. teneriffæ.* NECTRIA, n. g. *Nec. oculifera* = *Asterias oculifera* LAMARCK. NEPANTHIA, n. g. *Nep. tessellata. Nep. maculata.* MITHRODIA, n. g. *M. spinulosa* = *Asterias clavigera* LAMARCK = *Pentadactylosaster reticulatus* LINCK = *Ast. reticulata* BLAINVILLE MAN.? non LINN. nec LAM. UNIOPHORA, n. g. *Un. globifera.* ASTERINIDÆ. PALMIPES. *P. membranaceus* LINCK = *Asterias membranacea* RETZIUS & LAM. = *Ast. placenta* PENNANT = *Ast. cartilaginea* FLEMING = *Ast. rosacea* LAM. *P. stokesii.* FORANIA, n. g. *P. gibbosa* = *Asterias gibbosus* LEACH '17 = ? *Ast. equestris* THOMPSON = *Goniaster templetoni* FORBES '39· ASTERINA. *A. gibbosa* FORBES = *Asterias gibbosa* PENNANT = *Pentaceros plicatus et concavus* LINCK = *Asteriscus exigua* = ? *Asterias minuta* LINN. = *Asterias stellata obtusa ciliata* LINN. = ? *Asterina minuta* AGASSIZ = *Asterias pulchella* BLAINV. *A. hartonii. A. minuta* = ? *Asterias minuta* LINN. GMELIN = *Asterias exigua* LAM., var. 1 & 2. *A. krausii,* n. sp. *A. gunnii,* n. sp., 2 var. or 1 var. & 1 monstrosity. *A. calcar* = *Asterias calcar* LAM. PATIRIA. *P. coccinea.* SOCOMIA, n. g. *S. paradoxa.*

'47. Descriptions of Some New Genera and Species of Asteriadæ. Ann. Mag. Nat. Hist., vol. xx, p. 193-204.

CULCITA. *C. schmideliana* = *A. schmideliana* RETZIUS = *A. discoidea* LAM. *C. pentangularis.* RANDASIA, n. g. *R. granulata,* n. sp. *R. spinulosa,* n. sp. ASTERODISCUS. *A. elegans.* PENTACEROS *granulosus.* STELLASTER *incei. Stell. belcheri.* CALLIDERMA. *C. emma.* ANTHENEA. *A. tuberculosa. A. granulifera.* HOSIA *spinulosa.* ASTROGONIUM (restricted). *A. granularis* = *Asterias granularis* RETZIUS. *A. miliare. A. inæquale. A. tuberculatum. A. paxillosum.* PENTAGONASTER *dubeni.* TOSIA, n. g. *T. grandis. T. aurata. T. tubercularis. Tosia rubra. T. australis* GRAY '40· PETRICIA. *P. punctata.* PATIRIA. *P. coccinea* GRAY = *Asteriscus coccineus* M. T. *P. granifera* = ? *Asterias granifera* LAM.? var. à petits grains OUDART. *P. ocellifera* = *Asterias ocellifera* LAM. *P. obtusa. P.? crassa.* PTERASTER *capensis.* GANERIA. *G. falklandica.*

'66. Synopsis of the Species of Starfish in the British Museum. (With Figures of Some of the New Species.) 18 pp. ÷ 16 pls.

ASTERIADÆ. ASTERIAS. *A. aster. A. calamaria. A. tenuispina* = *A. glacialis* GRUBE = *A. spinosa* PENNANT = *A. savaresii* CHIAJE. *A. rustica* = *A gelatinosa* MEYER. *A echinata. A. glacialis* MÜLLER = *A. angulosa* MÜLLER = *Stellonia glacialis* NARDO = *S. angulosa* AGASSIZ. *A. holsatica* = *A. violacea* MÜLLER = *A. glacialis* JOHN. *A. rubens* = *Stellonia rubens. A. katherinæ. A. wilkinsonii. A. helianthus* = *Stellonia helianthus. A. cumingii. A. multiradiata.* UNIOPHORA. *U. globifera.* MARGARASTER. *M. graniferus* = *Asterias granifera* LINCK = ? *Asterias serrulata* E. [ncyclopédie] M. [éthodique] = *Ast. janthina* BRANDT. TONIA. *T. atlantica* = *Asterias aurantia* MEYER = *Stichaster striatus* M. T. '40 = *Asteracanthion aurantiacum* M. T. '42· MITHRODIA. *M. spinosa* = *Asterias clavigera* LAM. = *Pentadactylosaster reticulatus* LINCK = *Asterias reticulata* BLAINV. MAN. ? non LINN. nec LAM. = *Asterac. linckii* M. T. ASTROPECTINIDÆ. NAURICIA. *N. pulchella* = ? *Stellaster sulcatus* MÖBIUS. ASTROPECTEN. *A. polaris* = *Ctenodiscus polaris* M. T. = *Ct. crispatus* LÜTKEN = *Asterias crispata* RETZIUS = ? *Astrop. corniculatus* LINCK = *Ct. pygmæus* M. T. *A. stellaris* = *Archaster typicus* M. T. '40· *A. duplicatus* = *A. brasiliensis* M. T. *A. aurantiacus* = *Asterias aurantiaca* LINN. *A. stellatus* = *A. valenciennii* M. T. *A. armatus*. Ditto var. *pulcher*. *A. echinatus*. *A. marginatus* = ? *A. fimbriatus* LINCK. *A. regalis*. *A. erinaceus*. *A. mauritianus* = *Archaster angulatus* M. T. *A. mesodiscus*. *A. gracilis*. *A. irregularis* = *Asterias aurantiaca* MÜLLER = ? *Ast. johnstonii* CHIAJE. *A. dubius*. *A. regularis* = *Asterias petalodea* RETZ. *A. longipes*. LUIDIA. *L. fragilissima* FORBES '39 = *Asterias rubens* JOHNSTON. *L. savignii* = *Asterias savignii* AUDOUIN. *L. ciliaris*. PETALASTER = CHAETASTER M. T. *P. hardwickii*. *P. columbia*. SOLASTER = CROSSASTER M. T. *S. endeca* FORBES = *Asterias endeca* LINN. = *Ast. aspersa* MÜLLER. *S. papposus* FORBES = *Asterias papposa* LINN. = *Ast. stellata* RETZ. HENRICIA = LINCKIA FORBES non NARDO = ECHINASTER sp. M. T. *H. oculata* = *Asterias oculata* = *Ast. seposita* RETZ. = *Echinaster oculatus* M. T. = *Linckia oculata* FORBES. PENTACEROTIDÆ. CULCITA. *C. schmideliana* = *Asterias schmideliana* RETZ. = *Ast. discoidea* LAM. = *Ast. placenta* LAM. = *Cul. discoidea* AGASSIZ, 2 figs. ? *C. pentangularis*, 2 figs.? ASTERODISCUS. *A. elegans*, 4 figs. PENTACEROS. *P. grandis*. *P. reticulatus* = *Asterias reticulata* LINN. = *Oreaster reticulatus* M. T. *P. gibbus*. *P. cumingii*. *P. nodosus* = *Ast. nodosa a* LAM. *P. chinensis*. *P. franklinii*, 2 figs. *P. muricatus* = *Asterias linckii* BLAINV. = *Ast. nodosa* LAM. = *Oreaster turritus* M. T. *P. granulosus*, 4 figs. *P. modestus*, 2 figs. *P. nodosus* = *Asterias nodosa* GMELIN part. = *Oreaster nodosus* M. T. *P. aculeatus* = *Oreaster aculeatus* M. T. *P. armatus* = *Oreaster armatus* M. T. = *Goniodiscus conifer* MÖBIUS, 4 figs. STELLASTER. *S. childreni*, 2 figs. *S. belcheri*, 2 figs. *S. incei*, 2 figs. DORIGONA. *D. reevesii*, 2 figs. COMPTONIA. *C. elegans*. CALLIDERMA. *C. emma*, 2 figs. GYMNASTERIA. *G. spinosa* = *Asterias carinifera* LAM. = *Asteropsis carinifera* M. T. *G. inermis*. PAULIA. *P. horrida*, 1 var. RANDASIA. *R. granulata*, 2 figs. *R. spinulosa*, 2 figs. *R. luzonica*. ANTHENEA. *A. chinensis* = *Asterias chinensis* = *Ast. pentagonula* LAM. = *Goniodiscus pentagonula* M. T. = *Goniaster articulatus* AGASSIZ. *A. tuberculosa*, 2 figs. *A. granulifera*, 2 figs. HOSEA. *H. flavescens*. *H. spinulosa*, 3 figs. HIPPASTERIA. *Hipp. europæa* = *Asterias equestris* PENNANT = *Ast. phrygiana* PARELIUS = *Goniaster phrygianus* M. T. *Hipp. johnstoni* = *Asterias johnstoni*. *Hipp. plana* = *Pentaceros planus* LINCK = *Asterias equestris* GMELIN & LAMK. *Hipp. cornuta* = *Pent. longiorum cornuum* LINCK. CALLIASTER. *C. childreni*, 4 figs. ASTROGONIUM. *A. granulare*, 2 figs. *A. miliare*, 2 figs. *A. inæquale*. *A. tuberculatum*, 2 figs. *A. paxillosum*, 2 figs. GONIASTER. *G. cuspidatus* = *Astrogonium tessellatum* M. T. = ? *Asterias tessellata* BLAINV. = *Pentagonaster semilunatus cuspidatus* LINCK =

Asterias tessellata d. & c. LAM. *G. sebæ. G. regularis = Pentagonaster regularis* LINCK.
PENTAGONASTER. *P. pulchellus = Astrogonium pulchellum* M. T. = *Asterias mülleri* AGASSIZ
MS. = *Stephanaster elegans* AYRES = *Asterias pulchella*, 2 figs. *P. abnormalis*, 3 figs. *P.
dubeni*, 3 figs. TOSIA. *T. grandis*, 2 figs. *T. aurata*, 2 figs. *T. tubercularis*, 2 figs , var.
or young. *T. rubra*, 2 figs. *T. australis = Astrogonium australe*, 2 figs. ECHINASTER.
E. ellisii = Asterias echinus SOLANDER & ELLIS = *Echinaster solaris* M. T. = *Asterias
echinites* LAM. *E. solaris = Asterias solaris* (Naturforscher). OTHILIA = ECHINASTER sp.
M. T. *O. spinosa = Asterias spinosa* RETZ. = *Pentadactylosaster spinosus* LINCK = *Asterias echinophora* LAM. DON CHIAJE. = *Stellaria spinosa* NARDO AGASSIZ = *Echinaster
spinosus* M. T. *O. aculeata. O. multispina = Echinaster brasiliensis* M. T. *O. purpurea
= Echinaster fallax* M. T. *O. luzonica.* METRODIRA. *M. subulata = Scytaster subulatus.*
RHOPIA = STELLONIA (part.) AGASSIZ. *R. seposita* GRAY '40 = *Echinaster sepositus* M. T. =
Asterias sanguinolenta & *Ast. sagena* RETZ. = *Echinaster sanguinolentus* M. T. = *Aste-
rias seposita* RETZ = *Pentadactylosaster reticulatus* LINCK = *Stellonia seposita* NARDO
AGASSIZ. *R. mediterranea*, var.? FERDINA. *F. flavescens. F. cumingii.* DACTYLOSAS-
TER. *D. cylindricus = Asterias cylindrica* LAM. *D. gracilis.* TAMARIA. *T. fusca.*
CISTINA. *C. columbiæ.* OPHIDIASTER. *O. aurantius = Asterias ophidiana* LAMK. =
? *Ophidiaster ophidianus* AGASSIZ. *O. leachii = ? Asterias cylindricus* LAMK. = *Ophidiaster
cylindricus* M. T. *O. guildingii* GRAY, 2 var. *O. attenuatus. O. pyramidatus.* LIN-
CKIA = OPHIDIASTER M. T. *L. typus = Ophid. miliaris* M. T. = *Pentadactylosaster miliaris*
LINCK = *Asterias lævigata* LINN. = *Ast. cometa* BLAINV. *L. crassa. L. brownii. L.
leachii. L. guildingii. L. pacifica. L. columbiæ. L. unifascialis = ? Ophidiaster
suturalis* M. T. *L. bifascialis. L. pulchella. L. intermedia. L. erythræa.* FROMIA.
F. milleporella GRAY '40 = *Scytaster posterius* M. T. = *Asterias sebæ* BLAINV., 3 var.
GOMOPHIA. *G. egyptica.* NARDOA = SCYTASTER sp. M. T. *N. variolata* GRAY '40 =
Scytaster variolatus M. T. = *Asterias variolatus* RETZ. = *Pentadactylosaster variolatus*
LINCK = *Linckia variolata* NARDO. *N. agassizii*, 1 var., 3 monstrosities. *N. tuberculata.*
NARCISSIA. *N. teneriffæ.* NECTBIA. *N. ocellifera* GRAY '40 = *Asterias oculifera* LAMK.
= ? *Goniodiscus ocelliferus* M. T. NEPANTHIA. *N. tessellata. N. maculata.* ASTERINI-
DÆ. PALMIPES. *P. membranaceus = Asteriscus palmipes* M. T. = *Ast. membranacea* RETZ.
& LAMK. = *Ast. placenta* PENNANT = *Ast. cartilaginea* FLEMING = *Ast. rosacea* LAME. *P.
stokesii = Asterias rosacea* var. LAMK. = *Asteriscus rosaceus* M. T. PORANIA = ASTER-
OPSIS sp. MULLER. *P. gibbosa* GRAY '40 = *Asterias pulvillus* O. F. MÜLLER = *Asteropsis
pulvillus* M. T. = *Asterias gibbosus* LEACH = ? *Asterias equestris* THOMPSON = *Goniaster templetoni* FORBES '39. ASTERINA. *A. gibbosa = Asterias gibbosa* PENNANT = *Pentaceros plicatus
et concavus* LINCK = *Asteriscus exigua* PET. GAZ. = ? *As. minuta* LINN. = *Ast. stellata
obtusa ciliata* LINN. = ? *Asterina minuta* AGASSIZ = *Asterias pulchella* BLAINV. = *Asterias
vermiculata* M. T. *A. burtonii. A. minuta = Asteriscus minutus* M. T. = *Asterias minuta*
LINN. = *Asterias exigua* LAMK , 2 var., 2 monstrosities. *A. krausii. A. gunnii*, 2 var.

or 1 var. and 1 monstr. *A. calcar=Asterias calcar* Lamk. Petricia. *P. punctata*, 2 figs. Patiria. *P. coccinea=Asteriscus coccineus* M. T. *P. granifera=?Asterias granifera* Lam., var. à petits grains Oudart, 1 var. *P. ocellifera=?? Goniodiscus ocelliferus* M. T.=*Asterias ocellifera* Lam. *P. obtusa.* *P. ?crassa.* Pteraster. *Pt. miliaris* M. T.=*Asterias miliaris* O. F. Müller=*Asteriscus miliaris* M. T. *Pt. capensis.* Ganeria. *G. falklandica.* Socomia. *S. paradoxa.*

'72. List of Echinoderms collected by Robert M'Andrew, Esq., F. R. S., in the Gulf of Suez in the Red Sea. Ann. Mag. Nat. Hist., 4th. ser., vol. 10, p. 115–125.

Rhopia seposita Gray=*Asterias seposita* Retzius=*Cribrella seposita* Dujardin & Hupé *Linckia typus* Gray=*Asterias laevigata* Linné=*Ophidiaster miliaris* M. T. *Gomophia aegyptica* Gray. *Asteriscus pentagonus* Dujardin & Hupé=*Asterias calcar* Aud. *Asterina burtonii* Gray = *Asteriscus vermiculatus* M. T. = *Asterina gibbosa* Martens. *Pentaceros—?* = *Asterias mamillata* Aud. = *Oreaster mamillatus* M. T. ?*Pentaceros tuberculatus* M. T. *Goniaster sebae* Gray=*Goniodiscus sebae* M. T. *Astropecten polyacanthus* M. T. *Astropecten hemprichii* M. T. Luidia Forbes=Hemicnemus M. T. *Luidia savignii* Gray=*Asterias savignii*.

Grieg, J. A.

* '89. Undersögelser over dyrelivet i de vestlandske fjorde. II. Echinodermer, Annelider etc. fra Moster. Bergens Mus. Aarsberet. 1888. 11 pp.+1 pl. Cited after Ludwig.

'96· Om echinodermfaunaen i de vestlandske fjorde. Bergens Mus. Aarbog 1894–'95, No. 12. 13 pp.

Pontaster tenuispinus Düb. & Kor. *Plutonaster parelii* Düb. & Kor. *Astropecten irregularis* Penn. *Psilaster andromeda* Müll. & Trosch. *Luidia sarsii* Düb. & Kor. Pentagonaster Retz. *Pentagonaster hispidus* M. Sars. *Hippasteria phrygiana* Par. *Porania pulvillus* O. F. Müll. *Stichaster roseus* Müll. & Trosch. *Solaster papposus* Fabr. *Solaster endeca* Lin. *Solaster furcifer* Düb. & Kor. *Pedicellaster typicus* M. Sars. *Pteraster militaris* O. F. Mull. *Pteraster pulvillus* M. Sars. *Henricia sanguinolenta* O. F. Mull. *Asterias glacialis* Linn. *Asterias mulleri* M. Sars. *Asterias rubens* Linn.

* '97. Om Bukkenfjordens echinodermer og mollusker. Stavanger Mus. Aarsberetn. 1896, p. 34–46. Cited after Ludwig.

: 02. Oversigt over det nordlige Norges echinodermer. Bergens Museums Aarbog, p. 1–37, tav. 1.

Pontaster tenuispinus Düben & Koren. Plutonaster pareli Düben & Koren. Ctenodiscus crispatus Retzius. Leptoptychaster arcticus M. Sars. Astropecten irregularis Pennant. Psilaster andromeda M. T. Pentagonaster granularis Retzius. Hippasteria phrygiana Parelius. Poraniomorpha rosea Danielssen & Koren, 2 figs. Lasiaster hispidus M. Sars, 2 figs. Solaster papposus Linné. Solaster endea Retzius. Solaster syrtensis Verrill. Solaster furcifer Düben & Koren. Pteraster obscurus Perrier. Pteraster pulvillus M. Sars. Pteraster militaris O. F. Müller. Retaster multipes M. Sars. Cribrella sanguinolenta O. F. Müller. Pedicellaster typicus M. Sars. Stichaster roseus O. F. Müller. Stichaster arcticus Danielssen & Koren. Asterias glacialis Linné. Asterias mülleri M. Sars. Asterias lincki M. T. Asterias rubens Linné. Brisinga coronata G. O. Sars.

: 06. Echinodermen von dem norwegischen Fischereidampfer "Michael Sars" in den Jahren 1900–1903 gesammelt. III. Asteroidea. Bergens Museums Aarbog. 88 pp. + 2 pls. (To be concluded).

Pontaster tenuispinus Düben & Koren. Plutonaster pareli Düben & Koren. Plutonaster bifrons Wyv. Thomson. Ctenodiscus crispatus Retzius. Leptoptychaster arcticus M. Sars. Astropecten irregularis Pennant. Psilaster andromeda M. T. Bathybiaster vexillifer Wyv. Thomson. Luidia sarsi Düben & Koren. Luidia ciliaris Philippi Pentagonaster granularis Retzius. Hippasteria phrygiana Parelius. Tylaster willei Danielssen & Koren. Porania pulvillus O. F. Müller. Poraniomorpha (Rhegaster) tumida Stuxberg. Poraniomorpha (Lasiaster) hispida M. Sars. Stichaster roseus O. F. Müller. Zoroaster fulgens Wyv. Thomson. Solaster papposus Linck. Solaster squamatus Döderlein. Solaster endea Linné Solaster glacialis Danielssen & Koren. Solaster furcifer Düben & Koren. Korethraster hispidus Wyv. Thomson.

: 07. Echinodermer, samlede sommeren 1905 af "Belgica" i Nordhavet. Nyt Mag. Naturv. Kristiania, 45, p. 131–137. I am indebted to my friend, Mr. H. Ohshima, formerly of this Institute, for being enabled to consult this paper.

Asterias mulleri f. grönlandica Steenstrup Stichaster albulus Stimpson. Solaster papposus L. Hymenaster pellucidus Wyv. Thomson. Ctenodiscus crispatus Retz. Poraniomorpha, Rhegaster, tumida Stuxberg.

Hartlaub, C.

'92· Ueber die Arten und den Skelettbau von Culcita. Notes from the Leyden Museum, vol. xiv, p. 65–118, pl. 2. April.

C. novæ guineæ M. T. '42 = C. pulverulenta Valenciennes MS. C. plana, n. sp.. C. grex M. T. '42, 2 figs. C. coriacea M T. '42 = Randasia spinulosa Gray. C. arenosa

PERRIER '69· *C. aculispinosa* BELL '83· *C. veneris* PERRIER '79 ('80). *C. selenoele-liana* RETZIUS '05 = *C. discoidea* (LAM.) AGASSIZ, 4 text-figs.

HERDMAN, W. A.

'92. Notes on the Collections made during the Cruise of the S. Y. "Argo" up the West Coast of Norway. Proc. Trans. Liverpool Biol. Soc., vol. vi, p. 70–93.

Astropecten arcticus. Archaster parelii. Stichaster roseus.

03. Report to the Government of Ceylon on the Pearl Oyster Fisheries of the Gulf of Manaar. With Supplementary Reports upon the Marine Biology of Ceylon, by Other Naturalists. Part I. Pearl Oyster Report, p. 1–162 + 1 pl. Supplem. Reports, p. 163–307 + 26 pls.

Starfishes mentioned : *Asterias sp.* (p. 113). *Asterina cepheus* (p. 23, 51, 90, 118). *Astropecten* (p. 35, 41, 52). *Astropecten euryacanthus* (p. 32). *Astropecten hemprichii* (p. 20, 23, 26, 27, 32, 36, 49, 51, 59). *Astropecten polyacanthus* (p. 36, 49, 80). *Astropecten zebra ?* (p. 33). *Culcita schmideliana* (p. 113). *Echinaster purpureus* (p. 24–27, 84). *Linckia sp.* (p. 103, 108, 111). *Linckia diplax* (p. 33, 52). *Linckia multifora* (p. 33). *Linckia lævigata* (p. 27, 113, 115). *Luidia* (p. 121). *Luidia maculata* (p. 20, 23, 26, 106). *Nardoa tuberculata* (p. 115, 118). *Ophidiaster cylindricus* (p. 49). *Pen, taceros* (p. 104, 106, 108, 121). *Pentaceros lincki* (p. 22, 23, 25, 84, 102, 106, 107, 109. 115, 118). *Pentaceros nodosus* (p. 22, 51, 84). *Pteraster sp.* (p. 51). *Stellaster sp,* (p. 32, 49, 51, 52).

06. Report &c. Part V. Pearl Oyster Report, p. 1–136. Supplementary Reports, p. 137–452. 3 figs. of *Oreaster lincki.*

HERDMAN, W. A., HERDMAN, J. B., and BELL, F. J.

04. Report on the Echinoderma collected by Professor HERDMAN, at Ceylon, in 1902. By W. A. H. and J. B. H., with Notes and Additions by F. J. BELL. Report to the Government of Ceylon on the Pearl Oyster Fisheries of the Gulf of Manaar, by W. A. Herdman. With Supplementary Reports upon the Marine Biology of Ceylon, by Other Naturalists. Part II, p. 137–150.

ASTROPECTINIDÆ. *Astropecten hemprichi* M. T., 3 figs. *Astropecten euryacanthus* LÜTKEN. *Astropecten polyacanthus* M. T. *Astropecten indicus* DÖDERLEIN. *Astropecten zebra* SLADEN. *Luidia maculata* M. T. *Luidia hardwickii* (GRAY). PENTAGONASTERIDÆ. *Stellaster incei* (GRAY). ANTHENEIDÆ. *Anthenea sp. ?* *Goniodiscus sp. ?* PENTACERO-

TIDÆ. *Pentaceros linchi* DE BL., 3 figs. *Pentaceros mammillatus* (M. T.). *Pentaceros nodosus* (GRAY). *Pentaceros sp.* (young). *Culcita schmideliana* (RETZ.). ASTERINIDÆ. *Asterina cepheus* (M. T.). LINCKIIDÆ. *Ophidiaster cylindricus* (LAMK.). *Ophidiaster helicostichus* (SLADEN). *Linckia multiforis* (LAMK). *Linckia lævigata* (GMEL.). *Linckia pacifica* var. *diplax* (teste SLADEN). *Nardoa tuberculata* GRAY. *Metrodira subulata* (GRAY). PTERASTERIDÆ. *Retaster cribrosus* v. MARTENS. ECHINASTERIDÆ. *Echinaster purpureus* (GRAY).

Notes by F. J. B.: on the specific identity of *Astropecten hemprichi* and *Astropecten zebra* SLADEN and possibly also of *Astrop. notograptus*; on *Retaster cribrosus*.

HOFFMANN, C. K.

* '74. Recherches sur la faune de Madagascar et de ses dépendences, après les découvertes de F. P. L. POLLEN et D. C. VAN DAM, vol. 2. Echinodermes.

* '82· Die Echinodermen, gesammelt während der Fahrten des "Willem Barents" in den Jahren 1878 und 1879. Niederl. Arch. Zool., Suppl. Bd. i. Cited after LUDWIG.

HUTTON, F. W.

: 04. Index Faunæ Novæ Zealandiæ.

ASTROPECTINIDÆ. *Astropecten edwardsi* VERRILL. *Astrop. polyacanthus* MÜLL. & TROSCHEL. *Psilaster acuminatus* SLADEN. PENTAGONASTERIDÆ. *Astrogonium pulchellum* GRAY. *Astrog. abnormale* GRAY. *Gnathaster miliaris* GRAY. *Gnath. rugsous* HUTTON. *Goniodon dilatatus* PERRIER. GYMNASTERIDÆ. *Asteropsis imperialis* FARQUHAR. ASTERINIDÆ. *Asterina regularis* VERRILL. *Asterina novæ-zealandiæ* PERRIER. *Stegnaster inflatus* HUTTON. STICHASTERIDÆ. *Stichaster australis* VERRILL. *Stich. polyplax* MÜLL. & TROSCHEL. *Stichaster suteri* LORIOL. *Stich. suteri* var. *lævigatus* HUTTON. *Stich. insignis* FARQUHAR. LINCKIIDÆ. *Ophidiaster cylindricus* SLADEN. SOLASTERIDÆ. *Solaster torulatus* SLADEN. ECHINASTERIDÆ. *Cribrella ornata* PERRIER. *C. compacta* SLADEN. *C. sufflata* SLADEN. *C. lukinsii* FARQUHAR. *Echinaster polycnema* SLADEN. ASTERIIDÆ. *Asterias calamaria* GRAY. *A. scabra* HUTTON. *A. mollis* HUTTON. *A. fragilis* SLADEN. *A. rodolphi* PERRIER. *Uniofora granifera* LAMARCK.

IVES, J. E.

* '89· Catalogue of the Asteroidea and Ophiuroidea of the Academy of Natural Sciences of Philadelphia. Proc. Acad. Nat. Sc. Philad., p. 169-179. Cited after FISHER.

'91· Echinoderms and Arthropods from Japan. Proc. Acad. Nat. Sc. Philadelphia, p. 210–223, pl. vii.

Astropecten armatus. Astrop. japonicus, 5 figs. *Astropecten scoparius*, 4 figs. *Luidia quinaria*, 5 figs. *Asterina pectinifera*, 4 figs. *Nardoa semiregularis* var. *japonica*, 4 figs. *Cribrella sanguinolenta*, 4 figs. *Asterias amurensis*, 4 figs. *Asterias torquata*.

* '92· List of the Echinoderms and Crustaceans in the Cabinet of FRED. STEARNS. Cited after DE LORIOL.

JARZYNSKY, T.

* '85· Catalogus Echinodermatum inventorum in mari albo et in mari glaciali ad litus murmanicum anno 1869 et 1870. In N. WAGNER'S "Die Wirbellosen des Weissen Meeres," Bd. i. Cited after LUDWIG.

KALISCHEWSKIJ, M.

: 07. Zur Kenntnis der Echinodermenfauna des sibirischen Eismeeres. Résultats scient. Expéd. pol. russe en 1900–1903, sous la direction du Baron E. TOLL. Zool. vol. i, livr. 4. Mém. acad. imp. sci. St.-Pétersbourg, 8. sér., xviii, No. 4. 67 pp.+3 pls.

ARCHASTERIDÆ. *Pontaster tenuispinus* (DÜB. & KOREN)=*Astropecten tenuispinus* DÜB. & KOREN '46=*Archaster tenuispinus* MÖBIUS & BÜTSCHLI '74=*P. tenuispinus* var. *platynota* SLADEN '89=*P. hebitus* SLADEN '89=*P. limbatus* SLADEN '89=*P. marionis* PERRIER '94· PORCELLANASTERIDÆ. *Ctenodiscus crispatus* (RETZIUS)=*Ct. polaris* M. & TR. '42=*Ct. pygmæus* M. & TR. '42=*Ct. corniculatus* PERRIER '75=*Ct. krausei* LUD-WIG '86=*Ct. procurator* SLADEN '89· 9 figs. GYMNASTERIDÆ. *Rhegaster tumidus* (STUXBERG)=*Solaster tumidus* STUXBERG '78=*Asterina tumida* DANIELSSEN & KOREN '84=*Asterina tumida* var. *tuberculata* DAN. & KOREN '84=*R. tumidus* var. *tuberculatus* SLADEN '89· SOLASTERIDÆ. *Crossaster papposus* (L.)=*Asterias affinis* BRANDT '35= *Solaster papposus* M. & TR. '42=*S. papposa* STIMPSON '53=*Crossaster papposus* var. *affinis* D'URBAN '80=*Solaster affinis* DAN. & KOREN '84=*Cross. papposus* var. *septen-trionalis* SLADEN '89· *S. syrtensis* VERRILL=*S endeca* LEVINSEN '87=*S. endeca* (partim) LUDWIG : 00=*S. endeca* var. *syrtensis* ÖSTERGREN : 01. *Lophaster furcifer* (DÜB. & KOREN)=*Solaster furcifer* DUB. & KOREN '46· PTERASTERIDÆ. *Pt. militaris* (O. F. MÜLLER)=*Asteriscus militaris* M. & TR. '42=*Pt. militaris* var. *prolata* SLADEN '89· *Pt. pulvillus* M. SARS. *Hymenaster pellucidus* WYV. THOMSON '73· 1 fig. ECHINASTE-RIDÆ. *Cribrella sanguinolenta* (O. F. MÜLLER)=*Echinaster oculatus* M. & TR. '42= *Echinaster eschrichtii* M. & TR. '42=*Echin. sarsi* M. & TR. '44=*Linckia oculata* STIMP-SON '53=*Linckia pertusa* STIMPSON '53=*Cribrella oculata* M'ANDREW & BARRETT '57=

Echinaster sanguinolentus STUXBERG '78 = *Henricia oculata* BELL '92· ASTERIIDÆ. *Asterias mulleri* (M. SARS) = *Asterias* (*Leptasterias*) *mülleri* SLADEN '89 = *Asterias hyperborea* MICHAILOVSKIJ : 02, 3 figs. *A. groenlandica* (STEENSTRUP) = *A. gronlandica* var. *robusta* LEVINSEN '86, 3 figs. *A. groenlandica* (STEENSTRUP) var. *spitsbergensis* (DAN. & KOREN) = *Asterias spitsbergensis* DAN. & KOREN '84, 2 figs. *A. groenlandica* (STEENSTRUP) var. *longimana* n., 4 figs. *A. hyperborea* DAN. & KOREN '84 = ?*A. normani* DAN. & KOREN '84, 7 figs. *A. panopla* STUXBERG '78, 14 figs. *A. lincki* (M. & TR.) = *Asteracanthion linckii* M. & TR. '42 = *Asteracanthion stellionura* PERR. '69 = *A. gunneri* DAN. & KOREN '84, 10 figs.

KŒHLER, R.

'95· Catalogue raisonné des Échinodermes recueillis par M. KOROTNEV aux îles de la Sonde. Mém. soc. zool. France, p. 374–423.

Archaster typicus M. T. *Astropecten polyacanthus* M. T. = *Astrop. armatus* SLADEN '79· *Luidia maculata* M. T. *Asterina exigua* LAMARCK. *Culcita novœ-guineœ* M. T. = *C. schmideliana* SLUITER '89· *Goniodiscus articulatus* LINNÉ = *Asterias articulata* LINNÉ 1753 = *Artocreatis altera species* SEBA 1758 = *Astrogonium articulatum* VALENCIENNES = *Goniaster articulatus* LÜTKEN '64 = *Goniodiscus sebœ* part. PERRIER '75· *Iconaster longimanus* (MÖBIUS) = *Pentagonaster longimanus* PERRIER '75· *Stellaster equestris* (RETZIUS) = *Pentagonaster* (*Stell.*) *equestris* PERRIER '75· *Stellaster squamulosus* STUDER = *Stell. incei ?* SLADEN '89· *Pentaceros turritus* LINCK = *Oreaster nodosus* BELL '84· *Pentaceros lutkeni* BELL. *Pentaceros mammillatus* M. T. = *O. alveolatus* BELL '84· *Fromia major*, n. sp., 2 figs. *Ophidiaster helichostichus* SLADEN var. *inarmatus*, 2 figs. *Acanthaster echinites* ELLIS & SOLANDER.

: 07. Note préliminaire sur quelques astéries et ophiures provenant des campagnes de la Princesse-Alice. Bull. Inst. Océanogr. Monaco, No. 99, April. 47 pp.

Freyella recta, n. sp. *Frey. edwardsi* PERRIER. *Styracaster elongatus*, n. sp. *Hyphalaster fortis*, n. sp. *Hyph. gracilis*, n. sp. MAGDALENASTER, n. g. *Magd. arcticus*, n. sp. *Hymenaster roseus*, n. sp. *Pteraster reductus*, n. sp. *Dytaster parvulus*, n. sp. *Psilasteropsis humilis*, n. sp. *Astrogonium longobrachiale* (DAN. & KOREN). *Astrog. eminens*, n. sp. *Astrog. œquabile*, n. sp.

: 09. Échinodermes provenant des campagnes du yacht Princesse-Alice (Astéries, Ophiures, Échinides et Crinoïdes). Résultats des campagnes scientifiques accomplis sur son yacht par ALBERT Ier. Prince Souverain de Monaco. 34. fasc. 317 pp. + 32 pls. Asteroidea, p. 5–136, pl. 1–6, 8, 10–24.

ARCHASTERIDÆ. *Pontaster tenuispinus* (DÜBEN & KOREN), 5 figs, *Pont. tenuispinus* (type). *Pont. tenuispinus* var. *platynota*. *Pont. tenuispinus* var. *nitida*, n. var. *Pont. tenuispinus* var. *Marionis*. *Pontaster venustus* SLADEN '89, 1 fig. *Plutonaster bifrons* (WYV. THOMSON). *Plut. marginatus* SLADEN '89, 3 figs. *Plut. notatus* SLADEN '89, 7 figs. *Plut. rigidus* SLADEN '89 = *P. granulosus* PERRIER '96, 3 figs. *Plut subinermis* (PHILIPPI). *Dytaster agassizi* PERRIER '94, 2 figs. *Dyt. biserialis* SLADEN '89, *Dyt. parvulus* KŒHLER : 07. *Dyt. rigidus* PERRIER '94, 1 fig. PORCELLANASTERIDÆ. *Albatrossaster richardi*, n. sp., 3 figs. *Ctenodiscus corniculatus* (LINCK) = *Ct. crispatus* (RETZIUS). *Hyphalaster antonii* PERRIER '94, *Hyph. fortis* KŒHLER : 07, 2 figs. *Hyph. gracilis* KŒHLER : 07, 2 figs. *Styracaster elongatus* KŒHLER : 07, 8 figs. *Styracaster horridus* SLADEN '89, *Styrac. spinosus* PERRIER '94, *Thoracaster cylindratus* SLADEN '89 = *Pseudaster cordifer* PERRIER '94 = *Hyphalaster valdiviæ* CHUN : 00, 1 fig. ASTROPECTINIDÆ. *Astropecten aurantiacus* (LINNÉ). *Astrop. sphenoplax* J. BELL '92, 1 fig. *Astrop. irregularis* LINCK (incl. *Astrop. serratus* M. T. & *Astrop. pentacanthus* (DELLE CHIAJE), 25 figs. *Bathybiaster vexillifer* (WYV. THOMSON) = *Archaster vexillifer* W. THOMSON '73 = *Astropecten pallidus* DANIELSSEN & KOREN '76 = *Astropecten vexillifer* VIGUIER '78 = *Bathybiaster pallidus* DANIELSSEN & KOREN = *Ilyaster mirabilis* DANIELSSEN & KOREN '83 = *Archaster robustus* VERRILL '85 = *Phoxaster pumilus* SLADEN '89 = *Bathybiaster robustus* VERRILL '94, 3 figs. *Leptoptychaster arcticus* (M. SARS). *Luidia sarsi* (DÜBEN & KOREN). *Psilaster andromeda* (M. T.), 3 figs. *Psilasteropsis humilis* KŒHLER : 07, 3 figs. *Psilasteropsis patagiatus* (SLADEN '89) = *Psilaster andromeda* pars PERRIER '94, 3 figs. PENTAGONASTERIDÆ. ASTROGONIUM M. T. (emend. PERRIER). *Astrog. annectens* PERRIER '94 = *Pseudarchaster annectens* VERRILL '99, *Astrog. æquabile* KŒHLER : 07, 4 figs. *Astrog. eminens* KŒHLER : 06, 4 figs. *Astrog. fallax* PERRIER '84 = ?*Archaster parelii* pars VERRILL '85 = *Astrog. annectens* PERRIER '86 = *Pseudarchaster fallax* VERRILL '99, 1 fig *Astrog. marginatum*, n. sp., 4 figs. *Astrog. necator* PERRIER '94 = *Pseudarchaster necator* VERRILL '99, *Astrog. parelii* var. *longobrachiale* (DANIELSSEN & KOREN) = *Plutonaster parelii* var. *longobrachialis* DANIELSSEN & KOREN '84 = *Pseudarchaster tessellatus* var. *arcticus* SLUITER '95 = *Plutonaster parelii* LUDWIG : 00 = *Astrop. longobrachiale* KŒHLER : 07, 10 figs. *Dorigona arenata* PERRIER '95 = *Pentagonaster arenatus* PERRIER '84 = *Nymphaster arenatus* SLADEN '89, 1 fig. *Odontaster mediterraneus* MARENZELLER '95 = *Gnathaster mediterraneus* MARENZELLER '93, 1 fig. *Pentagonaster gosselini* PERRIER '94, 1 fig. *Pentag. granularis* RETZIUS, 3 figs. *Pentag. perrieri* SLADEN '89 = *Pentag. grandis* PERRIER '85 = *Pentag. concinnus* SLADEN '91, 1 fig. *Paragonaster subtilis* PERRIER '94 = *Goniopecten subtilis* PERRIER '84 = *Parag. elongatus* PERRIER '94, 1 fig. *Goniaster semilunatus* (LINCK) var. *africanus* VERRILL = *Goniaster africanus* VERRILL '71 = *Pentagonaster semilunatus* PERRIER '75 = *Phaneraster semilunatus* PERRIER '96, 1 fig. *Hippasteria plana* (LINCK) = *Hipp. phrygiana* PARELIUS, 1 fig. LINCKIIDÆ. *Chætaster longipes* BRUZELIUS. *Ilacelia attenuata* (GRAY), 2 figs. *Linckia boutieri* PERRIER '75, 2 f gs. *Linckia guildingi* GRAY. *Narcissia canariensis* (D'ORBIGNY) = *Asterias canariensis* D'ORBIGNY '35 = *Narcissia teneriffæ* GRAY '42 = *Scytaster canariensis* DUJARDIN & HUPÉ '62 = *Fromia narcissiæ* PERRIER '94. *Ophidiaster ophidianus* (LAMARCK). PTERASTERIDÆ. *Hexaster obscurus* PERRIER. *Hymenaster giboryi* PERRIER '94, 2 figs. *Hymenaster pellucidus* WYV. THOMSON, 1 fig. *Hymenaster roseus* KŒHLER : 07, 2 figs. *Pteraster militaris* (O. F. MÜLLER). *Pter. reductus* KŒHLER : 07, 3 figs. *Retaster multipes* (M. SARS), 2 figs. ASTERINIDÆ. *Asterina gibbosa* (PENNANT). *Palmipes membranaceus* LINCK. PORANIIDÆ. *Porania pulvillus* (O. F. MÜLLER). *Poraniomorpha hispida* (M. SARS) = *Goniaster hispidus* M. SARS '77 = *Porania spinulosa* VERRILL '77 = *Asterina borealis* VERRILL '78 = *Poraniomorpha rosea* DANIELSSEN & KOREN '81 = *Rhegaster murrayi* SLADEN '83 = *Pentagonaster hispidus* DANIELSSEN & KOREN '84 = *Poraniomorpha spinulosa* VERRILL '85 = *Poran. borealis* VERRILL '85 = *Lasiaster hispidus* SLADEN '89, 3 figs. ECHINASTERIDÆ. *Cribrella abyssalis* PERRIER '94. *Cribrella oculata* (LINCK) = *Cribrella sanguinolenta* (O. F. MÜLLER). *Echinaster sepositus* (GRAY). CRYASTERIDÆ. MAGDALENASTER KŒHLER : 07. *Magd. arcticus* KŒHLER : 07, 4 figs. STICHASTERIDÆ. *Stichaster roseus* (O. F. MÜLLER). *Nanaster (Stichaster) albulus* (STIMPSON). *Neomorphaster talismani* (PERRIER), 1 fig. ZOROASTERIDÆ. *Zoroaster trispinosus* KŒHLER '96, 5 figs. PEDICELLASTERIDÆ. *Pedicellaster sexradiatus* PERRIER '94, 1 fig. SOLASTERIDÆ. *Crossaster papposus* (LINCK), 3 figs. *Lophaster furcifer* (DÜBEN & KOREN). *Solaster endeca* var. *glacialis* DANIELSSEN & KOREN '84 = *Solaster echinatus* STORM '88 = *Sol. syrtensis* VERRILL '94 = *Sol. glacialis* GRIEG : 06, 2 figs. *Korethraster hispidus* WYV. THOMSON. ASTERIDÆ. *Asterias glacialis* LINNÉ. *Asterias grœnlandica* (STEENSTRUP), 1 fig. *Asterias hyperborea* DANIELSSEN & KOREN. *Asterias linckii* (M. T.) = *Asterias stellionura* PERRIER. *Asterias panopla* STUXBERG. *Asterias rubens* O. F. MÜLLER. *Asterias tenuispina* LAMARCK. BRISINGIDÆ. *Brisinga coronata* SARS. *Brisinga endecacnemos* ASBJØRNSEN '56, 1 fig. *Odinia robusta* (PERRIER) = *Brisinga robusta* PERRIER '85 *Freyella edwardsi* PERRIER '94, 2 figs. *Freyella recta* KŒHLER : 07, 5 figs. *Freyella sexradiata* PERRIER '94, 1 fig. *Stellosphæra mirabilis* KŒHLER & VANEY : 06, 10 figs.

: 09a.   An Account of the Deep-sea Asteroidea collected by the Royal Indian Marine Survey Ship Investigator. Astéries recueillies par l'Investigator dans l'Océan indien. I. Les Astéries de Mer profonde. 143 pp. + 13 pls.

JOHANNASTER, n. g. *J. superbus*, n. sp., 3 figs. *Pectinaster hispidus* (ALCOCK & WOOD-MASON) = *Pontaster hispidus* ALCOCK & WOOD-MASON '91. *Pontaster pilosus* ALCOCK '93. *Cheiraster snyderi* FISHER : 06. *Cheiraster inops* FISHER : 06. *Pararchaster indicus* , n. sp., 2 figs. *Persephonaster roulei*, n. sp., 3 figs. *Astropecten griegi*, **n. sp.**, 2 figs. PHIDIASTER, n. g. *Ph. agassizi*, n. sp., 3 figs. *Porcellanaster caulifer* SLADEN '89, 2 figs. *Caulaster dubius*, n. sp., 3 figs. SIDONASTER, n. g. *S. vaneyi*, n. sp., 3 figs. *S.*

batheri, n. sp., 3 figs. *Styracaster caroli* Ludwig : 07 = *St. horridus* Alcock & Wood-Mason '91, 2 figs. *Thoracaster alberti*, n. sp., 5 figs. *Astrogonium jordani* (Fisher) = *Pseudarchaster jordani* Fisher : 06. *Astrogonium roseum* (Alcock) = *Mediaster roseus* Alcock '93· *Astrogonium mozaicum* (Alcock & Wood-Mason) = *Pseudarchaster mosaicus* Alcock & Wood-Mason '91, 1 fig. *Dorigona ternalis* Perrier, 2 figs. *Dorigona belli*, n. sp., 3 figs. *Dorigona ludwigi*, n. sp., 2 figs. *Iconaster pentaphyllus* (Alcock '93). *Pentagonaster (Tosia) annandalei*, n. sp., 3 figs. *Pentagonaster (Tosia) cuenoti*, n. sp., 2 figs. *Pentagonaster döderleini*, n. sp., 3 figs. *Pentagonaster (Philonaster) mortenseni*, n. sp., 3 figs. *Mediaster ornatus* Fisher : 06, 1 fig. *Astroceramus fisheri*, n. sp., 3 figs. Circeaster, n. g. *C. marcelli*, n. sp., 3 figs. *C. magdalenœ*, n. sp., 4 figs. Lydiaster, n. g. *L. johannœ*, n. sp., 4 figs. *Evoplosoma augusti*, n. sp., 3 figs. *Palmipes ludovici*, n. sp., 4 figs. *Fromia andamanensis*, n. sp., 2 figs. *Zoroaster adami*, n. sp., 2 figs. *Hymenaster alcocki*, n. sp., 3 figs. *Hymenaster pentagonalis* Fisher 06. *Cribrella mutans*, n. sp., 2 figs. *Brisinga gracilis*, n. sp., 2 figs. *Brisinga panopla* Fisher : 06, 1 fig. *Brisinga parallela*, n. sp., 3 figs. *Odinia clarki*, n. sp., 3 figs. *Odinia austini*, n. sp., 3 figs. *Freyella indica*, n. sp., 3 figs.

: 09b. Échinodermes recueillis dans les mers arctiques par la Mission arctique française, commandée par M. Bénard. Bull. Mus. Hist. Natur., p. 121–123.

Ctenodiscus corniculatus (Linck). *Pteraster militaris* (O. F. Müll). *Cribrella oculata*. *Crossaster papposus* (Linck). *Stichaster roseus* (O. F. Müller). *Nanaster albulus* (Stimpson). *Asterias rubens* (O. F. Müll.). *Asterias grœnlandica* (Steenstrup).

: 10. An Account of the Shallow-water Asteroidea. Echinoderma of the Indian Museum, Part vi. Asteroidea (II). 191 pp. + 20 pls. June.

Archaster typicus M. T., 2 figs. *Craspidaster hesperus* (M. T.). *Crasp. glauconotus* Bedford. *Luidia limbata* (Sladen). *L. maculata* M. T. *L. savignyi* (Audouin), 2 figs. *L. mauritiensis*, n. sp., 2 figs. *L. integra*, n. sp., 4 figs. *L. denudata*, n. sp., 4 figs. *Astropecten andersoni* Sladen, 2 figs. *Astrop. javanicus* Lütken, 2 figs. *Astrop. indicus* Döderlein (= *Astrop. koehleri* Loriol), 8 figs. *Astrop. mauritianus* Gray, 3 figs. *Astrop. monacanthus* Sladen, 4 figs. *Astrop. polyacanthus* M. T. *Astrop. tamilicus* Döderlein, 3 figs. *Astrop. velitaris* Martens. *Astrop. zebra* Sladen. *Astrop. debilis*, n. sp., 3 figs. *Astrop. inutilis*, n. sp., 2 figs. *Astrop. nobilis*, n. sp., 3 figs. *Astrop. pugnax*, n. sp., 4 figs. *Dorigona confinis*, n. sp., 2 figs. *Goniodiscus forficulatus* Perrier, 5 figs. *Goniod. insignis*, n. sp., 3 figs. *Goniod. porosus*, n. sp., 3 figs. *Goniod. vallei*, n. sp., 2 figs. *Ogmaster capella* (M. T.), 1 fig. *Stellaster equestris* (Retzius). *Stell. incei* Gray. *Stell. squamulosus* Studer, 2 figs. *Anthenea regalis*, n. sp., 2 figs. *Anthen. rudis*, n. sp., 3 figs. *Anthen. sp. juv.*, 1 fig. *Pentaceros affinis*

(M. T.), 2 figs. *Pent. australis* (Lütken), 2 figs. *Pent. alveolatus* Perrier, 2 figs. *Pent. hedemanni* (Lütken,, 2 figs. *Pent. productus* (Bell), 2 figs. *Pent. regulus* (M. T.), 2 figs. *Pent. reinhardti* (Lutken), 2 figs. *Pent. westermanni* (Lütken), 2 figs. *Pent. novæ-caledoniæ,* n. sp., 3 figs. *Pent. mammillatus* var. *tuberculatus* (M. T.) (= *O. tuber-culatus*), 2 figs. *Pent. indicus,* n. sp., 4 figs. *Culcita novæ-guineæ* M. T., 3 figs. *C. schmideliana* Retzius. *Palmipes rosaceus* (Lamarck), 2 figs. *P. sarasini* de Loriol, 2 figs. *Asterina cephea* (M. T.). *Ast. exigua* (Lamarck), 2 figs. *Ast. lorioli,* n. sp. (=*Palmipes sarasini* de Loriol, p. p.', 4 figs. *Disasterina spinosa,* n. sp., 2 figs. *Nepanthia suffarcinata* Sladen. *Nepanthia brachiata,* n. sp., 2 figs. *Chætaster vestitus,* n. sp., 3 figs. *Fromia major* Kœhler. *Fr. armata,* n. sp., 2 figs. *Ferdina offreti,* n. sp., 4 figs. *Ophidiaster armatus* Kœhler. *Ophid. tuberifer* Sladen. *Ophid. hirsutus,* n. sp., 2 figs. *Ophid. ornatus,* n. sp., 2 figs. *Leiaster callipeplus* Fisher, 2 figs. *Linckia ehrenbergi* (M. T.). *L. miliaris* (Linck). *L. pacifica* Gray, 1 fig. *L. dubiosa,* n. sp., 2 figs. *Nardoa ægyptiaca* (Gray) (=*Scytaster ægyptiacus* Perrier), 2 figs. *Nard. frianti,* n. sp., 2 figs. *Nard. le monnieri,* n. sp.. 2 figs. *Nard. carinata,* n. sp., 3 figs. *Nard. squamulosa,* n. sp , 3 figs. *Metrodira subulata* Gray (=*Scaphaster humberti* de Loriol), 3 figs. *Echinaster callosus* Marenzeller. *Echin. eridanella* M. T. *Valvaster striatus* Perrier, 4 figs. *Sclerasterias nitida,* n. sp., 3 figs.

:11. Astéries et Ophiures des îles Aru et Kei. Abhandl. Senckenb. Naturf. Gesell., Bd. 33, p. 261–295, 3 pls.

Archaster typicus M. T. *Astropecten granulatus* M. T. *Astrop. polyacanthus* M. T. *Astrop. scoparius* Val. *Luidia maculata* M. T., 5 figs. *Craspidaster hesperus* (M. T.). *Iconaster longimanus* (Möbius). *Stellaster incei* Gray. *Anthenea mertoni,* n. sp., 2 figs. *Pentaceros rouxi,* n. sp., 2 figs. *Pentac. turritus* Linck. *Ophidiaster armatus,* n. sp . 2 figs. *Ophid. tumescens,* n. sp., 2 figs. *Linckia miliaris* Linck'. *Fromia major* Kœhler, 3 figs. *Metrodira subulata* Gray=*Scaphaster humberti* de Loriol '99' 4 figs. *Echinaster purpureus* (Gray). *Nepanthia suffarcinata* Sladen

Lamarck, J. B. P. A. d.

'16· Histoire naturelle des animaux sans vertébres, présentant les caractères généraux et particuliers de ces animaux, leur distribution, leurs classes, leurs familles, leurs genres, et la citation des principales espèces qui s'y rapportent; etc. Vol. i, March, 1815; vol. ii, 1816; vol. iii, Aug., 1816; vol. iv, March, 1817; vol. v, July, 1818. Astérie, vol. ii, p. 547–568.

Asterias tessellata = *Pentetagonaster regularis* Linck 1733. *Aster. punctata. Aster. cuspidata. Aster. pleyadella. Ast. ocellifera. Aster. vernicina. Aster. discoidea.*

Aster. exigua = *Pentaceros plicatus et concavus* LINCK 1733 = *Anasterias minuta* GMELIN
Aster. pentagonula.   *Aster. pulvillus.*   *Aster. penicillaris* = ?*Stella obtusangula* LINCK 1733.
Aster. equestris = *Pentaceros planus* LINCK 1733.   *Aster. carinifera.*   *Aster. obtusangula.*
Aster. reticulata = ditto LINNÉ.   *Aster. nodosa* = ditto LINNÉ.   *Aster. calcar.*   *Aster.*
membranacea = ditto RETZIUS GMELIN.   *Aster. rosacea.*   *Aster. helianthus.*   *Aster.*
echinites.   *Aster. papposa* = ditto LINNÉ   *Aster. endeca* = ditto LINNÉ.   *Aster. granifera.*
Aster. echinophora = *Pentadactylosaster spinosus* LINCK 1733.   *Aster. glacialis* = *Aster.*
angulosa O. F. MÜLLER 1776.   *Aster. tenuispina.*   *Ast. rubens.*   *Aster. clavigera.*   *Aster.*
seposita = *Pentadactylosaster reticulatus* LINCK.   *Aster. aranciaca* = ditto LINNÉ O.
F. MÜLLER.   *Aster. calcitrapa.*   *Aster. acuminata.*   *Aster. striata.*   *Aster. milleporella.*
Aster. variolata.   *Aster. multifora* = ?*An pentadactylosaster* LINCK.   *Aster. bicolor.*   *Aster.*
lævigata.   *Aster. arenata.*   *Aster. cylindrica.*   *Aster. senegalensis.*   *Aster. ophidiana.*
Aster. subulata.

LEIPOLDT, F.

'95.   Asteroidea der " Vettor-Pisani " Expedition (1882–1885).   Zeitsch. f.
wiss. Zool., Bd. 59, p. 545–654, Taf. xxxi, xxxii.   Published July 9.
HELIASTERIDÆ.   HELIASTER.   *H. helianthus* (LAM.) DUJ. & HUPÉ, 5 figs.   *H. cumingii*
(GRAY) VERRILL '67 = *Asterias* (*Heliaster*) *cumingii* GRAY '40, 1 fig.   *H. multiradiatus*
(GRAY) VERRILL = *Asterias* (*Heliaster*) *multiradiata* GRAY '40 = *H. kubiniji* XANTUS '60,
ASTERIIDÆ.   ASTERIAS.   *A.* (*Cosmasterias*) *tomidata* SLADEN '89,   *A.* (*Cosmas.*) *sulcifera*
(VALENC.) PERRIER '69 = *Asteracanthion sulciferus* VALENCIENNES = ?*Asterac. luridum*
PHILIPPI '58 = *Asterac. sulcifer* PERRIER '69 = *Asterac. clavatum* PHILIPPI '70 = *Asterac.*
fulvum, spectabile, mite PHILIPPI '70 = *A. clavata* PERRIER '78 = *A. mitis* PERRIER
'78 = *Stichaster polygrammus* SLADEN '89 = *Diplasterias sulcifera* PERRIER '91,   *A. rugi-*
spina STIMPSON '60 = ?*Asterac. antarcticum* LÜTKEN '56 = ?*Asterac. varium* PHILIPPI
'70 = ?*Asterac. fulgens* PHILIPPI '70 = *A. cunninghami* PERRIER '75 = *Anasterias minuta*
PERRIER '75 = *A. perrieri* SMITH '76 = (*A. rupicola* VERRILL '76) = ?*A. varia* PERRIER
'74 = (*A. spirabilis* BELL '81) = *A. hyndesi* PERRIER '86 (probably also = *A verrilli* BELL
'81 = ?*Calvasterias antipodum* BELL '82 = *Calvast. stolidota* SLADEN '89 = *Anasterias minuta*
var. *Asteroderma papillosum* PERRIER '91).   PYCNOPODIA.   *P. helianthoides* (BRANDT) STIMP-
SON '61 = *Asterias helianthoides* BRANDT '35,   STICHASTERIDÆ.   STICHASTER.   *St.*
aurantiacus (MEYEN) VERRILL.   ECHINASTERIDÆ.   CRIBRELLA.   *C. hyadesi* PERRIER '91,
ECHINASTER.   *E. panamensis,* n. sp., 4 figs.   PORANIIDÆ.   PORANIA.   *P. antarctica*
SMITH '76 = *P. magellanica* SLADEN '89,   PORANIOPSIS.   *P. echinasteroides* PERRIER '91,
1 fig.   ASTERINIDÆ.   ASTERINA.   *A. stellifer* MÖBIUS [= *marginata* (VAL.) PERRIER] var.
obtusa? = *A. marginata* RATHBUN '79,   *A. fimbriata* PERRIER '75.   *A. chilensis* LÜTKEN
'59 = *Asteriscus* (*Patiria*) *chilensis* VERRILL '67 = ?*A gayi* PERRIER '75, 5 figs.
GANERIIDÆ.   CYCETHRA.   *C. nitida* SLADEN '89, 1 fig.   *C. electilis* SLADEN '89,   ASTRO-

Pectinidæ. Luidia. *L. columbiæ* (Gray) Perrier '75 = *Petalaster columbiæ* Gray '40 = *L. tessellata* Lütken '58· *L. magellanica*, n. sp., 5 figs. Archasteridæ. Odontaster. *O. singularis* (M. T.) Bell '93 = *Goniodiscus singularis* M. T. '43 = *Pentagonaster* (*Astrogonium*) *singularis* Perrier '75 = *Gnathaster singularis* Sladen '89 = *Asterodon singularis* Bell '93· 3 figs. *O. meridionalis* (Smith) Bell = *Astrogonium meridionale* Smith '76 = *Pentagonaster meridionalis* Smith '79 = *Calliderma grayi* Bell '81 = *Gnathaster meridionalis* Sladen '89 = *Gn. pilulatus* Sladen '89 = *Gn. grayi* Sladen '89 = *Asterodon pedicellaris* Perrier '91 = *Asterod. grayi* Perrier '91 = *O. grayi* Bell '93 = *O. pedicellaris* Bell '93· 7 figs. Linckiidæ. Pharia. *P. pyramidata* Gray. Linckia. *L. miliaris* (Linck) v. Martens. Pentacerotidæ. Pentaceros. *P. reticulatus* Linck. *P. occidentalis* (Verrill) Sladen '89· Nidorellia. *N. armata* (Gray) Verrill '67 = *Pentaceros* (*N.*) *armatus* Gray '40 = *Goniodiscus armatus* Lütken '58 = *Goniod. conifer* Möbius '59 = *Goniod. stella* Verrill '67 = *Goniod. michelini* Perrier '69 = *N. michelini* Sladen '89· Culcita. *C. plana* Hartlaub '92· 2 figs. *C. coriacea* M. T., 1 fig. Appendix. Asteroidea collected by F. Orsini in the Red Sea. *Astropecten acanthifer* Sladen '83· *Astrop. polyacanthus* M. T. *Astrop. orsinii*, n. ·sp., 6 figs. *Linckia miliaris* (Linck) v. Martens. *L. ehrenbergii* (M. T.) Perrier. *Ogmaster capella* (M. T.) v. Martens = *Goniodiscus capella* M. T. '42 = *Goniaster* (*Ogmaster*) *capella* v. Martens '65 = *Dorigona reevesi* Gray '66 = *Goniaster mülleri* Lütken '71 = *Pentagonaster mulleri* Perrier '75· *Gymnasteria carinifera* (Lam.) v. Martens, 3 figs. *Pentaceros mammillatus* (Audouin) Perrier '76· *Culcita coriacea* M. T.

Levinsen, G. W. R.

* '86· Kara-Havets Echinodermata. In Lütken's "Dijmphna-Togtets zoologisk-botanisk Udbytte." Cited after Ludwig.

Loriol, P. de.

'84. Notes pour servir à l'étude des Échinodermes. Rec. zool. suis., t. i, p. 605–643, pl. 31–35.

Aspidaster Loriol '84· *Asp. delgadoi* Loriol '84· Goniodiscus *articulatus* (Linné) Lütken = *Asterias articulata* Linné 1753 = *Artocreatis altera species* Seba 1758 = *Astrogonium articulatum* Valenc. ? = *Goniaster articulatus* Lütken '64 = *Goniodiscus sebæ* pars Perrier '75·

'85· Catalogue raisonné des Échinodermes recueillis par M. V. de Robillard à l'île de Maurice. II. Stellérides. Mém. soc. d. phys. et d'hist. natur. d. Genève, t. xxix, no. 4. 84 pp. + 16 pls. (pl. vii–xxii).

Aster. calamaria Gray '40 = *Asteracanthion tenuispinus* Michelin '45 = *Coscinasterias*

714 S. GOTO :

muricata VERRILL '67 = *Cosc. muricata* HUTTON '72 = ? *Aster. jehennesii* PERRIER '75, 8 figs. *Acanthaster mauritiensis*, n. sp. = *Acanth. echinites* MŒBIUS '80, 19 figs. *Echinaster purpureus* (GRAY) v. MARTENS '67 = *Aster. sp.* SAVIGNY '04 = *Othilia purpurea* GRAY '40 = *Othilia luzonica* GRAY '40 = *Echinaster fallax* M. T. '42 = *Echin. sepositus* MICHELIN '45. *Valvaster striatus* (LAMARCK) PERRIER '75 = *Aster. striata* LAM. '16, 8 figs *Mithrodia clavigera* (LAMARCK) PERRIER '75 = *Aster. clavigera* LAM. = *Mithrodia spinulosa* GRAY '40 = *Heresaster papillosus* MICHELIN '45 = *Ophidiaster echinulatus* M. T. '42 = *Echinaster echinulatus* v. MARTENS '66, 5 figs. *Ophidiaster duncani*, n. sp., 10 figs. *Ophidiaster perrieri*, n. sp., 11 figs. *Ophidiaster cylindricus* (LAMARCK) M. T. '42 = *Asterias cylindrica* LAM. '16 = *Dactylosaster cylindricus* GRAY '40 = *Ophidiaster asperulus* LÜTKEN '71, 10 figs. *Ophidiaster purpureus* PERRIER '69, 9 figs. *Ophidiaster robillardi*, n. sp., 12 figs. *Linckia multifora* (LAMARCK) GRAY '40 = *Aster. multifora* LAMARCK '16 = ? *L. typus* GRAY '40 = ? *L. leachii* GRAY '40 = *Ophidiaster multiforis* M. T. '42, 20 figs. *Linckia ehrenbergii* (M. T.) '42 = *Ophidiaster ehrenbergii* M. T. '42, 17 figs. *Linckia marmorata* (MICHELIN) PERRIER '75 = *Ophidiaster marmoratus* MICHELIN '44, 6 figs. *Linckia miliaris* (LINCK) v. MARTENS '66 = *Pentadactylosaster asper* var. *miliaris* LINCK 1733 = *Aster. lœvigata* LINNÉ 1788 = *Linckia typus* GRAY '40 = *Ophidiaster miliaris* M. T. '42, *Linckia pacifica* GRAY '40 = *L. nicobarica* LÜTKEN '61, *Leiaster coriaceus* PETERS '52, 12 figs. *Leiaster leachii* GRAY '40 = *Ophidiaster leachii* GRAY '40 = ? *Ophid. ophidianus* MICHELIN '45, 11 figs. *Scytaster variolatus* (LINCK) M. T. '42 = *Nardoa variolata* GRAY '40, *Fromia milleporella* (LAMARCK) GRAY '40 = *Asterias milleporella* LAMARCK '16 = ? *Scytaster milleporellus* M. T. '42 = *Scytaster pistorius* M. T. '42 = *Scytaster milleporellus* MICHELIN '45, 6 figs. *Ferdina flavescens* GRAY '40, 4 figs. *Goniodiscus sebœ* M. T. '42 = *Hosea sebœ* GRAY '66 = *Goniaster sebœ* GRAY '72, 6 figs. *Goniodiscus studeri*, n. sp., 6 figs. *Pentagonaster spinulosus* (GRAY) PERRIER '75 = *Hosea spinulosa* GRAY '47 = *Goniaster spinulosus* v. MARTENS '66, 8 figs. *Pentaceros belli*, n. sp., 10 figs. *Pentaceros sladeni*, n. sp., 13 figs. *Pentaceros grayi* BELL '84 = *Stella marina orientalis* SEBA 1761 = *P. nodosus* GRAY '41, 9 figs. *Culcita schimideliana* (RETZIUS) GRAY '40 = ... BRUGUIÈRES 1791 = *Aster. schmideliana* RETZIUS '05 = *C. coriacea* STUDER '84, 13 figs. *Gymnasteria carinifera* (LAMARCK) v. MARTENS '66 = *Asterias carinifera* LAMARCK '16 = *Gymnasteria spinosa* GRAY '41 = *Gymnast. inermis* GRAY '41 = *Asteropsis carinifera* M. T. '42, 11 figs. *Asterina cephea* VALENCIENNES = ... SAVIGNY '09 = ? *Asterina burtonii* GRAY '40 = *Asteriscus cepheus* VAL. in M. T. '42, 10 figs. *Luidia savignyi* (AUDOUIN) GRAY '40 = *Asterias savignyi* AUDOUIN '24 = ? *Luidia maculata* PETERS '52, 2 figs. *Astropecten hemprichii* M. T. '42 = ? *Astropecten articulatus* MICHELIN '45 (non SAY) = *Astrop. mauritianus* MŒBIUS '81 (non GRAY), 8 figs. *Astropecten polyacanthus* M. T. '42 = ... SAVIGNY '03 = *Astrop. armatus* M. T. '42 = *Astrop. hystrix* M. T. '42 = *Astrop. vappa* M. T. '43. *Astropecten sp.*, 4 figs. *Archaster angulatus* M. T. '42 = *Arch. mauritianus* v. MARTENS '66, 5 figs.

'93. Échinodermes de la Baie d'Amboine. Rev. suis. zool. et Ann. d. Mus. d'hist. nat. Genève, t. i, p. 359–426, pl. xiii–xv.

Archaster typicus M. T. '40 = *Astropecten stellaris* GRAY '40 = *Arch. nicobaricus* BEHN in MÖBIUS '59· *Astropecten polyacanthus* M. T. '42 = *Astrop. armatus* M. T. '42 = *Astrop. hystrix* M. T. '42 = *Astrop. vappa* M. T. '43· *Luidia maculata* M. T. '42· *Pentaceros turritus* LINCK 1733 = *Asterias nodosa* (pars) LINNÉ 1788 = *Oreaster turritus* M. T. '42 = *Oreaster nodosus* BELL '84· *Pentaceropsis obtusatus* (BORY ST. VINCENT) SLADEN '89 = *Asterias obtusata* BORY ST. VINCENT '27 = *Oreaster obtusatus* M. T. '42 = *Pentaceros obtusatus* PERRIER '75· *Culcita grex* M. T. '42· *Asterina cepheus* VALENCIENNES = *Asteriscus cepheus* VAL. in M. T. '42 = *Asterina cephea* PERRIER '75· *Asterina exigua* LAMARCK = *Asterias exigua* LAMARCK '16 = *Asteriscus pentagonus* M. T. '42· *Linckia miliaris* (LINCK) v. MARTENS '66 = *Pentadactylosaster asper* var. *miliaris* LINCK 1733 = *Asterias lævigata* (pars) LINNÉ '66 = *Linckia typus* GRAY '40· *Ophidiaster purpureus* PERRIER '69· *Nardoa tuberculata* GRAY '40 = *Ophidiaster tuberculatus* M. T. '42 = *Linckia tuberculata* v. MARTENS '66 = *Scytaster tuberculatus* PERRIER '75· *Acanthaster echinites* (ELLIS & SOLANDER) LÜTKEN '71 = *Stella marina quindecim radiorum* RUMPH 1705 = *Asterias echinites* ELLIS & SOL. 1786 = *Echinaster solaris* M. T. '42 = *Echinaster* (*Heliaster*) *solaris* v. MARTENS '66· *Echinaster eridanella* VALENCIENNES in M. T. '42 = *Othilia eridanella* GRAY '66 = *Echinaster affinis* PERRIER '69·

'99. Notes pour servir à l'étude des Échinodermes. Mém. soc. d. phys. et d'hist. natur. d. Genève, t. xxxiii, 2me. partie, no. 1. 34 pp. + 3 pls.

Astrop. penangensis, n. sp., 13 figs. *Astrop. zebra* SLADEN '83 = *Astrop. coppingeri* BELL '84· *Astrop. verrilli*, n. sp., 9 figs. *Astrop. inermis*, n. sp., 8 figs. *Astrop. rubidus*, n. sp., 11 figs. *Astrop. kœhleri*, n. sp., 10 figs. *Astrop. ludwigi*, n. sp., 9 figs. *Astrop. kagoshimensis*, n. sp., 8 figs. SCAPHASTER, n g *Scaph. humberti*, n. sp., 9 figs.

LUDWIG, H.

'86. Echinodermen des Beringsmeeres. Zool. Jahrb., Bd. 1, p. 275–296. 1 Taf.

Asterias acervata STIMPS. *Asterias cribraria* STIMPS. *Asterias sp. Cribrella oculata* (LINCK) FORB. = *Echinaster eschrichtii* BRANDT '51 = *Echin. sanguinolentus* HOFFMANN '82 = *Echin. sanguinolentus* STUXBERG '83· *Ctenodiscus krausei*, n. sp , 4 figs.. *Pteraster aporus*, n. sp.

'97· Die Seesterne des Mittelmeeres. Fauna u. Flora des Golfes von Neapel u. d. angrenzenden Meeres-Abschnitte. 23. Monographie. x + 491 pp. + 12 pls.

716 S. GOTO :

ASTROPECTINIDÆ. ASTROPECTEN. *A. aurantiacus* (LINNÉ) = *Astropecten echinatus major* LINCK 1733 = *Astrop. stellatus* LINCK 1733 = *Asterias aranciaca* LINNÉ 1758 = *Asterias aurantiaca* TIEDEMANN '16 = *Stellaria aurantiaca* NARDO '34 = *Astrop. crenaster* DUJARDIN & HUPÉ '62, 7 figs. *A. bispinosus* (OTTO '23) = *Astropecten echinatus minor* LINCK 1733 = *Asterias bispinosa* OTTO '23 = *Stellaria bispinosa* NARDO '34 = *Asterias platyacantha* PHILIPPI '37 = *Astrop. echinatus* GRAY '40 = *Astropecten platyacanthus* M. T. '42 = *Astrop. myosurus* PERIER '69, 2 figs. *A. spinulosus* (PHILIPPI '37) = *Asterias spinulosa* PHILIPPI '37 = *Astrop. johnstoni* M. T. '42, 2 figs. *A. pentacanthus* (DELLE CHIAJE '25) = *Asterias pentacantha* DELLE CHIAJE '25, 2 figs. *A. pentacanthus* (DELLE CHIAJE) var. *serratus* (M. T. '42) = *Astrop. serratus* M. T. '42 = *Astrop. mülleri* MARION '83, *A. johnstoni* (DELLE CHIAJE '25) = *Asterias johnstoni* DELLE CHIAJE '25 = *Asterias aranciaca* var. *aculeis marginalibus minimis* DESHAYES in LAMARCK '40 = *Astrop. squamatus* M. T. '44 = *Astrop. aster* DE FILIPPI '59 = *Astrop. platyacanthus* PERR. '75, 2 figs. LUIDIA. *L. ciliaris* (PHILIPPI '37) = *Asterias tenuissima* RISSO '26 = *Asterias rubens* JOHNSTON '36 = *Asterias ciliaris* PHILIPPI '37 = *L. fragilissima* FORBES '39 = *Asterias pectinata* COUCH '40 = *Asterias imperati* DELLE CHIAJE '41 = *L. savignyi* M. T. '42, 14 figs. *L. sarsi* (DÜBEN & KOREN '45) = *Asterias* sp. n. M. SARS '35 = *L. fragilissima* FORBES '39 = *Luidya sarsii* DÜBEN & KOREN '45 = *L. savignyi* DÜB. & KOR. '46 = *Astrella simplex* PERRIER '82 = *L. ciliaris* COLOMBO '88 = *L. paucispina* v. MARENZELLER in STEINDACHNER '91, 13 figs. ARCHASTERIDÆ. PLUTONASTER. *P. subinermis* (PHILIPPI '37) = *Asterias subinermis* PHILIPPI '37 = *Astropecten subinermis* M. T. '42 = *Archaster subinermis* PERRIER '78 = *Goniopecten subinermis* PERRIER '85 = *Plut.* (subg. *Tethyaster*) *subinermis* SLADEN '89 = *Tethyaster subinermis* PERRIER '94, 27 figs. *P. bifrons* (WYV. THOMSON '73) = *Archaster bifrons* WYV. THOMSON '73 = *Goniopecten bifrons* PERRIER '85, ODONTASTER. *O. mediterraneus* (v. MARENZELLER '91) = *Gnathaster mediterraneus* v. MARENZELLER in STEINDACHNER '91, CHAETASTERIDÆ. CHAETASTER. *Ch. longipes* (RETZIUS '05) = *Asterias longipes* RETZIUS '05 = *Asterias subulata* LAMARCK '16 = *Asterias verrucosa* RISSO '26 = *Nepanthia tessellata* GRAY '40 = *Ch. subulatus* GAUDRY '51, 19 figs. PENTAGONASTERIDÆ. PENTAGONASTER. *P. placenta* (M. T. '42) = *Goniodiscus placenta* M. T. '42 = *Goniodiscus placentæformis* HELLER '63 = *Goniodiscus acutus* HELLER '63 = *Goniaster placentæformis* LÜTKEN '64 = *Goniaster placenta* v. MARENZELLER '75 = *P. mirabilis* PERRIER '75 = *P. acutus* PERRIER '78 = *Pent. minor* KŒHLER '96, 22 figs. *P. hystricis* v. MARENZ. in STENDACHNER. '91 = *P. kergroheni* KŒHLER '96 = *P. grecni* BELL '89 = *P. balteatus* SLADEN '91 = *P. concinnus* SLADEN '91, 1 fig. PORANIIDÆ. MARGINASTER. *M. capreensis* (GASCO) = *Asteropsis capreensis* GAS. '76 = *M. fimbriatus* SLADEN '89 = *Cheilaster fimbriatus* BELL '92, 11 figs. ASTERINIDÆ. ASTERINA. *A. gibbosa* (PENNANT 1777) = *Pentaceros gibbus plicatus et concavus* LINCK 1733 = *Asterias gibbosa* PENNANT 1777 = *Asterias minuta* OLIVI 1792 = *Asterias verruculata* RETZIUS '05 = *Asterias umbilicata* KONRAD '14 = *Asterias exigua* DELLE CHIAJE '25 = *Asterias membranacea* RISSO '26 = *Asterias pulchella* BLAINVILLE '34

= *Asteriscus verruculatus* M. T. '42 = *Asterias minima* VERANY '46 = *Asteriscus ciliatus* LORENZ· '60 = *Asteriscus pulchellus* PERRIER '69 = *Asteriscus pancerii* GASCO '70 = *Asteriscus arrecifiensis* GREEFF '72· 18 figs. PALMIPES. *P. membranaceus* LINCK 1733 = *Stella cartilaginea* ALDROVANDI 1638 = *Stella (Palmipes) membranacea* LINCK 1733 = *Asterias placenta* PENNANT 1777 = *Asterias membranacea* RETZIUS 1783 = *Asterias palmipes* OLIVI 1792 = *Asterias papyracea* KONRAD '14 = *Asterias rosacea, Stella rossa membranacea* DELLE CHIAJE '25 = *Asterias cartilaginea* FLEMING '28 = *Asterias (Palmasterias) membranacea* BLAINVILLE '30 = *Anseropoda membranacea* NARDO '34 = *Asteriscus palmipes* DUVERNOY '49 = *Asteriscus placenta* LÜTKEN '64 = *P. placenta* NORMAN '65· 17 figs. *P. lobianci* = *P. membranaceus* × *Asterina gibbosa*, 1 fig. LINCKIIDÆ. HACELIA. *H. attenuata* (GRAY '40) = *Asterias lævigata varietas* LAMARCK '16 = *Asterias variolata* RISSO '26 = *Asterias coriacea* GRUBE '40 = *Ophidiaster (Hacelia) attenuatus* GRAY '40 = *Ophidiaster ophidianus* LÜTKEN '64 = *Ophidiaster lessonæ* GASCO '76· 19 figs. OPHIDIASTER. *O. ophidianus* (LAMARCK '16) = *Asterias ophidiana* LAMARCK '16 = *Ophidiaster aurantius* GRAY '40 = *Ophidiaster canariensis* GREEFF '72· 15 figs. ECHINASTERIDÆ. ECHINASTER. *E. sepositus* (GRAY '40) = *Pentadactylosaster asper reticulatus digitis brevioribus* LINCK 1733 = *Asterias rubens* OLIVI 1792 = *Asterias sagena* RETZIUS '05 = *Asterias sanguinolenta* RETZIUS '05 = *Asterias seposita* LAMARCK '16 = *Stellonia seposita* NARDO '34 = *Rhopia mediterranea* GRAY '40 = *Rhopia seposita* GRAY '40 = *Echinaster sanguinolentus* M. T, '42 = *Asterias rosacea* VERANY '46 = *Cribrella seposita* DUJARDIN & HUPÉ '62 = *Cribrella oculata* RUSSO '93 = *E. sepositus* var. *mediterraneus* MARCHISIO '96· 20 figs. ASTERIIDÆ. ASTERIAS. *A. tenuispina* LAMARCK '16 = *Stella marina echinata* COLUMNA 1616 = *Asterias heptactis* KONRAD (MECKEL) '14 = *Asterias savaresi* DELLE CHIAJE '25 = *Asterias rubens* RISSO '26 = *Stellonia tenuispina* D'ORBIGNY '39 = *Asterias glacialis* var. *savaresi* GRUBE '40 = *Asterias glacialis* GRAY '40· 3 figs. *A. glacialis* L. 1758 = *Sol. echinatus cancellatus* LINCK 1733 = *A. spinosa* PENNANT 1777 = *A. angulosa* O. F. MÜLLER 1788 = *A. echinophora* DELLE CHIAJE '25 = *Stellonia glacialis* NARDO '34 = *Stellonia angulosa* L. AGASSIZ '35 = *Stellonia webbiana* D'ORBIGNY '39 = *Uraster glacialis* FORBES '41 = *Asteracanthion glacialis variatio profundus* LORENZ '60 = *A. madeirensis* STIMPSON = *Asteracanthion linckii* PERRIER '69 = *Marthasterias foliacea* JULLIEN '78 = *Asterias africana* GREEFF '82 = *Asterias (Stolasterias) glacialis* SLADEN = *Stolasterias glacialis* PERRIER '94· 19 figs. *A. edmundi* LUDWIG = *Stolasterias neglecta* PERRIER '91· 1 fig. *A. richardi* PERRIER '82 = *Hydrasterias richardi* PERRIER '94· 5 figs. BRISINGIDÆ. BRISINGA. *B. coronata* G. O. SARS '72 = *B. sp.* MARION '83 = *B. mediterranea* PERRIER '85·

'99· Echinodermen des Sansibargebietes. VOELTZKOW's Wissensch. Ergebn. d. Reisen in Madagaskar u. Ost-Afrika in den Jahren 1889-1895. Abhandl. Senckenb. naturforsch. Gesellsch., Bd. xxi, Hft. iv, p. 537–563.

ASTROPECTINIDÆ. *Astropecten hemprichii* M. T. *Astrop. polyacanthus* M. T. *Luidia savignyi* AUDOUIN. PENTAGONASTERIDÆ. *Pentagonaster semilunatus* LINCK. *Goniodiscus sebæ* M. T. *Goniod. sanderi* MEISSNER. *Pentaceros grayi* BELL. *Pent. hiulcus* LINCK. *Pentac. muricatus* LINCK. *Pentac. tuberculatus* M. T. *Pentac. turritus* LINCK. *Culcita coriacea* M. T. *Cul. pentangularis* GRAY. *Cul. schmideliana* RETZIUS var. *africana* DÖDERLEIN. GYMNASTERIIDÆ. *Gymnasteria carinifera* LAMARCK. ASTERINIDÆ. *Asterina cepheus* VALENCIENNES. *Aster. coccinea* M. T. LINCKIIDÆ. *Ferdina kuhlii* M. T. *Ophidiaster pustulatus* VON MARTENS + *purpureus* PERRIER. *Leiaster coriaceus* PETERS. *Leiaster glaber* PETERS. *Linckia ehrenbergii* M. T. *Linck. miliaris* LINCK. *Linck. multifora* LAMARCK. *Linckia pacifica* var. *diplax* M. T. *Nardoa variolata* LINCK. PTERASTERIDÆ. *Retaster cribrosus* VON MARTENS = *Pteraster cribrosus* v. MARTENS '67 = *Retaster insignis* SLADEN '89· ECHINASTERIDÆ. *Mithrodia clavigera* LAMARCK. *Echinaster purpureus* GRAY.

: 00. Arktische Seesterne. Fauna arctica. Eine Zusammenstellung der arktischen Tierformen, mit besonderer Berücksichtigung des Spitzbergen-Gebietes auf Grund der Ergebnisse der Deutschen Expedition in das nördische Eismeer im Jahre 1898. Unter Mitwirkung zahlreicher Fachgenossen herausgegeben von Dr. FRITZ RÖMER u. Dr. FRITZ SCHAUDINN. I. Bd, p. 445–502. A very complete list of synonyms is given under each species.

ARCHASTERIDÆ. *Pontaster tenuispinus* (DÜBEN & KOREN) = *Astropecten tenuispinus* DÜBEN & KOREN '46 = *Archaster tenuispinus* M. SARS '61 = *Pontaster tenuispinus* SLADEN '89 = *Pont. hebitus* SLADEN '89 = *Pont. limbatus* SLADEN '89 = *Pont. tenuispinus* BELL '89 = *Archaster tenuispinus* DALLA TORRE '89 = *Astropecten tenuispinus* RODGER '93 = *Pontaster marionis* PERRIER '94· *Plutonaster parelii* (DUBEN & KOREN) = *Asterias aranciata* var. PARELIUS 1768 = *Astropecten parelii* DÜBEN & KOREN '46 = *Archaster parelii* M. SARS '61 = *Pseudarchaster tessellatus* var. *arcticus* SLUITER '95· PORCELLANASTERIDÆ. *Ctenodiscus crispatus* (RETZIUS) = *Astropecten corniculatus* LINCK 1733 = *Asterias crispata* RETZIUS '05 = *Asterias polaris* SABINE '24 = *Astropecten polaris* GRAY '40 = *Ctenodiscus polaris* M. T. '42 = *Ctenodiscus pygmæus* M. T. '42 = *Ctenodiscus crispatus* DÜBEN & KOREN '46 = *Ctenodiscus corniculatus* PERRIER '75 = *Ctenodiscus krausei* LUDWIG '86· ASTROPECTINIDÆ. *Leptoptychaster arcticus* (M. SARS) = *Astropecten arcticus* M. SARS '50 = *Astropecten lutkeni* M'ANDREW & BARRETT '57 = *Archaster arcticus* VERRILL '73 = *Leptophycaster arcticus* NORMAN '93· *Astropecten irregularis* (PENNANT) = *Astropecten irregularis* LINCK 1733 = *Asterias aranciaca* O. F. MÜLLER 1776 = *Asterias irregularis* PENNANT 1777 = *Asterias aurantiaca* FORBES '39 = *Astropecten mülleri* M. T. '44 = *Astropecten echinulata* M. T. '44 = *Astropecten acicularis* NORMAN '65 = *Astropecten helgolandicus*

GREEFF '71· *Psilaster andromeda* (M. T.) = *Asterias aranciata* var. PARELIUS 1768 = *Astropecten andromeda* M. T. '42 = *Archaster andromeda* M. SARS '65 = *Archaster christi* PERRIER '75 = *Archaster floræ* VERRILL '78 = *Psilaster andromeda* SLADEN '91· *Bathybiaster pallidus* (DANIELSSEN & KOREN) = *Astropecten pallidus* DAN. & KOREN '76 = *Bathybiaster pallidus* DAN. & KOREN '82· *Iliaster mirabilis* DANIELSSEN & KOREN '83· PENTAGONASTERIDÆ. *Pentagonaster granularis* (RETZIUS) = *Asterias granularis* RETZIUS 1783 = *Astrogonium granulare* M. T. '42 = *Astrogonium boreale* M'ANDREW & BARRETT '57 = *Goniaster granularis* LÜTKEN '65 = *Pentagonaster granularis* PERRIER '75 = *Tosia* (*Ceramaster*) *granularis* VERRILL '99· ANTHENEIDÆ. *Hippasteria phrygiana* (PARELIUS) = *Pentaceros planus* LINCK 1733 = *Asterias phrygiana* PARELIUS 1768 = *Asterias equestris* FLEMING '28 = *Asterias johnstoni* (GRAY bei) JOHNSTON '36 = *Hippasteria plana, europæa, johnstoni* & *cornuta* GRAY '40 = *Asterias* (*Goniaster*) *equestris* GOULD '41 = *Goniaster equestris* FORBES '41 = *Astrogonium phrygianum* M. T. '42 = *Goniaster abbensis* FORBES '43 = *Hippasteria equestris* GRAY '48 = *Hippasteria abbensis* GRAY '48 = *Goniaster phrygiana* STIMPSON '53 = *Astrogonium aculeatum* M'ANDREW & BARRETT '57 = *Goniaster nidarosiensis* STORM '81· GYMNASTERIDÆ. *Tylaster willei* DANIELSSEN & KOREN '80· *Rhegaster tumidus* (STUXBERG) = *Solaster tumidus* STUXBERG '78 = *Asterina tumida* DANIELSSEN & KOREN '80 = *Rhegaster tumidus* SLADEN '83· *Poraniomorpha rosea* DANIELSSEN & KOREN '80· *Lasiaster hispidus* (M. SARS) = *Goniaster hispidus* M. SARS '72 = *Pentagonaster hispidus* DANIELSSEN & KOREN '84 = *Lasiaster hispidus* SLADEN '89 = *Lasiaster* (*Pentagonaster*) *hispidus* PFEFFER '94· SOLASTERIDÆ. *Crossaster papposus* (L.) = *Triskaidecactis papposa* LINCK 1733 = *Dodecactis reticulata* LINCK 1733 = *Asterias papposa* WALCH 1774 = *Asterias helianthemoides* PENNANT 1777 = *Asterias affinis* BRANDT '35 = *Solaster papposa* FORBES '39 = *Solaster* (*Polyaster*) *papposa* GRAY '40 = *Crossaster papposus* VERRILL '66 = *Crossaster affinis* DANIELSSEN & KOREN '76 = *Crossaster papposus* var. *affinis* D'URBAN '80 = *Crenaster papposus* HONEYMAN '88 = *Crossaster papposus* var. *septentrionalis* SLADEN '89· *Solaster endeca* (RETZIUS) = *Octactis dactyloides* LINCK 1733 = *Enneactin coriacea dentata* LINCK 1733 = *Asterias* (*quarta species*) PARELIUS 1768 = *Asterias aspera* O. F. MÜLLER 1776 = *Asterias endeca* RETZIUS 1783 = *Asterias endica* FLEMING '28 = *Asterias alboverrucosa* BRANDT '35 = *Solaster endeca* FORBES '39 = *Solaster* (*Endeca*) *endeca* GRAY '48 = *Solaster syrtensis* VERRILL '94 = *Sol. intermedius* SLUITER '95· *Solaster glacialis* DANIELSSEN & KOREN '84· *Lophaster furcifer* (DÜBEN & KOREN) = *Solaster furcifer* DÜBEN & KOREN '46 = *Lophaster furcifer* VERRILL '78· *Korethraster hispidus* WYV. THOMSON '73 = *Corethraster hispidus* v. MARENZELLER '77· PTERASTERIDÆ. *Hexaster obscurus* PERRIER '91 = *Pteraster* (*Temnaster*) *hexactis* VERRILL '94 = *Temnaster hexactis* VERRILL '95 = *Pteraster hexactis* DÖDERLEIN '99· *Pteraster militaris* (O. F. MÜLLER) = *Asterias militaris* O. F. MÜLLER 1776 = *Asteriscus militaris* M. T. '42 = *Pteraster militaris* M. T. '42 = *Pteraster sp.* WHITEAVES '74· *Pteraster pulvillus* M. SARS '61· *Retaster multipes* (M. SARS)

'66 = *Diplopteraster multipes* Verrill '80 = *Retaster multipes* Sladen '80 = *Diplopteraster multipes* Verrill '95 = *Pteraster multipes* Grieg '96· *Hymenaster pellucidus* Wyv. Thomson '73 = *Hymenaster sp.* Ruijs '97· Echinasterid.e. *Cribrella sanguinolenta* (O. F. Müller) = *Pentadactylosaster oculatus* Linck 1733 = *Asterias sanguinolenta* O. F. Müller 1776 = *Asterias pertusa* O. F. Müller 1776 = *Asterias oculata* Pennant 1777 = *Asterias spongiosa* Fabricius 1780 = *Asterias seposita* Retzius 1783 = *Linckia oculata* Forbes '39 = *Henricia oculata* Gray '40 = *Cribrella oculata* Forbes '41 = *Echinaster oculatus* M. T. '42 = *Echin. eschrichtii* M. T. '42 = *Echin. sanguinolentus* M. Sars '44 = *Echinaster sarsii* M. T. '44 = *Linckia pertusa* Stimpson '53 = *Cribrella sanguinolenta* Lütken '57· *Echinaster scrobiculatus* Danielssen & Koren '83· Pedicellasterid.e. *Pedicellaster typicus* M. Sars '71 = *Asteracanthion palaeocrystallus* Duncan & Sladen '77 = *Pedicellaster palaeocrystallus* Sladen '80· Stichasterid.e. *Stichaster roseus* (O. F. Müller) = *Asterias rosea* O. F. Müller 1776 = *Linckia rosea* Thomson '40 = *Cribrella rosea* Forbes '41 = *Asteracanthion roseus* (pars) M. T. '42 = *Henricia rosea* Gray '48 = *Stichaster roseus* M. Sars '61· *Stichaster arcticus* Danielssen & Koren '82· *Stichaster albulus* (Stimpson) = *Asteracanthion roseus* M. T. '42 = *Asteracanthion albulus* Stimpson '53 = *Asteracanthion problema* Steenstrup '55 = *Stichaster albulus* Verrill '66 = *Stephanasterias albula* Verrill '71· Asteriid.e. *Asterias glacialis* L. *Asterias mülleri* (M. Sars '44) = *Leptasterias mülleri* Verrill '66· *Asterias cribraria* Stimpson '62· *Asterias groenlandica* (Steenstrup) = *Ast. rubens* Phipps 1774 = ?*Uraster violacea* Forbes '52 = *Asteracanthion mülleri* var. Steenstrup '55 = *Leptasterias groenlandica* Verrill '79· *Asterias spitzbergensis* Danielssen & Koren '84· *Asterias hyperborea* Danielssen & Koren '82 = *Asterias normani* Danielssen & Koren '83· *Asterias polaris* (M. T. '42) = *Asterias rubens* Fabricius partim 1780 = *Ast. minuta* Fabricius 1780 = *Ast. violacea* Sabine '24 = *Ast. borealis* Perrier '75 = *Ast. douglasi* Perrier '75· *Asterias camtschatica* Brandt '35 = *Asteracanthium camtschaticum* Brandt '51 = *Asterias acervata* Stimpson '62· *Asterias panopla* Stuxberg '78· *Asterias linckii* (M. T.) '42 = *Asteracanthion stellionura* Perrier '69 = *Ast. gunneri* Danielssen & Koren '82· *Asterias rubens* L. = *Ast. violacea* Perrier '75· *Brisinga coronata* G. O. Sars.

: 03.   Seesterne. Résultats du Voyage du S. Y. Belgica en 1897–1898–1899 sous le commandement de A. DE Gerlache de Gomery. Rapports scientifiques publiés aux frais du gouvernement belge, sous la direction de la commission de la Belgica. Zoologie. 72 pp. + 7 pls.

Astropectinid.e. *Mimaster cognatus* Sladen, 7 figs. Archasterid.e. Subf. *Pararchasterinæ. Cheiraster gerlachei*, n. sp., 10 figs. (Synoptic key to *Cheiraster* species) Odontasterid.e. *Asterodon singularis* (M. T.) = *Asterodon granulosus* Perrier '91 = *Odontaster singularis* Leipoldt '95· *Odontaster cremeus*, n. sp. Poraniid.e. *Porania*

antarctica [E. A. SMITH, 3´ figs. SOLASTERIDÆ. *Solaster octoradiatus*, n. sp., 2 figs.
Lophaster stellans SLADEN, 2 figs. PTERASTERIDÆ. *Pteraster lebruni* PERRIER, 4 figs.
Hymenaster perspicuus, n. sp. ECHINASTERIDÆ. *Echinaster smithi*, n. sp. PEDICELLAS-
TERIDÆ. *Pedicellaster antarcticus*, n. sp. (Synoptic key to *Pedicellaster* species.)
ASTERIIDÆ. *Sporasterias antarctica* (LÜTKEN) = *Sporast. spirabilis* PERRIER '94 = *Ast.*
(*Sporast.*) *antarctica* MEISSNER '96 = *Ast. antarctica* BELL : 02. *Sporasterias antarctica*
(LÜTKEN) var. *rupicola* VERRILL. *Cosmasterias lurida* (PHILIPPI) = *Asteracanthion luridum*
PHILIPPI '58 = *Asterias* (*Cosmasterias*) *sulcifera* LEIPOLDT '95· *Diplasterias lütkeni*
PERRIER. *Stolasterias candicans*, n. sp. ANASTERIAS. (Synoptic key to *Anasterias*
species.) *Anast. chirophora*, n. sp., 10 figs. *Anast. lactea*, n. sp. *Anast. belgicæ*, n.
sp., 7 figs. (Starfishes that incubate their eggs: *Leptoptychaster kerguelensis* E. A.
SMITH, *Stichaster nutrix* STUDER, *Pteraster militaris* (O. F. MÜLLER), *Hexaster obscurus*
PERRIER, *Hymenaster nobilis* WYV. THOMSON, *Hymenaster præcoquis* SLADEN, *Cribrella*
sanguinolenta (O. F. MÜLLER), *Asterias mülleri* (M. SARS), *Asterias sp.* LUDWIG, *Asterias*
antarctica (LÜTKEN), *Asterias perrieri* E. A. SMITH, *Diplasterias steineni* (STUDER),
Diplasterias lütkeni PERRIER, *Anasterias studeri* PERRIER, *Anasterias chirophora* LUDWIG,
Anasterias belgicae LUDWIG.) BRISINGIDÆ. *Labidiaster radiosus* LUTKEN '71 = *Labid.*
lutkeni BELL '81· 1 fig. *Belgicella racovitzana*, n. g. n. sp , 12 figs.

: 05.   Asteroidea. Reports on an Exploration off the West Coasts of
Mexico, Central and South America, and off the Galapagos Islands,
in Charge of ALEXANDER AGASSIZ, by the U. S. Fish Commission
Steamer "Albatross," during 1891, Lieut. Commander Z. L. TANNER,
U.S.N., commanding. xxxv.
Reports on the Scientific Results of the Expedition to the Tropical
Pacific, in Charge of ALEXANDER AGASSIZ, on the U. S. Fish Com-
mission Steamer "Albatross," from August, 1899, to March, 1900,
Commander JEFFERSON F. MOSER, U.S.N., commanding. vii. Publish-
ed July. xii + 292 pp. + 35 pls. + 1 chart.

ARCHASTERIDÆ. *Cheiraster agassizii*, n. sp., 14 figs. *Pararchaster pectinifer*, n. sp., 9
figs. *Pararchaster pectinifer* juvenes, 5 figs. *Pararchaster cognatus*, n. sp., 1 fig.
Pararchaster spinuliger, n. sp., 1 fig. *Plutonaster abyssicola*, n. sp., 7 figs. *Per-*
sephonaster armiger, n. sp. *Dytaster demonstrans*, n. sp., 15 figs. *Archaster typicus*
M. T., 1 fig. ASTROPECTINIDÆ. *Psilaster sladeni*, n. sp. *Psilaster armatus*, n. sp., 3
figs. *Astropecten sulcatus*, n. sp., 2 figs. *Astropecten benthophilus*, n. sp., 2 figs.
Astropecten exiguus, n. sp., 3 figs. *Astropecten polyacanthus* M. T. *Parastropecten*
inermis, n. g. n. sp., 4 figs. LUIDIA. *L. ferruginea*, n. sp. *L. armata*, n. sp. POR-
CELLANASTERIDÆ. *Porcellanaster pacificus*, n. sp., 2 figs. *Porcellanaster waltharii*, n.

sp., 2 figs. *Albatrossia semimarginalis*, n. g. n. sp., 2 figs *Hyphalaster moseri*, n. sp., 2 figs. *Ctenodiscus crispatus* (RETZIUS) = *Ct. corniculatus* MICHAILOVSKY : 03 = *Ct. procurator* SLADEN '89, 2 figs. PENTAGONASTERIDÆ. *Pseudarchaster pectinifer*, n. sp., 2 figs. *Pseudarchaster pulcher*, n. sp., 5 figs. *Pseudarchaster verrilli*, n. sp., 5 figs. *Mediaster transfuga*, n. sp., 9 figs. *Mediaster elegans*, n. sp., 3 figs. *Mediaster elegans abyssi*, n. var. *Nymphaster diomedeæ*, n. sp., 6 figs. *Pentagonaster ernesti*, n. sp., 2 figs. ANTHENEIDÆ. *Hippasteria pacifica*, n. sp., 3 figs. PENTACEROTIDÆ. *Paulia horrida* GRAY *galapagensis*, n. var., 7 figs. *Pauliella ænigma*, n. g. n. sp., 7 figs. *Culcita novæ guineæ* M. T. GYMNASTERIDÆ. *Gymnasteria carinifera* (LAMARCK), 2 figs. ASTERINIDÆ. *Asterina cepheus* (M. T.) juv. LINKIIDÆ. *Linckia miliaris* (M. T.). *Linckia multifora* (LAMARCK). *Linckia pacifica* GRAY var. *diplax* (M. T.). *Ophidiaster cylindricus* (LAMARCK). ZOROASTERIDÆ. *Zoroaster magnificus*, n. sp., 5 figs. *Zoroaster nudus*, n. sp., 9 figs. *Zoroaster hirsutus*, n. sp., 3 figs. *Zoroaster sp.* juv., 3 figs. *Zoroaster longispinus*, n. sp., 6 figs. SOLASTERIDÆ. *Sarkaster validus*, n. g. n. sp., 9 figs. PTERASTERIDÆ. *Hymenaster platyacanthus*, n. sp., 1 fig. *Hymenaster purpureus*, n. sp. *Hymenaster violaceus*, n. sp. *Hymenaster gracilis*, n. sp., 4 figs. *Hymenaster sp.* *Retaster diaphanus*, n. sp. ECHINASTERIDÆ. *Cribrella gracilis*, n. sp., 4 figs. *Cribrella nana*, n. sp., 3 figs. *Alexandraster mirus*, n. g. n. sp., 6 figs. *Acanthaster echinites* ELLIS & SOLANDER. *Mithrodia clavigera* (LAMARCK). PEDICELLASTERIDÆ. *Pedicellaster improvisus*, n. sp., 6 figs ASTERIIDÆ. *Stolasterias alexandri*, n. sp., 3 figs. *Stolasterias robusta*, n. sp. *Sporasterias mariana*, n. sp., 5 figs. *Sporasterias cocosana*, n. sp., 3 figs. *Sporasterias galapagensis*, n. sp., 2 figs. *Hydrasterias diomedeæ*, n. sp., 2 figs. *Hydrasterias (?)* n. sp., 1 fig. BRISINGIDÆ. *Brisinga variispina*, n. sp. *Brisinga tenella*, n. sp. *Brisinga panamensis*, n. sp., 8 figs. *Freyella pacifica*, n. sp. *Freyella insignis*, n. sp., 2 figs *Freyella propinqua*, n. sp.

:05a. Asterien und Ophiuren der schwedischen Expedition nach den Magalhaensländern 1895–1897. Zeitsch. wiss. Zool., Bd. 82, p. 39–79, pl. v, vi.

Asterodon singularis M. T. = *Asterod. granulosus* PERRIER '91 = *Odontaster singularis* LEIPOLDT '95, 3 figs. *Odont. penicillatus* PHILIPPI = *Goniodiscus penicillatus* PHILIPPI '70 = *Gnathaster pilulatus* SLADEN '89 = *Asterodon grayi* PERRIER '91 = *Gnathaster grayi* PERRIER '94 = *Odont. meridionalis* LEIPOLDT '95, 2 figs. *Odont. grayi* BELL = *Calliderma grayi* BELL '81 = *Pentagonaster paxillosus* BELL '81 (non GRAY) = *Gnathaster grayi* SLADEN '89 = *Asterodon pedicellaris* PERRIER '91 = *Odont. grayi* BELL (partim) '93 = *Odont. pedicellaris* BELL '93 = *Gnathaster pedicellaris* PERRIER '94 = *Gnath. pedicellaris* VERRILL '99 = *Gnath. grayi* VERRILL '99, 4 figs. *Porania antarctica* E. A. SMITH, 1 fig. *Cycethra verrucosa* PHILIPPI = *Goniodiscus verrucosus* PHILIPPI '57 = *Cycethra nitida* LEIPOLDT '95 = *Cyc. electilis* LEIPOLDT '95 = *Cyc. simplex* MEISSNER '96 = *Cyc. simplex* forma *subelectilis* MEISS. '96, 2 figs. *Asterina fimbriata* ED. PERRIER '75, 6 figs. *Solaster australis* PERRIER = *Crossaster australis* PERRIER '91. *Retaster gibber* SLADEN, 2 figs. *Cribrella pagenstecheri* STUDER '85 = *Crib. obesa* SLADEN '89 = *Crib. hyalesi* PERRIER '91 = *Cribrella studeri* '91. *Cosmasterias lurida* PHILIPPI = *Asterias sulcifera* LEIPOLDT '95 = *Diplasterias lurida* MEISSNER : 04. *Sporasterias antarctica* LUTKEN = *Asterias antarctica* LUDWIG : 03 = *Sporasterias antarctica* var. *rupicola* LUDWIG :03 = *Asterias rugispina* LEIPOLDT '96. *Labidiaster radiosus* LÜTKEN.

:10. Notomyota, eine neue Ordnung der Seesterne. Sitzungsb. k. preus. Akad. Wiss., xxiii, p. 435–466. May 12.

1. Fam. CHEIRASTERIDÆ LUDW. PONTASTER SLADEN, s. str. *P. tenuispinus* DÜB. & KOREN '46 = *P. hebitus* & *P. limbatus* SLADEN '89 = *P.* (*Cremaster*) *marionis* PERRIER '85 = *P. perplexus* & *P. oligoporus* PERRIER '94. *P. planeta* SLADEN '89. PECTINASTER PERRIER, s. str. *P. echinulatus* = *Archaster echinulatus* PERRIER '75 = *Archaster sepitus* VERRILL '85. *P. filholi* PERRIER '85 = *Pontaster forcipatus* SLADEN '89 = *Pontaster venustus* SLADEN '89. *P. mimicus* = *Pontaster mimicus* SLADEN '89. *P. pristinus* = *Pontaster pristinus* SLADEN '89. *P. hispidus* = *Pontaster hispidus* ALCOCK & WOOD-MASON '91. *P. cribrellum* = *Pontaster cribrellum* ALCOCK '93. *P. agassizii* = *Cheiraster agassizii* LUDWIG '95. LUIDIASTER STUDER em. LUDWIG. *L. dawsoni* = *Archaster dawsoni* VERRILL '80 = *Pontaster oxyacanthus* SLADEN '89 = *Cheiraster horridus* FISHER : 06. *L. hirsutus* STUDER '84. *L. teres* = *Pontaster teres* SLADEN '89. *L. vincenti* = *Cheiraster vincenti* PERRIER '94. *L. gerlachei* = *Cheiraster gerlachei* LUDWIG : 03. CHEIRASTER STUDER, s. str. *Ch. gazellæ* STUDER '83 = *Ch. pedicellaris* STUDER '83. *Ch. trullipes* = *Pontaster trullipes* SLADEN '89. *Ch. subtuberculatus* = *Pontaster subtuberculatus* SLADEN '89. *Ch. pilosus* = *Pontaster pilosus* ALCOCK '93. *Ch. coronatus* PERRIER '94 = *Ch. mirabilis* PERRIER '85. *Ch. snyderi* FISHER : 06. *Ch. inops* FISHER : 06. *Ch. granulatus*, n. sp. *Ch. niasicus*, n. sp. MARCELASTER KŒHLER. *M. antarcticus* KŒHLER : 07. GAUSSASTER, n. g. *G. vanhöffeni*, n. sp. 2. Fam. BENTHOPECTINIDÆ VERRILL '99. PARARCHASTER SLADEN, s. str. *P. pedicifer* SLADEN '85. *P. folini* = *Cheiraster folini* PERRIER '85. *P. spinosissimus* SLADEN '89. *P. violaceus* ALCOCK '93. *P. fischeri* PERRIER '94. *P. spinuliger* LUDWIG : 05. *P. indicus* KŒHLER : 09. BENTHOPECTEN VERRILL, s. str. *B. simplex* = *Archaster simplex* PERRIER '81 = *Pararchaster armatus* SLADEN '89. *B. spinosus* VERRILL '84. *B. semisquamatus* = *Pararchaster semisquamatus* SLADEN '89. *B. antarcticus* = *Pararchaster antarcticus* SLADEN '80. *B. huddlestonii* = *Pararchaster huddlestonii* ALCOCK '93. *B. acanthonotus* FISHER : 05. *B. pectinifer* = *Pararchaster pectinifer* LUDWIG : 05. *B. cognatus* = *Pararchaster cognatus* LUDWIG : 05. *B. incertus*, n. sp.

:12. Uber die J. E. GRAY' sehen Gattungen Pentagonaster und Tosia. Zool. Jahrb., Suppl. xv, 1. Bd., p. 1–44.

PENTAGONASTER GRAY '40 = ASTROGONIUM (pars) MÜLL. & TROSCHEL '42 = STEPHANASTER
AYRES '51 = ASTROGONIUM (pars) + STEPHANASTER BATHER : 00. *P. pulchellus* GRAY '40
= *Astrogonium pulchellum* MULL. & TROSCHEL '42 = *Stephanaster elegans* AYRES '51 =
Stephanaster pulchellus PERRIER '94· *P. abnormalis* GRAY '66 = *P. pulchellus* var. *B.*
HUTTON '72 = *Astrogonium abnormale* SLADEN '89 = *Astrogonium pulchellum* var *B.*
FARQUHAR '95 = *Astrogonium huttoni* FARQUHAR '97· *P. crassimanus* MÖBIUS = *Astro-
gonium crassimanum* MÖBIUS '59 = *P. crassissimus* GRAY '66 = *Astrogonium pulchellum*
(pars) SLADEN '89· *P. dubeni* GRAY '47 = *Goniaster dubenii* LÜTKEN '71 = *P. gunnii*
PERRIER '75 = *Astrogonium dübeni* SLADEN '89 = *Astrogonium gunnii* SLADEN '89 =
Stephanaster dubeni PERRIER '94 = *Steph. gunnii* PERRIER '94· *Tosia* GRAY '40 =
= ASTROGONIUM (pars) MÜLL. & TROSCHEL '42 = PENTAGONASTER (pars) PERRIER '75 =
STEPHANASTER (pars) PERRIER '85· *T. australis* GRAY '40 = *Astrogonium geometricum*
MÜLL. & TROSCHEL '42 = *Astrogonium astrologorum* MÜLL. & TROSCHEL '42 = *Pentagon-
aster astrologorum* GRAY '66 = *Pentagon. australis* PERRIER '75 = *Pentagon. (Tosia)
astrologorum* PERRIER '78 = *Pentagon. (Tosia) australis* PERRIER '78 = *Stephanaster
astrologorum* PERRIER '94 = *Stephan. australis* PERRIER '94 = *Stephan. procyon* PERRIER
'94 = *Tosia astrologorum* VERRILL '99· *T. ornata* (MÜLL. & TROSCHEL) = *Astrogonium
ornatum* MÜLL. & TROSCHEL '42 = *Pentagonaster (Tosia) ornatus* PERRIER '78· *T. rubra*
GRAY '47 = *Pentagonaster (Tosia) ruber* PERRIER '78· *T. nobilis* (MULL. & TROSCHEL) .
= *Astrogonium nobile* MÜLL. & TROSCHEL '43 = *Tosia tubercularis* GRAY '66 = *Pentagonaster (Tosia) nobilis* PERRIER '78 = *Pentagonaster (Tosia) tubercularis* PERRIER '78 =
Tosia tubercularis VERRILL '99· *T. aurata* GRAY '47 = *Astrogonium australe* MULL. &
TROSCHEL '42 = *Astrogonium emilii* PERRIER '69 = *Pentagonaster auratus* PERRIER '75 =
Pentagon. astrologorum B PERRIER '75 = *Pentagon. (Tosia) auratus* PERRIER '78· *T.
magnifica* (MÜLL. & TROSCHEL) = *Astrogonium magnificum* MÜLL. & TROSCHEL '42 = *T.
grandis* GRAY '47 = *Pentagonaster (Astrogonium) magnificus* GRAY '66 = *Astrogonium
magnificum* PERRIER '69 = *Pentagon. (Tosia) magnificum* PERRIER '78 = *Pentagon. (Tosia)
grandis* PERRIER '78 = *Pentagon. magnificus* SLADEN '89 = *Pentagon. grandis* SLADEN '89.
Pentagon. minimus PERRIER '75·

LÜTKEN, C.

* '57· Oversigt over Grönlands Echinodermata. Cited after DUJARDIN and
HUPÉ.

'64· Kritiske Bemærkninger om forskjellige Søstjerner (Asterider), med
Beskrivelse af nogle nye Arter. Vidensk. Meddel. fra d. naturhist.
Foren. i Kjøbenhavn, p. 123–169. Résumé in French, 3 pp.
ASTROPECTINIDÆ. *Astropecten aculidatus* SAY. *Astro. ciliatus* *Astro. aster* DE FILIPPI. .
Astro. armatus M. T. *Luidia bellonæ*, n. sp. (End of *Astropectinidæ.) *Archaster*

nicobaricus BEHN = *Arch. typicus* M. T. *Stellaster sulcatus* MÖBIUS, an *Archaster. Asteriscus ciliatus* LORZ. = *Asterina gibbosa* PENNANT = *Asteriscus verruculatus* M. T GONIASTER AGASSIZ. *Astrogonium souleyeti. Astrogonium longimanum. Goniodiscus acutus* HELLER. *Goniodiscus placentæformis* HELLER. *Goniaster articulatus* (LINN.). OREASTER M. T. *Oreaster armatus* LÜTKEN = *Goniodiscus armatus* LÜTKEN. *O. nodosus* (GRAY) non LINNÉ nec LAMARCK. *O. forcipulosus*, n. sp. *O. linckii* (BLAINVILLE). *O. reinhardti*, n. sp. *O. dorsatus* (LINNÉ). *Oreaster gigas* (LIN.) = *O. reticulatus* (LIN) = *O. lapidarius* GRUBE = *O. tuberosus* BEHN-MÖBIUS. OPHIDIASTER & SCYTASTER. *Ophid.* (*Linckia*) *unifascialis* GRAY = *Ophid. bifascialis* GRAY ? = *Ophid. suturalis* M. T. ? *Scyaster galatheæ*, n. sp. *Oreaster desjardinsii* MICHELIN '45' a *Scytaster. Asterias canariensis* D'ORBIGNY '39'

'69. Choriaster granulatus, eine neue Gattung aus der Familie der Asteriden. Catalog iv des Museum Godeffroy, Hamburg.

'71. Fortsatte kritiske og beskrivende Bidrag til Kundskab om Søstjerne (Asteriderne). Tredie Række. Vidensk. Meddel. fra d. naturhist. Foren. i Kjøbenhavn, p. 227–304, 1 pl. Summary in French, 8 pp. ASTROPECTINIDÆ. *Luidia brevispina*, n. sp. *Astropecten euryacanthus*, n. sp. *Astropecten javanicus*, n. sp. *Astropecten velitaris* v. MARTENS. *Ctenodiscus australis* LOVÉN. *Archaster tenuispinus* DÜBEN & KOREN. (End of *Astropectinidæ*) ASTERINA = ASTERISCUS M. T. *A. cabbalistica*, n. sp., 1 fig. CHORIASTER LTK. *Ch. granulatus* LTK. GONIASTER (STELLASTER, ASTROGONIUM, GONIODISCUS). *G.* (*Stellaster*) *equestris* (RETZ) = *Asterias equestris* RETZIUS '05 = *Stellaster equestris* M. T. '42 = *St. childreni* M. T. '42 = *St. childreni* GRAY '66 = *Goniaster equestris* v. MARTENS '65' *G.* (*Stell.*) *incei* GRAY '47 = *Stell. gracilis* MÖBIUS '59' *G.* (*Stell.*) *tuberculosus* v. MARTENS '65' *G.* (*Stell.*) *belcheri* GRAY '66' 1 fig. *G. mülleri* v. MARTENS = *Stell.* (*Gon*) *mülleri* v. MARTENS '65 = *Dorigona reevesii* GRAY '66' *G. dubenii* GRAY, 1 fig, OREASTER. *O. australis*, n. sp. *O. hedemanni*, n. sp. *O. westermanni*, n. sp. *O. gracilis*, n. sp. OPHIDIASTER M. T. (LINCKIA, SCYTASTER, etc.). *Linckia nicobarica*, n. sp. *Ophidiaster asperulus*, n. sp., 1 fig. *O. granifer*, n sp. *O. cribrarius*, n. sp. *Scytaster subtilis*, n. sp , 1 fig. *Scyt. desjardinsii* (MICH.). ECHINASTER. *E. gracilis* M. T.? *E. brasiliensis* M. T. '42 = ?*Othilia multispina* GRAY '40 = *E. brasiliensis* LÜTKEN '59 = *Othilia braziliensis* AGASSIZ '69' *E. sentus* (SAY) '25 = *Othilia spinosa* GRAY '40 = *E. spinosus* M. T. '42' *E. spinulosus* VERRILL '69' *E crassispinus* VERRILL = *E. spinosus* LÜTKEN '59 = *E.* (*Othilia*) *crassispina* VERRILL '68' *E. serpentarius* VAL. M. T. '12 *E. tenuispinus* VERRILL. *E. cribrella*, n. sp.? LABIDIASTER, n. g. *L. radiosus* (LOV). ASTERIAS L = ASTERACANTHION M. T.). *A. amurensis*, n. sp. *A.* (*Leptast.*) *sp. A stellionura* VAL. PERRIER '69'

M'ANDREW R., & BARRETT, L.

* '57· List of the Echinodermata dredged between Drontheim and the North Cape. Ann. Mag. Nat. Hist., 2. ser., vol. xx, p. 43-46. Cited after LUDWIG.

MARENZELLER, E. v.

'78· Die Coelenteraten, Echinodermen und Würmer der k. k. öst.-ung. Nordpol-Expedition. Denkschr. k. Akad. Wissensch., mat.-naturw. Cl., Bd. 35, p. 357-398 + 4 pls.

Asterias albulus = Asteracanthion albulus STIMPSON '53 = Asteracanthion problema STEEN-strup '57· Corethraster hispidus WAV. THOMSON '73. Pteraster militaris M. & TR. '42 = Asterias militaris O. F. MULLER 1776. Archaster tenuispinus (DUB. & KOREN) SARS '61 = Astropecten tenuispinus DÜB. & KOREN '46· Ctenodiscus crispatus (RETZ.) LÜTK '57 = Asterias crispata RETZ. '05 = Ct. polaris M. & TR. '42·

MARTENS, E. von.

'65· Ueber ostasiatische Echinodermen, I. Archiv f. Naturgesch., p. 345-360.

Japanese Asteroiden Asterias rubens L. Linckia semiregularis var. japonica, n. Asterina pectinifera M. T. Astropecten armatus M. T. Astrop. scoparius VAL. M. T. Luidia maculata M. T. var. quinaria. Echinaster? sp. Stellaster childreni GRAY. Stellaster mülleri, n. sp. Archaster typicus M. T. Archaster hesperus M. T. = Stellaster sulcatus MÖBIUS. Astropecten japonicus M. T. Navricia pulchella GRAY (not known to v. M.). Chinese Asteroiden. Linckia (Subg. Scytaster M. T.) semiseriata, n. sp. Goniaster (Stellaster) equestris RETZIUS = Stellaster childreni GRAY? = St. gracilis MÖBIUS. Goniaster (Stell) tuberculosus, n. sp. Goniaster (Stell.) mülleri, n sp Subg OGMASTER, n. Goniaster (Ogm) capella M. T. Astropecten velutaris, n. sp.

'66. Ditto, II. Arch. f. Naturgesch., p. 57-88.

Asteroidea of the Indian Archipelago. Echinaster echinulatus M. T. = Ophidiaster echinulatus M. T. '42 = Heresaster papillosus MICHELIN '44· Echinaster fallax M. T. Linckia (Ophidiaster) tuberculata M. T. '42· Linckia pustulata, n. sp. Linckia rosenbergi, n. sp. Linckia miliaris (LINCK) M. T. = Stella marina I RUMPH = Pentadactylos-aster asper miliaris LINCK = Asterias laevigata LINNÉ et LAMARCK = Linckia typus NARDO '34 = Linckia brownii GRAY '40 = Ophidiaster miliaris M. T. '42· Linckia (Asterias) multiforis LAM. = Ophidiaster multiforis M. T. '42 = Linckia typus GRAY '40 = perhaps also Linckia leachii GRAY '40· Linckia paucifonis, n. sp L. milleporella LAM. sp. = Asterias

milleporella LAM. '16 = *Scytaster milleporellus* M. T. '42· *Leiaster speciosus*, n. sp.
Culcita discoidea LAM. M. T. '42· *Asterina gibbosa* PENN. sp. var. = *Asterina burtoni*
GRAY '40 = *Asteriscus verruculatus* M. T. '42· *Asterina coronata*, n. sp. = *Stella coriacea*
umbilicata LINCK. *Asterina pentagona* M. T. '42· *Asterina penicillaris* LAM. sp.
Gymnasterias carinifera LAM. sp. = *Asteropsis carinifera* M. T. '42· *Gymnasterias*
biserrata, n. sp. *Oreaster turritus* (LINCK) GRAY = *Stella marina quarta* RUMPH =
Pentaceros gibbus turritus LINCK = *Asterias nodosa* LINNÉ part. LAMARCK = *P. turritus*
GRAY '40· *O. muricatus* (LINCK) GRAY = *Asterias nodosa* LINNÉ part = *1st. nodosa* var.
3 LAM. = ? *O. tuberculatus* M. T. '42 = *O. castellum* GRUBE '65· *O. muricatus* var. *mul-*
tispina. *O. muricatus* var. *mutica* = *O. hiulcus* M. T. '42 = probably *Pentaceros hiulcus*
GRAY '40· *O. obtusatus* M. T. '42· *Goniaster elavatus*, n. sp. *Archaster typicus* M.
T. '42· *Luidia maculata* M. T. '42·

Mentioned as occurring in the Indian Archipelago according to authors : *Asterias*
tenuispina LAM., *Asterias calamaria* GRAY, *Echinaster oculatus* (LINCK) = *E. sepositus*
RETZIUS, *Echinaster crassus* M. T., *Solaster gracilis* GRUBE, *Chaetaster cylindratus* MÖBIUS,
Linckia pusilla M. T. '44' *Linckia suturalis* M. T., *Linckia cylindrica* LAM., *Linckia*
(Scytaster) pistorius M. T. = *Fromia milleporella* GRAY, *Linckia (Scytaster) semiregularis*
M. T., *Linckia (Scytaster) kuhlii* M. T., *Culcita novae guineae* M. T., *Asterina cepheus*
VAL., *Asterina trochiscus* RETZ., *Oreaster superbus* MÖBIUS, *Oreaster reticulatus* L.,
Oreaster aculeatus GRAY, *Goniaster (Astrogonium) semilunatus* LINCK = *cuspidatus* GRAY,
Goniaster inaequalis GRAY, *Goniaster (Goniodiscus) sebae* M. T., *Goniaster (Goniod.)*
pleyadella LAM., *Goniaster (Goniod.) cuspidatus* LAM. = *scaber* MÖBIUS, *Goniaster (Randa-*
sia) luzonicus GRAY, *Goniaster (Rand.) gracilis* GRAY, *Goniaster (Hosea) spinulosus* GRAY,
Goniaster (Longimani) longimanus MÖBIUS '60 ('59) = *Astrogonium longimanum* MÖBIUS
'60 ('59) = *Astrog. souleyeti* DUJARDIN & HUPÉ, *Goniaster (Stellaster) belcheri* GRAY,
Goniaster gracilis MÖBIUS = *equestris* RETZ., *Archaster angulatus* M. T., *Astropecten*
scoparius M. T., *Astropecten chinensis* GRUBE '65· *Astropecten umbrinus* GRUBE '65·

'67· Ditto, IV. Arch. f. Naturgesch., p. 106–119.

Echinaster purpureus GRAY = *Othilia purpurea* & *O. luzonica* GRAY '40 = *Echinaster*
fallax & *E. eridanella* (VAL.) M. T. '42· *Echinaster solaris* SCHMIDEL. *Linckia*
(Metrodira) subulata GRAY '40· *Pteraster cribrosus*, n. sp., 1 fig. *Gymnasteria carini-*
fera LAM. *Goniaster pentagonulus* LAM. *Goniaster (Stellaster) equestris* RETZIUS.
Archaster hesperus M. T.

* '69. Seesterne u. Seeigel. VON DER DECKEN's Reisen in Ostafrika, iii.
Bd., 1. Abth., p. 123–134, 1 Taf. Cited after PERRIER and
LUDWIG.

* '89· Echinodermen aus Neu Guinea. Sitzungsber. naturf. Gesellsch. Berlin, p. 183–185.

* : 04. Die Mollusken (Conchylien) und die übrigen Wirbellosen Thiere in RUMPH's Rariteitkamer. Rhumphius Gedenkboek, Kolon. Mus. Haarlem, p. 109–136. Cited after the Zoological Record.

MEISSNER, M.

'92· Asteriden gesammelt von Herrn Stabarzt Dr. SANDER auf der Reise S. M. S. "Prinz Adelbert." Arch. f. Naturgesch., 58, p. 183–190.

Asterias glacialis L. Asterias amurensis LÜTK. Asterias torquata SLADEN. Heliaster helianthus LM. Stichaster aurantiacus (MEYEN). Echinaster sepositus (LM.). Echinaster cylindricus, n. sp. Linckia multiforis (LM.). Goniodiscus sanderi, n. sp., 6 figs. Pentaceros muricatus (GRAY). Asterina penicillaris (LM.). Asterina pectinifera (M. T.). Asterina cepheus (M. T.). Asterina chilensis (LÜTK.). Astropecten scoparius (M. T.). Astropecten latespinosus, n. sp., 3 figs. Luidia bellonae LUTK. Luidia limbata SLADEN.

* : 04. Asteroideen. Ergebn. Hamburger Magalhaenischen Sammelreise, Lief. vii, no. 1, 28 pp. + 1 pl.

MICHAÏLOVSKIJ, M.

: 02. Zoologische Ergebnisse der Russischen Expedition nach Spitzbergen. Echinodermen (Holothurioidea, Echinoidea, Asteroidea, Ophiuroidea u. Crinoidea). Ann. d. Musée zool. Acad. imp. Sci. St.-Pétersbourg, vii, p. 460–546. Asteroidea p. 473–489. Mit einer Karte.

Asterias lincki (M. T.) = Pentadactylosaster reticulatus LINCK 1733 = Asteracanthion stellionura PERRIER '69 = Asterias gunneri DANIELSSEN & KOREN '84· Asterias panopla STUXBERG. Asterias grœnlandica (STEENSTRUP) = Asterias grœnlandica + var. robusta LEVINSEN '86· Asterias hyperborea DANIELSSEN & KOREN = Asterias normani DANIELSSEN & KOREN '84· Stichaster albulus (STIMPSON) = Asteracanthion roseus (pars) M. T. '42 = Asteracanthion albulus STIMPSON '53 = Asteracanthion problema LÜTKEN '57 = Stichaster albulus DUNCAN & SLADEN '77· Pedicellaster typicus M. SARS '61 = Asteracanthion palœocrystallus DUNCAN & SLADEN '77 = Pedicellaster palœocrystallus DUNCAN & SLADEN '81· Cribrella sanguinolenta (O. F. MULLER) = Pentadactylosaster oculatus LINCK 1733 = Asterias sanguinolenta O. F. MÜLLER 1776 = Asterias pertusa O. F. MÜLLER 1776 = Asterias spongiosa FABRICIUS 1780 = Henricia oculata GRAY '40 = Cribrella oculata FORBES '41 = Echinaster oculatus M.

T. '12 = *Echinaster eschrichtii* M. T. '42 = *Echinaster sarsii* M. T. '44 = *Linckia oculata* STIMPSON '53 = *Linckia pertusa* STIMPSON '53 = *Echinaster sanguinolentus* M. SARS '61 = *Cribrella sanguinolenta* DUJARDIN & HUPÉ '62 = *Cribrella eschrichtii* DUJARDIN & HUPÉ '62 = *Henricia sanguinolenta* BELL '92 (forma *lænior*, forma *scabrior*). *Hymenaster pellucidus* WYV. THOMSON '75 ('73). *Pteraster militaris* (O. F. MÜLLER) = *Asterias militaris* O. F. MÜLLER 1776 = *Asteriscus militaris* M. T. '42· *Pteraster obscurus* (E. PERRIER) = *Pt. pulvillus* M. SARS (pars) '61 = *Hexaster obscurus* PERRIER '91 = *Pteraster (Temnaster) hexactis* VERRILL. *Lophaster furcifer* (DÜBEN & KOREN) = *Solaster furcifer* DÜBEN & KOREN '16· *Solaster endeca* (RETZIUS) = *Oktactis dactyloides* LINCK 1733 = *Enneactis coriacea dentata* LINCK 1733 = *Asterias aspera* O. F. MULLER 1776 = *Asterias alboverrucosa* BRANDT '35· *Crossaster papposus* (LINCK) = *Triskaidekaktis papposa* LINCK 1733 = *Dodekaktis reticulata* LINCK 1733 = *Asterias papposa* O. F. MÜLLER 1776 = *Asterias affinis* BRANDT '35 = *Solaster papposa* GRAY '40 = *Solaster affinis* DANIELSSEN & KOREN = *Crossaster affinis* SLADEN '89· *Rhegaster tumidus* (STUXBERG) = *Solaster tumidus* STUXBERG '82 = *Asterina tumida* DANIELSSEN & KOREN '84 = *Asterina tumida* var. *tuberculata* DANIELSSEN & KOREN '84· *Lasiaster hispidus* (M. SARS) = *Goniaster hispidus* M. SARS '77 = *Pentagonaster hispidus* DANIELSSEN & KOREN. *Leptoptychaster arcticus* (M. SARS) = *Astropecten arcticus* M. SARS = *Astropecten lütkeni* BARRETT '57· *Ctenodiscus corniculatus* (LINCK) = *Astropecten corniculatus* LINCK 1733 = *Ctenodiscus polaris* M. T. '42 = *Ct. pygmæus* M. T. '42 = *Ct. crispatus* DÜBEN & KOREN '46 = *Ct. krausei* LUDWIG '86· *Pontaster tenuispinus* (DUBEN & KOREN) = *Astropecten tenuispinus* DÜB. & KOR. '46 = *Archaster tenuispinus* M. SARS. *Astropecten?* sp.

: 04. Die Echinodermen der zoologischen Ausbeute des Eisbrechers "Jermak" vom Sommer 1901. Ann. Mus. zool. Ac. imp. Sc. St.-Pétersbourg, ix, p. 157–188.

Asterias lincki (MULL. & TROSCH.). *Ast. panopla* STUXBERG. *Ast. mulleri* (M. SARS) = *Asteracanthion mulleri* M. SARS '46· *Stichaster albulus* (STIMPSON). *Cribrella sanguinolenta* (O. F. MULLER). *Hymenaster pellucidus* WYV. THOMSON. *Pteraster militaris* (O. F. MÜLLER). *Pteraster pulvillus* M. SARS. *Lophaster furcifer* (DÜBEN & KOREN) = *Solaster furcifer* MORTENSEN : 03. *Solaster endeca* (RETZIUS). *Crossaster papposus* (LINCK) = *Solaster papposus* MORTENSEN : 03. *Rhegaster tumidus* (STUXBERG) = *Solaster tumidus* '78· *Leptoptychaster arcticus* M. SARS. *Ctenodiscus corniculatus* (LINCK). *Pontaster tenuispinus* (DUBEN & KOREN). *Plutonaster parelii* (DÜBEN & KOREN) = *Astropecten parelii* DÜBEN & KOREN = *Archaster parelii* M. SARS '61 = *Plut. (Tethyaster) parelii* SLADEN '89 = *Plut. pareli* BELL '92.

MICHELIN, H.

* '45· Zoophytes, Échinodermes et Stellérides de l'ile Maurice. Mag. d.

Zool., d' Anat. comp. et d. Palæont. GUÉRIN-MÉNEVILLE. 27 pp.+ 6 pls. Cited after DE LORIOL and LUDWIG.

MÖBIUS, K.

'59. Neue Seesterne des Hamburger und Kieler Museums. 14 pp. + 4 pls.

Chætaster minutus, n. sp., 2 figs Chæt. cylindratus, n. sp., 2 figs. Asteriscus stellifer, n. sp. Oreaster superbus, n. sp., 2 figs Oreas tuberosus, n. sp Astrogonium longimanum, n sp., 2 figs Astrog. crassimanum, n. sp., 2 figs Goniodiscus stella, n. sp., 2 figs. Goniod. scaber, n. sp, 2 figs Goniod. conifer, n sp, 2 figs. Stellaster sulcatus, n. sp., 2 figs Stell. gracilis, n. sp, 2 figs Archaster nicobaricus, n sp.

* '80. Beiträge zur Meeresfauna der Insel Mauritius und der Seychellen. 1 Karte u. 22 Taf.

MÖBIUS, K., & BÜTSCHLI, O.

'75. Echinodermata der Nordsee. Jahresb. Komm. Untersuch. deutsch. Meere, Bd. II & III, p. 143–151.

Luidia savignyi AUDOUIN. Astropecten mülleri M. & T.= Asterias aranciaca O. F. MÜLL. Asteracanthion glacialis L. Asteracanthion mulleri. Asteracanthion rubens L. Pteraster militaris O. F. MULL. Cribrella sanguinolenta O. F. MULL. Astrogonium granulare O. F. MÜLL. Solaster papposus L. Solaster endeca L. Archaster tenuispinus DUB. & KOREN. Archaster andromeda M & TR. Archaster parelii DUB. & KOREN.

MORTENSEN, T.

: 10. Report on the Echinoderms collected by the Danmark-Expedition at North-east Greenland. Danmark-Ekspeditionen til Grønlands Nordøstkyst 1906–1908, B. v, Nr. 4, 239–302 + 10 pls.

Bathybiaster vexillifer (WYV. THOMSON) = Archaster vexillifer WYV. THOMSON '73 = Bathybiaster pallidus DAN. & KOREN '84 = Hyaster mirabilis DAN. & KOREN '84, 4 figs. Pontaster tenuispinus (DUB. & KOREN) = Astropecten tenuispinus DUB. & KOREN '44 = Archaster tenuispinus M. SARS '61. Ctenodiscus crispatus (RETZ.) = Ct. corniculatus DUNCAN & SLADEN '81 = Ct. krausei LUDWIG '86, 4 figs Poraniomorpha tumida (STUXBERG) = Solaster tumida STUXBERG '78 = Asterina tumida DAN. & KOREN '81 = Rhegaster tumidus DÖDERLEIN : 00 = Poraniomorpha (Rhegaster) tumida GRIEG : 06, 2 figs. Cribrella sanguinolenta (O. F. MÜLLER) = Echinaster oculatus DÜBEN & KOREN '44 = Echin. sanguinolentus SARS '61 = Cribrella oculata DUNCAN & SLADEN '81 = Echinaster

scrobiculatus DAN. & KOREN '84, forma *scabrior* MICHAIL, forma *lævior* MICHAIL
Pteraster militaris (O. F. MÜLLER), 3 figs. *Pteraster pulvillus* M. SARS, 2 figs. *Solaster*
glacialis DAN. & KOR. '84 = *Sol. endeca* LEVINSEN '86 = *Sol. syrtensis* DÖDERLEIN : 00
Solaster papposus (L.) = *Crossaster papposus* DUNCAN & SLADEN '81, *Pedicellaster*
palæocrystallus SLADEN, 5 figs. *Stichaster albulus* (STIMPSON) = *Asteracanthion problema*
STEENSTR. LUTKEN '57 = *Nanaster* (*Stichaster*) *albulus* KOEHLER : 09, 11 figs. *Asterias*
panopla STUXB. '78, 1 fig.

: 13. Conspectus faunæ groenlandicæ. Echinodermer.

> *Pontaster tenuispinus* (DÜBEN & KOREN). *Leptychaster arcticus* (M. SARS). *Bathybiaster*
> *vexillifer* WYV. THOMSON. *Psilaster andromeda* (M. TR). *Astrogonium parelii* (DUBEN
> & KOR) var. *longobrachiale* DANIELSSEN & KOREN. *Ceramaster granularis* (RETZIUS).
> *Ctenodiscus crispatus* (RETZIUS). *Hippasteria phrygiana* (PARELIUS) *Tylaster willei*
> DANIELSSEN & KOREN. *Poraniomorpha tumida* (STUXBERG). *Poraniom. hispida* (M. SARS)
> *Solaster papposus* (LINNÉ). *Sol. squamatus* DÖDERLEIN. *Sol. endeca* (LINNÉ). *Sol.*
> *glacialis* DANIELSSEN & KOREN. *Lophaster furcifer* (DÜBEN & KOREN). *Korethraster*
> *hispidus* WYV. THOMSON. *Pteraster militaris* (O. F. MULL.). *Pter. pulvillus* M. SARS
> *Pter. obscurus* (PERRIER). *Pt hastatus* MORTENS. *Diplopteraster multipes* (M. SARS).
> *Hymenaster pellucidus* WYV. THOMSON. *Henricia sanguinolenta* (O. F MULL.). *Pedi-*
> *cellaster typicus* M. SARS. *Pedicell. palæocrystallus* SLADEN. *Stichaster albulus* (STIMP-
> SON). [*Stich. roseus* (O. F. MULLER)]. *Asterias mulleri* SARS (incl. var. *grönlandica*
> STEENSTRUP). *Ast. polaris* (M. TR.). *Ast. linckii* (M TR). *Ast. panopla* STUXBERG.
> *Odinia semicoronata* PERRIER.

MULLER, J.

* '40. Ueber den Bau des Pentacrinus caput medusae. Ber. Akad. Wiss.
 Berlin, p. 89–106. Starfish, p. 99–106.

'54, Ueber den Bau der Echinodermen. 99 pp. + 9 pls. Publ. in
 Abhandl. Akad. Wiss. Berlin for 1853.

MÜLLER, J., & TROSCHEL, F. H.

*'40. Ueber die Gattungen der Asterien. Arch. f. Naturgesch., p. 318–
 326. Cited after LUDWIG.

'42. System der Asteriden. xx pp. + 134 pp. + 12 pls. Asteriæ, p. 7–78
 + p. 126–129 + p. 131–134 + Taf. i–vi + Taf. xi–xii.

> I Fam. ASTERACANTHION. *Asterac. glacialis* Nob. = *Sol echinatus cancellatus* LINCK

1733 = *Stellonia glacialis* NARDO '34 = *Stellonia angulosa* AGASSIZ '35· *Asterac. gelatinosus* Nob. = *Asterias rustica* GRAY '40· 2 figs. *Asterac. africanus*, n. sp. *Asterac. tenuispinus* Nob. = *Stella marina echinata* COLUMNA 1616 = *Asterias saveresii* DELLE CHIAJE '23–'29 = *Asterias glacialis* GRUBE '40 = ditto GRAY '40· 2 figs. *Asterac. violaceus* Nob. = ? *Stella marina holsatica* KADE bei LINCK 1733 = *Stellonia rubens* FORBES '39· *Asterac. polaris*, n. sp. *Asterac. rubens* Nob. = *Tetractis petaloides* LINCK 1733 = *Stella coriacea vulgaris Ludii* LINCK 1733 = *Stellonia rubens* AGASSIZ '35· 2 figs. *Asterac. roseus* Nob. *Asterac. loofts*, n. sp. *Asterac. linckii* Nob. = *Pentadactylosaster reticulatus* LINCK 1733 = *Mithrodia spinulosa* GRAY '40· *Asterac. striatus* Nob. *Asterac. aster* Nob. *Asterac. helianthus* Nob. = *Asterias cumingii* GRAY '40 = *Asterias multiradiata* GRAY. *Asterias echinata* GRAY '40· *Asterias katherinæ* GRAY '40· *Asterias wilkinsonii* GRAY '40· *Asterias calamaria* GRAY '40· *Asterac. graniferus* Nob = ? *Asterias serrulata* Encyclopédie 1792 = ? *Asterias ianthina* BRANDT '35· 2 figs. *Asterac. margaritifer* Nob. = ? *Asterias ochracea* BRANDT '35· *Asterac. aurantiacus* Nob = *Stichaster striatus* M. T. '40 = *Tonia atlantica* GRAY '40· 2 figs. II. Fam. ECHINASTER. *Echin. spinosus* Nob. = *Echinaster seu stella coriacea pentadactyla echinata* PETIVER 1711 = *Pentadactylosaster spinosus* LINCK 1733 = *Asterias echinophora* LAMARCK '16 = *Stellonia spinosa* NARDO AGASSIZ = *Othilia spinosa* GRAY '40 = *O. aculeata* GRAY '40· *Echin. brasiliensis* Nob. = *Othilia multispina* GRAY '40· 2 figs. *Echin. crassus*, n. sp. *Echin. gracilis*, n. sp. *Echin. sepositus* Nob. = *Asterias seposita* RETZIUS 1783 = *Stellonia seposita* NARDO AGASSIZ = *Rhopia seposita* GRAY '40 = *Rhopia mediterranea* GRAY '40· *Echin. fallax* Nob. = *Othilia purpurea* GRAY '40 = *Oth. luzonica* GRAY '40· *Echin. eridanella* VALENCIENNES, n. sp. *Echin. serpentarius* VALENCIENNES, n. sp. *Echin. oculatus* Nob. = *Pentadactylosaster oculatus* LINCK 1733 = *Asterias oculata* PENNANT 1777 = *Linckia oculata* FORBES '39 = *Henricia oculata* GRAY '40· *Echin. eschrichtii*, n. sp. *Echin. solaris* Nob. = *Stella pentekaidekaktis* COLUMNA 1616 = *Asterias solaris* Naturforscher Stück 1793 = *Asterias echinites* LAMARCK '16 = *Stellonia echinitis* AGASSIZ '35 = *Echin. ellisii* GRAY '40· SOLASTER. *Sol. papposus* FORBES '39 = *Stella marina tredecim radiis* BESLER 1616 = *Stella rubra coriacea 12 radiis* LUIDIUS bei LINCK 1733 = *Asterias papposa* FABRICIUS 1780, 4 figs. *Sol. endeca* FORBES '39 = *Asterias endeca* RETZIUS 1783. CHÆTASTER. *Chæt. subulatus* Nob. = *Asterias subulata* LAMARCK '16· 3 figs. *Chæt. hermanni*, n. sp. *Chæt. troschelii* VALENCIENNES, n. sp. *Nepanthia tessellata* GRAY '40· *Nep. maculata* GRAY '40· OPHIDIASTER. *Ophid. ophidianus* AGASSIZ '35 = *Asterias ophidiana* LAMARCK '16 = ? *Ophid. aurantius* GRAY '40· *Ophid. hemprichii*, n. sp. *Ophid. cylindricus* Nob. *Ophid. attenuatus* GRAY '40 = *Asterias coriacea* GRUBE '40· *Ophid. miliaris* Nob. = *Pentadactylosaster miliaris* LINCK 1733 = *Asterias lævigata* LINNÉ GMELIN 1788 = *Linckia typus* NARDO '34 = *Linckia brownii* GRAY '40· 2 figs. *Ophid. suturalis* Nob. = ? *Linckia unifascialis* GRAY '40 = ? *Linck. bifascialis* GRAY '40· *Ophid. diplax*, n. sp. *Ophid. ornithopus* VALENCIENNES, n. sp. *Ophid. multiforis* Nob = *Asterias multifora* LAMARCK

'16· *Ophid. ehrenbergii*, n. sp. *Ophid. tuberculatus*, n sp. *Ophid. echinulatus*, n. sp.
Ophid. pyramidatus GRAY '40· *Linckia guildingii* GRAY '40· DACTYLOSASTER. *Dactyl.*
cylindricus GRAY '40· *Dactyl. gracilis* GRAY '40· TAMARIA. *T. fusca* GRAY '40·
CISTINA. *C. columbiæ* GRAY '40· SCYTASTER. *Scyt. variolatus* Nob. = *Pentadactylosaster*
variolatus LINCK 1733 = *Asterias variolata* RETZIUS '05 = *Linckia variolata* AGASSIZ '35 =
Nardoa variolata GRAY '40 = *Nard. agassizii* GRAY '40· 2 figs. *Scyt. milleporellus* Nob.
= *Asterias milleporella* LAMARCK '16· *Scyt zodiacalis*, n sp *Scyt. pistorius* Nob. =
Fromia milleporella GRAY '40· *Scyt. semiregularis*, n sp. *Scyt. kuhlii*, n. sp. *Scyt.*
subulatus Nob. = *Metrodira subulata* GRAY '40· (*Linck. pulchella. Linck. intermedia*
Linck. erythræa. Gomophia ægyptica GRAY '40.) CULCITA. *Cul. discoidea* AGASSIZ '35
= *Asterias discoidea* LAMARCK '16 = *Cul. schmideliana* GRAY '40· *Cul. coriacea* Nob., 1
fig *Cul. novæ guineæ*, n. sp. *Cul. grex*, n sp. ASTERISCUS. *Asteris. palmipes* Nob.
= *Palmipes* LINCK 1733 = *Asterias placenta* PENNANT 1777 = *Asterias membranacea*
RETZIUS 1783 = *Palmipes membranaceus* AGASSIZ '35· *Asteris. rosaceus* Nob. = *Asterias*
rosacea var *lobis quindenis* LAMARCK '16 = *Palmipes stokesii* GRAY '40· *Asteris.*
pectinifer, n sp. *Asteris. verriculatus* Nob. = *Asterias verruculata* RETZIUS '05 = *Asterias*
exigua DELLE CHIAJE '23–'29 = *Asterias membranacea* GRUBE '40 = *Asterina gibbosa* GRAY
'40 = *Asterina burtoni* GRAY '40· *Asteriscus minutus* Nob. *Asteris. cepheus* VALENCIEN-
NES, n. sp. *Asteris. krausii* Nob. *Asteris. pentagonus* Nob. *Asteris. penicillaris* Nob
= *Asterias penicillaris* LAMARCK '16· 2 figs *Asteris. australis* Nob. = *Asterina calcar*
GRAY '40 = *Asterina gunnii* GRAY '40· *Asteris. diesingi*, n. sp. *Asteris. coccineus* Nob.
= *Patiria coccinea* GRAY '40· *Asteris. setaceus* VALENCIENNES, n. sp. *Asteris. militaris*
Nob. = *Asterias militaris* O. F. MÜLLER 1776. *Asteris. trochiscus* Nob. = *Asterias*
trochiscus RETZIUS '05· OREASTER. *O. reticulatus* Nob. = *Stella reticulata sive cancellata*
RONDELET 1555 = *Pentaceros gibbus et reticulatus* LINCK 1733 = *Pentaceros lentiginosus*
LINCK 1733 = *Asterias reticulata* SCHROETER 1777 = *Asterias pentascyphus* RETZIUS '05 =
Asterias sebæ BLAINVILLE '34 = *Pentaceros grandis* GRAY '40 = *Pentac. gibbus* GRAY '40·
1 fig. *O. affinis*, n. sp. *O. chinensis*, n. sp. *O. tuberculatus*, n. sp. *O. turritus* Nob
= *Asterias nodosa* LINNÉ GMELIN 1788 = *Pentaceros franklinii*. *O. hiulcus* Nob. =
Pentrac. hiulcus et gibbus LINCK 1733, 3 figs. *O. mammillatus* Nob. = *Asterias mam-*
millata AUDOUIN '09· *O. verrucosus*, n. sp. *O. elavatus* Nob. = *Asterias stellata* Mus.
Tessin. 1753. *O. carinatus*, n. sp. *O. aculeatus* Nob. *O. obtusatus* Nob = *Asterias*
obtusata Encyclopédie 1792. *O. obtusangulus* Nob. = *Asterias obtusangula* LAMARCK '16·
O. regulus VALENCIENNES, n. sp. *O. nodosus* Nob. *O. armatus* Nob. *O. orientalis* Nob
ASTROGONIUM *Astrog. phrygianum* Nob. = *Pentaceros planus* seu *oxyceros* LINCK 1733
= *Pentac. macroceros* LINCK 1733 = *Asterias phrygiana* PARELIUS 1770 = *Asterias equestris*
LAMARCK '16 = *Goniaster equestris* AGASSIZ = *Asterias johnstoni* GRAY in JOHNSTON '35 =
Hippasteria europæa GRAY '40 = *Hippas. johnstoni* GRAY '40 = *Hippas. plana* GRAY '40·
2 figs. *Astrog magnificum*, n. sp., 2 figs. *Astrog astrologorum*, n. sp. *Astrog.*

geometricum VALENCIENNES, n. sp. *Astrog. pulchellum* Nob. = *Pentagonaster pulchellus* GRAY '40· *Astrog. australe* Nob. = *Tosia australis* GRAY '40· *Astrog. ornatum*, n. sp. *Astrog. lamarckii*, n. sp. *Astrog. cuspidatum* Nob. = *Pentagonaster semilunatus* LINCK 1733 = ? *Asterias tessellata* BLAINVILLE '34 = *Goniaster cuspidatus* GRAY '40· *Astrog. granulare* Nob. = *Asterias granularis* O. F. MÜLLER 1776. GONIODISCUS. *G. pentagon-ulus* Nob. = *Asterias pentagonula* LAMARCK '16 = *Anthenea chinensis* GRAY '40· 2 figs. *G. sebae* Nob. = *Artocreas altera* SEBA. *G. placenta*, n. sp. *G. regularis* Nob = *Pentagon-aster regularis* LINCK 1733. *G. pleyadella* Nob = *Asterias pleyadella* LAMARCK '16· *G. ocelliferus* Nob. = *Asterias ocellifera* LAMARCK '16 = *Nectria ocellifera* GRAY '40· *G. cuspidatus* Nob. = *Asterias cuspidata* LAMARCK '16· *G. mammillatus* VALENCIENNES, n. sp. *G. capella*, n. sp. STELLASTER. *Stell. childreni* GRAY '40 = *Stell. equestris* Nob. = *Asterias equestris* RETZIUS '05· 3 figs. ASTEROPSIS. *Asterop. carinifera* Nob. = *Asterias carinifera* LAMARCK '16· 2 figs. *Asterop. ctenacantha* VALENCIENNES, n. sp. *Asterop. pulvillus* Nob. = *Asterias pulvillus* O. F. MÜLLER 1776 = *Goniaster templetoni* FORBES '39 = *Porania gibbosa* GRAY '40· 2 figs. *Asterop. vernicina* Nob. = *Asterias vernicina* LAMARCK '16· ARCHASTER. *Arch. typicus* Nob. = *Astropecten stellaris* GRAY '40· 4 figs. *Arch. hesperus* M. T. '40· *Arch. angulatus* Nob. = ? *Astropecten mauritianus* GRAY '40· III. Fam ASTROPECTEN. *Astrop. aurantiacus* Nob. = *Astrop. echinatus major* LINCK 1733 = *Asterias aurantiaca* PHILIPPI '37· *Astrop. brasiliensis*, n. sp = ? *Astrop. duplicatus* GRAY '40· *Astrop. valenciennii*, n. sp. = ? *Astrop. stellatus* GRAY '40· *Astrop. tiedemanni* n. sp. *Astrop. bispinosus* Nob. = *Astrop. echinatus minor* LINCK 1733 = *Asterias bispinosa* OTTO '23· *Astrop. polyacanthus*, n. sp., 2 figs. *Astrop. platyacanthus* Nob. = *Asterias platyacantha* PHILIPPI '37· *Astrop. hystrix* VAL., n. sp. *Astrop. armatus*, n. sp. *Astrop. scoparius* VAL., n. sp. *Astrop. hemprichii*, n. sp. *Astrop. articulatus* Nob. = *Asterias articulata* SAY. *Astrop. johnstoni* Nob. = *Asterias johnstoni* DELLE CHIAJE '23–'29. *Astrop. serratus* VAL., n. sp. *Astrop. spinulosus* Nob. = *Asterias spinulosa* PHIL. *Astrop. japon-icus*, n. sp. *Astrop. hispidus*, n. sp. *Astrop. longispinus*, n. sp. *Astrop. pentacanthus* Nob. = *Asterias pentacantha* DELLE CHIAJE '23 '29 = *Asterias aranciaca* JOHNSTON '36· *Astrop. subinermis* Nob. = *Asterias subinermis* PHILIPPI '37· *Astrop. marginatus*, n. sp. *Astrop. schoenleinii*, n. sp. *Astrop. granulatus*, n. sp. *Astrop. andromeda*, n. sp. CTENODISCUS. *Ct. polaris* Nob. = ? *Astrop. corniculatus* LINCK 1733 = *Asterias polaris* SABINE '24· 2 figs. *Ct. pygmaeus* Nob. LUIDIA. *L. savignii* Nob. = *Asterias savignyii* AUDOUIN '09 = *Asterias rubens* JOHNSTON '36 = *Asterias ciliaris* PHILIPPI '37 = *Luidia fragilissima* FORBES '39 = *L. ciliaris* GRAY '40· 1 fig. *L. maculata*, n. sp. *L. senegalensis* Nob. = *Stella marina* MARCGRAV 1648 = *Asterias senegalensis* LAMARCK '16· 2 figs. Addi-ton to II. Fam PTERASTER. *Pt. militaris* Nob. = *Asterias militaris* O. F. MÜLLER 1776, 2 figs.

* '43. Neue Beiträge zur Kenntniss der Asteriden. Arch. f. Naturgesch., p. 113–131.

NICHOLS, A. R.

: 03. A List of Irish Echinoderms. Proc. Roy. Irish Acad., vol. xxiv, sec. B., 1902–1904, p. 231–267.

ARCHASTERIDÆ. *Pontaster tenuispinus* (DÜB & KOR.) = *P. limbatus* SLADEN. *Plutonaster bifrons* (WYV. THOMSON). ASTROPECTINIDÆ. *Astropecten irregularis* (PENNANT) = *Asterias aurantiaca* FORBES. *Astropecten sphenoplax* BELL. *Psilaster andromeda* (M. & TR.). *Luidia ciliaris* (PHILIPPI) = *L. fragillissima* FORBES pars. *Luidia sarsi* DÜB. & KOR. = *L. fragillissima* FORBES pars. PENTAGONASTERIDÆ. *Pentagonaster granularis* (RETZ.) = *P. balteatus* SLADEN = *P. concinnus* SLADEN. *Pentagonaster greeni* (BELL). *Nymphaster subspinosus* (PERRIER) = *N. protentus* SLADEN. GYMNASTERIIDÆ. *Porania pulvillus* (O. F. MÜLLER) = *Goniaster templetoni* FORBES = *Asterias equestris* SOW. ? ASTERINIDÆ. *Asterina gibbosa* (PENNANT). *Palmipes placenta* (PENNANT) = *P. membranaceus* FORBES. STICHASTERIDÆ. *Stichaster roseus* (O. F. MÜLLER) = *Cribrella rosea* FORBES. *Neomorphaster talismani* (PERRIER) = *N. enstichus* BELL. *Zoroaster fulgens* WYV. THOMSON. SOLASTERIDÆ. *Solaster papposus* (FABR). *Solaster endeca* (LINNÉ). PTERASTERIDÆ. *Pteraster personatus* SLADEN. *Hymenaster giganteus* SLADEN. ECHINASTERIDÆ. *Henricia sanguinolenta* (O. F. MÜLLER) = *Cribrella oculata* FORBES (with var. *abyssicola*). ASTERIIDÆ. *Asterias glacialis* LINNÉ = *Uraster glacialis* FORBES. *Asterias rubens* LINNÉ = *Uraster rubens* & *Uraster violacea* FORBES. *Asterias murrayi* BELL. *Asterias hispida* PENNANT = *Uraster hispida* FORBES. BRISINGIDÆ. *Brisinga endecacnemos* ASBJ. *Brisinga coronata* G. O. SARS.

NORDGAARD, O.

'93· Enkelte træk af Beitstadfjordens evertebratfauna. Bergens Mus. Aarsberet. 1892, no. 2. 11 pp.

Asteracanthion rubens LIN. *Stichaster roseus* MÜLL. *Cribrella sanguinolenta* MÜLL. *Astropecten mülleri* MÜLL. & TROSCH. *Astropecten andromeda* MÜLL. & TROSCH. *Pteraster militaris* MÜLL. *Ctenodiscus crispatus* RETZ. *Solaster furcifer* DÜB. & KOR. *Solaster endeca* GMEL. *Astrogonium granulare* MÜLL. *Astrogonium phrygianum* PAB.

* :05. Hydrographical and Biological Investigations in Norwegian Fiords. Bergens Mus. Skrift. 256 pp. + 21 pls. Cited after the Zoological Record.

NORMAN, A. M.

'65. On the Genera and Species of British Echinodermata. I. Crinoidea, Ophiuroidea, Asteroidea. Ann. Mag. Nat. Hist., 3. ser., vol. 15, p. 98–129.

ASTROPECTINIDÆ. ASTROPECTEN. *A. irregularis* (PENNANT) = *Asterias aurantiaca* O F. MULLER 1776 = *Asterias irregularis* PENNANT 1777 = *Astrop. mulleri* MÜLL. & TROSCH. '44 = *Astrop. echinulatus* MULL. & TROSCH. '44· *A. acicularis*, n. sp. LUIDIA. *L. Savignii* (AUDOUIN) = *Asterias Savignii* AUDOUIN '28 = *L. fragilissima* FORBES '39 = *L. Savignii* MULL. & TROSCH '42· *L. sarsii* DUBEN & KOREN = *L. fragilissima* FORBES '39 = *L. Sarsii* DUBEN & KOREN '44 = *L. Savignii* DUBEN & KOREN '44· SOLASTERIDÆ ARCHASTER. *A. Parelii* (DÜBEN & KOREN) = *Asterias aurantiaca* var. PARELIUS 1768 = *Astropecten Parelii* DÜBEN & KOREN '44 = *Archaster Parelii* SARS '61· PALMIPES. *P. placenta* (PENNANT) = *Asterias placenta* PENNANT 1777 = *Asterias membranacea* RETZIUS 1783 = *P. membranaceus* FORBES '39· ASTERINA. *A. gibbosa* (PENNANT) = *Asterias gibbosa* PENNANT 1777 = *Asterias verruculenta* RETZIUS '05 = *Asterina gibbosa* FORBES '41 = *Asteriscus verruculentus* MÜLL. & TROSCH. '42 SOLASTER. *S. papposus* (LINNÆUS) = *S. papposa* FORBES '41· *S. endeca* (LINNÆUS) = *S. endeca* FORBES '41· FORANIA. *P. pulvillus* (O. F. MULLER) = *Asterias pulvillus* MÜLLER 1788 = *Goniaster Templetoni* FORBES '39 = *P. gibbosa* GRAY '41 = *Asteropsis pulvillus* MÜLL. & TROSCH. '42· GONIASTER. *G. phrygianus* (PARELIUS) = *Asterias phrygiana* PARELIUS 1768 = *Asterias equestris* GMELIN (?) 1788 = *G. equestris* FORBES '41 = *Hippasteria plana, Europæa,* & *Johnstoni* GRAY '41 = *Astrogonium phrygianum* MULL. & TROSCH. '42 = *Goniaster abbensis* FORBES '43 = *Astrogonium aculeatum* BARRETT '57 = *Astrogonium phrygianum* var. SARS '61 CRIBRELLA. *C. sanguinolenta* (O. F. MÜLLER) = *Asterias sanguinolenta* MULLER 1776 = *Asterias oculata* PENNANT 1777 = *C. oculata* FORBES '41 = *Henricia oculata* GRAY '41 = *Echinaster oculatus* MÜLL. & TROSCH. '42 = *Echinaster Eschrichtii* MULL. & TROSCH. '42 = *Echinaster sarsii* MÜLL. & TROSCH. '44 = *Linckia oculata* & *pertusa* STIMPSON '53 = *C. sanguinolenta* LÜTKEN '57· ASTERIADÆ. STICHASTER. *S. roseus* (O. F. MÜLLER) = *Asterias rosea* MÜLL. 1776 = *Cribrella rosea* FORBES '41 = *Asteracanthion roseus* MULL. & TROSCHEL '42 = *S. roseus* SARS '61· ASTERIAS. *A. glacialis* LINNÆUS = *Uraster glacialis* FORBES '41· *A mulleri* (SARS) = *Asteracanthion mulleri* SARS '46 = *A. rubens* (LINNÆUS) = *Uraster rubens* FORBES '41· *A. violacea* O. F. MÜLLER = *Uraster violacea* FORBES '41· *A. hispida* PENNANT = *Uraster hispida* FORBES '41·

'93· A Month on the Trondhjem Fiord. Ann. Mag. Nat. Hist., 6. ser., vol. 12, p. 341–367.

Pontaster tenuispinus. Plutonaster parelii.. Leptophychaster arcticus. Psilaster andromeda. Pentagonaster granularis. Lasiaster hispidus. Lophaster furcifer. Pteraster militaris. Pt. pulvillus. Retaster multipes. Brisinga endecacnemos. Br. coronata.

:03. Notes on the Natural History of East Finmark. Ann. Mag. Nat. Hist., 7. ser., vol. 12, p. 406–417.

Archaster tenuispinus DÜB. & KOREN. *Plutonaster parelii* DÜB. & KOREN, var. *longi-*

brachialis DAN. & KOREN. *Ctenodiscus crispatus* RETZ. *Leptoptychaster arcticus* M.
SARS. *Psilaster andromeda* MULL. & TROSCHEL *Pentagonaster granularis* RETZ.
Goniaster phrygianus PARELIUS. *Poraniomorpha rosea* DANIEL & KOREN. *Hexaster
obscurus* PERRIER. *Pteraster militaris* O. F. MULL. *Pt. pulvillus* M. SARS. *Crossaster
papposus* FABRICIUS. *Cros. affinis* BRANDT. *Solaster endeca* GMELIN. *Sol. syrtensis*
VERRILL. *Cribrella sanguinolenta* O. F. MULL. *Pedicellaster typicus* M. SARS. *Asterias
rubens* L. *Ast. mulleri* M. SARS. ?*Ast. glacialis* L. *Ast. linckii* MULL. & TROSCHEL =
Pentadactylosaster reticulatus digitis prælongatis LINCK 1733 = *Ast. linckii* MÜLL. &
TROSCHEL '42 = *Asteracanthion stellionura* PERRIER '69 = *Ast. gunneri* DANIEL. & KOREN
'82 = *Ast. stellionura* DANIEL. & KOREN '84· *Ast. panopla* STUXBERG '78·

PARELIUS, J. v. d. L.

* 1768. Beschreibung einiger Sternrochen oder Asterien. Kongl. norweg.
Gesellsch. Wissen. Drontheim, Schriften, 4. Tl. Cited after
LUDWIG.

PERRIER, E.

'69. Recherches sur les pédicellaires et les ambulacres des Astéries et
des Oursins. 188 pp. + 7 pls. Asteroidea, p. 5–110 + pl. i–ii.
ASTERACANTHION. *Asteracanthion glacialis*, 2 figs. *Asterac. gelatinosus* M. T., 1 fig.
Asterac. africanus M. T., 2 figs. *Asterac. tenuispinus* M. T., 3 figs. *Asterac. violaceus*
M. T. *Asterac. polaris?*, 2 figs. *Asterac. rubens* M. T., 7 figs. *Asterac. roseus* M. T.,
1 fig. *Asterac. linckii* M T. *Asterac. striatus* M. T. *Asterac. aurantiacus* M. T., 2 figs.
Heliaster helianthus, 2 figs. *Asteracanthion novæ boracensis* VAL., 2 figs. *Asterac.
sulcifer*, 4 figs. *Asterac. gemmifer* VAL., 1 fig. Sp. prox. *Asterac. gemmifer*, 3 figs.
Asterac. stellionura VAL., 4 figs. *Asterac. lacazii*, n. sp. = *Echinaster echinura* VAL. MS.
ECHINASTER. *Echinaster clouei* VAL. Coll. Mus. *Echinas. affinis*, n. sp. *Echinaster
(Cribrella) ornatus*, n. sp. SOLASTER. CHÆTASTER. OPHIDIASTER. *Ophidiaster atten-
uatus*, n. sp. *Ophid. diplax* M. T. *Ophid. irregularis*, n. sp. *Ophid. purpureus*, n. sp.
Ophidiaster (?) vestitus, n. sp. SCYTASTER. *Scytaster indicus*, n. sp. CULCITA. *Cul.
discoidea* AG., 3 figs. *Culcita grex (?)* M. T., 2 figs. *Cul. coriacea* M. T. *Cul. novæ
guineæ. Cul. arenosa* VAL., 1 fig. *Cul. pulverulenta* VAL., 3 figs. OREASTER. *Oreas.
reticulatus* M. T. *Oreas. turritus* M. T., 3 figs. *Oreas. muricatus* DUJARDIN &
HUPÉ = *Pent. muricatus* LINCK = *Asterias linckii* DE BLAINVILLE = *Oreas. linckii* VAL.
Coll. Mus, 2 figs. *Oreas. hiulcus* M. T., 2 figs. *Oreas. mammilatus* M. T. *Oreas.
mammosus* VAL. Coll. Mus. *Oreas. clouei*, n. sp., 3 figs. *Oreas. obtusangulus* M.
T. *Oreas. obtusatus.* ASTROGONIUM. *Astrog. phrygianum* M. T., 1 fig. *Astrog.
pulchellum* M. T. *Astrog. australe* M. T. *Astrog. emilii*, n. sp. *Astrog. dubium.*
GONIODISCUS. *Gonio. pleyadella* M. T. *Gonio. cuspidatus* M. T. *Gonio. articulatus*,

n. sp. = *Astrogonium articulatum* VAL. Coll. Mus.　*Gonio. acutus*, n. sp.　*Gonio. michelini*, n. sp.　NECTRIA.　STELLASTER.　*S. childreni* GRAY.　ASTEROPSIS *carinifera, vernicina, ctenacanthus, pulvillus, imbricata*.　ARCHASTER.　*Archas. typicus* M. T., 1 fig.　*Archas. angulatus* M. T. = *Arch. angulosus*, 2 figs.　ACANTHASTER.　*A. solaris* DUJARDIN & HUPÉ = *Echinaster solaris* M. T. = *Stellonia echinites* AGASSIZ, 1 fig.　ASTERISCUS.　*Asteris. marginatus* VAL., 1 fig.　*Asteris. verruculatus* M. T., 1 fig.　*Asteris. pulchellus* VAL.　*Asteris. calcaratus* VAL.　*Asteris. exiguus* VAL.　*Asteris. squamatus* VAL.　*Asteris. wega* VAL.　ASTROPECTEN.　*Astropecten perarmatus*, n. sp.　*Astrop. samoensis*, n. sp.　*Astrop. mulleri* VAL.　*Astrop. myosurus* VAL.　CTENODISCUS.　LUIDIA.　*Luid. savignyi* AUDOUIN, 1 fig.　*Luid. ciliaris* GRAY, 1 fig.　*Luid. granulosa* VAL., 1 fig.

'76· Révision des Stellérides du Muséum (d'histoire naturelle à Paris), II, III. Arch. zool. expér. gén., t. v, p. 1–104 + p. 209–304.

GONIASTERIDÆ.　NECTRIA.　*Nectria ocellifera* DUJARDIN & HUPÉ '62 = *Aster. ocellifera* LAMARCK (pars) '15 = *Goniodiscus ocelliferus* M. T. '42·　*N. ocellata* C. P. [1] = *A. ocellifera* LAMARCK (pars) '15 = *N. ocellifera* GRAY '40·　PENTAGONASTER.　*P. astrologorum* M. T. '42 = *Astrogonium astrologorum* M. T. '42·　*P. australis* = *Tosia australis* GRAY '40 = *Astrogonium geometricum* (VALENCIENNES MS) M. T. '42·　*P. pulchellus* GRAY '40 = *Astrogonium pulchellum* M. T. '42 = *Asterias mulleri* AGASSIZ MS. = *Goniodiscus mulleri* VALENCIENNES MS. = *Stephanaster elegans* AYRES '52·　*P. dubeni* GRAY '47 = *Goniaster dubeni*.　*P. gunnii*, n. sp.　*P. (Tosia) auratus* GRAY '47 = *Astrogonium australe* M. T. '42 = *Astrog. emilii* PERRIER '69 = *Goniodiscus sebæ* Coll, Mus.　*P. (Tosia) minimus* (E. P.).　*P. (Tosia) semilunatus* LINCK 1733 = *Asterias granularis* GMELIN (pars) 1788 = *Aster. tessellata* LAMARCK '15 = *Goniaster cuspidatus* GRAY '40 = *Goniast. sebæ* GRAY '40 = *Astrogonium cuspidatum* M. T. '42 = *Goniaster semilunatus* v. MARTENS '66 = *Astrogonium dubium* PERRIER '69 = *Goniaster americanus* VERRILL '71 = *Goniast. africanus* VERRILL '71·　*P. (Tosia) lamarckii* (M. T.) = *Astrogonium lamarckii* M. T. '42 = *Astrog. cuspidatum* Coll. Mus.　*P. (Calliaster) childreni* (E. P.) = *Calliaster childreni* GRAY '40·　*P. (Astrogonium) dilatatus*, n. sp.　*P. (Astrog.) spinulosus* = *Hosea spinulosa* GRAY '47 = *(Goniaster) Hosia spinulosus* v. MARTENS '66·　*P. (Astrog.) gibbosus*, n. sp.　*P. (Astrog.) miliaris* = *Astrog. miliare* GRAY '47·　*P. (Astrog.) paxillosus* = *Astrog. paxillosum* GRAY '47·　*P. (Astrog.) singularis* = *Goniodiscus singularis* M. T. '43·　*P. (Astrog.) tuberculatus* = *Astrog. tuberculatum* GRAY '47·　*P. (Astrog.) mammillatus* = *Goniodiscus mammillatus* VALENCIENNES Coll. Mus. = ditto M. T. '42 = *Hosea mammillata* GRAY '66·　*P. (Astrog.) granularis* = *Asterias granularis* O. F. MÜLLER 1788 = *Aster. tessellata* (var. *A*) LAMARCK '15 = *Astrog. granulare* M. T. '42·　*P. (Calliderma) emma* = *Calliderma emma* GRAY '47·　*P. (Stellaster) equestris* = *Asterias equestris* RETZIUS '20 = *Stellaster childreni* GRAY '40 =

---

1) Probably a misprint for E P.

Stellaster equestris M. T. '42 = *Goniaster (Stellaster) equestris* v. MARTENS '65· *P. (Stellast.)*
belcheri = *Stellaster belcheri* GRAY '47 = *Goniaster (Stellaster) belcheri* v. MARTENS '65· *P.*
(Stellast.) incei = *Stellaster incei* GRAY '47 = *Stellast. gracilis* MÖBIUS '59 = *Goniaster*
(Stellaster) incei v. MARTENS '65· *P. (Stellast.) granulosus*, n sp. *P. (Dorigona)*
mulleri = *Goniaster (Stellaster) mülleri* v. MARTENS '65 = *Dorigona reevesii* GRAY '66· *P.*
(Dorig.) longimanus = *Archaster lucifer* VALENCIENNES Museum label = *Astrogonium*
longimanum MÖBIUS '60 = *Astrog. souleyeti* DUJARDIN & HUPÉ '62 = *Goniaster longimanus*
LÜTKEN '64 & '71. GONIODISCUS. *G. cuspidatus* = *Aster. cuspidata* LAMARCK '15 =
Hosea cuspidata GRAY '65· *G. sebæ* = *Asteroceras altera* SEBA 1761 = *Hosea ? seba* GRAY
'65 = *Goniaster sebæ* v. MARTENS '66. *G. pleyadella* (M. T) = *Asterias pleyadella* LA-
MARCK '15 = *Hosea pleyadella* GRAY '65 = *Goniaster pleyadella* v. MARTENS '66· *G.*
rugosus, n. sp. *G. forficulatus*, n sp. *G. granuliferus* = *Anthenæa granulifera* GRAY
'47· PENTACEROS. *P. granulosus* GRAY '47· *P. nodulosus*, n sp. *P. nodosus* GRAY
'40· *P. muricatus* = *P. gibbus* var. *muricatus* LINCK 1733 = *P. gibbus & muricatus*
SEBA 1761 = *Asterias linckii* DE BLAINVILLE '34 = *Oreaster linckii* MS. Coll. Mus. =
Oreaster nodosus var. *muricatus* v. MARTENS '66 *P. turritus* = *Asterias nodosa* LINNÉ
1788 (GMELIN) = *P. franklinii* GRAY '40· *P. mammosus* PERRIER '69· *P. hiulcus* GRAY
'40 = *P. gibbus & hiulcus* LINCK 1733 = *P. nodosus* GRAY '66· *P. alveolatus*, n. sp.
P. dorsatus = *Asterias dorsata* LINNÉ 1785 = *Asterias stellata* 1753 = *Oreaster clavatus* M.
T. '42· *P. regulus* M. T. '42· *P. mammillatus* = *Asterias mammillata* AUDOUIN '24 =
Oreaster muricatus v. MARTENS (pars) '66· *P. gracilis* LÜTKEN '71· *P. reticulatus* =
Stella reticulata seu cancellata RONDELET 1554 = *Stella reticulata* ALDROVANDI 1602 =
Asterias secunda, major, pentadactyla, crassa et tuberculata BROWNE 1756 = *P. lentiginosus*
LINCK 1733 = *Asterias gigas* LINNÉ 1753 = *Aster. reticulata* P. L. S. MÜLLER = *Asterias*
pentascyphus & Ast. reticulata RETZIUS '05 = *Aster. sebæ* DE BLAINVILLE '33 = *P. granlis,*
P. gibbus & P. reticulatus GRAY '40 = *Oreaster coronatus* VALENCIENNES MS. Coll Mus.
= *Oreas. lapidarius & O. gigas* GRUBE '57 = *Oreas. tuberosus* BELM '62 = *Oreast. gigas*
LÜTKEN '64 = *P. grandis, P. gibbosus & P. reticulatus* GRAY '66 = *Oreast. aculeatus*
VERRILL '67· *P. obtusatus* = *Asterias obtusata* BORY DE SAINT-VINCENT. NIDORELLIA.
Nidor. armata = *P. (Nidorellia) armatus* GRAY '40 = *Goniodiscus armatus* LÜTKEN '59 =
Goniod. conifer MÖBIUS '60 = *Goniodiscus stella* VERRILL '67· *Nidor. michelini* =
Goniodiscus michelini PERRIER '69 = *Oreast. armatus* LÜTKEN '71. *Nidor. horrida* =
Paulia horrida GRAY '40 = *Goniodiscus (?) horridus* M. T. '42· CHORIASTER. *Ch.*
granulatus LUTKEN '69· ASTERODISCUS. *A. elegans* GRAY '47· CULCITA. *C. schmideliana*
= *Asterias schmideliana* RETZIUS '05 = *Aster. discoidea* LAMARCK '15 = *C. discoidea*
AGASSIZ '35· *Culcita coriacea* M. T. '42· *C. grex* M. T. '42 = *Culc. novæ guineæ* Mus.
label = ditto PERRIER '69· *C. novæ guineæ* M. T. '42 = *C. pulverulenta* VALENCIENNES
MS. Coll. Mus. *Culc. pentangularis* GRAY '66 = *C. grex* Mus. = *Randasia granulata* GRAY

'47 = *C. grex* Perrier '69· *Cule arenosa* Perrier '69· Randasia. *Rand. spinulosa* Gray '47· *R. granulata* Gray '47· Goniaster. *G. obtusangulus = Aster. obtusangula* Lam. '15 = *Oreaster obtusangulus* M. T. '42 = *Pentac. obtusangula* Gray '65· Hippasteria *H. plana = Pentac. planus* (oxyceros & macroceros) Linck 1733 = *Aster. phrygiana* Parelius 1770 = *Asterias equestris* Pennant 1777 = *Goniaster phrygianus* Agassiz '35 = *Aster. johnstoni* Gray = *Aster. (Goniaster) equestris* Gould '41 = *Astrogonium phrygianum* M. T. '42 = *Goniaster abbensis* Forbes '43 = *Astrog. asculeatum* Barrett '57· Anthenea. *A. tuberculosa* Gray '47· *A. articulata = Astrogonium articulatum* Valenciennes Coll. Mus. *A. pentagonula = Asterias pentagonula* Lamarck '15 = *Goniaster articulatus* Agassiz '33 Mus. Paris = *A. chinensis* Gray '40 = *Goniodiscus pentagonulus* M. T. '42 = *Astrog. articulatum* Valenciennes Coll. Mus. = *Goniaster articulatus* Lütken '64 = *Goniaster pentagonulus* v. Martens '65 = *Goniodiscus articulatus* Perrier '69· *A. acuta = Goniodiscus acutus* Perrier '69 *A. flavescens = Hosia flavescens* Gray '40· *A. grayi*, n. sp. Torania. *Por. pulvillus = Asterias pulvillus* O. F. Müller 1788 = *Aster. gibbosa* Leach '17 = *Goniaster templetoni* Forbes '39 = *Porania gibbosa* Gray '40 = *Asteropsis pulvillus & Ast. ctenacantha* M. T. '42 = *Asteropsis pulvillus* Lütken '58 Asteropsis *Astrop. verrucina = Asterias verrucina* Lamarck '15 = *Petricia punctata* Gray '47· Dermasterias, n. g *D. inermis =* ? *Gymnasteria inermis* Verrill '40 = ? *Asteropsis imbricata* Grube = ? ditto Verrill '67· Gymnasteria. *G. valvulata*, n sp. *G. carinifera = Asterias carinifera* Lamarck '15 = *Gymnast. spinosa & Gymnast. inermis* Gray '40 = *Asteropsis carinifera* M. T. '42· Asterinidæ. Disasterina. *D. abnormalis.* Palmipes. *P. rosaceus = Asterias rosacea* var. *lobis quindenis* Lamarck '15 = *P. stokesii* Gray '40 = *Asteriscus rosaceus* M. T. '42· *P. membranaceus* Linck 1733 = *Stella cartilaginea* Aldrovandi 1638 = *Asterias placenta* Pennant 1777 = *Asterias membranacea* Retzius 1783 = *Aster. cartilaginea* Fleming '23 = *Asteriscus palmipes* M. T. '42 = *P. placenta* Norman '65· *P. inflatus = Pteraster inflatus* Hutton '72· Asterina. *A. gibbosa* Forbes '39 = *Stellula hibernica glabra* Petiver 1709 = *Asteriscus seu Stella pentadactyla exigua, canis marini corio utrinque munita* Petiver 1709 = *Pentaceros plicatus & concavus* Linck 1733 = *Asterias gibbosa* Pennant 1777 = *Asterias verruculata* Retzius '05 = *Asterias exigua* Delle Chiaje '23 = *Asterias pulchella* de Blainville '34 = *Asterina minuta* Nardo '34 = *Asteriscus verruculata* M. T. '42· *A. calcar* Gray '40 = *Asterias calcar* var. *c. octogona* Lamarck '15 = *Asteria calcar* Oudart = *Asteriscus australis* M. T. '42· *A. gunnii* Gray '40 = *Asterias calcar* var. *b.* Lamarck '15 = *Asteriscus australis* M. T. (pars) '42 = ? *Asteriscus diesingii* M. T. '42 = *Asteriscus zelandicus* Valenciennes Mus. label = *Asteriscus exiguus* Valenciennes (pars) Mus. label = *Asteriscus calcar* Dujardin & Hupé '62· *A. regularis* Verrill '67 = *Ast. gunnii* var. *pentagonale* Gray '40 = *Asteriscus australis* M. T. (pars) '42 = *Ast. gunnii* Dujardin & Hupé (pars) '62 = *Asterina (Asteriscus) regularis* Verrill '67 = *Ast. cabbalistica* Lütken '71· *A. marginata = Asteriscus minutus* M. T. '42. = *Asteriscus. marginatus*

VALENCIENNES MS. Mus. Coll. = *Asteriscus stellifer* MÖBIUS '59 = *Asteriscus brasiliensis*
LÜTKEN '59 = *Ast. stellifera* LÜTKEN '71· *A. calcarata* = *Asteriscus calcaratus* VALEN-
CIENNES (pars) MS. Mus. Coll. = *Asteriscus calcaratus* VALENCIENNES in Cl GAY '54·
Ast. chilensis LÜTKEN '59 = *Patiria chilensis* VERRILL. *A exiqua* = ? *Asterias minuta*
GMELIN 1788 = *Asterias exiqua* LAMARCK '15 = *Asterias minuta* DE BLAINVILLE '34 =
Asterina minuta NARDO '34 = *Ast. kraussii* GRAY '40 = *Asteriscus pentagonus* M. T. '42 =
Asterina pentagona v. MARTENS '66 *A. gayi*, n. sp = *Asteriscus calcaratus* VALEN-
CIENNES (pars) MS. Mus. Coll. *A. pusilla*, n. sp = *Asteriscus calcaratus* VALENCIENNES
(pars) MS. Mus. Coll. *A. fimbriata*, n. sp. *A. novæ-zelandiæ*, n. sp. *A minuta*
GRAY '40 = *A. folium* LÜTKEN '59 = *Asteriscus folium* VERRILL '67· *A. wesveli*, n. sp.
A. granulosa, n. sp. *Ast. squamata*, n. sp. *A. stellaris*, n. sp. *A. pectinifera* =
Asteriscus pectinifer M. T. '42 = *Asterina pectinifera* v. MARTENS '65· *A. coccinea* =
Asteriscus coccineus M. T. '42 = *Patiria coccinea* GRAY '47 = *Asteriscus coccineus*
DUJARDIN & HUPÉ '62 *A. cephea* = *Asterias calcar* var. *a* AUDOUIN '25 = *Asteriscus*
cepheus VALENCIENNES MS. Mus Coll. = *Asterina burtonii* GRAY '40 = *Asteriscus cepheus*
M. T. '42 = ? *Asteriscus verruculatus* & *Asteriscus cepheus* PETERS '52 = ? *Asterina*
gibbosa v. MARTENS '66 = *Asterina cepheus* v. MARTENS '66· *Ast. wega* = *Asteriscus wega*
VALENCIENNES MS. Mus. Coll. *A. setacea* = *Asteriscus setaceus* VALENCIENNES MS Mus.
Coll. = ditto M. T. '42 *A. obtusa* = *Patiria obtusa* GRAY '47 *A granifera* = *Patiria*
granifera GRAY '47· *A. (Nepanthia) belcheri*, n. sp. *A. (Nep.) brevis*, n. sp. *A.*
(Nep.) maculata = *Nepanthia maculata* GRAY '40 = *Chætaster (?) maculatus* M. T. '42·
PATIRIA. *P. ocellifera* GRAY '47· *P. crassa* GRAY '47· GANERIA. *Gan. falklandica*
GRAY '47· ASTROPECTINIDÆ. CHAETASTER. *Ch. longipes* = *Asterias longipes* RETZIUS '05
= *Asterias subulata* LAMARCK '15 = *Nepanthia tessellata* GRAY '40 = *Ch. subulata* M. T.
'42 = *Ch. longipes* SARS '57· *Ch. nodosus*, n. sp. LUIDIA. *L. hardwickii* = *Petalaster*
hardwickii GRAY '40· *L clathrata* LÜTKEN '59 = *Asterias clathrata* SAY '25 = *L. gemmacea*
VALENCIENNES MS. Mus. Coll. *L. colombiæ* = *Petalaster colombiæ* GRAY '40 = *L. tessellata*
LÜTKEN '59· *L alternata* = *Asterias alternata* SAY '25 = *L. granulosa* VALENCIENNES MS.
Mus Coll. = *L. granulosa* PERRIER '69· *L. elegans*, n sp. *L. variegata*, n. sp. *L.*
brevispina LÜTKEN '71· *L. bellonæ* LÜTKEN '64· *L. maculata* M. T. '42· *L. savignyi*
= *Asterias savignyi* AUD. '09 = *Asterias rubens* JOHNSTON '36 = *Asterias ciliaris* PHILIPPI
'37 = *L. fragilissima* FORBES '39 = *Asterias pectinata* COUCH '40 = *L. sarsii* DUBEN & KOREN
'44· *L senegalensis* M. T. '42 = ? *Stella marina* MARCGRAV 1648 = *Asterias senegalensis*
LAMARCK '16 = *L. marcgravii* STEENSTRUP in LÜTKEN '59· ARCHASTER. *A. typicus* M.
T. '40 = *Astropecten stellaris* GRAY '40 = *Arch. nicobaricus* BEHN in MOBIUS '59· *Arch.*
angulatus M. T. '42 = *Arch. angulosus* DUJARDIN & HUPÉ '62 = *Arch. mauritianus* v.
MARTENS '66 *A. hesperus* M. T. '40 = *Stellaster sulcatus* MÖBIUS '59· *A. christi* =
Asterias christi DUBEN & KOREN '34 = *Astropecten andromeda* M. T. '42 *A. parelii*
SARS '61 = *Asterias aurantiaca* var. *parelius* 1768 = *Astropecten parelii* DUBEN & KOREN

'44· *A. tenuispinus* SARS '61 = *Astropecten tenuispinus* DÜBEN & KOREN '44· *A. echinulatus*, n sp. ASTROPECTEN. *A. aurantiacus* GRAY '40 = *Stella marina major* BESLER 1616 = *Stella pectinata* ALDROVANDI 1638 = *Crenaster* LUIDIUS in LINCK 1699 = *Stella marina major, spinosior, fusca* BARELIERI 1714 = *Astrop. echinatus major* LINCK 1733 = *Asterias aurantiaca* LINNÉ '35 = *Asterias aurantiaca* LAMARCK '16 = *Astrop. crenaster* DUJARDIN & HUPÉ '62 = *Astrop. perarmatus* PERRIER '69· *A. duplicatus* GRAY '40 = *A. valenciennii* M. T. '42 = *A. variabilis* LÜTKEN '58· *A. bispinosus* M. T. '42 = *A. echinatus minor* LINCK 1733 = *Asterias bispinosa* OTTO '23 = *A. echinatus* GRAY '40 = ?? *A. myosurus* VALENCIENNES MS. · Mus. Coll. *A. platyacanthus* M. T. '42 = *Asterias platyacantha* PHILIPPI '37 = *Astrop. aster* PHILIPPI '59· *A. polyacanthus* M. T. '42 = *A. hystrix* VALENCIENNES MS. Mus. Coll. = *A. hystrix* M. T. '42 = *A. armatus* M. T. '42 = *A. rappa* M. T. '43 = *A. aster* LÜTKEN '64· *A. erinaceus* GRAY '40 = *A. armatus* GRAY '40 = *A. oerstedii* LÜTKEN '59· *A. scoparius* M. T. '42 = ditto VALENCIENNES MS. Mus. Coll *A. mauritianus* GRAY '40· *A. johnstoni* M. T. '42 = *Asterias johnstoni* DELLE CHIAJE '22 = ? *A. irregularis* GRAY '40· *A. serratus* M. T. '42 = ditto VALENCIENNES MS. Mus. Coll. = *A. aranciaca* FISCHER '72· *A. spinulosus* M. T. '42 = *Asterias spinulosa* PHILIPPI '37 = *A. archimedis* VALENCIENNES Mus. label. *A. antillensis* LÜTKEN '59· *A. javanicus* LÜTKEN '71 = *A. armatus* Mus. label. *A. arenarius* = ditto VALENCIENNES MS. Mus. Coll. *Astrop. samoensis* PERRIER '69· *A. brasiliensis* M. T. '42· *A. irregularis* LINCK 1733 = *Asterias aurantiaca* O. F. MÜLLER 1776 = *Asterias irregularis* PENNANT 1777 = *A. aurantiaca* FORBES '41 = *A. mulleri* M. T. '42· *A. subinermis* M. T. '42 = *Asterias subinermis* PHILIPPI '37· *A. pentacanthus* M. T. '42 = *Asterias pentacantha* DELLE CHIAJE '25 = *Asterias aranciaca* JOHNSTON '36· *A. articulatus* M. T. '42 = *Asterias articulata* SAY '25 = *Astrop. dubius* GRAY '40 = *Asterias aranciaca* GOULD '41· *A. dussumieri* = ditto VALENCIENNES Mus. label *A. richardi* = ditto VALENCIENNES MS. Mus. Coll. *A. alatus*, n. sp. *A. spatuliger*, n. sp. *A. latiradiatus* = *Platasterias latiradiata* GRAY '71· CTENODISCUS. *Ct. corniculatus* = *Astrop. corniculatus* LINCK 1733 = *Asterias crispata* RETZIUS '05 = *Asterias polaris* SABINE '24 = *Asterias aurantiaca* DEWHURST = *Astrop. polaris* GRAY '40 = *Ct. polaris* M. T. '42 = *Ct. pygmaus* M. T. '42 = *Anodiscus crispatus* Mus. label. PTERASTERIDÆ. PTERASTER. *Pt. militaris* M. T. '42 = *Asteriscus militaris* M. T. '42· *Pt. capensis* GRAY '40· *Pt. cribrosus* V. MARTENS '67·

* '78. Étude sur la répartition géographique des Astérides. Nouv. Arch. Mus. Hist. nat., 2. sér., t. 1, p. 1–108. Cited after LUDWIG.

'21· Description sommaire des espèces nouvelles d'Astéries. Bull. Mus. Comp. Zool. Harv. Coll., ix, No. 1, p. 1–31.

Asterias contorta (E. PERR). *Asterias fascicularis*, n. sp. *Asterias linearis*, n. sp. *Asterias angulosa*, n. sp. *Asterias gracilis*, n. sp. *Zoroastr sigsbeei*, n. sp. *Zoroas-*

ter ackleyi, n. sp. *Pedicellaster pourtalesi* (E. PERR.). *Echinaster modestus* (E. PERR.). *Cribrella antillarum* (E. PERR.). *Crib. sex-radiata* (E. PERR.). *Ophidiaster floridæ,* n. sp. *Ophid. agassizii,* n. sp. *Korethraster palmatus,* n. sp. *Kor. radians* (E. PERR.). *Pteraster caribbæus.* n. sp. *Fromia japonica,* n. sp. *Asterina lymani,* n. sp. *Asterina pilosa,* n. sp. *Marginaster pectinatus* (E. PERR.). *Marg. echinulatus* (E. PERR.). *Radiaster elegans,* n. sp. *Ctenaster spectabilis,* n. sp. *Pentagonaster (Tosia) parvus* (E. PERR.). *Pent. grenadensis* (E. PERR.). *Pent. ternalis* (E. PERR.). *Pent. subspinosus,* n. sp. *Pent. arenatus,* n. sp. *Pent. alexandri,* n. sp. *Goniodiscus pedicellaris,* n. sp. *Anthenoides peircei* (E. PERR.). *Goniopecten demonstrans,* n. sp. *Goniop. intermedius,* n. sp. *Goniop. subtilis,* n. sp. *Archaster pulcher,* n. sp. *Arch. mirabilis* (E. PERR). *Arch. simplex,* n. sp. *Blakiaster conicus* (E. PERR.). *Luidia barbadensis,* n. sp. *Luid. convexiuscula* (E. PERR.). *Astropecten alligator,* n. sp.

84. Mémoire sur les Étoiles de mer recueillies dans la mer des Antilles et le Golfe du Mexique durant les expéditions de dragage faites sous la direction de M. ALEXANDER AGASSIZ. Nouv. Arch. d. Mus. d'Hist. nat., 2. sér., t. 6, p. 127–216 + 10 pls.

General results. Classification of Asteroidea. Morphological significance of pedicellariæ, skeletal pieces, odontophore. Relations between the families; characters of secondary importance furnished by teeth and ambulacral tubes.

BRISINGIDÆ. HYMENODISCUS, n. g. *H. agassizii,* n. sp., 3 figs. PEDICELLASTERIDÆ. PEDICELLASTER LOVÉN. *P. pourtalesi* E. PERR., 1 fig. ZOROASTER WYV. THOM. *Z. sigsbeei,* n. sp., 1 fig. *Z. ackleyi,* n. sp., 1 fig. ASTERIADÆ. ASTERIAS LIN. *A. contora* E. PERR. *A. fascicularis,* n. sp., 1 fig. *A. linearis,* n, sp., 1 fig. *A. angulosa,* n. sp. *A. gracilis,* n. sp. ECHINASTERIDÆ. ECHINASTER M. TR. *E. modestus* E. PERR., 1 fig. CRIBRELLA AG. *Cr. antillarum* E. PERR., 1 fig. *Cr. sexradiata* E. PERR., 1 fig. SOLASTERIDÆ. KORETHRASTER WYV. THOM. *K. palmatus,* 2 figs. *K. hispidus,* n. sp. RADIASTER, n. g. *R. elegans,* n. sp., 1 fig. CTENASTER, n. g. *Ct. spectabilis,* n. sp , 2 figs. PTERASTERIDÆ. PTERASTER M. TR. *Pt. caribbæus.* ASTERINIDÆ. ASTERINA NARDO. *A. lymani. A. pilosa,* n. sp., 1 fig. *A. wesseli* LÜTK. LINCKIADÆ. OPHIDIASTER M. TR. *O. floridæ,* n. sp., 1 fig. *O. agassizii,* n. sp. LINCKIA GRAY. *L. nodosa* E. PERR. FROMIA GRAY. *F. japonica,* n. sp., 1 fig. GYMNASTERIDÆ. MARGINASTER, n. g. *M. pectinatus* E. PERR., 2 figs. *M. echinulatus* E. PERR., 2 figs. PENTAGONASTERIDÆ. PENTAGONASTER. *P. parvus* E. PERR., 2 figs. *P. grenadensis* E. PERR., 1 fig. *P. ternalis* E. PERR., 1 fig. *P. subspinosus,* n. sp., 1 fig. *P. arenatus,* n. sp. 2 figs. *P. alexandri,* n. sp., 6 figs. *P. dentatus* E. PERR., 1 fig. *P. affinis* E. PERR., 1 fig. *P. intermedius,* 1 fig. GONIODISCUS M. TR. *G. pedicellaris,* n. sp., 1 fig. ANTHENOIDES, n. g. *A. peircei* E. PERR., 1 fig. ARCHASTERIDÆ. GONIOPECTEN, n. g. *G. demonstrans,* n. sp., 1 fig. *G. intermedius,* n. sp., 3 figs. *G. subtilis,* n. sp., 2 figs. ARCHASTER, M.

Tr. *A. pulcher*, n sp, 1 fig *A. efflorescens*, n sp *A. insignis*, n. sp., 1 fig. *A. mirabilis*, n sp, 5 figs. *A. coronatus*, n. sp. *A. echinulatus* E. Perr., 1 fig *A. simplex*, n. sp., 1 fig Blakiaster, n. g *B. conicus*, n. sp.. 1 fig. Astropectinide Luidia Forbes. *L. barbalaevis*, n sp, 1 fig *L. convexiuscula*, n. sp., 1 fig. *L. elegans* E Perr. *L. alternata* Say. *L. clathrata* Say. Astropecten Linck. *A. articulatus* Say. *A. alligator*, n. sp.

'94. Échinodermes. Expéditions scientifiques du Travailleur et du Talisman pendant les années 1880, 1881, 1882, 1883. 431 pp. + 26 pls.

Brisingidæ. Brisinga. *B. endecacnemos* Asbjörnsen '56, 1 fig. *B. hirsuta*, n. sp, 3 figs *B. coronata* O. Sars '71. *B. mediterranea* Perrier '81. Odinia. *O. elegans* Perrier = *Brisinga elegans* Perrier '85, 1 fig. *O. semi-coronata* Perrier '85, 4 figs. *O. robusta* Perrier = *Brisinga robusta* Perrier '85, 5 figs. Freyella. *F. edwardsi* Perrier = *Brisinga edwardsi* Perrier '82, 1 fig. *F. spinosa* Perrier, 7 figs. *Freyella sexradiata* Perrier '85, 1 fig. Pedicellasteridæ. Coronaster. *C. parfaiti* Perrier, 1 fig. *C. antonii* Perrier., 1 fig. Lytaster, n g *L. inæqualis* Perrier, 1 fig. Pedicellaster. *P. sexradiatus* Perrier '82, 1 fig Gastraster, n. g. *G. margaritaceus* Perrier = *Pedicellaster margaritaceus* Perrier '82, 1 fig. Asteriidæ. Stolasterias. *St. glacialis* Linck. Hydrasterias. *H. richardi* Perrier, 1 fig. Zoroasteridæ. Zoroaster. *Z. fulgens* Jeffreys. *Z. ackleyi* Perrier '81, Prognaster, n. g *P. longicauda*, n. sp., 1 fig. Mammaster, n. g *M. sigsbeei* Perrier '81, Stichasteridæ. Neomorphaster Sladen = Glyptaster Sladen '85 = Stichaster (part) Perrier '85, *N. talismani* = *Stichaster talismani* Perrier '85, 1 fig. Echinasteridæ Cribrella. *C. abyssalis*, n. sp., 1 fig. Echinaster. *E. sepositus* Retzius. Solasteridæ. (Trib. Solasterinæ.) Ctenaster. *Ct. spectabilis* Perrier '81, (Trib. Korethrasterinæ.) Korethraster. *K. setosus*, n. sp., 1 fig. *K. (Remaster) palmatus* Perrier '84, Poraniidæ. Marginaster. *M. pentagonus* Perrier '82, 1 fig. *M. pectinatus* Perrier '84, *M. echinulatus* Perrier '84, Ganeriidæ. Radiaster. *R. elegans* Perrier '81, Myxasteridæ. Myxaster, n. g. *M. sol*, n. sp., 1 fig. Pterasteridæ. Pteraster. *Pt. sordidus*, n. sp., 1 fig. *Pt. alveolatus*, n. sp., 1 fig. Hymenaster. *H. rex*, n. sp, 1 fig. *H. giboryi*, n. sp, 1 fig, Cryptaster, n. g. *C. personatus*, n. sp., 1 fig. Astropectinidæ. (Trib. Luidiinæ.) Astrella. *A. simplex* Perrier '82, 1 fig. Luidia. *L. sarsii* Düben & Koren. (Trib. Astropectinæ.) Psilaster. *Ps. andromeda* M. T. '42, Astropecten. *A. ibericus*, n. sp., 1 fig. Porcellanasteridæ. Caulaster. *C. pedunculatus* Perrier, 1 fig. *C. sladeni* Perrier, 1 fig. Porcellanaster. *P. inermis* Perrier, 1 fig *P. granulosus* Perrier, 1 fig. Styracaster. *St. edwardsi* Perrier, 1 fig. *St. spinosus*, n. sp, 1 fig. Hyphalaster. *H. parfaiti* Perrier, 1 fig *H. antonii* Perrier, 1 fig. Pseudaster. *Ps. cordifer* Perrier, 1 fig.

ARCHASTERIDÆ. (Trib. *Pararchasterinæ*) PARARCHASTER *P. simplex* = *Archaster simplex* PERRIER '81· *P. folini* PERRIER = *Cheiraster folini* PERRIER '85· 2 figs. *P. fischeri*, n sp, 1 fig. CHEIRASTER *Ch. coronatus* PERRIER = *Archaster mirabilis* (part) PERRIER '81· *Ch. vincenti* PERRIER = *Archaster mirabilis* (part) PERRIER '81· *Ch. mirabilis* PERRIER '81· *Ch. echinulatus* = *Archaster echinulatus* PERRIER '82· PECTIN-ASTER. *P. filholi* PERRIER '85· 1 fig PONTASTER *P. venustus* SLADEN '88 *Pontaster perplexus*, n sp, 1 fig *P. marionis* PERRIER = *Crenaster marionis* PERRIER '85· 1 fig *P. oligoporus* PERRIER = *Archaster mirabilis* (part) PERRIER '81 GONIOPECTEN. *G. demonstrans* PERRIER '81· DYTASTER *D. insignis* = *Archaster insignis* PERRIER '84· *D. agassizii*, n sp. = *Pectinaster insignis* (part) PERRIER '85· 1 fig. *D. rigidus*, n. sp. = *Archaster rigidus* PERRIER '84 (in FILHOL: La vie au fond des mers). CRENASTER. *C. semispinosus*, n sp. *C. spinulosus*, n. sp. *C. mollis* PERRIER '85· 1 fig. PLUTONAS-TER. *P. edwardsi* PERRIER = *Goniopecten edwardsi* PERRIER '82· *P. bifrons* = *Archaster bifrons* WYV. THOMSON '73· *P. intermedius* = *Goniopecten intermedius* PERRIER '81· *P. notatus* SLADEN '89· *P. inermis* PERRIER = *Goniopecten inermis* PERRIER '85· *P. efflorescens* PERRIER = *Archaster efflorescens* PERRIER '84· *P. pulcher* = *Archaster pulcher* PERRIER '84· TETHYASTER *T. subinermis* SLADEN '89 = *Asterias subinermis* PHILIPPI '37 = *Astropecten subinermis* M. T. '42· (Trib. *Gnathasterinæ*.) HOPLASTER. *H. spinosus* PERRIER '82· 1 fig LINCKIIDÆ. CHÆTASTER *Ch. longipes* M. T. OPHIDIASTER. *O. ophidianus* LAMARCK. NARCISSIA. *N. canariensis* = *Asterias canariensis* D'ORBIGNY '39 = *N. teneriffæ* GRAY '40· FROMIA. *F. narcissia*, n. sp. PENTAGONASTERIDÆ. (Trib. *Astrogoninæ*.) ASTROGONIUM. *A. annectens* PERRIER, 2 figs. *A. hystrix* PERRIER, 2 figs. *A fallax* PERRIER '85· 2 figs. *A. necator* PERRIER, 1 fig. PARAGONASTER SLA-DEN '85 = GONIOPECTEN (part.) PERRIER '81 = ASTROGONIUM (part.) PERRIER '85· *P. subtilis* = *Goniopecten subtilis* PERRIER '81· 2 figs. *P. elongatus* = *Astrogonium elongatum* PERRIER '85· 2 figs. *P. strictus* PERRIER, 1 fig. DORIGONA GRAY '66 = PENTAGONASTER (subg DORIGONA) PERRIER '82 = NYMPHASTER SLADEN '85 *D. ternalis* = *Pentagonaster ternalis* PERRIER '81· *D. subspinosa* = *Pentagonaster subspinosa* PERRIER '81· *D. arenata* PERRIER = *Pentagonaster arenatus* PERRIER '81· 1 fig *D. jacqueti*, 2 figs. ROSASTER, n. g. *R alexandri* = *Pentagonaster alexandri* PERRIER '81· PHANERASTER, n. g. *Ph. semilunatus* LINCK = *Pentagonaster semilunatus* LINCK 1733 = *Asterias tessellata* LAMARCK '16 = *Goniaster cuspidatus* GRAY '40 = *Astrogonium cuspidatum* M. T. '42· PENTAGONAS-TER. *P. perrieri* SLADEN = *P. grandis* PERRIER '85· *P. vincenti*, n sp, 1 fig. *P. hæsitans*, n. sp, 2 figs *P. gosselini*, n. sp, 1 fig *P. granularis* var. *deplasi* = *Pentagonaster deplasi* PERRIER '85· 1 fig. STEPHANASTER *St. bourgeti*, n. sp., 1 fig. PENTACEROTIDÆ. *P. dorsatus* (LINNÉ) PERRIER

PETERS, W.

* '52· Uebersicht der Seesterne von Mosambique. Monatsber. Akad. Berl, p. 177–178. Cited after PERRIER, DE LORIOL & LUDWIG [BRONN.]

PFEFFER, G.

'94. Fische, Mollusken und Echinodermen von Spitzbergen, gesammelt von Herrn Prof. W. KÜKENTHAL im Jahre 1886. Zool. Jahrb., Abth. Syst. Geogr. Biol., Bd. viii, p. 91–99.

Ctenodiscus corniculatus LINCK. *Stichaster albulus* STIMPSON. *Crossaster papposus* FABRICIUS. *Solaster endeca* RETZIUS.

'94a. Echinodermen von Ost-Spitzbergen nach der Ausbeute der Herren Prof. W. KÜKENTHAL und Dr. ALFR. WALTER im Jahre 1889. Zool. Jahrb., Abth. Syst. Geogr. Biol., Bd. viii, p. 100–127.

Pontaster tenuispinus DUB. & KOREN = *P. tenuispinus* var. *platynotus* SLADEN = *P. hebitus* SLADEN = *P. limbatus* SLADEN. *Ctenodiscus corniculatus* LINCK = *C. crispatus* RETZIUS. *Rhegaster tumidus* STUXBERG. *Stichaster albulus* STIMPSON = *Asterias problema* STEEN-STRUP. *Crossaster affinis* BRANDT. *Solaster endeca* RETZIUS. *Lophaster furcifer* DÜB. & KOREN. *Pteraster militaris* O. F. MÜLL. *Pteraster pulvillus* M. SARS. *Hymenaster pellucidus* WYV. THOMSON. *Cribrella oculata* LINCK = *C. sanguinolenta* O. F. MÜLL. *Asterias groenlandica* LÜTKEN. *Asterias stellionura* PERRIER. *Asterias gunneri* DANIEL. & KOREN.

* '96· Ostafrikanische Echiniden, Asteriden und Ophiuriden, gesammelt von Herrn F. STUHLMANN im Jahre 1888 und 1889. Mitteil. Mus. Hamburg, xiii, p. 43–48. Cited after LUDWIG.

: 00. Echinodermen von Ternate. Echiniden, Asteriden, Ophiuriden u. Comatuliden. Ergebn. zoolog. Forschungsreise in d. Molukken u. Borneo, W. KÜKENTHAL. Abh. Senckenberg. Ges., xxv, p. 81–86. Holothurien von v. MARENZELLER.

Archaster typicus M. T. *Pentaceros turritus* LINCK. *Goniodiscus pleiadella* LAM. *Asterina cepheus* VAL. *Asterina exigua* LAM. *Fromia variolaris* LINCK. *Linckia miliaris* LINCK. *L. multiforis* LAM. *Nardoa tuberculata* GRAY. *Echinaster eridanella* VAL.

REIN, J. J.

: 05. Japan nach Reisen und Studien im Auftrage der Köngl. preuss. Regierung dargestellt. I. Bd. Natur u. Volk des Mikadoreiches. 2. neu bearb. Aufl.

RETZIUS, A. J.

* '05· Dissertatio sistens species cognitas Asteriarum. Cited after MÜLLER and TROSCHEL.

Ruijs, J. M.

* '87· Zoologische Bijdragen tot de kennis der Karazee (Nederl. Pool-Exp. 1882–1883). Bijdragen tot de Dierkunde, Afl. 14. Cited after Ludwig.

Russo, A.

* '94· Echinodermi raccolti nel Mar Rosso dagli Ufficiali della R. marina italiana. Boll. Soc. Napoli, vii, p. 159–163.

Sabine.

* '24· Supplement to the Appendix of Captain Parry's Voyage.

Sars, M.

* '50· Beretning om en i Sommeren 1849 foretagen zoologisk Reise i Lofoten og Finmarken. Nyt Mag. for Naturvidensk., Bd. 6, Hft. 2, p. 121–211. Cited after Ludwig.

* '56. Fauna littoralis Norvegiæ. 2. Hft. Cited after Sladen.

* '61. Oversigt af Norges Echinodermer. 16 pls. Cited after Duncan and Sladen.

* '65· Bemærkninger over det dyriske Livs Udbredning i Havets Dybder. Forhandl. Vidensk. Selsk. Christiania, Aar 1846, p. 53–68. Cited after Ludwig.

* '69· Fortsatte Bemærkninger over det dyriske Livs Udbredning i Havets Dybder. Forhandl. Vidensk. Selsk. Christiania, Aar 1868, p. 246–275. Cited after Ludwig.

Savigny, J. C. de.

* '03· Planches d'Échinodermes d'Egypt. Cited after de Loriol.

Schmidt, J.

* : 05. Fiskeriundersøgelser ved Island og Færøerne i Sommeren 1903. Skrift. Komm. f. Havundersøgelser, no. 1. vi+148 pp.+10 charts.

SIMPSON, J. J., & BROWN, R.N.R.

: 10.   Asteroidea of Portuguese East Africa, collected by JAS. J. SIMPSON, M. A., B. Sc. (1907–1908). Proc. Roy. Phys. Soc. Edinburgh, vol. 18, p. 45–60. Feb. 18.

Archaster angulatus M. & T.   *Astropecten hemprichii* M. & T.   *Astropecten polyacanthus* M. & T.   *Luidia maculata* M & T.   *Luidia aspera* SLADEN.   *Stellaster incei* GRAY. *Anthenea sp.?* *Pentaceros linrki* DE BLAINV. *Pentaceros superbus* MÖBIUS. *Pentaceros gracilis* LÜTKEN. *Pentaceros sp.* *Pentaceros sp.* *Culcita schmideliana* (RETZ) *Culcita sp.* *Ophidiaster cylindricus* (LAM.) M. & T.   *Linckia diplax* (M. & T.). *Linckia marmorata* (MICHELIN), 4 figs   *Nardoa variolata* GRAY   *Retaster cribrosus* (v. MART.).

SLADEN, W. P.

'79.   On the Asteroidea and Echinoidea of the Korean Seas. Journ. Linn. Soc. London, vol. xiv, p. 424–445, pl. 8.

Astropecten formosus, n. sp , 4 figs.   *Astrop. japonicus* M. T.   *Astrop. polyacanthus* (of *A. armatus*-type) M. T.= *Astrop. hystrix* VALEN. MS.= *Astrop. armatus* M. T. '42 = *Astrop. vappa* M. T. '43.   *Stellaster belcheri* GRAY '47= *Goniaster* (*Stellaster*) *belcheri* v. MARTENS '66= *Pentagonaster* (*Stellaster*) *belcheri* '76.   *Cribrella densispina*, n. sp., 5 figs. *Asteracanthion rubens* LINNÉ var. *migratum* mihi.

'83.   The Asteroidea of H.M.S. 'Challenger' Expedition. Part II. Journ. Linn. Soc. London, Zool., vol. xvii, p. 214–269.

ASTROPECTINIDÆ.   Subfam. *Porcellanasteridæ.* PORCELLANASTER. *P. cœruleus*, WYV. THOMSON. *P. caulifer*, n. sp. *P. tuberosus*, n. sp. *P. crassus*, n. sp. *P. gracilis*, n. sp. STYRACASTER, n. g. *St. horridus*, n. sp. *St. armatus*, n. sp. HYPHALASTER, n. g. *H. hyalinus*, n. sp. *H. diadematus*, n. sp. *H. inermis*, n. sp. *H. planus*, n. sp. THORACASTER, n. g *Th. cylindratus*, n. sp. *Astropectinidæ.* ASTROPECTEN LINCK. *A. brasiliensis*, M. T. *A. brevispinus*, n. sp. *A. polyacanthus* M. T. *A. pectinatus*, n. sp. *A. acantháfer*, n. sp. *A. japonicus* M. T. *A. imbellis*, n. sp. *A. hermatophilus*. n. sp. *A. pontoporæus*, n. sp. *A. zebra*, n. sp. *A. zebra* var. *rosea*. *A. velitaris* v. MARTENS. *A. monacanthus*, n. sp. *A. cingulatus*, n sp. *A. mesactus*, n. sp.

'89.   Report on the Asteroidea collected by H. S. M. Challenger during the Years 1873–1876. Report on the Scientific Results of the Voyage of H. M. S. Challenger during the Years 1873–76 under the Command of Capt. GEORGE S. NARES, R. N., F. R. S. and the late Capt. FRANK TOURLE THOMSON, R. N., prepared under the

Superintendence of the late Sir C. WYVILLE THOMSON, Knt., F.R.S., &c. and now of JOHN MURRAY, LL.D., Ph. D., &c. Zoology, vol. xxx. xlii + 893 pp. + cxvii pls. + 1 chart.

ARCHASTERIDÆ. Subf. *Pararchasterinæ.* PARARCHASTER *P. semisquamatus,* n. sp, 4 figs. *P. semisquamatus* var. *occidentalis,* n. *P. antarcticus,* n. sp. *P. spinosissimus,* n. sp, 4 figs. *P. pedicifer,* n. sp., 4 figs *P. armatus,* n. sp., 4 figs. PONTASTER. *P. tenuispinus* DÜBEN & KOREN = *Astropecten tenuispinus* DÜBEN & KOREN '46 = *Archaster tenuispinus* SARS '61· *P. tenuispinus* var. *platynota,* n , 3 figs. *P. planeta,* n. sp , 4 figs. *P. hebitus,* n. sp , 4 figs. *P. limbatus,* n sp , 4 figs. *P. oxyacanthus,* n. sp.. 4 figs. *P. teres,* n. sp., 4 figs. *P. forcipatus,* n sp., 4 figs. *P. forcipatus* var *echinata,* n. *P. mimicus,* n. sp., 4 figs. *P. pristinus,* n. sp , 4 figs. *P. venustus,* n. sp., 4 figs. *P. venustus* var. *robusta,* n. *P. trullipes.* n sp., 4 figs. *P. subtuberculatus,* n. sp., 4 figs. Subf. *Plutonasterinæ.* DYTASTER. *D spinosus,* n. sp., 4 figs. *D. exilis,* n. sp., 4 figs. *D. exilis* var. *gracilis* n., 4 figs. *D. exilis* var. *carinata,* n. *D. madreporifer,* n. sp., 4 figs. *D. nobilis,* n. sp., 4 figs. *D. æquivocus,* n. sp., 5 figs. *D. biserialis,* n. sp., 4 figs. *D. inermis,* n, sp., 4 figs. PLUTONASTER. *P. bifrons* WYV. THOMSON = *Archaster bifrons* WYV. THOMSON '73· 6 figs. *P. marginatus,* n. sp.. 4 figs *P. rigidus,* n. sp., 4 figs. *P. rigidus* var. *semiarmata,* n., 1 fig. *P. ambiguus,* n. sp , figs *P. notatus,* n. sp., 4 figs. *P. abbreviatus,* n sp. Subg. Tethyaster, n. *T. subinermis* PHILIPPI = *Asterias subinermis* PHILIPPI '37 = *Astropecten subinermis* M. T. '42 = *Archaster subinermis* PERRIER '78 *T. parelii* DÜBEN & KOREN = *Astropecten parelii* DÜBEN & KOREN '46 = *Archaster parelii* SARS '61· LONCHOTASTER. *L. tartareus,* n. sp., 5 figs. *L. forcipifer,* n. sp., 4 figs. Subf. *Pseudarchasterinæ.* PSEUDARCHASTER. *P. discus,* n. sp., 4. figs. *P. tessellatus,* n sp., 4 figs. *P. intermedius,* n. sp , 4 figs. APHRODITASTER. *A. gracilis,* n. sp., 4 figs. Subf. *Archasterinæ.* ARCHASTER. *A. typicus* M. T. '40 (April) = *Astropecten stellaris* GRAY '40 (Nov.) = *A. nicobaricus* MÖBIUS (BEHN) '59· PORCELLANASTERIDÆ. Subf. *Porcellanasterinæ.* PORCELLANASTER. *P. caruleus* WYV. THOMSON '77· 7 figs. *P. caulifer* SLADEN '83· 10 figs. *P. tuberosus* SLADEN '83· 8 figs. *P. crassus* SLADEN '83· 8 figs. *P. gracilis* SLADEN '83· 7 figs. *P. eremicus,* n. sp., 1 fig. STYRACASTER. *S. horridus* SLADEN '83· 7 figs. *S. armatus* SLADEN '83· 8 figs HYPHALASTER. *H. hyalinus* SLADEN '83· 10 figs. *H. diadematus* SLADEN '83· 8 figs. *H. inermis* SLADEN '83· 7 figs. *H. planus* SLADEN '83· 7 figs. THORACASTER. *T. cylindratus* SLADEN '83· 6 figs. Subf. *Ctenodiscinæ.* CTENODISCUS. *Ct. corniculatus* (LINCK) PERRIER '76 = *Astropecten corniculatus* LINCK 1733 = *Asterias crispata* RETZIUS '05 = *Asterias polaris* SABINE '21 = *Asterias arancia* DEWHURST '34 = *Astropecten polaris* GRAY '40 = *Ctenodiscus polaris* M. T. '42 = *Ct. pygmæus* M. T. '42 = *Ct. crispatus* DÜBEN & KOREN '46 = *Anodiscus crispatus* (? VAL. MS.) PERRIER '69· *Ct. australis* LÜTKEN (LOVÉN MS.) '71· 7 figs. *Ct. procurator,* n. sp , 6 figs. ASTROPECTINIDÆ. Subf. *Astropectininæ.* CRASPIDASTER, n. g *C. hesperus* M. T. = *Archaster*

hesperus M. T. '40 = *Stellaster sulcatus* Mōbius '59, 7 figs. LEPTOPTYCHASTER = LEPTYCH-
ASTER SMITH '76· *L. kerguelensis* SMITH '76 (Feb.) = *Archaster excavatus* WYV. THOM-
SON '76 (Dec.), 4 figs. *L. arcticus* SARS = *Astropecten arcticus* SARS '51 = *Astropecten
lütkeni* BARRETT '57 = *Archaster arcticus* PERRIER '78· *L. arcticus* var. *elongata*, n. *L.
antarcticus*, n. sp., 4 figs. MOIRASTER, n. g. *M. magnificus* BELL = *Archaster magnificus*
BELL '81· ASTROPECTEN = STELLARIA NARDO '34 = ASTERIAS AGASSIZ '35 = CRENASTER
D'ORBIGNY '50· *A. brasiliensis* M. T. '42· *Astropecten brevispinus* SLADEN '83· 5 figs.
A. polyacanthus M. T. '42 = *A. hystrix* M. T. '42 = *A. armatus* M. T. '42 = *A. vappa*
M. T. '43 = *A. chinensis* GRUBE '65 = *A. ensifer* GRUBE '65· *A. pectinatus* SLADEN '83·
5 figs. *A. acanthifer* SLADEN '83· 5 figs. *A. japonicus* M. T. '42· *A. imbellis* SLADEN
'83· 5 figs. *A. hermatophilus* SLADEN '83· 5 figs. *A. irregularis* LINCK 1733 = *Asterias
aranciaca* O. F. MÜLLER 1776 = *A. mulleri* M. T. '44 = *A. echinulatus* M. T. '44· *A.
pontoporaus* SLADEN '83· 5 figs. *A. zebra* SLADEN '83· 5 figs. *A. velitaris* VON MARTENS
'65· *A. granulatus* M. T. '42· 5 figs. *A. monacanthus* SLADEN '83· 5 figs. *A. cingulatus* SLADEN '83· 5 figs. *A. mesactus* SLADEN '83· 5 figs. PSILASTER. *Ps. andromeda*
M. T. '42 = *Asterias aranciata (varietas)* PARELIUS 1768 = *Astropecten christi* DÜBEN &
KOREN '44 = *Archaster andromeda* MÖBIUS & BÜTSCHLI '75 = *Archaster christi* PERRIER
'76 = *Goniopecten christi* PERRIER '85· *Ps. acuminatus*, n. sp., 4 figs. *Ps. cassiope*, n.
sp, 4 figs. *Ps. gracilis*, n. sp., 5 figs. *Ps. patagiatus*, n. sp., 4 figs. PHOXASTER.
Ph. pumilus, n. sp., 9 figs. BATHYBIASTER. *B. loripes*, n. sp., 4 figs. *B. loripes* var.
obesa, n. *B. vexillifer* WYV. THOMSON = *Archaster vexillifer* WYV. THOMSON '73· Subf.
Luidiinæ. LUIDIA. *L. aspera*, n, sp., 4 figs. *L. limbata*, n. sp., 4 figs. *L. clathrata*
(SAY) LÜTKEN '60 = *Asterias clathrata* SAY '25· *L. ciliaris* (PHILIPPI) GRAY '40 = *Asterias
rubens* JOHNSTON '36 = *Asterias ciliaris* PHILIPPI '37 = *L. fragilissima* (part) FORBES '39
= *Asterias pectinata* COUCH '40 = *Asterias imperati* DELLE CHIAJE '41· *L. longispina*, n.
sp, 4 figs. *L. africana*, n. sp., 4 figs. *L. sarsii* DÜBEN & KOREN '44 = *Asterias* n. sp.
SARS '35 = *L. fragilissima* (part) FORBES '41 = *Luydia savignyi* (part) DÜBEN & KOREN
'46· *L. forficifer*, n. sp., 4 figs. PENTAGONASTERIDÆ. Subf. *Pentagonasterinæ*. PENTA-
GONASTER LINCK 1733 = GONIASTER (part) AGASSIZ '35 = ASTROGONIUM (part) M T.
'42 = GONIODISCUS (part) M. T. '42 = HOSIA (part) GRAY '40 = TOSIA GRAY '40· *P.
semilunatus* LINCK 1733 = *Asterias granularis* (part) GMELIN 1789 = *Asterias tessellata*
(part) LAMARCK '16 = *Goniaster cuspidatus* GRAY '40 = *Astrogonium cuspidatum* M. T.
'42 = *Goniaster semilunatus* v. MARTENS '66 = *Astrogonium dubium* PERRIER '69 = *Goniaster
americanus* VERRILL '71 = *Goniaster africanus* VERRILL '71 = *Astrogonium semilunatum*
PERRIER '85· *P. granularis* RETZIUS = *Asterias granularis* RETZIUS 1783 = *Asterias
granularis* ABILDGAARD 1789 = *Asterias tessellata* (part) LAMARCK '16 = *Astrogonium
granulare* M. T. '42 = *Astrogonium boreale* BARRETT '57 = *Goniaster granularis* LÜTKEN
'65 = *Pentagonaster (Astrogonium) granularis* PERRIER '76· *P. astrologorum* (M. T.)
PERRIER '76 = *Astrogonium astrologorum* M. T. '42· *P. patagonicus*, n. sp , 4 figs.

P. japonicus, n. sp., 4 figs. *P. lepidus*, n. sp , 4 figs. *P. arcuatus*, n. sp , 4 figs. CALLIASTER. *C. baccatus*, n. sp , 4 figs. CHITONASTER *Ch. cataphractus*, n. sp., 3 figs. GNATHASTER, n. g. *Gn. meridionalis* SMITH = *Astrogonium meridionale* SMITH '76 = *Pentagonaster meridionalis* SMITH '79, 6 figs. *Gn. elongatus*, n. sp., 12 figs. *Gn. pilulatus*, n. sp., 3 figs. NYMPHASTER. *N. symbolicus*, n. sp., 4 figs. *N. bipunctus*, n. sp., 4 figs. *N. protentus*, n. sp., 4 figs. *N. albidus*, n. sp., 4 figs. *N. basilicus*, n. sp , 2 figs PARAGONASTER. *P. ctenipes*, n. sp., 4 figs. *P. cylindratus*, n. sp., 4 figs. NECTRIA GRAY '40 = GONIODISCUS M. T. (part) '42. *N. ocellifera* (LAMARCK) GRAY '40 = *Asterias ocellifera* LAMARCK '16 = *Goniodiscus ocelliferus* M. T. '42 = *Chætaster munitus* MÖBIUS '59, 7 figs. Subf. *Goniodiscinæ.* STELLASTER GRAY '40 = GONIASTER (subg. STELLASTER) v. MARTENS '65 = PENTAGONASTER (subg. STELLASTER) PERRIER '76. *St. incei* GRAY '47 = *St. belcheri* GRAY '47 = *St. gracilis* MÖBIUS '59. *St. princeps*, n. sp., 2 figs. LEPTOGONASTER. *L. cristatus*, n. sp., 7 figs. Subf. *Mimasterinæ.* MIMASTER. *M. tizardi* SLADEN '82. *M. cognatus*, n. sp., 4 figs. ANTHENEIDÆ. ANTHENEA GRAY '40 = HOSIA (part) GRAY '40 = GONIODISCUS (part) M. T. '42 = GONIASTER (part) v. MARTENS '65. *A. acuta* PERRIER '76 = *Goniodiscus acutus* PERRIER '69. *A. tuberculosa* GRAY '47. *A. tuberculosa* GRAY (?) juv., 4 figs. HIPPASTERIA GRAY = GONIASTER (part) AGASSIZ '35 = ASTROGONIUM (part) M. T. '42. *H. plana* (LINCK) GRAY '40 = *Pentaceros planus* LINCK 1733 = *Asterias equestris* (part) LINNÉ 1766 = *Asterias phrygiana* PARELIUS 1768 = *Asterias johnstoni* GRAY (in JOHNSTON) '36 = *H. europæa* GRAY '40 = *H. johnstoni* GRAY '40 = *H. cornuta* GRAY '40 = *Goniaster equestris* FORBES '41 = *Astrogonium phrygianum* M. T. '42 = *Goniaster abbensis* FORBES '43 = *Astrogonium aculeatum* BARRETT '57 = *Goniaster phrygianus* NORMAN '65 = *Hippasteria phrygiana* VERRILL '85. PENTACEROTIDÆ. PENTACEROS LINCK 1733 = GONIASTER (part) AGASSIZ '35 = OREASTER M. T. '42. *P. dorsatus* (LINNÉ) PERRIER '76 = *Asterias dorsatus* LINNÉ 1753 = *Asterias nodosa* (part) LINNÉ 1766 = *Oreaster clavatus* M. T. '42 = *O. dorsatus* LÜTKEN '65. *P. turritus* LINCK 1733 = *Asterias nodosa* LINNÉ (part) 1766 = *P. franklinii* GRAY '40 = *O. turritus* M. T. '42 = *P. modestus* GRAY '66. *P. productus* BELL '84 var. *tuberata*, n. *P. callimorphus*, n. sp. PENTACEROPSIS, n. g. *P. obtusatus* BORY DE SAINT VINCENT = *Asterias obtusatus* B. S. V. '27 = *O. obtusatus* M. T. '42. CULCITA. *C. novæ-guineæ* M. T. '42 = *C. pulverulenta* (VALENCIENNES MS.) PERRIER '69. ASTERODISCUS. *A. elegaus* GRAY '47. CHORIASTER. *Ch. granulatus* LÜTKEN '69. GYMNASTERIIDÆ. GYMNASTERIA. *G. carinifera* (LAMARCK) VON MARTENS '67 = *Asterias carinifera* LAMARCK '16 = *Asterope carinifera* M. T. '40 = *G. spinosa* GRAY '40 = *G. inermis* GRAY '40 = *Asteropsis carinifera* M. T. '42 = ? *G. biserrata* v. MARTENS '66. PORANIA GRAY '40 = GONIASTER (part) FORBES '39 = ASTEROPSIS (part) M. T. '42. *P. pulvillus* (O. F. MULLER) NORMAN '65 = *Asterias pulvillus* O. F. MÜLLER 1788 = *Goniaster templetoni* FORBES '39 = *Porania gibbosa* GRAY '40 = *Asteropsis pulvillus* M. T. '42 = *Asteropsis ctenicantha* M. T. '42. *P. antarctica* SMITH '76, 1 fig. *P. glaber*, n. sp , 2 figs. *P. spiculata*, n. sp., 1 fig. *P. magellanica*

Studer '76 =? *P. patagonica* Perrier '78 = *P. nanjelharnica* Studer '84· Marginaster *M. fimbriatus*, n. sp . 3 figs Rhegaster. *R. murrayi* Sladen '83· Lasiaster, n. g *L. villosus*, n. sp., 4 figs. Asterinidæ. Subf. *Ganeriinæ*. Cycethra *C. electilis*, n sp., 4 figs. *C. nitida*, n. sp , 4 figs. *C. pinguis*, n. sp., 4 figs. Ganeria. *G. falklandica* Gray '47· 4 figs. Subf. *Asterininæ*. Patiria. *P. bellula*, n. sp., 4 figs. Nepanthia Gray '40 = Asterina (subg. Nepanthia) Perrier '76· *N. brevis* Perrier = *Asterina (Nep.) brevis* Perrier '76· 3 figs. *N. maculata* Gray '40 = *Chætaster*(?) *maculatus* M.T. '42 = ?*Chætaster cylindratus* Möbius '59 = *Asterina (Nep) maculata* Perrier '76· Asterina. *A. regularis* Verrill '71 = *A. gunnii* var. Gray '40 = *Asteriscus australis* (part) M. T. '42 = *A. (Asteriscus) regularis* Verrill '71 = *A. cabbalistica* Lütken '71· *A. exigua* (Lamarck) Perrier '76 = *Asterias exigua* Lamarck '16 = *Asterias minuta* de Blainville '34 = *A. kraussii* Gray '40 = *Asteriscus pentagonus* M. T. '42 = *A. pentagona* v. Martens '66· *A. gunnii* Gray '40 = *Asterias calcar* var. *b* Lamarck '16 = *Asteriscus australis* (part) M. T. '42 = *Asteris. diesingi* M. T. '42 = *Asteris. calcar* Dujardin & Hupé '62 = *Asteris. exiguus* Perrier '69· *A. pectinifera* (M. T.) v. Martens '65· *A. folium* (Lütken '60) Agassiz '77· *A. cepheus* (M. T. '42) v. Martens '66 = *A. burtonii* Gray '40· *A. penicillaris* Lamarck v. Martens '66 = *Asterias penicillaris* Lamarck '16· Subf. *Palmipedinæ*: Palmipes Linck 1733 = Anseropoda Nardo '34 = Asteriscus (part) M. T. '40· *P. membranaceus* Linck 1733 = *Stella cartilaginea* Aldrovandus 1602 = *Asterias placenta* Pennant ,1777 = *Asterias membranacea* Retzius 1783 = *Asterias cartilaginea* Fleming '28 = *Anseropoda membranacea* Nardo '34 = *Asteriscus palmipes* M. T. '42 = *P. placenta* Norman '65· *P. diaphanus*, n. sp. Linckiidæ. Subf. *Chætasterinæ*. Chætaster M. T. '40 = ?Astropus (subg.) Gray '40 = Nepanthia (part) Gray '40· *Ch. longipes* (Retzius) Sars '57 = *Asterias longipes* Retzius '05 = *Asterias subulata* Lamarck '16 = *Chætaster subulata* M. T. '40 = ?*Astropus (Astropus) longipes* Gray '40 = *Nepanthia tessellata* Gray '40· Subf. *Linckiinæ*. Fromia = Linckia (part) M. T. '40 = Scytaster (part) M. T. '42· *F. milleporella* (Lamarck) Gray '40 = *Asterias milleporella* Lamarck '16 = *L. milleporella* M. T. '40 = *Scytaster pistorius* M. T. '42 = *Scyt. milleporellus* Michelin '45 = *Linckia (Scytaster) milleporella* v. Martens '66 = *Linckia pistoria* v. Martens '69· Ophidiaster Agassiz '35 = Dactylosaster Gray '40 = Tamaria Gray '40 = ?Cistina Gray '40 = Hacelia (subg) Gray '40 = Linckia (part) v. Martens '65· *O. attenuatus* Gray '40 = *O. (Hacelia) attenuatus* Gray '40 = *Asterias coriacea* Grube '40· *O. ophidianus* (Lamarck) Agassiz '35 = *Asterias ophidianus* Lamarck '16 = *O. aurantius* Gray '40· *O. cylindricus* (Lamarck) M. T. '40 = *Asterias cylindrica* (part) Lamarck '16 = *Dactylosaster cylindricus* Gray '40 = *Linckia cylindrica* v. Martens '66 = *O. asperulus* Lütken '71· *O. tuberifer*, n. sp., 4 figs. *O. helicostichus* n. sp., 3 figs. Leiaster Peters '52 = Lepidaster Verrill '71 = Ophidiaster (part) Perrier '75· *L. speciosus* v. Martens '66· Linckia Nardo '34 = Ophidiaster (part) M. T. '40 = Acalia (subg.) Gray '40· *L. guildingii* Gray '40· = *Ophidiaster ornithopus*

M. T. '42 = *Scytaster stella* DUCHASSAING '50 = *L. ornithopus* VERRILL '71· *L. miliaris* (LINCK) v. MARTENS '66 = *Pentadactylosaster miliaris* LINCK 1733 = *Asterias lævigata* (part) LINCK 1766 = *L. typus* NARDO '34 = *Ophidiaster lævigata* M. T. '40 = *L. brownii* GRAY '40 = *L. crassa* GRAY '40 = *Ophidiaster milium* is M. T. '42 = *Ophidiaster clathrata* GRUBE '64 = *L. lævigata* LÜTKEN '71· *L. pacifica* GRAY '40 var. *diplax* (M. T.) = = *Ophidiaster diplax* M. T. '42 = ? *Ophidiaster irregularis* PERRIER '69 = *L. nicobarica* LÜTKEN '71· NARDOA GRAY '40 = LINCKIA (part) M. T. '40 = GOMOPHIA GRAY '40 = SCYTASTER (part) M. T. '42 = OPHIDIASTER (part) M. T. '42· *N. tuberculata* GRAY '40 = *Ophidiaster tuberculatus* M. T. '42 = *Scytaster tuberculatus* LÜTKEN '65 = *Linckia tuberculata* v. MARTENS '66· NARCISSIA GRAY '40 = SCYTASTER (part) DUJARDIN & HUPÉ '62· *N. canariensis* D'ORBIGNY = *Asterias canariensis* D'ORBIGNY '39 = *N. teneriffæ* GRAY '40 = *Scytaster canariensis* DUJARDIN & HUPÉ '62· *N. trigonaria*, n. sp, 4 figs. Subf. *Metrodirinæ*. METRODIRA GRAY '40 = SCYTASTER (part) M. T. '42· *M. subulata* GRAY '40 = *Scytaster subulatus* M. T. '42· ZOROASTERIDÆ. ZOROASTER. *Z. fulgens* WYV. THOMSON '73, 4 figs. *Z. tenuis*, n. sp, 4 figs. CNEMIDASTER, n. g. *Cn. wyvillii*, n. sp., 4 figs. PHOLIDASTER. *Ph. squamatus*, n. sp., 4 figs. *Ph. distinctus*, n. sp., 1 fig STICHASTERIDÆ. STICHASTER M. T. '40 = TONIA GRAY '40 = CŒLASTERIAS VERRILL '71 = STEPHANASTERIAS VERRILL '72. *St. aurantiacus* (MEYEN) VERRILL '71 = *Asterias aurantiaca* MEYEN '34 = *St. striatus* M. T. '40 = *Tonia atlantica* GRAY '40 = *Asteracanthion aurantiacus* M. T. '42· *St. roseus* (O. F. MÜLLER) SARS '61 = *Asterias rosea* O. F. MÜLLER 1788 = *Linckia rosea* THOMPSON '40 = *Cribrella rosea* FORBES '41 = *Asteracanthion roseus* (part) M. T. '42 = *St. albulus* (STIMPSON) VERRILL '66 = *Asteracanthion roseus* (part) M. T. '42 = *Asterac. albulus* STIMPSON '53 = *Asterac. problema* STEENSTRUP '55 = *Asterias albua* STIMPSON '63 = *Stephanasterias albula* VERRILL '72· *St. polyplax* M. T. = *Asteracanthion polyplax* M. T. '44· *St. felipes*, n. sp., 4 figs. *St. polygrammus*, n. sp., 5 figs. NEOMORPHASTER, n. g. = GLYPTASTER SLADEN '85· *N. eustichus*, n. sp., 4 figs. TABSASTER, n. g. *T. stoichodes*, n. sp., 4 figs. SOLASTERIDÆ. Subf. *Solasterinæ*. CROSSASTER M. T. '40· *C. papposus* (LINCK) M. T. '40 = *Triskaidecactis papposus* LINCK 1733 = *Asterias helianthemoides* PENNANT 1777 = *Asterias papposa* FABRICIUS 1780 = *Asterias* (*Solasterias*) *papposus* BLAINVILLE '34 = *Stellonia papposa* AGASSIZ '35 = *Solaster papposa* FORBES '39 = *Solaster* (*Polyaster*) *papposa* GRAY '40· *C. papposus* var. *septentrionalis* SLADEN '82· *C. penicillatus*, n. sp., 3 figs. RHIPIDASTER, n. g. *R. vannipes*, n. sp., 4 figs. SOLASTER FORBES '39 = STELLONIA (part) AGASSIZ '35 = CROSSASTER (part) M. T. '40 = SOLASTER (subg. ENDECA) GRAY '40· *S. endeca* (RETZIUS) FORBES '39 = *Asterias aspera* O. F. MÜLLER 1776 = *Asterias endeca* RETZIUS 1783 = *Asterias* (*Solasterias*) *endeca* BLAINVILLE '34 = *Stellonia endeca* AGASSIZ '35 = *Solaster* (*Endeca*) *endeca* GRAY '40· *S. regularis*, n. sp., 3 figs. *S. subarcuatus*, n. sp., 3 figs. *S. torulatus*, n. sp., 4 figs. LOPHASTER. *L. furcifer* (DÜBEN & KOREN) VERRILL '78 = *Chætaster borealis* DÜBEN '44 = *Solaster furcifer* DÜBEN & KOREN '46· *L. stellans*, n. sp., 4 figs. Subf. *Korethras-*

terin.r. KORETHRASTER. *K. hispidus* WYV. THOMSON '73, 4 figs. PERIBOLASTER. *P. folliculatus,* n. sp , 4 figs. PTERASTERIDÆ. Subf. *Pterasterinæ.* PTERASTER. *Pt. militaris* M. T. '42 = *Asteriscus militaris* M. T. '42 = *Asterias militaris* O. F. MÜLLER 1766. *Pt militaris* var. *prolata* SLADEN. *Pt. affinis* SMITH '76· *Pt. rugatus* SLADEN '82, 4 figs. *Pt. stellifer* SLADEN '82, 4 figs *Pt. semireticulatus* SLADEN '82, 4 figs. RETASTER PERRIER '78 = DIPLOPTERASTER VERRILL '80· *R. verrucosus* SLADEN '82, 4 figs. *R peregrinator* SLADEN '82, 4 figs. *R. gibber* SLADEN '82, 4 figs. *R. insignis* SLADEN '82, 4 figs. MARSIPASTER. *M. spinosissimus* SLADEN '82, 5 figs. *M. hirsutus* SLADEN '82, 5 figs. CALYPTRASTER. *C. coa* SLADEN '82, 5 figs HYMENASTER. *H. nobilis* WYV. THOMSON '76, 3 figs. *H formosus* SLADEN '82, 5 figs. *H. pergamentaceus* SLADEN '82, 5 figs. *H. sacculatus* SLADEN '82, 6 figs *H. echinulatus* SLADEN '82, 5 figs. *H. carnosus* SLADEN '82, 5 figs. *H. glaucus* SLADEN '82, 5 figs. *H. vicarius* SLADEN '82, 5 figs. *H. pellucidus* WYV. THOMSON '73, 5 figs. *H. infernalis* SLADEN '82, 4 figs. *H. caelatus* SLADEN '82, 5 figs. *H. crucifer* SLADEN '82, 5 figs. *H anomalus* SLADEN '82, 5 figs. *H. latebrosus* SLADEN '82, 5 figs. *H. porosissimus* SLADEN '82, 5 figs. *H. graniferus* SLADEN '82, 5 figs *H. geometricus* SLADEN '82, 5 figs. *H. pullatus* SLADEN '82, 4 figs. *H. membranaceus* WYV THOMSON '77, 5 figs. *H. coccinatus* SLADEN '82, 5 figs. *H. praecoquis* SLADEN '82, 5 figs. BENTHASTER. *B. wyville-thomsoni* SLADEN '82, 4 figs. *B. penicillatus* SLADEN '82, 4 figs. Subf. *Pythonasterinæ.* PYTHONASTER. *P. murrayi,* n. sp., 5 figs. ECHINASTERIDÆ. Subf. *Acanthasterinæ.* ACANTHASTER GERVAIS '41 = STELLONIA (part) AGASSIZ '35 = ECHINASTER GRAY '40 = ECHINITES M. T. '44· *A. echinites* (ELLIS & SOLANDER) LÜTKEN '71 = *Asterias echinites* ELLIS & SOLANDER 1786 = *Stellonia echinites* AGASSIZ '35 = *Echinaster ellisii* (? part) GRAY '40 = *A. echinus* GERVAIS '41 = *Echinaster solaris* M. T. '42 Subf. *Mithrodiinæ.* MITHRODIA GRAY '40 = HERESASTER MICHELIN '44· *M. clavigera* (LAMARCK) PERRIER '75 = *Asterias clavigera* LAMARCK '16 = *M. spinulosa* GRAY '40 = *Ophidiaster echinulatus* M. T. '42 = *Heresaster papillosus* MICHELIN '44 = *Echinaster echinulatus* v. MARTENS '66 = *M. echinulata* LÜTKEN '71. Subf. *Echinasterinæ.* CRIBRELLA (AGASSIZ) FORBES = PENTADACTYLOSASTER (part) LINCK 1733 = CRIBRELLA (part) AGASSIZ '35 = LINCKIA FORBES '39 = HENRICIA GRAY '40 = CRIBRELLA (part) FORBES '41 = ECHINASTER (part) M. T. '42· *C. oculata* (LINCK) FORBES '41 = *Pentadactylosaster oculatus* LINCK 1733 = *Asterias sanguinolenta* O. F. MÜLLER 1776 = *Asterias pertusa* O. F. MÜLLER 1776 = *Asterias oculata* PENNANT 1777 = *Asterias spongiosa* FABRICIUS 1780 = *Asterias seposita* RETZIUS 1783 = *Linckia oculata* FORBES '39 = *Henricia oculata* GRAY '40 = *Echinaster oculatus* M. T. '42 = *Echin. eschrichtii* M. T. '42 = *Echin. sanguinolentus* SARS '44 = *Echin. sarsii* M. T. '44 = *Linckia pertusa* STIMPSON '53 = *Cribrella sanguinolenta* LÜTKEN '57 = *Crib. eschrichtii* DUJARDIN & HUPÉ '62 = *Crib. sarsii* PERRIER '78· *C. ornata* PERRIER '75 = *Echinaster (Cribrella) ornatus* PERRIER '69· *C. compacta,* n. sp., 4 figs. *C. obesa,* n. sp., 4 figs. *C. praestans,* n. sp., 3 figs. *C. simplex,* n. sp., 4 figs. *C. simplex* var. *granulosa,* n. *C. sufflata,* n. sp., 4 figs. PERKNASTER, n. g. *P. fuscus,* n. sp., 3 figs.

P. densus, n. sp., 4 figs. ECHINASTER M. T. '40 = ECHINASTER LLHUYD 1703 = STELLONIA
(part.) NARDO '34 = OTHILIA GRAY '40 = RHOPIA GRAY '40· *E. spinosus* (RETZIUS) M.
T. '42 = *Pentalactylosaster spinosus regularis* LINCK 1733 = *Asterias spinosa* RETZIUS
'05 = *Asterias echinophora* LAMARCK '16 = *Othilia spinosa* GRAY '40 = *E (Othilia) crassispina*
VERRILL '71 = *E. crassispina* LÜTKEN '71 = *E. echinophorus* PERRIER '75· *E eridanella*
M. T. '42 = *E. affinis* PERRIER '69· *E. spinulifer* SMITH '79 = *Othilia spinulifer* SMITH '76·
HELIASTERIDÆ. HELIASTER. *H. helianthus* (LAMARCK) DUJARDIN & HUPÉ '62 = *Asterias*
helianthus LAMARCK '16 = *Stellonia helianthus* AGASSIZ '35 = *Asteracanthion helianthus* M.
T. '42· PEDICELLASTERIDÆ. PEDICELLASTER *P. scaber* SMITH '76· *P. hypernotius*, n.
sp., 3 figs. ASTERIIDÆ. ASTERIAS LINNÉ 1766 = STELLONIA (part.) NARDO '34 =
URASTER (AGASSIZ) FORBES '39 = ASTERACANTHION (part) M. T. '40 = LEPTASTERIAS
VERRILL '66 = COSCINASTERIAS VERRILL '71 = MARGARASTER HUTTON '72 = MARTHASTERIAS
JULLIEN '78· (*Asterias vera.*) *A. vesiculosa*, n sp., 4 figs *A. meridionalis* PERRIER '75·
A. perrieri SMITH '76 = *Othilia sexradiata* STUDER '76· *A. torquata*, n. sp, 4 figs. *A.*
glomerata, n. sp., 4 figs. *A. rubens* LINNÉ 1766 = *A. glacialis* PENNANT 1777 = *A.*
clathrata PENNANT 1777 = *A. holsatica* RETZIUS '05 = *A. minuta* RETZIUS '05 = *Stellonia*
rubens NARDO '34 = *Uraster rubens* FORBES '41· *A. versicolor*. n. sp., 4 figs. *A. amurensis*
LÜTKEN '76· *A. cunninghami* PERRIER '75· Subf. *Cosmasterias*, n. *A. tomidata*,
n. sp., 3 figs. *A. sulcifera* (VALENCIENNES MS.) PERRIER '69 = *Asterac. sulcifer*
(VALENCIENNES MS.) PERRIER '69 = *A. sulcifer* PERRIER '75· Subg *Smilasterias*, n. *A.*
scalprifera, n. sp , 5 figs. A. *triremis*, n. sp., 4 figs. Subg. *Hydrasterias*, n. *A. ophidion*
n. sp., 4 figs. Subg. *Leptasterias*. *A. mulleri* SARS '44 = *Leptasterias mülleri* VERRILL
'66· *A. compta* STIMPSON '61 = *Leptasterias compta* VERRILL '66· Subg *Stolasterias*, n.
A. gemmifera (VALENCIENNES MS.) PERRIER '69 = *Asterac. gemmifer* PERRIER '69· *A*
tenuispina LAMARCK '16 = *A. savaresi* DELLE CHIAJE '25 = *Echinaster doriæ* FILIPPI '59
= *Echin. tribulus* FILIPPI '59 = A. *glacialis* GRUBE '40 = ? *A. atlantica* VERRILL '71 *A.*
calamaria GRAY '40 = *Coscinasterias muricata* VERRILL '71 = *Asterac. australis* PERRIER '69
= *A. jehensenii* (VALENCIENNES MS) PERRIER '75· *A. colsellata*, n. sp., 4 figs *A stichantha*,
n. sp , 4 figs. *A. eustyla*, n. sp., 4 figs. *A glacialis* O. F. MÜLLER 1776 = *A. spinosa*
1777 = *A. angulosa* O. F MULLER 1788 = *A. echinophora* DELLE CHIAJE '25 = *Stellonia*
glacialis NARDO '34 = *Stellonia angulosa* AGASSIZ '35 = *Stellonia webbiana* D'ORBIGNY '39 =
Uraster glacialis FORBES '41 = *A. madeirensis* STIMPSON '62 = *Marthasterias foliacea*
JULLIEN '78· *A. africana* (M. T. '42) PERRIER '75· CALVASTERIAS *C. stolidota*, n. sp.,
4 figs. BRISINGIDÆ. LABIDIASTER. *L. radiosus* (LOVÉN MS) LÜTKEN '71· 1 fig *L.*
annulatus, n. sp., 1 fig. ODINIA. *O. pandina*, n sp. = *Brisinga coronata* (part) WYV.
THOMSON '73· 5 figs. BRISINGA. *B endecacnemos* ASBJØRNSEN '56 *B coronata* SARS
'71· *B. verticillata*, n. sp., 3 figs. *B. cricophora*, n sp , 3 figs. *B. armillata*, n. sp.,
3 figs. *B. membranacea*, n sp , 2 figs. *B. discincta*, n sp., 4 figs FREYELLA. *F.*
pennata, n. sp., 4 figs. *F. polycnema*, n. sp., 6 figs *F. echinata*, n. sp., 5 figs. *F.*

fragilissima, n. sp, 4 figs   *F. bracteata*, n. sp, 4 figs.   *F. dimorpha*, n. sp, 5 figs.
F. remex, n. sp, 3 figs   *F. tuberculata*, n. sp., 3 figs.   *F. benthophila*, n. sp., 4 figs.
F. heroina, n. sp, 4 figs. *F. attenuata*, n. sp, 3 figs.   COLPASTER, n g   *C. scutigerula*.
n sp, 4 figs

'89a.   On the Asteroidea of the Mergui Archipelago, collected for the
Trustees of the Indian Museum, Calcutta, by Dr. JOHN ANDERSON,
F.R.S., Superintendent of the Museum. Journ. Linn. Soc. London,
vol. xxi, p. 319–331, pl. 28.

ARCHASTERIDÆ. Subf. *Archasterinæ*. ARCHASTER. *A. typicus* M. T. '40 = *Astropecten stellaris* GRAY '40 = *Archaster nicobaricus* MÖBIUS '59· ASTROPECTINIDÆ. Subf. *Astropectininæ*. ASTROPECTEN. *A. andersoni*, n. sp, 4 figs. *A. hemprichii* M. T. '42 = ?*A. articulatus* MICHELIN '45 = *A. mauritianus* MÖBIUS '81· *A. notograptus*, n. sp., 4 figs. Subf. *Luidiinæ*. LUIDIA. *L. forficifer* SLADEN '89· *L. maculata* M. T. PENTAGONASTERIDÆ. Subf. *Goniodiscinæ*. GONIODISCUS. *G. articulatus* (LINNÉ) DE LORIOL '84 = *Asterias articulata* LINNÉ 1753 = *Artocreatis altera* species SEBA 1758 = *Goniaster articulatus* LÜTKEN '64· ASTERINIDÆ. Subf. *Asterininæ* NEPANTHIA. *N. suffarcinata*, n. sp., 4 figs. ASTERINA. *A. cepheus* (M. T) v. MARTENS = *A. burtonii* GRAY '40·

'97.   Notes on Rockall Island and Bank, with an Account of the Petro-
logy of Rockall, and of its Winds, Currents, etc.: with Reports on
the Ornithology, the Invertebrate Fauna of the Bank, and on its
Previous History. Trans. Roy. Irish Acad., xxxi, 1896–1901, p. 39
–98, pl. ix–xiv. Echinodermata, by S., p. 78.

Pontaster tenuispinus (DÜB. & KOREN) SLADEN. *Plutonaster (Tethyaster) parelii* (DÜB. & KOREN) SLADEN. *Astropecten irregularis* LINCK. *Luidia sarsii* DÜB. & KOREN. *Hippasteria plana* (LINCK) GRAY. *Stichaster roseus* (O. F. MÜLLER) SARS. *Cribrella sanguinolenta* (O. F. MULLER) LUTKEN.

SLUITER, C. P.

'89·   Die Evertebraten aus der Sammlung des Königl. naturwissenschaft-
lichen Vereins in Niederländisch Indien in Batavia. Zugleich eine
Skizze der Fauna des Java-Meeres, mit Beschreibung der neuen
Arten. Naturk. Tijdschrift voor Nederlandsch-Indië, uitgeg. door de
Koninkl. Natuurkund. Vereenig. in Nederlandsch-Indië, 8te. ser.,
deel ix, p. 285–313. Asteroidea, p. 297–313.

ECHINASTERIDÆ. *Echinaster eridanella* (VAL.). *Echin. purpureus* (GRAY) oder *E. fallax*

(M. T.). *Mithrodia claviqera* (LAM). LINCKIADÆ. *Ophidiaster cylindricus* (M. T.). *Ophid. robillardi* (P. DE LORIOL). *Linckia miliaris* (M. T.). *Linck. ehrenbergii* (M. T.). *Linckia multifora* (GRAY). *Scytaster variolatus* (M. T.). *Sc. tuberculatus* (GRAY). *Fromia milleporella* (GRAY). GONIASTERIDÆ. *Pentagonaster* (*Stellaster*) *belcheri* (GRAY). *Pent.* (*Astrogonium* GRAY) *gibbosus* PERRIER *Pentaceros nodosus* (LINN.) = *P. turritus* LINCK. *Pent. hiulcus* (M. T.). *Pent. mulleri* (BELL). *Pent. troscheli* (BELL.) *Pent. grayi* (BELL) = *P. nodosus* (GRAY). *Culcita schmideliana* (GRAY) = *C. discoidea* (AG.). *Gymnasterias carinifera* (LAM.). ASTERINIDÆ. *Asterina cephea* (VALENCIENNES). *Archaster typicus* (M. T.). *Astropecten squamosus*, n. sp. *Astrop javanicus* (LUTKEN). *Astrop. polyacanthus* (M. T.). *Astrop. pusillus*, n. sp. *Astrop. macer*, n. sp. *Luidia maculata* (M. T.). *Luid. hardwickii* (GRAY).

'95. Die Asteriden Sammlung des Museums zu Amsterdam. Bijd. tot de Dierk., Koninkl. Zoöl. Genoot. "Natura Artis Magistra," Amsterdam, Afl. 17, p. 51–64.

ARCHASTERIDÆ. *Pontaster tenuispinus* (DUB. & KOR.) SLADEN. *Plutonaster* (subg. *Tethyaster*) *parelii* (DUB. & KOR.) SLADEN. *Pseudarchaster tessellatus* SLADEN var. *arcticus*, n. var. *Archaster typicus* M. T. PORCELLANASTERIDÆ *Ctenodiscus corniculatus* (LINCK) PERRIER. ASTROPECTINIDÆ. *Craspidaster hesperus* (M. T.) SLADEN. *Leptoptych-aster arcticus* (SARS) SLADEN. *Astropecten irregularis* LINCK. *Astrop. aurantiacus* (LINNÉ) GRAY. *Astrop. polyacanthus* M. T. *Astrop. abatus* PERRIER. *Astrop. javanicus* LÜTKEN. *Astrop. ternatensus*, n. sp. *Astrop. ornans*, n sp. *Luidia columbiæ* (GRAY) PERRIER. *Luid. hardwickii* (GRAY) PERRIER. *Luid. maculata* M T. PENTAGONASTERIDÆ. *Pentagonaster astrologorum* (M. T) PERRIER. *Pentag. granularis* (RETZIUS) PERRIER. *Nectria ocellifera* (LAM.) GRAY. *Stellaster incei* GRAY. *Goniodiscus sebæ* M. T. ANTHENEIDÆ. *Anthenea pentagonula* (LAM) PERRIER. *Anthenea flavescens* (GRAY) PERRIER. *Hippasteria plana* (LINCK) GRAY. PENTACEROTIDÆ. *Pentaceros affinis* M. T. *Pentac. grayi* BELL. *Pentac. hiulcus* M. T. *Pentac. muricatus* LINCK. *Pentac. pro-ductus* BELL. *Pentac. reinhardti* LUTKEN. *Pentac. reticulatus* LINCK. *Pentac. sladeni* DE LORIOL. *Pentac. tuberculatus* M. T. *Pentac. turritus* LINCK. *Pentaceropsis euphues*, n. sp. *Nidorellia michelini* PERRIER. *Culcita coriacea* M. T. *Cul. novae guineae* M. T. *C. niassensis*, n. sp. GYMNASTERIIDÆ. *Gymnasterias carinifera* (LAM) v. MARTENS. *Porania pulvillus* (O. F. MÜLLER) NORMAN. *Rhegaster tumidus* var. *tuberculata* (DAN. & KOR) SLADEN. *Poraniomorpha rosea* DAN. & KOR. ASTERINIDÆ. *Asterina cephea* (M. T.) v. MARTENS. *Asterina calcar* (LAM) GRAY. *Asterin. gibbosa* (PENNANT) FORBES. *Asterina exigua* (LAM) PERRIER. *Asterin. marginata* (VAL.) PERRIER. *Asterin. regularis* VERRILL. *Asterin. coronata* v. MARTENS. *Asteria pectinifera* M. T. *Palmipes mem-branaceus* LINCK. LINCKIIDÆ. *Fromia milleporella* (LAM.) GRAY. *Ophidiaster cylindricus* M. T. *Ophid. ophidianus* AG. *Ophid. purpureus* PERRIER *Ophid. pusillus* M. T. *Leiaster lenchii* GRAY. *Linckia ehrenbergii* M. T. *Linck. multifora* (LAM.) v. MARTENS. *Linck. pacifica* GRAY. *Linckia miliaris* (LINCK) v. MARTENS. *Linck. guildingii* GRAY. *Nardoa novae caledoniae* PERRIER. *Nardoa pauciforis* (v. MARTENS) SLADEN. *Nardoa variolata* (RETZ.) GRAY. *Nard. tuberculata* GRAY. *Nard. semiregularis* M. T. STICHASTERIDÆ *Stichaster aurandiacus* (MEYEN) VERRILL. *Stich. roesus* (O. F. MULLER) SARS. SOLASTERIDÆ. *Crossaster affinis* (BRDT.) SLADEN. *Cross. papposus* M. T. *Solaster glacialis* DAN. & KOR. *Sol. endeca* (RETZ.) FORBES *Sol. intermedius*, n. sp. *Lophaster furcifer* (DUB. & KOR.) VERRILL. PTERASTERIDÆ *Pteraster militaris* M. T. *Pt. pulvillus* SARS. *Hymenaster pellucidus* W. THOM. ECHINASTERIDÆ. *Acanthaster echinites* LUTKEN. *Mithrodia clavigera* (LAM.) PERRIER. *Cribrella oculata* (LINCK) FORBES. *Echinaster spinosus* (RETZ.) M. T. *Echin. purpureus* (GRAY) BELL. *Echin. eridanella* M. T. *Echin. sepositus* M. T. HELIASTERIDÆ. *Heliaster helianthus* (LAM.) DUJ. & HUPÉ. ASTERIDÆ. *Asterias mulleri* SARS. *Ast. gelatinosa* MEYEN. *Ast. polaris* M. T. *Ast. rubens* L. *Ast. glacialis* O. F. MULLER. *Ast. knckii* M. T. *Ast. stellionura* PERRIER. *Ast. gunneri* DAN. & KOR. *Ast. panopla* STUXB. *Ast. groenlandica* (LUTK.) STIMPS. *Ast. gemmifera* PERRIER. *Ast. tenera* STIMPSON. *Ast. forbesi* (DESOR) VERRILL.

STIMPSON, W.

* '53· Synopsis of the Marine Invertebrata of Grand Manan. 3 pls. Cited after DUNCAN & SLADEN.

STORM, V.

* '78· Beretning om Selskabets zoologiske Samling i Aaret 1877. Kongel. Norske Vidensk. Selskabs Skrifter, Bd. viii, p. 223-261. Cited after LUDWIG.

* '79. Bidrag til Kundskab om Throndhjemsfjordens Fauna. Kong. Nors. Vidensk. Selsk. Skrifter (1878), p. 9-36.

* '80. Bidrag til Kundskab om Throndhjemsfjordens Fauna. Kong. Norske Vidensk. Selsk. Skrifter (1879), p. 109-125· Cited after LUDWIG.

STUDER, T.

'84· Verzeichnis der Während der Reise S. M. S. Gazelle um die Erde 1874-1876 gesammelten Asteriden und Euryaliden. Abhandl. kngl. preuss. Akad. Wiss. Berlin v. J. 1884, p. 1-64, Taf. i-v.

Asteriadæ. Asterias. *Ast. perrieri* Smith '76· *Ast. rugispina* Stimpson '62· *Ast. rupicola* Verrill '76· *Ast. antarctica* Lütken '56· *Ast. calamaria* Gray. *Ast. meridionalis* Perrier '75. *Ast. studeri* Bell '81 = *Ast. mollis* Studer '77, 3 figs. *Ast. harttii* Rathbun '79 =? *Ast. gracilis* Perrier '81· *Ast. sulcifera* Valenciennes Perrier '69· *Ast. fragilis*, n. sp., 4 figs. *Ast. belli*, n. sp , 2 figs. Heliasteridæ vacat. Brisingidæ. Gymnobrisinga, n. g. *Gymnobr. sarsii*, n. sp., 1 fig. Labidiaster. *Lab. radiosus* Lovén, 25 figs. Echinasteridæ. Echinaster. *Echin. fallax* M. T. *Echin. spinulifer* Smith '76 = *Othilia spinulifer* Smith '76· Cribrella. *Crib. antillarum* Perrier '81· Mithrodia. *M. clavigera* Lam. Acanthaster. *Ac. echinites* Ell. *Ac. ellisi* Gray. Linckiadæ. Linckia. *Linck. miliaris* Linck. *Linck. pacifica* Gray. *Linck. diplax* M. T. *L. multiforis* Lam. *Linck. bouvieri* Perrier '76· Chætaster. *Ch. longipes* Betz. *Ch. nodosus* Perrier '76· Ophidiaster. *Ophid. pustulatus* v. Martens = *Linckia pustulata* v. Martens '66· *Ophid. fuscus* Gray = *Tamaria fusea* Gray '66· *Ophid. cylindricus* Lam. = *Dactyloaster* Gray. Scytaster. *Scyt. variolatus* M. T. Goniasteridæ. *Pentagonasterinæ.* Fromia. *Fr. monilis* Val. = *Scyt. milleporellus* M. T. *Fr. milleporella* Lam. = *Scytaster pistorius* M. T. Metrodira. *M. subulata* Gray '40· Ferdina. *Ferd. flavescens* Gray. Pentagonaster. *Pent. spinulosus* Gray '66· *Pent. belli*, n. sp. *Pent. meridionalis* Smith '76· *Pent. tuberculatus* Gray = *Astrogonium tuberculatum* Gray '66· *Pent. (Stellaster* Gray) *squamulosus*, n. sp., 3 figs. *Pent. (Dorigona* Gray) *mæbii*, n. sp. *Goniasterinæ.* Anthenea. *Anth. pentagonula* Lam Culcita. *Cul. schmideliana* Retz. *Cul. coriacea* M. T. *Cul. novæ guineæ* M. T. Pentaceros. *Pentac. hiulcus* Gray. *Pent. turritus* Linck. *Pent. orientalis* M. T. Gymnasteria. *Gymn. carinifera* Lam. Asterinidæ. Cycethra. *Cyc. simplex* Bell '81· Asterina. *Aster. cepheus* Val. *Aster. exigua* Lam. *Aster. fimbriata* Perr. *Aster. (Nepanthia* Gray) *brevis* Perr. '76· *Aster. (Nephanthia) maculata* Gray. Forania. *P. magelhanica* Studer '76 = *P. patagonica* Perr. '78· Astropectinidæ. Ctenodiscus. *Ct. australis* Lütken. Leptoptychaster. *Lept. kerguelensis* Smith '79 = *Archaster excavatus* Wyv. Thomson '76 = *Leptychaster kerguelensis* Smith '76· Luidia. *L. sarsii* Dub. & Kor. *L. ciliaris.* Astropecten. *Astrop. aurantiacus* L. = *Astrop. antarcticus* Studer '76· *Astrop. polyacanthus* M. T. *Astrop. velitaris* v. Martens '65. *Astrop. pentacanthus* Müll. *Astrop. irregularis* Linck. *Astrop. capensis*, n. sp. *Astrop. mesactus* Sladen '83· *Astrop. subinermis* Phil. Luidiaster. *Luid. hirsutus* Studer '83· 4 figs. Archasteridæ. Archaster. *Arch. typicus* M. T. *Arch. angulatus* M. T. *Arch. christii* Dub. & Kor. '44· Cheiraster. *Cheir. gazellæ* Studer '83· 3 figs. *Cheir. pedicellaris* Studer '83· 5 figs.

* '89. Die Forschungsreise S. M. S. Gazelle in den Jahren 1874-76. III. Teil. Zoologie. Geologie. Berlin. vi + 322 pp. + 33 pls.

STUXBERG, A.

'78· Echinodermer från Novaja Semljas haf samlade under Norden-
skiöldska expeditionerna 1875 og 1876. Oefversigt K. Vet. Akad.
Förhandlinger, no. 3. I have been enabled to consult this paper
through the kindness of Mr. H. OHSHIMA, formerly of this Institute.
Ctenodiscus crispatus (RETZ.) = *Asterias crispata* RETZ. *Archaster tenuispinus* DÜB. &
KOR = *Astropecten tenuispinus* DÜB. & KOREN. *Pteraster militaris* (O. F. MÜLL.) =
Asterias militaris O. F. MÜLL. *Pteraster pulvillus* M. SARS. *Solaster tumidus*, n. sp., 6
figs. *Solaster papposus* (L.) = *Asterias papposa* L. *Solaster furcifer* DÜB. & KOR.
Echinaster sanguinolentus (O. F. MÜLL.) = *Asterias sanguinolenta* O. F. MÜLL. *Asterias
lincki* (M. & TR). *Asterias panopla*, n. sp. *Asterias problema* (STEENST.). *Asterias
grönlandica* (STEENST.). *Pedicellaster typicus* M. SARS.

* '80. Evertebratfaunan i Sibiriens Ishaf. Bihang Kongl. Svenska Vet.-
Akad. Handlinger, v. Cited after LUDWIG.

* '86. Fauna på och kring Novaja Semlja. Vega Expeditionens Veten-
skapliga Jakttagelser, v. Cited after LUDWIG.

THOMSON, Wyv.

'72· On the Echinidea of the 'Porcupine' Deep-sea Dredging-Expedi-
tions. Ann. Mag. Nat. Hist., 4. ser., vol. 12, p. 300–306.

SÜSSBACH, S., & BRECKNER, A.

: 10. Die Seeigel, Seesterne und Schlangensterne der Nord- und Ostsee.
Wissenschaftl. Meeresunters. herausg. v. d. Kommiss. z. Untersuch.
d. deutsch. Meere i. Kiel u. d. Biolog. Anst. a. Helgoland, N. F.,
Bd. xii, Abteil. Kiel, p. 169–300, Taf. I–III. August. I owe my
access to this paper to the kindness of my colleague, Prof. HARA
of the College of Agriculture.
PHANEROZONIA. CRYPTOZONIA. *Pontaster tenuispinus* (DÜBEN & KOREN) = *Astropecten
tenuispinus* DÜB. & KOREN '45 = *Archaster tenuispinus* SARS '61· *Tethyaster parelii*
(DÜBEN & KOREN) = *Astropecten pareli* DÜB. & KOREN '45· *Astropecten irregularis*
(PENNANT) = *Asterias irregularis* PENNANT 1777 = *Asterias aranciaca* O. F. MÜLL. =
Astrop. mülleri M. & TR. '44 = *Astrop. echinulatus* M. & TR. '44 = *Astrop. acicularis*
NORMAN '65 = *Astrop. helgolandicus* GREEFF '71· *Astrop. pentacanthus* (DELLE CHIAJE)
var. *serratus* (M. & TR) = *Astrop. serratus* M. & TR. '42 = *Astrop. mulleri* MARION '83·

Psilaster andromeda M. & Tr. = *Astrop. andromeda* M. & Tr. = *Astrop. christii* Düben
& Koren '44 = *Archaster florae* Verrill '78 = *Archaster andromeda* Möbius & Bütschli
'73 = *Psilaster cassiope* Sladen '89· *Luidia ciliaris* (Philippi). *Luidia sarsii* (Düben
& Koren) = *Luydia sarsii* Düb. & Koren '44 = *Luydia savignyi* Düb. & Koren '46 part.
= *Luid. savignyi* Möbius & Bütcshli '72 = *Luid. ciliaris* (Phil.) var. *sarsi* Meissner &
Collin '96· *Pentagonaster granularis* (Retzius) = *Astrogonium granulare* M. & Tr. '42·
Hippasteria phrygiana (Parelius) = *Goniaster phrygianus* Petersen '89· *Culcita borealis*,
n sp., 3 figs. *Porania pulvillus* (O. F. Müller) = *Asterias pulvillus* O. F. Müll. 1788
= *Goniaster templetoni* Forbes '40 = *Porania gibbosa* Gray '40· *Lasiaster hispidus* (M.
Sars) = *Goniaster hispidus* M. Sars '71 = *Poraniomorpha rosea* Danielssen & Koren '84 =
Rhegaster murrayi Sladen '83 = *Lasiaster villosus* Sladen '89 = *Poraniomorpha spinulosus*
Verrill '95 = *Poraniom. borealis* Verrill '78 *Palmipes placenta* (Pennant) = *Asterias
placenta* Pennant 1777 = *Asterias membranacea* Retzius 1783 = *Asterias cartilaginea*
Fleming '28 = *Palm. membranaceus* Linck. *Solaster endeca* (Retzius). *Solaster papposus*
L. = *Crossaster papposus* Ludwig. *Cribrella sanguinolenta* (O. F. Müll.) = *Henricia
sanguinolenta*. *Pteraster militaris* O. F. Müll. *Pteraster pulvillus* M. Sars. *Retaster
multipes* (M. Sars). *Stichaster roseus* (O. F. Müll). *Asterias rubens* Linné. *Asterias
mülleri* M. Sars. *Asterias glacialis* L.

Vanhöffen, E.

* '97. Die Fauna und Flora Grönlands. In E. v. Drygalski's "Grön-
land-Expedition der Gesellschaft f. Erdkunde zu Berlin 1891-1893,"
Bd. ii, p. 1-383 + 8 pls. Cited after Ludwig.

Verrill, A. G.

* '66. On the Polyps and Echinoderms of New England with Description
of New Species. Proc. Boston Soc. Nat. Hist., vol. x, p. 333-357.
Cited after Perrier.

'67-'71· Notes on the Radiata in the Museum of Yale College, with Descriptions of New Genera and Species. Trans. Connect. Acad. Arts
and Sciences, vol. 1, pt. 2, p. 247-613, pl. iv-x. Issue of the part
following p. 503 considerably delayed owing to its destruction by
fire.

No. 1. Descriptions of New Starfishes from New Zealand.

Cœlasterias, n. g. *C. australis*, n. sp. Coscinasterias, n g *C. muricata*, n. sp.
Asterina (*Asteriscus*) *regularis*, n. sp. *Astropecten Edwardsii*, n sp.

No. 2   Notes on the Echinoderms of Panama and West Coast of America, with Descriptions of New Genera and Species

Luidia tessellata LÜTKEN '59· *Astropecten fragilis*, n sp *Astropecten regalis* GRAY '40· *Astropecten Örstedii* LÜTKEN '59· *Astropecten parvimanus*, n. sp. *Patiria obtusa* GRAY '47· *Asterina (Asteriscus) modesta*, n sp. *Oreaster occidentalis*, n. sp. NIDRELLIA. *N. armata* GRAY = *Pentaceros (Nidorellia) armatus* GRAY '40 = *Oreaster armatus* M. T. '42 = *Goniodiscus armatus* LÜTKEN '59· LINCKIA   *L. unifascialis* GRAY = *L. (Phataria) unifascialis* GRAY '40 = *Ophidiaster (Linckia) unifascialis* LÜTKEN '64 = ? *Ophidiaster suturalis* M. T. '42· *Ophidiaster pyramidatus* GRAY = *O. (Phania) pyramidatus* GRAY '40 = *O. porosissimus* LÜTKEN '59· *Mithrodia Bradleyi*, n sp. *Heliaster helianthus* GRAY = *Asterias helianthus* LAMARCK = *Asterias (Heliaster) helianthus* GRAY '40· *Heliaster microbrachia* XANTUS '60· *Heliaster Cumingii* GRAY. *Heliaster Kubiniji* XANTUS '60 *Stichaster aurantiacus* VERRILL = *Asterias aurantiacus* MEYEN '34 = *Stichaster striatus* M. T. '40 = *Tonia atlantica* GRAY '40 = *Asteracanthion aurantiacus* M. T. '42   *Luidia bellona* LÜTKEN '64·

No. 3. On the Geographical Distribution of the Echinoderms of the West Coast of America. Publ. July, 1867.

List of Species found at Sitcha. *Asteropsis imbricata* GRUBE. *Patiria miniata (Asterias miniata* BRANDT) *Solaster decemradiatus* (BRANDT sp) STIMPSON. *Asterias ochracea* BRANDT. *Asterias epichlora*.

List of Species found in Puget Sound and along the Coast to Cape Mendocino, Cal. *Mediaster æqualis* STIMP. *Patiria miniata* (BRANDT sp.). *Cribrella leviuscula (Linckia leviuscula* STIMP). *Pycnopodia helianthoides* STIMP. *Asterias epichlora* BRANDT (*A. Katherinæ* GRAY). *A. ochracea* BRANDT. *A. conferta* STIMP. *A. fissispina* STIMP. *A. Lutkenii* STIMP. *A. paucispina* STIMP. *A. Troschelii* STIMP. *A hexactis* STIMP.

List of Species found between Cape Mendocino and San Diego. Cal. *Mediaster æqualis* STIMPSON. *Patiria miniata* (BRANDT sp). *Pycnopodia helianthoides* STIMP. *Asterias gigantea* STIMP. *A. brevispina* STIMP·. *A. ochracea* BRANDT. *A. capitata* STIMP. *A. æqualis* STIMP. *?Chætaster californicus* GRUBE.

List of Species found at Margarita Bay and Cape St. Lucas. *Astropecten Örstedii* LÜTKEN. *Nidorellia armata* GRAY *Oreaster occidentalis* VERRILL. *Linckia unifascialis* GRAY. Ditto var. *bifascialis* GRAY. *Ophidiaster pyramidatus* GRAY. *Heliaster microbrachia* XANTUS. *H. Kubiniji* KANTUS. *Asterias sertulifera* XANTUS.

List of Species found at Acapulco, Mazatlan, and in the Gulf of California. *Nidorellia armata* GRAY. *Linckia unifascialis* GRAY. *Heliaster* sp. = (*H. Kubiniji ?*).

List of Species of the West Coast of Central America and the Bay of Panama. *Luidia tessellata* LÜTK. (?) *Petalaster columbiæ* GRAY. *Astropecten fragilis* VERRILL. *A. regalis* GRAY. *A. Örstedii* LÜTK. *Patiria obtusa* GRAY. *Asteriscus modestus* VERRILL. *Gymnasteria spinosa* GRAY. *G. inermis* GRAY. *Nidorellia armata* GRAY.

Oreaster occidentalis VERRILL. *Luichia unifascialis* GRAY. *Ophidiaster pyramidatus* GRAY. *Mithrodia Bradleyi* VERRILL. *Echinaster aculeatus* (GRAY sp.) LÜTE. *Heliaster microbrachia* KANTUS.

List of Species from the West Coast of Ecuador and Southern Part of New Granada. *Luidia bellonæ* LÜTKEN. *Astropecten armatus* GRAY (non MÜLLER & TR.). *A. erinaceus* GRAY. *A. fragilis* VERRILL. *Paulia horrida* GRAY *Nidorellia armata* GRAY. *Oreaster Cumingii* (GRAY sp.) LÜTE. *Ophidiaster pyramidatus* GRAY. *Linckia unifascialis* GRAY. *L. Columbiæ* GRAY. *Cistina Columbiæ* GRAY. *Dactylosaster gracilis* GRAY *Ferdina Cumingii* GRAY. (?) *Acanthaster Ellisii* (*Echinaster Ellisii* GRAY).

List of Species found at Zorritos, Peru. *Astropecten fragilis* VERRILL *Nidorellia armata* GRAY. *Linckia unifascialis* GRAY. *Ophidiaster pyramidatus* GRAY. *Heliaster Cumingii* GRAY.

List of Species recorded from the Galapago Islands (?*Culcita Schmideliana* GRAY. *Heliaster Cumingii* GRAY. (?) *Acanthaster Ellisii* GRAY.

List of Species found on the Coast of Peru, at Paita and southward. *Luidia bellonæ* LÜTK *Astropecten Peruanus* VERRILL. *Asteriscus* (*Patiria*) *Chilensis* LÜTK. *Stichaster aurantiacus* V. (MEYEN sp.). *Heliaster helianthus* GRAY. *H. Cumingii* GRAY.

List of Species from the Coast of Chili. *Goniodiscus verrucosus* PHIL *G. singularis* M. T. *Astrogonium Fonki* PHIL. *Asteriscus* (*Patiria*) *Chilensis* LÜTK. *Heilaster helianthus* GRAY. *Stichaster aurantiacus* V. (MEYEN sp.). *Asterias gelatinosa* MEYEN (*A. rustica* GRAY). *Asterias echinata* GRAY. *Asterias Germanii* nobis (PHILIPPI sp.). *Asterias lurida* nobis (PHILIPPI sp.).

List of Species from the Southern Extremity of South America, and the Neighbouring Islands. *Ganeria Falklandica* GRAY *Asterias arturetica* (LÜTK sp.). *Asterias rugispina* STIMP.

Comparison of the Tropical Echinoderm Faunæ of the East and West Coasts of America.

No 4. Notice of the Corals and Echinoderms collected by Prof. C. F. HARTT at the Abrolhos Reefs, Province of Bahia, Brazil, 1867.

Oreaster gigas (LINN. sp) LUTKEN = *Pentaceros reticulatus* GRAY '40· *Linckia ornithopus* LÜTKEN = *Ophidiaster ornithopus* M. T. =? *Linckia Guildingii* GRAY. *Echinaster* (*Othilia*) *crassispina*, n. sp., 1 fig. *Asterias atlantica*, n. sp.

No. 5. Notice of a Collection of Echinoderms from La Paz, Lower California, with Descriptions of a New Genus. Publ. April, 1868.

Linckia unifascialis GRAY (var. *bifascialis*) VERRILL. *Nidorellia armata* GRAY = *Pentaceros* (*Nidorellia*) *armatus* GRAY = *Oreaster armatus* M. T. '42. = *Goniodiscus conifer* MÖBIUS '59 = *Goniodiscus armatus* LÜTKEN '59 = *Nidorellia armata* VERRILL '67 = *Goniodiscus stella* VERRILL '67 (young) non MÖBIUS. AMPHIASTER, n g *A. insignis* VERRILL, n. sp, 1 fig. *Oreaster occidentalis* VERRILL.

No 6. (Corals and Polyps only.)

No. 7. (Distribution of Polyps.)

No. 8. Additional Observations on Echinoderms, chiefly from the Pacific Coast of America. Presented January, 1871.

Pteraster Danœ VERRILL '69, 2 figs. *Oreaster occidentalis* VERRILL (wrongly called *Pentaceros occidentalis* '70). *Nidorellia armata* GRAY. *Gymnasteria spinosa* GRAY. *Mithrodia Bradleyi* VERRILL. *Acanthaster Ellisii* VERRILL '69 = *Echinaster Ellisii* GRAY '10 = *Acanthaster solaris* (pars) DUJ. & HUPÉ '62 *Echinaster tenuispina*, n. sp. *Ophidiaster pyramidatus* GRAY. LEPIDASTER, n. g *L. teres*, n. sp. *Heliaster Kubiniji* XANTUS.

No. 9. The Echinoderm-Fauna of the Gulf of California and Cape St. Lucas.

Astropecten Örstedii LÜTKEN. *Luidia brevispina* LTK *Gymnasteria spinosa* GRAY. *Amphiaster insignis* VERRILL. *Nidorellia armata* GRAY. *Oreaster occidentalis* VERRILL. *Acanthaster Ellisii* VERRILL. *Mithrodia Bradleyi* VERRILL. *Echinaster tenuispina* VERRILL. *Lepidaster teres* VERRILL. *Linckia unifascialis* GRAY. Ditto var. *bifascialis* GRAY. *Ophidiaster pyramidatus* GRAY. *Heliaster microbrachia* XANTUS. *H. Kubiniji* XANTUS. *Asterias sertulifera* XANTUS.

* '73.   Results of Recent Dredging Operations on the Coast of New England. Amer. Journ. Sc. Arts, vol. v, p. 1–16, vol. vi, p. 435–441. Cited after LUDWIG.

* '80.   List of Marine Invertebrata from the New England Coast. Proc. U. S. Nat. Museum, vol. ii, p. 227–232. Cited after LUDWIG.

'82.   Notice of the Remarkable Marine Fauna occupying the Outer Banks off the Southern Coast of New England, no. 4. Amer. Journ. Sc., 3. ser., vol. 23, p. 135–142, 216–225.

Asterias vulgaris (STIMPS.) VERRILL. *Asterias tanneri* VERR. *Stephanasterias albua* (STIMPS.) VERR. *Cribrella sanguinolenta* (MÜLLER) LÜTKEN. *Diplopteraster multipes* (SARS) VERR *Porania grandis* VERR. *Porania spinulosa* VERR. *Porania borealis* VERR. = *Asterina borealis* V. *Odontaster hispidus* VERR. *Archaster florœ* VERR. *Archaster americanus* VERR. *Archaster agassizii* VERR. *Archaster parelii* DÜB. & KOREN. *Archaster tenuispinus* DÜB. & KOREN. *Archaster mirabilis* (?) PERRIER. *Archaster arcticus* M. SARS. *Archaster bairdii* VERR., n. sp. *Luidia elegans* PER. *Ctenodiscus crispatus* DÜB. & KOREN.

'84.   Notice of the Remarkable Marine Fauna occupying the Outer Banks off the Southern Coast of New England, no. 9.   Brief Contributions to Zoology from the Museum of Yale College. No. LV. Amer. Journ. Sci., 3. ser., vol., 28, p. 213–220.

Zoroaster diomedeæ, n. sp.   *Archaster grandis*, n. sp.   *Benthopecten spinosus*, n. sp.

'85.   Results of the Explorations made by the Steamer Albatross off the Northern Coast of the United States in 1883. Ann. Rep. Commiss. Fish and Fisher. for 1883, p. 503–601, pls. 1–44.

Asterias forbesii DESOR.   *Asterias vulgaris* ST. = *A. rubens* L. (?)   *Asterias tanneri* VERR.
Ast. briareus VERR.   *Leptasterias compta* (STIMP.) VERR.   *Stephanasterias albua* (STIMP.)
VERR. = *Stichaster albulus* VERR.   *Zoroaster diomedeæ* VERR.   *Brisinga elegans* VERR.
Bris. costata VERR.   *Solaster abyssicola* VERR. = *Sol. earlii* VERR.   *Solaster endeca* FORBES
Crossaster papposus M. & TR.   *Lophaster furcifer* (DUB. & KOR.) VERR. = *Solaster
furcifer* auth.   *Pteraster militaris* M. & TR.   *Diplopteraster multipes* (SARS.) VERR.
Hymenaster modestus VERR.   *Porania grandis* VERR.   *Poraniomorpha spinulosa* VERR.
= *Porania spinulosa* VERR.   *Astrogonium granulare* M. & TR.   *Hippasteria phrygiana*
GRAY = *H. plana* auth.   *Odontaster hispidus* VERR.   *Astropecten articulatus* SAY.   *Archas-
ter arcticus* M. SARS.   *Archaster americanus* VERR.   *Archaster floræ* VERR.   *Archaster
robustus* VERR.   *Archaster grandis* VERR.   *Archaster agassizii* VERR.   *Archaster parelii*
DÜBEN & KOREN.   *Archaster formosus* VERR.   *Archaster tenuispinus* DÜBEN & KOREN.
Archaster sepitus VERR.   *Benthopecten spinosus* VERR.   *Luidia elegans* VERR.   *Luidia
clathrata* (SAY).   *Porcellanaster cœruleus* W. THOMSON.   *Poraniomorpha borealis* VERRILL.
Archaster bairdii VERR.   *Ctenodiscus crispatus* DÜBEN & KOREN.   *Zoroaster fulgens* W.
THOMSON.   *Pteraster pulvillus* SARS.

Figures of the following species: *Pteraster militaris* 1.   *Archaster floræ* 1.   *Archaster
parelii* 1.   *Archaster tenuispinus* 1.   *Luidia elegans* 1.   *Asterias tanneri* 2.   *Solaster
earlii* 3.   *Porcellanster cœruleus* 2.   *Diplopteraster multipes* 1.   *Porania grandis* 4.   *Lophaster furcifer* 2.   *Hippasteria phrygiana* 1.   *Brisinga americana* 3.   *Asterina borealis* 2
Astrogonium granulare 2.   *Tremaster mirabilis* 1.

'94.   Descriptions of New Species of Starfishes and Ophiurans, with a Revision of Certain Species formerly described; mostly from the Collections made by the U. S. Commission of Fish and Fisheries. Proc. U. S. National Mus., xvii, p. 246–297.

ARCHASTERIDÆ.   *Benthopectininæ*, n.   *Benthopecten spinosus* VER. = *Pararchaster semisquamatus*, var. *occidentalis* SLADEN = *Pararch. armatus*.   *Pontasterinæ*, n.   *Pontaster
hebitus* SLADEN = *Archaster tenuispinus* VER.   *Pontaster forcipatus* SLADEN = *Archaster
tenuispinus* VER.   *Pont. sepitus* VER.   *Plutonasterinæ*.   *Dytaster grandis* VER. = *Archaster
grandis* VER. = *Dytaster madreporifer* SLADEN.   *Plutonaster agassizii* VER. = *Archaster
agassizii* VER. = *Plutonaster rigidus* SLADEN = *Pluton. bifrons* (part) SLADEN.   *Pseudarch-*

asterina. *Pseudarchaster intermedius* Sladen = *Archaster parelii* Ver. *Pseudarch. concinnus*, n sp.

'95. Distribution of the Echinoderms of North-eastern America. Amer. Journ. Sc., 3. ser., vol. 49, p. 127-141, 199-212.

In a foot-note the author says that in following Sladen in the arrangement and nomenclature of the species he by no means approves of the changes of names made by that writer, but that in several cases he declines to follow him in the re-surrection of the ante-Linnæan names given by Linck.

Archasteridæ. *Benthopecten spinosus* Verr. = *Pararchaster semisquamatus* var. *occidentalis* Slad. = *Pararchaster armatus* Slad. *Pontaster hebitus* Slad. = *Archaster tenuispinus* Verr. *Pont. forcipatus* Slad. *Pont. sepitus* Verr. = *Archaster sepitus* Verr. *Dytaster grandis* Verr. = *Archaster grandis* Verr. = *Dytaster madreporifer* Slad. *Plutonaster agassizii* Verr. = *Archaster agassizii* Verr. = *Pluton. rigidus* Slad. (also var. *semiarmata*) = *Pluton. bifrons* (part) Slad. *Pseudarchaster intermedius* Slad. = *Archaster parelii* Verr. & n. var. *insignis*. *Pseudarch. concinnus* Verr. Porcellanasteridæ. *Ctenodiscus crispatus* Düb. & Koren = *Asterias crispatus* Retzius = *Ctenodiscus polaris* M. & Tr = *Ct. corniculatus* Perrier. ("This is an instance in which certain writers have resurrected Linck's ante-binomial names to displace those given under the Linnæan system. In this I cannot concur.") *Porcellanaster cæruleus* Thoms. Astropectinidæ. *Astropecten americanus* Verr. = *Archaster americanus* Verr. *Astropecten vestitus* Lütken = *Asterias vestita* Say. *Astrop. articulatus* M. & Tr. *Leptoptychaster arcticus* Sladen = *Astropecten arcticus* M. Sars = *Archaster arcticus* Verr. = *Leptoptychaster arcticus* var. *elongatus* Sladen. *Psilaster floræ* Verr. = *Archaster floræ* Verr. *Bathybiaster robustus* Verr. = *Archaster robustus* Verr. = *Phoxaster pumilus* Sladen. *Luidia clathrata* (Say) = *Asterias clathrata* (Say). *Luidia elegans* Perrier. Goniasteridæ or Pentagonasteridæ. *Pentagonaster eximius* Verr. *Pentagonaster granularis* Perrier = *Asterias granularis* Retzius = *Astrogonium granulare* M. & Tr. = *Goniaster granularis* Lütken. *Pentagonaster simplex*, n. sp. *Pentagonaster planus*, n. sp. *Odontaster hispidus* Verr. *Isaster bairdii* Verr. = *Archaster bairdii* Verr. *Paragonaster formosus* Verr. = *Archaster formosus* Verr. = ? *Paragonaster cylindratus* Sladen. *Hippasteria phrygiana* Ag. = *Asterias phrygiana* Parelius = *Asterias equestris* Pennant = *Hippasteria plana* Gray = *Goniaster equestris* Forbes = *Astrogonium phrygianum* M. & Tr. = *Goniaster phrygianus* Norman. Gymnasteridæ. *Porania (Chondraster) grandis* Verr. *Porania insignis*, n. sp. = *Porania grandis* (pars) Verr. *Poraniomorpha spinulosa* Verr. = *Porania spinulosa* Verr. with var. *rudis* & *inermis*. *Poraniomorpha borealis* Verr. = *Asterina borealis* Verr. = *Porania borealis* Verr *Rhegaster abyssicola*, n. sp.

ASTERINIDÆ. *Asterina pygmaea* VERR. *Tremaster mirabilis* VERR. SOLASTERIDÆ. *Solaster enleca* FORBES = *Asterias enleca* RETZIUS. *Solaster syrtensis* VERR. *Solaster abyssicola* VERR. *Solaster benedicti* VERR. *Solaster earlii* VERR. *Crossaster papposus* M. & TR. = *Asterias papposa* FABRICIUS = *Solaster papposus* FORBES. *Crossaster helianthus* VERR. *Lophaster furcifer* VERR. = *Solaster furcifer* DUB. & KOREN. PTERASTERIDÆ. *Pteraster pulvillus* M. SARS. *Pteraster militaris* M. & TR. = *Asterias militaris* MÜLL. *Temnaster hexactis* VERR. = *Pteraster* (*Temnaster*) *hexactis* VERR. *Diplopteraster multipes* VERR. = *Pteraster multipes* M. SARS. *Retaster multipes* SL. (*Diplopteraster verrucosus* V. = *Retaster verrucosus* SL. *D. peregrinator* V. = *Retaster peregrinator* SL) LOPHOPTER- ASTER, n. g. *Lophopteraster abyssorum*, n. sp. *Hymenaster modestus* VERR. *Hymen- aster regalis*, n. sp. ECHINASTERIDÆ. *Cribrella pectinata* VERR. *Cribrella sanguinolenta* LÜTKEN = *Asterias sanguinolenta* MÜLL. = *Asterias oculata* PENNANT = *Asterias spongiosa* FABRICIUS = *Linckia oculata* FORBES = *Cribrella oculata* FORBES = *Echinaster oculatus* M. & TR. = *Linckia pertusa* STIMPSON = *Echinaster sanguinolentus* SARS. PEDICELLASTERIDÆ. *Pedicellaster typicus* M. SARS = *Pedicellaster palæocrystallus* DUNCAN & SLAD. ZOROAS- TERIDÆ. *Zoroaster diomedeæ* VERR. = ? *Z. fulgens* (pars) SLADEN. STICHASTERIDÆ. *Neomorphaster forcipatus* VERR. *Stichaster albulus* VERR = *Asteracanthion albulus* STIMP. = *Asteracanthion problema* STEENSTRUP = *Stephanasterias albula* VERR. ASTERIDÆ. *As- terias forbesii* VERR. = *Asteracanthion forbesii* DESOR = *Asterias arenicola* STIMPS. = *Asteracanthion berylinus* AG. MSS. *Asterias vulgaris* STIMPS. MSS. = *Asterias stimpsoni* (pars) VERR. = *Asteracanthion pallidus* AG. MSS. = *Asterias pallida* & *Ast. fabricii* PER- RIER. *Asterias stellionura* PERR. = *Asterac. stellionura* PERR. *Asterias enopla*, n. sp. *Asterias polaris* VERR. = *Asterac. polaris* M. & TR. = *Ast. borealis* PERR. *Asterias tanneri* VERR. *Ast. briareus* VERR. *Leptasterias compta* VERR = *Asterias compta* STIMPS. *Leptasterias tenera* VERR. = *Asterias tenera* STIMPS. *Leptasterias grœnlandica* VERR. = *Asterac. grœnlandicus* LÜTKEN = *Ast. grœnlandica* VERR. *Leptasterias littoralis* VERR. = *Asterac. littoralis* STIMPS. = *Ast. littoralis* VERR. *Hydrasterias ophidion* SLAD. = *Asterias* (*Hydrasterias*) *ophidion* SLAD. BRISINGIDÆ. *Odinia americana* VERR. = *Brisinga ameri- cana* VERR. = *Freyella americana* SLAD. *Brisinga costata* VERR. *Brisinga multicostata* VERR. *Brisinga verticillata* SLADEN. *Freyella elegans* SLADEN = *Brisinga elegans* = *Freyella bracteata* SLADEN. *Freyella aspera* VERR.

'99. Revision of Certain Genera and Species of Starfishes with Descrip- tions of New Forms. Trans. Connect. Acad. Arts and Sciences, vol. x, p. 145–234, pls. xxiv–xxx (8 pls.).

GONIASTERIDÆ (= PENTACEROTIDÆ GRAY '66 part. = PENTAGONASTERIDÆ PERRIER '84). GONIASTER (AGASSIZ) GRAY (restr.) = GONIASTER AGASSIZ '36 part. = PENTAGONASTER PERRIER '76 part. = ditto SLADEN '89 = ASTROGONIUM M. T. '42 part. = PHANERASTER PERRIER '94· *Goniaster americanus* VERRILL '71 = *Pentagonaster semilunatus* PERRIER

'76 part. = *Phaneraster semilunatus* PERRIER '91 part = *Pentagonaster parvus* PERRIER '84, 8 figs. *Goniaster africanus* VERRILL '71, 2 figs. PENTAGONASTER GRAY '40 = STEPHANASTER AYRES '51 = PENTAGONASTER (Sect. A, *a*, pars) PERRIER '76 = ASTROGONIUM SLADEN '89. TOSIA GRAY '40 = ASTROGONIUM (pars) M. T. '42 = PENTAGONASTER (Sec. A, *b*, pars) PERRIER '76 = PENTAGONASTER (pars) SLADEN '89. Section A—Typical. Section B—PLINTHASTER. Section C—CERAMASTER. *T. granularis* (RETZIUS) = *Asterias granularis* RETZIUS 1783 = *Astrogonium granulare* M. T. '42 = *Goniaster granularis* LÜTKEN '65 = *Pentagonaster granularis* PERRIER '76 = *Pentagon. balteatus* SLADEN '91 = *Pentagon. concinnus* SLADEN '91. *T.* (*Plinthaster*) *compta*, n. sp., 1 fig. *T.* (*Plinthaster*) *nitida*, n. sp., 3 figs. PYRENASTER, n. g. *P. dentatus* PERRIER = *Pentagonaster dentatus* PERRIER '84, 3 figs. *P. affinis* PERRIER = *Pentagonaster affinis* PERRIER '81. PELTASTER, n. g. *P. hebes*, n. sp., 1 fig. *P. planus* VERRILL = *Pentagonaster planus* VERRILL '85. 2 figs. LITONOTASTER, n. g. *L. intermedius* PERRIER = *Pentagon. intermedius* PERRIER '84, 3 figs. EUGONIASTER, n. g. *E. investigatoris* ALCOCK = *Pentagonaster investigatoris* ALCOCK '93. ANTHENIASTER, n. g. *A. sarissa* ALCOCK = *Anthenioides sarissa* ALCOCK '93. Subf. *Hippasteriinæ*, n. *Hippasteria caribæa*, n. sp., 2 figs. CLADASTER, n. g. *C. rudis*, n. sp., 4 figs. Subf. *Mediasterinæ*, n. MEDIASTER STIMPSON '57 = ISASTER VERRILL '94. *M. æqualis* STIMPSON '57, 3 figs. *M. bairdii* VERRILL = *Archaster bairdii* VERRILL '82 = *Isaster bairdii* VERRILL '94 = *Mediaster stellatus* PERRIER '91, 11 figs. *M. agassizii*, n. sp. *M.* (?) *pedicellaris* VERRILL = *Goniodiscus pedicellaris* PERRIER '89. *M. arcuatus* (SLADEN) = *Pentagonaster arcuatus* SLADEN '89. *M. japonicus* (SLADEN) = *Pentagonaster japonicus* SLADEN '89. *M. patagonicus* (SLADEN) = *Pentagonaster patagonicus* SLADEN '89. NYMPHASTER SLADEN '85 = PENTAGONASTER (pars) PERRIER '84 = DORIGONA PERRIER '94 non GRAY '66 nec PERRIER '76. *N. ternalis* (PERRIER) = *Pentagonaster ternalis* PERRIER '81 = *N.* (?) *ternalis* SLADEN '89 = *Dorigona ternalis* PERRIER '94, 1 fig *N. subspinosus* (PERRIER) = *Pentagonaster subspinosus* PERRIER '81 = *Nymphaster* (?) *subspinosus* SLADEN '89 = *Dorigona subspinosa* PERRIER '74. *N. arenatus* (PERRIER) = *Pentagonaster arenatus* PERRIER '81 = *Nymphaster* (?) *arenatus* SLADEN '89 = *Dorigona arenata* PERRIER '94. *N. jacqueti* (PERRIER) = *Dorigona jacqueti* PERRIER '94 = *Dorigona prehensilis* PERRIER '85 = *N.* (?) *prehensilis* SLADEN '89. *Nymphaster protentus* SLADEN '89. *N. albidus* SLADEN '89. *N. basilicus* SLADEN '89. NEREIDASTER, n. *N. symbolicus* (SLADEN) = *Nymph. symbolicus* SLADEN '89. *N. bipunctus* (SLADEN) = *Ny. bipunctus* SLADEN '89. Subf. *Pseudarchasterinæ* = *Pseudarchasterinæ* SLADEN '89 = *Astrogoniinæ* (pars) PERRIER '94. PSEUDARCHASTER SLADEN = ASTROGONIUM (pars) PERRIER '94 non M. T. nec GRAY. *Ps. intermedius* SLADEN '89 = *Archaster parelii* (pars) VERRILL '74 non DÜBEN & KOREN, 3 figs. *Ps. fallax* PERRIER = *Astrogonium fallax* PERRIER '85 = *Archaster parelii* (pars) VERRILL '83, 2 figs. *Ps.* (?) *hispidus* VERRILL, n. sp., 1 fig. *Ps. granuliferus*, n. sp., 2 figs. *Ps. concinnus* VERRILL '94, 3 figs. *Ps. ordinatus*, n. sp., 3 figs. *Ps. annectens* (PERRIER) = *Astrogonium annectens* PERRIER '94 = *Ps. hystrix*

PERRIER. *Ps. necator* (PERRIER) = *Astrogonium necator* PERRIER '94· *Ps. aphrodite* (PERRIER) = *Astrogonium aphrodite* PERRIER '94· APHRODITASTER *gracilis* SLADEN '89 = *Astrogonium gracile* PERRIER '94· *Ps. tessellatus* SLADEN. *Ps. patagonicus* (PERRIER). *Ps. discus* SLADEN. *Ps. mosaicus* ALCOCK & WOOD-MASON. *Ps. roseus* = *Mediaster roseus* ALCOCK '93· PARAGONASTER *subtilis* PERRIER '94 = *Goniopecten subtilis* PERRIER '81· *P. formosus* (VERRILL '84). *P. strictus* PERRIER. *P. elongatus* (PERRIER.'85). *P. cylindratus* SLADEN '89· ROSASTER *alexandri* PERRIER '94 = *Pentagonaster alexandri* PER. '81· Incerta sedes. HOPLASTER PERRIER '82· *H. spinosus* PERRIER '82· *H. lepidus* (SLADEN) = *Pentagonaster lepidus* SLADEN '89· LASIASTER SLADEN. *L. hispidus* (SARS) SLADEN '89 = *Goniaster hispidus* M. SARS '77 = *Pentagonaster hispidus* PERRIER '78·

Revision of the Classification of the Orders Valvata and Paxillosa of PERRIER, and especially of the ARCHASTERIDÆ. ARCHASTERIDÆ SLADEN '89· Order *Phanerozonia* SLADEN (rest.). Subord. I. *Valvata* PERRIER (sens. ext.). Fam. I. LINCKIIDÆ PERRIER. Fam. II. PENTACEROTIDÆ GRAY (restr.). Fam. III. ANTHENEIDÆ PERRIER (restr.). Fam. IV. GONIASTERIDÆ FORBES (restr.). Subf. I. *Goniasterinæ* VERRILL, n. = *Pentagonasterinæ* SLADEN (pars). Subf. II. *Goniodiscinæ* SLADEN. Subf. III. *Mediasterinæ* VERRILL, n. Subf. IV. *Pseudarchasterinæ* SLADEN. Subf. V. *Hippasterinæ* VERRILL, n. Fam. V. ODONTASTERIDÆ VERRILL, n. = *Gnathasterinæ* PERRIER (pars). Fam. VI. PLUTONASTERIDÆ VERRILL. Subf. I. *Mimasterinæ* SLADEN. Subf. II. *Plutonasterinæ* SLADEN. Subf. III. *Pontasterinæ* VERRILL '94· Fam. VII. GONIOPECTINIDÆ VERRILL, n. Fam. VIII. BENTHOPECTINIDÆ VERRILL = *Benthopectininæ* VERRILL '94· Subord. II. *Paxillosa* PERRIER (sens. restr.). Fam. IX. PORCELLANASTERIDÆ SLADEN. Fam. X. ARCHASTERIDÆ VIGUIER (restr. to ARCHASTER). Fam. XI. ASTROPECTINIDÆ GRAY (restr.). Fam. XII. LUIDIIDÆ VIGUIER, n. = *Luidiinæ* SLADEN.

ODONTASTERIDÆ VERRILL, n. = *Gnathasterinæ* (pars) PERRIER '94· ACODONTASTER VERRILL, n. g. = GNATHASTER (pars) SLADEN '89 = ODONTASTER (pars) BELL '93· GNATHASTER SLADEN (restr.) = GNATHASTER (pars) SLADEN '89 = ODONTASTER (pars) BELL '93· ODONTASTER VERRILL '80 = GNATHASTER SLADEN (pars) '89 = ODONTASTER BELL (pars) '93· *O. hispidus* VERRILL '80· 2 figs. *O. setosus*, n. sp., 5 figs. *O. robustus*, n. sp., 2 figs. PLUTONASTERIDÆ = *Plutonasterinæ* (subf.) SLADEN '89· PLUTONASTER *agassizii* VERRILL '94 = *Archaster agassizii* VERRILL '80 = *P. rigidus* SLADEN '89 (also var. *semiarmatus*) = *P. bifrons* (pars) SLADEN '89· 1 fig. *P. efflorescens* PERRIER '94 = *Archaster efflorescens* PERRIER '87· GONIOPECTINIDÆ VERRILL, n. GONIOPECTEN *demonstrans* PERRIER '81· 1 fig PRIONASTER, n. g. *P. elegans*, n. sp., 4 figs. BENTHOPECTINIDÆ VERRILL, n. = *Benthopectininæ* VERRILL '94· BENTHOPECTEN *spinosus* VERRILL '84 = *Pararchaster semisquamatus* var. *occidentalis* SLADEN '89 = *Pararchaster armatus* SLADEN '89· 2 figs. ASTROPECTINIDÆ GRAY. BLAKIASTR *conicus* PERRIER '81 = *Leptoptychaster conicus* PERRIER '94· 1 fig. SIDERIASTER, n. g. *S. grandis*, n. sp., 3 figs. PTERASTERIDÆ PERRIER. HEXASTER *obscurus* PERRIER '91 = *Pteraster* (*Temnus-*

ter) hexactis VERRILL '94 = *Temnaster hexactis* VERRILL '95· HYMENASTER *regalis*
VERRILL '94 var. *agassizii*, n. ASTERINIDÆ GRAY. MARGINASTER *austerus*, n. sp.
STICHASTERIDÆ PERRIER '85· STEPHANASTERIAS VERRILL '71 = NANASTER PERRIER '94 =
STICHASTER (pars) VERRILL '66·

: 09.   Descriptions of New Genera and Species of Starfishes from the
North Pacific Coast of America. Americ. Journ. Sc., 4. ser , vol.
28, p. 59–70. July.

Solaster galaxides, n. sp., 2 figs. *Solaster constellatus*, n. sp., 2 figs. *Pteraster octaster*,
n. sp., 1 fig. *Pteraster hebes*, n. sp. *Hippasteria spinosa*, n. sp. *Tosia arctica*, n. sp.,
2 figs. *Asterias (Pisaster) papulosa*, n. sp. ALLASTERIAS, n. g. *Allasterias rathbuni*,
n. sp., 2 figs.; var. *anomala*, n.; var. *nortonensis*, n. *Asterias (Urasterias) forcipulata*,
n. sp. *Asterias polythela*, n. sp. *Asterias victoriana*, n. sp.

: 09a. Remarkable Development of Starfishes on the Northwest American
Coast; Hybridism; Multiplicity of Rays; Teratology; Problems in
Evolution; Geographical Distribution. Amer. Nat., 43, p. 542-555.
September.

VIGUIER, C.

'78· Anatomie comparée du squelette des Stellérides. Arch. zool. expér.
gén., t. 7, p. 33–250, pl. 5–16.

ASTERIADÆ. ASTERIAS LINNÉ 1735 = STELLA CORIACEA, SOL, PENTADACTYLOSASTER
(pars), HEXAKTIN, HEPTAKTIN LINCK 1733 = STELLONIA NARDO '34 = STELLONIA &
URASTER AGASSIZ '34 = ASTERACANTHION (pars) MÜLL. & TROSCHEL. Chiefly *A. glacialis*,
10 figs. ANASTERIAS PERRIER '75· *A. minuta.* STICHASTER MÜLL. & TROSCHEL '40 =
ASTERACANTHION (pars) MÜLL. & TROSCHEL '42 = STEPHANASTERIAS '61 & CÆLASTERIAS
'72 VERRILL = TONIA GRAY '65· *St. aurantiacus*, 5 figs. CALVASTERIAS PERRIER '78·
C. asterinoides. PYCNOPODIA STIMPSON '61· *P. helianthoides*, 2 figs. HELIASTERIDÆ.
HELIASTER GRAY '40· *H. helianthus*, 1 fig. *H. kubiniji. H. microbrachia*, 8 figs. BRI-
SINGIDÆ. BRISINGA SARS. LABIDIASTER LÜTKEN. PEDICELLASTER SARS. ECHINASTERI-
DÆ. *Echinasterinæ.* ECHINASTER M. & TR. '40 = STELLONIA (pars) NARDO '34 = OTHILIA
& RHOPIA GRAY '40· *E. sepositus*, 7 figs. CRIBRELLA AGASSIZ '35 (pars) = HENRICIA
GRAY '40 = ECHINASTER (pars) MÜLL. & TROSCHEL '42· *C. oculata*, 8 figs. *Mithrodinæ.*
MITHRODIA GRAY '40 = HERESASTER MICHELIN '44· *M. clavigera*, 6 figs. *Valvasterinæ.*
VALVASTER. *Solasterinæ.* SOLASTER FORBES '33 = STELLONIA (pars) NARDO '34 = CROS-
SASTER MÜLL. & TROSCHEL '40· *S. papposus*, 6 figs. *S. endeca*, 1 fig. ACANTHASTER
P. GERVAIS '41 = STELLONIA (pars) AGASSIZ '35 = ECHINASTER GRAY '40 = ECHINITES

MÜLL. & TROSCH. '44· *A. echinites*, 6 figs. LINCKIADÆ. LINKIA NARDO '34 emend., 1 fig. *L. miliaris*, 7 figs. *L. diplax*, 1 fig. CHÆTASTER MÜLL. & TROSCH. '40 = NEPANTHIA GRAY '40· *Ch. longipes*, 6 figs. OPHIDIASTER AGASSIZ '34 emend. = LINCKIA (pars) NARDO '34 = DACTYLOSASTER, TAMARIA, CISTINA, OPHIDIASTER GRAY '40· *O. pyramidatus*, 7 figs. *O. ophidianus. O. germani*, 3 figs. SCYTASTER LÜTKEN '64 = NARDOA, GOMOPHIA, NARCISSIA GRAY '40· *Sc. novæ-caledoniæ*, 8 figs. GONIASTERIDÆ. *Pentagonasterinæ.* FROMIA GRAY '40 = SCYTASTER MÜLL. & TROSCH. '42 = LINCKIA v. MARTENS '66· *F. milleporella*, 6 figs. METRODIRA GRAY '40 = SCYTASTER MULL. & TROSCH. '42· *M. subulata.* FERDINA GRAY '40· *F. flavescens.* PENTAGONASTER LINCK 1733 = GONIASTER (pars) AGASSIZ '36 = ASTROGONIUM, STELLASTER & GONIODISCUS (pars) MÜLL. & TROSCH. '42 = STELLASTER, HOSEA (pars), CALLIASTER, ASTROGONIUM, PENTAGON-ASTER, TOSIA GRAY. *P. astrologorum*, 6 figs. HIPPASTERIA GRAY '40 = PENTACEROS (pars) LINCK 1733 = ASTROGONIUM (pars) M. & TR. '44· *H. plana. Goniasterinæ.* ANTHENEA GRAY '40 = GONIASTER (pars) AGASSIZ '35 = GONIODISCUS (pars) MÜLL. & TROSCH. '42· *Anthenea articulata*, 5 figs. GONIASTER (s. n) PERRIER '75 = OREASTER (pars) MÜLL. & TROSCH. '42 = PENTACEROS (pars) GRAY '40· *G. obtusangulus.* GONIODISCUS (pars) MÜLL. & TROSCH. '42 = HOSEA (pars) GRAY = GONIASTER (pars) v. MARTENS '66· *G. sebæ.* NECTRIA GRAY '40 = GONIODISCUS (pars) MÜLL. & TROSCH. '42· ASTERODISCUS GRAY '47· *A. elegans.* CULCITA AGASSIZ '35· 1 fig. *C. schmideliana*, 6 figs. CHORIAS-TER LÜTKEN '69· *Ch. granulatus.* NIDORELLIA GRAY '40 (subg.) = PAULIA GRAY '40 = GONIODISCUS (pars) MÜLL. & TROSCH. '42· PENTACEROS LINCK 1733 = GONIASTER (pars) AGASSIZ '33 = OREASTER MÜLL. & TROSCH. *P. reticulatus*, 5 figs. *P. muricatus*, 6 figs. GYMNASTERIA GRAY '40 = ASTEROPSIS (pars) MÜLL. & TROSCH. '42 = GYMNASTRIAS v. MARTENS '66· *G. carinifera*, 6 figs. ASTERINIDÆ. PATIRIA GRAY '40· ASTERINA NARDO '34· *A. gibbosa*, 5 figs. *A. calcar*, 1 fig. PALMIPES LINCK 1733. *P. membranaceus*, 5 figs. *P. inflatus*, 2 figs. DISASTERINA E. PERRIER '75· *D. abnormalis.* ASTEROPSIS MÜLL. & TROSCHEL '42· DERMASTERIAS PERRIER '75· FORANIA GRAY '40· *P. pulvillus*, 5 figs. GANERIA GRAY '47· *G. falklandica.* PTERASTERIDÆ. ASTROPECTINIDÆ. CTENO-DISCUS MÜLL. & TROSCHEL '42· *Ct. corniculatus*, 6 figs. LUIDIA FORBES '39· *L. clathrata*, 7 figs. ASTROPECTEN LINCK 1733. *A. aurantiacus*, 6 figs. ARCHASTERIDÆ. ARCHASTER MÜLL. & TROSCHEL '40· *A. typicus*, 6 figs. *A. angulatus*, 1 fig.

WALTER, A.

'85· Ceylons Echinodermen. Jen. Zeitschr., Bd. xviii, p. 365–384.

Linckia sp. Luidia maculata M. & TR. *Astropecten armatus* M. & TR. = *hystrix* = *polyacanthus* M. & TR.

List of starfishes known from the coasts of Ceylon: *Linckia sp., Scytaster variolatus, Scytaster novæ-caledoniæ, Fromia milleporella, Luidia maculata, Astropecten armatus* et var. *hystrix, Astropecten sp.*

WHITEAVES, J. F.

'72· Notes on a Deep-sea Dredging-Expedition round the Island of Anticosti, in the Gulf of St. Lawrence. Ann. Mag. Nat. Hist., 4. ser., vol. x, p. 341–354.

" *Calveria hystrix* (singular asterid allied to *Pteraster*)." *Ctenodiscus crispatus.*

* '74· On Recent Deep-sea Dredging Operations in the Gulf of St. Lawrence. Amer. Journ. Sc. Arts, vol. vii, p. 210–219. Cited after LUDWIG.

* : 01. Catalogue of the Marine Invertebrata of Eastern Canada. Geol. Surv. Canada. iv + 272 pp.

WOOD-MASON, J., & ALCOCK, A.

'91· Natural History Notes from H. M. Indian Marine Survey Steamer 'Investigator,' Commander R. F. HOSKYN, R. N., commanding.— Series II., no. 1. On the Results of Deep-sea Dredging during the Season 1890–91. Ann. Mag. Nat. Hist., 6. ser., vol. 8, p. 427–452. Asteroidea p. 427–440.

ARCHASTERIDÆ. PARARCHASTER. *P. semisquamatus* SLADEN. FONTASTER. *P. hispidus,* n. sp. DYTASTER. *D. exilis* SLADEN. *D. anacanthus,* n. sp. PERSEPHONASTER, n. g. *P. croceus,* n. sp. = *Plutonaster sp.* WOOD-MASON & ALCOCK '91· *P. rhodopeplus,* n. sp. PSEUDARCHASTER. *P. mosaicus,* n. sp. PORCELLANASTERIDÆ. PORCELLANASTER. *P. caruleus* WYV. THOMSON. *P. sp.* prox. *caruleus* WYV. THOMSON. STYRACASTER. *S. horridus* SLADEN. *S. clavipes,* n. sp. HYPHALASTER. *H. tara,* n. sp. PENTAGONASTERIDÆ. PARAGONASTER. *P. sp.* prox. *ctenipes* SLADEN. *P. sp.* ZOROASTERIDÆ. ZOROASTER. *Z. sp.* ASTERIADÆ. ASTERIAS. *A. mazophorus,* n. sp. PTERASTERIDÆ. MARSIPASTER. *M. hirsutus* SLADEN. HYMENASTER. *H. nobilis* WYV. THOMSON. ECHINASTERIDÆ. DICTYASTER, n. g. *D. xenophilus,* n. sp. = *Plectaster sp.* WOOD-MASON & ALCOCK '91· BRISINGIDÆ. BRISINGA. *B. insularum,* n. sp. *B. bengalensis,* n. sp. *B. andamanica,* n. sp. FREYELLA. *F. benthophila* SLADEN.

INDEX

OF

SPECIFIC GENERIC AND FAMILY NAMES.

Acalia   752.
Acantharchaster   693, 738.
    aciculosus   694.
    dawsoni   44, 694.
    echinites   707.
    intermedius   694.
    solaris   738.
    variabilis   694.
    variabilis pedicellaris   694.
Acanthaster   682, 754, 759, 770.
    echinites 670, 674, 678, 679, 714, 715, 722,
        754, 758, 759, 770.
    echinus   682, 754.
    ellisii   676, 759, 763, 764.
    mauritiensis   714.
    solaris   682, 764.
Acodontaster   769.
Albatrossaster richardi   708.
Albatrossia semimarginalis   722.
Alexandraster inflatus   695.
    mirus   ˙722˙
Allasterias   770.
    rathbuni   770.
    rathbuni var. anomala   770.
    rathbuni var. nortonensis   770.
Amphiaster   763.
    insignis   677, 763, 764.
An pentadactylosaster   712.
Anasterias   692, 721, 770.
    belgicæ   721.
    chirophora   721.
    lactea   721.
    minuta   712, 770.
    studeri   721.
Anodiscus crispatus   55, 688, 742, 749.
Anseropoda   656, 672, 683, 692, 693, 697.
    insignis   692.
    membranacea   672, 717, 752.
Anseropodidæ   692.

Anthenea   440, 570, 581, 584, 618, 689, 698,
        699, 700, 740, 751, 759, 771.
    acuta   740, 751.
    articulata   443, 740, 771.
    chinensis   440, 685, 698, 700, 734, 740.
    flavescens   668, 669, 676, 740, 757.
    granulifera   699, 700, 739.
    grayi   740.
    mertoni   711.
    pentagonula 439, 443, 444, 584, 670, 676,
                740, 757, 759.
    regalis   710˙
    rudis   710.
    sp.   704, 710, 748.
    tuberculosa   679, 699, 700, 740, 751.
Antheneidæ 569, 581, 704, 719, 722, 751, 737, 769.
Antheniaster   691, 768.
    epixanthus   691.
    sarissa   768.
Anthenoides   667, 697, 743.
    granulosus   697.
    lithosurus   697.
    peircei   743.
    rugulosus   697.
    sarissa   667, 768.
Anthosticte   178.
Aphroditaster   749.
    gracilis   749, 769.
    microceramus   696.
Archaster   18, 24, 25, 228, 229, 260, 262, 264,
            265, 372, 373, 388, 400, 437, 685,
            691, 734, 736, 738, 741, 743, 749,
            756, 759, 771.
    agassizii   764, 765, 766, 769.
    americanus   764, 765, 766.
    andromeda   719, 730, 750, 761.
    angulatus 4, 9, 10, 11, 14, 16, 17, 22, 265,
            714, 727, 734, 738, 741, 748, 759,
        771.

angulosus 685, 738, 741.
arcticus 228, 229, 671, 680, 694, 718, 750, 764, 765, 766.
bairdii 764, 765, 766, 768.
bifrons 671, 688, 716, 745, 749.
christi 719, 741, 750, 759.
coronatus 744.
dawsoni 44, 694, 723.
echinulatus 723, 742, 744, 745.
efflorescens 744, 745, 769.
excavatus 750, 759.
floræ 719, 761, 764, 765, 766.
formosus 765, 766.
grandis 765, 766.
hesperus 257, 258, 259, 262, 265, 272, 437, 438, 685, 726, 734, 741, 750.
hystrix 257.
insignis 744, 745.
lucifer 739.
lütkeni 671.
magnificus 750.
mauritianus 9, 714, 741.
mirabilis 743, 744, 745, 764.
nicobaricus 5, 6, 15, 715, 725, 730, 741, 749, 756.
parelii 62, 261, 371, 372, 373, 374, 375, 376, 379, 383, 387, 388, 391, 400, 667, 668, 671, 676, 694, 708, 718, 729, 730, 736, 741, 749, 764, 765, 766, 768.
parelii var. longobrachialis 374, 376, 383, 386, 387, 391, 678, 694.
pulcher 743, 744, 745.
rigidus 745.
robustus 708, 765, 766.
sepitus 723, 765, 766.
simplex 35, 38, 40, 723, 743, 744, 745.
subinermis 716, 749.
sulcatus 259, 725.
tenuis 24, 673, 674.
tenuispinus 375, 671, 677, 680, 688, 706, 718, 726, 729, 730, 736, 742, 749, 760, 764, 765, 766.
typicus 3, 4, 6, 7, 8, 9, 10, 13, 14, 16, 17, 22, 24, 262, 263, 265, 668, 670, 673, 674, 675, 676, 678, 685, 691, 700, 707, 710, 711, 715, 721, 725, 726, 727, 734, 738, 741, 746, 749, 756, 757, 759, 771.
typicus multispina 14, 17.
vexillifer 671, 708, 730, 750.
Archasteridæ 3, 360, 410, 666, 667, 671, 691, 706, 708, 713, 716, 718, 720, 721, 735, 743, 745, 749, 756, 757, 759, 765, 766, 769, 771, 772.
Arthraster 687.
dixoni 687.
Artocreatis altera species 707, 713, 756.
Artoceras altera 685.
Artocreas altera 734.
Aspidaster 713.
delgadoi 713.
Asteracanthion 615, 672, 673, 681, 725, 731, 737, 755, 770.
africanus 681, 732, 737.
albulus 687, 720, 726, 728, 753, 767.
americanus 681.
antarcticum 712.
aster 681, 732.
aurantiacum 700.
aurantiacus 681, 732, 737, 753, 762.
australis 689, 755.
berylinus 767.
bootes 681.
bootfs 732.
calamaria 681.
camtschaticum 681, 720.
clavatum 712.
constellatus 681.
desmoulinsii 681.
echinata 681.
epichlora 681.
fulgens 712.
fulvum 712.
forbesii 681, 767.
gaveyi 681.
gemmifer 737, 755.
gelatinosa 681.
gelatinosus 419, 732, 737.
germanii 681.
glacialis 671, 681, 730, 731, 737.
glacialis variatio profundus 717.
globiferus 681.
graniferus 681, 689, 732.
grœnlandicum 679, 683.
grœnlandicus 681, 688, 767.
helianthoides 681.

helianthus   681, 732, 755.
hispidum   681·
janthina   681.
katherinæ   681·
lacazii   737.
lanceolata   681.
linckii   679, 681, 700, 707, 717, 732, 737.
littoralis   767.
lumbricalis   681.
luridum   681, 712, 721.
margaritifer   681, 732.
matutina   681.
miniatum   681.
mite   712.
mülleri   676, 681, 729, 730, 736.
novæ boracensis   737.
ochotense   687.
ochraceus   681.
palæocrystallus   687, 720, 728.
pallidus .   767.
pectinatum   681.
polare   687.
polaris   681, 732, 737, 767.
polyplax   681, 689, 753.
problema   681, 687, 720, 726, 728, 731,
               753, 767.
roseus   672, 681, 687, 720, 728, 732, 736,
              737, 753.
rubens   676, 681, 687, 730, 732, 735, 737.
rubens var. migratum   748.
spectabile   712.
stellionura   679, 707, 720, 728, 737, 767.
striatus   681, 732, 737.
sulcifer   712, 737, 755.
sulciferus   712.
tenuiradiatus   681.
tenuispinus   681, 713, 732, 737.
varium   712.
violaceus   681, 732, 737.
webbianus   673, 681.
wilkinsonii   681.
yvaryana   681.
Asterias 610, 667, 671, 673, 680, 681, 682, 683,
          685, 686, 697, 699, 712, 717, 725, 736,
          743, 750, 755, 759, 770, 772.
    acervata   715, 720.
    acicularis   671.
    acuminata   .   675, 682, 712.
    æqualis   762·

affinis   682, 687, 695, 706, 719, 729.
africana   717, 755.
albula   681, 726, 753.
alboverrucosa   682, 687, 695, 719, 729.
alternata   741.
americana   687.
amurensis   680, 706, 725, 728, 755.
angulosa   673, 675, 681, 699, 712,
             717, 742,743, 755.
antarctica   721, 723, 759, 763.
antiqua   683.
arancia   749.
aranciaca 671, 673, 686, 712, 716, 718,
           730, 734, 742, 750, 760.
aranciata   371, 718, 719, 750.
arenata   675, 712.
arenicola   686, 687, 767.
articulata   707, 713, 734, 742, 756.
aspera   672, 682, 688, 695, 698, 700, 719,
          729, 753.
aster   697, 699.
atlantica   755, 763.
attenuata   682.
auranciaca   675, 688.
aurantia   699.
aurantiaca 52, 371, 671, 685, 686, 698, 700,
            716, 718, 734, 735, 736, 741, 742,
            753, 762.
belli   759.
bicolor   675, 683, 712.
bispinosa   675, 686, 716, 734, 742.
borealis   720, 767.
brevispina   762.
briareus   765, 767.
calamaria 669, 680, 681, 689, 697, 699, 705,
           713, 727, 732, 755, 759.
calcar   675, 683, 699, 702, 712, 740, 741,
          752.
calcitrapa   675, 686, 712.
camtschatica   720.
canariensis   683, 709, 725, 745, 753.
capitata   762.
carinifera   610, 611, 618, 624, 675, 679,
             692, 700, 712, 714, 734, 740,
             751.
cartilaginea 672, 699, 701, 717, 740, 752,
             761.
chilopora   685.
chinensis   440, 442, 698, 700.

christi 741.
ciliaris 671, 686, 698, 716, 734, 741, 750.
clathrata 673, 675, 681, 741, 755, 766.
clavata 712.
clavigera 675, 679, 683, 699, 700, 712, 714 754.
cometa 675, 701.
compta 755, 767.
conferta 762.
constellata 687.
contorta 742, 743.
coriacea 682, 717, 732, 752.
cribraria 715, 720.
crispata 52, 53, 54, 55, 671, 680, 686, 688 694, 700, 718, 726, 742, 749, 760, 766.
cumingii 681, 697, 699, 712, 732.
cunninghami 712, 755.
cuspidata 675, 685, 711, 734, 739.
cylindrica 675, 682, 698, 701, 712, 714, 752.
discoidea 675, 698, 699, 700, 711, 733, 739.
distichum 681.
dorsata 669, 739, 751.
douglasi 720.
dunkeri 684.
echinata 681, 697, 699, 732, 763.
echinites 675, 679, 682, 698, 712, 715, 732, 754.
echinophora 673, 681, 682, 698, 701, 712, 717, 732, 755.
echinulata 671.
echinus 682, 698, 701.
edmundi 717.
endeca 672, 675, 680, 682, 688, 695, 698, 700, 712, 719, 732, 753, 767.
endica 672, 688, 719.
enopla 767.
epichlora 762.
equestris 411, 412, 413, 414, 611, 618, 663, 671, 675, 684, 685, 698, 699, 700, 701, 712, 719, 725, 733, 734, 735, 736, 738, 740, 751.
eustyla 755.
exigua 672, 683, 699, 701, 712, 716, 733, 740, 741, 752.
fabricii 767.
fascicularis 742, 743.
fimbriata 675, 686.
fissispina 762.

forbesii 690, 758, 765, 767.
forcipulata 770.
forreri 677.
fragilis 689, 705, 759.
franklinii 683.
gelatinosa 699, 758, 763.
gemmifera 755, 758.
germannii 763.
gibbosa 672, 675, 683, 685, 699, 701, 716, 736, 740.
gigantea 762.
gigas 739.
glacialis 482, 667, 673, 675, 677, 681, 690, 697, 699, 702, 703, 709, 712, 717, 720, 728, 732, 735, 736, 737, 755, 758, 762, 770.
glomerata 755.
gracilis 742, 743, 759.
granifera 675, 689, 699, 702, 712.
granularis 671, 675, 684, 699, 719, 734, 738, 750, 766, 768.
grœnlandica 679, 707, 709, 710, 720, 728, 731, 746, 758, 760, 767.
gunneri 679, 707, 720, 728, 737, 746, 758.
harttii 759.
helgolandica 673.
helianthemoides 687, 695, 719, 753.
helianthoides 672, 712.
helianthosus 682.
helianthus 675, 681, 697, 699, 712, 755, 762.
heptactis 717.
hexactis 762.
hispida 673, 681, 735, 736.
holsatica 673, 681, 697, 699, 755.
hyndesi 712.
hyperborea 679, 707, 709, 720, 728.
ianthina 699, 732.
imbricata 685, 762.
imperati 671, 716, 750.
irregularis 671, 675, 718, 736, 742, 760.
janthina 699, 732.
japonica 680.
jehennesii 689, 714, 755.
johnstoni 671, 684, 686, 698, 700, 716, 719, 733, 734, 740, 742, 751.
jurensis 684.
katherinæ 681, 697, 699, 732, 762.
lævigata 675, 682, 698, 701, 702, 712, 715.

717, 726, 732, 753.
linearis 742, 743.
linckii 494, 495, 500, 505, 669, 670, 675, 678, 679, 683, 698, 700, 703, 707, 709, 720, 728, 729, 731, 737, 739, 758, 760.
littoralis 767.
longipes 682, 716, 741, 752.
lütkeni 762.
luna 675.
lurida 763.
madeirensis 673, 717, 755.
major 739.
mammillata 445, 684, 702, 733, 739.
mandelslohi 686, 687.
matutina 687.
mazophorus 667, 772.
membranacea 662, 672, 675, 683, 699, 701, 712, 716, 717, 733, 736, 740, 752, 761.
meridionalis 755, 759.
miliaris 702, 733.
militaris 663, 672, 680, 688, 696, 719, 726, 729, 734, 754, 760, 767.
milleporella 670, 675, 683, 699, 712, 714, 727, 733, 752.
miniata 695, 762.
minima 717.
minuta 673, 675, 681, 687, 699, 701, 716, 720, 741, 752, 755.
mitis 712.
mollis 689, 705, 759.
mulleri 667, 671, 673, 675, 685, 687, 701, 702, 703, 707, 720, 721, 729, 731, 736, 737, 738, 755, 758, 762.
multifora 675, 679, 682, 692, 712, 714, 732.
multiforis 726.
multiradiata 681, 697, 699, 712, 732.
murrayi 673, 735.
nautarum 668.
nipon 680.
nobilis 675.
nodosa 468, 470, 494, 500, 669, 675, 683, 698, 700, 712, 715, 727, 733, 739, 751.
normani 707, 720, 728.
obtusa 687.
obtusangula 445, 675, 684, 712, 733, 740.
obtusata 445, 679, 684, 715, 733, 739, 751.

ocellifera 675, 685, 699, 702, 711, 734, 738, 751.
ochotense 681.
ochracea 732, 762.
oculata 672, 675, 680, 682, 687, 695, 698, 700, 720, 732, 736, 754, 767.
oculifera 699, 701.
ophidiana 675, 682, 701, 712, 717, 732, 752.
ophidion 755, 767.
pallida 767.
palmipes 717.
panopla 679, 707, 709, 720, 728, 729, 731, 737, 758, 760.
papposa 672, 675, 679, 682, 687, 695, 698, 700, 712, 719, 729, 732, 753, 760, 767.
papulosa 770.
papyracea 717.
paucispina 762.
pectinata 671, 716, 741, 750.
penicillaris 683, 712, 733, 752.
pentacantha 686, 716, 734, 742.
pentadactyla 739.
pentagonula 442, 439, 675, 685, 700, 712, 734, 740.
pentascyphus 683, 733, 739.
perrieri 712, 721, 755, 759.
pertusa 672, 675, 682, 687, 695, 720, 728, 754.
petalodea 700.
phrygiana 671, 675, 680, 684, 700, 719, 733, 736, 740, 751.
placenta 672, 683, 698, 699, 700, 701, 717, 733, 736, 740, 752, 761.
platyacantha 716, 734, 742.
pleyadella 675, 679, 685, 711, 734, 739.
polaris 52, 54, 671, 680, 686, 688, 690, 697, 718, 720, 731, 734, 742, 749, 758, 767.
polyplax 669.
polythela 770.
prisca 686.
problema 746, 760.
propinquus 686.
pulchella 672, 675, 683, 698, 699, 701, 716, 740.
pulvillus 663, 672, 675, 685, 701, 712, 734, 736, 740, 751, 761.
punctata 675, 684, 711.
quinqueloba 685.
rectus 686.

regularis 675.
reticulata 669, 675, 683, 698, 699, 700, 712, 733, 739.
richardi 717.
rodolphi 688, 705.
rollestoni 626, 677, 680.
rosacea 662, 663, 664, 675, 683, 699, 701, 712, 717, 733, 740.
rosea 672, 720, 736, 753.
rubens 667, 668, 671, 673, 675, 678, 679, 681, 687, 690, 697, 698, 699, 700, 702, 703, 709, 710, 712, 716, 717, 720, 726, 734, 735, 736, 737, 741, 750, 755, 758, 762, 765.
rugispina 712, 723, 759, 763.
rupicola 689, 712, 759.
rustica 681, 697, 699, 732, 763.
sagena 701, 717.
sanguinolenta 672, 680, 682, 687, 695, 701, 717, 720, 728, 736, 754, 760, 767.
satsumana 680.
saveresii 681, 699, 717, 732, 755.
savignyi 686, 698, 700, 702, 714, 734, 736, 741.
scabra 689, 705.
scalpifera 755.
schmideliana 698, 699, 700, 714, 739.
schultzii colta 685.
scutata 684.
sebæ 675, 698, 701, 733, 739.
secunda 739.
senegalensis 675, 686, 712, 734, 741.
seposita 672, 675, 682, 687, 695, 698, 700, 701, 702, 712, 717, 720, 732, 754.
serrulata 681, 699, 732.
sertulifera 762, 764.
similispinus 677.
solaris 682, 698, 701, 732.
sp. 675, 704, 714, 715, 716, 719, 721, 725, 750.
sphenoplax 671.
spinosa 673, 675, 681, 697, 698, 699, 701, 717, 755.
spinulosa 686, 716, 734, 742.
spirabilis 712.
spitzbergensis 707, 720.
spongiosa 672, 675, 682, 687, 695, 720, 728, 754, 767.

stellata 445, 682, 698, 700, 733, 739.
stellata obtusa ciliata 699, 701.
stellifera 684.
stellionura 675, 679, 688, 709, 725, 737, 746, 758, 767.
stichantha 755.
stimpsoni 767.
stratifera 685.
striata 675, 692, 712, 714.
studeri 759.
subinermis 686, 716, 734, 745, 749.
subulata 675, 682, 712, 716, 732, 741, 752.
sulcifera 712, 721, 723, 755, 759.
tabulata 684.
tanneri 764, 765, 767.
tenera 758, 767.
tenuiradiata 687.
tenuispina (-is, -us) 673, 675, 689, 699, 709, 712, 717, 725, 727, 755.
tenuissima 716.
tessellata 440, 671, 675, 684, 685, 700, 701, 711, 734, 738, 745, 750.
tomidata 712, 755.
torquata 680, 706, 728, 755.
triremis 755.
troschelii 762.
trochiscus 733.
tuberculata 739.
umbilicata 716.
vappa 152, 153, 164, 165.
varia 712.
variolata 670, 675, 699, 701, 712, 717, 733.
vermiculata 701.
vernicina 675, 685, 711, 734, 740.
verrilli 712.
verrucosa 716.
verruculata 672, 683, 716, 733, 740.
verruculenta 736.
versicolor 680, 755.
vesiculosa 755.
vestita 766.
victoriana 770.
violacea 673, 675, 687, 697, 699, 720, 736.
volsellata 673, 674, 755.
volsatella 680.
vulgaris 690, 764, 765, 767.
wilkinsonii 681, 687, 697, 699, 732.
Asteriidæ (-adæ) 667, 673, 689, 690, 692, 697, 699, 705, 707, 709, 712, 717,

720, 721, 722, 735, 736, 743,
744, 755, 758, 759, 767, 770,
772.

Asterina 494, 653, 672, 683, 692, 693, 695,
697, 699, 712, 716, 725, 736, 740,
743, 752, 756, 759, 771.

  batheri 647, 648, 651.
  belcheri 669, 741.
  borealis 690, 709, 764, 765, 766.
  brevis 669, 679, 741, 752, 759.
  burtonii 670, 683, 701, 702, 714, 727,
    733, 741, 752, 756.
  cabbalistica 644, 645, 689, 725, 740, 752.
  calcar 669, 683, 699, 702, 733, 740,
    757, 771.
  calcarata 645, 741.
  cephea 647, 669, 670, 674, 676, 678,
    679, 704, 705, 711, 714, 715,
    718, 722, 727, 728, 741, 746,
    752, 756, 757, 759.
  chilensis 712, 728, 741.
  coccinea 718, 741.
  coronata 727, 757.
  exigua 674, 676, 707, 711, 715, 741,
    746, 752, 757, 759.
  fimbriata 712, 723, 741, 759.
  folium 741, 752.
  gayi 712, 741.
  gibbosa 635, 647, 672, 677, 683, 699, 701,
    702, 709, 716, 717, 725, 727, 733,
    735, 736, 740, 741, 757, 771.
  granifera 741.
  granulosa 692, 741.
  gunnii 643, 644, 669, 683, 689, 699,
    701, 733, 740, 752.
  krausii 683, 699, 701, 741, 752.
  lorioli 711.
  lincki 679.
  lymani 743.
  maculata 741, 752, 759.
  marginata 673, 712, 740, 757.
  miniata 636, 637, 695.
  minuta 672, 683, 699, 701, 740, 741.
  modesta 762.
  neozelanica 645, 646, 689.
  novæ-zelandiæ 643, 644, 647, 649, 689,
    705, 741.
  obtusa 741.
  pectinifera 626, 634, 635, 636, 637, 653,

676, 680, 706, 726, 728, 741,
752, 757.

  penicillaris 651, 652, 727, 728, 752.
  pentagona 727, 741, 752.
  pilosa 743.
  pusilla 741.
  pygmæa 767.
  regularis 644, 646, 647, 669, 688, 689,
    705, 740, 752, 757, 762.
  setacea 741.
  squamata 741.
  stellaris 741.
  stellifera 673, 712, 741.
  trochiscus 727.
  tumida 680, 706, 719, 729, 730.
  verruculata 635.
  wega 741.
  wesseli 741, 743.

Asterinidæ 634, 667, 672, 674, 689, 692, 695,
699, 701, 705, 709, 712, 716, 718,
722, 735, 740, 743, 752, 756, 757,
759, 767, 770, 771.

Asteriscus 635, 672, 683, 725, 733, 738, 740.
  antiquus 683.
  arrecifensis 717.
  australis 683, 689, 733, 740, 752.
  brasiliensis 673, 741.
  calcar 683, 740, 752.
  calcaratus 738, 741.
  cepheus 683, 714, 715, 733, 741.
  chilensis 712, 763.
  ciliatus 717, 725.
  coccineus 683, 699, 702, 733, 741.
  diesingii 684, 733, 740, 752.
  exiguus 699, 701, 738, 740, 752.
  folium 741.
  gibbosus 672.
  krausii 683, 733.
  marginatus 673, 738, 740.
  miliaris 702.
  militaris 672, 687, 688, 696, 706, 719,
    729, 733, 742, 754.
  miniatus 695.
  minutus 673, 683, 701, 733, 740.
  modestus 762.
  palmipes 672, 683, 701, 717, 733, 740, 752.
  panceri 717.
  pectinifer 634, 635, 683, 733, 741.
  penicillaris 683, 733.

pentagonus   683, 702, 715, 733, 741, 752.
placenta   717.
pulchellus   717, 738.
regularis   644, 740, 752, 762.
rosaceus 663, 664, 668, 683, 701, 733, 740.
setaceus   683, 733, 741.
squamatus   738.
stellifer   730, 741.
trochiscus   683, 733.
vermiculatus   702.
verriculatus   733.
verruculatus   683, 717, 725, 727, 738.
                740, 741.
wega   738, 741.
zelandicus   740.
Asteroceras altera   739.
Asteroderma papillosum   712.
Asterodiscus   691, 699, 700, 739, 751, 771.
elegans   699, 700, 739, 751, 771.
tuberculosus   691.
Asterodon granulosus   720, 722.
grayi   713, 722.
pedicellaris   713, 722.
singularis   713, 720, 722.
Asterope   693.
carinifera   610, 751.
Asteropidæ   695.
Asteropsis   615, 672, 685, 692, 693, 701,
             734, 740, 751, 771.
capreensis   716.
carinifera   610, 611, 612, 613, 615, 616,
             618, 627, 685, 700, 714, 727,
             734, 738, 740, 751.
ctenacantha 613, 615, 685, 734, 738, 740,
             751.
imbricata   615, 695, 738, 740.
imperialis   688, 705.
pulvillus   615, 672, 676, 685, 701, 734,
             736, 738, 740, 751.
vernicina   615, 734, 738, 740.
Asthenactis   692.
papyraceus   692.
Astrella   744.
simplex   716, 744.
Astroceramus   691.
callimorphus   691.
fisheri   710.
lionotus   696.
sphæriostictus   696.

Astrogonium 367, 383, 388, 400, 410, 415, 417,
             671, 684, 691, 699, 700, 708, 724,
             725, 733, 737, 745, 750, 751, 767,
             768, 771.
abbensis   684.
abnormale   688, 705, 724.
aculeatum   671, 684, 719, 736, 751.
æquabile   707, 708.
angustatum   684.
annectens   708, 745, 768.
aphrodite   367, 769.
articulatum   441, 442, 444, 707, 713,
                738, 740.
asculentum   740.
astrologorum   684, 724, 733, 738, 750.
australe   684, 701, 724, 734, 737, 738.
boreale   671, 684, 719.
bowerbanki   684.
capella   415.
chilopora   685.
combii   684.
compactum   684.
costatum   685.
couloni   684.
crassimanum   688, 724, 730.
cuspidatum   684, 734, 738, 745, 750.
dilatatum   688.
dubium   737, 738, 750.
dübeni   724.
dutempleanum   684.
dutemplei   685.
elongatum   745.
emilii   724, 737, 738.
eminens   707, 708.
fallax   708, 745, 768.
fleuriausum   684.
fonki   684, 763.
geometricum   684, 724, 734, 738.
gibbosum   738, 757.
gracile   769.
granulare 671, 676, 684, 694, 699, 700,
           719, 724, 730, 734, 735, 738,
           750, 761, 765, 766, 768.
hunteri   684.
huttoni   724.
hystrix   745.
inæquale   699, 700.
jordani   710.
jurense   . 684.

lamarckii 684, 734. 738.
latum 684.
longimanum 436, 437, 668, 678, 725, 727, 730, 739.
longobrachiale 387, 388, 410, 707.
lunatum 685.
magnificum 684, 724, 733.
malbosii 684.
mammillatum 738.
mantellii 684.
marginatum 685, 708.
meridionale 713, 751.
miliare 688, 699, 700, 738.
moulinsii 685.
mosaicum 684, 710.
necator 708, 745, 769.
nobile 684, 724.
ornatum 684, 724, 734.
parelii 388, 390, 391, 394, 395, 396, 397, 399, 400, 410, 690, 694, 708, 731.
parkinsoni 684.
paxillosum 699, 700, 738.
phrygianum 611, 671, 676, 684, 719, 733, 735, 736, 737, 740, 751, 766.
porosum 684.
pretiosum 366, 680.
pulchellum 684, 688, 701, 705, 724, 734, 737, 738.
punctatum 684.
quinquelobum 685.
roseum 710.
rugatum 684.
rugosum 688.
scutatum 684.
semilunatum 727, 750.
singulare 713, 738.
smithii 684.
souleyetii 417, 684, 725, 727, 739.
sp. 688.
spinulosum 738.
stelliferum 684.
stokesii 685.
stratiferum 685.
sublunatum 684.
tabulatum 684.
tesselatum 700.
tuberculatum 685, 699, 700, 738. 759.
uncatum 685.

variabile 684.
Astromesites 178, 696.
compactus 696.
Astropecten 4, 8, 10, 11, 12, 13, 23, 25, 52, 55, 57, 58, 59, 66, 76, 94, 119, 122, 134, 148, 164, 178, 201, 208, 214, 219, 221, 222, 228, 229, 241, 249, 250, 260, 270, 273, 278, 368, 370, 372, 373, 391, 400, 667, 671, 680, 685, 691, 692, 693, 694, 697, 700, 704, 716, 718, 734, 735, 736, 742, 744, 748, 750, 756, 759, 771.
acanthifer 119, 696, 713, 748, 750.
acicularis 718, 736, 760.
adriaticus 686.
alatus 263, 742, 757.
alligator 743, 744.
americanus 766.
andersoni 668, 710, 756.
andromeda 261, 372, 375, 668, 671, 675, 686, 719, 734, 735, 741, 761.
antarcticus 759.
antillensis 227, 742.
archimedis 742.
arcticus (-um) 227, 228, 229, 231, 238, 671, 680, 686, 694, 718, 729, 750, 766.
aranciaca 742.
arenarius 742.
arenicolus 686.
armatus 145, 147, 148, 149, 150, 152, 153, 155, 157, 159, 162, 164, 165, 166, 670, 680, 686, 688, 694, 698, 700, 706, 707, 714, 715, 724, 726, 734, 742, 748, 750, 763, 771.
articulatus 119, 121, 129, 686, 714, 724, 734, 742, 744, 756, 765, 766.
aster 150, 152, 716, 724, 742.
aurantiacus 134, 227, 685, 698, 700, 708, 716, 734, 742, 757, 759, 771.
benthophilus 721.
bispinosus 188, 685, 716, 734, 742.
brasiliensis 227, 685, 700, 734, 742, 748, 750.
brevispinus 215, 221, 222, 224, 227, 748, 750.
buschii 686.
calcitrapa 686.

californicus 694.
callistus 691.
capensis 759.
castellensis 686.
chinensis 157, 159, 164, 688, 727, 750.
ciliatus 686, 724.
cingulatus 748, 750.
christii 671, 683, 750, 761.
colei 686.
coppingeri 669, 678, 715.
corniculatus 52, 54, 55, 57, 277, 686, 688, 697, 700, 718, 729, 734, 742, 749.
cottaldinus 686.
cotteswoldiæ 686.
crenaster 685, 716, 742.
crispatus 686.
ctenophorus 691.
debilis 710.
dubius 686, 698, 700, 742.
duplicatus 227, 685, 697, 700, 734, 742.
dussumieri 742.
echinatus 685, 686, 698, 700, 716, 734, 742.
echinulata 202, 686, 718, 736, 750, 760.
edwardsii 686, 705, 762.
ensifer 157, 159, 164, 688, 750.
eremicus 696.
erinaceus 145, 227, 677, 686, 694, 698, 700, 742, 763.
euryacanthus 125, 670, 704, 725.
exiguus 721.
fimbriatus 686, 698, 700.
formosus 124, 200, 748.
fragilis 762, 763.
gracilis 258, 273, 686, 698, 700.
granulatus 678, 686, 711, 734, 750.
griegi 709.
hastingiæ 686.
helgolandicus 718, 760.
hemprichii 119, 120, 121, 123, 125, 126, 127, 128, 129, 130, 131, 142, 143, 670, 674, 686, 702, 704, 705, 714, 718, 734, 748, 756.
hermatophilus 748, 750.
hestrix see hystrix.
hispidus 686, 734.
hystrix 144, 146, 150, 151, 152, 153, 155, 164, 166, 167, 670, 686, 714, 715, 734, 742, 748, 750, 771.

ibericus 744.
imbellis 748, 750.
indicus 704, 710.
inermis 715.
inutilis 710.
irregularis 202, 667, 668, 671, 685, 689, 698, 700, 702, 703, 708, 718, 735, 736, 742, 750, 756, 757, 759, 760.
japonicus 119, 120, 122, 123, 124, 125, 128, 129, 142, 143, 174, 176, 177, 680, 686, 706, 726, 734, 748, 750.
javanicus 187, 373, 668, 710, 725, 742, 757.
johnstoni 152, 686, 716, 734, 742.
kagoshimensis 129, 185, 187, 188, 680, 715.
kœhleri 710, 715.
lævis 686.
latespinosus 193, 194, 680, 728.
latiradiatus 742.
longipes 686, 698, 700, 752.
longispinus 686, 734.
longobrachiale 708.
ludwigi 174, 177, 178, 217, 680, 715.
lutkeni 227, 228, 686, 694, 718, 729.
luzonicus 696.
macer 263, 272, 757.
mandelslohi 686.
marginatus 686, 698, 700, 734.
mauritianus 125, 127, 128, 129, 675, 685, 698, 700, 710, 714, 734, 742, 756.
mesactus 748, 750, 759.
mesodiscus 686, 698, 700.
monacanthus 674, 710, 748, 750.
mülleri 202, 676, 718, 730, 735, 736, 738, 742, 750, 760.
myosurus 716, 738, 742.
nobilis 710.
nodotianus 686.
notograptus 131, 705, 756.
orion 686.
ornans 757.
ornatissimus 692, 694.
orsinii 187, 713.
órstedii 694, 742, 762, 764.
pallidus 678, 708.
pareh i 371, 372, 376, 407, 408, 671, 686, 694, 718, 729, 736, 741, 749, 760.

pectinatus 748, 750.
pedicellaris 696.
penangensis 715.
pentacanthus 123, 708, 716, 734, 742, 759, 760.
perarmatus 738, 742.
peruanus 763.
peruvianus 762.
philipsii 686.
platyacanthus 686, 716, 734, 742.
pleiacanthus 188, 668.
polaris 52, 54, 671, 686, 688, 697, 700, 718, 742, 749.
polyacanthus 143, 144, 145, 149, 150, 151, 152, 153, 155, 156, 157, 158, 159, 160, 162, 163, 164, 165, 166, 167, 182, 183, 184, 185, 669, 670, 674, 675, 678, 680, 686, 688, 691, 702, 704, 705, 707, 710, 711, 713, 714, 715, 718, 721, 734, 742, 748, 750, 757, 759, 771.
pontoporæus 175, 177, 748, 750.
poritoides 686.
pressii 686.
priscus 686.
productus 691.
propinquus 686.
pugnax 710.
pusillulus 691.
pusillus 757.
rectus 686.
regalis 194, 698, 700, 762.
regularis 686, 698, 700.
richardi 263, 742.
rubidus 715.
rupellensis 686.
samoensis 738, 742.
schœnleini 674, 686, 734.
scoparius 119, 121, 123, 128, 129, 130, 132, 142, 143, 168, 170, 172, 173, 177, 182, 183, 184, 185, 187, 188, 191, 195, 680, 686, 706, 711, 726, 727, 728, 734, 742.
serratus 686, 708, 734, 742, 760.
sp. 127, 139, 667, 670, 714, 729, 771.
spatuliger 263, 742.
sphenoplax 708, 735.
spinulosus 686, 716, 734.

squamatus 686, 716.
squamosus 757.
stellaris 3, 4, 6, 9, 15, 262, 685, 697, 700, 715, 734, 741, 749, 756.
stellatus 685, 698, 700, 716, 716, 734.
subinermis 686, 716, 734, 742, 745, 749, 759.
sulcatus 721.
tamalicus 187, 678, 710.
tenellus 696.
tenuispinus 671, 680, 686, 706, 718, 726, 729, 730, 742, 749, 760.
ternatensis 757.
tiedemanni 685, 734.
triseriatus 686.
umbrinus 727.
valenciennii 685, 700, 734, 742.
vappa 145, 147, 149, 150, 153, 157, 158, 164, 166, 686, 688, 714, 715, 742, 748, 750.
variabilis 742.
velitaris 130, 158, 678, 691, 710, 725, 726, 748, 750, 759.
verrilli 715.
vestitus 766.
vexillifer 671, 708.
zebra 119, 130, 674, 675, 678, 704, 705, 710, 715, 748, 750.
Astropectinidæ 119, 122, 264, 277, 667, 671, 674, 688, 689, 690, 691, 692, 693, 694, 696, 697, 700, 704, 705, 708, 713, 716, 718, 720, 721, 724, 725, 735, 736, 741, 744, 748, 749, 756, 757, 759, 766, 769, 771.
Astropus 752.
longipes 752.
Astrothauma 696.
euphylacteum 697.
Bathybiaster 178, 671, 690, 750.
loripes 750.
pallidus 708, 719, 730.
pectinatus 690, 694.
robustus 708, 766.
vexillifer 671, 703, 708, 730, 731, 750.
Bdellacoma 687.
colvini 687.
cygnipus 687.
marstoni 687.

pyrotechnicum 687.
vermiformis 687.
Belgicella
racovitzana 721.
Benthaster 692.
eritimus 692.
penicillatus 754.
wyville-thomsoni 754.
Benthopecten 25, 37, 690, 693, 694, 723.
acanthonotus 39, 690, 693, 694, 723.
antarcticus 37, 39, 723.
armatus 38, 39, 40, 689.
claviger 693, 694.
cognatus 723.
huddlestonii 723.
incertus 723.
mutabilis 693, 694.
pectinifer 723.
pedicifer 37.
semisquamatus 37, 38, 39, 723.
simplex 37, 38, 40, 723.
spinosissimus 37.
spinosus 24, 25, 36, 38, 39, 40, 723,
765, 766, 769.
Benthopectinidæ 24, 689, 690, 691, 693,
694, 723, 769.
Blakiaster 178, 744.
conicus 743, 744, 769.
Brisinga 667, 673, 687, 691, 692, 717,
744, 755, 770, 772.
alberti 692.
americana 765, 767.
andamanica 667, 772.
armillata 755.
bengalensis 667, 772.
coronata 673, 703, 709, 717, 720, 735,
736, 744, 755.
costata 765, 767.
cricophora 755.
distincta 755.
edwardsi 744.
elegans 744, 765, 767.
endecacnemos 668, 673, 676, 687, 690, 709,
735, 736, 744, 755.
evermanni 692.
exilis 691.
fragilis 692.
gracilis 710.
gunnii 667.

hirsuta 744.
insularum 667, 772.
mediterranea 717, 744.
membranacea 755.
multicostata 767.
panamensis 677, 722.
panopla 692, 710.
parallela 710.
robusta 709, 744.
sp. 717.
tenella 722.
variispina 722.
verticillata 755, 767.
Brisingidæ 667, 673, 690, 691, 692, 709, 717,
721, 722, 735, 743, 744, 755, 759,
767, 770, 772.
Bunaster
ritteri 679.

Calliaster 429, 431, 570, 572, 667, 691,
698, 700, 751, 771.
baccatus 751.
childreni 429, 431, 698, 700, 738.
corynetes 696.
mammillifer 667.
pedicellaris 691.
Calliderma 431, 433, 691, 699, 700.
dixoni 432.
emma 431, 699, 700, 738.
grayi 713, 722.
spectabilis 433, 691.
Calveria
hystrix 772.
Calvasterias 755, 770.
asterinoides 770.
stolidata 755.
Calyptraster 754.
coa 754.
Caulaster 744.
dubius 709.
pedunculatus 744.
sladeni 744.
Ceramaster 312, 691, 694, 768.
arcticus 312, 695.
clarki 312, 693, 694.
granularis 321, 322, 323, 719, 731.
japonicus 312, 317, 320, 321, 322, 323, 694.
leptoceramus 312, 319, 321, 323, 677, 694.
micropelta 311.

patagonicus 312, 677, 694.
smithi 696.
Chætaster 667, 682, 700, 716, 732, 737, 741, 745, 752, 771.
   borealis 672, 688, 753.
   californicus 695, 762.
   cylindratus 689, 727, 730, 752.
   hermanni 682, 732.
   longipes 682, 708, 716, 741, 745, 752, 759, 771.
   maculatus 682, 689, 741, 752.
   moorei 673, 674.
   munitus 730, 751.
   nodosus 741, 759.
   sp. 667.
   subulatus 682, 716, 732, 741, 752.
   tessellatus 682.
   troschelii 682, 732.
   vestitus 711.
Chætasteridæ 716.
Cheilaster 672.
   fimbriatus 672, 716.
Cheiraster 44, 49, 51, 691, 720, 723, 759.
   agassizi 693, 694, 721, 723.
   coronatus 723, 745.
   echinulatus 745.
   folini 723, 745.
   gazellæ 723, 759.
   gerlachei 720, 723.
   granulatus 723.
   horridus 44, 691, 723.
   inops 691, 709, 723.
   mirabilis 723, 745.
   niasicus 723.
   oxyacanthus 40, 51.
   pedicellaris 723, 759.
   pilosus 723.
   snyderi 691, 709, 723.
   subtuberculatus 723.
   trullipes 723.
   vincenti 723, 745.
   yodomiensis 45.
Cheirasteridæ 44, 689, 723.
Chitonaster 751.
   cataphractus 751.
Chœtaster see Chætaster.
Chondraster grandis 766.
Choriaster 605, 725, 739, 751, 771.
   granulatus 604, 605, 607, 689, 725, 739, 751, 771.

Cidaris
   variabilis 684.
Circeaster 710.
   magdalenæ 710.
   marcelli 710.
Cistina 698, 701, 733, 752, 771.
   columbiæ 683, 698, 701, 733, 763.
Cladaster 695, 768.
   rudis 768.
   validus 693, 695.
Cnemidaster 753.
   wyvillii 753.
Cœlaster 687, 697.
   americanus 681.
   constellata 681.
   couloni 687.
   mandelslohi 686.
   matutina 681.
   tenuiradiatus 681.
Cœlasterias 672, 753, 762, 770.
   australis 689, 762.
Colpaster 756.
   scutigerula 756.
Comptonia 685, 698, 700.
   elegans 685, 698, 700.
Corethraster see Korethraster.
Corethrasteridæ see Korethrasteridæ.
Coronaster 744.
   antoni 744.
   parfaiti 744.
Coscinasterias 692, 755, 762.
   euplecta 692.
   muricata 689, 714, 755, 762.
Cosmasterias 692, 755.
   lurida 721, 723.
   sulcifera 712, 721.
   tomidata 712.
Craspidaster 264, 749.
   glauconotus 668, 710.
   hesperus 257, 265, 272, 273, 277, 278, 675, 710, 711, 749, 757.
Crenaster 671, 685, 692, 742, 745, 750.
   adriatica 686.
   arenicolus 686.
   armatus 686.
   castellanensis 686.
   cottaldina 686.
   cotteswoldiæ 686.
   hastingiæ 686.

lævis   686.
luidius   685.
mandelslohi   686.
marionis   723, 745.
mollis   745.
nodotiana   686.
orion   686.
papposus   719.
philipsii   686.
poritoides   686.
propinquus   686.
rectus   686.
rupellensis   686.
semispinosus   745.
spinulosus   745.
Cribrella   667, 672, 682, 692, 693, 697, 712,
           736, 743, 744, 754, 759, 770.
    abyssalis   709, 744.
    antillarum   743, 759.
    brasiliensis   682.
    compacta   689, 705, 754.
    densispina   680, 748.
    eschrichtii   682, 687, 729, 754.
    fallax   682, 689.
    gracilis   722.
    hyadesi   712, 723.
    leviuscula   695, 762.
    lukinsii   689, 705.
    mutans   710.
    nana   722.
    obesa   723, 754.
    oculata 672, 673, 679, 680, 682, 687, 695,
           706, 709, 710, 715, 717, 720, 728,
           730, 735, 736, 746, 754, 758, 767,
           770.
    ornata   689, 705, 737, 754.
    pagenstecheri   723.
    pectinata   767.
    præstans   667, 754.
    rosea   672, 681, 720, 735, 736, 753.
    sanguinolenta 672, 673, 678, 680, 682, 687,
           690, 695, 703, 706, 709, 720,
           721, 728, 729, 730, 735, 736,
           737, 746, 754, 756, 761, 764,
           767.
    sarsii   754.
    seposita   682, 702, 717.
    sexradiata   743.
    simplex   754.

    spiculifera   695.
    studeri   723.
    sufflata   705, 754.
Crossaster   672, 682, 692, 700, 753, 770.
    affinis   680, 719, 729, 737, 746, 758.
    alternatus   692, 696.
    australis   723.
    borealis   692, 695.
    helianthus   679, 767.
    papposus   672, 678, 679, 680, 687, 688,
           690, 695, 706, 709, 710, 719,
           729, 731, 737, 746, 753, 758,
           761, 765, 767.
    penicillatus   753.
    vancouverensis   695.
Cryasteridæ   709.
Cryptaster   744.
    personatus   744.
Cryptopeltaster   690, 695.
    lepidonotus   690, 695.
Ctenaster   692, 743, 744.
    spectabilis   743, 744.
Ctenodiscus 52, 54, 55, 57, 58, 59, 64, 67, 71,
           113, 116, 671, 686, 694, 734, 738,
           742, 749, 759, 771.
    australis 56, 57, 64, 65, 69, 70, 71, 81, 89,
           90, 93, 97, 99, 100, 694, 725, 749,
           759.
    corniculatus 55, 56, 57, 59, 61, 63, 64, 65,
           66, 67, 68, 69, 70, 71, 72, 73,
           77, 79, 88, 89, 671, 679, 680,
           688, 706, 708, 710, 718, 722,
           729, 730, 742, 746, 749, 757,
           766, 771.
    crispatus 52, 53, 54, 55, 59, 66, 69, 73, 75,
           76, 77, 79, 80, 81, 82, 83, 84, 85,
           86, 87, 88, 90, 91, 93, 97, 99, 100,
           277, 671, 675, 677, 678, 680, 686,
           688, 690, 694, 700, 703, 706, 708,
           718, 722, 726, 729, 730, 731, 735,
           737, 746, 749, 760, 764, 765, 766,
           772.
    krausei 64, 65, 66, 68, 73, 75, 76, 93, 100,
           679, 680, 694, 706, 715, 718, 729,
           730.
    orientalis   696.
    polaris 52, 53, 54, 66, 680, 686, 688, 700,
           706, 718, 726, 729, 734, 742, 749,
           766.

procurator 69, 70, 80, 81, 89, 93, 97, 99, 100, 102, 694, 706, 722, 749.

pygmæus 53, 54, 55, 61, 671, 686, 688, 700, 706, 718, 729, 734, 742, 749.

Ctenophoraster 178, 691.

diploctenius 696.

hawaiiensis 208, 691.

Ctenopleura ⸍ 178, 696·

astropectinides 178, 696.

Culcita 515, 521, 531, 537, 539, 558, 559, 561, 567, 579, 580, 581, 607, 622, 683, 691, 698, 699, 700, 713, 733, 737, 739, 751, 759.

acutispinosa 517, 533, 550, 556, 561, 563, 565, 583, 668, 704.

arenosa 517, 525, 530, 531, 548, 549, 553, 556, 557, 558, 561, 562, 563, 565, 566, 568, 582, 587, 583, 668, 691, 703, 740, 737.

borealis 518, 690, 761.

coriacea 518, 519, 533, 538, 541, 547, 548, 549, 550, 567, 582, 679, 683, 703, 713, 714, 718, 733, 737, 739, 757, 759.

discoidea 526, 538, 541, 678, 683, 700, 704, 727, 733, 737, 739, 757.

grex 517, 524, 526. 529, 530, 531, 532, 537, 541, 549, 552, 557, 565, 567, 670, 683, 703, 715, 733, 737, 739, 740.

niassensis 518, 757.

novæ-guineæ 517, 519, 520, 522, 523, 524, 526, 528, 529, 530, 537, 538, 540, 541, 545, 546, 547, 550, 551, 552, 556, 557, 558, 559, 560, 561, 562, 563, 564, 565, 566, 567, 568, 582, 586, 587, 588, 596, 668, 670, 676, 679, 683, 691, 703, 707, 711, 722, 727, 733, 737, 739, 751, 757, 759.

pentangularis 520, 523, 529, 532, 538, 540, 546, 551, 556, 557, 560, 563, 582, 583, 674, 699, 700, 718, 739.

plana 517, 546, 553, 556, 557, 558, 559, 560, 561, 563, 564, 565, 566, 568, 582, 587, 679, 703, 713.

pulverulenta 526, 527, 528, 529, 538, 703, 737, 739, 751.

schmideliana 516, 521, 536, 537, 538, 540, 541, 543, 552, 556, 560, 563, 564, 566, 567, 568, 571, 578, 579, 580, 582, 588, 596, 597, 601, 670, 676, 678, 679, 698, 699, 700, 704, 705, 707, 711, 714, 718, 733, 739, 748, 757, 759, 763, 771.

sp. 670, 673, 674, 748.

veneris 518, 567, 704.

Cycethra 712, 752, 759.

electilis 712, 722, 752.

nitida 712, 722, 752.

pinguis 752.

simplex 722, 723, 759.

verrucosa 722.

Dactylosaster 698, 701, 733, 752, 759, 771.

cylindricus 683, 698, 701, 714, 733, 752.

gracilis 683, 698, 701, 733, 763.

miliaris 682.

Decactis 697.

Dermaster imbricatus 695.

Dermasterias 695, 740, 771.

imbricata 695.

inermis 620, 695, 740.

Dictyaster 667, 772.

xenophilus 667, 772.

Diplasterias 673, 692.

lütkeni 721.

lurida 723.

steineni 721.

sulcifera 712.

Diplodontias 693.

Diplopteraster 696, 754.

multipes 696, 720, 731, 764, 765, 767.

peregrinator 767.

verrucosus 767.

Dispsacaster 178, 245, 251, 667, 690, 691, 693, 694.

anoplus 251, 693, 694.

borealis 251, 693, 694.

diaphorus 696.

eximius 251, 690, 693, 694.

grandissimus 252.

laetmophilus 251, 693, 694.

nesiotes 251, 691.

pentagonalis 251, 667.

sladeni 251, 667.

Disasterina   740, 771.
  abnormalis   740, 771.
  ceylanica   678.
  spinosa   711.
Distolasterias   692.
Dodecactis   697.
  reticulata   687, 719, 729.
Dorigona   420, 435, 437, 667, 700, 768.
  arenata   690, 708, 745, 768.
  belli   710.
  confinis   710.
  jacqueti   745, 768.
  longimana   436, 437, 668, 669, 739.
  ludwigi   710.
  mœbii   759.
  mülleri   438, 739.
  pentaphylla   667.
  prehensilis   768.
  pulchellus   688.
  reevesii 420, 435, 436, 700, 713, 725, 739.
  subspinosa   745, 768.
  ternalis   710, 745, 768.
Dytaster   666, 690, 692, 694, 696, 745, 749.
  æquivocus   749.
  agassizi   708, 745.
  anacanthus   666, 772.
  biserialis   708, 749.
  demonstrans   721.
  exilis   666, 749, 772.
  evaulus   696.
  gilberti   690, 694.
  grandis   765, 766.
  inermis   749.
  insignis   745.
  madreporifer   749, 765, 766.
  nobilis ·   749.
  parvulus   707, 708.
  rigidus   708, 745.
  spinosus   749.

Echinaster   672, 681, 682, 692, 698, 700, 701,
              717, 725, 732, 737, 743, 744, 754,
              755, 759, 770.
  aculeatus   763.
  affinis   715, 737, 755.
  brasiliensis   682, 701, 725, 732.
  callosus   711.
  clouei   737.
  crassispinus   725, 755, 763.

  crassus   682, 727, 732.
  cribrella   725.
  cylindricus   728.
  decanus   682.
  deplanatus   682.
  doriæ   755.
  echinophorus   755.
  echinulatus   714, 726, 754.
  echinus   737.
  ellisii   682, 698, 701, 732, 754, 763, 764.
  eridanella   674, 677, 679, 682, 711, 715,
                 727, 732, 746, 755, 756, 758.
  eschrichtii   673, 682, 687, 706, 715, 720,
                 729, 732, 736, 754.
  fallax   669, 682, 689, 701, 714, 726,
                 727, 732, 756, 759.
  gracilis   682, 725, 732.
  lacunosus   682.
  modestus   743.
  oculatus 673, 682, 687, 695, 700, 706, 720,
           727, 728, 730, 732, 736, 754, 767.
  ornatus   689, 737, 754.
  panamensis   712.
  polycnema   705.
  purpureus   669, 670, 674, 676, 689, 704,
                 705, 711, 714, 718, 727, 756,
                 758.
  rigidus   682.
  sanguinolentus   672, 676, 682, 687, 695,
                     701, 707, 715, 717, 720,
                     729, 730, 754, 760, 767.
  sarsii   673, 682, 687, 695, 706,
                 720, 729, 736, 754.
  scrobiculatus   720, 731.
  sepositus   682, 701, 709, 714, 717, 727,
                 728, 732, 744, 758, 770.
  serpentarius   682, 725, 732.
  smithi   721.
  solaris   678, 682, 698, 701, 715,
                 727, 732, 738, 754.
  sp.   692, 726.
  spinosus   681, 701, 725, 732, 755, 758.
  spinulifer   755, 759.
  spinulosus   725.
  tenuispinus   677, 725, 764.
  tribulus   755.
Echinasteridæ   667, 672, 674, 689, 690, 692,
                 693, 695, 705, 706, 709, 712,
                 717, 718, 720, 721, 722, 735,

743, 744, 754, 756, 758, 759, 767, 770, 772.

Echinites 682, 770.
  solaris 682.
Endeca
  endeca 672, 688, 695, 719.
Enneactis (-n) 697.
  coriacea dentata 719, 729.
Eremicaster 690, 694.
  pacificus 677, 694.
  tenebrarius 96, 677, 694.
  waltharii 694.
Eugoniaster 768.
  investigatoris 768.
Evoplosoma 691.
  augusti 710.
  forcipifera 691.

Ferdina : 698, 701, 759, 771.
  cumingii 682, 698, 701, 763.
  flavescens 682, 698, 701, 714, 759, 771.
  kuhlii 718.
  offreti 711.
Freyella : 667, 691, 772.
  americana 767.
  aspera 767.
  attenuata 756.
  benthophila 667, 756, 772.
  bracteata 756, 767.
  dimorpha 756.
  edwardsi 707, 709, 744.
  echinata 755.
  elegans 767.
  fecunda 691.
  fragilissima 756.
  heroina 756.
  indica 710.
  insignis 722.
  pacifica 722.
  pennata 755.
  polycnema 755.
  propinqua 722.
  recta 707, 709.
  remex 756.
  sexradiata 709, 744.
  spinosa 744.
  tuberculata 667, 756.
Fromia 683, 698, 701, 743, 759, 771.

andamanensis 710.
armata 711.
indica 670.
japonica 743.
major 707, 711·
milleporella 670, 674, 676, 678, 683, 698, 701, 714, 727, 752, 757, 759, 771.
monilis 759.
narcissiæ 709, 745.
tumida 670, 678.
variolaris 746.

Ganeria 699, 702, 741, 752, 771.
  falklandica 699, 702, 741, 752, 763, 771.
Ganeriidæ 712, 744.
Gastraster 744.
  margaritaceus 744.
Gaussaster 723.
  vanhöffeni 723.
Gephyreaster 694.
  swifti 249, 694.
Gilbertaster 691.
  anacanthus 691.
Glyphaster 242, 250.
  anomalus 694.
Glyptaster 672, 744, 753.
Gnathaster 751, 769.
  dilatatus 688.
  elongatus 751.
  grayi 713, 722.
  mediterraneus 708, 716.
  meridionalis 713, 751.
  miliaris 688, 705.
  pedicellaris 722.
  pilulatus 713, 722, 751.
  rugosus 688, 705.
  singularis 713.
Gomophia 683, 692, 699, 701, 753, 771.
  ægyptica 683, 692, 699, 701, 702, 733.
Goniaster 262, 415, 417, 420, 437, 605, 671, 672, 683, 684, 693, 697, 698, 700, 725, 736, 740, 750, 751, 767, 771.
  abbensis 671, 719, 736, 740, 751.
  africanus 708, 738, 750, 768.
  americanus 738, 750, 767.
  articulatus 442, 528, 673, 700, 707, 713, 725, 740, 756.

belcheri 416, 436, 678, 727, 725, 748.
boworbankii 684.
capella 415, 435, 713, 726.
clavatus ' 727.
compactus 684.
couloni 684.
cuspidatus 519, 684, 698, 700, 727, 734, 738, 745, 750.
dübenii 417, 724, 725, 738.
equestris 414, 417, 418, 436, 437, 671, 684, 719, 725, 726, 727, 733, 736, 739, 740, 751, 766.
gracilis 727.
granularis 671, 719, 750, 766, 768.
hispidus 709, 719, 729, 761, 769.
hunteri 684.
inæqualis 727.
incei 416, 417, 725, 739.
jurensis 684.
longimanus 437, 727, 739.
luzonicus 727.
mantellii 684.
mülleri 416, 417, 420, 435, 436, 437, 438, 713, 725 726.
nidaroensis 719.
obtusangulus 669, 740, 771.
parkinsoni 684.
pentagonulus 727, 740.
placenta 716.
placentæformis 716.
pleyadella 727, 739.
phrygianus 671, 700, 719, 736, 737, 740, 751, 761, 766.
porosus 684.
rectilineus 685.
regularis 685, 698, 701.
rugatus 684.
scaber 727.
sebæ 519, 523, 524, 527, 685, 698, 701, 702, 714, 727, 738, 739.
semilunatus 708, 727, 738, 750.
spinulosus 524, 714, 727, 738.
sublunatus 684.
templetoni 672, 685, 699, 701, 734, 735, 736, 740, 751, 761.
tuberculosus 416, 417, 725, 726.
uncatus 685.
Goniasteridæ 12, 360, 410, 622, 690, 691, 693, 694, 696, 738, 757, 759.

766, 767, 769, 771.
Gonioliscides 583, 691.
    sebæ 584, 691.
Gonioliscus 415, 417, 420, 440, 536, 569, 570, 580, 581, 583, 584, 588, 595, 685, 725, 734, 737, 739, 743, 750, 751, 756, 771.
    acutus 716, 725, 738, 740, 751.
    armatus 713, 725, 739, 762, 763.
    articulatus 442, 534, 668, 673, 676, 707, 713, 737, 740, 756.
    capella 420, 433, 435, 436, 584, 685, 713, 734.
    conifer 700, 713, 730, 739, 763.
    cuspidatus 584, 685, 727, 734, 737, 739.
    flavescens 685.
    forficulatus 710, 739.
    gracilis 673
    granuliferus 739.
    horridus 685, 739.
    insignis 710.
    luzonicus 685.
    mammillatus 584, 685, 734, 738.
    michelini 713, 738, 739.
    mülleri 738.
    ocelliferus 584, 685, 701, 734, 738, 751.
    pedicellaris 743, 768.
    penicillatus 722.
    pentagonulus 440, 441, 442, 584, 685, 700, 734, 740.
    placenta 584, 685, 716, 734.
    placentæformis 716, 725.
    pleyadella 584, 678, 685, 727, 734, 737, 739, 746.
    porosus 710.
    regularis 584, 685, 734.
    rugosus 674, 739.
    sanderi 718, 728.
    scaber 442, 668, 730.
    sebæ 519, 522, 524, 527, 528, 533, 534, 537, 551, 558, 560, 568, 569, 570, 571, 572, 577, 578, 579, 580, 581, 583, 584, 586, 587, 588, 597, 673, 678, 679, 685, 702, 707, 713, 714, 718, 727, 738, 739, 757, 771.
    seriatus 685.
    singularis 685, 713, 738, 763.
    sp. 673, 674, 704.
    stella 713, 730, 739, 763.

studeri 588, 714.
vallei 710.
verrucosus 685, 722, 763.
Goniodon 693.
   dilatatus 705.
Goniopecten 696, 743, 745.
   asiaticus 696.
   bifrons 716.
   christi 750.
   demonstrans 743, 745, 769.
   edwardsi 745.
   inermis 745.
   intermedius 743, 745.
   subinermis 716.
   subtilis 708, 743, 745, 769.
Goniopectinidæ 696, 769.
Gymnasteria 610, 611, 615, 622, 630, 685, 692, 693, 698, 700, 740, 751, 759, 771.
   biserrata 613, 614, 615, 624, 625, 626, 630, 727, 751.
   carinifera 610, 613, 614, 615, 619, 620, 624, 625, 626, 627, 628, 629, 630, 670, 674, 676, 677, 678, 679, 692, 713, 714, 718, 722, 727, 740, 751, 757, 759, 771.
   inermis 611, 613, 615, 619, 620, 685, 698, 700, 714, 740, 751, 762.
   spinosa 611, 613, 615, 616, 617, 624, 685, 698, 700, 714, 740, 751, 762, 764.
   valvulata 629, 673, 676, 740.
Gymnasteriidæ 569, 581, 610, 672, 690, 692, 705, 706, 718, 719, 722, 735, 743, 751, 757, 766.
Gymnobrisinga 759.
   sarsii 759.
Hacelia 717, 752.
   attenuata 717, 708, 752.
Heliaster 681, 692, 712, 755, 770.
   cumingii 676, 712, 762, 763.
   helianthus 681, 712, 728, 737, 755, 758, · 762, 763, 770.
   kubiniji 677, 712, 762, 764, 770.
   microbrachia 762, 763, 764, 770.
   multiradiatus (-a) 676, 692, 712.
   solaris 715.
   sp. 715, 762.
Heliasteridæ 692, 712, 755, 758, 770.
Hemicnemis (-us) 671, 686, 702.
Henricia 672, 681, 692, 693, 695, 698, 700, 770.

abyssicola 690.
annectens 693.
aspera 692, 693, 695.
asthenactis 693, 695.
clarki 677, 693, 695.
leviuscula 677, 693, 695.
longispina 693, 695.
multispina 693.
oculata 672, 687, 689, 695, 698, 700, 707, 720, 728, 732, 736, 754.
pauperrima 692.
polyacantha 692, 693, 695.
robusta 692.
rosea 672, 681, 720.
sanguinolenta 672, 673, 675, 690, 693, 695, 702, 729, 731, 735, 762.
spiculifera 695.
tumida 695.
Heresaster 754, 770.
   papillosus 714, 726, 754.
Heptaktin 770.
Heterozonias 696.
   alternatus 677, 696.
Hexaktin 770.
Hexaster
   obscurus 680, 696, 709, 719, 721, 729, 737, 769.
Hippasteria 312, 344, 429, 431, 671, 684, 690, 695, 698, 740, 751, 771.
   abbensis 719.
   californica 351, 353, 677, 690, 695.
   caribæa 768.
   cornuta 700, 719, 751.
   equestris 684, 719.
   europæa 671, 684, 698, 700, 719, 733, 736, 751.
   heathi 690, 695.
   imperialis 338.
   johnstoni 671, 684, 698, 700, 719, 733, 736, 751.
   leiopelta 348, 353, 354, 693, 695.
   nozawai 344.
   pacifica 722.
   plana 611, 618, 624, 671, 680, 700, 708, 719, 733, 736, 740, 751, 756, 757, 765, 771.
   phrygiana 349, 350, 351, 671, 680, 690, 702, 703, 708, 719, 731, 751, 761, 765, 766.

spinosa 319, 352, 353, 354, 677, 693, 770.
Hoplaster   745, 769.
  lepidus   769.
  spinosus   745, 769.
Hosia (Hosea)   440, 581, 685, 698, 700, 750,
              751, 771.
  cuspidata   739.
  flavescens 522, 581, 668, 685, 698, 700, 740.
  mammillata   738.
  pleyadella   739.
  sebæ   714, 739.
  spinulosa 436, 522, 523, 524, 527, 570, 581,
        587, 689, 700, 714, 727, 738,
Hydrasterias   692, 744, 755.
  diomedeæ   722.
  ophidion   767.
  richardi   717, 744.
  sp.   722.
  verrilli   692.
Hymenaster 667, 672, 691, 692, 744, 754, 772.
  alcocki   710.
  anomalus   754.
  cœlatus   754.
  carnosus   754.
  coccinatus   754.
  crucifer   754.
  echinulatus   754.
  formosus   754.
  geometricus   754.
  giboryi   709, 744.
  giganteus   672, 690, 735.
  glaucus   754.
  gracilis   722.
  graniferus   754.
  infernalis   754.
  kœhleri   696.
  latebrosus   754.
  membranaceus   754.
  modestus   765, 767.
  nobilis   667, 721, 754, 772.
  pellucidus   672, 678, 680, 703, 706, 709,
       720, 729, 731, 746, 754, 758.
  pentagonalis   692, 710.
  pergamentaceus   754.
  perissonotus   677, 696.
  perspicuus   721.
  platyacanthus   722.
  porosissimus   754.
  præcoquis   721, 754.

  pullatus   754.
  purpureus   722.
  quadrispinosus   677, 691, 696.
  regalis   767, 770.
  rex   744.
  roseus   707, 709.
  sacculatus   754.
  sp.   720, 722.
  vicarius   754.
  violaceus   722.
Hymenodiscus   743.
  agassizii   743.
Hyphalaster   667, 744, 748, 749, 772.
  antonii   704, 744.
  diadematus   748, 749.
  fortis   707, 708.
  gracilis   707, 708.
  hyalinus   748, 749.
  inermis   112, 113, 115, 118, 748, 749.
  moseri   722.
  parfaiti   744.
  planus   748, 749.
  tara   667, 772.
  valdiviæ   708.

Iconaster
  longimanus   417, 668, 674, 678, 707, 711.
  pentaphyllus   710.
  perierctus   696.
Ilyaster
  mirabilis   708, 719, 730.
Isaster   768.
  bairdii   766, 768.

Johannaster   359, 360, 709.
  giganteus   361.
  superbus   709.

Koremaster   696.
Korethraster   672, 743, 744, 754.
  hispidus   672, 703, 709, 719, 726, 731,
       743, 754..
  palmatus   743, 744.
  radians   743.
  setosus   744.
  sp.   690.
Korethrasteridæ   672, 691, 696.

Labidiaster   725, 755, 759, 770.

annulatus 755.
lütkeni 721.
radiosus 721, 723, 725, 755, 759.
Lætmaster 692.
Lasiaster 672, 752, 769.
  hispidus 703, 709, 719, 729, 736, 761, 769.
  villosus 672, 690, 752, 761.
Leiaster 691, 752.
  callipeplus 691, 711.
  coriaceus 674, 714, 718.
  glaber 718.
  leachii 674, 714, 758.
  speciosus 674, 727, 752.
Lepidaster 687, 752, 764.
  grayi 687.
  teres 764.
Leptasterias 755.
  compta 755, 765, 767.
  grœnlandica 767.
  littoralis 767.
  mülleri 673, 707, 755.
  sp. 725.
  tenera 690, 767.
Leptogonaster 751.
  cristatus 751.
Leptophycaster see Leptychaster.
Leptoptychaster see Leptychaster.
Leptychaster 178, 229, 242, 249, 250, 671,
    692, 694, 750, 759.
  anomalus 236, 238, 239, 241, 242, 246,
    247, 250, 692, 694.
  antarcticus 750.
  arcticus 227, 229, 230, 231, 233, 234, 235,
    237, 242, 249, 385, 671, 680, 694,
    703, 708, 718, 729, 731, 736, 737,
    750, 757, 766.
  conicus 769.
  inermis 250, 677.
  kerguelensis 242, 249, 721, 750, 759.
  pacificus 236, 237, 246, 249, 250, 692, 694.
  propinquus 249, 250, 693, 694.
Linckia 672, 682, 683, 691, 692, 693, 695,
    697, 698, 700, 701, 713, 725, 743,
    752, 753, 759, 762, 771.
  bifascialis 682, 698, 701, 732.
  bouvieri 709, 759.
  brownii 682, 698, 701, 726, 732, 753.
  columbiæ 677, 683, 695, 698, 701.
  crassa 683, 698, 701, 753.

cylindrica 727, 752.
diplax 670, 692, 695, 704, 748, 759, 771.
dubiosa 711.
ehrenbergii 674, 711, 713, 714, 718, 757, 758.
erythræa 683, 698, 701, 733.
guildingii 683, 698, 701, 709, 733,
    752, 758, 763.
intermedia 683, 698, 701, 733.
lævigata 669, 670, 676, 704, 705, 753.
leachii 683, 692, 698, 701, 714, 726.
leviuscula 695, 762.
marmorata 669, 674, 714, 748.
megaloplax 669, 674.
miliaris 679, 711, 713, 714, 715, 718, 722,
    726, 746, 753, 757, 758, 759, 771.
milleporella 726, 752.
multifora (-is) 670, 674, 678, 679, 692,
    704, 705, 714, 718, 722,
    726, 728, 746, 757, 758,
    759.
nicobarica 714, 725, 753.
nodosa 669, 743.
oculata 672, 682, 687, 695, 700, 706, 720,
    729, 732, 736, 754, 767.
ornithopus 695, 753, 763.
pacifica 670, 683, 692, 695, 698, 701, 705,
    711, 714, 718, 722, 753, 758, 759.
pauciforis 669, 726.
pertusa 682, 687, 695, 706, 720, 729,
    736, 754, 767.
pistoria 727, 752.
pulchella 683, 698, 701, 733.
pusilla 727.
pustulata 679, 726, 759.
rosea 672, 720, 753.
rosenbergi 726.
semiregularis 726, 727.
semiseriata 726.
sp. 669, 670, 704, 771.
subulata 727.
suturalis 727.
tuberculata 726, 753.
typus 682, 692, 698, 701, 702, 714,
    715, 726, 732, 753.
unifascialis 682, 698, 701, 725, 732,
    762, 763, 764.
variolata 701, 733.
variolosa 699.
Linckiidæ 667, 674, 689, 691, 695, 705, 708,

713, 717, 718, 722, 743, 745, 752,
757, 759, 769, 771
Litonotaster   768.
  intermedius   768.
Lonchotaster   178, 749.
  forcipifer   749.
  tartareus   749.
Lophaster   672, 691, 696, 753.
  furcifer 672, 680, 688, 706, 709, 719, 729,
    731, 736, 746, 753, 758, 765, 767.
  furcilliger   677, 691, 693, 696.
  stellans   721, 753.
Lophopteraster   767.
  abyssorum   767.
Luidia 12, 57, 58, 297, 300, 671, 686, 691, 692,
    694, 698, 704, 713, 716, 734, 736, 738,
    741, 744, 750, 756, 759, 771.
  africana   750.
  alternata   741, 744.
  armata   721.
  aspera   301, 305, 674, 748, 750.
  asthenosoma   692, 694.
  barbadensis   743, 744.
  bellonæ   676, 724, 728, 741, 762, 763.
  brevispina   691, 725, 741, 764.
  chefuensis   296, 297, 311.
  ciliaris   279, 671, 677, 686, 689, 698,
    700, 703, 716, 734, 735, 738,
    750, 759, 761.
  clathrata   741, 744, 750, 765, 766, 771.
  columbiæ   676, 713, 741, 757, 763.
  convexiuscula   743, 744.
  denudata   710.
  elegans   741, 744, 764, 765, 766.
  ferruginea   721.
  foliata   694.
  foliolata   694.
  forficifer   673, 674, 676, 750, 756.
  fragilissima 671, 686, 698, 700, 716, 734,
    735, 736, 741, 750.
  gemmacea   741.
  granulosa   738, 741.
  hardwickii   670, 674, 704, 741, 757.
  hystrix   301, 305, 691.
  integra   305, 710.
  limbata   294, 297, 680, 710, 728, 750.
  longispina (·is)   668, 673, 674, 750.
  ludwigi   692, 694.
  maculata 278, 279, 280, 281, 282, 283, 284,

285, 293, 294, 297, 310, 668, 670,
  673, 674, 676, 678, 680, 686, 704,
  707, 710, 711, 714, 715, 726, 727,
  734, 741, 748, 756, 757, 771.
  magellanica   713.
  magnifica   285, 301, 305, 691.
  maregravii   741.
  mauritiensis   710.
  moroisoana   301.
  murchisoni   687.
  paucispina   716.
  penangensis   668.
  phragma   677.
  quinaria   279, 293, 294, 297, 306, 308,
    310, 680, 706.
  sarsii   671, 676, 686, 689, 716, 702, 703,
    708, 735, 736, 741, 744, 750, 756,
    759, 761.
  savignyi 279, 671, 674, 686, 698, 702, 710,
    714, 716, 718, 730, 734, 736, 738,
    741, 750, 761.
  senegalensis   280, 686, 734, 741.
  singaporensis   297.
  sp.   669, 670, 673, 691.
  tessellata   713, 741, 762.
  variegata   741.
  williamsoni   687.
  yesoensis   306.
Luidiaster   44, 51, 694, 723, 759.
  dawsoni   44, 694, 723.
  gerlachei   44, 723.
  hirsutus   44, 723, 759.
  teres   44, 723.
  vincenti   44, 723.
  yodomiensis   51.
Luidiidæ   278, 689, 691, 692, 769.
Lydiaster
  johannæ   710.
Lytaster   744.
  inæqualis   744.
Lysasterias   692.

Magdalenaster   707, 709.
  arcticus   707, 709.
Mammaster   744.
  sigsbeei   744.
Marcellaster   693, 723.
  aciculosus   693.
  antarcticus   · 723.

dawsoni 693.
intermedius 693.
variabilis 693.
Margaraster 699, 755.
grauiferus 689, 699.
scaber 689.
Marginaster 672, 716, 743, 744, 752.
austerus 770.
capreensis 716.
echinulatus 743, 744.
fimbriatus 672, 716, 752.
pectinatus 743, 744.
pentagonus 744.
Marsipaster 667, 754, 772.
hirsutus 667, 754, 772.
spinosissimus 754.
Marthasterias 775.
foliacea 673, 717, 755.
Mediaster 311, 312, 317, 318, 329, 330, 332, 336, 667, 690, 691, 694, 768.
æqualis 311, 329, 683, 694, 762, 768.
agassizii 768.
arcuatus 317, 768.
báirdii 768.
brachiatus 331, 354.
elegans 358, 722,
japonicus 317, 694, 768.
ornatus 311, 691, 710.
patagonieus 694, 768.
pedicellaris 768.
roseus 667, 710, 769.
stellatus 768.
tenellus 358, 690, 694.
transfuga 722.
Metopaster 584.
Metrodira 683, 698, 701, 753, 759, 771.
subulata 669, 674, 683, 689, 698, 701, 705, 711, 727, 733, 753, 759, 771.
Milteliphaster 667.
wood-masoni 667.
Mimaster 249, 672, 690, 751.
cognatus 249, 720, 751.
notabilis 696.
swifti 690, 694.
tizardi 672, 690, 751.
Mithrodia 683, 692, 699, 700, 754, 759, 770.
bradleyi 692, 762, 763, 764.
clavigera 674, 679, 683, 714, 718, 722, 754, 757, 758, 759, 770.
echinulata 754.
spinulosa 681, 699, 700, 714, 732, 754.
Mithrodiidæ 692.
Moiraster 750.
magnificus 750.
Myonotus 694.
intermedius 694.
Myxaster 744.
sol 744.
Myxasteridæ 692, 744.
Myxoderma 691.
Nanaster 770.
albulus 709, 710, 731.
Narcissia 683, 699, 745, 753, 771.
canariensis 709, 745, 753.
teneriffæ 683, 699, 701, 709, 745, 753.
trigonaria 753.
Nardoa 683, 692, 699, 701, 753, 771.
ægyptica 692, 711.
agassizii 699, 701, 733.
carinata 711.
frianti 711.
le monnieri 711.
novæ caledoniæ 758.
pauciforis 758.
semiregularis 680, 706, 758.
squamulosa 711.
tuberculata 674, 677, 679, 699, 701, 704, 705, 715, 746, 753, 758.
variolata 674, 699, 701, 714, 718, 733, 748, 758.
Nauricia 277, 697, 700.
pulchella 277, 278, 686, 697, 700, 726.
Nearchaster 694.
aciculosus 677, 694.
variabilis 694.
Nectria 584, 615, 685, 699, 701, 738, 751, 771.
ocellata 738.
ocellifera 584, 685, 701, 734, 738, 751, 757.
oculifera 699.
Neomorphaster 672, 744, 753.
eustichus 672, 735, 753.
forcipatus 767.
talismani 709, 735, 744.
Nepanthia 699, 701, 752, 756, 771.
belcheri 741.
brachiata 711.
brevis 669, 679, 741, 752, 759.

maculata 682, 680, 699, 701, 732, 741, 752, 759.
  suffarcinata 711, 756.
  tessellata 682, 699, 701, 716, 732, 741, 752.
Nereidaster 360, 691, 768.
  bipunctus 768.
  bowersi 360, 691.
  symbolicus 768.
Nidorellia 683, 691, 713, 739, 762, 771.
  armata 669, 676, 691, 713, 739, 762, 763, 764.
  horrida 739.
  michelini 669, 713, 739, 757.
Nymphaster 329, 360, 439, 667, 671, 696, 745, 751, 768.
  albidus 751, 768.
  arenatus 690, 708, 768.
  arthrocnemis 696.
  atopus 696.
  basilicus 667, 751, 768.
  bipunctus 751, 768.
  diomedeæ 722.
  dyscrita 696.
  euryplax 696.
  florifer 667.
  habrotatus 696.
  jacqueti 768.
  leptodomus 696.
  meseres 696.
  moluccanus 696.
  mucronatus 696.
  nora 667.
  prehensilis 768.
  protentus 667, 671, 690, 735, 751, 768.
  subspinosus 671, 690, 735, 768.
  symbolicus 360, 751, 768.
  ternalis 768.

Octactis 697.
  dactyloides 719, 729.
Odinia 673, 692, 744, 755.
  americana 767.
  austini 710.
  clarki 710.
  elegans 744.
  pacifica 692.
  pandina 673, 755.
  robusta 709, 744.
  semicoronata 731, 744.

Odontaster 344, 690, 694, 713, 716, 769.
  cremeus 720.
  crassus 690, 694.
  grayi 344, 713, 722.
  hispidus 764, 765, 766, 769.
  mediterraneus 708, 716.
  meridionalis 713, 722.
  pedicellaris 713, 722.
  penicillatus 722.
  robustus 769.
  setosus 769.
  singularis 713, 720, 722.
Odontasteridæ 690, 694, 720, 769.
Ogmaster 262, 420, 435, 726.
  capella 420, 433, 439, 584, 710, 713, 726.
Oktactis see Octactis.
Ophidiaster 682, 691, 692, 697, 698, 701, 725, 732, 737, 743, 745, 752, 753, 759, 771.
  agassizii 743.
  arenatus 683.
  armatus 711.
  attenuatus 682, 698, 701, 717, 732, 737, 752.
  asperulus 714, 725, 752.
  aurantius 682, 698, 701, 717, 732, 752.
  bicolor 683.
  bifascialis 725.
  canariensis 717.
  clathrata 753.
  colombiæ 683.
  cribrarius 725.
  cylindricus 682, 701, 704, 705, 714, 722, 732, 748, 752, 757, 759.
  diplax 682, 692, 732, 737, 753.
  duncani 714.
  echinulatus 683, 714, 726, 733, 754.
  ehrenbergii 683, 714, 733.
  floridæ 743.
  fuscus 674, 759.
  germani 771.
  granifer 725.
  guildingii 683, 698, 701.
  helicostichus 673, 674, 705, 707, 752.
  hemprichii 682, 732.
  hirsutus 711.
  irregularis 737, 753.
  lævigata 753.
  leachii 682, 698, 701, 714.
  lessoniæ 717.

lorioli   584, 691.
marmoratus   669, 714.
miliaris 670, 682, 701, 702, 726, 732, 753.
multiforis   679, 682, 692, 714, 726, 732.
ophidianus   682, 701, 709, 714, 717, 732,
                    745, 752, 757, 771.
ornatus   711.
ornithopus   682, 732, 752, 763.
pacifica   683.
perrieri   714.
porosissimus   762.
purpureus   679, 714, 715, 718, 737, 757.
pusillus   682, 757.
pustulatus   679, 718, 759.
pyramidatus   683, 698, 701, 733, 762,
                        763, 764, 771.
rhabdotus   691.
robillardi   714, 757.
sclerodermus   691.
sp.   688.
squameus   691.
suturalis   682, 701, 725, 732, 762.
tenellus   691.
triseriatus   691.
tuberculatus   683, 715, 726, 733, 753.
tuberifer   679, 711, 752.
tumescens   711.
unifascialis   725, 762.
vestitus   737.
Ophiura
arenata   683.
Oreaster 415, 443, 445, 493, 494, 505, 605, 607,
            668, 683, 693, 725, 733, 737, 771.
aculeatus   683, 700, 727, 733, 739.
affinis   445, 669, 683, 733.
alveolatus   513, 669, 707.
armatus 669, 683, 700, 725, 733, 762, 763.
australis   669, 725.
boysii   683.
bulbiferus   683.
carinatus   445, 669, 683, 733.
castellum   500, 727.
chinensis   445, 463, 669, 683, 733.
clavatus   445, 669, 683, 733, 739, 751.
clouei   468, 669, 737.
coronatus   683, 739.
cumingii   763.
decipiens   463, 669.
desjardinsii   683, 725.

dœderleini   451.
dorsatus   506, 669, 725, 751.
forcipulosus   669, 725.
franklini   468, 669.
gigas   445, 669, 725, 739, 763.
gracilis   669, 725.
granulosus   669.
grayi   469, 669.
hedemanni   669, 725.
hiulcus   474, 475, 502, 504, 506, 513,
             669, 683, 727, 733, 737.
lapidarius   669, 683, 725, 739.
lincki   482, 487, 493, 495, 505, 506,
            510, 511, 513, 669, 670, 704,
            725, 737, 739.
lütkeni   669.
magnificus   453, 457.
mammilatus   445, 504, 506, 669, 683,
             702, 733, 737.
mammosus   468, 474, 477, 478, 480,
             484, 669, 737.
modestus   444, 463, 465.
mülleri   669.
muricatus   494, 495, 500, 502, 503, 504,
             505, 506, 683, 727, 737, 739.
muticus   669.
nahensis   463.
nodosus 444, 445, 463, 465, 468, 469, 482,
             488, 494, 513, 514, 669, 683, 676,
             700, 707, 715, 725, 733, 739.
nodulosus   480, 668, 669.
obtusangulus 445, 669, 683, 733, 737, 740.
obtusatus 445, 683, 715, 727, 733, 737, 751.
obtusus   683.
ocellatus   684.
occidentalis   669, 677, 762, 763, 764.
orientalis   445, 669, 683, 733.
pistilliformis   684.
productus   669.
regulus   445, 669, 683, 733.
reinhardti   463, 669, 670, 725.
reticulatus 669, 683, 725, 727, 733, 737.
rouxi   463.
sebæ   669.
sp.   669.
superbus   669, 670, 727, 730.
squamatus   684.
thurstoni   670.
troscheli   669.

tuberculatus 445, 471, 500, 503, 669, 683,
      711, 727, 733.
  tuberosus   669, 725, 730, 739.
  turritus 468, 470, 471, 472, 473, 474, 478,
      500, 504, 506, 513, 669, 683, 700,
      715, 727, 733, 737, 751.
  valvulatus   669, 683.
  verrucosus   445, 669, 683, 733.
  westermanni   669, 670, 725.
Oreasteridæ   444.
Oreillers   683.
Othilia   681, 682, 698, 701, 755, 770.
  aculeata   682, 698, 701, 732.
  braziliensis   725.
  crassispina   725, 763.
  eridanella   715.
  luzonica 682, 698, 701, 714, 727, 732.
  multispina   682, 698, 701, 725, 732.
  purpurea 669, 682, 689, 698, 701, 714,
      727, 732..
  sexradiata   755.
  spinosa   682, 698, 701, 725, 732, 755.
  spinulifer   755, 759.
Oxyceros   684.

Palæaster   687.
  asperrimus   687.
  coronella   687.
  hirudo   687.
  niagarensis   687.
  obtusus   687.
  ruthveni   687.
Palasterina   687.
  primæva   687.
Palæocoma   687.
  colvini   687.
  cygnipus   687.
  marstoni   687.
  pyrotechnicum   687.
  vermiformis   687.
Palmasterias   697.
  membranacea   717.
Palmipes   634, 656, 667, 672, 683, 692,
      693, 697, 699, 701, 717, 733,
      736, 740, 752, 771.
  antiquus   683.
  diaphanus   752.
  inflatus   689, 740, 771.
  lobianci   717.

  lorioli   711.
  ludovici   710.
  membranaceus   659, 662, 672, 683, 699,
      701, 709, 717, 733, 735,
      736, 740, 752, 757, 761,
      771.
  pellucidus   667.
  petaloides   659.
  placenta   672, 690, 717, 735, 736, 740,
      752, 761.
  rosaceus   662, 663, 665, 668, 676, 683,
      711, 740.
  sarasini   711.
  stokesii   635, 663, 664, 683, 699, 701,
      733, 740.
  tenuis   656, 659.
Paragonaster   667, 696, 745, 751, 772.
  ctenipes   696, 751, 772.
  cylindratus   751, 766, 769.
  elongatus   708, 745, 769.
  formosus   766, 769.
  sp.   667, 772.
  stenostichus   696.
  strictus   745, 769.
  subtilis   708, 745, 769.
  tenuiradiis   667.
Pararchaster 37, 666, 693, 723, 745, 749, 772.
  antarcticus   29, 723, 749.
  armatus   31, 35, 36, 38, 39, 40, 689, 723,
      749, 765, 766, 769.
  cognatus   721, 723.
  fischeri   723, 745.
  folini   723, 745.
  huddlestonii   666, 723.
  indicus   709, 723.
  pectinifer   721, 723.
  pedicifer   723, 749.
  semisquamatus 24, 26, 29, 30, 36, 38, 666,
      723, 749, 765, 766, 769, 772.
  simplex   745.
  spinosissimus   723, 749.
  spinuliger   721, 723.
  violaceus   666, 723.
Parasterina   693.
Parastropecten   249, 250.
  inermis 239, 241, 242, 249, 250, 721.
Pataginaster   178, 691.
  nuttingi   691.
  sphærioplax   . 696.

Patiria 635, 693, 699, 702, 752, 771.
  bellula 752.
  briareus 673, 674.
  chilensis 712, 741, 763.
  coccinea 635, 699, 702, 733, 741.
  crassa 669, 699, 702, 741.
  granifera 699, 702, 741.
  miniata 762.
  obtusa 699, 702, 741, 762.
  ocellifera 699, 702, 741.
Paulia 440, 685, 698, 700, 771.
  horrida 434, 676, 685, 698, 700, 722, 739, 763.
Pauliella
  aenigma 722.
Pectinaster 694, 723, 745.
  agassizii 694, 677, 723.
  cribellum 723.
  echinulatus 723.
  filholi 723, 745.
  hispidus 709, 723.
  insignis 745.
  mimicus 723.
  pristinus 723.
Pedicellaster 667, 721, 743, 744, 755, 770.
  antarcticus 721.
  atratus 667.
  hypernotius 755.
  hyperoncus 677.
  improvisus 677, 722.
  margaritaceus 744.
  palaeocrystallus 687, 720, 728, 731, 767.
  pourtalesi 743.
  scaber 755.
  sexradiatus 744.
  typicus 702, 703, 720, 728, 731, 737, 760, 767.
Pedicellasteridae 667, 709, 720, 721, 722, 743, 744, 755, 767.
Peltaster 768.
  cycloplax 696.
  hebes 768.
  planus 768.
Pentaceropsis 751.
  euphues 757.
  obtusatus 674, 679, 715, 751.
Pentacoros 478, 487, 507, 508, 509, 511, 521, 585, 607, 683, 691, 693, 697, 698, 700, 704, 713, 739, 751, 759, 771.

aculeatus 669, 698, 700.
affinis 710, 757.
alveolatus 487, 711, 739.
armatus 669, 691, 698, 700, 713, 739, 762, 763.
australis 711.
belli 714.
callimorphus 751.
chinensis 698, 700.
concavus 699, 701, 712, 740.
cumingii 698, 700.
dorsatus 739, 745, 751.
franklini 444, 469, 470, 477, 698, 700, 733, 739, 751.
gibbosus 739.
gibbus 669, 683, 698, 700, 716, 733, 739.
gracilis 515, 676, 739, 748.
grandis 669, 683, 698, 700, 733, 739.
granulosus 668, 676, 699, 700, 739.
grayi 714, 718, 757.
hawaiiensis 691.
hedemanni 711.
hiuculus 469.
hiulcus 469, 487, 502, 503, 507, 674, 698, 718, 727, 733, 739, 757, 759.
indicus 711.
lentiginosus 683, 733, 739.
lincki 486, 513, 514, 515, 674, 676, 704, 705, 748.
longiorum cornuum 700.
lütkeni 707.
macroceros 684, 733, 740.
mammillatus 507, 705, 707, 711, 713, 739.
mammosus 478, 739.
modestus 445, 477, 485, 700, 751.
mülleri 757.
muricatus 479, 487, 494, 495, 500, 505, 506, 511, 513, 622, 669, 670, 676, 698, 700, 718, 728, 737, 739, 757, 771.
nodosus 468, 469, 476, 484, 486, 488, 669, 674, 698, 700, 704, 705, 714, 739, 757.
nodulosus 475, 480, 674, 739.
novae-caledoniae 711.
obtusangula 740.
obtusatus 715, 739.
occidentalis 676, 713, 761.
orientalis 759.

oxyceros   733, 740.
planus   684, 700, 712, 719, 733, 740, 751.
plicatus   699, 701, 712, 740.
productus   711, 751, 757.
regulus   711, 739.
reinhardti   711, 757.
reticulatus   507, 508, 509, 510, 511, 669,
              683, 698, 700, 713, 733, 739,
              757, 763, 771.
rouxi   489, 711.
sladeni   714, 757.
sp.   678, 702, 705, 748.
superbus   515, 676, 748.
troscheli   757.
tuberculatus   702, 718, 757.
turritus 470, 477, 479, 480, 483, 484, 485,
          486, 487, 489, 507, 511, 668, 669,
          679, 707, 711, 715, 718, 727, 739,
          746, 751, 757, 759.
westermanni   711.
Pentacerotidæ   569, 572, 581, 689, 690, 691,
                698, 700, 705, 713, 722, 745,
                750, 751, 757, 767, 769.
Pentadactyla
echinata   682.
Pentadactylosaster   697, 754, 770.
asper   714, 715, 717, 726.
miliaris   698, 701, 732, 753.
oculatus   682, 687, 720, 728, 732, 754.
reticulatus   681, 698, 699, 700, 701, 712,
              728, 732, 737.
spinosus   681, 682, 698, 701, 732, 755.
variolatus   699, 701, 733.
Pentagonaster   252, 311, 312, 329, 420, 429,
                431, 433, 570, 582, 667, 671,
                684, 691, 693, 697, 698, 701,
                702, 716, 723, 724, 738, 743,
                745, 750, 759, 767, 768, 771.
abnormalis   312, 701, 724.
acutus   716.
affinis   743, 768.
alexandri   743, 745, 769.
ammophilus   691.
annandalei   710.
arcuatus   311, 312, 326, 329, 330, 360,
            667, 751, 768.
arenatus   690, 708, 743, 745, 768.
astrologorum   724, 738, 750, 757, 771.
auratus   724, 738.

australis   528, 724, 738.
balteatus   671, 689, 716, 735, 768.
belcheri   678, 739, 748, 757.
belli   759.
bowerbankii   684.
childreni   429, 430, 738.
chilopora   685.
compactus   684.
concinnus   671, 690, 708, 716, 735, 768.
coppingeri   669.
couloni   684.
crassimanus   312, 724.
cuenoti   311, 329, 330, 710.
dentatus   690, 743, 768.
deplasi   745.
dilatatus   688.
döderleini   710.
dübeni   312, 417, 699, 701, 724, 738.
emma   738.
equestris   419, 707, 738.
ernesti   311, 722.
eximius   766.
gibbosus   738, 757.
gosselini   689, 708, 745.
grandis   690, 708, 724, 745.
granularis   667, 668, 671, 689, 690, 703,
              708, 719, 735, 736, 737, 738,
              745, 750, 757, 761, 766, 768.
granulosus   739.
greeni   671, 716, 735.
grenadensis   743.
gunnii   724, 738.
hæsitans   745.
hispidus   702, 709, 719, 729, 769.
hiulcus   507.
hunteri   684.
hystricis   716.
incei   678, 739.
intermedius   329, 667, 743, 768.
investigatoris   667, 768.
japonicus   311, 313, 316, 694, 751, 768.
kergroheni   716.
lamarckii   429, 738.
lepidus   751, 769.
longimanus   438, 678, 707, 739.
magnificus   724·
mammillatus   507.
mantelli   684.
meridionalis   713, 751, 759.

miliaris 688, 738.
minimus 724, 738.
minor 716.
misakiensis 311, 332, 360.
mirabilis 337, 716.
moebii 759.
mortenseni 311, 710.
mülleri 438, 713, 739.
nobilis 724.
ornatus 724.
parkinsoni 684.
parvus 743, 768.
patagonicus 311, 316, 694, 750, 768.
paxillosus 722, 738.
perrieri 690, 708, 745.
placenta 716.
planus 766, 768.
porosus 684.
pulchellus 312, 684, 688, 698, 701, 724, 734, 738.
pulvillus 330.
pulvinus 336, 337, 667.
regularis 685, 698, 701, 711, 734.
ruber 724.
rugatus 684.
schultzii 685.
semilunatus 430, 684, 698, 700, 708, 718, 734, 738, 745, 750, 767.
simplex 766.
singularis 713, 738.
spinulosus 527, 535, 588, 591, 597, 714, 738, 759.
squamulosus 759.
stratifera 685.
sublunatus 685.
subspinosus 671, 743, 745, 768.
ternalis 743, 745, 768.
tubercularis 724.
tuberculatus 738, 759.
turritus 478, 507.
uncatus 685.
validus 669.
vincenti 745.
Pentagonasteridæ 311, 360, 410, 420, 569, 570, 581, 667, 671, 674, 688, 689, 704, 705, 708, 716, 718, 719, 722, 735, 743, 745, 756, 757, 766, 767, 772.
Pentasterias 697.

Pentetagonaster
   chilopora 685.
   costata 685.
   dutempleanus 684.
   dutemplei 685.
   fleuriausa 684.
   jurensis 684.
   malbosii 684.
   moulinsii 685
   quinqueloba 685.
   stellifer 684.
   stratifera 685.
   tabulatus 684.
   variabilis 684.
Peribolaster 691, 696, 754.
   biserialis 677, 691, 696.
   folliculatus 754.
Perissogonaster 696.
   insignis 696.
Perknaster 754.
   densus 755.
   fuscus 754
Persephonaster 178, 208, 209, 215, 221, 666, 690, 696, 772.
   anchistus 696.
   armiger 721.
   asper 202.
   brevispinus 220, 221, 227.
   cœlochilus 667.
   croceus 666, 772.
   euryactis 696.
   habrogenys 696.
   luzonicus 696.
   misakiensis 202, 203, 208, 216, 217, 218, 219.
   monostœchus 696.
   multicinctus 696.
   œdiplax 696.
   penicillatus 690, 694.
   rhodopeplus 666, 772.
   roulei 709.
   suluensis 696.
   tenuis 696.
   triacanthus 202, 203, 211, 215, 220, 221.
Petalaster 671, 698, 700.
   columbiæ (-a) 687, 698, 700, 713, 741, 762.
   hardwickii (-a) 687, 698, 700, 741.
Petricia 699, 702.
   punctata 699, 702, 740.

Phaneraster   693, 745, 767.
  semilunatus   708, 745, 768.
Pharia   713.
  pyramidata   676, 713, 762.
Phataria
  fascialis   695.
  unifascialis   676, 677, 695, 762.
Phidiaster   709.
  agassizi   709.
Philonaster
  mortenseni   710.
Pholidaster   753.
  distinctus   753.
  squamatus   753.
Phoxaster   750.
  pumilus   708, 750, 766.
Pisaster   692.
  ochraceus   677.
  papulosus   770.
Platasterias   697.
  latiradiata   742.
Plectaster sp.   772.
Pleuraster   687.
  arenicola   687.
  obtusa   687.
Plinthaster   691, 768.
  compta   768.
  nitida   768.
Plutonaster   178, 208, 376, 388, 400, 408, 410,
    671, 716, 745, 749.
  abbreviatus   749.
  abyssicola   268, 694, 721.
  agassizii   765, 766, 769.
  ambiguus   749.
  bifrons   671, 689, 703, 708, 716, 735,
    715, 749, 765, 766, 769.
  edwardsi   745.
  efflorescens   745, 769.
  granulosus   708.
  inermis   745.
  intermedius   745.
  marginatus   708, 749.
  notatus   708, 745, 749.
  parelii   379, 380, 382, 383, 384, 385, 386,
    388, 671, 690, 694, 702, 703, 708,
    718, 729, 736, 749, 756, 757.
  pulcher   745.
  rigidus   708, 749, 765, 766, 769.
  sp.   772.

  subinermis   208, 359, 708, 716, 749.
Plutonasteridæ   359, 360, 689, 690, 769.
Polyaster   682.
  papposa   672, 687, 695, 719, 753.
Pontaster   666, 671, 691, 723, 745, 749, 772.
  cribrellum   666, 723.
  forcipatus   723, 749, 765, 766.
  hebitus   671, 706, 718, 723, 746, 749,
    765, 766.
  hispidus   666, 709, 723, 772.
  limbatus 671, 706, 718, 723, 735, 746, 749.
  marionis   706, 718, 723, 745.
  mimicus   666, 723, 749.
  oligoporus   723, 745.
  oxyacanthus   40, 44, 723, 749.
  perplexus   723, 745.
  pilosus   666, 709, 723.
  planeta   723, 749.
  pristinus   723, 749.
  sepitus   765, 766.
  subtuberculatus   723, 749.
  tenuispinus 671, 680, 689, 702, 703, 706,
    708, 718, 723, 729, 730, 731,
    735, 736, 746, 749, 756, 757,
    760.
  teres   723, 749.
  trullipes   723, 749.
  venustus   708, 723, 745, 749.
Porania   622, 672, 685, 699, 701, 712,
    736, 740, 751, 759, 771.
  antarctica   712, 721, 722, 751.
  borealis   764, 766.
  gibbosa   672, 685, 699, 701, 734, 736,
    740, 751, 761.
  glaber   751.
  grandis   764, 765, 766.
  insignis   766.
  magelhænica   752, 759.
  magellanica   712, 751.
  patagonica   752, 759.
  pulvillus 663, 672, 690, 702, 703, 709, 735,
    736, 740, 751, 757, 761, 771.
  spiculata   751.
  spinulosa   709, 764, 765, 766.
Poraniidæ   709, 712, 716, 720, 744.
Poraniomorpha
  borealis   709, 761, 765, 766.
  hispida   703, 709, 731.
  rosea   703, 709, 719, 737, 757, 761.

spinulosa · 709, 761, 765, 766.
tumida : . 703, 730, 731.
villosa 690.
Poraniopsis 695, 712.
  echinasteroides 712.
  inflata 693, 695.
Porcellanaster 96, 112, 667, 690, 744, 748, 749, 772.
  cæruleus 667, 748, 749, 765, 766, 772.
  caulifer 709, 748, 749.
  crassus 748, 749.
  eremicus 749.
  gracilis 748, 749.
  granulosus 744.
  inermis 744.
  pacificus 694, 721.
  sp. 772.
  tenebrarius 690, 694.
  tuberosus 108, 110, 748, 749.
  waltharii 82, 694, 721.
Porcellanasteridæ 52, 667, 671, 690, 694, 696,
  706, 708, 718, 721, 744, 749,
  757, 766, 769, 772.
Prionaster 696.
  analogus 696.
  elegans 769.
  gracilis 696.
  megaloplax 696.
Prognaster 744.
  longicauda 744.
Pseudarchaster 383, 400, 408, 667, 690, 691,
  749, 768, 772.
  alascensis 690, 694.
  annectens 708, 768.
  aphrodite 769.
  concinnus 766, 768.
  discus 378, 381, 408, 749, 769.
  dissonus 693, 694.
  fallax 400, 408, 708, 768.
  granuliferus 768.
  hispidus 768.
  hystrix 768.
  intermedius 373, 376, 377, 380, 381, 383,
  384, 400, 407, 408, 694, 749,
  766, 768.
  jordani 691, 710.
  mosaicus 667, 710, 769, 772.
  myobrachius 691.
  necator 708, 769.
  oligoporus 696.

ordinatus 768.
parelii 371, 380, 402, 406, 408, 690, 694.
patagonicus 769.
pectinifer 677, 722.
pretiosus 366.
pulcher 407, 722.
pusillus 247, 677, 690, 694.
roseus 769.
tessellatus 378, 379, 381, 382, 384, 394,
  400, 694, 708, 718, 749, 757,
  769.
verrilli 722.
Pseudarchasteridæ 690, 691.
Pseudaster 744.
  cordifer 708, 744.
Psilaster 92, 178, 671, 691, 694, 750.
  acuminatus 688, 705, 750.
  andromeda 385, 671, 689, 702, 703, 708,
  719, 731, 735, 736, 737, 744,
  750, 761.
  armatus 721.
  attenuatus 691.
  cassiope 750, 761.
  floræ 766.
  gotoi 696.
  gracilis 274, 750.
  patagiatus 689, 750.
  pectinatus 677, 694.
  robustus 696.
  sladeni 721.
Psilasteropsis 691.
  cingulata 691.
  humilis 707, 708.
  patagiatus 689, 708.
Pteraster 672, 677, 687, 691, 692, 696, 702,
  734, 742, 743, 744, 754, 772.
  affinis 754.
  alveolatus 744.
  aporus 696, 715.
  capensis 687, 699, 702, 742.
  caribbæus 743.
  coscinopeplus 696.
  cribrosus 668, 678, 679, 718, 727, 742.
  danæ 764.
  gracilis 696.
  hastatus 731.
  hebes 770.
  hexactis 679, 680, 696, 719, 729, 767, 770.
  inflatus 689, 740.

jordani 677, 691, 696.
lebruni 721.
marsippus 696.
miliaris 702.
militaris 663, 672, 675, 676, 678, 680, 687, 688, 690, 696, 702, 703, 706, 709, 710, 719, 721, 726, 729, 730, 731, 734, 735, 736, 737, 742, 746, 754, 758, 760, 762, 765, 767.
multipes 696, 720, 767.
multiporus 677.
multispinus 696.
obesus 677.
obscurus 680, 696, 703, 729, 731.
octaster 696, 770.
personatus 672, 690, 735.
pulvillus 675, 676, 678, 680, 696, 702, 703, 706, 719, 729, 731, 736, 737, 746, 758, 760, 762, 765, 767.
reductus 707, 709.
reticulatus 692.
rugatus 754.
semireticulatus 754.
sordidus 744.
sp. 704, 719.
stellifer 754.
temnochiton 696.
tesselatus 696.
trigonodon 696.
Pterasteridæ 667, 672, 674, 690, 691, 692, 696, 705, 706, 709, 719, 721, 722, 735, 742, 743, 744, 754, 758, 767, 769, 771, 772.
Pycnopodia 712, 770.
helianthoides 712, 762, 770.
Pycnopodidæ 692.
Pyrenaster 768.
affinis 768.
dentatus 768.
Pythonaster 754.
murrayi 754.

Radiaster 743, 744.
elegans 743, 744.
Randasia 440, 521, 531, 536, 537, 685, 698, 699, 700, 740.
gracilis 727.
granulata 521, 523, 529, 530, 532, 537, 588, 593, 670, 699, 700, 739, 740

luzonica 685, 698, 700, 727.
spinulosa 521, 532, 588, 699, 700, 703, 740.
Rathbunaster 692.
californicus 692.
Remaster palmatus 744.
Retaster 754.
cribrosus 668, 674, 676, 679, 705, 718, 748.
diaphanus 722.
gibber 723, 754.
gracilis 696.
insignis 668, 670, 679, 718, 754.
multipes 696, 703, 709, 719, 720, 736, 762, 767.
peregrinator 754, 767.
verrucosus 754, 767.
Rhegaster 672, 752.
abyssicola 766.
murrayi 672, 709, 752, 761.
tumidus 678, 680, 703, 706, 719, 729, 730, 746, 757.
Rhipidaster 753.
vannipes 673, 674, 753.
Rhopia 681, 698, 701, 755, 770.
mediterranea 682, 698, 701, 717, 732.
seposita 682, 698, 701, 702, 717, 732.
Ripaster 179.
Rosaster 696, 745.
alexandri 745, 769.
mamillatus 696.
mimicus 696.
nanuus 696.

Sarkaster validus 722.
Scaphaster humberti 711, 715.
Sclemsterias nitida 711.
Scutasteries 697.
Scytaster 683, 692, 701, 725, 733, 737, 752, 753, 759, 771.
ægyptiacus (-icus) 670, 678, 692, 711.
canariensis 683, 709, 753.
cancellatus 683.
desjardinsii 725.
galatheæ 670, 725.
indicus 737.
kuhlii 683, 727, 733.
milleporellus 683, 714, 727, 733, 752, 759.
novæ-caledoniæ 670, 678, 771.
pistorius 683, 714, 727, 733, 752, 759.
posterius . 701.

semiregularis 683, 727, 733.
semiseriatus 726.
stella 753.
subtilis 725.
subulatus 683, 689, 701, 733, 753.
tuberculatus 715, 753, 757.
variolatus 670, 683, 701, 714, 733, 757, 759, 771.
zodiacalis 678, 683, 733.
Sideriaster 769.
grandis 769.
Sidonaster 709.
batheri 710.
psilonotus 696.
vaneyi 709.
Smilasterias 755.
Socomia 699, 702.
paradoxa 699, 702.
Sol 770.
coriaceus planus 681.
echinatus cancellatus 681, 717, 731.
Solaster 672, 682, 695, 697, 698, 700, 732, 736, 737, 753.
abyssicola 765, 767.
affinis 677, 679, 682, 690, 706, 729.
alboverrucosa 682.
australis 723.
benedicti 767.
borealis 677, 695.
constellatus 770.
dawsoni 695.
decanus 682.
decemradiatus 695, 762.
earlii 765, 767.
echinatus 709.
endeca 667, 672, 675, 676, 678, 679, 680, 682, 688, 690, 695, 698, 700, 702, 703, 706, 709, 719, 729, 730. 731, 732, 735, 736, 737, 746, 753, 758, 761, 765, 767, 770.
exiguus 693, 695.
furcifer 672, 676, 680, 682, 702, 703, 706, 719, 729, 735, 753, 760, 765, 767.
galaxides 695, 770.
glacialis 703, 709, 719, 731, 758.
gracilis 727.
helianthus 679.
hypothrissus 693, 695.
ntermedius 719, 758.

japonicus 696.
moretonis 682.
octoradiatus 721.
papposus 672, 675, 676, 679, 680, 682, 687, 690, 695, 698, 700, 702, 703, 706, 719, 729, 730, 731, 732, 735, 736, 753, 760, 761, 767, 770.
paxillatus 677, 695.
regularis 753.
squamatus 703, 731.
stimpsoni 695.
subarcuatus 753.
syrtensis 679, 680, 703, 706, 709, 719, 731, 737, 767.
torulatus 705, 753.
tumidus 680, 706, 719, 729, 730, 760.
Solasterias (-es) 682, 697.
endeca 682, 688, 753.
papposa 687, 753.
Solasteridæ 672, 690, 692, 693, 695, 705, 706, 709, 719, 721, 722, 735, 736, 743, 744, 753, 758, 767.
Sphæriodiscus scrotocryptus 696.
Sporasterias
antarctica 721, 723.
cocosana 722.
galapagensis 722.
mariana 722.
spirabilis 721.
Stegnaster
inflatus 689, 705.
Stella
cancellata 683, 733, 739.
cartilaginea 717, 740, 752.
coriacea 682, 697, 770.
coriacea acutangula hispida 681.
coriacea pentadactyla echinata 732.
coriacea umbilicata 727.
coriacea vulgaris 732.
coriacea vulgaris luidii 681.
hibernica echinata 681.
marina etc. 681, 685, 714, 715, 717, 726, 732, 734, 741, 742.
membranacea 717.
obtusangula 712.
pectinata 742.
pentadactyla exigua, etc. 740.
pentekaidekaktis 682, 732.
reticulata 683, 733, 739.

rossa membrancea   717.
rubra coriacea, etc.   682, 732.
Stellaria   671, 685, 697, 750.
   aurantiaca   716.
   bispinosa   716.
   spinosa   701.
Stellaster   261, 262, 411, 415, 417, 420, 432,
      437, 438, 615, 685, 698, 700, 725,
      734, 738, 751, 771.
   belcheri   413, 416, 420, 436, 668, 669,
      678, 699, 700, 725, 727, 739,
      748, 751, 757.
   childreni 411, 412, 413, 414, 415, 416, 419,
      421,685,698,700,725,726,734,738.
   comptoni   685.
   elegans   685.
   equestris 411, 412, 415, 416, 417, 419, 420,
      433, 665, 685, 707, 710, 725, 726,
      727, 734, 738, 739.
   gracilis   414, 415, 416, 417, 420, 668. 725,
      726, 730, 739, 751.
   granulosus   739.
   hesperus   727.
   incei   413, 416, 419, 668, 669, 674, 676,
      678, 699, 700, 704, 707, 710, 711,
      725, 739, 748, 751, 757.
   mülleri   416, 420, 438, 725, 726, 739.
   princeps   420, 751.
   schultzii   685.
   sp.   704.
   squamulosus   668, 707, 710, 759.
   sulcatus 257, 259, 262, 265, 272, 278, 437,
      700, 725, 726, 730, 741, 750.
   tuberculosus   416, 417, 436, 725, 726.
Stellonia   673, 681, 682, 697, 698, 699, 701,
      753, 754, 755, 770.
   angulosa   673, 681, 699, 717, 732, 755.
   echinites   682, 732, 738, 754.
   endeca   672, 682, 688, 695, 753.
   glacialis   673, 681, 699, 717, 732, 755.
   helianthus   681, 699, 755.
   hispida   673, 681.
   papposa   672, 682, 687, 695, 753.
   rubens   681, 699, 732, 755.
   seposita   698, 701, 717, 732.
   spinosa   682, 732.
   tenuispina   717.
   webbiana   673, 681, 717, 755.
Stellosphæra mirabilis   709.

Stellula hiberniea glabra   740.
Stephanaster   693, 721, 745, 768.
   astrologorum   724.
   australis   724.
   bourgeti   745.
   dübeni   724.
   elegans   684, 688, 701, 721, 738.
   gunnii   724.
   procyon   724.
   pulchellus   724.
Stephanasterias   672, 753, 770.
   albula   720, 753, 764, 765, 767.
Stichaster   672, 709, 712, 736, 744, 753, 770.
   albulus   687, 703, 709, 720, 728, 729, 731,
      746, 753, 765, 767.
   arcticus   703. 720.
   aurantiacus   712, 728, 753, 758, 762, 763,
      770.
   australis   689, 705.
   felipes   753.
   insignis   689, 705.
   littoralis   689.
   nutrix   721.
   polygrammus   712, 753.
   polyplax   688, 689, 705, 753.
   roseus   667, 668, 672, 676, 690, 702, 703,
      709, 710, 720, 731, 735, 736, 753,
      756, 758, 762.
   striatus   681, 700, 732, 753, 762.
   suteri   688, 689, 705.
   talismani   744.
Stichasteridæ 672, 689, 690, 705, 709, 712, 720,
      735, 744, 753, 758, 767, 770.
Stolasterias   667, 744, 755.
   alexandri   722.
   candicans   721.
   glacialis   673, 717, 744.
   neglecta   717.
   robusta   722.
Styracaster   667, 744, 748, 749, 772.
   armatus   667, 748, 749.
   clavipes   667, 772.
   caroli   710.
   edwardsi   744.
   elongatus   707, 708.
   horridus   667, 708, 710, 748, 749, 772.
   spinosus   708, 744.

Tamaria   698, 701, 733, 752, 771.

fusca 683, 698, 701, 733, 759.
Tarsaster 753.
　neozealanicus 689.
　stoichodes 753.
Temnaster hexactis 680, 696, 719, 729, 767, 770.
Tethyaster 178, 376, 745, 749.
　parelii 376, 382, 383, 400, 401, 408, 690, 694, 729, 756, 757, 760.
　subinermis 716, 745, 749.
Tetractis petaloides 681, 732.
Thoracaster 748, 749.
　alberti 710.
　cylindratus 708, 748, 749.
Thrissacanthias 178, 208, 694.
　penicillatus 677, 694.
Tonia 672, 697, 753.
　atlantica 681, 697, 732, 753, 762.
Tosia 311, 312, 336, 431, 432, 582, 584, 684, 690, 698, 699, 701, 723, 724, 750, 768, 771.
　annandalei 710.
　arctica 695, 770.
　astrologorum 724.
　atlantica 699.
　aurata 312, 699, 701, 724, 738.
　australis 312, 522, 684, 698, 699, 701, 724, 734, 738.
　ceramoidea 691.
　compta 768.
　cuenoti 710.
　dübeni 417.
　grandis 699, 701, 724.
　granularis 719, 768.
　lamarckii 738.
　leptocerama 311, 690, 694.
　magnifica 312, 724.
　mammillata 584.
　micropelta 311, 691.
　minima 738.
　nitida 768.
　nobilis 312, 724.
　ornata 312, 724.
　parva 743.
　placenta 584.
　rubra 312, 699, 701, 724.
　semilunata 738.
　tubercularis 699, 701, 724.
Tremaster mirabilis 765, 767.
Triskaidecactis 697.

papposa 687, 719, 729, 753.
Tritonaster 178, 691.
　craspedotus 691.
　evorus 696.
Tropidaster 687.
　pectinatus 687.
Tylaster willei 703, 719, 731.

Uniophora 699.
　globifera 681, 699.
　granifera 689, 705.
Uraster 673, 681, 687, 755, 770.
　gaveyi 681.
　glacialis 673, 681, 717, 735, 736, 755.
　hirudo 687.
　hispida 673, 735, 736.
　obtusus 687.
　primæva 687.
　rubens 673, 681, 735, 736, 755.
　ruthveni 687.
　violacea (-us) 673, 681, 687, 735, 736.
Urasterias forcipulata 770.

Valvaster 692, 770.
　striatus 692, 711, 714.

Zoroaster 667, 672, 691, 692, 743, 744, 753, 772.
　ackleyi 743, 744.
　adami 710.
　alfredi 667.
　angulatus 667.
　barathri 667.
　carinatus 667.
　diomedeæ 690, 765, 767.
　evermanni 677, 691.
　fulgens 672, 690, 703, 735, 744, 753, 765, 767.
　gilesii 667.
　hirsutus 722.
　longispinus 722.
　magnificus 722.
　nudus 722.
　ophiurus 677, 691.
　planus 667.
　platyacanthus 677.
　sacculatus 691.
　sigsbeei 742, 743.
　sp. 722, 772.
　spinulosus 692.

squameus	667.	zen	667.
tenuis	753.	Zoronsteridæ	667, 691, 692, 709, 722, 744,
trispinosus	690, 709.		753, 767, 772.

ADDENDUM.

To p. 354, after line 12 (*Hipp. spinosa*) :

CLARK describes a young specimen from lower California [:13, p. 194]:

" A specimen with R only 9 mm. seems to be undoubtedly the young of this species, although it was taken at a considerably greater depth than has been hitherto known for *spinosa*. There are only four marginal plates in each series. These carry conspicuous thick spines; if there are two or three on a plate, one (the median of three) is notably larger than the others. The abactinal plates are each bordered with spiniform granules from four to twelve in number according to the size of the plate. The primary plates are conspicuous and each carries a central spinelet. Actinally the furrow and subambulacral spines are conspicuous, but the spiniform granules of the actinal intermediate plates are very small. No pedicellariæ are to be seen anywhere actinally but five or six on the abactinal surface are very conspicuous; there are none on the marginal plates.

" Station 5693. Northwest of San Nicolas Island, California, 451 fms."

ERRATA.

P. 381, l. 20 :	*for* " some "	*read* " same."
„ 385, „ 24 :	„ " de thos "	„ " det bos."
„ 489, „ 20 :	„ " échantillons "	„ " échantillon."
„ 622, „ 13 :	„ " Goniasterianæ "	„ " Goniasterinæ."
„ 659, „ 18 :	„ " Palmipes- "	„ " Palmipes."
„ 674, „ 11 :	„ " (juv.),"	, " (juv.)."
„ 685, „ 11 from below :	„ " Gymnasteria "	„ " Gymnasteria."
„ 687, „ 7 :	„ " PALÆASTRINA "	„ " PALASTERINA."
„ 691, „ 4 from below :	„ " LINKIIDÆ "	„ " LINCKIIDÆ."
„ 716, „ 7 :	„ " PERIER "	„ " PERRIER."
„ 722, „ 10 :	„ " LINKIIDÆ "	„ " LINCKIIDÆ."
„ 725, „ 10 :	„ " Scyaster "	„ " Scytaster."
„ 734, „ 7 from below :	„ " Asterias,"	„ " Asterias."
„ 751, „ 9 „ :	„ " elegaus "	„ " elegans."
„ 755, „ 18 :	„ " subf."	„ " subg."
„ „ „ 28 :	„ " jehensenii "	„ " jehennesii."
„ 763, „ 1 :	„ " Luickia "	„ " Linckia."

Published December 17th, 1914.

S. GOTO :

A DESCRIPTIVE MONOGRAPH OF JAPANESE ASTEROIDEA.

PLATE I.

PLATE I.

Cheiraster yodomiensis.

1. Abactinal view. × 0.66.
2. Actinal view. × 0.66.
3. Mouth-plates. R=170 mm. × 3.3.
4. Two adambulacral plates. The under side is the adcentral end. R=170 mm. × 3.5.
5. Two somewhat large ventrolateral plates with pectinate pedicellaria. × 4.
6. Four ventrolateral plates. × 7.
7. Abactinal plates, with three pores. R=170 mm. × 8.

Persephonaster asper.

8. Actinal view. × 0.5.
9. Abactinal view. × 0.5.
10. Mouth-plates. R=205 mm. × 2.8.
11. Sixth adambulacral plate. The under side is the adcentral end. R=205 mm. × 4.
12. Paxilla. R=205 mm. × 4.5.
13. Madreporite. × 4.5. The under side is the outer.

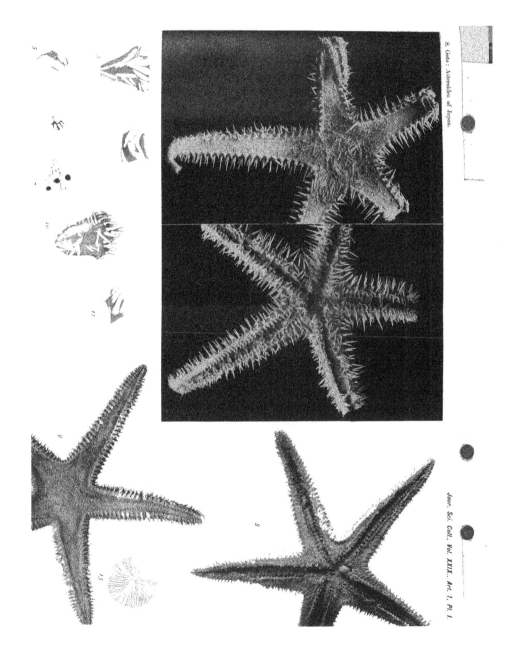

S. Goto : Asteroidea of Japan.

Jour. Sci. Coll., Vol. XXIX, Art. 1, Pl. I.

PLATE II.

PLATE II.

Persephonaster triacanthus.

14. Abactinal view. × 1.
15. Actinal view. × 1.
16. Actinal interradius, denuded. × 1.
17. Mouth-plates and adjoining adambulacrals and ventrolaterals, denuded. × 5.
18. Mouth-plates and adjoining adambulacrals. × 4.
19. Third inferomarginal and adambulacral plates. The under side is the adcentral end. × 5.5.
20. Eleventh superomarginal plate. The under side is the adcentral, the right side the inner. × 5.
21. Paxilla. Three different views. × 8.
22. Madreporite. The under side is the outer. × 7.

Persephonaster misakiensis.

23. Abactinal view. × 0.8.
24. Actinal view. × 0.8.
25. Actinal interradius. × 1.
26. Third inferomarginal with adambulacral and ventrolateral plates. The under side is the adcentral end. × 6.
27. Tenth and eleventh adambulacral plates, denuded. The under side is the abcentral end.
28. Tenth superomarginal plate. The under side is the inner, the left side the abcentral end. × 6.5.
29. Mouth-plates. × 4.
30. Ditto ; side view. × 5.5.
31. Paxilla of somewhat small size. × 40.
32. Abactinal skeleton. × 7.
33. Madreporite. × 8.

Goto del. Ucheyama photo.

14-22 *Persephonaster triacanthus*.

23-33 *Persephonaster misakiensis*.

S. GOTO:

A DESCRIPTIVE MONOGRAPH OF JAPANESE ASTEROIDEA.

PLATE III.

PLATE III.

Astropecten scoparius.

34. Abactinal view. × 1.
35. Actinal view. × 1.
36. Actinal interradius. × 6.
37. Mouth-plates. R = 55 mm. × 4.5.
38. Fifth inferomarginal plate and corresponding adambulacral plates. The under side is the abcentral end. × 5.5.
39. Superomarginal plate, about one-third from the tip of an arm. The under side is the abcentral end. × 5.5.
40. Madreporite. The upper side is the inner. × 5.
41. Paxilla, from middle of the base of an arm. × 7.

Astropecten polyacanthus.

42. Abactinal view. × 1.
43. Actinal view. × 1.
44. Actinal interradius, denuded. × 3.
45. Mouth-plates. × 4.
46. Fourth inferomarginal and corresponding adambulacral plates. R = 59 mm. The under side is the adcentral end. × 4.5.
47. First armed superomarginal plate of an arm. The right side is the inner. × 6.
48. First superomarginal spine of an arm. × 5.5.
49. Paxilla, from middle of the base of an arm.
50. Madreporite, covered with granules. × 5.
51. Ditto, denuded. The upper side is the inner. × 8.

Astropecten latespinosus.

52. Abactinal view. × 1.
53. Actinal view. × 1.
54. Abactinal view of a small specimen. × 1.
55. Actinal view of same. × 1.
56. Actinal interradius, denuded. R = 36 mm. × 4.5.
57. Mouth-plates. × 5.
58. Fifth inferomarginal plate and corresponding adambulacrals. The upper side is the abcentral end. R = 36 mm. × 5.
59. Fifth superomarginal and inferomarginal plates. × 5.
60. Paxilla. × 9.
61. Madreporite. The upper side is the inner. × 8.

34-41 Astropecten seguarius.   42-51 Astrop. polyacanthus.   52-61 Astrop. latespinosus.

PLATE IV.

PLATE IV.

Astropecten kagoshimensis.

62. Abactinal view. × 1.

63. Actinal view. × 1.

64. Actinal interradius, denuded. R=50 mm. × 3.5.

65. Mouth-plates. R=49 mm. × 6.

66. Sixth inferomarginal plate and corresponding adambulacrals.
R=49 mm. × 4.

67. Fifth superomarginal plate. R=50 mm. The upper side is
the abcentral end, the right the outer. × 9.

68. Paxilla. R=50 mm. × 10.

69. Madreporite. The upper side is the inner. R=50 mm. × 9.

Astropecten ludwigi.

70. Abactinal view. × 1.

71. Actinal view. × 1.

72. Actinal interradius, denuded. R=62 mm. × 1.

73. Month-plates and first adambulacral plates, viewed slightly
obliquely to the surface of the plates. R=102 mm. × 4.

74. Mouth-plates and adjoining adambulacrals, denuded. R=62
mm. × 4.

75. Fifth inferomarginal plate and corresponding adambulacral.
R=102 mm. × 6.

76. Tenth superomarginal plate. The upper side is the abcentral
end, the right the inner. R=102 mm. × 4.

77. Paxilla near madreporite. R=102 mm. × 5.

78. Madreporite. The upper side is the outer. R=102 mm. × 6.

79. Ditto, denuded. The upper side is the outer. R=62 mm.
× 10.

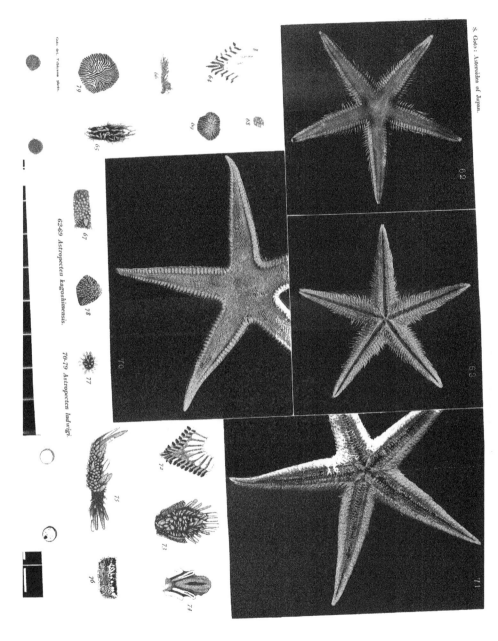

S. Goto : Asteroidea of Japan.

62-69 Astropecten kagoshimensis.   70-79 Astropecten ludwigi.

Goto del. T.Okuyama photo.

S. GOTO:

A DESCRIPTIVE MONOGRAPH OF JAPANESE ASTEROIDEA.

PLATE V.

PLATE V.

Luidia maculata.

80. Abactinal view. × 0.5.
81. Actinal view. × 0.5.
82. Mouth-plates. R=184 mm. × 4.
83. Adambulacral and inferomarginal plates, 25 mm. from the base of an arm. R=184 mm. × 4.5.
84. Pedicellaria from an adambulacral plate, from near the base of an arm. R=184 mm. × 45.
85. One of the three valves of the pedicellaria figured in 84. × 45.
86. Another valve of same. × 45.
87. Paxilla from about the middle of an arm. × 45.
88. Madreporite. R=263 mm. The upper side is the outer. × 6.

Luidia yesoensis.

89. Abactinal view. × 1.
90. Actinal view. × 1.

80-88 Luidia maculata.

89-90 Luidia yesoensis.

S. GOTO:

A DESCRIPTIVE MONOGRAPH OF JAPANESE ASTEROIDEA.

PLATE VI.

PLATE VI.

Luidia yesoensis.

91. Mouth-plates. R=47 mm. × 7.

92. Thirteenth adambulacral and corresponding inferomarginal plates. R=47 mm. × 7.

93. a–d, adambulacral spines from the innermost to the outermost in the order of the alphabet. e, adambulacral pedicellaria. R=47 mm. × 45.

94. Madreporite. The upper side is the outer. R=47 mm. × 5.

Luidia moroisoana.

95. Abactinal view. × 0.8.

96. Actinal view. × 0.8.

97. Mouth-plates. R=194 mm. × 3.5.

98. Seventeenth adambulacral and corresponding inferomarginal plates. R=194 mm. × 5.

99. Fifteenth adambulacral and corresponding inferomarginal plates. R=194 mm. The right side is the outer. × 6.

100. Pedicellaria close to the third adambulacral spine. × 45.

101. Furrow pedicellaria. × 45.

102. Pedicellaria from a superomarginal plate. × 45.

103. Superomarginal plate corresponding to the seventeenth inferomarginal. R=194 mm. × 7.

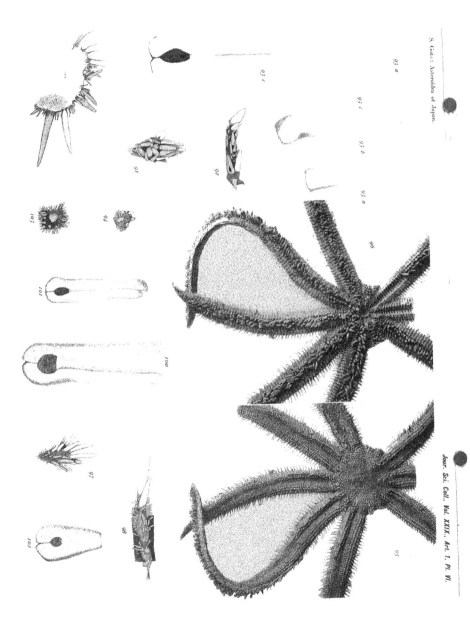

S. GOTO:

A DESCRIPTIVE MONOGRAPH OF JAPANESE ASTEROIDEA.

PLATE VII.

PLATE VII.

Luidia quinaria.

104. Abactinal view. × 1.
105. Actinal view. × 1.
106. Mouth-plates. R=78 mm. × 6.
107. Twelfth adambulacral and corresponding inferomarginal plates. R=78 mm. The upper side is the abcentral end. × 6.
108. Twelfth inferomarginal and superomarginal plates. R=78 mm. × 7.
109. Oral pedicellaria. × 45.
110. Adambulacral pedicellaria. × 45.
111. Paxilla. R=78 mm. × 7.
112. Madreporite. R=78 mm. The upper side is the outer. × 7.

Ctenodiscus crispatus.

113. Abactinal view of a Misaki specimen. × 1.
114. Actinal view of same. × 1.
115. Mouth-plates and portion of interradius. R=27 mm. Misaki specimen. × 6.
116. Paxillæ in profile, from near the margin of the disk; after treatment with caustic potash. Misaki specimen. × 45.
117. Paxillæ viewed from inside, with papular pores between. Misaki specimen. × 45.
118. Flattened calcareous rod from fasciolar groove. Misaki specimen. × 130.
119. Madreporite. Misaki specimen. R=27 mm. The upper side is the inner. × 5.

104-112 Luidia quinaria.

113-119 Ctenodiscus crispatus.

Goto del. Echiyama photo.

S. GOTO:

A DESCRIPTIVE MONOGRAPH OF JAPANESE ASTEROIDEA.

PLATE VIII.

PLATE VIII.

Ctenodiscus crispatus.

120. Abactinal view of a specimen from the Japan Sea. × 1.

121. Actinal view of same. × 1.

122. Part of an actinal interradius. Japan Sea specimen. R=19 mm. × 5.

123. Paxilla in profile. Japan Sea specimen. R=23 mm. × 45.

124. Paxillæ viewed from inside, with papular pores between. Japan Sea specimen. × 45.

125. Flattened calcareous rod from fasciolar groove. Japan Sea specimen. × 130.

126. Madreporite. Japan Sea specimen. R=19 mm. × 9.

Pseudarchaster pretiosus.

127. Abactinal view. × 1.

128. Actinal view. × 1.

129. Abactinal view of a denuded specimen. × 1.

130. Actinal view of same. × 1.

131. Mouth-plates and adjoining plates. R=32.5 mm. × 6.

132. Ditto, denuded. R=39.5 mm. × 6.

133. Fourth inferomarginal and corresponding adambulacral plates. R=32.5 mm. × 8.

134. Paxilla, from the base of an arm. × 7.

135. Madreporite. R=32.5 mm. × 5.

Dipsacaster grandissimus.

136. Mouth-plates. R=115 mm. × 4.

137. Ditto, denuded. × 4.7.

138. Fifth adambulacral plate and the adjoining ventrolateral. × 4.5.

139. Paxilla in profile, with 45 coronal spinelets. × 8.

Jour. Sci. Coll., Vol. XXIX., Art. 1, Pl. VIII.

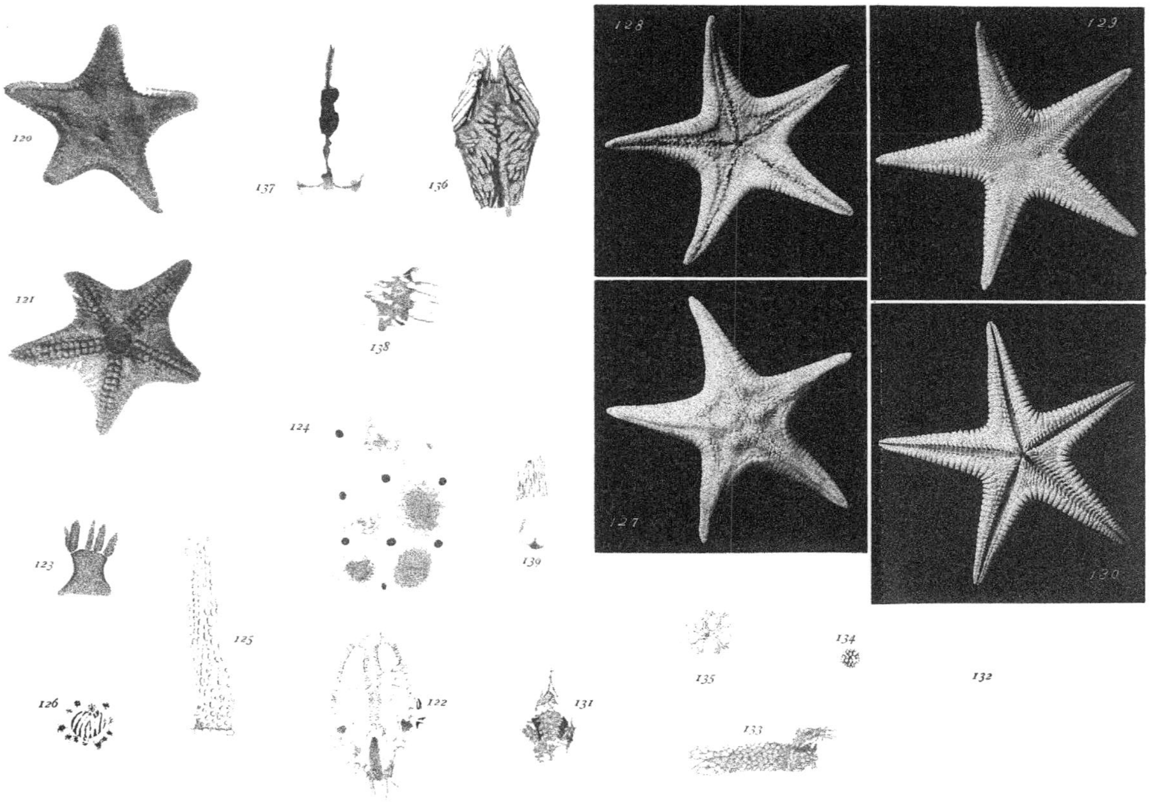

S. GOTO:

A DESCRIPTIVE MONOGRAPH OF JAPANESE ASTEROIDEA.

PLATE IX.

PLATE IX.

Dipsacaster grandissimus.

140. Abactinal view. × 0.6.
141. Actinal view. × 0.6.

Mediaster brachiatus.

142. Abactinal view. × 1.
143. Actinal view. × 1.
144. Abactinal view of a smaller specimen. × 1.
145. Actinal view of same. × 1.
146. Mouth-plates and adjoining parts of the larger specimen. × 5.
147. Mouth-plates and adjoining parts of the smaller specimen. × 6.
148. Eighth and ninth adambulacral plates with adjoining ventro-
     laterals. × 5.
149. Ventrolateral pedicellaria. × 50.
150. Pedicellaria valve from a ventrolateral plate. × 130.
151. Pedicellaria valve from a paxilla. × 130.
152. Pedicellaria on a paxilla. × 90.

140-141 Dipsacaster grandissimus.

142-152 Mediaster brachiatus.

PLATE X.

PLATE X.

Mediaster brachiatus.

153. End of an arm of the larger specimen, denuded. × 6.
154. Paxilla. × 7.
155. Madreporite. The upper side is the outer. × 7.

Johannaster giganteus.

156. Abactinal view. × 0.3.
157. Actinal view. × 0.3.
158. Mouth-plates. $R = 338$ mm. × 3.5.
159. Twelfth adambulacral and adjoining ventrolateral plates. $R = 338$ mm. The upper side is the adcentral end. × 4.
160. Valve of a large pedicellaria from the adcentral end of an adambulacral plate. $R = 338$ mm. × 90.
161. Pedicellaria valve from a ventrolateral plate adjoining an adambulacral. $R = 338$ mm. × 90.
162. Pedicellaria valve from an abactinal plate in papular area. $R = 338$ mm. × 90.
163. Ditto. $R = 338$ mm. × 90.

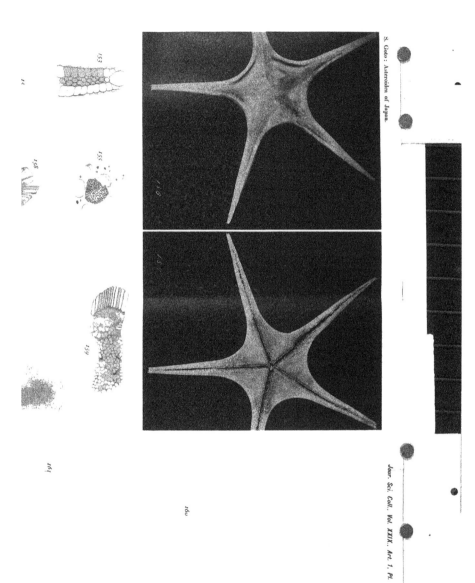

S. GOTO:

A DESCRIPTIVE MONOGRAPH OF JAPANESE ASTEROIDEA.

PLATE XI.

PLATE XI.

Pentagonaster japonicus.

164. Abactinal view. × 1.
165. Actinal view. × 1.
166. Mouth-plates. R=74 mm. × 4.5.
167. Ninth and tenth adambulacral with adjoining ventrolateral plates. R=74 mm. The left side is the adcentral end. × 7.
168. Pedicellaria valve from an adambulacral plate. × 90.
169. Abactinal plate. R=74 mm. × 8.
170. Madreporite. R=74 mm. The right side is the outer. × 5.

Pentagonaster arcuatus.

171. Abactinal view. × 1.
172. Actinal view. × 1.
173. Mouth-plates. R=56 mm. × 5.5.
174. Third adambulacral plate and the adjoining ventrolateral. R= 58 mm. The upper side is the adcentral end. × 5.5.
175. Paxilla from near the madreporite. R=58 mm. × 7.5.
176. Paxilla. R=58 mm. × 9.
177. Madreporite. R=56 mm. The upper side is the outer. × 7.5.

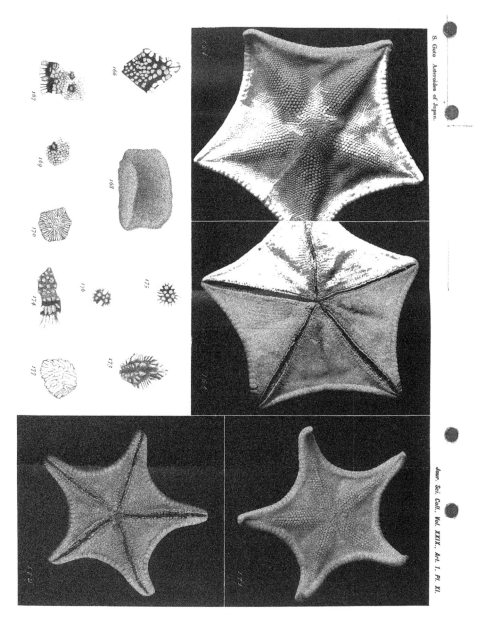

PLATE XII.

PLATE XII.

Hippasteria imperialis.

178. Abactinal view. × 0.5.
179. Actinal view. × 0.5.
180. Abactinal view of a small specimen. × 1.
181. Actinal view of same. × 1.
182. Mouth-plates. × 2.
183. Ninth adambulacral plate. The upper side is the adcentral end. × 6.
184. Marginals. The upper row is the superomarginal, the cleft at the middle the interradial line. × 2.
185. Ventrolateral plate somewhat near the border of disk and the ambulacral furrow. × 4.5.
186. Ventrolateral. × 4.5.
187. Abactinal plate. × 5.
188. Ditto. × 5.
189. Ditto. × 5.
190. Valves of an adambulacral pedicellaria. × 50.
191. Valve of a ventrolateral pedicellaria. × 50.
192. Valve of an abactinal pedicellaria. × 50.
193. Madreporite. The right side is the outer. × 7.

Jour. Sci. Coll., Vol. XXIX., Art. 1, Pl. X

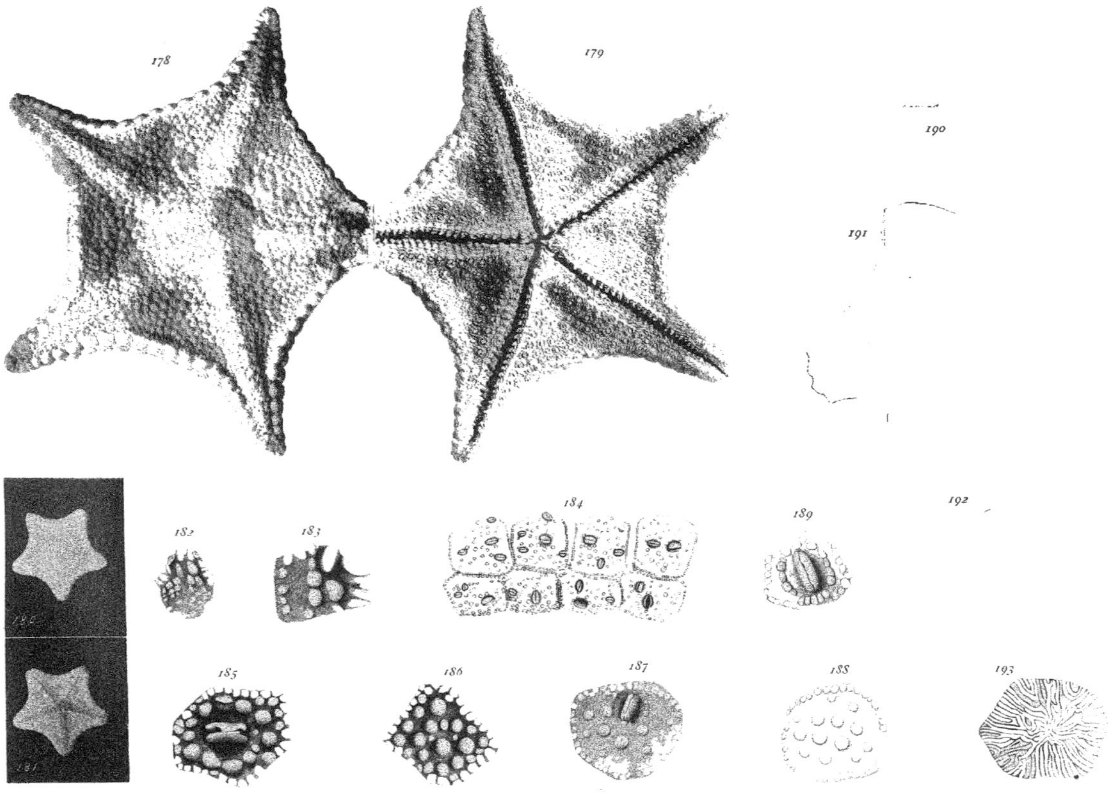

S. GOTO:

A DESCRIPTIVE MONOGRAPH OF JAPANESE ASTEROIDEA.

PLATE XIII.

PLATE XIII.

Pentagonaster misakiensis.

194. Abactinal view. × 1.

195. Actinal view. × 1.

196. Mouth-plates. × 5.

167. Ninth and tenth adambulacral plates. The lower end is the adcentral side. × 5.

198.
199.
200. } Abactinal plales. × 8.
201.

202. Madreporite. The right side is the outer. × 6.5.

Hippasteria nozawai.

203. Abactinal view. × 1.

204. Actinal view. × 1.

205. Mouth-plates. × 5.

206. Adambulacral plates with adjoining ventrolaterals. The right side is the month end. × 5.

207. Two superomarginals and adjoining abactinal plates. × 5.

208. Two inferomarginal and adjoining ventrolateral plates. × 5.

209. Ventrolateral pedicellaria with three valves. × 7.

210. Valve of an abactinal pedicellaria. × 50.

211. Madreporite and adjoining abactinal plates. The right side is the outer. × 5.

Stellaster equestris

212. Mouth-plates. R=62 mm. × 5.

213. Seventh and eighth adambulacral plates. The upper side is the adcentral. R=62 mm. × 5.

214. Valves of an adambulacral pedicellaria. R=62 mm. × 45.

215. Abactinal plates. R=62 mm. × 6.

216. Ventrolateral plate. R=62 mm. × 6.

217. Valve of a small ventrolateral pedicellaria. R=62 mm. × 45.

218. Madreporite. R=62 mm. × 6.5.

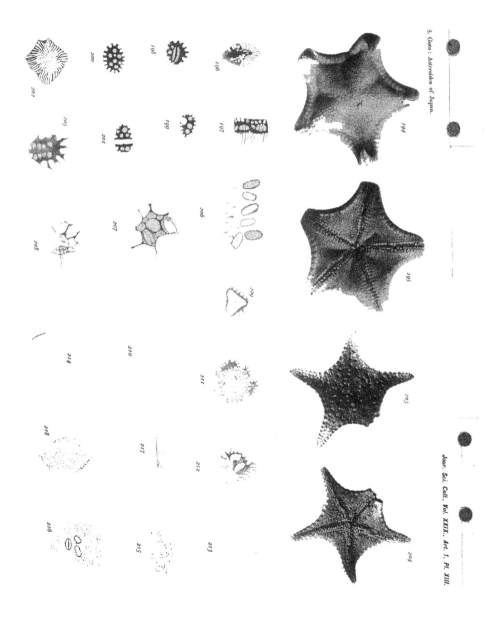

S. GOTO:

A DESCRIPTIVE MONOGRAPH OF JAPANESE ASTEROIDEA.

PLATE XIV.

PLATE XIV.

Stellaster equestris.

219. Abactinal view. × 1.
220. Actinal view. × 1.

Oreaster nodosus=turritus.

221. Abactinal view. × 0.5.
222. Actinal view. × 0.5.
223. Three adambulacral plates, viewed from the furrow side. The
      upper side is the adcentral. × 4.
224. Two adambulacral and adjoining ventrolateral plates, viewed
      from the actinal side. × 3.7.
225. Adambulacral pedicellaria, after treatment with caustic potash.
      Two views of one and the same pedicellaria. × 45.
226. Pedicellaria from the abactinal side. Two views at right
      angles to each other. × 90.
227. Small portion of an actinal interradius. × 4.

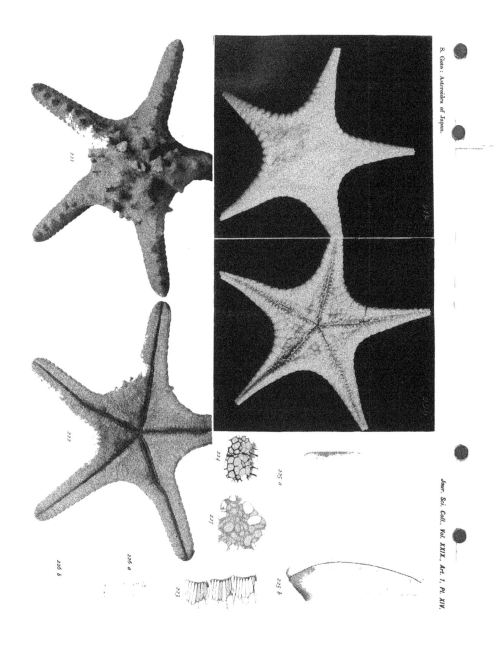

S. GOTO:

A DESCRIPTIVE MONOGRAPH OF JAPANESE ASTEROIDEA.

PLATE XV.

PLATE XV.

Oreaster modestus.

228. Abactinal view. × 0.8.

229. Actinal view. × 0.8.

230. Three adambulacral plates, viewed from the furrow side. The upper side is the adcentral. × 4.5.

231. Fourth, fifth and sixth adambulacral plates with adjoining ventrolateral plates, viewed from the actinal side. The under side is the adcentral. × 4.

232. Small portion of an actinal interradius, near the mouth. × 8.

233. Adambulacral pedicellaria, after treatment with caustic potash. × 45.

234. Adambulacral pedicellaria with three jaws, after treatment with caustic potash. × 45.

235. Abactinal pedicellaria. × 45.

236. Abactinal pedicellaria (valvate) with surrounding granules. × 5.

Oreaster magnificus.

237. Abactinal view. × 0.4.

238. Actinal view. × 0.4.

239. Mouth angle. × 1.5.

240. Small portion of abactinal surface, carinal part. × 3.

228-236 *Oreaster modestus.*

237-240 *Oreaster magnificus.*

S. GOTO:

A DESCRIPTIVE MONOGRAPH OF JAPANESE ASTEROIDEA

PLATE XVI.

PLATE XVI.

Oreaster nahensis.

241. Abactinal view. × 0.6.
242. Actinal view. × 0.6.
243. Valve of an abactinal pedicellaria. × 90.

Oreaster dœderleini.

244. Abactinal view. × 0.5.
245. Actinal view. × 0.5.
246. Valve of a pedicellaria close to the mouth. × 45.
247.
248. } Valves of ventrolateral pedicellariæ. × 45.
249. Abactinal pedicellaria. × 45.
250. Abactinal pedicellaria (valvate). × 45.
251. Another form of abactinal pedicellaria. × 45.

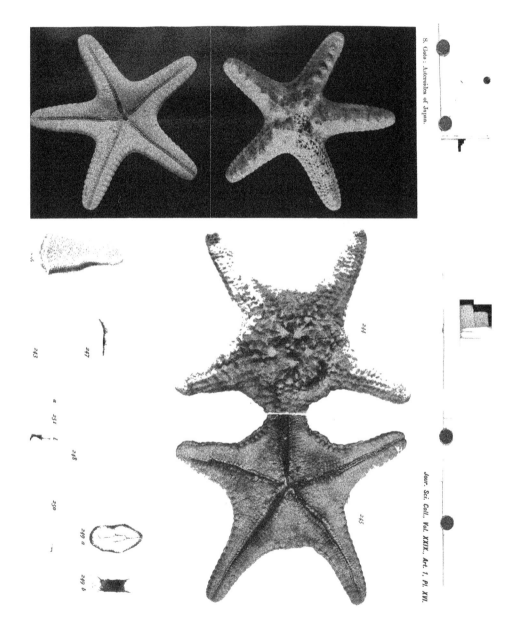

PLATE XVII.

PLATE XVII.

Culcita novæ-guineæ.

252. Abactinal view. Slightly reduced.

253. Actinal view. Slightly reduced.

254. Mouth angle. × 3.

255. Portion fo furrow margin. The right side is the adcentral. × 5.

256. Adambulacral pedicellaria, after treatment with caustic potash. · × 90.

257.⎫
258.⎬ Valves of actinal pedicellariæ. × 90.

259.⎫
260.⎬ Valve of an abactinal pedicellaria. × 90.

261.⎫
262.⎬ Two views of a flattened specimen. Slightly reduced.

Choriaster granulatus.

263. Adambulacral pedicellaria. × 90.

PLATE XVIII.

PLATE XVIII.

Choriaster granulatus.

264. Abactinal view. × 0.9.

265. Actinal view. × 0.9.

266. Mouth angle. × 4.

267. Three adambulacral plates and adjoining parts. × 4.

268. Jaw of an adambulacral pedicellaria, after treatment with caustic potash. × 90.

269. Madreporite. The right side is the outer. × 4.

Gymnasteria carinifera.

270. Abactinal view. Slightly reduced.

271. Actinal view. Slightly reduced.

Asterina pectinifera.

272. Abactinal view. × 1.

273. Actinal view. × 1.

PLATE XIX.

PLATE XIX.

Asterina pectinifera.

274. Mouth-plates and adjacent parts. R=55 mm. ×3.

Asterina batheri.

275. Abactinal view. × 1.
276. Actinal view. × 1.
277. Mouth-plates. R=24 mm. × 6.
278. Four adambulacral plates and adjoining parts, from about the middle of an arm. R=31 mm. × 7.

Asterina novæ-zelandiæ.
(Name misprinted in plate.)

279. Abactinal view. ×|1.
280. Actinal view. × 1.
281. Mouth angle. R=23 mm. × 4.

Palmipes tenuis.

282. Abactinal view. × 1.
283. Actinal view. × 1.

Palmipes petaloides.

284. Abactinal view. × 1.
285. Actinal view. × 1.